ORTHOGONAL POLYNOMIALS OF SEVERAL VARIABLES

Serving both as an introduction to the subject and as a reference, this book presents the theory in elegant form and with modern concepts and notation. It covers the general theory and emphasizes the classical types of orthogonal polynomials whose weight functions are supported on standard domains. The approach is a blend of classical analysis and symmetry-group-theoretic methods. Finite reflection groups are used to motivate and classify the symmetries of weight functions and the associated polynomials.

This revised edition has been updated throughout to reflect recent developments in the field. It contains 25 percent new material including two brand new chapters, on orthogonal polynomials in two variables, which will be especially useful for applications, and on orthogonal polynomials on the unit sphere. The most modern and complete treatment of the subject available, it will be useful to a wide audience of mathematicians and applied scientists, including physicists, chemists and engineers.

Encyclopedia of Mathematics and Its Applications

This series is devoted to significant topics or themes that have wide application in mathematics or mathematical science and for which a detailed development of the abstract theory is less important than a thorough and concrete exploration of the implications and applications.

Books in the **Encyclopedia of Mathematics and Its Applications** cover their subjects comprehensively. Less important results may be summarized as exercises at the ends of chapters. For technicalities, readers can be referred to the bibliography, which is expected to be comprehensive. As a result, volumes are encyclopedic references or manageable guides to major subjects.

All the titles listed below can be obtained from good booksellers or from Cambridge University Press. For a complete series listing visit www.cambridge.org/mathematics.

106 A. Markoe *Analytic Tomography*
107 P. A. Martin *Multiple Scattering*
108 R. A. Brualdi *Combinatorial Matrix Classes*
109 J. M. Borwein and J. D. Vanderwerff *Convex Functions*
110 M.-J. Lai and L. L. Schumaker *Spline Functions on Triangulations*
111 R. T. Curtis *Symmetric Generation of Groups*
112 H. Salzmann *et al. The Classical Fields*
113 S. Peszat and J. Zabczyk *Stochastic Partial Differential Equations with Lévy Noise*
114 J. Beck *Combinatorial Games*
115 L. Barreira and Y. Pesin *Nonuniform Hyperbolicity*
116 D. Z. Arov and H. Dym *J-Contractive Matrix Valued Functions and Related Topics*
117 R.Glowinski, J.-L.Lions and J. He *Exact and Approximate Controllability for Distributed Parameter Systems*
118 A. A. Borovkov and K. A. Borovkov *Asymptotic Analysis of Random Walks*
119 M. Deza and M. Dutour Sikirić *Geometry of Chemical Graphs*
120 T. Nishiura *Absolute Measurable Spaces*
121 M. Prest *Purity, Spectra and Localisation*
122 S. Khrushchev *Orthogonal Polynomials and Continued Fractions*
123 H. Nagamochi and T. Ibaraki *Algorithmic Aspects of Graph Connectivity*
124 F. W. King *Hilbert Transforms I*
125 F. W. King *Hilbert Transforms II*
126 O. Calin and D.-C. Chang *Sub-Riemannian Geometry*
127 M. Grabisch *et al. Aggregation Functions*
128 L.W. Beineke and R. J. Wilson (eds.) with J. L. Gross and T. W. Tucker *Topics in Topological Graph Theory*
129 J. Berstel, D. Perrin and C. Reutenauer *Codes and Automata*
130 T. G. Faticoni *Modules over Endomorphism Rings*
131 H. Morimoto *Stochastic Control and Mathematical Modeling*
132 G. Schmidt *Relational Mathematics*
133 P. Kornerup and D. W. Matula *Finite Precision Number Systems and Arithmetic*
134 Y. Crama and P. L. Hammer (eds.) *Boolean Models and Methods in Mathematics, Computer Science, and Engineering*
135 V. Berthé and M. Rigo (eds.) *Combinatorics, Automata and Number Theory*
136 A. Kristály, V. D. Rădulescu and C. Varga *Variational Principles in Mathematical Physics, Geometry, and Economics*
137 J. Berstel and C. Reutenauer *Noncommutative Rational Series with Applications*
138 B. Courcelle and J. Engelfriet *Graph Structure and Monadic Second-Order Logic*
139 M. Fiedler *Matrices and Graphs in Geometry*
140 N. Vakil *Real Analysis through Modern Infinitesimals*
141 R. B. Paris *Hadamard Expansions and Hyperasymptotic Evaluation*
142 Y. Crama and P. L. Hammer *Boolean Functions*
143 A. Arapostathis, V. S. Borkar and M. K. Ghosh *Ergodic Control of Diffusion Processes*
144 N. Caspard, B. Leclerc and B. Monjardet *Finite Ordered Sets*
145 D. Z. Arov and H. Dym *Bitangential Direct and Inverse Problems for Systems of Integral and Differential Equations*
146 G. Dassios *Ellipsoidal Harmonics*
147 L. W. Beineke and R. J. Wilson (eds.) with O. R. Oellermann *Topics in Structural Graph Theory*
148 L. Berlyand, A. G. Kolpakov and A. Novikov *Introduction to the Network Approximation Method for Materials Modeling*
149 M. Baake and U. Grimm *Aperiodic Order I: A Mathematical Invitation*
150 J. Borwein *et al. Lattice Sums Then and Now*
151 R. Schneider *Convex Bodies: The Brunn–Minkowski Theory (Second Edition)*
152 G. Da Prato and J. Zabczyk *Stochastic Equations in Infinite Dimensions (Second Edition)*
153 D. Hofmann, G. J. Seal and W. Tholen (eds.) *Monoidal Topology*
154 M. Cabrera García and Á. Rodríguez Palacios *Non-Associative Normed Algebras I: The Vidav–Palmer and Gelfand–Naimark Theorems*
155 C. F. Dunkl and Y. Xu *Orthogonal Polynomials of Several Variables (Second Edition)*

ENCYCLOPEDIA OF MATHEMATICS AND ITS APPLICATIONS

Orthogonal Polynomials of Several Variables

Second Edition

CHARLES F. DUNKL
University of Virginia

YUAN XU
University of Oregon

CAMBRIDGE
UNIVERSITY PRESS

CAMBRIDGE
UNIVERSITY PRESS

University Printing House, Cambridge CB2 8BS, United Kingdom

Cambridge University Press is part of the University of Cambridge.

It furthers the University's mission by disseminating knowledge in the pursuit of education, learning and research at the highest international levels of excellence.

www.cambridge.org
Information on this title: www.cambridge.org/9781107071896

First published 2001
Second edition 2014

Printed in the United Kingdom by CPI Group Ltd, Croydon CR0 4YY

A catalogue record for this publication is available from the British Library

Library of Congress Cataloguing in Publication data
Dunkl, Charles F., 1941–
Orthogonal polynomials of several variables / Charles F. Dunkl, University of Virginia, Yuan Xu, University of Oregon. – Second edition.
pages cm. – (Encyclopedia of mathematics and its applications; 155)
Includes bibliographical references and indexes.
ISBN 978-1-107-07189-6
1. Orthogonal polynomials. 2. Functions of several real variables.
I. Xu, Yuan, 1957– II. Title.
QA404.5.D86 2014
515′.55–dc23

2014001846

ISBN 978-1-107-07189-6 Hardback

To our wives
Philomena and Litian
with deep appreciation

Contents

Preface to the Second Edition

In this second edition, several major changes have been made to the structure of the book. A new chapter on orthogonal polynomials in two variables has been added to provide a more convenient source of information for readers concerned with this topic. The chapter collects results previously scattered in the book, specializing results in several variables to two variables whenever necessary, and incorporates further results not covered in the first edition. We have also added a new chapter on orthogonal polynomials on the unit sphere, which consolidates relevant results in the first edition and adds further results on the topic. Since the publication of the first edition in 2001, considerable progress has been made in this research area. We have incorporated several new developments, updated the references and, accordingly, edited the notes at the ends of relevant chapters. In particular, Chapter 5, "Examples of Orthogonal Polynomials in Several Variables", has been completely rewritten and substantially expanded. New materials have also been added to several other chapters. An index of symbols is given at the end of the book.

Another change worth mentioning is that orthogonal polynomials have been renormalized. Some families of orthogonal polynomials in several variables have expressions in terms of classical orthogonal polynomials in one variable. To provide neater expressions without constants in square roots they are now given in the form of orthogonal rather than orthonormal polynomials as in the first edition. The L^2 norms have been recomputed accordingly.

The second author gratefully acknowledges support from the National Science Foundation under grant DMS-1106113.

Charles F. Dunkl
Yuan Xu

Preface to the First Edition

The study of orthogonal polynomials of several variables goes back at least as far as Hermite. There have been only a few books on the subject since: Appell and de Fériet [1926] and Erdélyi *et al.* [1953]. Twenty-five years have gone by since Koornwinder's survey article [1975]. A number of individuals who need techniques from this topic have approached us and suggested (even asked) that we write a book accessible to a general mathematical audience.

It is our goal to present the developments of very recent research to a readership trained in classical analysis. We include applied mathematicians and physicists, and even chemists and mathematical biologists, in this category.

While there is some material about the general theory, the emphasis is on classical types, by which we mean families of polynomials whose weight functions are supported on standard domains such as the simplex and the ball, or Gaussian types, which satisfy differential–difference equations and for which fairly explicit formulae exist. The term "difference" refers to operators associated with reflections in hyperplanes. The most desirable situation occurs when there is a set of commuting self-adjoint operators whose simultaneous eigenfunctions form an orthogonal basis of polynomials. As will be seen, this is still an open area of research for some families.

With the intention of making this book useful to a wide audience, for both reference and instruction, we use familiar and standard notation for the analysis on Euclidean space and assume a basic knowledge of Fourier and functional analysis, matrix theory and elementary group theory. We have been influenced by the important books of Bailey [1935], Szegő [1975] and Lebedev [1972] in style and taste.

Here is an overview of the contents. Chapter 1 is a summary of the key one-variable methods and definitions: gamma and beta functions, the classical and related orthogonal polynomials and their structure constants, and hypergeometric and Lauricella series. The multivariable analysis begins in Chapter 2 with some examples of orthogonal polynomials and spherical harmonics and specific two-variable examples such as Jacobi polynomials on various domains and disk polynomials. There is a discussion of the moment problem, general properties of orthogonal polynomials of several variables and matrix three-term recurrences in Chapter 3. Coxeter groups are treated systematically in a self-contained way, in a style suitable for the analyst, in Chapter 4 (a knowledge of representation theory is not necessary). The chapter goes on to introduce differential–difference operators, the intertwining operator and the analogue of the exponential function and concludes with the construction of invariant differential operators. Chapter 5 is a presentation of h-harmonics, the analogue of harmonic homogeneous polynomials associated with reflection groups; there are some examples of specific reflection groups as well as an application to proving the isometric properties of the generalized Fourier transform. This transform uses an analogue of the exponential function. It contains the classical Hankel transform as a special case. Chapter 6 is a detailed treatment of orthogonal polynomials on the simplex, the ball and of Hermite type. Then, summability theorems for expansions in terms of these polynomials are presented in Chapter 7; the main method is Cesàro (C,δ) summation, and there are precise results on which values of δ give positive or bounded linear operators. Nonsymmetric Jack polynomials appear in Chapter 8; this chapter contains all necessary details for their derivation, formulae for norms, hook-length products and computations of the structure constants. There is a proof of the Macdonald–Mehta–Selberg integral formula. Finally, Chapter 9 shows how to use the nonsymmetric Jack polynomials to produce bases associated with the octahedral groups. This chapter has a short discussion of how these polynomials and related operators are used to solve the Schrödinger equations of Calogero–Sutherland systems; these are exactly solvable models of quantum mechanics involving identical particles in a one-dimensional space. Both Chapters 8 and 9 discuss orthogonal polynomials on the torus and of Hermite type.

The bibliography is intended to be reasonably comprehensive into the near past; the reader is referred to Erdélyi *et al.* [1953] for older papers, and Internet databases for the newest articles. There are occasions in the book where we suggest some algorithms for possible symbolic algebra use; the reader is encouraged to implement them in his/her favorite computer algebra system but again the reader is referred to the Internet for specific published software.

There are several areas of related current research that we have deliberately avoided: the role of special functions in the representation theory of Lie groups (see Dieudonné [1980], Hua [1963], Vilenkin [1968], Vilenkin and Klimyk [1991a, b, c, 1995]), basic hypergeometric series and orthogonal polynomials of q-type (see Gasper and Rahman [1990], Andrews, Askey and Roy [1999]), quantum groups (Koornwinder [1992], Noumi [1996],

Koelink [1996] and Stokman [1997]), Macdonald symmetric polynomials (a generalization of the q-type) (see Macdonald [1995, 1998]). These topics touch on algebra, combinatorics and analysis; and some classical results can be obtained as limiting cases for $q \to 1$. Nevertheless, the material in this book can stand alone and 'q' is not needed in the proofs.

We gratefully acknowledge support from the National Science Foundation over the years for our original research, some of which is described in this book. Also we are grateful to the mathematics departments of the University of Oregon for granting sabbatical leave and the University of Virginia for inviting Y. X. to visit for a year, which provided the opportunity for this collaboration.

Charles F. Dunkl
Yuan Xu

1

Background

The theory of orthogonal polynomials of several variables, especially those of classical type, uses a significant amount of analysis in one variable. In this chapter we give concise descriptions of the needed tools.

1.1 The Gamma and Beta Functions

It is our opinion that the most interesting and amenable objects of consideration have expressions which are rational functions of the underlying parameters. This leads us immediately to a discussion of the gamma function and its relatives.

Definition 1.1.1 The gamma function is defined for $\operatorname{Re} x > 0$ by the integral

$$\Gamma(x) = \int_0^\infty t^{x-1} \mathrm{e}^{-t}\, \mathrm{d}t.$$

It is directly related to the beta function:

Definition 1.1.2 The beta function is defined for $\operatorname{Re} x > 0$ and $\operatorname{Re} y > 0$ by

$$B(x,y) = \int_0^1 t^{x-1}(1-t)^{y-1} \mathrm{d}t.$$

By making the change of variables $s = uv$ and $t = (1-u)v$ in the integral $\Gamma(x)\Gamma(y) = \int_0^\infty \int_0^\infty s^{x-1} t^{y-1} \mathrm{e}^{-(s+t)} \mathrm{d}s\, \mathrm{d}t$, one obtains

$$\Gamma(x)\Gamma(y) = \Gamma(x+y)B(x,y).$$

This leads to several useful definite integrals, valid for $\operatorname{Re} x > 0$ and $\operatorname{Re} y > 0$:

1. $\displaystyle\int_0^{\pi/2} \sin^{x-1}\theta \cos^{y-1}\theta\, \mathrm{d}\theta = \tfrac{1}{2} B\left(\tfrac{x}{2}, \tfrac{y}{2}\right) = \frac{\tfrac{1}{2}\Gamma\left(\tfrac{x}{2}\right)\Gamma\left(\tfrac{y}{2}\right)}{\Gamma\left(\tfrac{x+y}{2}\right)};$

2. $\Gamma\left(\frac{1}{2}\right) = \sqrt{\pi}$ (set $x = y = 1$ in the previous integral);
3. $\displaystyle\int_0^\infty t^{x-1} \exp(-at^2)\,\mathrm{d}t = \tfrac{1}{2} a^{-x/2} \Gamma\left(\tfrac{x}{2}\right), \quad \text{for } a > 0;$
4. $\displaystyle\int_0^1 t^{x-1}(1-t^2)^{y-1}\mathrm{d}t = \tfrac{1}{2} B\left(\tfrac{x}{2}, y\right) = \frac{\frac{1}{2}\Gamma\left(\frac{x}{2}\right)\Gamma(y)}{\Gamma\left(\frac{x}{2}+y\right)};$
5. $\displaystyle\Gamma(x)\Gamma(1-x) = B(x, 1-x) = \frac{\pi}{\sin \pi x}.$

The last equation can be proven by restricting x to $0 < x < 1$, making the substitution $s = t/(1-t)$ in the beta integral $\int_0^1 [t/(1-t)]^{x-1}(1-t)^{-1}\mathrm{d}t$ and computing the resulting integral by residues. Of course, one of the fundamental properties of the gamma function is the recurrence formula (obtained from integration by parts)

$$\Gamma(x+1) = x\Gamma(x),$$

which leads to the fact that Γ can be analytically continued to a meromorphic function on the complex plane; also, $1/\Gamma$ is entire, with (simple) zeros exactly at $\{0, -1, -2, \ldots\}$. Note that Γ interpolates the factorial; indeed, $\Gamma(n+1) = n!$ for $n = 0, 1, 2, \ldots$

Definition 1.1.3 The Pochhammer symbol, also called the shifted factorial, is defined for all x by

$$(x)_0 = 1, \quad (x)_n = \prod_{i=1}^{n}(x+i-1) \qquad \text{for } n = 1, 2, 3, \ldots$$

Alternatively, one can recursively define $(x)_n$ by

$$(x)_0 = 1 \quad \text{and} \quad (x)_{n+1} = (x)_n(x+n) \qquad \text{for } n = 1, 2, 3, \ldots$$

Here are some important consequences of Definition 1.1.3:

1. $(x)_{m+n} = (x)_m(x+m)_n$ for $m, n \in \mathbb{N}_0$;
2. $(x)_n = (-1)^n(1-n-x)_n$ (writing the product in reverse order);
3. $(x)_{n-i} = (x)_n(-1)^i/(1-n-x)_i$.

The Pochhammer symbol incorporates binomial-coefficient and factorial notation:

1. $(1)_n = n!, \quad 2^n\left(\frac{1}{2}\right)_n = 1 \times 3 \times 5 \times \cdots \times (2n-1)$;
2. $(n+m)! = n!(n+1)_m$;
3. $\displaystyle\binom{n}{i} = (-1)^i \frac{(-n)_i}{i!}$, where $\binom{n}{i}$ is the binomial coefficient;
4. $(x)_{2n} = 2^{2n}\left(\frac{x}{2}\right)_n\left(\frac{x+1}{2}\right)_n$, the duplication formula.

The last property includes the formula for the gamma function:

$$\Gamma(2x) = \frac{2^{2x-1}}{\sqrt{\pi}}\Gamma(x+\tfrac{1}{2})\Gamma(x).$$

For appropriate values of x and n, the formula $\Gamma(x+n)/\Gamma(x) = (x)_n$ holds, and this can be used to extend the definition of the Pochhammer symbol to values of $n \notin \mathbb{N}_0$.

1.2 Hypergeometric Series

The two most common types of hypergeometric series are (they are convergent for $|x| < 1$)

$$ {}_1F_0(a;x) = \sum_{n=0}^{\infty} \frac{(a)_n}{n!} x^n, $$

$$ {}_2F_1\left(\begin{matrix} a,b \\ c \end{matrix};x\right) = \sum_{n=0}^{\infty} \frac{(a)_n (b)_n}{(c)_n n!} x^n, $$

where a and b are the "numerator" parameters and c is the "denominator" parameter. Later we will also use ${}_3F_2$ series (with a corresponding definition). The ${}_2F_1$ series is the unique solution analytic at $x = 0$ and satisfying $f(0) = 1$ of

$$x(1-x)\frac{\mathrm{d}^2}{\mathrm{d}x^2}f(x) + [c-(a+b+1)x]\frac{\mathrm{d}}{\mathrm{d}x}f(x) - abf(x) = 0.$$

Generally, classical orthogonal polynomials can be expressed as hypergeometric polynomials, which are terminating hypergeometric series for which a numerator parameter has a value in $-\mathbb{N}_0$. The two series can be represented in closed form. Obviously ${}_1F_0(a;x) = (1-x)^{-a}$; this is the branch analytic in $\{x \in \mathbb{C} : |x| < 1\}$, which has the value 1 at $x = 0$. The Gauss integral formula for ${}_2F_1$ is as follows.

Proposition 1.2.1 *For* $\mathrm{Re}(c-b) > 0$, $\mathrm{Re}\, b > 0$ *and* $|x| < 1$,

$$ {}_2F_1\left(\begin{matrix} a,b \\ c \end{matrix};x\right) = \frac{\Gamma(c)}{\Gamma(b)\Gamma(c-b)} \int_0^1 t^{b-1}(1-t)^{c-b-1}(1-xt)^{-a}\,\mathrm{d}t. $$

Proof Use the ${}_1F_0$ series in the integral and integrate term by term to obtain a multiple of

$$\sum_{n=0}^{\infty} \frac{(a)_n}{n!} x^n \int_0^1 t^{b+n-1}(1-t)^{c-b-1}\mathrm{d}t = \sum_{n=0}^{\infty} \frac{(a)_n \Gamma(b+n)\Gamma(c-b)}{n!\Gamma(c+n)} x^n,$$

from which the stated formula follows. □

Corollary 1.2.2 *For* $\operatorname{Re} c > \operatorname{Re}(a+b)$ *and* $\operatorname{Re} b > 0$ *the Gauss summation formula is*

$$ {}_2F_1\left(\begin{matrix} a,b \\ c \end{matrix};1\right) = \frac{\Gamma(c)\Gamma(c-b-a)}{\Gamma(c-a)\Gamma(c-b)}. $$

The terminating case of this formula, known as the Chu–Vandermonde sum, is valid for a more general range of parameters.

Proposition 1.2.3 *For* $n \in \mathbb{N}_0$, *any* a,b, *and* $c \neq 0,1,\ldots,n-1$ *the following hold:*

$$ \sum_{i=0}^{n} \frac{(a)_{n-i}(b)_i}{(n-i)!i!} = \frac{(a+b)_n}{n!} \quad \text{and} \quad {}_2F_1\left(\begin{matrix} -n,b \\ c \end{matrix};1\right) = \frac{(c-b)_n}{(c)_n}. $$

Proof The first formula is deduced from the coefficient of x^n in the expression $(1-x)^{-a}(1-x)^{-b} = (1-x)^{-(a+b)}$. The left-hand side can be written as

$$ \frac{(a)_n}{n!} \sum_{i=0}^{n} \frac{(-n)_i(b)_i}{(1-n-a)_i i!}. $$

Now let $a = 1-n-c$; simple computations involving reversals such as $(1-n-c)_n = (-1)^n(c)_n$ finish the proof. □

The following transformation often occurs:

Proposition 1.2.4 *For* $|x| < 1$,

$$ {}_2F_1\left(\begin{matrix} a,b \\ c \end{matrix};x\right) = (1-x)^{-a} {}_2F_1\left(\begin{matrix} a,c-b \\ c \end{matrix};\frac{x}{x-1}\right). $$

Proof Temporarily assume that $\operatorname{Re} c > \operatorname{Re} b > 0$; then from the Gauss integral we have

$$ \begin{aligned} {}_2F_1\left(\begin{matrix} a,b \\ c \end{matrix};x\right) &= \frac{\Gamma(c)}{\Gamma(b)\Gamma(c-b)} \int_0^1 t^{b-1}(1-t)^{c-b-1}(1-xt)^{-a}\,\mathrm{d}t \\ &= \frac{\Gamma(c)}{\Gamma(b)\Gamma(c-b)} \int_0^1 s^{c-b-1}(1-s)^{b-1}(1-x)^{-a}\left(1-\frac{xs}{x-1}\right)^{-a} \mathrm{d}s, \end{aligned} $$

where one makes the change of variable $t = 1-s$. The formula follows from another application of the Gauss integral. Analytic continuation in the parameters extends the validity to all values of a,b,c excluding $c \in \mathbb{N}_0$. For this purpose we tacitly consider the modified series

$$ \sum_{n=0}^{\infty} \frac{(a)_n(b)_n}{\Gamma(c+n)n!} x^n, $$

which is an entire function in a,b,c. □

Corollary 1.2.5 *For $|x| < 1$,*

$$ {}_2F_1\left(\begin{matrix} a,b \\ c \end{matrix};x\right) = (1-x)^{c-a-b}\,{}_2F_1\left(\begin{matrix} c-a,c-b \\ c \end{matrix};x\right). $$

Proof Using Proposition 1.2.4 twice,

$$ \begin{aligned} {}_2F_1\left(\begin{matrix} a,b \\ c \end{matrix};x\right) &= (1-x)^{-a}\,{}_2F_1\left(\begin{matrix} a,c-b \\ c \end{matrix};\frac{x}{x-1}\right) \\ &= (1-x)^{-a}\left(1-\frac{x}{x-1}\right)^{b-c}{}_2F_1\left(\begin{matrix} c-a,c-b \\ c \end{matrix};x\right) \end{aligned} $$

and $1-x/(x-1) = (1-x)^{-1}$. □

Equating coefficients of x^n on the two sides of the formula in Corollary 1.2.5 proves the *Saalschütz summation formula* for a balanced terminating ${}_3F_2$ series.

Proposition 1.2.6 *For $n = 0,1,2,\ldots$ and $c,d \neq 0,-1,-2,\ldots,-n$ with $-n+a+b+1 = c+d$ (the "balanced" condition), we have*

$$ {}_3F_2\left(\begin{matrix} -n,a,b \\ c,d \end{matrix};1\right) = \frac{(c-a)_n(c-b)_n}{(c)_n(c-a-b)_n} = \frac{(c-a)_n(d-a)_n}{(c)_n(d)_n}. $$

Proof Considering the coefficient of x^n in the equation

$$ (1-x)^{a+b-c}\,{}_2F_1\left(\begin{matrix} a,b \\ c \end{matrix};x\right) = {}_2F_1\left(\begin{matrix} c-a,c-b \\ c \end{matrix};x\right) $$

yields

$$ \sum_{j=0}^{n} \frac{(c-a-b)_{n-j}(a)_j(b)_j}{(n-j)!\,j!\,(c)_j} = \frac{(c-a)_n(c-b)_n}{n!\,(c)_n}, $$

but

$$ \frac{(c-a-b)_{n-j}}{(n-j)!} = \frac{(c-a-b)_n(-n)_j}{n!\,(1-n-c+a+b)_j}; $$

this proves the first formula with $d = -n+a+b-c+1$. Further, $(c-b)_n = (-1)^n(1-n-c+b)_n = (-1)^n(d-a)_n$ and $(-1)^n(c-a-b)_n = (1-n-c+a+b)_n = (d)_n$, which proves the second formula. □

1.2.1 Lauricella series

There are many ways to define multivariable analogues of the hypergeometric series. One straightforward and useful approach consists of the Lauricella generalizations of the ${}_2F_1$ series; see Exton [1976]. Fix $d = 1,2,3,\ldots$ vector parameters $\mathbf{a},\mathbf{b},\mathbf{c} \in \mathbb{R}^d$, scalar parameters α,γ and the variable $x \in \mathbb{R}^d$. For concise formulation we use the following: let $\mathbf{m} \in \mathbb{N}_0^d$, $\mathbf{m}! = \prod_{j=1}^d (m_j)!$, $|\mathbf{m}| = \sum_{j=1}^d m_j$, $(\mathbf{a})_{\mathbf{m}} = \prod_{j=1}^d (a_j)_{m_j}$ and $x^{\mathbf{m}} = \prod_{j=1}^d x_j^{m_j}$.

The four types of Lauricella functions are (with summations over $\mathbf{m} \in \mathbb{N}_0^d$):

1. $F_A(\alpha, \mathbf{b}; \mathbf{c}; x) = \sum_{\mathbf{m}} \frac{(\alpha)_{|\mathbf{m}|} (\mathbf{b})_{\mathbf{m}}}{(\mathbf{c})_{\mathbf{m}} \mathbf{m}!} x^{\mathbf{m}}$, convergent for $\sum_{j=1}^d |x_j| < 1$;
2. $F_B(\mathbf{a}, \mathbf{b}; \gamma; x) = \sum_{\mathbf{m}} \frac{(\mathbf{a})_{\mathbf{m}} (\mathbf{b})_{\mathbf{m}}}{(\gamma)_{|\mathbf{m}|} \mathbf{m}!} x^{\mathbf{m}}$, convergent for $\max_j |x_j| < 1$;
3. $F_C(\alpha, \beta; \mathbf{c}; x) = \sum_{\mathbf{m}} \frac{(\alpha)_{|\mathbf{m}|} (\beta)_{|\mathbf{m}|}}{(\mathbf{c})_{\mathbf{m}} \mathbf{m}!} x^{\mathbf{m}}$, convergent for $\sum_{j=1}^d |x_j|^{1/2} < 1$;
4. $F_D(\alpha, \mathbf{b}; \gamma; x) = \sum_{\mathbf{m}} \frac{(\alpha)_{|\mathbf{m}|} (\mathbf{b})_{\mathbf{m}}}{(\gamma)_{|\mathbf{m}|} \mathbf{m}!} x^{\mathbf{m}}$, convergent for $\max_j |x_j| < 1$.

There are integral representations of Euler type (the following are subject to obvious convergence conditions; further, any argument of a gamma function must have a positive real part):

1. $$F_A(\alpha, \mathbf{b}; \mathbf{c}; x) = \prod_{j=1}^d \frac{\Gamma(c_j)}{\Gamma(b_j)\Gamma(c_j - b_j)} \times \int_{[0,1]^d} \prod_{j=1}^d \left(u_j^{b_j - 1} (1 - u_j)^{c_j - b_j - 1} \right) \left(1 - \sum_{j=1}^d u_j x_j \right)^{-\alpha} du;$$
2. $$F_B(\mathbf{a}, \mathbf{b}; \gamma; x) = \prod_{j=1}^d \Gamma(a_j)^{-1} \frac{\Gamma(\gamma)}{\Gamma(\delta)} \times \int_{T^d} \prod_{j=1}^d \left(u_j^{a_j - 1} (1 - u_j x_j)^{-b_j} \right) \left(1 - \sum_{j=1}^d u_j \right)^{\delta - 1} du,$$
   where $\delta = \gamma - \sum_{j=1}^d a_j$ and T^d is the simplex $\{u \in \mathbb{R}^d : u_j \geq 0$ for all j, and $\sum_{j=1}^d u_j \leq 1\}$;
3. $$F_D(\alpha, \mathbf{b}; \gamma; x) = \frac{\Gamma(\gamma)}{\Gamma(\alpha)\Gamma(\gamma - \alpha)} \times \int_0^1 u^{\alpha - 1} (1 - u)^{\gamma - \alpha - 1} \prod_{j=1}^d (1 - u x_j)^{-b_j} du,$$
   a single integral.

1.3 Orthogonal Polynomials of One Variable

1.3.1 General properties

We start with a determinant approach to the Gram–Schmidt process, a method for producing orthogonal bases of functions given a linearly (totally) ordered basis. Suppose that X is a region in $\mathbb{R}^d$ (for $d \geq 1$), μ is a probability measure on X

and $\{f_i(x) : i = 1,2,3\ldots\}$ is a set of functions linearly independent in $L^2(X,\mu)$. Denote the inner product $\int_X fg\,\mathrm{d}\mu$ as $\langle f,g\rangle$ and the elements of the Gram matrix $\langle f_i, f_j\rangle$ as g_{ij}, $i,j \in \mathbb{N}$.

Definition 1.3.1 For $n \in \mathbb{N}$ let $d_n = \det(g_{ij})_{i,j=1}^n$ and

$$D_n(x) = \det\begin{bmatrix} g_{11} & g_{12} & \cdots & g_{1n} \\ \vdots & \vdots & \cdots & \vdots \\ g_{n-1,1} & g_{n-1,2} & \cdots & g_{n-1,n} \\ f_1(x) & f_2(x) & \cdots & f_n(x) \end{bmatrix}.$$

Proposition 1.3.2 *The functions $\{D_n(x) : n \geq 1\}$ are orthogonal in $L^2(X,\mu)$, $\operatorname{span}\{D_j(x) : 1 \leq j \leq n\} = \operatorname{span}\{f_j(x) : 1 \leq j \leq n\}$ and $\langle D_n, D_n\rangle = d_{n-1}d_n$.*

Proof By linear independence, $d_n > 0$ for all n; thus $D_n(x) = d_{n-1}f_n(x) + \sum_{j<n} c_j f_j(x)$ for some coefficients c_j (where $d_0 = 1$) and $\operatorname{span}\{D_j : j \leq n\} = \operatorname{span}\{f_j : j \leq n\}$. The inner product $\langle f_j, D_n\rangle$ is the determinant of the matrix in the definition of D_n with the last row replaced by $(g_{j1}, g_{j2}, \ldots, g_{jn})$ and hence is zero for $j < n$. Thus $\langle D_j, D_n\rangle = 0$ for $j < n$ and $\langle D_n, D_n\rangle = d_{n-1}\langle f_n, D_n\rangle = d_{n-1}d_n$. □

There are integral formulae for d_n and $D_n(x)$ which are interesting foreshadowings of multivariable weight functions P_n involving the discriminant, as follows.

Definition 1.3.3 For $n \in \mathbb{N}$ and $x_1, x_2, \ldots, x_n \in X$ let

$$P_n(x_1, x_2, \ldots, x_n) = \det(f_j(x_i))_{i,j=1}^n.$$

Proposition 1.3.4 *For $n \in \mathbb{N}$ and $x_1, x_2, \ldots, x_n \in X$,*

$$\int_{X^n} P_n(x_1, x_2, \ldots, x_n)^2\,\mathrm{d}\mu(x_1)\cdots\mathrm{d}\mu(x_n) = n!d_n,$$

and

$$\int_{X^n} P_n(x_1, x_2, \ldots, x_n)P_{n+1}(x_1, x_2, \ldots, x_n, x)\,\mathrm{d}\mu(x_1)\cdots\mathrm{d}\mu(x_n) = n!D_{n+1}(x).$$

Proof In the first integral, make the expansion

$$P_n(x_1, x_2, \ldots, x_n)^2 = \sum_\sigma \sum_\tau \varepsilon_\sigma \varepsilon_\tau \prod_{i=1}^n f_{\sigma i}(x_i) f_{\tau i}(x_i),$$

where the summations are over the symmetric group S_n (on n objects); $\varepsilon_\sigma, \sigma i$ denote the sign of the permutation σ and the action of σ on i, respectively. Integrating over X^n gives the sum $\sum_\sigma \sum_\tau \varepsilon_\sigma \varepsilon_\tau \prod_{i=1}^n g_{\sigma i, \tau i} = n! \sum_\tau \varepsilon_\tau \prod_{i=1}^n g_{\tau i, i} = n!d_n$.

The summation over σ is done by first fixing σ and then replacing i by $\sigma^{-1}i$ and τ by $\tau\sigma^{-1}$. Similarly,

$$P_n(x_1,x_2,\ldots,x_n)P_{n+1}(x_1,x_2,\ldots,x_n,x)$$
$$=\sum_{\sigma}\sum_{\tau}\varepsilon_\sigma\varepsilon_\tau\prod_{i=1}^{n} f_{\sigma i}(x_i)f_{\tau i}(x_i)f_{\tau(n+1)}(x),$$

and the τ-sum is over S_{n+1}. As before, the integral has the value

$$\sum_{\sigma}\sum_{\tau}\varepsilon_\sigma\varepsilon_\tau\prod_{i=1}^{n} g_{\sigma i,\tau i}f_{\tau(n+1)}(x),$$

which reduces to the expression $n!\sum_\tau \varepsilon_\tau \prod_{i=1}^{n} g_{\tau i,i} f_{\tau(n+1)}(x) = n!D_{n+1}(x)$. □

We now specialize to orthogonal polynomials; let μ be a probability measure supported on a (possibly infinite) interval $[a,b]$ such that $\int_a^b |x|^n \,\mathrm{d}\mu < \infty$ for all n. We may as well assume that μ is not a finite discrete measure, so that $\{1,x,x^2,x^3,\ldots\}$ is linearly independent in $L^2(\mu)$; it is not difficult to modify the results to the situation where $L^2(\mu)$ is of finite dimension. We apply Proposition 1.3.2 to the basis $f_j(x) = x^{j-1}$; the Gram matrix has the form of a Hankel matrix $g_{ij} = c_{i+j-2}$, where the nth moment of μ is

$$c_n = \int_a^b x^n \,\mathrm{d}\mu(x)$$

and the orthonormal polynomials $\{p_n(x) : n \geq 0\}$ are defined by

$$p_n(x) = (d_{n+1}d_n)^{-1/2} D_{n+1}(x);$$

they satisfy $\int_a^b p_m(x)p_n(x)\,\mathrm{d}\mu(x) = \delta_{mn}$, and the leading coefficient of p_n is $(d_n/d_{n+1})^{1/2} > 0$. Of course this implies that $\int_a^b p_n(x)q(x)\,\mathrm{d}\mu(x) = 0$ for any polynomial $q(x)$ of degree $\leq n-1$. The determinant P_n in Definition 1.3.3 is exactly the Vandermonde determinant $\det(x_i^{j-1})_{i,j=1}^{n} = \prod_{1\leq i<j\leq n}(x_j - x_i)$.

Proposition 1.3.5 *For $n \geq 0$,*

$$\int_{[a,b]^n} \prod_{1\leq i<j\leq n}(x_j - x_i)^2 \,\mathrm{d}\mu(x_1)\cdots\mathrm{d}\mu(x_n) = n!d_n,$$

$$\int_{[a,b]^n} \prod_{i=1}^{n}(x-x_i) \prod_{1\leq i<j\leq n}(x_j - x_i)^2 \,\mathrm{d}\mu(x_1)\cdots\mathrm{d}\mu(x_n) = n!(d_n d_{n+1})^{1/2} p_n(x).$$

It is a basic fact that $p_n(x)$ has n distinct (and simple) zeros in $[a,b]$.

Proposition 1.3.6 *For $n \geq 1$, the polynomial $p_n(x)$ has n distinct zeros in the open interval (a,b).*

Proof Suppose that $p_n(x)$ changes sign at $t_1,\ldots,t_m$ in (a,b). Then it follows that $\varepsilon p_n(x)\prod_{i=1}^m(x-t_i) \geq 0$ on $[a,b]$ for $\varepsilon = 1$ or -1. If $m < n$ then $\int_a^b p_n(x)$ $\prod_{i=1}^m(x-t_i)\,\mathrm{d}\mu(x) = 0$, which implies that the integrand is zero on the support of μ, a contradiction. □

In many applications one uses orthogonal, rather than orthonormal, polynomials (by reason of their neater notation, generating function and normalized values at an end point, for example). This means that we have a family of nonzero polynomials $\{P_n(x) : n \geq 0\}$ with $P_n(x)$ of exact degree n and for which $\int_a^b P_n(x)x^j\,\mathrm{d}\mu(x) = 0$ for $j < n$.

We say that the squared norm $\int_a^b P_n(x)^2\,\mathrm{d}\mu(x) = h_n$ is a *structural constant*. Further, $p_n(x) = \pm h_n^{-1/2}P_n(x)$; the sign depends on the leading coefficient of $P_n(x)$.

1.3.2 Three-term recurrence

Besides the Gram matrix of moments there is another important matrix associated with a family of orthogonal polynomials, the Jacobi matrix. The principal minors of this tridiagonal matrix provide an interpretation of the three-term recurrence relations. For $n \geq 0$ the polynomial $xP_n(x)$ is of degree $n+1$ and can be expressed in terms of $\{P_j : j \leq n+1\}$, but more is true.

Proposition 1.3.7 *There exist sequences $\{A_n\}_{n\geq 0}, \{B_n\}_{n\geq 0}, \{C_n\}_{n\geq 1}$ such that*

$$P_{n+1}(x) = (A_n x + B_n)P_n(x) - C_n P_{n-1}(x),$$

where

$$A_n = \frac{k_{n+1}}{k_n}, \qquad C_n = \frac{k_{n+1}k_{n-1}h_n}{k_n^2 h_{n-1}}, \qquad B_n = -\frac{k_{n+1}}{k_n h_n}\int_a^b xP_n(x)^2\,\mathrm{d}\mu(x),$$

and k_n is the leading coefficient of $P_n(x)$.

Proof Expanding $xP_n(x)$ in terms of polynomials P_j gives $\sum_{j=0}^{n+1} a_j P_j(x)$ with $a_j = h_j^{-1}\int_a^b xP_n(x)P_j(x)\,\mathrm{d}\mu(x)$. By the orthogonality property, $a_j = 0$ unless $|n-j| \leq 1$. The value of $a_{n+1} = A_n^{-1}$ is obtained by matching the coefficients of x^{n+1}. Shifting the label gives the value of C_n. □

Corollary 1.3.8 *For the special case of monic orthogonal polynomials, the three-term recurrence is*

$$P_{n+1}(x) = (x + B_n)P_n(x) - C_n P_{n-1}(x),$$

where

$$C_n = \frac{d_{n+1}d_{n-1}}{d_n^2} \quad \text{and} \quad B_n = -\frac{d_n}{d_{n+1}}\int_a^b xP_n(x)^2\,\mathrm{d}\mu(x).$$

Proof In the notation from the end of the last subsection, the structure constant for the monic case is $h_n = d_{n+1}/d_n$. □

It is convenient to restate the recurrence, and some other, relations for orthogonal polynomials with arbitrary leading coefficients in terms of the moment determinants d_n (see Definition 1.3.1).

Proposition 1.3.9 *Suppose that the leading coefficient of $P_n(x)$ is k_n, and let $b_n = \int_a^b xp_n(x)^2\,\mathrm{d}\mu(x)$; then*

$$h_n = k_n^2\frac{d_{n+1}}{d_n},$$

$$xP_n(x) = \frac{k_n}{k_{n+1}}P_{n+1}(x) + b_nP_n(x) + \frac{k_{n-1}h_n}{k_nh_{n-1}}P_{n-1}(x).$$

Corollary 1.3.10 *For the case of orthonormal polynomial p_n,*

$$xp_n(x) = a_np_{n+1}(x) + b_np_n(x) + a_{n-1}p_{n-1}(x),$$

where $a_n = k_n/k_{n+1} = (d_nd_{n+2}/d_{n+1}^2)^{1/2}$.

With these formulae one can easily find the reproducing kernel for polynomials of degree $\leq n$, the Christoffel–Darboux formula:

Proposition 1.3.11 *For $n \geq 1$, if k_n is the leading coefficient of p_n then we have*

$$\sum_{j=0}^{n} p_j(x)p_j(y) = \frac{k_n}{k_{n+1}}\frac{p_{n+1}(x)p_n(y) - p_n(x)p_{n+1}(y)}{x-y},$$

$$\sum_{j=0}^{n} p_j(x)^2 = \frac{k_n}{k_{n+1}}\left[p_{n+1}'(x)p_n(x) - p_n'(x)p_{n+1}(x)\right].$$

Proof By the recurrence in the previous proposition, for $j \geq 0$,

$$(x-y)p_j(x)p_j(y) = \frac{k_j}{k_{j+1}}\left[p_{j+1}(x)p_j(y) - p_j(x)p_{j+1}(y)\right] + \frac{k_{j-1}}{k_j}\left[p_{j-1}(x)p_j(y) - p_j(x)p_{j-1}(y)\right].$$

The terms involving b_j arising from setting $n = j$ in Corollary 1.3.10 cancel out. Now sum these equations over $0 \leq j \leq n$; the terms telescope (and note that the case $j = 0$ is special, with $p_{-1} = 0$). This proves the first formula in the proposition; the second follows from L'Hospital's rule. □

The Jacobi tridiagonal matrix associated with the three-term recurrence is the semi-infinite matrix

$$J = \begin{bmatrix} b_0 & a_0 & & & \bigcirc \\ a_0 & b_1 & a_1 & & \\ & a_1 & b_2 & a_2 & \\ & & a_2 & b_3 & \ddots \\ \bigcirc & & & \ddots & \ddots \end{bmatrix}$$

where $a_{n-1} = \sqrt{C_n} = \sqrt{d_{n+1}d_{n-1}}/d_n$ (a_{n-1} is the same as the coefficient a_n in Corollary 1.3.10 orthonormal polynomials). Let J_n denote the left upper submatrix of J of size $n \times n$, and let I_n be the $n \times n$ identity matrix. Then $\det(xI_{n+1} - J_{n+1}) = (x - b_n)\det(xI_n - J_n) - a_{n-1}^2 \det(xI_{n-1} - J_{n-1})$, a simple matrix identity; hence $P_n(x) = \det(xI_n - J_n)$. Note that the fundamental result regarding the eigenvalues of symmetric tridiagonal matrices with all superdiagonal entries nonzero shows again that the zeros of P_n are simple and real. The eigenvectors of J_n have elegant expressions in terms of p_j and provide an explanation for the Gaussian quadrature.

Let $\lambda_1, \lambda_2, \ldots, \lambda_n$ be the zeros of $p_n(x)$.

Theorem 1.3.12 *For $1 \le j \le n$ let*

$$\mathbf{v}^{(j)} = (p_0(\lambda_j), p_1(\lambda_j), \ldots, p_{n-1}(\lambda_j))^{\mathrm{T}};$$

then $J_n \mathbf{v}^{(j)} = \lambda_j \mathbf{v}^{(j)}$. Further,

$$\sum_{j=1}^{n} \gamma_j p_r(\lambda_j) p_s(\lambda_j) = \delta_{rs}$$

where $0 \le r, s \le n-1$ and $\gamma_j = \left(\sum_{i=0}^{n-1} p_i(\lambda_j)^2\right)^{-1}$.

Proof We want to show that the ith entry of $(J_n - \lambda_j I_n)\mathbf{v}^{(j)}$ is zero. The typical equation for this entry is

$$\begin{aligned} & a_{i-2}p_{i-2}(\lambda_j) + (b_{i-1} - \lambda_j)p_{i-1}(\lambda_j) + a_{i-1}p_i(\lambda_j) \\ &= \left(a_{i-2} - \frac{k_{i-2}}{k_{i-1}}\right) p_{i-2}(\lambda_j) + \left(a_{i-1} - \frac{k_{i-1}}{k_i}\right) p_i(\lambda_j) \\ &= 0, \end{aligned}$$

because $k_i = (d_i/d_{i+1})^{1/2}$ and $a_{i-1} = (d_{i+1}d_{i-1}/d_i^2)^{1/2} = k_{i-1}/k_i$ for $i \ge 1$, where k_i is the leading coefficient of p_i. This uses the recurrence in Proposition 1.3.9; when $i = 1$ the term $p_{-1}(\lambda_j) = 0$ and for $i = n$ the fact that $p_n(\lambda_j) = 0$ is crucial. Since J_n is symmetric, the eigenvectors are pairwise orthogonal, that is, $\sum_{i=1}^{n} p_{i-1}(\lambda_j)p_{i-1}(\lambda_r) = \delta_{jr}/\gamma_j$ for $1 \le j, r \le n$. Hence the matrix

$\left(\gamma_j^{1/2} p_{i-1}(\lambda_j)\right)_{i,j=1}^n$ is orthogonal. The orthonormality of the rows of the matrix is exactly the second part of the theorem. □

This theorem has some interesting implications. First, the set $\{p_i(x) : 0 \le i \le n-1\}$ comprises the orthonormal polynomials for the discrete measure $\sum_{j=1}^n \gamma_j \delta(\lambda_j)$, where $\delta(x)$ denotes a unit mass at x. This measure has total mass 1 because $p_0 = 1$. Second, the theorem implies that $\sum_{j=1}^n \gamma_j q(\lambda_j) = \int_a^b q\,\mathrm{d}\mu$ for any polynomial $q(x)$ of degree $\le n-1$, because this relation is valid for every basis element p_i for $0 \le i \le n-1$. In fact, this is a description of Gaussian quadrature. The Christoffel–Darboux formula, Proposition 1.3.11, leads to another expression (perhaps more concise) for the weights γ_j.

Theorem 1.3.13 *For $n \ge 1$ let $\lambda_1, \lambda_2, \ldots, \lambda_n$ be the zeros of $p_n(x)$ and let $\gamma_j = (\sum_{i=0}^{n-1} p_i(\lambda_j)^2)^{-1}$. Then*

$$\gamma_j = \frac{k_n}{k_{n-1} p_n'(\lambda_j) p_{n-1}(\lambda_j)},$$

and for any polynomial $q(x)$ of degree $\le 2n-1$ the following holds:

$$\int_a^b q\,\mathrm{d}\mu = \sum_{j=1}^n \gamma_j q(\lambda_j).$$

Proof In the Christoffel–Darboux formula for $y = x$, set $x = \lambda_j$ (thus $p_n(\lambda_j) = 0$). This proves the first part of the theorem (note that $0 < \gamma_j \le 1$ because $p_0 = 1$). Suppose that $q(x)$ is a polynomial of degree $2n-1$; then, by synthetic division, $q(x) = p_n(x)g(x) + r(x)$ where $g(x), r(x)$ are polynomials of degree $\le n-1$. Thus $r(\lambda_j) = q(\lambda_j)$ for each j and, by orthogonality,

$$\int_a^b q\,\mathrm{d}\mu = \int_a^b p_n g\,\mathrm{d}\mu + \int_a^b r\,\mathrm{d}\mu = \int_a^b r\,\mathrm{d}\mu = \sum_{j=1}^n \gamma_j r(\lambda_j).$$

This completes the proof. □

One notes that the leading coefficients k_i (for orthonormal polynomials, see Corollary 1.3.10) can be recovered from the Jacobi matrix J; indeed $k_i = \prod_{j=1}^i c_j^{-1}$ and $k_0 = 1$, that is, $p_0 = 1$. This provides the normalization of the associated discrete measures $\sum_{j=1}^n \gamma_j \delta(\lambda_j)$. The moment determinants are calculated from $d_i = \prod_{j=1}^{i-1} k_j^2$.

If the measure μ is finite with point masses at $\{t_1, t_2, \ldots, t_n\}$ then the Jacobi matrix has the entry $c_n = 0$ and J_n produces the orthogonal polynomials $\{p_0, p_1, \ldots, p_{n-1}\}$ for μ, and the eigenvalues of J_n are $t_1, t_2, \ldots, t_n$.

1.4 Classical Orthogonal Polynomials

This section is concerned with the Hermite, Laguerre, Chebyshev, Legendre, Gegenbauer and Jacobi polynomials. For each family the orthogonality measure has the form $d\mu(x) = cw(x)\,dx$, with *weight function* $w(x) > 0$ on an interval and *normalization constant* $c = [\int_{\mathbb{R}} w(x)\,dx]^{-1}$. Typically the polynomials are defined by a Rodrigues relation which easily displays the required orthogonality property. Then some computations are needed to determine the other structures: h_n (which is always stated with respect to a normalized μ), the three-term recurrences, expressions in terms of hypergeometric series, differential equations and so forth. When the weight function is even, so that $w(-x) = w(x)$, the coefficients b_n in Proposition 1.3.9 are zero for all n and the orthogonal polynomials satisfy $P_n(-x) = (-1)^n P_n(x)$.

1.4.1 Hermite polynomials

For the Hermite polynomials $H_n(x)$, the weight function $w(x)$ is e^{-x^2} on $x \in \mathbb{R}$ and the normalization constant c is $\Gamma(\frac{1}{2})^{-1} = \pi^{-1/2}$.

Definition 1.4.1 For $n \geq 0$ let $H_n(x) = (-1)^n e^{x^2} \left(\frac{d}{dx}\right)^n e^{-x^2}$.

Proposition 1.4.2 *For* $0 \leq m < n$,

$$\int_{\mathbb{R}} x^m H_n(x) e^{-x^2}\,dx = 0$$

and

$$h_n = \pi^{-1/2} \int_{\mathbb{R}} [H_n(x)]^2 e^{-x^2}\,dx = 2^n n!$$

Proof For any polynomial $q(x)$ we have

$$\int_{\mathbb{R}} q(x) H_n(x) e^{-x^2}\,dx = \int_{\mathbb{R}} \left(\frac{d}{dx}\right)^n q(x) e^{-x^2}\,dx$$

using n-times-repeated integration by parts. From the definition it is clear that the leading coefficient $k_n = 2^n$, and thus $h_n = \pi^{-1/2} \int_{\mathbb{R}} 2^n n! e^{-x^2}\,dx = 2^n n!$ □

Proposition 1.4.3 *For* $x, r \in \mathbb{R}$, *the generating function is*

$$\exp(2rx - r^2) = \sum_{n=0}^{\infty} H_n(x) \frac{r^n}{n!}$$

and

$$H_n(x) = \sum_{j \leq n/2} \frac{n!}{(n-2j)!\,j!} (-1)^j (2x)^{n-2j},$$

$$\frac{d}{dx} H_n(x) = 2n H_{n-1}(x).$$

Proof Expand $f(r) = \exp[-(x-r)^2]$ in a Taylor series at $r = 0$ for fixed x. Write the generating function as

$$\sum_{m=0}^{\infty} \frac{1}{m!} \sum_{j=0}^{m} \binom{m}{j} (2x)^{m-j} r^{m+j} (-1)^j,$$

set $m = n + j$ and collect the coefficient of $r^n/n!$ Now differentiate the generating function and compare the coefficients of r^n. □

The three-term recurrence (see Proposition 1.3.7) is easily computed using $k_n = 2^n$, $h_n = 2^n n!$ and $b_n = 0$:

$$H_{n+1}(x) = 2xH_n(x) - 2nH_{n-1}(x)$$

with $a_n = \sqrt{\frac{1}{2}(n+1)}$. This together with the fact that $H_n'(x) = 2nH_{n-1}(x)$, used twice, establishes the differential equation

$$H_n''(x) - 2xH_n'(x) + 2nH_n(x) = 0.$$

1.4.2 Laguerre polynomials

For a parameter $\alpha > -1$, the weight function for the Laguerre polynomials is $x^\alpha e^{-x}$ on $\mathbb{R}_+ = \{x : x \geq 0\}$ with constant $\Gamma(\alpha+1)^{-1}$.

Definition 1.4.4 For $n \geq 0$ let $L_n^\alpha(x) = \frac{1}{n!} x^{-\alpha} e^x \left(\frac{d}{dx}\right)^n (x^{n+\alpha} e^{-x})$.

Proposition 1.4.5 *For $n \geq 0$,*

$$L_n^\alpha(x) = \frac{(\alpha+1)_n}{n!} \sum_{j=0}^{n} \frac{(-n)_j}{(\alpha+1)_j} \frac{x^j}{j!}.$$

Proof Expand the definition to

$$L_n^\alpha(x) = \frac{1}{n!} \sum\nolimits_{j=0}^{n} \binom{n}{j} (-n-\alpha)_{n-j} (-1)^n x^j$$

(with the n-fold product rule) and write $(-n-\alpha)_{n-j} = (-1)^{n-j}(\alpha+1)_n / (\alpha+1)_j$. □

Arguing as for Hermite polynomials, the following hold for the Laguerre polynomials:

1. $\displaystyle\int_0^\infty q(x) L_n^\alpha(x) x^\alpha e^{-x}\, dx = \frac{(-1)^n}{n!} \int_0^\infty \left(\frac{d}{dx}\right)^n q(x) x^{\alpha+n} e^{-x}\, dx$

   for any polynomial $q(x)$;
2. $\displaystyle\int_0^\infty x^m L_n^\alpha(x) x^\alpha e^{-x}\, dx = 0$ for $0 \leq m < n$;

3. the leading coefficient is $k_n = \dfrac{(-1)^n}{n!}$;
4. $h_n = \dfrac{1}{\Gamma(\alpha+1)} \displaystyle\int_0^\infty [L_n^\alpha(x)]^2 x^\alpha e^{-x}\, dx$
$= \dfrac{1}{\Gamma(\alpha+1)n!} \displaystyle\int_0^\infty x^{\alpha+n} e^{-x}\, dx = \dfrac{(\alpha+1)_n}{n!}$.

Proposition 1.4.6 *The three-term recurrence for the Laguerre polynomials is*

$$L_{n+1}^\alpha(x) = \frac{1}{n+1}(-x+2n+1+\alpha)L_n^\alpha(x) - \frac{n+\alpha}{n+1} L_{n-1}^\alpha(x).$$

Proof The coefficients A_n and C_n are computed from the values of h_n and k_n (see Proposition 1.3.7). To compute B_n we have that

$$L_n^\alpha(x) = \frac{(-1)^n}{n!}\left[x^n - n(\alpha+n)x^{n-1}\right] + \text{terms of lower degree;}$$

hence

$$\begin{aligned}
&\frac{1}{\Gamma(\alpha+1)} \int_0^\infty x L_n^\alpha(x)^2 x^\alpha e^{-x}\, dx \\
&\quad = \frac{1}{n!\Gamma(\alpha+1)} \int_0^\infty [(n+1)x - n(\alpha+n)] x^{\alpha+n} e^{-x}\, dx \\
&\quad = \frac{1}{n!\Gamma(\alpha+1)}[(n+1)\Gamma(\alpha+n+2) - n(\alpha+n)\Gamma(\alpha+n+2)] \\
&\quad = \frac{1}{n!}(\alpha+1)_n(2n+1+\alpha).
\end{aligned}$$

The value of B_n is obtained by multiplying this equation by $-k_{n+1}/(k_n h_n) = n!/[(n+1)(\alpha+1)_n]$. □

The coefficient $a_n = \sqrt{(n+1)(n+1+\alpha)}$. The generating function is

$$(1-r)^{-\alpha-1} \exp\left(\frac{-xr}{1-r}\right) = \sum_{n=0}^\infty L_n^\alpha(x) r^n, \qquad |r| < 1.$$

This follows from expanding the left-hand side as

$$\sum_{j=0}^\infty \frac{(-xr)^j}{j!}(1-r)^{-\alpha-j-1} = \sum_{j=0}^\infty \frac{(-xr)^j}{j!} \sum_{m=0}^\infty \frac{(\alpha+1+j)_m}{m!} r^m,$$

and then changing the index of summation to $m = n - j$ and extracting the coefficient of r^n; note that $(\alpha+1+j)_{n-j} = (\alpha+1)_n/(\alpha+1)_j$. From the ${}_1F_1$ series for $L_n^\alpha(x)$ the following hold:

1. $L_n^\alpha(0) = \dfrac{(\alpha+1)_n}{n!}$;

2. $\dfrac{d}{dx}L_n^\alpha(x) = -L_{n-1}^{\alpha+1}(x)$;
3. $x\dfrac{d^2}{dx^2}L_n^\alpha(x) + (\alpha+1-x)\dfrac{d}{dx}L_n^\alpha(x) + nL_n^\alpha(x) = 0.$

The Laguerre polynomials will be used in the several-variable theory, as functions of $|x|^2$. The one-variable version of this appears for the weight function $|x|^{2\mu}e^{-x^2}$ for $x \in \mathbb{R}$, with constant $\Gamma\left(\mu+\frac{1}{2}\right)^{-1}$. Orthogonal polynomials for this weight are given by $P_{2n}(x) = L_n^{\mu-1/2}(x^2)$ and $P_{2n+1}(x) = xL_n^{\mu+1/2}(x^2)$. This family is discussed in more detail in what follows. In particular, when $\mu = 0$ there is a relation to the Hermite polynomials (to see this, match up the leading coefficients): $H_{2n}(x) = (-1)^n 2^{2n} n! L_n^{-1/2}(x^2)$ and $H_{2n+1}(x) = (-1)^n 2^{2n+1} n! x L_n^{1/2}(x^2)$.

1.4.3 Gegenbauer polynomials

The Gegenbauer polynomials are also called ultraspherical polynomials. For a parameter $\lambda > -\frac{1}{2}$ their weight function is $(1-x^2)^{\lambda-1/2}$ on $-1 < x < 1$; the constant is $B(\frac{1}{2}, \lambda+\frac{1}{2})^{-1}$. The special cases $\lambda = 0, 1, \frac{1}{2}$ correspond to the Chebyshev polynomials of the first and second kind and the Legendre polynomials, respectively. The usual generating function does not work for $\lambda = 0$, neither does it obviously imply orthogonality; thus we begin with a different choice of normalization.

Definition 1.4.7 For $n \geq 0$, let

$$P_n^\lambda(x) = \frac{(-1)^n}{2^n(\lambda+\frac{1}{2})_n}(1-x^2)^{1/2-\lambda}\frac{d^n}{dx^n}(1-x^2)^{n+\lambda-1/2}.$$

Proposition 1.4.8 *For $n \geq 0$,*

$$P_n^\lambda(x) = \left(\frac{1+x}{2}\right)^n {}_2F_1\left(\begin{matrix}-n, -n-\lambda+\frac{1}{2}\\ \lambda+\frac{1}{2}\end{matrix}; \frac{x-1}{x+1}\right)$$
$$= {}_2F_1\left(\begin{matrix}-n, n+2\lambda\\ \lambda+\frac{1}{2}\end{matrix}; \frac{1-x}{2}\right).$$

Proof Expand the formula in the definition with the product rule to obtain

$$P_n^\lambda(x) = \frac{(-1)^n}{2^n(\lambda+\frac{1}{2})_n}\sum_{j=0}^{n}\left[\binom{n}{j}(-n-\lambda+\tfrac{1}{2})_{n-j}(-n-\lambda+\tfrac{1}{2})_j \times (-1)^j(1-x)^j(1+x)^{n-j}\right].$$

This produces the first ${}_2F_1$ formula; note that $(-n-\lambda+\frac{1}{2})_{n-j} = (-1)^{n-j}(\lambda+\frac{1}{2})_n/(\lambda+\frac{1}{2})_j$. The transformation in Proposition 1.2.4 gives the second expression. □

As before, this leads to the orthogonality relations and structure constants.

1. The leading coefficient is $k_n = \dfrac{(n+2\lambda)_n}{2^n(\lambda+1/2)_n} = \dfrac{2^{n-1}(\lambda+1)_{n-1}}{(2\lambda+1)_{n-1}}$, where the second equality holds for $n \geq 1$;
2. $\displaystyle\int_{-1}^{1} q(x) P_n^\lambda(x)(1-x^2)^{\lambda-1/2}\,\mathrm{d}x = \frac{2^{-n}}{(\lambda+\frac{1}{2})_n}\int_{-1}^{1}\frac{\mathrm{d}^n}{\mathrm{d}x^n} q(x)(1-x^2)^{n+\lambda-1/2}\mathrm{d}x,$

   for any polynomial $q(x)$;
3. $\displaystyle\int_{-1}^{1} x^m P_n^\lambda(x)(1-x^2)^{\lambda-1/2}\,\mathrm{d}x = 0 \quad \text{for } 0 \leq m < n;$
4. $\displaystyle h_n = \frac{1}{B(\frac{1}{2},\lambda+\frac{1}{2})}\int_{-1}^{1}[P_n^\lambda(x)]^2(1-x^2)^{\lambda-1/2}\,\mathrm{d}x = \frac{n!(n+2\lambda)}{2(2\lambda+1)_n(n+\lambda)};$
5. $\displaystyle P_{n+1}^\lambda(x) = \frac{2(n+\lambda)}{n+2\lambda} x P_n^\lambda(x) - \frac{n}{n+2\lambda} P_{n-1}^\lambda(x),$

   where the coefficients are calculated from k_n and h_n;
6. $\displaystyle a_n = \left(\frac{(n+1)(n+2\lambda)}{4(n+\lambda)(n+1+\lambda)}\right)^{1/2}.$

Proposition 1.4.9 *For* $n \geq 1$,

$$\frac{\mathrm{d}}{\mathrm{d}x} P_n^\lambda(x) = \frac{n(n+2\lambda)}{1+2\lambda} P_{n-1}^{\lambda+1}(x)$$

and $P_n^\lambda(x)$ *is the unique polynomial solution of*

$$(1-x^2) f''(x) - (2\lambda+1) x f'(x) + n(n+2\lambda) f(x) = 0, \quad f(1) = 1.$$

Proof The ${}_2F_1$ series satisfies the differential equation

$$\frac{\mathrm{d}}{\mathrm{d}x} {}_2F_1\left(\begin{matrix} a,b \\ c \end{matrix}; x\right) = \frac{ab}{c} {}_2F_1\left(\begin{matrix} a+1,b+1 \\ c+1 \end{matrix}; x\right);$$

applying this to $P_n^\lambda(x)$ proves the first identity. Similarly, the differential equation satisfied by ${}_2F_1\left(\begin{matrix} -n,n+2\lambda \\ \lambda+\frac{1}{2} \end{matrix}; t\right)$ leads to the stated equation, using $t = \frac{1}{2}(1-x)$, and $\mathrm{d}/\mathrm{d}t = -2\,\mathrm{d}/\mathrm{d}x$. □

There is a generating function which works for $\lambda > 0$ and, because of its elegant form, it is used to define Gegenbauer polynomials with a different normalization. This also explains the choice of parameter.

Definition 1.4.10 For $n \geq 0$, the polynomial $C_n^\lambda(x)$ is defined by

$$(1-2rx+r^2)^{-\lambda} = \sum_{n=0}^{\infty} C_n^\lambda(x) r^n, \quad |r| < 1, \quad |x| \leq 1.$$

Proposition 1.4.11 *For $n \geq 0$ and $\lambda > 0$,* $C_n^\lambda(x) = \frac{(2\lambda)_n}{n!} P_n^\lambda(x)$ *and*

$$C_n^\lambda(x) = \frac{(\lambda)_n 2^n}{n!} x^n {}_2F_1\left(\begin{matrix} -\frac{n}{2}, \frac{1-n}{2} \\ 1-n-\lambda \end{matrix}; \frac{1}{x^2}\right).$$

Proof First, expand the generating function as follows:

$$\begin{aligned}
[1+2r(1-x)-2r+r^2]^{-\lambda} &= (1-r)^{-2\lambda}\left(1-\frac{2r(x-1)}{(1-r)^2}\right)^{-\lambda} \\
&= \sum_{j=0}^{\infty} \frac{(\lambda)_j}{j!} 2^j r^j (x-1)^j (1-r)^{-2\lambda-2j} \\
&= \sum_{i,j=0}^{\infty} \frac{(\lambda)_j (2\lambda+2j)_i}{j!\,i!} 2^j r^{j+i} (x-1)^j \\
&= \sum_{n=0}^{\infty} r^n \frac{(2\lambda)_n}{n!} \sum_{j=0}^{n} \frac{(-n)_j (2\lambda+n)_j}{(\lambda+\frac{1}{2})_j\, j!} \left(\frac{1-x}{2}\right)^j,
\end{aligned}$$

where in the last line i has been replaced by $n-j$, and $(2\lambda+2j)_{n-j} = (2\lambda)_{2j}(2\lambda+2j)_{n-j}/(2\lambda)_{2j} = (2\lambda)_n(2\lambda+n)_j/[2^{2j}(\lambda)_j(\lambda+\frac{1}{2})_j]$.

Second, expand the generating function as follows:

$$\begin{aligned}
[1-(2rx-r^2)]^{-\lambda} &= \sum_{j=0}^{\infty} \frac{(\lambda)_j}{j!}(2rx-r^2)^j \\
&= \sum_{j=0}^{\infty}\sum_{i=0}^{j} \frac{(\lambda)_j}{(j-i)!\,i!}(-1)^j (2x)^{j-i} r^{i+j} \\
&= \sum_{n=0}^{\infty} r^n \sum_{i\leq n/2} \frac{(\lambda)_{n-i}}{i!(n-2i)!}(-1)^i (2x)^{n-2i},
\end{aligned}$$

where in the last line j has been replaced by $n-i$. The latter sum equals the stated formula. □

We now list the various structure constants for $C_n^\lambda(x)$ and use primes to distinguish them from the values for $P_n^\lambda(x)$:

1. $k_n' = \dfrac{(\lambda)_n 2^n}{n!}$;
2. $h_n' = \dfrac{\lambda(2\lambda)_n}{(n+\lambda)n!} = \dfrac{\lambda}{n+\lambda} C_n^\lambda(1)$;
3. $C_{n+1}^\lambda(x) = \dfrac{2(n+\lambda)}{n+1} x C_n^\lambda(x) - \dfrac{n+2\lambda-1}{n+1} C_{n-1}^\lambda(x)$;
4. $\dfrac{\mathrm{d}}{\mathrm{d}x} C_n^\lambda(x) = 2\lambda C_{n-1}^{\lambda+1}(x)$;
5. $C_n^\lambda(1) = \dfrac{(2\lambda)_n}{n!}$.

There is another expansion resembling the generating function which is used in the Poisson kernel for spherical harmonics.

Proposition 1.4.12 *For* $|x| \le 1$ *and* $|r| < 1$,

$$\frac{1-r^2}{(1-2xr+r^2)^{\lambda+1}} = \sum_{n=0}^{\infty} \frac{n+\lambda}{\lambda} C_n^{\lambda}(x) r^n$$
$$= \sum_{n=0}^{\infty} \frac{1}{h_n} P_n^{\lambda}(x) r^n.$$

Proof Expand the left-hand side as $\sum_{n=0}^{\infty} \left[C_n^{\lambda+1}(x) - C_{n-2}^{\lambda+1}(x)\right] r^n$. Then

$$C_n^{\lambda+1}(x) - C_{n-2}^{\lambda+1}(x) = \sum_{i \le n/2} \frac{(\lambda+1)_{n-i}}{i!(n-2i)!}(-1)^i(2x)^{n-2i}$$
$$- \sum_{j \le n/2-1} \frac{(\lambda+1)_{n-2-j}}{j!(n-2-2j)!}(-1)^j(2x)^{n-2-2j}$$
$$= \sum_{i \le n/2} \frac{(\lambda+1)_{n-1-i}}{i!(n-2i)!}(-1)^i(2x)^{n-2i}(\lambda+n).$$

In the sum for $C_{n-2}^{\lambda+1}(x)$ replace j by $i-1$, and combine the two sums. The end result is exactly $[(n+\lambda/\lambda)]C_n^{\lambda}(x)$. Further,

$$\frac{(n+\lambda)(2\lambda)_n}{\lambda n!} = h_n^{-1}$$

(the form for h_n given in the list before Proposition 1.4.9 avoids the singularity at $\lambda = 0$). □

Legendre and Chebyshev polynomials

The special case $\lambda = \frac{1}{2}$ corresponds to the Legendre polynomials, often denoted by $P_n(x) = P_n^{1/2}(x) = C_n^{1/2}(x)$ (in later discussions we will avoid this notation in order to reserve P_n for general polynomials). Observe that these polynomials are orthogonal for dx on $-1 \le x \le 1$. Related formulae are for the Legendre polynomials $h_n = 1/(2n+1)$ and

$$(n+1)P_{n+1}(x) = (2n+1)xP_n(x) - nP_{n-1}(x).$$

The case $\lambda = 0$ corresponds to the Chebyshev polynomials of the first kind denoted by $T_n(x) = P_n^0(x)$. Here $h_0 = 1$ and $h_n = \frac{1}{2}$ for $n > 0$. The three-term recurrence for $n \ge 1$ is

$$T_{n+1}(x) = 2xT_n(x) - T_{n-1}(x)$$

and $T_1(x) = xT_0$. Of course, together with the identity $\cos(n+1)\theta = 2\cos\theta\cos n\theta - \cos(n-1)\theta$ for $n \geq 1$, this shows that $T_n(\cos\theta) = \cos n\theta$. Accordingly the zeros of T_n are at

$$\cos\frac{(2j-1)\pi}{2n}$$

for $j = 1, 2, \ldots, n$. Also, the associated Gaussian quadrature has equal weights at the nodes (a simple but amusing calculation).

The case $\lambda = 1$ corresponds to the Chebyshev polynomials of the second kind denoted by $U_n(x) = (n+1)P_n^1(x) = C_n^1(x)$. For this family $h_n = 1$ and, for $n \geq 0$,

$$U_{n+1}(x) = 2xU_n(x) - U_{n-1}(x).$$

This relation, together with

$$\frac{\sin(n+2)\theta}{\sin\theta} = 2\cos\theta\frac{\sin(n+1)\theta}{\sin\theta} - \frac{\sin n\theta}{\sin\theta}$$

for $n \geq 0$, implies that

$$U_n(\cos\theta) = \frac{\sin(n+1)\theta}{\sin\theta}.$$

1.4.4 Jacobi polynomials

For parameters $\alpha, \beta > -1$ the weight function for the Jacobi polynomials is $(1-x)^\alpha(1+x)^\beta$ on $-1 < x < 1$; the constant c is $2^{-\alpha-\beta-1}B(\alpha+1, \beta+1)^{-1}$.

Definition 1.4.13 For $n \geq 0$, let

$$P_n^{(\alpha,\beta)}(x) = \frac{(-1)^n}{2^n n!}(1-x)^{-\alpha}(1+x)^{-\beta}\frac{d^n}{dx^n}\left[(1-x)^{\alpha+n}(1+x)^{\beta+n}\right].$$

Proposition 1.4.14 *For $n \geq 0$,*

$$\begin{aligned} P_n^{(\alpha,\beta)}(x) &= \frac{(\alpha+1)_n}{n!}\left(\frac{1+x}{2}\right)^n {}_2F_1\left(\begin{matrix}-n, -n-\beta\\ \alpha+1\end{matrix}; \frac{x-1}{x+1}\right)\\ &= \frac{(\alpha+1)_n}{n!}{}_2F_1\left(\begin{matrix}-n, n+\alpha+\beta+1\\ \alpha+1\end{matrix}; \frac{1-x}{2}\right).\end{aligned}$$

Proof Expand the formula in Definition 1.4.13 with the product rule to obtain

$$\begin{aligned} P_n^{(\alpha,\beta)}(x) = \frac{(-1)^n}{2^n n!}\sum_{j=0}^{n}\Bigg[\binom{n}{j}&(-n-\alpha)_{n-j}(-n-\beta)_j(-1)^j\\ &\times(1-x)^j(1+x)^{n-j}\Bigg].\end{aligned}$$

This produces the first ${}_2F_1$ formula (note that $(-n-\alpha)_{n-j} = (-1)^{n-j}(\alpha+1)_n/(\alpha+1)_j$. The transformation in Proposition 1.2.4 gives the second expression. □

The orthogonality relations and structure constants for the Jacobi polynomials are found in the same way as for the Gegenbauer polynomials.

1. The leading coefficient is $k_n = \dfrac{(n+\alpha+\beta+1)_n}{2^n n!}$;
2. $\displaystyle\int_{-1}^{1} q(x)P_n^{(\alpha,\beta)}(x)(1-x)^\alpha(1+x)^\beta \mathrm{d}x$
$$= \frac{1}{2^n n!}\int_{-1}^{1}\left(\frac{\mathrm{d}}{\mathrm{d}x}\right)^n q(x)(1-x)^{\alpha+n}(1+x)^{\beta+n}\,\mathrm{d}x$$
for any polynomial $q(x)$;
3. $\displaystyle\int_{-1}^{1} x^m P_n^{(\alpha,\beta)}(x)(1-x)^\alpha(1+x)^\beta \mathrm{d}x = 0 \quad \text{for } 0 \le m < n;$
4. $\displaystyle h_n = \frac{1}{2^{\alpha+\beta+1}B(\alpha+1,\beta+1)}\int_{-1}^{1} P_n^{(\alpha,\beta)}(x)^2(1-x)^\alpha(1+x)^\beta\,\mathrm{d}x$
$$= \frac{(\alpha+1)_n(\beta+1)_n(\alpha+\beta+n+1)}{n!(\alpha+\beta+2)_n(\alpha+\beta+2n+1)};$$
5. $P_n^{(\alpha,\beta)}(-x) = (-1)^n P_n^{(\beta,\alpha)}(x)$ and $P_n^{(\alpha,\beta)}(-1) = (-1)^n(\beta+1)_n/n!$.

The differential relations follow easily from the hypergeometric series.

Proposition 1.4.15 *For $n \ge 1$,*

$$\frac{\mathrm{d}}{\mathrm{d}x}P_n^{(\alpha,\beta)}(x) = \frac{n+\alpha+\beta+1}{2}P_{n-1}^{(\alpha+1,\beta+1)}(x)$$

where $P_n^{(\alpha,\beta)}(x)$ is the unique polynomial solution of

$$(1-x^2)f''(x) - (\alpha-\beta+(\alpha+\beta+2)x)f'(x) + n(n+\alpha+\beta+1)f(x) = 0$$

with $f(1) = (\alpha+1)_n/n!$.

The three-term recurrence calculation requires extra work because the weight lacks symmetry.

Proposition 1.4.16 *For $n \ge 0$,*

$$\begin{aligned} P_{n+1}^{(\alpha,\beta)}(x) &= \frac{(2n+\alpha+\beta+1)(2n+\alpha+\beta+2)}{2(n+1)(n+\alpha+\beta+1)}xP_n^{(\alpha,\beta)}(x) \\ &\quad + \frac{(2n+\alpha+\beta+1)(\alpha^2-\beta^2)}{2(n+1)(n+\alpha+\beta+1)(2n+\alpha+\beta)}P_n^{(\alpha,\beta)}(x) \\ &\quad - \frac{(\alpha+n)(\beta+n)(2n+\alpha+\beta+2)}{(n+1)(n+\alpha+\beta+1)(2n+\alpha+\beta)}P_{n-1}^{(\alpha,\beta)}(x). \end{aligned}$$

Proof The values of $A_n = k_{n+1}/k_n$ and $C_n = k_{n+1}k_{n-1}h_n/k_n^2 h_{n-1}$ (see Proposition 1.3.7) can be computed directly. To compute the B_n term, suppose that

$$q(x) = v_0\left(\frac{1-x}{2}\right)^n + v_1\left(\frac{1-x}{2}\right)^{n-1};$$

then

$$\begin{aligned}
&2^{-\alpha-\beta-1}B(\alpha+1,\beta+1)^{-1}\int_{-1}^{1} xq(x)P_n^{(\alpha,\beta)}(x)(1-x)^{\alpha}(1+x)^{\beta}\,\mathrm{d}x\\
&= (-1)^n 2^{-\alpha-\beta-2n-1}\frac{1}{n!}B(\alpha+1,\beta+1)^{-1}\\
&\quad\times\int_{-1}^{1}[-(n+1)v_0(1-x)+(v_0-2v_1)](1-x)^{n+\alpha}(1+x)^{n+\beta}\,\mathrm{d}x\\
&= \frac{(-1)^n(\alpha+1)_n(\beta+1)_n}{(\alpha+\beta+2)_{2n}}v_0\left(\frac{-2(n+1)(\alpha+n+1)}{\alpha+\beta+2n+2}+1-2\frac{v_1}{v_0}\right).
\end{aligned}$$

From the series for $P_n^{(\alpha,\beta)}(x)$ we find that

$$v_0 = (-2)^n k_n, \qquad \frac{v_1}{v_0} = -\frac{n(\alpha+n)}{\alpha+\beta+2n}$$

and the factor within large parentheses evaluates to

$$\frac{\beta^2-\alpha^2}{(\alpha+\beta+2n)(\alpha+\beta+2n+2)}.$$

Multiply the integral by $-k_{n+1}/(k_n h_n)$ to get the stated result. □

Corollary 1.4.17 *The coefficient for the corresponding orthonormal polynomials (and $n \geq 0$) is*

$$a_n = \frac{2}{\alpha+\beta+2n+2}\left(\frac{(\alpha+n+1)(\beta+n+1)(n+1)(\alpha+\beta+n+1)}{(\alpha+\beta+2n+1)(\alpha+\beta+2n+3)}\right)^{1/2}.$$

Of course, the Jacobi weight includes the Gegenbauer weight as a special case. In terms of the usual notation the relation is

$$C_n^{\lambda}(x) = \frac{(2\lambda)_n}{\left(\lambda+\frac{1}{2}\right)_n}P_n^{(\lambda-1/2,\lambda-1/2)}(x).$$

In the next section we will use the Jacobi polynomials to find the orthogonal polynomials for the weight $|x|^{2\mu}(1-x^2)^{\lambda-1/2}$ on $-1 < x < 1$.

1.5 Modified Classical Polynomials

The weights for the Hermite and Gegenbauer polynomials can be modified by multiplying them by $|x|^{2\mu}$. The resulting orthogonal polynomials can be expressed in terms of the Laguerre and Jacobi polynomials. Further, there is an integral transform which maps the original polynomials to the modified ones. We also present in this section a limiting relation between the Gegenbauer and Hermite polynomials.

Note that the even- and odd-degree polynomials in the modified families have separate definitions. For reasons to be explained later, the integral transform is denoted by V (it is an implicit function of μ).

Definition 1.5.1 For $\mu \geq 0$, let $c_\mu = \left[2^{2\mu} B(\mu, \mu+1)\right]^{-1}$ (see Definition 1.1.2) and, for any polynomial $q(x)$, let the integral transform V be defined by

$$Vq(x) = c_\mu \int_{-1}^{1} q(tx)(1-t)^{\mu-1}(1+t)^{\mu}\, dt.$$

Note that

$$\int_{-1}^{1} (1-t)^{\mu-1}(1+t)^{\mu}\, dt = 2\int_{0}^{1} (1-t^2)^{\mu-1} dt = B(\tfrac{1}{2}, \mu);$$

thus c_μ can also be written as $B\left(\frac{1}{2}, \mu\right)^{-1}$ (and this proves the duplication formula for Γ, given in item 5 of the list after Definition 1.1.2). The limit

$$\lim_{\lambda \to 0} c_\lambda \int_{-1}^{1} f(x)(1-x^2)^{\lambda-1} dx = \tfrac{1}{2}[f(1) + f(-1)] \tag{1.5.1}$$

shows that V is the identity operator when $\mu = 0$.

Lemma 1.5.2 *For $n \geq 0$,*

$$V(x^{2n}) = \frac{\left(\frac{1}{2}\right)_n}{\left(\mu+\frac{1}{2}\right)_n} x^{2n} \quad \textit{and} \quad V(x^{2n+1}) = \frac{\left(\frac{1}{2}\right)_{n+1}}{\left(\mu+\frac{1}{2}\right)_{n+1}} x^{2n+1}.$$

Proof For the even-index case

$$\begin{aligned} c_\mu \int_{-1}^{1} t^{2n}(1-t)^{\mu-1}(1+t)^{\mu} dt &= 2c_\mu \int_{0}^{1} t^{2n}(1-t^2)^{\mu-1} dt \\ &= \frac{B\left(n+\frac{1}{2}, \mu\right)}{B\left(\frac{1}{2}, \mu\right)} = \frac{\left(\frac{1}{2}\right)_n}{\left(\mu+\frac{1}{2}\right)_n}, \end{aligned}$$

and for the odd-index case

$$\begin{aligned} c_\mu \int_{-1}^{1} t^{2n+1}(1-t)^{\mu-1}(1+t)^{\mu} dt &= 2c_\mu \int_{0}^{1} t^{2n+2}(1-t^2)^{\mu-1} dt \\ &= \frac{B\left(n+\frac{3}{2}, \mu\right)}{B\left(\frac{1}{2}, \mu\right)} = \frac{\left(\frac{1}{2}\right)_{n+1}}{\left(\mu+\frac{1}{2}\right)_{n+1}}. \end{aligned}$$

□

1.5.1 Generalized Hermite polynomials

For the generalized Hermite polynomials H_n^μ, the weight function is $|x|^{2\mu}\mathrm{e}^{-x^2}$ on $x \in \mathbb{R}$ and the constant is $\Gamma\left(\mu+\frac{1}{2}\right)^{-1}$. From the formula

$$\int_R f(x^2)|x|^{2\mu}\mathrm{e}^{-x^2}\mathrm{d}x = \int_0^\infty f(t)t^{\mu-1/2}\mathrm{e}^{-t}\mathrm{d}t,$$

we see that the polynomials $\{P_n(x) : n \geq 0\}$ given by $P_{2n}(x) = L_n^{\mu-1/2}(x^2)$ and $P_{2n+1}(x) = xL_n^{\mu+1/2}(x^2)$ form an orthogonal family. The normalization is chosen so that the leading coefficient becomes 2^n.

Definition 1.5.3 For $\mu \geq 0$ the generalized Hermite polynomials $H_n^\mu(x)$ are defined by

$$H_{2n}^\mu(x) = (-1)^n 2^{2n} n! L_n^{\mu-1/2}(x^2),$$
$$H_{2n+1}^\mu(x) = (-1)^n 2^{2n+1} n! x L_n^{\mu+1/2}(x^2).$$

The usual computations show the following (for $n \geq 0$):

1. $k_n = 2^n$;
2. $h_{2n} = 2^{4n} n!\left(\mu+\frac{1}{2}\right)_n$ and $h_{2n+1} = 2^{4n+2} n!\left(\mu+\frac{1}{2}\right)_{n+1}$;
3. $H_{2n+1}^\mu(x) = 2xH_{2n}^\mu(x) - 4nH_{2n-1}^\mu(x)$,
   $H_{2n+2}^\mu(x) = 2xH_{2n+1}^\mu(x) - 2(2n+1+2\mu)H_{2n}^\mu(x)$.

The integral transform V maps the Hermite polynomials H_{2n} and H_{2n+1} onto this family:

Theorem 1.5.4 *For $n \geq 0$ and $\mu > 0$,*

$$VH_{2n} = \frac{\left(\frac{1}{2}\right)_n}{\left(\mu+\frac{1}{2}\right)_n} H_{2n}^\mu \quad \textit{and} \quad VH_{2n+1} = \frac{\left(\frac{1}{2}\right)_{n+1}}{\left(\mu+\frac{1}{2}\right)_{n+1}} H_{2n+1}^\mu.$$

Proof The series for the Hermite polynomials can now be used. For the even-index case,

$$\begin{aligned} VH_{2n}(x) &= \sum_{j=0}^{n} \frac{(2n)!}{(2n-2j)!j!}(-1)^j(2x)^{2n-2j}\frac{\left(\frac{1}{2}\right)_{n-j}}{\left(\mu+\frac{1}{2}\right)_{n-j}} \\ &= 2^{2n}(-1)^n\left(\tfrac{1}{2}\right)\sum_{ni=0}^{n}\frac{(-n)_i}{i!\left(\mu+\frac{1}{2}\right)_i}x^{2i} \\ &= 2^{2n}(-1)^n\frac{\left(\frac{1}{2}\right)_n n!}{\left(\mu+\frac{1}{2}\right)_n}L_n^{\mu-1/2}(x^2). \end{aligned}$$

For the odd-index case,

$$\begin{aligned}VH_{2n+1}(x) &= \sum_{j=0}^{n} \frac{(2n+1)!}{(2n+1-2j)!j!}(-1)^j(2x)^{2n+1-2j}\frac{\left(\frac{1}{2}\right)_{n-j+1}}{\left(\mu+\frac{1}{2}\right)_{n-j+1}} \\ &= (-1)^n\frac{(2n+1)!}{n!}x\sum_{i=0}^{n}\frac{(-n)_i}{i!\left(\mu+\frac{1}{2}\right)_{i+1}}x^{2i} \\ &= (-1)^n\frac{(2n+1)!}{\left(\mu+\frac{1}{2}\right)_{n+1}}xL_n^{\mu+1/2}(x^2),\end{aligned}$$

where we have changed the index of summation by setting $i = n - j$ and used the relations $(2n)! = 2^{2n}n!\left(\frac{1}{2}\right)_n$ and $(2n+1)! = 2^{2n+1}n!\left(\frac{1}{2}\right)_{n+1}$. □

1.5.2 Generalized Gegenbauer polynomials

For parameters λ, μ with $\lambda > -\frac{1}{2}$ and $\mu \geq 0$ the weight function is $|x|^{2\mu}(1-x^2)^{\lambda-1/2}$ on $-1 < x < 1$ and the constant c is $B\left(\mu+\frac{1}{2}, \lambda+\frac{1}{2}\right)^{-1}$. Suppose that f and g are polynomials; then

$$\begin{aligned}&\int_{-1}^{1} f(x^2)g(x^2)|x|^{2\mu}(1-x^2)^{\lambda-1/2}\mathrm{d}x \\ &\quad = 2^{-\mu-\lambda}\int_{-1}^{1} f\left(\frac{1+t}{2}\right)g\left(\frac{1+t}{2}\right)(1+t)^{\mu-1/2}(1-t)^{\lambda-1/2}\,\mathrm{d}t\end{aligned}$$

and

$$\begin{aligned}&\int_{-1}^{1}(xf(x^2))(xg(x^2))|x|^{2\mu}(1-x^2)^{\lambda-1/2}\,\mathrm{d}x \\ &\quad = 2^{-\mu-\lambda-1}\int_{-1}^{1} f\left(\frac{1+t}{2}\right)g\left(\frac{1+t}{2}\right)(1+t)^{\mu+1/2}(1-t)^{\lambda-1/2}\,\mathrm{d}t.\end{aligned}$$

Accordingly, the orthogonal polynomials $\{P_n : n \geq 0\}$ for the above weight function are given by the formulae

$$P_{2n}(x) = P_n^{(\lambda-1/2,\mu-1/2)}(2x^2-1) \quad \text{and} \quad P_{2n+1}(x) = xP_n^{(\lambda-1/2,\mu+1/2)}(2x^2-1).$$

As in the Hermite polynomial case, we choose a normalization which corresponds in a natural way with V.

Definition 1.5.5 For $\lambda > -\frac{1}{2}$, $\mu \geq 0$, $n \geq 0$, the generalized Gegenbauer polynomials $C_n^{(\lambda,\mu)}(x)$ are defined by

$$C_{2n}^{(\lambda,\mu)}(x) = \frac{(\lambda+\mu)_n}{\left(\mu+\frac{1}{2}\right)_n}P_n^{(\lambda-1/2,\mu-1/2)}(2x^2-1),$$

$$C_{2n+1}^{(\lambda,\mu)}(x) = \frac{(\lambda+\mu)_{n+1}}{\left(\mu+\frac{1}{2}\right)_{n+1}} x P_n^{(\lambda-1/2,\mu+1/2)}(2x^2-1).$$

The structural constants are as follows (for $n \geq 0$):

1. $k_{2n}^{(\lambda,\mu)} = \dfrac{(\lambda+\mu)_{2n}}{\left(\mu+\frac{1}{2}\right)_n n!}$ and $k_{2n+1}^{(\lambda,\mu)} = \dfrac{(\lambda+\mu)_{2n+1}}{\left(\mu+\frac{1}{2}\right)_{n+1} n!}$;
2. $h_{2n}^{(\lambda,\mu)} = \dfrac{\left(\lambda+\frac{1}{2}\right)_n (\lambda+\mu)_n (\lambda+\mu)}{n!\left(\mu+\frac{1}{2}\right)_n (\lambda+\mu+2n)}$,

   $h_{2n+1}^{(\lambda,\mu)} = \dfrac{\left(\lambda+\frac{1}{2}\right)_n (\lambda+\mu)_{n+1} (\lambda+\mu)}{n!\left(\mu+\frac{1}{2}\right)_{n+1} (\lambda+\mu+2n+1)}$;
3. $C_{2n+1}^{(\lambda,\mu)}(x) = \dfrac{2(\lambda+\mu+2n)}{2\mu+2n+1} x C_{2n}^{(\lambda,\mu)}(x) - \dfrac{2\lambda+2n-1}{2\mu+2n+1} C_{2n-1}^{(\lambda,\mu)}(x)$,

   $C_{2n+2}^{(\lambda,\mu)}(x) = \dfrac{\lambda+\mu+2n+1}{n+1} x C_{2n+1}^{(\lambda,\mu)}(x) - \dfrac{\lambda+\mu+n}{n+1} C_{2n}^{(\lambda,\mu)}(x)$;
4. $C_n^{(\lambda,\mu)}(1) = \dfrac{n+\lambda+\mu}{\lambda+\mu} h_n^{(\lambda,\mu)}$.

The modified polynomials $C_n^{(\lambda,\mu)}$ are transforms of the ordinary Gegenbauer polynomials $C_n^{(\lambda+\mu)}$.

Theorem 1.5.6 *For $\lambda > -\frac{1}{2}$, $\mu \geq 0$, $n \geq 0$,*

$$VC_n^{\lambda+\mu}(x) = C_n^{(\lambda,\mu)}(x).$$

Proof We use the series found in Subsection 1.4.3. In the even-index case,

$$\begin{aligned}
VC_{2n}^{\lambda+\mu}(x) &= \sum_{j=0}^{n} \frac{(\lambda+\mu)_{2n-j}}{j!(2n-2j)!}(-1)^j (2x)^{2n-2j} \frac{\left(\frac{1}{2}\right)_{n-j}}{\left(\mu+\frac{1}{2}\right)_{n-j}} \\
&= (-1)^n \frac{(\lambda+\mu)_n}{n!} {}_2F_1\left(\begin{matrix}-n, n+\lambda+\mu \\ \mu+\frac{1}{2}\end{matrix}; x^2\right) \\
&= \frac{(-1)^n (\lambda+\mu)_n}{\left(\mu+\frac{1}{2}\right)_n} P_n^{(\mu-1/2,\lambda-1/2)}(1-2x^2) \\
&= \frac{(\lambda+\mu)_n}{\left(\mu+\frac{1}{2}\right)_n} P_n^{(\lambda-1/2,\mu-1/2)}(2x^2-1),
\end{aligned}$$

and in the odd-index case

$$\begin{aligned}
VC_{2n+1}^{\lambda+\mu}(x) &= \sum_{j=0}^{n} \frac{(\lambda+\mu)_{2n+1-j}}{j!(2n+1-2j)!}(-1)^j(2x)^{2n+1-2j}\frac{\left(\frac{1}{2}\right)_{n+1-j}}{\left(\mu+\frac{1}{2}\right)_{n+1-j}} \\
&= (-1)^n x \frac{(\lambda+\mu)_{n+1}}{n!\left(\mu+\frac{1}{2}\right)} {}_2F_1\left(\begin{matrix}-n, n+\lambda+\mu+1 \\ \mu+\frac{3}{2}\end{matrix}; x^2\right) \\
&= \frac{x(-1)^n(\lambda+\mu)_{n+1}}{\left(\mu+\frac{1}{2}\right)_{n+1}} P_n^{(\mu+1/2,\lambda-1/2)}(1-2x^2) \\
&= \frac{x(\lambda+\mu)_{n+1}}{\left(\mu+\frac{1}{2}\right)_{n+1}} P_n^{(\lambda-1/2,\mu+1/2)}(2x^2-1).
\end{aligned}$$

Note that the index of summation was changed to $i = n - j$ in both cases. □

1.5.3 A limiting relation

With suitable renormalization the weight $\left(1 - x^2/\lambda\right)^{\lambda-1/2}$ tends to e^{-x^2}. The effect on the corresponding orthogonal polynomials is as follows.

Proposition 1.5.7 *For* $\mu, n \geq 0$, $\lim_{\lambda\to\infty} \lambda^{-n/2} C_n^{(\lambda,\mu)}(x/\sqrt{\lambda}) = s_n H_n^{\mu}(x)$, *where* $s_{2n} = \left[2^{2n} n! \left(\mu+\frac{1}{2}\right)_n\right]^{-1}$, $s_{2n+1} = \left[2^{2n+1} n! \left(\mu+\frac{1}{2}\right)_{n+1}\right]^{-1}$.

Proof Replace x by $x/\sqrt{\lambda}$ in the ${}_2F_1$ series expressions (with argument x^2) found above. Observe that $\lim_{\lambda\to\infty} \lambda^{-j}(\lambda+a)_j = 1$ for any fixed a and any $j = 1, 2, \ldots$ □

By specializing to $\mu = 0$ the following is obtained.

Corollary 1.5.8 *For* $n \geq 0$, $\lim_{\lambda\to\infty} \lambda^{-n/2} C_n^{\lambda}\left(\frac{x}{\sqrt{\lambda}}\right) = \frac{1}{n!} H_n(x)$.

1.6 Notes

The standard reference on the hypergeometric functions is the book of Bailey [1935]. It also contains the Lauricella series of two variables. See also the books Andrews, Askey and Roy [1999], Appell and de Fériet [1926], and Erdélyi, Magnus, Oberhettinger and Tricomi [1953]. For multivariable Lauricella series, see Exton [1976].

The standard reference for orthogonal polynomials in one variable is the book of Szegő [1975]. See also Chihara [1978], Freud [1966] and Ismail [2005].

2

Orthogonal Polynomials in Two Variables

We start with several examples of families of orthogonal polynomials of two variables. Apart from a brief introduction in the first section, the general properties and theory will be deferred until the next chapter where they will be given in d variables for all $d \geq 2$. Here we focus on explicit constructions and concrete examples, most of which will be given explicitly in terms of classical orthogonal polynomials of one variable.

2.1 Introduction

A polynomial of two variables in the variables x, y is a finite linear combination of the monomials $x^j y^k$. The total degree of $x^j y^k$ is defined as $j+k$ and the total degree of a polynomial is defined as the highest degree of the monomials that it contains. Let $\Pi^2 = \mathbb{R}[x,y]$ be the space of polynomials in two variables. For $n \in \mathbb{N}_0$, the space of homogeneous polynomials of degree n is denoted by

$$\mathscr{P}_n^2 := \operatorname{span}\{x^j y^k : j+k=n, j,k \in \mathbb{N}_0\}$$

and the space of polynomials of total degree n is denoted by

$$\Pi_n^2 := \operatorname{span}\{x^j y^k : j+k \leq n, j,k \in \mathbb{N}_0\}.$$

Evidently, Π_n^2 is a direct sum of $\mathscr{P}_m^2$ for $m = 0, 1, \ldots, n$. Furthermore,

$$\dim \mathscr{P}_n^2 = n+1 \qquad \text{and} \qquad \dim \Pi_n^2 = \binom{n+2}{2}.$$

Let $\langle \cdot, \cdot \rangle$ be an inner product defined on the space of polynomials of two variables. An example of such an inner product is given by

$$\langle f, g \rangle_\mu = \int_{\mathbb{R}^2} f(x) g(x) \, \mathrm{d}\mu(x),$$

where $\mathrm{d}\mu$ is a positive Borel measure on $\mathbb{R}^2$ such that the integral is well defined on all polynomials. Mostly we will work with $\mathrm{d}\mu = W(x,y)\,\mathrm{d}x\mathrm{d}y$, where W is a nonnegative function called the weight function.

Definition 2.1.1 A polynomial P is an *orthogonal polynomial* of degree n with respect to an inner product $\langle\cdot,\cdot\rangle$ if $P \in \Pi_n^2$ and

$$\langle P, Q\rangle = 0 \qquad \forall Q \in \Pi_{n-1}^2. \tag{2.1.1}$$

When the inner product is defined via a weight function W, we say that P is orthogonal with respect to W.

According to the definition, P is an orthogonal polynomial if it is orthogonal to all polynomials of lower degree; orthogonality to other polynomials of the same degree is not required. Given an inner product one can apply the Gram–Schmidt process to generate an orthogonal basis, which exists under certain conditions to be discussed in the next chapter. In the present chapter we deal with only specific weight functions for which explicit bases can be constructed and verified directly.

We denote by $\mathscr{V}_n^2$ the space of orthogonal polynomials of degree n:

$$\mathscr{V}_n^2 := \operatorname{span}\{P \in \Pi_n^2 : \langle P, Q\rangle = 0, \forall Q \in \Pi_{n-1}^2\}.$$

When the inner product is defined via a weight function W, we sometimes write $\mathscr{V}_n^2(W)$. The dimension of $\mathscr{V}_n^2$ is the same as that of $\mathscr{P}_n^2$:

$$\dim \mathscr{V}_n^2 = \dim \mathscr{P}_n^2 = n+1.$$

A basis of $\mathscr{V}_n^2$ is often denoted by $\{P_k^n : 0 \le k \le n\}$. If, additionally, $\langle P_k^n, P_j^n\rangle = 0$ for $j \ne k$ then the basis is said to be mutually orthogonal and if, further, $\langle P_k^n, P_k^n\rangle = 1$ for $0 \le k \le n$ then the basis is said to be orthonormal.

In contrast with the case of one variable, there can be many distinct bases for the space $\mathscr{V}_n^2$. In fact let $\mathbb{P}_n := \{P_k^n : 0 \le k \le n\}$ denote a basis of $\mathscr{V}_n^2$, where we regard $\mathbb{P}_n$ both as a set and as a column vector. Then, for any nonsingular matrix $M \in \mathbb{R}^{n+1,n+1}$, $M\mathbb{P}_n$ is also a basis of $\mathscr{V}_n^2$.

2.2 Product Orthogonal Polynomials

Let W be the product weight function defined by $W(x,y) = w_1(x)w_2(y)$, where w_1 and w_2 are two weight functions of one variable.

Proposition 2.2.1 *Let $\{p_k\}_{k=0}^\infty$ and $\{q_k\}_{k=0}^\infty$ be sequences of orthogonal polynomials with respect to w_1 and w_2, respectively. Then a mutually orthogonal basis of $\mathscr{V}_n^2$ with respect to W is given by*

$$P_k^n(x,y) = p_k(x)q_{n-k}(y), \qquad 0 \le k \le n.$$

Furthermore, if $\{p_k\}$ and $\{q_k\}$ are orthonormal then so is $\{P_k^n\}$.

Proof Verification follows from writing

$$\begin{aligned}\int_{\mathbb{R}^2} & P_k^n(x,y)P_j^m(x,y)W(x,y)\,\mathrm{d}x\,\mathrm{d}y \\ &= \int_{\mathbb{R}} p_k(x)p_j(x)w_1(x)\,\mathrm{d}x \int_{\mathbb{R}} q_{n-k}(y)q_{m-j}(y)w_2(y)\,\mathrm{d}y\end{aligned}$$

and using the orthogonality of $\{p_n\}$ and $\{q_n\}$. □

Below are several examples of weight functions and the corresponding classical orthogonal polynomials.

Product Hermite polynomials

$$W(x,y) = e^{-x^2-y^2}$$

and

$$P_k^n(x,y) = H_k(x)H_{n-k}(y), \qquad 0 \le k \le n. \tag{2.2.1}$$

Product Laguerre polynomials

$$W(x,y) = x^{\alpha}y^{\beta}e^{-x-y},$$

and

$$P_k^n(x,y) = L_k^{\alpha}(x)L_{n-k}^{\beta}(y), \qquad 0 \le k \le n. \tag{2.2.2}$$

Product Jacobi polynomials

$$W(x,y) = (1-x)^{\alpha}(1+x)^{\beta}(1-y)^{\gamma}(1+y)^{\delta},$$

and

$$P_k^n(x,y) = P_k^{(\alpha,\beta)}(x)P_{n-k}^{(\gamma,\delta)}(y), \qquad 0 \le k \le n. \tag{2.2.3}$$

There are also mixed-product orthogonal polynomials, such as the product of the Hermite polynomials and the Laguerre polynomials.

2.3 Orthogonal Polynomials on the Unit Disk

The unit disk of $\mathbb{R}^2$ is defined as $B^2 := \{(x,y) : x^2+y^2 \le 1\}$, on which we consider a weight function defined as follows:

$$W_{\mu}(x,y) := \frac{\mu+\frac{1}{2}}{\pi}(1-x^2-y^2)^{\mu-1/2}, \qquad \mu > -\tfrac{1}{2},$$

which is normalized in such a way that its integral over B^2 is 1. Let

$$\langle f,g\rangle_{\mu} := \int_{B^2} f(x,y)g(x,y)W_{\mu}(x,y)\,\mathrm{d}x\,\mathrm{d}y$$

in this section. There are several distinct explicit orthogonal bases.

First orthonormal basis This basis is given in terms of the Gegenbauer polynomials.

Proposition 2.3.1 *Let $\mu > 0$. For $0 \le k \le n$ define the polynomials*

$$P_k^n(x,y) = C_{n-k}^{k+\mu+1/2}(x)(1-x^2)^{k/2} C_k^{\mu}\left(\frac{y}{\sqrt{1-x^2}}\right), \tag{2.3.1}$$

which holds in the limit $\lim_{\mu\to 0} \mu^{-1} C_k^{\mu}(x) = (2/k)T_k(x)$ if $\mu = 0$. Then $\{P_k^n : 0 \le k \le n\}$ is a mutually orthogonal basis for $\mathcal{V}_n^2(W_\mu)$. Moreover,

$$\langle P_k^n, P_k^n \rangle_\mu = \frac{(2k+2\mu+1)_{n-k}(2\mu)_k(\mu)_k(\mu+\frac{1}{2})}{(n-k)!k!(\mu+\frac{1}{2})_k(n+\mu+\frac{1}{2})}, \tag{2.3.2}$$

where, when $\mu = 0$, we multiply (2.3.2) by μ^{-2} and set $\mu \to 0$ if $k \ge 1$ and multiply by an additional 2 if $k = 0$.

Proof Since the Gegenbauer polynomial C_k^{μ} is even if k is even and odd if k is odd, it follows readily that $P_{k,n}$ is a polynomial of degree at most n. Using the integral relation

$$\begin{aligned}\int_{B^2} f(x,y)\,\mathrm{d}x\,\mathrm{d}y &= \int_{-1}^{1}\int_{-\sqrt{1-x^2}}^{\sqrt{1-x^2}} f(x,y)\,\mathrm{d}x\,\mathrm{d}y \\ &= \int_{-1}^{1}\int_{-1}^{1} f\left(x, t\sqrt{1-x^2}\right)\mathrm{d}t\sqrt{1-x^2}\,\mathrm{d}x\end{aligned}$$

and the orthogonality of the Gegenbauer polynomials, we obtain

$$\langle P_j^m, P_k^n \rangle_\mu = \frac{\mu+\frac{1}{2}}{\pi} h_{n-k}^{k+\mu+1/2} h_k^{\mu} \delta_{n,m}\delta_{j,k}, \qquad 0 \le j \le m, \quad 0 \le k \le n,$$

where $h_k^{\mu} = \int_{-1}^{1} \left[C_k^{\mu}(x)\right]^2 (1-x^2)^{\mu-1/2}\,\mathrm{d}x$, from which the constant in (2.3.2) can be easily verified. □

In the case $\mu = 0$, the above proposition holds under the limit relation $\lim_{\mu\to 0} \mu^{-1} C_k^{\mu}(x) = (2/k)T_k(x)$, where T_k is the Chebyshev polynomial of the first kind. We state this case as a corollary.

Corollary 2.3.2 *For $0 \le k \le n$, define the polynomials*

$$P_k^n(x,y) = C_{n-k}^{k+1/2}(x)(1-x^2)^{k/2} T_k\left(\frac{y}{\sqrt{1-x^2}}\right). \tag{2.3.3}$$

Then $\{P_k^n : 0 \le k \le n\}$ is a mutually orthogonal basis for $\mathcal{V}_n^2(W_0)$. Moreover, $\langle P_0^n, P_0^n \rangle_0 = 1/(2n+1)$ and, for $0 < k \le n$,

$$\langle P_k^n, P_k^n \rangle_0 = \frac{(2k+1)_{n-k}k!}{2(n-k)!(\frac{1}{2})_k(2n+1)}. \tag{2.3.4}$$

Second orthonormal basis The basis is given in terms of the Jacobi polynomials in the polar coordinates $(x,y) = (r\cos\theta, r\sin\theta)$, $0 \le r \le 1$ and $0 \le \theta \le 2\pi$.

Proposition 2.3.3 *For $1 \le j \le \frac{n}{2}$, define*

$$\begin{aligned} P_{j,1}(x,y) &= [h_{j,n}]^{-1} P_j^{(\mu-1/2,n-2j)}(2r^2-1) r^{n-2j} \cos(n-2j)\theta, \\ P_{j,2}(x,y) &= [h_{j,n}]^{-1} P_j^{(\mu-1/2,n-2j)}(2r^2-1) r^{n-2j} \sin(n-2j)\theta, \end{aligned} \tag{2.3.5}$$

where the constant is given by

$$[h_{j,n}]^2 = \frac{(\mu+\frac{1}{2})_j (n-j)!(n-j+\mu+\frac{1}{2})}{j!(\mu+\frac{3}{2})_{n-j}(n+\mu+\frac{1}{2})} \begin{cases} \times 2 & n \ne 2j, \\ \times 1 & n = 2j. \end{cases} \tag{2.3.6}$$

Then $\{P_{j,1} : 0 \le j \le \frac{n}{2}\} \cup \{P_{j,2} : 0 \le j < \frac{n}{2}\}$ is an orthonormal basis of the space $\mathscr{V}_n^2(W(\mu)$.

Proof From Euler's formula $(x+\mathrm{i}y)^m = r^m(\cos m\theta + \mathrm{i}\sin m\theta)$, it follows readily that $r^m \cos m\theta$ and $r^m \sin m\theta$ are both polynomials of degree m in (x,y). Consequently, $P_{j,1}$ and $P_{j,2}$ are both polynomials of degree at most n in (x,y). Using the formula

$$\int_{B^2} f(x,y)\,\mathrm{d}x\,\mathrm{d}y = \int_0^1 r \int_0^{2\pi} f(r\cos\theta, r\sin\theta)\,\mathrm{d}r\,\mathrm{d}\theta,$$

that $P_{j,1}$ and $P_{k,2}$ are orthogonal to each other follows immediately from the fact that $\sin(n-2k)\theta$ and $\cos(n-2k)\theta)$ are orthogonal in $L^2([0,2\pi])$. Further, we have

$$\begin{aligned} \langle P_{j,1}, P_{k,1}\rangle_\mu = A_j \frac{\mu+\frac{1}{2}}{\pi} [h_{j,n}]^{-2} \int_0^1 r^{2n-2j+1} P_j^{(\mu-1/2,n-2j)}(2r^2-1) \\ \times P_k^{(\mu-1/2,n-2j)}(2r^2-1)(1-r^2)^{\mu-1/2}\,\mathrm{d}r, \end{aligned}$$

where $A_j = \pi$ if $2j \ne n$ and 2π if $2j = n$. Make the change of variables $t \mapsto 2r^2-1$, the orthogonality of $P_{j,1}$ and $P_{k,1}$ follows from that of the Jacobi polynomials. The orthogonality of $P_{j,2}$ and $P_{k,2}$ follows similarly. □

Orthogonal polynomials via the Rodrigues formula Classical orthogonal polynomials of one variable all satisfy Rodrigues' formula. For polynomials in two variables, an orthogonal basis can be defined by an analogue of the Rodrigues formula.

Proposition 2.3.4 *For $0 \le k \le n$, define*

$$U_{k,n}^\mu(x,y) = (1-x^2-y^2)^{-\mu+1/2} \frac{\partial^n}{\partial x^k \partial y^{n-k}} \left[(1-x^2-y^2)^{n+\mu-1/2}\right].$$

Then $\{U_{k,n}^\mu : 0 \le k \le n\}$ is a basis for $\mathscr{V}_n^2(W_\mu)$.

Proof Let $Q \in \Pi^2_{n-1}$. Directly from the definition,

$$\langle U^\mu_{k,n}, Q\rangle_\mu = \frac{\mu+\frac{1}{2}}{\pi}\int_{B^2} \frac{\partial^n}{\partial x^k \partial y^{n-k}}\left[(1-x^2-y^2)^{n+\mu-1/2}\right] Q(x,y)\,\mathrm{d}x\,\mathrm{d}y,$$

which is zero after integration by parts n times and using the fact that the nth-order derivative of Q is zero. □

This basis is not mutually orthogonal, that is, $\langle U^\mu_{k,n}, U^\mu_{j,n}\rangle_\mu \neq 0$ for $j \neq k$. Let $V^\mu_{m,n}$ be defined by

$$V^\mu_{m,n}(x,y) = \sum_{i=0}^{\lfloor m/2\rfloor}\sum_{j=0}^{\lfloor n/2\rfloor} \frac{(-m)_{2i}(-n)_{2j}(\mu+\frac{1}{2})_{m+n-i-j}}{2^{2i+2j}i!\,j!\,(\mu+\frac{1}{2})_{m+n}} x^{m-2i}y^{n-2j}.$$

Then $V_{m,n}$ is the orthogonal projection of $x^m y^n$ in $L^2(B^2, W_\mu)$ and the two families of polynomials are biorthogonal.

Proposition 2.3.5 *The set* $\{V^\mu_{k,n-k} : 0 \le k \le n\}$ *is a basis of* $\mathscr{V}^2_n(W_\mu)$ *and, for* $0 \le j,k \le n$,

$$\langle U^\mu_{k,n}, V^\mu_{j,n-j}\rangle_\mu = \frac{(\mu+\frac{1}{2})k!(n-k)!}{n+\mu+\frac{1}{2}}\delta_{j,k}.$$

The proof of this proposition will be given in Chapter 4 as a special case of orthogonal polynomials on a d-dimensional ball.

An orthogonal basis for constant weight In the case of a constant weight function, say $W_{1/2}(x) = 1/\pi$, an orthonormal basis can be given in terms of the Chebyshev polynomials U_n of the second kind; this has the distinction of being related to the Radon transform. Let ℓ denote the line $\ell(\theta,t) = \{(x,y) : x\cos\theta + y\sin\theta = t\}$ for $-1 \le t \le 1$, which is perpendicular to the direction $(\cos\theta, \sin\theta)$, $|t|$ being the distance between the line and the origin. Let $I(\theta,t) = \ell(\theta,t)\cap B^2$, the line segment of ℓ inside B^2. The Radon projection $\mathscr{R}_\theta(f;t)$ of a function f in the direction θ with parameter $t \in [-1,1]$ is defined by

$$\begin{aligned}\mathscr{R}_\theta(f;t) &:= \int_{I(\theta,t)} f(x,y)\,\mathrm{d}\ell \\ &= \int_{-\sqrt{1-t^2}}^{\sqrt{1-t^2}} f(t\cos\theta - s\sin\theta, t\sin\theta + s\cos\theta)\,\mathrm{d}s, \qquad (2.3.7)\end{aligned}$$

where $\mathrm{d}\ell$ denotes the Lebesgue measure on the line $I(\theta,t)$.

Proposition 2.3.6 *For* $0 \le k \le n$, *define*

$$P^n_k(x,y) = U_n\left(x\cos\frac{k\pi}{n+1} + y\sin\frac{k\pi}{n+1}\right). \qquad (2.3.8)$$

Then $\{P^n_k : 0 \le k \le n\}$ *is an orthonormal basis of* $\mathscr{V}^2_n(W_{1/2})$.

Proof For $\theta \in [0,\pi]$ and $g : \mathbb{R} \mapsto \mathbb{R}$, define $g_\theta(x,y) := g(x\cos\theta + y\sin\theta)$. The change of variables $t = x\cos\phi + y\sin\phi$ and $s = -x\sin\phi + y\cos\phi$ amounts to a rotation and leads to

$$\langle f, g_\theta \rangle := \frac{1}{\pi}\int_{B^2} f(x,y) g_\theta(x,y)\,\mathrm{d}x\,\mathrm{d}y = \frac{1}{\pi}\int_{-1}^{1} \mathscr{R}_\theta(f;t) g(t)\,\mathrm{d}t. \tag{2.3.9}$$

Now, a change of variable in (2.3.7) gives

$$\mathscr{R}_\theta(f;t) = \sqrt{1-t^2} \times \int_{-1}^{1} f\left(t\cos\theta - s\sqrt{1-t^2}\sin\theta,\quad t\sin\theta + s\sqrt{1-t^2}\cos\theta\right)\mathrm{d}s.$$

If f is a polynomial of degree m then the last integral is a polynomial in t of the same degree, since an odd power of $\sqrt{1-t}$ in the integrand is always companied by an odd power of s, which has integral zero. Therefore, $Q(t) := \mathscr{R}_\theta(f;t)/\sqrt{1-t^2}$ is a polynomial of degree m in t for every θ. Further, the integral also shows that $Q(1) = 2f(\cos\theta, \sin\theta)$. Consequently, by the orthogonality of U_k on $[-1,1]$, (2.3.9) shows that $\langle f, P_k^n \rangle = 0$ for all $f \in \Pi_m^2$ whenever $m < n$. Hence $P_k^n \in \mathscr{V}_n^2(W_{1/2})$.

In order to prove orthonormality, we consider $\mathscr{R}_\theta(P_j^n;t)$. By (2.3.9) and the fact that $P_j^n \in \mathscr{V}_n^2(W_{1/2})$,

$$\int_{-1}^{1} \frac{\mathscr{R}_\theta(P_j^n;t)}{\sqrt{1-t^2}} U_m(t)\sqrt{1-t^2}\,\mathrm{d}t = \int_{B^2} P_j^n(x,y)(U_m)_\theta(x,y)\mathrm{d}x\mathrm{d}y = 0,$$

for $m = 0,1,\ldots,n-1$. Thus the polynomial $Q(t) = \mathscr{R}_\theta(P_j^n;t)/\sqrt{1-t^2}$ is orthogonal to U_m for $0 \le m \le n-1$. Since Q is a polynomial of degree n, it must be a multiple of U_n; thus $Q(t) = cU_n(t)$ for some constant independent of t. Setting $t = 1$ and using the fact that $U_n(1) = n+1$, we have

$$c = \frac{P_j^n(\cos\,\theta,\,\sin\,\theta)}{n+1} = \frac{2U_n(\cos\,\theta - \frac{j\pi}{n+1})}{(n+1)}.$$

Consequently, we conclude that

$$\langle P_k^n, P_j^n \rangle = \frac{2}{\pi}\frac{U_n\left(\cos\frac{(k-j)\pi}{n+1}\right)}{n+1}\int_{-1}^{1}[U_n(t)]^2\sqrt{1-t^2}\,\mathrm{d}t = \delta_{k,j},$$

using the fact that $U_n\left(\cos\frac{(k-j)\pi}{n+1}\right) = \sin(k-j)\pi/\sin\frac{(k-j)\pi}{n+1} = 0$ when $k \ne j$. □

Since P_j^n is a basis of $\mathscr{V}^2(W_{1/2})$, the proof of the proposition immediately implies the following corollary.

Corollary 2.3.7 *If $P \in \mathscr{V}_n^2(W_{1/2})$ then for each $t \in (-1,1)$, $0 \le \theta \le 2\pi$,*

$$\mathscr{R}_\theta(P;t) = \frac{2}{n+1}\sqrt{1-t^2}\,U_n(t)P(\cos\theta, \sin\theta).$$

2.4 Orthogonal Polynomials on the Triangle

We now consider the triangle $T^2 := \{(x,y) : 0 \le x, y, x+y \le 1\}$. For $\alpha, \beta, \gamma > -1$, define the Jacobi weight function on the triangle,

$$W_{\alpha,\beta,\gamma}(x,y) = \frac{\Gamma(\alpha+\beta+\gamma+3)}{\Gamma(\alpha+1)\Gamma(\beta+1)\Gamma(\gamma+1)} x^\alpha y^\beta (1-x-y)^\gamma, \tag{2.4.1}$$

normalized so that its integral over T^2 is 1. Define, in this section,

$$\langle f, g \rangle_{\alpha,\beta,\gamma} = \int_{T^2} f(x,y) g(x,y) W_{\alpha,\beta,\gamma}(x,y)\, \mathrm{d}x\, \mathrm{d}y.$$

An orthonormal basis An orthonormal basis can be given in the Jacobi polynomials.

Proposition 2.4.1 *For $0 \le k \le n$, define*

$$P_k^n(x,y) = P_{n-k}^{(2k+\beta+\gamma+1,\alpha)}(2x-1)(1-x)^k P_k^{(\gamma,\beta)}\left(\frac{2y}{1-x} - 1\right). \tag{2.4.2}$$

Then $\{P_k^n : 0 \le k \le n\}$ is a mutually orthogonal basis of $\mathscr{V}_n^d(W_{\alpha,\beta,\gamma})$ and, further.

$$\begin{aligned} \langle P_k^n, P_k^n \rangle_{\alpha,\beta,\gamma} = & \frac{(\alpha+1)_{n-k}(\beta+1)_k(\gamma+1)_k(\beta+\gamma+2)_{n+k}}{(n-k)!k!(\beta+\gamma+2)_k(\alpha+\beta+\gamma+3)_{n+k}} \\ & \times \frac{(n+k+\alpha+\beta+\gamma+2)(k+\beta+\gamma+1)}{(2n+\alpha+\beta+\gamma+2)(2k+\beta+\gamma+1)}. \end{aligned} \tag{2.4.3}$$

Proof It is evident that $P_k^n \in \Pi_n^2$. We need the integral relation

$$\begin{aligned} \int_{T^2} f(x,y) dx dy &= \int_0^1 \int_0^{1-x} f(x,y) \mathrm{d}x \mathrm{d}y \\ &= \int_0^1 \int_0^1 f(x,(1-x)t))(1-x) \mathrm{d}t \mathrm{d}x. \end{aligned} \tag{2.4.4}$$

Since $P_k^n(x,(1-x)y) = P_{n-k}^{(2k+\beta+\gamma+1,\alpha)}(2x-1)(1-x)^k P_k^{(\gamma,\beta)}(2y-1)$ and, moreover, $W_{\alpha,\beta,\gamma}(x,(1-x)y) = c_{\alpha,\beta,\gamma} x^\alpha (1-x)^{\beta+\gamma} y^\beta (1-y)^\gamma$, where $c_{\alpha,\beta,\gamma}$ denotes the normalization constant of $W_{\alpha,\beta,\gamma}$, we obtain from the orthogonality of the Jacobi polynomials $P_n^{(a,b)}(2u-1)$ on the interval $[0,1]$ that

$$\langle P_j^m, P_k^n \rangle_{\alpha,\beta,\gamma} = c_{\alpha,\beta,\gamma} h_{n-k}^{(2k+\beta+\gamma+1,\alpha)} h_k^{(\gamma,\beta)} \delta_{j,k} \delta_{m,n},$$

where $h_k^{(a,b)} = \int_0^1 |P_k^{(a,b)}(2t-1)|^2 (1-t)^a t^b \,\mathrm{d}t$ and the constant (2.4.3) can be easily verified. □

The double integral over the triangle T^2 can be expressed as an iterated integral in different ways. This leads to two other orthonormal bases of $\mathscr{V}^2(W_{\alpha,\beta,\gamma})$.

Proposition 2.4.2 *Denote by $P_{k,n}^{\alpha,\beta,\gamma}$ the polynomials defined in* (2.4.2) *and by $h_{k,n}^{\alpha,\beta,\gamma}$ the constant* (2.4.3). *Define the polynomials*

$$Q_k^n(x,y) := P_{k,n}^{\beta,\alpha,\gamma}(y,x) \quad \text{and} \quad R_k^n(x,y) := P_{k,n}^{\gamma,\beta,\alpha}(1-x-y,\, y).$$

Then $\{Q_k^n : 0 \le k \le n\}$and $\{R_k^n : 0 \le k \le n\}$ are also mutually orthogonal bases of $\mathscr{V}_n^d(W_{\alpha,\beta,\gamma})$, and, furthermore,

$$\langle Q_k^n, Q_k^n \rangle_{\alpha,\beta,\gamma} = h_{k,n}^{\beta,\alpha,\gamma} \quad \text{and} \quad \langle R_k^n, R_k^n \rangle_{\alpha,\beta,\gamma} = h_{k,n}^{\gamma,\beta,\alpha}.$$

Proof For Q_k^n, we exchange x and y in the integral (2.4.4) and the same proof then applies. For R_k^n, we make the change of variables $(x,y) \mapsto (s,t)$ with $x = (1-s)(1-t)$ and $y = (1-s)t$ in the integral over the triangle T^2 to obtain

$$\int_{T^2} f(x,y)\,\mathrm{d}x\,\mathrm{d}y = \int_0^1 \int_0^1 f((1-s)(1-t),(1-s)t)\,(1-s)\,\mathrm{d}s\,\mathrm{d}t.$$

Since $P_{k,n}^{\gamma,\beta,\alpha}(1-x-y,y) = P_{k,n}^{\gamma,\beta,\alpha}(s,(1-s)t)$ and $W_{\alpha,\beta,\gamma}(x,y) = c_{\alpha,\beta,\gamma}(1-s)^{\alpha+\beta}s^{\gamma}(1-t)^{\alpha}t^{\beta}$, the proof follows, again from the orthogonality of the Jacobi polynomials. □

These bases illustrate the symmetry of the triangle T^2, which is invariant under the permutations of $(x,y,1-x-y)$.

Orthogonal polynomials via the Rodrigues formula The Jacobi polynomials can be expressed in the Rodrigues formula. There is an analogue for the triangle.

Proposition 2.4.3 *For $0 \le k \le n$, define*

$$U_{k,n}^{\alpha,\beta,\gamma}(x,y) := \left[W_{\alpha,\beta,\gamma}(x)\right]^{-1} \frac{\partial^n}{\partial x^k \partial y^{n-k}} \left[x^{k+\alpha} y^{n-k+\beta}(1-x-y)^{n+\gamma}\right].$$

Then the set $\{U_{k,n}^{\alpha,\beta,\gamma}(x,y) : 0 \le k \le n\}$ is a basis of $\mathscr{V}_n^2(W_{\alpha,\beta,\gamma})$.

Proof For $Q \in \Pi_{n-1}^2$, the proof of $\langle U_{k^n}, Q \rangle_{\alpha,\beta,\gamma} = 0$ follows evidently from integration by parts and the fact that the nth derivative of Q is zero. □

Just as for the disk, this basis is not mutually orthogonal. We can again define a biorthogonal basis by considering the orthogonal projection of $x^m y^n$ onto $\mathscr{V}_n^2(W_{\alpha,\beta,\gamma})$. Let $V_{m,n}^{\alpha,\beta,\gamma}$ be defined by

$$V_{m,n}^{(\alpha,\beta,\gamma)}(x,y) = \sum_{i=0}^{m} \sum_{j=0}^{n} (-1)^{n+m+i+j} \binom{m}{i} \binom{n}{j} \times \frac{(\alpha+\frac12)_m(\beta+\frac12)_n(\alpha+\beta+\gamma+\frac12)_{n+m+i+j}}{(\alpha+\frac12)_i(\beta+\frac12)_j(\alpha+\beta+\gamma+\frac12)_{2n+2m}} x^i y^j.$$

Proposition 2.4.4 *The set* $\{V^{\alpha,\beta,\gamma}_{k,n-k} : 0 \le k \le n\}$ *is a basis of* $\mathscr{V}^2_n(W_{\alpha,\beta,\gamma})$ *and, for* $0 \le j,k \le n$,

$$\langle U^{\alpha,\beta,\gamma}_{k,n}, V^{\alpha,\beta,\gamma}_{j,n-j}\rangle_{\alpha,\beta,\gamma} = \frac{(\alpha+\frac{1}{2})_k(\beta+\frac{1}{2})_{n-k}(\gamma+\frac{1}{2})_n k!(n-k)!}{(\alpha+\beta+\gamma+\frac{3}{2})_{2n}}\delta_{k,j}.$$

The proof of this proposition will be given in Chapter 4, as a special case of orthogonal polynomials on a d-dimensional simplex.

2.5 Orthogonal Polynomials and Differential Equations

The classical orthogonal polynomials in one variable, the Hermite, Laguerre and Jacobi polynomials, are eigenfunctions of second-order linear differential operators. A similar result also holds in the present situation.

Definition 2.5.1 A linear second-order partial differential operator L defined by

$$Lv := A(x,y)v_{xx} + 2B(x,y)v_{xy} + C(x,y)v_{yy} + D(x,y)v_x + E(x,y)v_y,$$

where $v_x := \partial v/\partial x$ and $v_{xx} := \partial^2 v/\partial x^2$ etc., is called *admissible* if for each nonnegative integer n there exists a number λ_n such that the equation

$$Lv = \lambda_n v$$

has $n+1$ linearly independent solutions that are polynomials of degree n and no nonzero solutions that are polynomials of degree less than n.

If a system of orthogonal polynomials satisfies an admissible equation then all orthogonal polynomials of degree n are eigenfunctions of the admissible differential operator for the same eigenvalue λ_n. In other words, the eigenfunction space for each eigenvalue is $\mathscr{V}^2_n$. This requirement excludes, for example, the product Jacobi polynomial $P^{(\alpha,\beta)}_k(x)P^{(\gamma,\mathrm{d})}_{n-k}$, which satisfies a second-order equation of the form $Lu = \lambda_{k,n}u$, where $\lambda_{k,n}$ depends on both k and n.

Upon considering lower-degree monomials, it is easy to see that for L to be admissible it is necessary that

$$\begin{aligned} A(x,y) &= Ax^2 + a_1x + b_1y + c_1, & D(x,y) &= Bx + d_1, \\ B(x,y) &= Axy + a_2x + b_2y + c_2, & E(x,y) &= By + d_2, \\ C(x,y) &= Ay^2 + a_3x + b_3y + c_3, \end{aligned}$$

and, furthermore, for each $n = 0, 1, 2, \ldots,$

$$nA + B \ne 0, \qquad \text{and} \qquad \lambda_n = -n[A(n-1)+B].$$

A classification of admissible equations is to be found in Krall and Sheffer [1967] and is given below without proof. Up to affine transformations, there are only nine equations. The first five are satisfied by orthogonal polynomials that we have already encountered. They are as follows:

1. $v_{xx} + v_{yy} - (xv_x + yv_y) = -nv$;
2. $xv_{xx} + yv_{yy} + (1+\alpha - x)v_x + (1+\beta - y)v_y = -nv$;
3. $v_{xx} + yv_{yy} - xv_x + (1+\alpha - y)v_y = -nv$;
4. $(1-x^2)v_{xx} - 2xyv_{xy} + (1-y^2)v_{yy} - (2\mu+1)(xv_x + yv_y) = -n(n+2\mu+2)v$;
5. $x(1-x)v_{xx} - 2xyv_{xy} + y(1-y)v_{yy} - \left[(\alpha+\beta+\gamma+\frac{3}{2})x - (\alpha+\frac{1}{2})\right] v_x$
$- \left[(\alpha+\beta+\gamma+\frac{3}{2})y - (\beta+\frac{1}{2})\right] v_y = -n\left(n+\alpha+\beta+\gamma+\frac{1}{2}\right) v.$

These are the admissible equations for, respectively, product Hermite polynomials, product Laguerre polynomials, product Hermite–Laguerre polynomials, orthogonal polynomials on the disk $\mathcal{V}_n^2(W_\mu)$ and orthogonal polynomials on the triangle $\mathcal{V}_n^2(W_{\alpha,\beta,,\gamma})$. The first three can be easily verified directly from the differential equations of one variable satisfied by the Hermite and Laguerre polynomials. The fourth and the fifth equations will be proved in Chapter 7 as special cases of the d-dimensional ball and simplex.

The other four admissible equations are listed below:

6. $3yv_{xx} + 2v_{xy} - xv_x - yv_y = \lambda v$;
7. $(x^2+y+1)v_{xx} + (2xy+2x)v_{xy} + (y^2+2y+1)v_{yy} + g(xv_x + yv_y) = \lambda v$;
8. $x^2 v_{xx} + 2xyv_{xy} + (y^2 - y)v_{yy} + g[(x-1)v_x + (y-\alpha)v_y] = \lambda v$;
9. $(x+\alpha)v_{xx} + 2(y+1)v_{yy} + xv_x + yv_y = \lambda v.$

The solutions for these last four equations are weak orthogonal polynomials, in the sense that the polynomials, are orthogonal with respect to a linear functional that is not necessarily positive definite.

2.6 Generating Orthogonal Polynomials of Two Variables

The examples of orthogonal polynomials that we have described so far are given in terms of orthogonal polynomials of one variable. In this section, we consider two methods that can be used to generate orthogonal polynomials of two variables from those of one variable.

2.6.1 A method for generating orthogonal polynomials

Let w_1 and w_2 be weight functions defined on the intervals (a,b) and (d,c), respectively. Let ρ be a positive function defined on (a,b) such that

Case I ρ is a polynomial of degree 1;

Case II ρ is the square root of a nonnegative polynomial of degree at most 2; we assume that $c = -d > 0$ and that w_2 is an even function on $(-c,c)$.

For each $k \in \mathbb{N}_0$ let $\{p_{n,k}\}_{n=0}^{\infty}$ denote the system of orthonormal polynomials with respect to the weight function $\rho^{2k+1}(x)w_1(x)$, and let $\{q_n\}$ be the system of orthonormal polynomials with respect to the weight function $w_2(x)$.

Proposition 2.6.1 *Define polynomials P_k^n of two variables by*

$$P_k^n(x,y) = p_{n-k,k}(x)[\rho(x)]^k q_k\left(\frac{y}{\rho(x)}\right), \qquad 0 \le k \le n. \tag{2.6.1}$$

Then $\{P_k^n : 0 \le k \le n\}$ is a mutually orthogonal basis of $\mathcal{V}_n^d(W)$ for the weight function

$$W(x,y) = w_1(x)w_2(\rho^{-1}(x)y), \qquad (x,y) \in R, \tag{2.6.2}$$

where the domain R is defined by

$$R = \{(x,y) : a < x < b,\ d\rho(x) < y < c\rho(x)\}. \tag{2.6.3}$$

Proof That the P_k^n are polynomials of degree n is evident in case 1; in case 2 it follows from the fact that q_k has the same parity as k, since w_2 is assumed to be even. The orthogonality of P_k^n can be verified by a change of variables in the integral

$$\begin{aligned}
&\iint_R P_k^n(x,y)P_j^m(x,y)W(x,y)\,\mathrm{d}x\,\mathrm{d}y \\
&\quad = \int_a^b p_{n-k,k}(x)p_{m-j,j}(x)\rho^{k+j+1}(x)w_1(x)\,\mathrm{d}x \int_c^d q_k(y)q_j(y)w_2(y)\,\mathrm{d}y \\
&\quad = h_{k,n}\delta_{n,m}\delta_{k,j}
\end{aligned}$$

for $0 \le j \le m$ and $0 \le k \le n$. □

Let us give several special cases that can be regarded as extensions of Jacobi polynomials in two variables.

Jacobi polynomials on the square Let $w_1(x) = (1-x)^{\alpha}(1+x)^{\beta}$ and $w_2(x) = (1-x)^{\gamma}(1+x)^{\delta}$ on $[-1,1]$ and let $\rho(x) = 1$. Then the weight function (2.6.2) becomes

$$W(x,y) = (1-x)^{\alpha}(1+x)^{\beta}(1-y)^{\gamma}(1+y)^{\delta}, \qquad \alpha,\beta,\gamma,\delta > -1,$$

on the domain $[-1,1]^2$ and the orthogonal polynomials P_k^n in (2.6.1) are exactly those in (2.2.3).

Jacobi polynomials on the disk Let $w_1(x) = w_2(x) = (1-x^2)^{\mu-1/2}$ on $[-1,1]$ and let $\rho(x) = (1-x^2)^{1/2}$. Then the weight function (2.6.2) becomes

$$W_\mu(x,y) = (1-x^2-y^2)^{\mu-1/2}, \quad \mu > -\frac{1}{2},$$

on the domain $B^2 = \{(x,y) : x^2+y^2 \le 1\}$, and the orthogonal polynomials P_k^n in (2.6.1) are exactly those in (2.3.1).

Jacobi polynomials on the triangle Let $w_1(x) = x^\alpha(1-x)^{\beta+\gamma}$ and $w_2(x) = x^\beta(1-x)^\gamma$, both defined on the interval $(0,1)$, and $\rho(x) = 1-x$. Then the weight function (2.6.2) becomes

$$W_{\alpha,\beta,\gamma}(x,y) = x^\alpha y^\beta (1-x-y)^\gamma, \qquad \alpha,\beta,\gamma > -1,$$

on the triangle $T^2 = \{(x,y) : x \ge 0, y \ge 0, 1-x-y \ge 0\}$, and the orthogonal polynomials P_k^n in (2.6.1) are exactly those in (2.4.2).

Orthogonal polynomials on a parabolic domain Let $w_1(x) = x^a(1-x)^b$ on $[0,1]$, $w_2(x) = (1-x^2)^a$ on $[-1,1]$, and $\rho(x) = \sqrt{x}$. Then the weight function (2.6.2) becomes

$$W_{a,b}(x,y) = (1-x)^b(x-y^2)^a, \quad y^2 < x < 1, \qquad \alpha,\beta > -1.$$

The domain $R = \{(x,y) : y^2 < x < 1\}$ is bounded by a straight line and a parabola. The mutually orthogonal polynomials P_k^n in (2.6.1) are given by

$$P_k^n(x,y) = p_{n-k}^{(a,b+k+1/2)}(2x-1)x^{k/2}p_k^{(b,b)}(y/\sqrt{x}), \qquad n \ge k \ge 0. \tag{2.6.4}$$

The set of polynomials $\{P_{k,n} : 0 \le k \le n\}$ is a mutually orthogonal basis of $\mathcal{V}_n^d(W_{a,b})$.

2.6.2 Orthogonal polynomials for a radial weight

A weight function W is called radial if it is of the form $W(x,y) = w(r)$, where $r = \sqrt{x^2+y^2}$. For such a weight function, an orthonormal basis can be given in polar coordinates $(x,y) = (r\cos\theta, r\sin\theta)$.

Proposition 2.6.2 *Let $p_m^{(k)}$ denote the orthogonal polynomial of degree m with respect to the weight function $r^{k+1}w(r)$ on $[0,\infty)$. Define*

$$\begin{aligned} P_{j,1}(x,y) &= p_{2j}^{(2n-4j)}(r)r^{n-2j}\cos(n-2j)\theta, \qquad 0 \le j \le \frac{n}{2}, \\ P_{j,2}(x,y) &= p_{2j}^{(2n-4j)}(r)r^{n-2j}\sin(n-2j)\theta, \qquad 0 \le j < \frac{n}{2}. \end{aligned} \tag{2.6.5}$$

Then $\{P_{j,1} : 0 \le j \le n/2\} \cup \{P_{j,2} : 0 \le j < n/2\}$ is a mutually orthogonal basis of $\mathcal{V}_n^2(W)$ with $W(x,y) = w(\sqrt{x^2+y^2})$.

The proof of this proposition follows the same line as that of Proposition 2.3.3 with an obvious modification. In the case of $w(r) = (1-r^2)^{\mu-1/2}$, the basis

(2.6.5) is, up to a normalization constant, that given in (2.3.5). Another example gives the second orthogonal basis for the product Hermite weight function $W(x,y) = e^{-x^2-y^2}$, for which an orthogonal basis is already given in (2.2.1).

Product Hermite weight $w(r) = e^{-r^2}$ The basis (2.6.5) is given by

$$\begin{aligned} P_{j,1}(x,y) &= L_j^{n-2j}(r^2) r^{(n-2j)} \cos(n-2j)\theta, \qquad 0 \le j \le \frac{n}{2}, \\ P_{j,2}(x,y) &= L_{2j}^{n-2j}(r^2) r^{(n-2j)} \sin(n-2j)\theta, \qquad 0 \le j < \frac{n}{2} \end{aligned} \tag{2.6.6}$$

in terms of the Laguerre polynomials of one variable.

Likewise, we call a weight function ℓ_1-radial if it is of the form $W(x,y) = w(s)$ with $s = x+y$. For such a weight function, a basis can be given in terms of orthogonal polynomials of one variable.

Proposition 2.6.3 *Let $p_m^{(k)}$ denote the orthogonal polynomial of degree m with respect to the weight function $s^{k+1}w(s)$ on $[0,\infty)$. Define*

$$P_k^n(x,y) = p_{n-k}^{(2k)}(x+y)(x+y)^k P_k\left(2\frac{x}{x+y} - 1\right), \qquad 0 \le k \le n, \tag{2.6.7}$$

where P_k is the Legendre polynomial of degree k. Then $\{P_k^n : 0 \le k \le n\}$ is a mutually orthogonal basis of $\mathscr{V}_n^2(W)$ with $W(x,y) = w(x+y)$ on $\mathbb{R}_+^2$.

Proof It is evident that P_k^n is a polynomial of degree n in (x,y). To verify the orthogonality we use the integral formula

$$\int_{\mathbb{R}_+^2} f(x,y)\,\mathrm{d}x\,\mathrm{d}y = \int_0^\infty \left(\int_0^1 f(st, s(1-t))\,\mathrm{d}t \right) s\,\mathrm{d}s$$

and the orthogonality of $p_m^{(2k)}$ and P_k. □

One example of proposition 2.6.3 is given by the Jacobi weight (2.4.1) on the triangle for $\alpha = \beta = 0$, for which $w(r) = (1-r)^\gamma$. Another example is given by the product Laguerre weight function $W(x,y) = e^{-x-y}$, for which a basis was already given in (2.2.2).

Product Laguerre weight $w(r) = e^{-r}$ The basis (2.6.7) is given by

$$P_k^n(x,y) = L_{n-k}^{2k+1}(x+y)(x+y)^k P_k\left(2\frac{x}{x+y} - 1\right), \qquad 0 \le k \le n, \tag{2.6.8}$$

in terms of the Laguerre polynomials and Legendre polynomials of one variable.

2.6.3 Orthogonal polynomials in complex variables

For a real-valued weight function W defined on $\Omega \subset \mathbb{R}^2$, orthogonal polynomials of two variables with respect to W can be given in complex variables z and $\bar{z}$.

For this purpose, we identify $\mathbb{R}^2$ with the complex plane $\mathbb{C}$ by setting $z = x + \mathrm{i}y$ and regard Ω as a subset of $\mathbb{C}$. We then consider polynomials in z and $\bar{z}$ that are orthogonal with respect to the inner product

$$\langle f, g \rangle_W^{\mathbb{C}} := \int_\Omega f(z,\bar{z})\overline{g(z,\bar{z})}w(z)\,\mathrm{d}x\,\mathrm{d}y, \tag{2.6.9}$$

where $w(z) = W(x,y)$. Let $\mathscr{V}_n^2(W,\mathbb{C})$ denote the space of orthogonal polynomials in z and $\bar{z}$ with respect to the inner product (2.6.9). In this subsection we denote by $P_{k,n}(x,y)$ real orthogonal polynomials with respect to W and denote by $Q_{k,n}(z,\bar{z})$ orthogonal polynomials in $\mathscr{V}_n^2(W,\mathbb{C})$.

Proposition 2.6.4 *The space $\mathscr{V}_n^2(W,\mathbb{C})$ has a basis $Q_{k,n}$ that satisfies*

$$Q_{k,n}(z,\bar{z}) = \overline{Q_{n-k,n}(z,\bar{z})}, \qquad 0 \le k \le n. \tag{2.6.10}$$

Proof Let $\{P_{k,n}(x,y) : 0 \le k \le n\}$ be a basis of $\mathscr{V}_n^2(W)$. We make the following definitions:

$$\begin{aligned} Q_{k,n}(z,\bar{z}) &:= \frac{1}{\sqrt{2}}\left[P_{k,n}(x,y) - \mathrm{i}P_{n-k,n}(x,y)\right], \qquad 0 \le k \le \left\lfloor \frac{n-1}{2} \right\rfloor, \\ Q_{k,n}(z,\bar{z}) &:= \frac{1}{\sqrt{2}}\left[P_{n-k,n}(x,y) + \mathrm{i}P_{k,n}(x,y)\right], \qquad \left\lfloor \frac{n+1}{2} \right\rfloor < k \le n, \\ Q_{n/2,n}(z,\bar{z}) &:= \frac{1}{\sqrt{2}}P_{n/2,n}(x,y) \quad \text{if } n \text{ is even.} \end{aligned} \tag{2.6.11}$$

If f and g are real-valued polynomials then $\langle f,g\rangle_W^{\mathbb{C}} = \langle f,g\rangle_W$. Hence, it is easy to see that $\{Q_{k,n} : 0 \le k \le n\}$ is a basis of $\mathscr{V}_n^2(W,\mathbb{C})$ and that this basis satisfies (2.6.10). □

Conversely, given $\{Q_{k,n} : 0 \le k \le n\} \in \mathscr{V}_n^2(W,\mathbb{C})$ that satisfies (2.6.10), we can define

$$\begin{aligned} P_{k,n}(x,y) &:= \frac{1}{\sqrt{2}}\left[Q_{k,n}(z,\bar{z}) + Q_{n-k,n}(z,\bar{z})\right], \qquad 0 \le k \le \frac{n}{2}, \\ P_{k,n}(x,y) &:= \frac{1}{\sqrt{2}\mathrm{i}}\left[Q_{k,n}(z,\bar{z}) - Q_{k-k,n}(z,\bar{z})\right], \qquad \frac{n}{2} < k \le n. \end{aligned} \tag{2.6.12}$$

It is easy to see that this is exactly the converse of (2.6.11). The relation (2.6.10) implies that the $P_{k,n}$ are real polynomials. In fact $P_{k,n}(x,y) = \mathrm{Re}\{Q_{k,n}(z,\bar{z})\}$ for $0 \le k \le n/2$ and $P_{k,n}(x,y) = \mathrm{Im}\{Q_k^n(z,\bar{z})\}$ for $n/2 < k \le n$. We summarize the relation below.

Theorem 2.6.5 *Let $P_{k,n}$ and $Q_{k,n}$ be related by* (2.6.11) *or, equivalently, by* (2.6.12). *Then:*

(i) $\{Q_{k,n} : 0 \le k \le n\}$ *is a basis of* $\mathcal{V}_n^2(W,\mathbb{C})$ *that satisfies* (2.6.10) *if and only if* $\{P_{k,n} : 0 \le k \le n\}$ *is a basis of* $\mathcal{V}_n^2(W)$;

(ii) $\{Q_{k,n} : 0 \le k \le n\}$ *is an orthonormal basis of* $\mathcal{V}_n^2(W,\mathbb{C})$ *if and only if* $\{P_{k,n} : 0 \le k \le n\}$ *is an orthonormal basis of* $\mathcal{V}_n^2(W)$.

Expressing orthogonal polynomials in complex variables can often lead to formulas that are more symmetric and easier to work with, as can be seen from the following two examples.

Complex Hermite polynomials

These are orthogonal with respect to the product Hermite weight

$$w_H(z) = \mathrm{e}^{-x^2-y^2} = \mathrm{e}^{-|z|^2}, \qquad z = x + \mathrm{i}y \in \mathbb{C}.$$

We define these polynomials by

$$H_{k,j}(z,\bar{z}) := (-1)^{k+j} e^{z\bar{z}} \left(\frac{\partial}{\partial z}\right)^k \left(\frac{\partial}{\partial \bar{z}}\right)^j \mathrm{e}^{-z\bar{z}}, \qquad k,j = 0,1,\ldots,$$

where, with $z = x + \mathrm{i}y$,

$$\frac{\partial}{\partial z} = \frac{1}{2}\left(\frac{\partial}{\partial x} - \mathrm{i}\frac{\partial}{\partial y}\right) \quad \text{and} \quad \frac{\partial}{\partial \bar{z}} = \frac{1}{2}\left(\frac{\partial}{\partial x} + \mathrm{i}\frac{\partial}{\partial y}\right).$$

By induction, it is not difficult to see that the $H_{k,j}$ satisfy an explicit formula:

$$H_{k,j}(z,\bar{z}) = k!j! \sum_{\nu=0}^{\min\{k,j\}} \frac{(-1)^\nu}{\nu!} \frac{z^{k-\nu}\bar{z}^{j-\nu}}{(k-\nu)!(j-\nu)!} = z^k \bar{z}^j {}_2F_0\left(-k,-j;\frac{1}{z\bar{z}}\right)$$

from which it follows immediately that (2.6.10) holds, that is,

$$H_{k,j}(z,\bar{z}) = \overline{H_{j,k}(z,\bar{z})}. \tag{2.6.13}$$

Working with the above explicit formula by rewriting the summation in ${}_2F_0$ in reverse order, it is easy to deduce from $(x)_{j-i} = (-1)^i (x)_j/(1-j-x)_i$ that $H_{k,j}$ can be written as a summation in ${}_1F_1$, which leads to

$$H_{k,j}(z,\bar{z}) = (-1)^j j! z^{k-j} L_j^{k-j}(|z|^2), \qquad k \ge j, \tag{2.6.14}$$

where L_j^α is a Laguerre polynomial.

Proposition 2.6.6 *The complex Hermite polynomials satisfy the following properties:*

(i) $\dfrac{\partial}{\partial z} H_{k,j} = \bar{z} H_{k,j} - H_{k,j+1}, \quad \dfrac{\partial}{\partial \bar{z}} H_{k,j} = z H_{k,j} - H_{k+1,j}$;

(ii) $z H_{k,j} = H_{k+1,j} + j H_{k,j-1}, \quad \bar{z} H_{k,j} = H_{k,j+1} + k H_{k-1,j}$;

(iii) $\displaystyle\int_{\mathbb{C}} H_{k,j}(z,\bar{z}) \overline{H_{m,l}(z,\bar{z})} w_H(z)\, \mathrm{d}x\,\mathrm{d}y = j!k!\delta_{k,m}\delta_{j,l}$.

Proof Part (i) follows from directly from the above definition of the polynomials. Part (ii) follows from (2.6.14) and the identity $L_n^\alpha(x) = L_n^{\alpha+1}(x) - L_{n-1}^{\alpha+1}(x)$. The orthogonal relation (iii) follows from writing the integral in polar coordinates and applying the orthogonality of the Laguerre polynomials. □

Disk polynomials

Let $\mathrm{d}\sigma_\lambda(z) := W_{\lambda+1/2}(x,y)\,\mathrm{d}x\,\mathrm{d}y$ and define

$$P_{k,j}^\lambda(z,\bar{z}) = \frac{j!}{(\lambda+1)_j} P_j^{(\lambda,k-j)}(2|z|^2-1)z^{k-j}, \qquad k > j, \tag{2.6.15}$$

which is normalized by $P_{k,j}^\lambda(1,1) = 1$. For $k \le j$, we use

$$P_{k,j}^\lambda(z,\bar{z}) = \overline{P_{j,k}^\lambda(\bar{z},z)}. \tag{2.6.16}$$

Written in terms of hypergeometric functions, these polynomials take a more symmetric form:

$$P_{k,j}^\lambda(z,\bar{z}) = \frac{(\lambda+1)_{k+j}}{(\lambda+1)_k(\lambda+1)_j} z^k \bar{z}^j {}_2F_1\left(\begin{matrix}-k,-j\\ -\lambda-k-j\end{matrix}; \frac{1}{z\bar{z}}\right), \qquad k,j \ge 0.$$

To see this, write the summation in the ${}_2F_1$ sum in reverse order and use $(x)_{j-i} = (-1)^i (x)_j/(1-j-x)_i$, to get

$$P_{k,j}^\lambda(z,\bar{z}) = \frac{(k-j+1)_j}{(\lambda+1)_j}(-1)^j {}_2F_1\left(\begin{matrix}-j,k+\lambda+1\\ k-j+1\end{matrix}; |z|^2\right) z^{k-j},$$

which is (2.6.15) by Proposition 1.4.14 and $P_n^{(\beta,\alpha)}(t) = (-1)^n P_n^{(\alpha,\beta)}(t)$.

Proposition 2.6.7 *The disk polynomials satisfy the following properties:*

(i) $|P_{k,j}^\lambda(z,\bar{z})| \le 1$ *for* $|z| \le 1$ *and* $\lambda \ge 0$;

(ii) $zP_{k,j}^\lambda(z,\bar{z}) = \dfrac{\lambda+k+1}{\lambda+k+j+1} P_{k+1,j}^\lambda(z,\bar{z}) + \dfrac{j}{\lambda+k+j+1} P_{k,j-1}^\lambda(z,\bar{z})$

with a similar relation for $\bar{z}P_{k,j}^\lambda(z)$ *upon using* (2.6.16);

(iii) $\displaystyle\int_D P_{k,j}^\lambda \overline{P_{m,l}^\lambda}\,\mathrm{d}\sigma_\lambda(z) = \frac{\lambda+1}{\lambda+k+j+1}\frac{k!\,j!}{(\lambda+1)_k(\lambda+1)_j}\delta_{k,m}\delta_{j,l}$.

Proof Part (i) follows on taking absolute values inside the ${}_2F_1$ sum and using the fact that $|P_n^{(\alpha,\beta)}(-1)| = (\beta+1)_n/n!$; the condition $\lambda \ge 0$ is used to ensure that the coefficients in the ${}_2F_1$ sum are positive. The recursive relation (ii) is proved using the following formula for Jacobi polynomials:

$$(2n+\alpha+\beta+1)P_n^{(\alpha,\beta)}(t) = (n+\alpha+\beta+1)P_n^{(\alpha,\beta+1)}(t) + (n+\alpha)P_{n-1}^{(\alpha,\beta+1)}(t),$$

which can be verified using the ${}_2F_1$ expansion of Jacobi polynomials in Proposition 1.4.14. Finally, the orthogonal relation (iii) is established using the polar coordinates $z = re^{i\theta}$ and the structure constant of Jacobi polynomials. □

2.7 First Family of Koornwinder Polynomials

The Koornwinder polynomials are based on symmetric polynomials. Throughout this section let w be a weight function defined on $[-1,1]$. For $\gamma > -1$, let us define a weight function of two variables,

$$B_\gamma(x,y) := w(x)w(y)|x-y|^{2\gamma+1}, \qquad (x,y) \in [-1,1]^2. \tag{2.7.1}$$

Since B_γ is evidently symmetric in x, y, we need only consider its restriction on the triangular domain $\triangle$ defined by

$$\triangle := \{(x,y) : -1 < x < y < 1\}.$$

Let Ω be the image of $\triangle$ under the mapping $(x,y) \mapsto (u,v)$ defined by

$$u = x+y, \quad v = xy. \tag{2.7.2}$$

This mapping is a bijection between $\triangle$ and Ω since the Jacobian of the change of variables is $\mathrm{d}u\,\mathrm{d}v = |x-y|\,\mathrm{d}x\,\mathrm{d}y$. The domain Ω is given by

$$\Omega := \{(u,v) : 1+u+v > 0, 1-u+v > 0, u^2 > 4v\} \tag{2.7.3}$$

and is depicted in Figure 2.1. Under the mapping (2.7.2), the weight function B_γ becomes a weight function defined on the domain Ω by

$$W_\gamma(u,v) := w(x)w(y)(u^2-4v)^\gamma, \quad (u,v) \in \Omega, \tag{2.7.4}$$

where the variables (x,y) and (u,v) are related by (2.7.2).

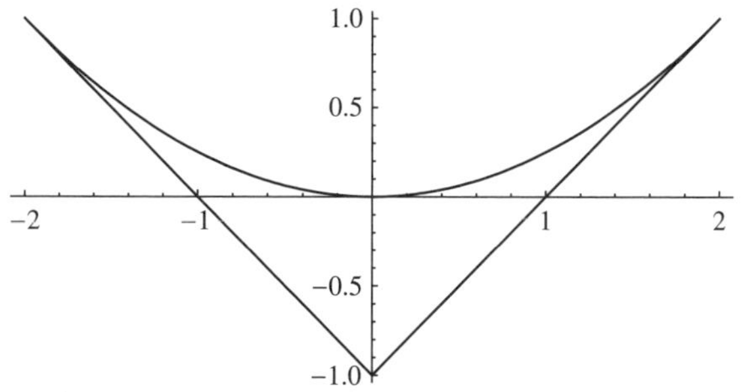

Figure 2.1 Domains for Koornwinder orthogonal polynomials.

Proposition 2.7.1 *Let $\mathcal{N} = \{(k,n) : 0 \le k \le n\}$. In $\mathcal{N}$ define an order $\prec$ as follows: $(j,m) \prec (k,n)$ if $m < n$ or $m = n$ and $j \le k$. Define monic polynomials $P_{k,n}^{(\gamma)}$ under the order $\prec$,*

$$P_{k,n}^{(\gamma)}(u,v) = u^{n-k}v^k + \sum_{(j,m)\prec(k,n)} a_{j,m} u^{m-j} v^j, \tag{2.7.5}$$

that satisfy the orthogonality condition

$$\int_\Omega P_{k,n}^{(\gamma)}(u,v) u^{m-j} v^j W_\gamma(u,v)\, du\, dv = 0, \qquad \forall (j,m) \prec (k,n). \tag{2.7.6}$$

Then these polynomials are uniquely determined and are mutually orthogonal with respect to W_γ.

Proof It is easy to see that $\prec$ is a total order, so that the monomials can be ordered linearly with this order. Applying the Gram–Schmidt orthogonalization process to the monomials so ordered, the uniqueness follows from the fact that $P_{k,n}^{(\gamma)}$ has leading coefficient 1. □

In the case $\gamma = \pm\frac{1}{2}$, the orthogonal polynomials $P_{k,n}^{(\gamma)}$ can be given explicitly in terms of orthogonal polynomials of one variable. Let $\{p_n\}_{m=0}^{\infty}$ be the sequence of monic orthogonal polynomials $p_n(x) = x^n +$ lower-degree terms, with respect to w.

Proposition 2.7.2 *The polynomials $P_{k,n}^{(\gamma)}$ for $\gamma = \pm\frac{1}{2}$ are given by*

$$P_{k,n}^{(-1/2)}(u,v) = \begin{cases} p_n(x)p_k(y) + p_n(y)p_k(x), & k < n, \\ p_n(x)p_n(y), & k = n \end{cases} \tag{2.7.7}$$

and

$$P_{k,n}^{(1/2)}(u,v) = \frac{p_{n+1}(x)p_k(y) - p_{n+1}(y)p_k(x)}{x-y}, \tag{2.7.8}$$

where (u,v) are related to (x,y) by $x = u+v$, $y = uv$.

Proof By the fundamental theorem of symmetric polynomials, each symmetric polynomial in x,y can be written as a polynomial in the elementary symmetric polynomials $x+y$ and xy. A quick computation shows that

$$x^n y^k + y^n x^k = u^{n-k} v^k + \text{lower-degree terms in } (u,v),$$

so that the right-hand sides of both (2.7.7) and (2.7.8) are of the form (2.7.5). Furthermore, the change of variables $(x,y) \mapsto (u,v)$ shows that

$$\int_\Omega f(u,v) W_\gamma(u,v)\,\mathrm{d}u\,\mathrm{d}v = \int_\triangle f(x+y,xy) B_\gamma(x,y)\,\mathrm{d}x\,\mathrm{d}y$$
$$= \frac{1}{2}\int_{[-1,1]^2} f(x+y,xy) B_\gamma(x,y)\,\mathrm{d}x\,\mathrm{d}y, \qquad (2.7.9)$$

where the second line follows since the integrand is a symmetric function of x and y and where $[-1,1]^2$ is the union of $\triangle$ and its image under $(x,y) \mapsto (y,x)$. Using (2.7.9), the orthogonality of $P_{k,n}^{(-1/2)}$ and $P_{k,n}^{(1/2)}$ in the sense of (2.7.6) can be verified directly from the orthogonality of p_n. The proof is then complete upon using Proposition 2.7.1. □

Of particular interest is the case when w is the Jacobi weight function $w(x) = (1-x)^\alpha(1+x)^\beta$, for which the weight function W_γ becomes

$$W_{\alpha,\beta,\gamma}(u,v) = (1-u+v)^\alpha(1+u+v)^\beta(u^2-4v)^\gamma, \qquad (2.7.10)$$

where $\alpha,\beta,\gamma > -1$, $\alpha+\gamma+\frac{3}{2} > 0$ and $\beta+\gamma+\frac{3}{2} > 0$; these conditions guarantee that $W_{\alpha,\beta,\gamma}$ is integrable. The polynomials p_n in (2.7.7) and (2.7.8) are replaced by monic Jacobi polynomials $P_n^{(\alpha,\beta)}$, and we denote the orthogonal polynomials $P_{k,n}^{(\gamma)}$ in Proposition 2.7.1 by $P_{k,n}^{\alpha,\beta,\gamma}$. This is the case originally studied by Koornwinder. These polynomials are eigenfunctions of a second-order differential operator.

Proposition 2.7.3 *Consider the differential operator*

$$\begin{aligned} L_{\alpha,\beta,\gamma} g &= (-u^2+2v+2)g_{uu} + 2u(1-v)g_{uv} + (u^2-2v^2-2v)g_{vv} \\ &\quad + [-(\alpha+\beta+2\gamma+3)u + 2(\beta-\alpha)]\,g_u \\ &\quad + [-(\beta-\alpha)u - (2\alpha+2\beta+2\gamma+5)v - (2\gamma+1)]\,g_v. \end{aligned}$$

Then, for $(k,n) \in \mathcal{N}$,

$$L_{\alpha,\beta,\gamma} P_{k,n}^{\alpha,\beta,\gamma} = -\lambda_{k,n}^{\alpha,\beta,\gamma} P_{k,n}^{\alpha,\beta,\gamma}, \qquad (2.7.11)$$

where $\lambda_{k,n}^{\alpha,\beta,\gamma} := n(n+\alpha+\beta+2\gamma+2) - k(k+\alpha+\beta+1)$.

Proof A straightforward computation shows that, for $0 \le k \le n$,

$$L_{\alpha,\beta,\gamma} u^{n-k}v^k = \lambda_{k,n}^{\alpha,\beta,\gamma} u^{n-k}v^k + \sum_{(j,m)\prec(k,n)} b_{j,m} u^{m-j} v^j \qquad (2.7.12)$$

for some $b_{j,m}$, which implies that $L_{\alpha,\beta,\gamma} P_{k,n}^{\alpha,\beta,\gamma}$ is a polynomial of the same form and degree. From the definition of $L_{\alpha,\beta,\gamma}$ it is easy to verify directly that $L_{\alpha,\beta,\gamma}$ is self-adjoint in $L^2(\Omega, W_{\alpha,\beta,\gamma})$ so that, for $(j,m) \in \mathcal{N}$ with $(j,m) \prec (k,n)$,

$$\int_\Omega \left(L_{\alpha,\beta,\gamma} P_{k,n}^{\alpha,\beta,\gamma}\right) P_{j,m}^{\alpha,\beta,\gamma} W_{\alpha,\beta,\gamma} \mathrm{d}u\,\mathrm{d}v$$
$$= \int_\Omega P_{k,n}^{\alpha,\beta,\gamma} \left(L_{\alpha,\beta,\gamma} P_{j,m}^{\alpha,\beta,\gamma}\right) W_{\alpha,\beta,\gamma} \mathrm{d}u\,\mathrm{d}v = 0,$$

by (2.7.6). By the uniqueness referred to in Proposition 2.7.1, we conclude that $L_{\alpha,\beta,\gamma} P_{k,n}^{\alpha,\beta,\gamma}$ is a constant multiple of $P_{k,n}^{\alpha,\beta,\gamma}$ and that the constant is exactly $\lambda_{k,n}^{\alpha,\beta,\gamma}$, as can be seen from (2.7.12). □

It should be noted that although $L_{\alpha,\beta,\gamma}$ is of the same type as the differential operator in Definition 2.5.1 it is not admissible, since the eigenvalues $\lambda_{k,m}^{\alpha,\beta,\gamma}$ depend on both k and n.

2.8 A Related Family of Orthogonal Polynomials

We consider the family of weight functions defined by

$$\mathscr{W}_{\alpha,\beta,\gamma}(x,y) := |x-y|^{2\alpha+1}|x+y|^{2\beta+1}(1-x^2)^\gamma(1-y^2)^\gamma \tag{2.8.1}$$

for $(x,y) \in [-1,1]^2$, where $\alpha,\beta,\gamma > -1$, $\alpha+\gamma+\frac{3}{2} > 0$ and $\beta+\gamma+\frac{3}{2} > 0$. These weight functions are related to the $W_{\alpha,\beta,\gamma}$ defined in (2.7.10). Indeed, let Ω in (2.7.3) be the domain of $W_{\alpha,\beta,\gamma}$ and define Ω^* by

$$\Omega^* := \{(x,y) : -1 < -y < x < y < 1\},$$

which is a quadrant of $[-1,1]^2$. We have the following relation:

Proposition 2.8.1 *The mapping $(x,y) \mapsto (2xy, x^2+y^2-1)$ is a bijection from Ω^* onto Ω, and*

$$\mathscr{W}_{\alpha,\beta,\gamma}(x,y) = 4^\gamma |x^2-y^2| W_{\alpha,\beta,\gamma}(2xy, x^2+y^2-1) \tag{2.8.2}$$

for $(x,y) \in [-1,1]^2$. Further,

$$4^\gamma \int_\Omega f(u,v) W_{\alpha,\beta,\gamma}(u,v)\,\mathrm{d}u\,\mathrm{d}v$$
$$= \int_{[-1,1]^2} f(2xy, x^2+y^2-1)\mathscr{W}_{\alpha,\beta,\gamma}(x,y)\,\mathrm{d}x\,\mathrm{d}y. \tag{2.8.3}$$

Proof For $(x,y) \in [-1,1]^2$, let us write $x = \cos\theta$ and $y = \cos\phi$, $0 \le \theta,\phi \le \pi$. Then it is easy to see that

$$2xy = \cos(\theta-\phi) + \cos(\theta+\phi), \qquad x^2+y^2-1 = \cos(\theta-\phi)\cos(\theta+\phi),$$

from which it follows readily that $(2xy, x^2+y^2-1) \in \Omega$. The identity (2.8.2) follows from a straightforward verification. For the change of variable $u = 2xy$ and $v = x^2+y^2-1$, we have $\mathrm{d}u\,\mathrm{d}v = 4|x^2-y^2|\,\mathrm{d}x\,\mathrm{d}y$, from which the integration relation (2.8.3) follows. □

An orthogonal basis for $\mathscr{W}$ can be given accordingly. Let $P_{k,n}^{\alpha,\beta,\gamma}$, $0 \le k \le n$, denote a basis of mutually orthogonal polynomials of degree n for $W_{\alpha,\beta,\gamma}$.

Theorem 2.8.2 *For $n \in \mathbb{N}_0$, a mutually orthogonal basis of $\mathscr{V}_{2n}(\mathscr{W}_{\alpha,\beta,\gamma})$ is given by*

$$ {}_1Q_{k,2n}^{\alpha,\beta,\gamma}(x,y) := P_{k,n}^{\alpha,\beta,\gamma}(2xy, x^2+y^2-1), \qquad 0 \le k \le n, $$
$$ {}_2Q_{k,2n}^{\alpha,\beta,\gamma}(x,y) := (x^2-y^2)P_{k,n-1}^{\alpha+1,\beta+1,\gamma}(2xy, x^2+y^2-1), \qquad 0 \le k \le n-1 $$

and a mutually orthogonal basis of $\mathscr{V}_{2n+1}(\mathscr{W}_{\alpha,\beta,\gamma})$ is given by

$$ {}_1Q_{k,2n+1}^{\alpha,\beta,\gamma}(x,y) := (x+y)P_{k,n}^{\alpha,\beta+1,\gamma}(2xy, x^2+y^2-1), \qquad 0 \le k \le n, $$
$$ {}_2Q_{k,2n+1}^{\alpha,\beta,\gamma}(x,y) := (x-y)P_{k,n}^{\alpha+1,\beta,\gamma}(2xy, x^2+y^2-1), \qquad 0 \le k \le n. $$

In particular, when $\gamma = \pm\frac{1}{2}$ the basis can be given in terms of the Jacobi polynomials of one variable, upon using (2.7.7) *and* (2.7.8).

Proof These polynomials evidently form a basis if they are orthogonal. Let us denote by $\langle \cdot,\cdot \rangle_{W_{\alpha,\beta,\gamma}}$ and $\langle \cdot,\cdot \rangle_{\mathscr{W}_{\alpha,\beta,\gamma}}$ the inner product in $L^2(\Omega, W_{\alpha,\beta,\gamma})$ and in $L^2([-1,1]^2, \mathscr{W}_{\alpha,\beta,\gamma})$, respectively. By (2.8.3), for $0 \le j \le m$ and $0 \le k \le n$,

$$ \left\langle {}_1Q_{k,2n}^{\alpha,\beta,\gamma}, {}_1Q_{j,2m}^{\alpha,\beta,\gamma} \right\rangle_{\mathscr{W}_{\alpha,\beta,\gamma}} = \langle P_{k,n}^{\alpha,\beta,\gamma}, P_{j,m}^{\alpha,\beta,\gamma} \rangle_{W_{\alpha,\beta,\gamma}} = 0 $$

for $(j,2m) \prec (k,2n)$ and, furthermore,

$$ \left\langle {}_1Q_{k,2n}^{\alpha,\beta,\gamma}, {}_2Q_{j,2m}^{\alpha,\beta,\gamma} \right\rangle_{\mathscr{W}_{\alpha,\beta,\gamma}} = \int_{[-1,1]^2} (x^2-y^2)P_{k,n}^{\alpha,\beta,\gamma}(2xy, x^2+y^2-1) \times P_{k-1,n}^{\alpha+1,\beta+1,\gamma}(2xy, x^2+y^2-1)\mathscr{W}_{\alpha,\beta,\gamma}(x,y)\,dx\,dy. $$

The right-hand side of the above equation changes sign under the change of variables $(x,y) \mapsto (y,x)$, which shows that $\left\langle {}_1Q_{k,2n}^{\alpha,\beta,\gamma}, {}_2Q_{j,2m}^{\alpha,\beta,\gamma} \right\rangle_{\mathscr{W}_{\alpha,\beta,\gamma}} = 0$. Moreover, since $(x^2-y^2)^2\mathscr{W}_{\alpha,\beta,\gamma}(x,y)\,dx\,dy$ is equal to a constant multiple of $W_{\alpha+1,\beta+1,\gamma}(u,v)\,du\,dv$, we see that

$$ \left\langle {}_2Q_{k,2n}^{\alpha,\beta,\gamma}, {}_2Q_{j,2m}^{\alpha,\beta,\gamma} \right\rangle_{\mathscr{W}_{\alpha,\beta,\gamma}} = \left\langle P_{k,n-1}^{\alpha+1,\beta+1,\gamma}, P_{j,m-1}^{\alpha+1,\beta+1,\gamma} \right\rangle_{W_{\alpha+1,\beta+1,\gamma}} = 0 $$

for $(k,2n) \ne (j,2m)$. Furthermore, setting $F = P_{k,n}^{(\gamma)}P_{k,n}^{(\gamma),0,1}$, we obtain

$$ \left\langle {}_1Q_{k,2n}^{\alpha,\beta,\gamma}, {}_1Q_{j,2m+1}^{\alpha,\beta,\gamma} \right\rangle_{\mathscr{W}_{\alpha,\beta,\gamma}} = \int_{[-1,1]^2} (x+y)P_{k,n}^{\alpha,\beta,\gamma}(2xy, x^2+y^2-1) \times P_{k,n}^{\alpha,\beta+1,\gamma}(2xy, x^2+y^2-1)\mathscr{W}_{\alpha,\beta,\gamma}(x,y)\,dx\,dy, $$

which is equal to zero since the right-hand side changes sign under $(x,y) \mapsto (-x,-y)$. The same proof shows also that $\left\langle {}_1Q_{k,2n}^{\alpha,\beta,\gamma}, {}_2Q_{j,2m+1}^{\alpha,\beta,\gamma} \right\rangle_{\mathscr{W}_{\alpha,\beta,\gamma}} = 0$. Together, these facts prove the orthogonality of ${}_1Q_{k,2n}^{\alpha,\beta,\gamma}$ and ${}_2Q_{k,2n}^{\alpha,\beta,\gamma}$.

Since $(x-y)(x+y) = x^2 - y^2$ changes sign under $(x,y) \mapsto (y,x)$, the same consideration shows that $\left\langle {}_1Q_{k,2n+1}^{\alpha,\beta,\gamma}, {}_2Q_{j,2m+1}^{\alpha,\beta,\gamma} \right\rangle_{\mathscr{W}_{\alpha,\beta,\gamma}} = 0$. Finally,

$$\left\langle {}_1Q_{k,2n+1}^{\alpha,\beta,\gamma}, {}_1Q_{j,2m+1}^{\alpha,\beta,\gamma} \right\rangle_{\mathscr{W}_{\alpha,\beta,\gamma}} = \left\langle P_{k,n}^{\alpha,\beta+1,\gamma}, P_{j,m}^{\alpha,\beta+1,\gamma} \right\rangle_{W_{\alpha,\beta,\gamma+1}} = 0,$$

$$\left\langle {}_2Q_{k,2n+1}^{\alpha,\beta,\gamma}, {}_2Q_{j,2m+1}^{\alpha,\beta,\gamma} \right\rangle_{\mathscr{W}_{\alpha,\beta,\gamma}} = \left\langle P_{k,n}^{\alpha+1,\beta,\gamma}, P_{j,m}^{\alpha+1,\beta,\gamma} \right\rangle_{W_{\alpha+1,\beta,\gamma}} = 0$$

for $(k,n) \neq (j,m)$, proving the orthogonality of ${}_iQ_{k,2n+1}^{\alpha,\beta,\gamma}$, $i = 1,2$. □

The structure of the basis in Theorem 2.8.2 can also be extended to the more general weight function $\mathscr{W}_\gamma(x,y) = |x^2 - y^2| W_\gamma(2xy, x^2 + y^2 - 1)$, with W_γ as in (2.7.4).

2.9 Second Family of Koornwinder Polynomials

The second family of Koornwinder polynomials is based on orthogonal exponentials over a regular hexagonal domain. It is convenient to use homogeneous coordinates in

$$\mathbb{R}_H^3 := \{\mathbf{t} = (t_1, t_2, t_3) \in \mathbb{R}^3 : t_1 + t_2 + t_3 = 0\},$$

in which the hexagonal domain becomes

$$\Omega := \{\mathbf{t} \in \mathbb{R}_H^3 : -1 \le t_1, t_2, t_3 \le 1\},$$

as seen in Figure 2.2. In this section we adopt the convention of using bold face letters, such as $\mathbf{t}$ and $\mathbf{k}$, to denote homogeneous coordinates. For $\mathbf{t} \in \mathbb{R}_H^3$ and $\mathbf{k} \in \mathbb{R}_H^3 \cap \mathbb{Z}^3$, define the exponential function

$$\phi_{\mathbf{k}}(\mathbf{t}) := \mathrm{e}^{2\pi \mathrm{i} \mathbf{k} \cdot \mathbf{t}/3} = \mathrm{e}^{2\pi \mathrm{i}(k_1 t_1 + k_2 t_2 + k_3 t_3)/3}.$$

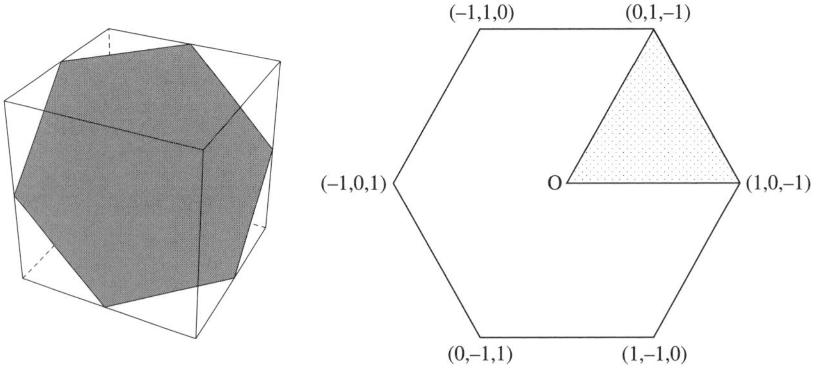

Figure 2.2 A hexagonal domain in homogeneous coordinates.

Proposition 2.9.1 *For* $\mathbf{k},\mathbf{j} \in \mathbb{R}^3_H \cap \mathbb{Z}^3$,

$$\tfrac{1}{3}\int_\Omega \phi_\mathbf{k}(\mathbf{t})\overline{\phi_\mathbf{j}(\mathbf{t})}\,\mathrm{d}\mathbf{t} = \delta_{\mathbf{k},\mathbf{j}}.$$

Proof Using homogeneous coordinates, it is easy to see that

$$\int_\Omega f(\mathbf{t})\,\mathrm{d}\mathbf{t} = \int_0^1 \mathrm{d}t_1 \int_{-1}^0 f(\mathbf{t})\,\mathrm{d}t_2 + \int_0^1 \mathrm{d}t_2 \int_{-1}^0 f(\mathbf{t})\,\mathrm{d}t_3 + \int_0^1 \mathrm{d}t_3 \int_{-1}^0 f(\mathbf{t})\,\mathrm{d}t_1,$$

from which the orthogonality can be easily verified using the homogeneity of $\mathbf{t}$, $\mathbf{k}$ and $\mathbf{j}$. □

Evidently the hexagon (see Figure 2.2) is invariant under the reflection group $\mathscr{A}_2$ generated by the reflections in planes through the edges of the shaded equilateral triangle and perpendicular to its plane. In homogeneous coordinates the three reflections $\sigma_1, \sigma_2, \sigma_3$ are defined by

$$\mathbf{t}\sigma_1 := -(t_1,t_3,t_2), \qquad \mathbf{t}\sigma_2 := -(t_2,t_1,t_3), \qquad \mathbf{t}\sigma_3 := -(t_3,t_2,t_1).$$

Because of the relations $\sigma_3 = \sigma_1\sigma_2\sigma_1 = \sigma_2\sigma_1\sigma_2$, the group $\mathscr{G}$ is given by

$$\mathscr{G} = \{1, \sigma_1, \sigma_2, \sigma_3, \sigma_1\sigma_2, \sigma_2\sigma_1\}.$$

For a function f in homogeneous coordinates, the action of the group $\mathscr{G}$ on f is defined by $\sigma f(\mathbf{t}) = f(\mathbf{t}\sigma)$, $\sigma \in \mathscr{G}$. We define functions

$$\begin{aligned}
\mathsf{TC}_\mathbf{k}(\mathbf{t}) :=& \tfrac{1}{6}\left[\phi_{k_1,k_2,k_3}(\mathbf{t}) + \phi_{k_2,k_3,k_1}(\mathbf{t}) + \phi_{k_3,k_1,k_2}(\mathbf{t})\right.\\
&\left. + \phi_{-k_1,-k_3,-k_2}(\mathbf{t}) + \phi_{-k_2,-k_1,-k_3}(\mathbf{t}) + \phi_{-k_3,-k_2,-k_1}(\mathbf{t})\right],\\
\mathsf{TS}_\mathbf{k}(\mathbf{t}) :=& \tfrac{-1}{6}\mathrm{i}\left[\phi_{k_1,k_2,k_3}(\mathbf{t}) + \phi_{k_2,k_3,k_1}(\mathbf{t}) + \phi_{k_3,k_1,k_2}(\mathbf{t})\right.\\
&\left. - \phi_{-k_1,-k_3,-k_2}(\mathbf{t}) - \phi_{-k_2,-k_1,-k_3}(\mathbf{t}) - \phi_{-k_3,-k_2,-k_1}(\mathbf{t})\right].
\end{aligned}$$

Then $\mathsf{TC}_\mathbf{k}$ is invariant under $\mathscr{A}_2$ and $\mathsf{TS}_\mathbf{k}$ is anti-invariant under $\mathscr{A}_2$; these functions resemble the cosine and sine functions. Because of their invariance properties, we can restrict them to one of the six congruent equilateral triangles inside the hexagon. We will choose the shaded triangle in Figure 2.2:

$$\triangle := \{(t_1,t_2,t_3) : t_1 + t_2 + t_3 = 0, 0 \le t_1, t_2, -t_3 \le 1\}. \qquad (2.9.1)$$

Since $\mathsf{TC}_{\mathbf{k}\sigma}(\mathbf{t}) = \mathsf{TC}_\mathbf{k}(\mathbf{t}\sigma) = \mathsf{TC}_\mathbf{k}(\mathbf{t})$ for $\sigma \in \mathscr{A}_2$ and $\mathbf{t} \in \triangle$, we can restrict the index of $\mathsf{TC}_\mathbf{k}$ to the index set

$$\Lambda := \{\mathbf{k} \in \mathbb{H} : k_1 \ge 0, k_2 \ge 0, k_3 \le 0\}. \qquad (2.9.2)$$

A direct verification shows that $\mathsf{TS}_\mathbf{k}(\mathbf{t}) = 0$ whenever $\mathbf{k}$ has a zero component; hence, $\mathsf{TS}_\mathbf{k}(\mathbf{t})$ is defined only for $\mathbf{k} \in \Lambda^\circ$, where

$$\Lambda^\circ := \{\mathbf{k} \in \mathbb{H} : k_1 > 0, k_2 > 0, k_3 < 0\},$$

the set of the interior points of Λ.

Proposition 2.9.2 *For* $\mathbf{k},\mathbf{j} \in \Lambda$,

$$2\int_{\triangle} \mathsf{TC}_{\mathbf{k}}(\mathbf{t})\overline{\mathsf{TC}_{\mathbf{j}}(\mathbf{t})}\,\mathrm{d}\mathbf{t} = \delta_{\mathbf{k},\mathbf{j}} \begin{cases} 1, & \mathbf{k}=0, \\ \frac{1}{3}, & \mathbf{k}\in\Lambda\setminus\Lambda^\circ,\ \mathbf{k}\neq 0, \\ \frac{1}{6}, & \mathbf{k}\in\Lambda^\circ \end{cases} \tag{2.9.3}$$

and, for $\mathbf{k},\mathbf{j}\in\Lambda^\circ$,

$$2\int_{\triangle} \mathsf{TS}_{\mathbf{k}}(\mathbf{t})\overline{\mathsf{TS}_{\mathbf{j}}(\mathbf{t})}\,\mathrm{d}\mathbf{t} = \tfrac{1}{6}\,\delta_{\mathbf{k},\mathbf{j}}. \tag{2.9.4}$$

Proof If $f\bar{g}$ is invariant under $\mathscr{A}_2$ then

$$\tfrac{1}{3}\int_{\Omega} f(\mathbf{t})\overline{g(\mathbf{t})}\,\mathrm{d}\mathbf{t} = 2\int_{\triangle} f(\mathbf{t})\overline{g(\mathbf{t})}\,\mathrm{d}\mathbf{t},$$

from which (2.9.3) and (2.9.4) follow from Proposition 2.9.1. □

We can use these generalized sine and cosine functions to define analogues of the Chebyshev polynomials of the first and the second kind. To see the polynomial structure among the generalized trigonometric functions, we make a change of variables. Denote

$$z := \mathsf{TC}_{0,1,-1}(\mathbf{t}) = \tfrac{1}{3}[\phi_{0,1,-1}(\mathbf{t}) + \phi_{1,-1,0}(\mathbf{t}) + \phi_{-1,0,1}(\mathbf{t})]. \tag{2.9.5}$$

Let $z = x+iy$. The change of variables $(t_1,t_2)\mapsto(x,y)$ has Jacobian

$$\begin{aligned} \left|\frac{\partial(x,y)}{\partial(t_1,t_2)}\right| &= \frac{16}{27}\pi^2 \sin\pi t_1 \sin\pi t_2 \sin\pi(t_1+t_2) \\ &= \frac{2}{3\sqrt{3}}\pi\left[-3(x^2+y^2+1)^2 + 8(x^3-3xy^2)+4\right]^{1/2}, \end{aligned} \tag{2.9.6}$$

and the region $\triangle$ is mapped onto the region $\triangle^*$ bounded by $-3(x^2+y^2+1)^2 + 8(x^3-3xy^2)+4 = 0$, which is called Steiner's hypocycloid. This three-cusped region is depicted in Figure 2.3.

Definition 2.9.3 Under the change of variables (2.9.5), define the generalized Chebyshev polynomials

$$\begin{aligned} T_k^m(z,\bar{z}) &:= \mathsf{TC}_{k,m-k,-m}(\mathbf{t}), && 0\le k\le m, \\ U_k^m(z,\bar{z}) &:= \frac{\mathsf{TS}_{k+1,m-k+1,-m-2}(\mathbf{t})}{\mathsf{TS}_{1,1,-2}(\mathbf{t})}, && 0\le k\le m. \end{aligned}$$

Proposition 2.9.4 *Let P_k^m denote either T_k^m or U_k^m. Then P_k^m is a polynomial of total degree m in z and $\bar{z}$. Moreover*

$$P_{m-k}^m(z,\bar{z}) = \overline{P_k^m(z,\bar{z})}, \qquad 0\le k\le m, \tag{2.9.7}$$

and the P_k^m satisfy the recursion relation

$$P_k^{m+1}(z,\bar{z}) = 3zP_k^m(z,\bar{z}) - P_{k+1}^m(z,\bar{z}) - P_{k-1}^{m-1}(z,\bar{z}), \tag{2.9.8}$$

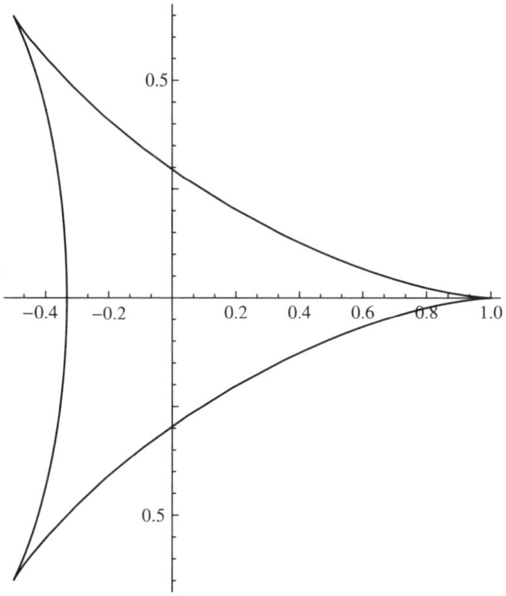

Figure 2.3 The domain for the Koornwinder orthogonal polynomials of the second type.

for $0 \le k \le m$ *and* $m \ge 1$*, where we use*

$$T_{-1}^{m}(z,\bar{z}) = T_{1}^{m+1}(z,\bar{z}), \quad T_{m+1}^{m}(z,\bar{z}) = T_{m}^{m+1}(z,\bar{z}),$$
$$U_{-1}^{m}(z,\bar{z}) = 0, \quad U_{m}^{m-1}(z,\bar{z}) = 0.$$

In particular, we have

$$T_0^0(z,\bar{z}) = 1, \quad T_0^1(z,\bar{z}) = z, \quad T_1^1(z,\bar{z}) = \bar{z},$$
$$U_0^0(z,\bar{z}) = 1, \quad U_0^1(z,\bar{z}) = 3z, \quad U_1^1(z,\bar{z}) = 3\bar{z}.$$

Proof Both (2.9.7) and (2.9.8) follow from a straightforward computation. Together they determine all P_k^m recursively, which shows that both T_k^m and U_k^m are polynomials of degree m in z and $\bar{z}$. □

The polynomials T_k^m and U_k^m are analogues of the Chebyshev polynomials of the first and the second kind. Furthermore, each family inherits an orthogonal relation from the generalized trigonometric functions, so that they are orthogonal polynomials of two variables.

Proposition 2.9.5 *Let* $w_\alpha(x,y) = \left|\frac{\partial(x,y)}{\partial(t_1,t_2)}\right|^{2\alpha}$ *as given in* (2.9.6)*. Define*

$$\langle f,g\rangle_{w_\alpha} = c_\alpha \int_{\Delta^*} f(x,y)\overline{g(x,y)}w_\alpha(x,y)\mathrm{d}x\mathrm{d}y,$$

where c_α *is a normalization constant;* $c_\alpha := 1/\int_\Delta w_\alpha(x,y)\,dx\,dy$. *Then*

$$\langle T_k^m, T_j^m \rangle_{w_{-1/2}} = \delta_{k,j} \begin{cases} 1, & m=k=0, \\ \frac{1}{3}, & k=0 \text{ or } k=m,\ m>0, \\ \frac{1}{6}, & 1 \le k \le m-1, m>0 \end{cases}$$

and

$$\langle U_k^m, U_j^m \rangle_{w_{1/2}} = \tfrac{1}{6}\delta_{k,j}, \qquad 0 \le j,k \le m.$$

In particular, $\{T_k^m : 0 \le k \le m\}$ *and* $\{U_k^m : 0 \le k \le m\}$ *are mutually orthogonal bases of* $\mathcal{V}_m^2(w_{-1/2})$ *and* $\mathcal{V}_m^2(w_{1/2})$*, respectively.*

Proof The change of variables (2.9.5) implies immediately that

$$2\int_\Delta f(\mathbf{t})\,d\mathbf{t} = c_{-1/2}\int_{\Delta^*} f(x,y) w_{-1/2}(x,y)\,dx\,dy.$$

As a result, the orthogonality of the T_k^m follows from that of $\mathsf{TC_j}$ in Proposition 2.9.2. Further, using

$$\mathsf{TS}_{1,1,-2}(\mathbf{t}) = \tfrac{4}{3}\sin\pi t_1 \sin\pi t_2 \sin\pi t_3$$

it is easily seen that the orthogonality of U_k^m follows from that of $\mathsf{TS_j}$. □

By taking the real and complex parts of T_k^m or U_k^m and using the relation (2.9.7), we can also obtain a real orthogonal basis for $\mathcal{V}_m^2$.

2.10 Notes

The book by Appell and de Fériet [1926] contains several examples of orthogonal polynomials in two variables. Many examples up to 1950 can be found in the collection Erdélyi, Magnus, Oberhettinger and Tricomi [1953]. An influential survey is that due to Koornwinder [1975]. For the general properties of orthogonal polynomials in two variables, see the notes in Chapter 4.

Section 2.3 The disk is a special case ($d = 2$) of the unit ball in $\mathbb{R}^d$. Consequently, some further properties of orthogonal polynomials on the disk can be found in Chapter 4, including a compact formula for the reproducing kernel.

The first orthonormal basis goes back as far as Hermite and was studied in Appell and de Fériet [1926]. Biorthogonal polynomials were considered in detail in Appell and de Fériet [1926]; see also Erdélyi *et al.* [1953]. The basis (2.3.8) was first discovered in Logan and Shepp [1975], and it plays an important role in computer tomography; see Marr [1974] and Xu [2006a]. For further studies on orthogonal polynomials on the disk, see Waldron [2008] and Wünsche [2005] as well as the references on orthogonal polynomials on the unit ball B^d for $d \ge 2$ given in Chapters 5 and 8.

Section 2.4 The triangle is a special case of the simplex in $\mathbb{R}^d$. Some further properties of orthogonal polynomials on the triangle can be found in Chapter 5, including a compact formula for the reproducing kernel.

The orthogonal polynomials (2.4.2) were first introduced in Proriol [1957]; the case $\alpha = \beta = \gamma = 0$ became known as Dubiner's polynomials in the finite elements community after Dubiner [1991], which was apparently unaware that they had appeared in the literature much earlier. The basis U_k^n was studied in detail in Appell and de Fériet [1926], and the biorthogonal basis appeared in Fackerell and Littler [1974]. The change of basis matrix connecting the three bases in Proposition 2.4.2 has Racah–Wilson ${}_4F_3$ polynomials as entries; see Dunkl [1984b].

Section 2.5 The classification of all admissible equations that have orthogonal polynomials as eigenfunctions was studied first by Krall and Sheffer [1967], as summarized in Section 2.5. The classification in Suetin [1999], based on Engelis [1974], listed 15 cases; some of these are equivalent under the affine transforms in Krall and Sheffer [1967] but are treated separately because of other considerations. The orthogonality of cases (6) and (7) listed in this section is determined in Krall and Sheffer [1967]; cases (8) and (9) are determined in Berens, Schmid and Xu [1995a]. For further results, including solutions of the cases (6)–(9) and further discussion on the impact of affine transformations, see Littlejohn [1988], Lyskova [1991], Kim, Kwon and Lee [1998] and references therein. Classical orthogonal polynomials in two variables were studied in the context of hypergroups in Connett and Schwartz [1995].

Fernández, Pérez and Piñar [2011] considered examples of orthogonal polynomials of two variables that satisfy fourth-order differential equations. The product Jacobi polynomials, and other classical orthogonal polynomials of two variables, satisfy a second-order matrix differential equation; see Fernández, Pérez and Piñar [2005] and the references therein. They also satisfy a matrix form of Rodrigues type formula, as seen in de Álvarez, Fernández, Pérez, and Piñar [2009].

Section 2.6 The method in the first subsection first appeared in Larcher [1959] and was used in Agahanov [1965] for certain special cases. It was presented systematically in Koornwinder [1975], where the two cases of ρ were stated. For further examples of explicit bases constructed in various domains, such as $\{(x,y) : x^2 + y^2 \le 1, -a \le y \le b\}$, $0 < a,b < 1$, see Suetin [1999]. The product formula for polynomials in (2.6.4) was given in Koornwinder and Schwartz [1997]; it generates a convolution structure for $L^2(W_{\alpha,\beta})$ and was used to study the convergence of orthogonal expansions in zu Castell, Filbir and Xu [2009]. The complex Hermite polynomials were introduced by Itô [1952]. They have been widely studied and used by many authors; see Ghanmi [2008, 2013], Intissar and Intissar [2006], Ismail [2013] and Ismail and Simeonov [2013] for some recent studies and their references. Disk polynomials were introduced in Zernike and Brinkman [1935] to evaluate the point image of an aberrated optical system taking into account the effects of diffraction (see the Wikipedia article on optical aberration); our normalization follows Dunkl [1982]. They were used in Folland [1975] to expand the Poisson–Szegő kernel for the ball in $\mathbb{C}^d$. A Banach algebra related to disk polynomials was studied in Kanjin [1985]. For further properties of disk polynomials, including the fact that for $\lambda = d-2$, $d = 2,3,\ldots$, they are spherical functions for $U(d)/U(d-1)$, see Ikeda [1967], Koornwinder [1975], Vilenkin and Klimyk [1991a, b, c], and Wünsche [2005]. The structure of complex orthogonal polynomials of two variables and its connection to and contrast with its real counterpart are studied in Xu [2013].

Section 2.7 The polynomials $P_{k,n}^{\alpha,\beta,\gamma}$ were studied in Koornwinder [1974a], which contains another differential operator, of fourth order, that has the orthogonal polynomials as eigenfunctions. See Koornwinder and Sprinkhuizen-Kuyper [1978] and Sprinkhuizen-Kuyper [1976] for further results on these polynomials, including an explicit formula for $P_{k,n}^{\alpha,\beta,\gamma}$ in terms of power series, and Rodrigues type formulas; in Xu [2012] an explicit formula for the reproducing kernel in the case of W_γ in (2.7.4) with $\gamma = \pm\frac{1}{2}$ was given in terms of the reproducing kernels of the orthogonal polynomials of one variable. The case $W_{\pm 1/2}$ for general w was considered first in Schmid and Xu [1994] in connection with Gaussian cubature rules.

Section 2.8 The result in this section was developed recently in Xu [2012], where further results, including a compact formula for the reproducing kernel and convergence of orthogonal expansions, can be found.

Section 2.9 The generalized Chebyshev polynomials in this section were first studied in Koornwinder [1974b]. We follow the approach in Li, Sun and Xu [2008], which makes a connection with lattice tiling. In Koornwinder [1974b] the orthogonal polynomials $P_{k,n}^{\alpha}$ for the weight function w_α, $\alpha > -\frac{5}{6}$, were shown to be eigenfunctions of a differential operator of third order. The special case $P_{0,n}^{\alpha}$ was also studied in Lidl [1975], which includes an interesting generating function. For further results, see Shishkin [1997], Suetin [1999], Li, Sun and Xu [2008] and Xu [2010].

Other orthogonal polynomials of two variables The Berstein–Szegő two-variable weight function is of the form

$$W(x,y) = \frac{4}{\pi^2}\frac{\sqrt{1-x^2}\sqrt{1-y^2}}{|h(z,y)|^2}, \qquad x = \frac{1}{2}\left(z + \frac{1}{z}\right).$$

Here, for a given integer m and $-1 \le y \le 1$,

$$h(z,y) = \sum_{i=0}^{N} h_i(y) z^i, \qquad z \in \mathbb{C},$$

is nonzero for any $|z| \le 1$ in which $h_0(y) = 1$ and, for $1 \le i \le m$, $h_i(y)$ are polynomials in y with real coefficients of degree at most $\frac{m}{2} - |\frac{m}{2} - i|$. For $m \le 2$, complete orthogonal bases for $\mathscr{V}_n^2(W)$ are constructed for all n in Delgado, Geronimo, Iliev and Xu [2009].

Orthogonal polynomials with respect to the area measure on the regular hexagon were studied in Dunkl [1987], where an algorithm for generating an orthogonal basis was given.

3

General Properties of Orthogonal Polynomials in Several Variables

In this chapter we present the general properties of orthogonal polynomials in several variables, that is, those properties that hold for orthogonal polynomials associated with weight functions that satisfy some mild conditions but are not any more specific than that.

This direction of study started with the classical work of Jackson [1936] on orthogonal polynomials in two variables. It was realized even then that the proper definition of orthogonality is in terms of polynomials of lower degree and that orthogonal bases are not unique. Most subsequent early work was focused on understanding the structure and theory in two variables. In Erdélyi *et al.* [1953], which documents the work up to 1950, one finds little reference to the general properties of orthogonal polynomials in more than two variables, other than (Vol. II, p. 265): "There does not seem to be an extensive general theory of orthogonal polynomials in several variables." It was remarked there that the difficulty lies in the fact that there is no unique orthogonal system, owing to the many possible orderings of multiple sequences. And it was also pointed out that since a particular ordering usually destroys the symmetry, it is often preferable to construct biorthogonal systems. Krall and Sheffer [1967] studied and classified two-dimensional analogues of classical orthogonal polynomials as solutions of partial differential equations of the second order. Their classification is based on the following observation: while the orthogonal bases of $\mathcal{V}_n^d$, the set of orthogonal polynomials of degree n defined in Section 3.1, are not unique, if the results can be stated "in terms of $\mathcal{V}_0^d, \mathcal{V}_1^d, \ldots, \mathcal{V}_n^d, \ldots$ rather than in terms of a particular basis in each $\mathcal{V}_n^d$, a degree of uniqueness is restored." This is the point of view that we shall adopt in much of this chapter.

The advantage of this viewpoint is that the result obtained will be basis independent, which allows us to derive a proper analogy of the three-term relation for orthogonal polynomials in several variables, to define block Jacobi matrices and study them as self-adjoint operators and to investigate common zeros of orthogonal polynomials in several variables, among other things. This approach is also natural for studying the Fourier series of orthogonal polynomial expansions, since it restores a certain uniqueness in the expansion in several variables.

3.1 Notation and Preliminaries

Throughout this book we will use the standard multi-index notation, as follows. We denote by $\mathbb{N}_0$ the set of nonnegative integers. A *multi-index* is usually denoted by α, $\alpha = (\alpha_1, \dots, \alpha_d) \in \mathbb{N}_0^d$. Whenever α appears with a subscript, it denotes the component of a multi-index. In this spirit we define, for example, $\alpha! = \alpha_1! \cdots \alpha_d!$ and $|\alpha| = \alpha_1 + \cdots + \alpha_d$, and if $\alpha, \beta \in \mathbb{N}_0^d$ then we define $\delta_{\alpha,\beta} = \delta_{\alpha_1,\beta_1} \cdots \delta_{\alpha_d,\beta_d}$.

For $\alpha \in \mathbb{N}_0^d$ and $x = (x_1, \dots, x_d)$, a *monomial* in the variables $x_1, \dots, x_d$ is a product

$$x^\alpha = x_1^{\alpha_1} \cdots x_d^{\alpha_d}.$$

The number $|\alpha|$ is called the *total degree* of x^α. A polynomial P in d variables is a linear combination of monomials,

$$P(x) = \sum_\alpha c_\alpha x^\alpha,$$

where the coefficients c_α lie in a field k, usually the rational numbers $\mathbb{Q}$, the real numbers $\mathbb{R}$ or the complex numbers $\mathbb{C}$. The degree of a polynomial is defined as the highest total degree of its monomials. The *collection* of all polynomials in x with coefficients in a field k is denoted by $k[x_1, \dots, x_d]$, which has the structure of a commutative ring. We will use the abbreviation Π^d to denote $k[x_1, \dots, x_d]$. We also denote the space of polynomials of degree at most n by Π_n^d. When $d = 1$, we will drop the superscript and write Π and Π_n instead.

A polynomial is called *homogeneous* if all the monomials appearing in it have the same total degree. Denote the space of homogeneous polynomials of degree n in d variables by $\mathscr{P}_n^d$, that is,

$$\mathscr{P}_n^d = \left\{ P : P(x) = \sum_{|\alpha|=n} c_\alpha x^\alpha \right\}.$$

Every polynomial in Π^d can be written as a linear combination of homogeneous polynomials; for $P \in \Pi_n^d$,

$$P(x) = \sum_{k=0}^{n} \sum_{|\alpha|=k} c_\alpha x^\alpha.$$

Denote by r_n^d the dimension of $\mathscr{P}_n^d$. Evidently $\{x^\alpha : |\alpha| = n\}$ is a basis of $\mathscr{P}_n^d$; hence, $r_n^d = \#\{\alpha \in \mathbb{N}_0^d : |\alpha| = n\}$. It is easy to see that

$$\frac{1}{(1-t)^d} = \prod_{i=1}^{d} \sum_{\alpha_i=0}^{\infty} t^{\alpha_i} = \sum_{n=0}^{\infty} \left(\sum_{|\alpha|=n} 1 \right) t^n = \sum_{n=0}^{\infty} r_n^d t^n.$$

Therefore, recalling an elementary infinite series

$$\frac{1}{(1-t)^d} = \sum_{n=0}^{\infty} \frac{(d)_n}{n!} t^n = \sum_{n=0}^{\infty} \binom{n+d-1}{n} t^n,$$

and comparing the coefficients of t^n in the two series, we conclude that

$$r_n^d = \dim \mathscr{P}_n^d = \binom{n+d-1}{n}.$$

Since the cardinalities of the sets $\{\alpha \in \mathbb{N}_0^d : |\alpha| \le n\}$ and $\{\alpha \in \mathbb{N}_0^{d+1} : |\alpha| = n\}$ are the same, as each α in the first set becomes an element of the second set upon adding $\alpha_{d+1} = n - |\alpha|$ and as this is a one-to-one correspondence, it follows that

$$\dim \Pi_n^d = \binom{n+d}{n}.$$

One essential difference between polynomials in one variable and polynomials in several variables is the lack of an obvious natural order in the latter. The natural order for monomials of one variable is the degree order, that is, we order monomials in Π according to their degree, as $1, x, x^2, \ldots$ For polynomials in several variables, there are many choices of well-defined total order. Two are described below.

Lexicographic order We say that $\alpha \succ_L \beta$ if the first nonzero entry in the difference $\alpha - \beta = (\alpha_1 - \beta_1, \ldots, \alpha_d - \beta_d)$ is positive.

Lexicographic order does not respect the total degree of the polynomials. For example, α with $\alpha = (3,0,0)$ of degree 3 is ordered so that it lies in front of β with $\beta = (2,2,2)$ of degree 6, while γ with $\gamma = (0,0,3)$ of degree 3 comes after β. The following order does respect the polynomial degree.

Graded lexicographic order We say that $\alpha \succ_{\text{glex}} \beta$ if $|\alpha| > |\beta|$ or if $|\alpha| = |\beta|$ and the first nonzero entry in the difference $\alpha - \beta$ is positive.

In the case $d = 2$ we can write $\alpha = (n-k, k)$, and the lexicographic order among $\{\alpha : |\alpha| = n\}$ is the same as the order $k = 0, 1, \ldots, n$. There are various other orders for polynomials of several variables; some will be discussed in later chapters.

Let $\langle \cdot, \cdot \rangle$ be a bilinear form defined on Π^d. Two polynomials P and Q are said to be orthogonal to each other with respect to the bilinear form if $\langle P, Q \rangle = 0$. A polynomial P is called an *orthogonal polynomial* if it is orthogonal to all polynomials of lower degree, that is, if

$$\langle P, Q \rangle = 0, \quad \forall Q \in \Pi^d \qquad \text{with } \deg Q < \deg P.$$

If the bilinear form is given in terms of a weight function W,

$$\langle P, Q \rangle = \int_{\Omega} P(x)Q(x)W(x)\,\mathrm{d}x \tag{3.1.1}$$

where Ω is a domain in $\mathbb{R}^d$ (this implies that Ω has a nonempty interior), we say that the orthogonal polynomials are orthogonal with respect to the weight function W.

For arbitrary bilinear forms there may or may not exist corresponding orthogonal polynomials. In the following we hypothesize their existence, and later we will derive necessary and sufficient conditions on the bilinear forms or linear functionals for the existence of such orthogonal polynomial systems.

Definition 3.1.1 Assume that orthogonal polynomials exist for a particular bilinear form. We denote by $\mathscr{V}_n^d$ the space of orthogonal polynomials of degree exactly n, that is,

$$\mathscr{V}_n^d = \{P \in \Pi_n^d : \langle P, Q \rangle = 0 \quad \forall Q \in \Pi_{n-1}^d\}. \tag{3.1.2}$$

When the dimension of $\mathscr{V}_n^d$ is the same as that of $\mathscr{P}_n^d$, it is natural to use a multi-index to index the elements of an orthogonal basis of $\mathscr{V}_n^d$. Thus, we shall denote the elements of such a basis by P_α, $|\alpha| = n$. We will sometimes use the notation P_α^n, in which the superscript indicates the degree of the polynomial. For orthogonal polynomials of two variables, instead of P_α with $\alpha = (k, n-k)$ the more convenient notation P_k or P_k^n for $0 \le k \le n$ is often used, as in Chapter 2.

3.2 Moment Functionals and Orthogonal Polynomials in Several Variables

We will use the standard multi-index notation as in the above section. Throughout this section we write r_n instead of r_n^d whenever r_n^d would appear as a subscript or superscript.

3.2.1 Definition of orthogonal polynomials

A multi-sequence $s : \mathbb{N}_0^d \mapsto \mathbb{R}$ is written in the form $s = \{s_\alpha\}_{\alpha \in \mathbb{N}_0^d}$. For each multi-sequence $s = \{s_\alpha\}_{\alpha \in \mathbb{N}_0^d}$, let $\mathscr{L}_s$ be the linear functional defined on Π^d by

$$\mathscr{L}_s(x^\alpha) = s_\alpha, \qquad \alpha \in \mathbb{N}_0^d; \tag{3.2.1}$$

call $\mathscr{L}_s$ the *moment functional* defined by the sequence s.

For convenience, we introduce a vector notation. Let the elements of the set $\{\alpha \in \mathbb{N}_0^d : |\alpha| = n\}$ be arranged as $\alpha^{(1)}, \alpha^{(2)}, \ldots, \alpha^{(r_n)}$ according to lexicographical order. For each $n \in \mathbb{N}_0$ let $\mathbf{x}^n$ denote the column vector

$$\mathbf{x}^n = (x^\alpha)_{|\alpha|=n} = (x^{\alpha^{(j)}})_{j=1}^{r_n}.$$

That is, $\mathbf{x}^n$ is a vector whose elements are the monomials x^α for $|\alpha| = n$, arranged in lexicographical order. Also define, for $k, j \in \mathbb{N}_0$ vectors of moments $\mathbf{s}_k$ and matrices of moments $\mathbf{s}_{\{k\}+\{j\}}$ by

$$\mathbf{s}_k = \mathscr{L}_s(\mathbf{x}^k) \qquad \text{and} \qquad \mathbf{s}_{\{k\}+\{j\}} = \mathscr{L}_s(\mathbf{x}^k(\mathbf{x}^j)^{\mathrm{T}}). \tag{3.2.2}$$

By definition $\mathbf{s}_{\{k\}+\{j\}}$ is a matrix of size $r_k^d \times r_j^d$ and its elements are $\mathscr{L}_s(x^{\alpha+\beta})$ for $|\alpha| = k$ and $|\beta| = j$. Finally, for each $n \in \mathbb{N}_0$, the $\mathbf{s}_{\{k\}+\{j\}}$ are used as building blocks to define a matrix

$$M_{n,d} = (\mathbf{s}_{\{k\}+\{j\}})_{k,j=0}^n \quad \text{with} \quad \Delta_{n,d} = \det M_{n,d}. \tag{3.2.3}$$

We call $M_{n,d}$ a moment matrix; its elements are $\mathscr{L}_s(x^{\alpha+\beta})$ for $|\alpha| \le n$ and $|\beta| \le n$.

In the following we will write $\mathscr{L}$ for $\mathscr{L}_s$ whenever s is not explicit. If $\mathscr{L}$ is a moment functional then $\mathscr{L}(PQ)$ is a bilinear form, so that we can define orthogonal polynomials with respect to $\mathscr{L}$. In particular, P is an orthogonal polynomial of degree n if $\mathscr{L}(PQ) = 0$ for every polynomial of degree $n-1$ or less. More generally, we can put the monomials in the basis $\{x^\alpha : \alpha \in \mathbb{N}_0^d\}$ of Π^d in a linear order and use Gram–Schmidt orthogonalization to generate a new basis whose elements are mutually orthogonal with respect to $\mathscr{L}$. However, since there is no natural order for the set $\{\alpha \in \mathbb{N}_0^d : |\alpha| = n\}$, we may just as well consider orthogonality only with respect to polynomials of different degree, that is, we take polynomials of the same degree to be orthogonal to polynomials of lower degree but not necessarily orthogonal among themselves. To make this precise, let us introduce the following notation. If $\{P_\alpha^n\}_{|\alpha|=n}$ is a sequence of polynomials in Π_n^d, denote by $\mathbb{P}_n$ the (column) polynomial vector

$$\mathbb{P}_n = (P_\alpha^n)_{|\alpha|=n} = (P_{\alpha^{(1)}}^n \quad \cdots \quad P_{\alpha^{(r_n)}}^n)^{\mathrm{T}}, \tag{3.2.4}$$

where $\alpha^{(1)}, \ldots, \alpha^{(r_n)}$ is the arrangement of elements in $\{\alpha \in \mathbb{N}_0^d : |\alpha| = n\}$ according to lexicographical order. Sometimes we use the notation $\mathbb{P}_n$ to indicate the set of polynomials $\{P_\alpha^n\}$.

Definition 3.2.1 Let $\mathscr{L}$ be a moment functional. A sequence of polynomials $\{P_\alpha^n : |\alpha| = n, n \in \mathbb{N}_0\}$, $P_\alpha^n \in \Pi_n^d$, is said to be orthogonal with respect to $\mathscr{L}$ if

$$\mathscr{L}(\mathbf{x}^m \mathbb{P}_n^{\mathrm{T}}) = 0, \quad n > m, \qquad \text{and} \quad \mathscr{L}(\mathbf{x}^n \mathbb{P}_n^{\mathrm{T}}) = S_n, \tag{3.2.5}$$

where S_n is an invertible matrix of size $r_n^d \times r_n^d$ (assuming for now that $\mathscr{L}$ permits the existence of such P_α^n).

We may also call $\mathbb{P}_n$ an orthogonal polynomial. We note that this definition agrees with our usual notion of orthogonal polynomials since, by definition,

$$\mathscr{L}(\mathbf{x}^m \mathbb{P}_n^{\mathrm{T}}) = 0 \iff \mathscr{L}(x^\beta P_\alpha^n) = 0, \qquad |\alpha| = n, \quad |\beta| = m;$$

thus (3.2.5) implies that the P^n_α are orthogonal to polynomials of lower degree. Some immediate consequences of this definition are as follows.

Proposition 3.2.2 *Let $\mathscr{L}$ be a moment functional and let $\mathbb{P}_n$ be the orthogonal polynomial defined above. Then $\{\mathbb{P}_0, \ldots, \mathbb{P}_n\}$ forms a basis for Π^d_n.*

Proof Consider the sum $\mathbf{a}_0^{\mathrm{T}}\mathbb{P}_0 + \cdots + \mathbf{a}_n^{\mathrm{T}}\mathbb{P}_n$, where $\mathbf{a}_i \in \mathbb{R}^{r_i}$. Multiplying the sum from the right by the row vector $\mathbb{P}_k^{\mathrm{T}}$ and applying $\mathscr{L}$, it follows from the orthogonality that $\mathbf{a}_k^{\mathrm{T}}\mathscr{L}(\mathbb{P}_k\mathbb{P}_k^{\mathrm{T}}) = 0$ for $0 \le k \le n$, and this shows, by the orthogonality of $\mathbb{P}_k$ to lower-degree polynomials, that $\mathbf{a}_k^{\mathrm{T}}S_k^{\mathrm{T}} = 0$. Thus, $\mathbf{a}_k = 0$ since S_k is invertible. Therefore $\{P^n_\alpha\}_{|\alpha|\le n}$ is linearly independent and forms a basis for Π^d_n. □

Using the vector notation, the orthogonal polynomial $\mathbb{P}_n$ can be written as

$$\mathbb{P}_n = G_{n,n}\mathbf{x}^n + G_{n,n-1}\mathbf{x}^{n-1} + \cdots + G_{n,0}\mathbf{x}^0, \tag{3.2.6}$$

where $G_n = G_{n,n}$ is called the leading coefficient of $\mathbb{P}_n$ and is a matrix of size $r^d_n \times r^d_n$.

Proposition 3.2.3 *Let $\mathbb{P}_n$ be as in the previous proposition. Then the leading-coefficient matrix G_n is invertible.*

Proof The previous proposition implies that there exists a matrix G'_n such that

$$\mathbf{x}^n = G'_n\mathbb{P}_n + \mathbb{Q}_{n-1},$$

where $\mathbb{Q}_{n-1}$ is a vector whose components belong to Π^d_{n-1}. Comparing the coefficients of $\mathbf{x}^n$ gives $G'_nG_n = I$, which implies that G_n is invertible. □

Proposition 3.2.4 *Let $\mathbb{P}_n$ be as in the previous proposition. Then the matrix $H_n = \mathscr{L}(\mathbb{P}_n\mathbb{P}_n^{\mathrm{T}})$ is invertible.*

Proof Since $H_n = \mathscr{L}(\mathbb{P}_n\mathbb{P}_n^{\mathrm{T}}) = G_n\mathscr{L}(\mathbf{x}^n\mathbb{P}_n^{\mathrm{T}}) = G_nS_n$, it is invertible by Proposition 3.2.3. □

Lemma 3.2.5 *Let $\mathscr{L}$ be a moment functional and let $\mathbb{P}_n$ be an orthogonal polynomial with respect to $\mathscr{L}$. Then $\mathbb{P}_n$ is uniquely determined by the matrix S_n.*

Proof Suppose, otherwise, that there exist $\mathbb{P}_n$ and $\mathbb{P}_n^*$ both satisfying the orthogonality conditions of (3.2.5) with the same matrices S_n. Let G_n and G_n^* denote the leading-coefficient matrices of $\mathbb{P}_n$ and $\mathbb{P}_n^*$, respectively. By Proposition 3.2.2, $\{\mathbb{P}_0, \ldots, \mathbb{P}_n\}$ forms a basis of Π^d_n; write the elements of $\mathbb{P}_n^*$ in terms of this basis. That is, there exist matrices $C_k : r^d_n \times r^d_k$ such that

$$\mathbb{P}_n^* = C_n\mathbb{P}_n + C_{n-1}\mathbb{P}_{n-1} + \cdots + C_0\mathbb{P}_0.$$

Multiplying the above equation by $\mathbb{P}_k^{\mathrm{T}}$ and applying the moment functional $\mathscr{L}$, we conclude that $C_k\mathscr{L}(\mathbb{P}_k\mathbb{P}_k^{\mathrm{T}}) = 0$, $0 \le k \le n-1$, by orthogonality. It follows from Proposition 3.2.4 that $C_k = 0$ for $0 \le k \le n-1$; hence $\mathbb{P}_n^* = C_n\mathbb{P}_n$. Comparing the coefficients of $\mathbf{x}^n$ leads to $G_n^* = C_nG_n$. That is, $C_n = G_n^*G_n^{-1}$ and $\mathbb{P}_n = G_nG_n^{*-1}\mathbb{P}_n^*$, which implies by (3.2.5) that $S_n = \mathscr{L}(\mathbf{x}^n\mathbb{P}_n^{\mathrm{T}}) = \mathscr{L}(\mathbf{x}^n\mathbb{P}_n^{*\mathrm{T}})(G_nG_n^{*-1})^{\mathrm{T}} = S_n(G_nG_n^{*-1})^{\mathrm{T}}$. Therefore $G_nG_n^{*-1} = I$ and so $G_n = G_n^*$ and $\mathbb{P}_n = \mathbb{P}_n^*$. □

Theorem 3.2.6 *Let $\mathscr{L}$ be a moment functional. A system of orthogonal polynomials in several variables exists if and only if*

$$\Delta_{n,d} \neq 0, \qquad n \in \mathbb{N}_0.$$

Proof Using the monomial expression of $\mathbb{P}_n$ in (3.2.6) and the notation $\mathbf{s}_k$ in (3.2.2),

$$\begin{aligned}\mathscr{L}(\mathbf{x}^k\mathbb{P}_n^{\mathrm{T}}) &= \mathscr{L}(\mathbf{x}^k(G_n\mathbf{x}^n + G_{n,n-1}\mathbf{x}^{n-1} + \cdots + G_{n,0}\mathbf{x}^0)^{\mathrm{T}}) \\ &= \mathbf{s}_{\{k\}+\{n\}}G_n^{\mathrm{T}} + \mathbf{s}_{\{k\}+\{n-1\}}G_{n,n-1}^{\mathrm{T}} + \cdots + \mathbf{s}_{\{k\}+\{0\}}G_{n,0}^{\mathrm{T}}.\end{aligned}$$

From the definition of $M_{n,d}$ in (3.2.3), it follows that the orthogonality condition (3.2.5) is equivalent to the following linear system of equations:

$$M_{n,d}\begin{bmatrix} G_{n,0}^{\mathrm{T}} \\ \vdots \\ G_{n,n}^{\mathrm{T}} \end{bmatrix} = \begin{bmatrix} 0 \\ \vdots \\ 0 \\ S_n^{\mathrm{T}} \end{bmatrix}. \tag{3.2.7}$$

If an orthogonal polynomial system exists then for each S_n there exists exactly one $\mathbb{P}_n$. Therefore, the system of equations (3.2.7) has a unique solution, which implies that $M_{n,d}$ is invertible; thus $\Delta_{n,d} \neq 0$.

However, if $\Delta_{n,d} \neq 0$ then, for each invertible matrix S_n, (3.2.7) has a unique solution $(G_{0,n} \;\; \cdots \;\; G_{n,n})^{\mathrm{T}}$. Let $\mathbb{P}_n = \sum G_{k,n}\mathbf{x}^k$; then (3.2.7) is equivalent to $\mathscr{L}(\mathbf{x}^k\mathbb{P}_n^{\mathrm{T}}) = 0$, $k < n$, and $\mathscr{L}(\mathbf{x}^n\mathbb{P}_n^{\mathrm{T}}) = S_n$. □

Definition 3.2.7 A moment linear functional $\mathscr{L}_s$ is said to be positive definite if

$$\mathscr{L}_s(p^2) > 0 \qquad \forall p \in \Pi^d, \quad p \neq 0.$$

We also say that $\{s_\alpha\}$ is positive definite when $\mathscr{L}_s$ is positive definite.

If $p = \sum a_\alpha x^\alpha$ is a polynomial in Π^d then $\mathscr{L}_s(p) = \sum a_\alpha s_\alpha$. In terms of the sequence s, the positive definiteness of $\mathscr{L}_s$ amounts to the requirement that

$$\sum_{\alpha,\beta} a_\alpha a_\beta s_{\alpha+\beta} > 0,$$

where $\alpha+\beta=(\alpha_1+\beta_1,\ldots,\alpha_d+\beta_d)$, for every sequence $a=\{a_\alpha\}_{\alpha\in\mathbb{N}_0^d}$ in which $a_\alpha=0$ for all but finitely many α. Hence it is evident that $\mathscr{L}_s$ is positive definite if and only if for every tuple $(\beta^{(1)},\ldots,\beta^{(r)})$ of distinct multi-indices $\beta^{(j)}\in\mathbb{N}_0^d$, $1\le j\le r$, the matrix $(s_{\beta^{(i)}+\beta^{(j)}})_{i,j=1,\ldots,r}$ has a positive determinant.

Lemma 3.2.8 *If $\mathscr{L}_s$ is positive definite then the determinant $\Delta_{n,d}>0$.*

Proof Assume that $\mathscr{L}_s$ is positive definite. Let $\mathbf{a}$ be an eigenvector of the matrix $M_{n,d}$ corresponding to eigenvalue λ. Then, on the one hand, $\mathbf{a}^{\mathrm{T}}M_{n,d}\mathbf{a}=\lambda\|\mathbf{a}\|^2$. On the other hand, $\mathbf{a}^{\mathrm{T}}M_{n,d}\mathbf{a}=\mathscr{L}_s(p^2)>0$, where $p(x)=\sum_{j=0}^n\mathbf{a}_j^{\mathrm{T}}\mathbf{x}^j$. It follows that $\lambda>0$. Since all the eigenvalues are positive, $\Delta_{n,d}=\det M_{n,d}>0$. □

Corollary 3.2.9 *If $\mathscr{L}$ is a positive definite moment functional then there exists a system of orthogonal polynomials with respect to $\mathscr{L}$.*

Definition 3.2.10 Let $\mathscr{L}$ be a moment functional. A sequence of polynomials $\{P_\alpha^n:|\alpha|=n,n\in\mathbb{N}_0\}$, $P_\alpha^n\in\Pi_n^d$, is said to be orthonormal with respect to $\mathscr{L}$ if

$$\mathscr{L}(P_\alpha^nP_\beta^m)=\delta_{\alpha,\beta};$$

in the vector notation, the above equations become

$$\mathscr{L}(\mathbb{P}_m\mathbb{P}_n^{\mathrm{T}})=0,\quad n\ne m,\qquad\text{and}\quad\mathscr{L}(\mathbb{P}_n\mathbb{P}_n^{\mathrm{T}})=I_{r_n},\tag{3.2.8}$$

where I_k denotes the identity matrix of size $k\times k$.

Theorem 3.2.11 *If $\mathscr{L}$ is a positive definite moment functional then there exists an orthonormal basis with respect to $\mathscr{L}$.*

Proof By the previous corollary, there is a basis of orthogonal polynomials $\mathbb{P}_n$ with respect to $\mathscr{L}$. Let $H_n=\mathscr{L}(\mathbb{P}_n\mathbb{P}_n^{\mathrm{T}})$. For any nonzero vector $\mathbf{a}$, $P=\mathbf{a}^{\mathrm{T}}\mathbb{P}_n$ is a nonzero polynomial by Proposition 3.2.2. Hence $\mathbf{a}H_n\mathbf{a}^{\mathrm{T}}=\mathscr{L}(P^2)>0$, which shows that H_n is a positive definite matrix. Let $H_n^{1/2}$ be the positive square root of H (the unique positive definite matrix with the same eigenvectors as H and $H_n=H_n^{1/2}H_n^{1/2}$). Define $\mathbb{Q}_n=(H_n^{1/2})^{-1}\mathbb{P}_n$. Then

$$\mathscr{L}(\mathbb{Q}_n\mathbb{Q}_n^{\mathrm{T}})=(H_n^{1/2})^{-1}\mathscr{L}(\mathbb{P}_n\mathbb{P}_n^{\mathrm{T}})(H_n^{1/2})^{-1}=(H_n^{1/2})^{-1}H_n(H_n^{1/2})^{-1}=I,$$

which shows that the elements of $\mathbb{Q}_n$ consist of an orthonormal basis with respect to $\mathscr{L}$. □

3.2.2 Orthogonal polynomials and moment matrices

Let $\mathscr{L}$ be a positive definite moment functional. For each $n\in\mathbb{N}_0$, let $M_{n,d}$ be the moment matrix of $\mathscr{L}$ as defined in (3.2.3). Then $M_{n,d}$ has a positive determinant by Lemma 3.2.8. For $\alpha \in \mathbb{N}_0^d$ we denote by $\mathbf{s}_{k,\alpha}$ the column vector $\mathbf{s}_{\alpha,k} := \mathscr{L}(x^\alpha \mathbf{x}^k)$; in particular, $\mathbf{s}_{\alpha,0} = s_\alpha$. We will define monic orthogonal polynomials in terms of the moment matrix. For $\alpha \in \{\beta : |\beta| = n\}$, define the polynomial P_α^n by

$$P_\alpha^n(x) := \frac{1}{\Delta_{n-1,d}} \det \left[\begin{array}{c|c} & \mathbf{s}_{\alpha,0} \\ M_{n-1,d} & \mathbf{s}_{\alpha,1} \\ & \vdots \\ & \mathbf{s}_{\alpha,n-1} \\ \hline 1 \ \ \mathbf{x}^{\mathrm{T}} \cdots (\mathbf{x}^{n-1})^{\mathrm{T}} & x^\alpha \end{array}\right]. \tag{3.2.9}$$

It is evident that the P_α^n are of degree n. Furthermore, they are orthogonal.

Theorem 3.2.12 *The polynomials P_α^n defined above are monomial orthogonal polynomials.*

Proof Expanding the determinant in (3.2.9) by its last row, we see that P_α^n is a monomial polynomial, $P_\alpha(x) = x^\alpha + \cdots$. Moreover multiplying $P_\alpha(x)$ by x^β, $|\beta| \le n-1$, and applying the linear functional $\mathscr{L}$, the last row of the determinant $\det \mathscr{L}[x^\beta M_\alpha(x)]$ coincides with one of the rows above; consequently, $\mathscr{L}(x^\beta P_\alpha^n) = 0$. □

For $d = 2$, this definition appeared in Jackson [1936]; see also Suetin [1999]. It is also possible to define a sequence of orthonormal bases in terms of moments. For this purpose let $\mathscr{L}$ be a positive definite moment functional and define a different matrix $\widetilde{M}_\alpha(x)$ as follows.

The rows of the matrix $M_{n,d}$ are indexed by $\{\alpha : |\alpha| = n\}$. Moreover, the row indexed by α is

$$\begin{pmatrix} \mathbf{s}_{\alpha,0}^{\mathrm{T}} & \mathbf{s}_{\alpha,1}^{\mathrm{T}} & \cdots & \mathbf{s}_{\alpha,n}^{\mathrm{T}} \end{pmatrix} = \mathscr{L}\begin{pmatrix} x^\alpha & x^\alpha \mathbf{x}^{\mathrm{T}} & \cdots & x^\alpha (\mathbf{x}^n)^{\mathrm{T}} \end{pmatrix}.$$

Let $\widetilde{M}_\alpha(x)$ be the matrix obtained from $M_{n,d}$ by replacing the above row, with index α, by $\begin{pmatrix} x^\alpha & x^\alpha \mathbf{x}^{\mathrm{T}} & \cdots & x^\alpha (\mathbf{x}^n)^{\mathrm{T}} \end{pmatrix}$. Define

$$\widetilde{P}_\alpha(x) := \frac{1}{\Delta_{n,d}} \det \widetilde{M}_\alpha(x), \qquad |\alpha| = n, \quad \alpha \in \mathbb{N}_0^d. \tag{3.2.10}$$

Multiplying the polynomial $\widetilde{P}_\alpha^n$ by x^β and applying the linear functional $\mathscr{L}$, we obtain immediately

$$\mathscr{L}(x^\beta \widetilde{P}_\alpha^n) = \begin{cases} 0 & \text{if } |\beta| \le n-1, \\ \delta_{\alpha,\beta} & \text{if } |\beta| = n. \end{cases} \tag{3.2.11}$$

Thus $\{\widetilde{P}_\alpha^n\}_{|\alpha|=n_{n=0}}^{\infty}$ is a system of orthogonal polynomials with respect to $\mathscr{L}$.

Let $\operatorname{adj} M_{n,d}$ denote the adjoint matrix of $M_{n,d}$; that is, its elements are the cofactors of $M_{n,d}$ (see, for example, p. 20 of Horn and Johnson [1985]). A theorem of linear algebra states that

$$M_{n,d}^{-1} = \frac{1}{\Delta_{n,d}} \operatorname{adj} M_{n,d}.$$

Let $M_{n,d}^{-1}(|\alpha| = n)$ denote the submatrix of $M_{n,d}^{-1} = (m_{\alpha,\beta})_{|\alpha|,|\beta| \le n}$ which consists of those $m_{\alpha,\beta}$ whose indices satisfy the condition $|\alpha| = |\beta| = n$; in other words, $M_{n,d}^{-1}(|\alpha| = n)$ is the principal minor of $M_{n,d}^{-1}$ of size $r_n^d \times r_n^d$ at the lower right corner. The elements of $M_{n,d}^{-1}(|\alpha| = n)$ are the cofactors of the elements of $\mathscr{L}(\mathbf{x}^n(\mathbf{x}^n)^{\mathrm{T}})$ in $M_{n,d}$. Since $M_{n,d}$ is positive definite, so is $M_{n,d}^{-1}$; it follows that $M_{n,d}^{-1}(|\alpha| = n)$ is also positive definite.

Theorem 3.2.13 *Let $\mathscr{L}$ be a positive definite moment functional. Then the polynomials*

$$\mathbb{P}_n = (M_{n,d}^{-1}(|\alpha| = n))^{-1/2}\widetilde{\mathbb{P}}_n = G_n \mathbf{x}^n + \cdots \qquad (3.2.12)$$

are orthonormal polynomials with respect to $\mathscr{L}$, and G_n is positive definite.

Proof By the definition of $\widetilde{M}_\alpha$, expand the determinant along the row $(1 \quad \mathbf{x}^{\mathrm{T}} \quad \cdots \quad (\mathbf{x}^n)^{\mathrm{T}})$; then

$$\det \widetilde{M}_\alpha(x) = \sum_{|\beta|=n} C_{\alpha,\beta} x^\beta + Q_\alpha^{n-1}, \quad Q_\alpha^{n-1} \in \Pi_{n-1}^d,$$

where $C_{\alpha,\beta}$ is the cofactor of the element $\mathscr{L}(x^\alpha x^\beta)$ in $M_{n,d}$; that is, $\operatorname{adj} M_{n,d} = (C_{\beta,\alpha})_{|\alpha|,|\beta|\le n}$. Therefore, since $\operatorname{adj} M_{n,d} = \Delta_{n,d} M_{n,d}^{-1}$, we can write

$$\begin{aligned}\widetilde{\mathbb{P}}_n &= \frac{1}{\Delta_{n,d}} \Delta_{n,d} M_{n,d}^{-1}(|\alpha| = n)\mathbf{x}^n + \mathbb{Q}_{n-1} \\ &= M_{n,d}^{-1}(|\alpha| = n)\mathbf{x}^n + \mathbb{Q}_{n-1}.\end{aligned}$$

From (3.2.11) we have that $\mathscr{L}(\mathbf{x}^n \widetilde{\mathbb{P}}_n^{\mathrm{T}}) = I_{r_n}$ and $\mathscr{L}(\mathbb{Q}_{n-1}\widetilde{\mathbb{P}}_n^{\mathrm{T}}) = 0$; hence,

$$\mathscr{L}(\widetilde{\mathbb{P}}_n \widetilde{\mathbb{P}}_n^{\mathrm{T}}) = M_{n,d}^{-1}(|\alpha| = n).$$

It then follows immediately that $\mathscr{L}(\mathbb{P}_n \mathbb{P}_n^{\mathrm{T}}) = I$; thus $\mathbb{P}_n$ is orthonormal. Moreover,

$$G_n = [M_{n,d}^{-1}(|\alpha| = n)]^{1/2}. \qquad (3.2.13)$$

Clearly, G_n is positive definite. □

As is evident from the examples in the previous chapter, systems of orthonormal polynomials with respect to $\mathscr{L}$ are not unique. In fact, if $\mathbb{P}_n$ is orthonormal then, for any orthogonal matrix O_n of size $r_n^d \times r_n^d$, the polynomial components in $\mathbb{P}_n^* = O_n \mathbb{P}_n$ are also orthonormal. Moreover, it is easily seen that if the components of $\mathbb{P}_n$ and $\mathbb{P}_n^*$ are each a collection of orthonormal polynomials then multiplication by an orthogonal matrix connects $\mathbb{P}_n$ and $\mathbb{P}_n^*$.

Theorem 3.2.14 *Let $\mathscr{L}$ be positive definite and let $\{Q_\alpha^n\}$ be a sequence of orthonormal polynomials with respect to $\mathscr{L}$. Then there is an orthogonal matrix O_n such that $\mathbb{Q}_n = O_n\mathbb{P}_n$, where $\mathbb{P}_n$ are the orthonormal polynomials defined in* (3.2.12). *Moreover, the leading-coefficient matrix G_n^* of $\mathbb{Q}_n$ satisfies $G_n^* = O_n G_n$, where G_n is the positive definite matrix in* (3.2.13). *Furthermore,*

$$(\det G_n)^2 = \frac{\Delta_{n-1,d}}{\Delta_{n,d}}. \tag{3.2.14}$$

Proof By Theorem 3.2.13, $\mathbb{Q}_n = O_n\mathbb{P}_n + O_{n,n-1}\mathbb{P}_{n-1} + \cdots + O_{n,0}\mathbb{P}_0$. Multiplying the equation by $\mathbb{P}_k$, $0 \le k \le n-1$, it follows that $O_{n,k} = 0$ for $k = 0, 1, \ldots, n-1$. Hence, $\mathbb{Q}_n = O_n\mathbb{P}_n$. Since both $\mathbb{P}_n$ and $\mathbb{Q}_n$ are orthonormal,

$$I = \mathscr{L}(\mathbb{Q}_n\mathbb{Q}_n^{\mathrm{T}}) = O_n\mathscr{L}(\mathbb{P}_n\mathbb{P}_n^{\mathrm{T}})O_n^{\mathrm{T}} = O_nO_n^{\mathrm{T}},$$

which shows that O_n is orthonormal. Comparing the leading coefficients of $\mathbb{Q}_n = O_n\mathbb{P}_n$ leads to $G_n^* = O_nG_n$. To verify equation (3.2.14), use (3.2.13) and the formula for the determinants of minors (see, for example, p. 21 of Horn and Johnson [1985]):

$$\det M_{n,d}^{-1}(|\alpha| = n) = \frac{\det M_{n,d}(1 \le |\alpha| \le n-1)}{\Delta_{n,d}} = \frac{\Delta_{n-1,d}}{\Delta_{n,d}},$$

from which the desired result follows immediately. □

Equation (3.2.14) can be viewed as an analogue of the relation $k_n^2 = d_n/d_{n+1}$ for orthogonal polynomials of one variable, given in Section 1.4.

3.2.3 The moment problem

Let $\mathscr{M} = \mathscr{M}(\mathbb{R}^d)$ denote the set of nonnegative Borel measures on $\mathbb{R}^d$ having moments of all orders, that is, if $\mu \in \mathscr{M}$ then

$$\int_{\mathbb{R}^d} |x^\alpha|\,\mathrm{d}\mu(x) < \infty \qquad \forall \alpha \in \mathbb{N}_0^d.$$

For $\mu \in \mathscr{M}$, as in the previous section its *moments* are defined by

$$s_\alpha = \int_{\mathbb{R}^d} x^\alpha\,\mathrm{d}\mu, \qquad \alpha \in \mathbb{N}_0^d. \tag{3.2.15}$$

However, if $\{s_\alpha\}$ is a multi-sequence and there is a measure $\mu \in \mathscr{M}$ such that (3.2.15) holds then s_α is called a *moment sequence*. Two measures in $\mathscr{M}$ are called equivalent if they have the same moments. The measure μ is called *determinate* if μ is unique in the equivalence class of measures. If a sequence is a moment sequence of a determinate measure, we say that the sequence is determinate.

Evidently, if $\mu \in \mathscr{M}$ then the moment functional $\mathscr{L}$ defined by its moments is exactly the linear functional

$$\mathscr{L}(P) = \int_{\mathbb{R}^d} P(x)\,\mathrm{d}\mu(x), \qquad P \in \Pi^d.$$

Examples include any linear functional expressible as an integral with respect to a nonnegative weight function W, that is for which $\mathrm{d}\mu(x) = W(x)\,\mathrm{d}x$ and $\int |x^\alpha| W(x)\,\mathrm{d}x < \infty$ for all α. Whenever W is supported on a domain with nonempty interior, it is positive definite in the sense of Definition 3.2.7, that is, $\mathscr{L}(P^2) > 0$ whenever $P \neq 0$.

The moment problem asks the question when is a sequence a moment sequence and, if so, when is the measure determinate? Like many other problems involving polynomials in higher dimensions, the moment problem in several variables is much more difficult than its one-variable counterpart. The problem is still not completely solved. A theorem due to Haviland gives the following characterization (see Haviland [1935]).

Theorem 3.2.15 *A moment sequence s can be given in the form* (3.2.15) *if and only if the moment functional $\mathscr{L}_s$ is nonnegative on the set of nonnegative polynomials $\Pi^d_+ = \{P \in \Pi^d : p(x) \geq 0, x \in \mathbb{R}^d\}$.*

The linear functional $\mathscr{L}$ is called positive if $\mathscr{L}(P) \geq 0$ on Π^d_+. In one variable, $\mathscr{L}$ being positive is equivalent to its being nonnegative definite, that is, to $\mathscr{L}(p^2) \geq 0$, since every positive polynomial on $\mathbb{R}$ can be written as a sum of the squares of two polynomials. In several variables, however, this is no longer the case: there exist positive polynomials that cannot be written as a sum of the squared polynomials. Thus, Hamburger's famous theorem that a sequence is a moment sequence if and only if it is positive definite does not hold in several variables (see Fuglede [1983], Berg, Christensen and Ressel [1984], Berg [1987] and Schmüdgen [1990]). There are several sufficient conditions for a sequence to be determinate. For example, the following result of Nussbaum [1966] extends a classical result of Carleman in one variable (see, for example, Shohat and Tamarkin [1943]).

Theorem 3.2.16 *If $\{s_\alpha\}$ is a positive definite sequence and if Carleman's condition for a sequence $\{a_n\}$,*

$$\sum_{n=0}^{\infty} (a_{2n})^{-1/2n} = +\infty, \tag{3.2.16}$$

is satisfied by each of the marginal sequences $s_{\{n,0,\ldots,0\}}, s_{\{0,n,\ldots,0\}}, \ldots, s_{\{0,0,\ldots,n\}}$ then $\{s_\alpha\}$ is a determinate moment sequence.

This theorem allows us to extend some results for the moment problem of one variable to several variables. For example, it follows that the proof of Theorem 5.2 in Freud [1966] in one variable can be extended easily to the following theorem.

Theorem 3.2.17 *If $\mu \in \mathscr{M}$ satisfies*

$$\int_{\mathbb{R}^d} \mathrm{e}^{c\|x\|}\,\mathrm{d}\mu(x) < \infty \tag{3.2.17}$$

for some constant $c > 0$ then μ is a determinate measure.

Evidently, we can replace $\|x\|$ by $|x|_1$. In particular, if $\mu \in \mathscr{M}$ has compact support then it is determinate. If K is a domain defined in $\mathbb{R}^d$ then one can consider the K-moment problem, which asks when a given sequence is a moment sequence of a measure supported on K. There are various sufficient conditions and results for special domains; see the references cited above.

A related question is the density of polynomials in $L^2(\mathrm{d}\mu)$. In one variable it is known that if μ is determinate then polynomials in $L^2(\mathrm{d}\mu)$ will be dense. This, however, is not true in several variables; see Berg and Thill [1991], where it is proved that there exist rotation-invariant measures μ on $\mathbb{R}^d$ which are determinate but for which the polynomials are not dense in $L^2(\mathrm{d}\mu)$. For the study of the convergence of orthogonal polynomial expansions in several variables in later chapters, the following theorem is useful.

Theorem 3.2.18 *If $\mu \in \mathscr{M}$ satisfies the condition* (3.2.17) *for some constant $c > 0$ then the space of polynomials Π^d is dense in the space $L^2(\mathrm{d}\mu)$.*

Proof The assumption implies that polynomials are elements of $L^2(\mathrm{d}\mu)$. Indeed, for each $\alpha \in \mathbb{N}_0^d$ and for every $c > 0$ there exists a constant A such that $|x^\alpha|^2 < A\mathrm{e}^{c\|x\|}$ for all sufficiently large $\|x\|$.

If polynomials were not dense in $L^2(\mathrm{d}\mu)$, there would be a function f, not almost everywhere zero, in $L^2(\mathrm{d}\mu)$ such that f would be orthogonal to all polynomials:

$$\int_{\mathbb{R}^d} f(x)x^\alpha \mathrm{d}\mu(x) = 0, \qquad \alpha \in \mathbb{N}_0^d.$$

Let $\mathrm{d}\nu = f(x)\,\mathrm{d}\mu$. Then, since $(\int |f|\,\mathrm{d}\mu)^2 \le \int |f|^2\,\mathrm{d}\mu \int \mathrm{d}\mu$, we can take the Fourier–Stieltjes transform of $\mathrm{d}\nu$, which is, by definition,

$$\hat{\nu}(z) = \int_{\mathbb{R}^d} \mathrm{e}^{-\mathrm{i}\langle z,x\rangle}\mathrm{d}\nu(x), \qquad z \in \mathbb{C}^d.$$

Let $y = \operatorname{Im} z$. By the Cauchy–Schwarz inequality, for $\|y\| \le c/2$ we have

$$|\hat{\nu}(z)|^2 = \left|\int_{\mathbb{R}^d} \mathrm{e}^{-\mathrm{i}\langle z,x\rangle}\mathrm{d}\nu\right|^2 \le \int_{\mathbb{R}^d} \mathrm{e}^{2\|y\|\,\|x\|}\mathrm{d}\mu \int_{\mathbb{R}^d} |f(x)|^2\,\mathrm{d}\mu < \infty,$$

since $|\langle y, x\rangle| \le \|y\|\,\|x\|$. Similarly,

$$\left|\frac{\partial \hat{v}(z)}{\partial z_i}\right|^2 \le \int_{\mathbb{R}^d} |x_i|^2 e^{2\|y\|\,\|x\|} d\mu \int_{\mathbb{R}^d} |f(x)|^2 d\mu < \infty$$

for $\|y\| \le c/4$. Consequently $\hat{v}(z)$ is analytic in the set $\{z \in \mathbb{C}^d : |\operatorname{Im} z| \le c/4\}$. For small $\|z\|$, expanding $\hat{v}(z)$ in a power series and integrating by parts gives

$$\hat{v}(z) = \sum_{n=0}^{\infty} \frac{(-\mathrm{i})^n}{n!} \int_{\mathbb{R}^d} f(x)\langle x, z\rangle^n d\mu = 0.$$

Therefore, by the uniqueness principle for analytic functions, $\hat{v}(z) = 0$ for $\|y\| \le c/4$. In particular $\hat{v}(z) = 0$ for $z \in \mathbb{R}^d$. By the uniqueness of the Fourier–Stieltjes transform, we conclude that $f(x) = 0$ almost everywhere, which is a contradiction. □

This proof is a straightforward extension of the one-variable proof on p. 31 of Higgins [1977].

3.3 The Three-Term Relation

3.3.1 Definition and basic properties

The three-term relation plays an essential role in understanding the structure of orthogonal polynomials in one variable, as indicated in Subsection 1.3.2. For orthogonal polynomials in several variables, the three-term relation takes a vector–matrix form. Let $\mathscr{L}$ be a moment functional. Throughout this subsection, we use the notation $\tilde{\mathbb{P}}_n$ for orthogonal polynomials with respect to $\mathscr{L}$ and the notation H_n for the matrix

$$H_n = \mathscr{L}(\tilde{\mathbb{P}}_n \tilde{\mathbb{P}}_n^{\mathrm{T}}).$$

When $H_n = I$, the identity matrix, $\tilde{\mathbb{P}}_n$ becomes orthonormal and the notation $\mathbb{P}_n$ is used to denote orthonormal polynomials. Note that H_n is invertible by Proposition 3.2.4.

Theorem 3.3.1 *For $n \ge 0$, there exist unique matrices $A_{n,i} : r_n^d \times r_{n+1}^d$, $B_{n,i} : r_n^d \times r_n^d$ and $C_{n,i}^{\mathrm{T}} : r_n^d \times r_{n-1}^d$ such that*

$$x_i \tilde{\mathbb{P}}_n = A_{n,i}\tilde{\mathbb{P}}_{n+1} + B_{n,i}\tilde{\mathbb{P}}_n + C_{n,i}\tilde{\mathbb{P}}_{n-1}, \quad 1 \le i \le d, \tag{3.3.1}$$

where we define $\mathbb{P}_{-1} = 0$ and $C_{-1,i} = 0$; moreover,

$$\begin{aligned} A_{n,i}H_{n+1} &= \mathscr{L}(x_i \tilde{\mathbb{P}}_n \tilde{\mathbb{P}}_{n+1}^{\mathrm{T}}), \\ B_{n,i}H_n &= \mathscr{L}(x_i \tilde{\mathbb{P}}_n \tilde{\mathbb{P}}_n^{\mathrm{T}}), \\ A_{n,i}H_{n+1} &= H_n C_{n+1,i}^{\mathrm{T}}. \end{aligned} \tag{3.3.2}$$

Proof Since the components of $x_i\tilde{\mathbb{P}}_n$ are polynomials of degree $n+1$, they can be written as linear combinations of orthogonal polynomials of degree $n+1$ and less by Proposition 3.2.2. Hence, in vector notation, there exist matrices $M_{k,i}$ such that

$$x_i\tilde{\mathbb{P}}_n = M_{n,i}\tilde{\mathbb{P}}_{n+1} + M_{n-1,i}\tilde{\mathbb{P}}_n + M_{n-2,i}\tilde{\mathbb{P}}_{n-1} + \cdots.$$

However, if we multiply the above equation by $\tilde{\mathbb{P}}_k^{\mathrm{T}}$ from the right and apply the linear functional $\mathscr{L}$ then $M_{k-1,i}H_k = \mathscr{L}(x_i\tilde{\mathbb{P}}_n\tilde{\mathbb{P}}_k^{\mathrm{T}})$. By the orthogonality of $\tilde{\mathbb{P}}_k$ and the fact that H_k is invertible, $M_{k,i} = 0$ for $k \le n-2$. Hence, the three-term relation holds and (3.3.2) follows. □

For orthonormal polynomials $\mathbb{P}_n$, $H_n = I$; the three-term relation takes a simpler form:

Theorem 3.3.2 *For $n \ge 0$, there exist matrices $A_{n,i} : r_n^d \times r_{n+1}^d$ and $B_{n,i} : r_n^d \times r_n^d$ such that*

$$x_i\mathbb{P}_n = A_{n,i}\mathbb{P}_{n+1} + B_{n,i}\mathbb{P}_n + A_{n-1,i}^{\mathrm{T}}\mathbb{P}_{n-1}, \quad 1 \le i \le d, \tag{3.3.3}$$

where we define $\mathbb{P}_{-1} = 0$ and $A_{-1,i} = 0$. Moreover, each $B_{n,i}$ is symmetric.

The coefficients of the three-term relation satisfy several properties. First, recall that the leading-coefficient matrix of $\tilde{\mathbb{P}}_n$ (or $\mathbb{P}_n$) is denoted by G_n. Comparing the highest-coefficient matrices on each side of (3.3.3), it follows that

$$A_{n,i}G_{n+1} = G_nL_{n,i}, \quad 1 \le i \le d, \tag{3.3.4}$$

where the $L_{n,i}$ are matrices of size $r_n^d \times r_{n+1}^d$, defined by

$$L_{n,i}\mathbf{x}^{n+1} = x_i\mathbf{x}^n, \quad 1 \le i \le d. \tag{3.3.5}$$

For example, for $d = 2$,

$$L_{n,1} = \begin{bmatrix} 1 & & \bigcirc & 0 \\ & \ddots & & \vdots \\ \bigcirc & & 1 & 0 \end{bmatrix} \quad \text{and} \quad L_{n,2} = \begin{bmatrix} 0 & 1 & & \bigcirc \\ \vdots & & \ddots & \\ 0 & \bigcirc & & 1 \end{bmatrix}.$$

We now adopt the following notation: if $M_1, \ldots, M_d$ are matrices of the same size $p \times q$ then we define their *joint matrix* M by

$$M = (M_1^{\mathrm{T}} \quad \cdots \quad M_d^{\mathrm{T}})^{\mathrm{T}}, \qquad M : dp \times q \tag{3.3.6}$$

(it is better to write M as a column matrix of $M_1, \ldots, M_d$). In particular, both L_n and A_n are joint matrices of size $dr_n^d \times r_{n+1}^d$. The following proposition collects the properties of $L_{n,i}$.

Proposition 3.3.3 *For each i, $1 \le i \le d$, the matrix $L_{n,i}$ satisfies the equation $L_{n,i}L_{n,i}^{\mathrm{T}} = I$. Moreover,*

$$\operatorname{rank} L_{n,i} = r_n^d, \quad 1 \le i \le d, \qquad \textit{and} \quad \operatorname{rank} L_n = r_{n+1}^d.$$

Proof By definition, each row of $L_{n,i}$ contains exactly one element equal to 1; the rest of its elements are 0. Hence $L_{n,i}L_{n,i}^{\mathrm{T}} = I$, which also implies that $\operatorname{rank} L_{n,i} = r_n^d$. Moreover, let

$$\mathcal{N}_n = \{\alpha \in \mathbb{N}_0^d : |\alpha| = n\} \quad \text{and} \quad \mathcal{N}_{n,i} = \{\alpha \in \mathbb{N}_0^d : |\alpha| = n, \alpha_i \ne 0\},$$

and let $\mathbf{a} = (a_\alpha)_{\alpha \in \mathcal{N}_n}$ be a vector of size $r_{n+1}^d \times 1$; then $L_{n,i}$ can be considered as a mapping which projects $\mathbf{a}$ onto its restriction on $\mathcal{N}_{n+1,i}$, that is, $L_{n,i}\mathbf{a} = \mathbf{a}|_{\mathcal{N}_{n+1,i}}$. To prove that L_n has full rank, we show that if $L_n\mathbf{a} = 0$ then $\mathbf{a} = 0$. By definition, $L_n\mathbf{a} = 0$ implies that $L_{n,i}\mathbf{a} = 0$, $1 \le i \le d$. Evidently $\bigcup_i \mathcal{N}_{n,i} = \mathcal{N}_n$. Hence $L_{n,i}\mathbf{a} = 0$ implies that $\mathbf{a} = 0$. □

Theorem 3.3.4 *For $n \ge 0$ and $1 \le i \le d$,*

$$\operatorname{rank} A_{n,i} = \operatorname{rank} C_{n+1,i} = r_n^d. \tag{3.3.7}$$

Moreover, for the joint matrix A_n of $A_{n,i}$ and the joint matrix C_n^{T} of $C_{n,i}^{\mathrm{T}}$,

$$\operatorname{rank} A_n = r_{n+1}^d \quad \textit{and} \quad \operatorname{rank} C_{n+1}^{\mathrm{T}} = r_{n+1}^d. \tag{3.3.8}$$

Proof From the relation (3.3.4) and the fact that G_n is invertible, $\operatorname{rank} A_{n,i} = r_n^d$ follows from Proposition 3.3.3. Since H_n is invertible, it follows from the third equation of (3.3.2) that $\operatorname{rank} C_{n+1,i} = r_n^d$. In order to prove (3.3.8) note that (3.3.4) implies that $A_nG_{n+1} = \operatorname{diag}\{G_n, \ldots, G_n\}L_n$, since G_n being invertible implies that the matrix $\operatorname{diag}\{G_n, \ldots, G_n\}$ is invertible. It follows from Proposition 3.3.3 that $\operatorname{rank} A_n = r_{n+1}^d$. Furthermore, the third equation of (3.3.2) implies that $A_nH_{n+1} = \operatorname{diag}\{H_n, \ldots, H_n\}C_{n+1}^{\mathrm{T}}$, and H_n is invertible; consequently, $\operatorname{rank} C_{n+1} = \operatorname{rank} A_n$. □

Since the matrix A_n has full rank, it has a generalized inverse D_n^{T}, which we write as follows:

$$D_n^{\mathrm{T}} = (D_{n,1}^{\mathrm{T}} \quad \cdots \quad D_{n,i}^{\mathrm{T}}), \qquad r_{n+1}^d \times dr_n^d,$$

where $D_{n,i}^{\mathrm{T}} : r_{n+1}^d \times r_n^d$, that is,

$$D_n^{\mathrm{T}} A_n = \sum_{i=1}^d D_{n,i}^{\mathrm{T}} A_{n,i} = I. \tag{3.3.9}$$

We note that the matrix D_n that satisfies (3.3.9) is not unique. In fact, let A be a matrix of size $s \times t$, $s > t$, with $\operatorname{rank} A = t$, and let the singular-value decomposition of A be given by

$$A = W^{\mathrm{T}} \begin{bmatrix} \Lambda \\ 0 \end{bmatrix} U,$$

where $\Lambda : t \times t$ is an invertible diagonal matrix and $W : s \times s$ and $U : t \times t$ are unitary matrices. Then the matrix D^{T} defined by

$$D^{\mathrm{T}} = U^{\mathrm{T}}(\Lambda^{-1}, \Lambda_1)W,$$

where $\Lambda_1 : s \times t$ can be any matrix, satisfies $D^{\mathrm{T}}A = I$. The matrix D^{T} is known as the *generalized inverse* of A. If $\Lambda_1 = 0$ then D^{T} is the unique Moore–Penrose generalized inverse, which is often denoted by A^+.

Using the generalized matrix D_n^{T}, we can deduce a recursive formula for orthogonal polynomials in several variables.

Theorem 3.3.5 *Let D_n^{T} be a generalized inverse of A_n. There exist matrices $E_n : r_{n+1}^d \times r_n^d$ and $F_n : r_{n+1}^d \times r_{n-1}^d$ such that*

$$\tilde{\mathbb{P}}_{n+1} = \sum_{i=1}^{d} x_i D_{n,i}^{\mathrm{T}} \tilde{\mathbb{P}}_n + E_n \tilde{\mathbb{P}}_n + F_n \tilde{\mathbb{P}}_{n-1}, \tag{3.3.10}$$

where

$$\sum_{i=1}^{d} D_{n,i}^{\mathrm{T}} B_{n,i} = -E_n, \qquad \sum_{i=1}^{d} D_{n,i}^{\mathrm{T}} C_{n,i}^{\mathrm{T}} = -F_n.$$

Proof Multiplying the three-term relation (3.3.3) by $D_{n,i}^{\mathrm{T}}$ and summing over $1 \le i \le d$, we find that the desired equality follows from (3.3.9). □

Equation (3.3.10) is an analogue of the recursive formula for orthogonal polynomials in one variable. However, unlike its one-variable counterpart, (3.3.10) does not imply the three-term relation; that is, if we choose matrices $A_{n,i}$, $B_{n,i}$ and $C_{n,i}$, and use (3.3.10) to generate a sequence of polynomials, in general the polynomials do not satisfy the three-term relation. We will discuss the condition under which equivalence does hold in Section 3.5.

3.3.2 Favard's theorem

Next we prove an analogy of Favard's theorem which states roughly that any sequence of polynomials mich satisfies a three-term relation (3.3.1) whose coefficient matrices satisfy the rank conditions (3.3.7) and (3.3.8) is necessarily a sequence of orthogonal polynomials. We need a definition.

Definition 3.3.6 A linear functional $\mathscr{L}$ defined on Π^d is called quasi-definite if there is a basis B of Π^d such that, for any $P, Q \in B$,

$$\mathscr{L}(PQ) = 0 \quad \text{if} \quad P \neq Q \qquad \text{and} \qquad \mathscr{L}(P^2) \neq 0.$$

From the discussion in Section 3.2 it is clear that if $\mathscr{L}$ is positive definite then it is quasi-definite. However, quasi-definiteness may not imply positive definiteness.

Theorem 3.3.7 *Let $\{\mathbb{P}_n\}_{n=0}^{\infty} = \{P_\alpha^n : |\alpha| = n, n \in \mathbb{N}_0\}$, $\mathbb{P}_0 = 1$, be an arbitrary sequence in Π^d. Then the following statements are equivalent.*

(i) *There exists a linear functional $\mathscr{L}$ which defines a quasi-definite linear functional on Π^d and which makes $\{\mathbb{P}_n\}_{n=0}^{\infty}$ an orthogonal basis in Π^d.*

(ii) *For $n \ge 0$, $1 \le i \le d$, there exist matrices $A_{n,i}$, $B_{n,i}$ and $C_{n,i}$ such that*

  (a) *the polynomials $\mathbb{P}_n$ satisfy the three-term relation* (3.3.1),

  (b) *the matrices in the relation satisfy the rank conditions* (3.3.7) *and* (3.3.8).

Proof That statement (i) implies statement (ii) is contained in Theorems 3.3.1 and 3.3.4. To prove the other direction, we first prove that the polynomial sequence $\{\mathbb{P}_n\}$ forms a basis of Π^d. It suffices to prove that the leading coefficient, G_n, of $\mathbb{P}_n$ is invertible.

The matrix $\operatorname{diag}\{G_n, \ldots, G_n\}$ is of size $dr_n^d \times dr_n^d$ and has copies of the matrix G_n as diagonal entries in the sense of block matrices. Since the polynomial $\mathbb{P}_n$ satisfies the three-term relation, comparing the coefficient matrices of $\mathbf{x}^{n+1}$ on each side of the relation leads to $A_{n,i}G_{n+1} = G_n L_{n,i}$, $1 \le i \le d$, from which it follows that $A_n G_{n+1} = \operatorname{diag}\{G_n, \ldots, G_n\} L_n$. We now proceed with induction in n to prove that each G_n is invertible. That $\mathbb{P}_0 = 1$ implies $G_0 = 1$. Suppose that G_n has been proved to be invertible. Then $\operatorname{diag}\{G_n, \ldots, G_n\}$ is invertible and from the relation $\operatorname{rank} L_n = r_{n+1}^d$ we have

$$\operatorname{rank} A_n G_{n+1} = \operatorname{rank} \operatorname{diag}\{G_n, \ldots, G_n\} L_n = r_{n+1}^d.$$

Therefore, by (3.3.8) and the rank inequality of product matrices (see, for example, p. 13 of Horn and Johnson [1985]),

$$\operatorname{rank} G_{n+1} \ge \operatorname{rank} A_n G_{n+1} \ge \operatorname{rank} A_n + \operatorname{rank} G_{n+1} - r_{n+1}^d = \operatorname{rank} G_{n+1}.$$

Thus, it follows that $\operatorname{rank} G_{n+1} = \operatorname{rank} A_n G_{n+1} = r_{n+1}^d$. Hence G_{n+1} is invertible and the induction is complete.

Since $\{P_k^n\}$ is a basis of Π^d, the linear functional $\mathscr{L}$ defined on Π^d by

$$\mathscr{L}(1) = 1, \quad \mathscr{L}(\mathbb{P}_n) = 0, \qquad n \ge 1,$$

is well defined. We now use induction to prove that

$$\mathscr{L}(\mathbb{P}_k \mathbb{P}_j^{\mathrm{T}}) = 0, \quad k \ne j. \tag{3.3.11}$$

Let $n \ge 0$ be an integer. Assume that (3.3.11) hold for every k, j such that $0 \le k \le n$ and $j > k$. Since the proof of (3.3.10) used only the three-term relation and the fact that rank $A_n = r_{n+1}^d$, it follows from (iia) and (iib) that (3.3.10) holds for $\mathbb{P}_n$ under these considerations. Therefore, for $\ell > n+1$,

$$\begin{aligned}\mathscr{L}(\mathbb{P}_{n+1}\mathbb{P}_\ell^{\mathrm{T}}) &= \mathscr{L}\left(\sum_{i=1}^d x_i D_{n,i}\mathbb{P}_n\mathbb{P}_\ell^{\mathrm{T}}\right)\\ &= \mathscr{L}\left(\sum_{i=1}^d D_{n,i}\mathbb{P}_n(A_{\ell,i}\mathbb{P}_{\ell+1} + B_{\ell,i}\mathbb{P}_\ell + C_{\ell,i}\mathbb{P}_{\ell-1})^{\mathrm{T}}\right) = 0.\end{aligned}$$

The induction is complete. Next we prove that $H_n = \mathscr{L}(\mathbb{P}_n\mathbb{P}_n^{\mathrm{T}})$ is invertible. Clearly, H_n is symmetric. From (iib) and (3.3.11), $A_{n,i}H_{n+1} = H_n C_{n+1,i}^{\mathrm{T}}$; thus

$$A_n H_{n+1} = \operatorname{diag}\{H_n, \ldots, H_n\} C_{n+1}^{\mathrm{T}}. \tag{3.3.12}$$

Since $\mathscr{L}(1) = 1$, it follows that $H_0 = \mathscr{L}(\mathbb{P}_0\mathbb{P}_0^{\mathrm{T}}) = 1$. Thus H_0 is invertible. Suppose that H_n is invertible. Then $\operatorname{diag}\{H_n, \ldots, H_n\}$ is invertible and, by $\operatorname{rank} C_{n+1} = r_{n+1}^d$,

$$\operatorname{rank} A_n H_{n+1} = \operatorname{rank}\big(\operatorname{diag}\{H_n, \ldots, H_n\} C_{n+1}^{\mathrm{T}}\big) = r_{n+1}^d.$$

However, rank $A_n = r_{n+1}^d$ by (iia) and $A_n : dr_n^d \times r_{n+1}^d$, $H_{n+1} : r_{n+1}^d \times r_{n+1}^d$; we then have

$$\operatorname{rank} H_{n+1} \geq \operatorname{rank}(A_n H_{n+1}) \geq \operatorname{rank} A_n + \operatorname{rank} H_{n+1} - r_{n+1}^d = \operatorname{rank} H_{n+1}.$$

Therefore rank $H_{n+1} = \operatorname{rank} A_n H_{n+1}^{\mathrm{T}} = r_{n+1}^d$, which implies that H_{n+1} is invertible. By induction, we have proved that H_n is invertible for each $n \geq 0$. Since H_n is symmetric and invertible, there exist invertible matrices S_n and Λ_n such that $H_n = S_n \Lambda_n S_n^{\mathrm{T}}$ and Λ_n is diagonal. For $\mathbb{Q}_n = S_n^{-1}\mathbb{P}_n$ it then follows that

$$\mathscr{L}(\mathbb{Q}_n\mathbb{Q}_n^{\mathrm{T}}) = S_n^{-1}\mathscr{L}(\mathbb{P}_n\mathbb{P}_n^{\mathrm{T}})(S_n^{-1})^{\mathrm{T}} = S_n^{-1}H_n(S_n^{-1})^{\mathrm{T}} = \Lambda_k.$$

This proves that $\mathscr{L}$ defines a quasi-definite linear functional in Π^d; $\mathscr{L}$ makes $\{\mathbb{P}_n\}_{n=0}^\infty$ an orthogonal basis. The proof is complete. □

If the polynomials in Theorem 3.3.7 satisfy (3.3.3) instead of (3.3.1) then they will be orthogonal with respect to a positive definite linear functional instead of a quasi-definite linear functional.

Theorem 3.3.8 *Let $\{\mathbb{P}_n\}_{n=0}^\infty = \{P_\alpha^n : |\alpha| = n, n \in \mathbb{N}_0\}$, $\mathbb{P}_0 = 1$, be an arbitrary sequence in Π^d. Then the following statements are equivalent.*

(i) *There exists a linear function $\mathscr{L}$ which defines a positive definite linear functional on Π^d and which makes $\{\mathbb{P}_n\}_{n=0}^\infty$ an orthonormal basis in Π^d.*
(ii) *For $n \geq 0$, $1 \leq i \leq d$, there exist matrices $A_{n,i}$ and $B_{n,i}$ such that*
  (a) *the polynomials $\mathbb{P}_n$ satisfy the three-term relation* (3.3.3),
  (b) *the matrices in the three-term relation satisfy the rank conditions* (3.3.7) *and* (3.3.8).

Proof As in the proof of the previous theorem, we only need to prove that statement (ii) implies statement (i). By Theorem 3.3.7 there exists a quasi-definite linear functional $\mathscr{L}$ which makes $\{\mathbb{P}_n\}_{n=0}^{\infty}$ orthogonal. We will prove that $\mathscr{L}$ is positive definite. It is sufficient to show that the matrix $H_n = \mathscr{L}(\mathbb{P}_n\mathbb{P}_n^{\mathrm{T}})$ is the identity matrix for every $n \in \mathbb{N}_0$. Indeed, since every nonzero polynomial P of degree n can be written as $P = \mathbf{a}_n^{\mathrm{T}}\mathbb{P}_n + \cdots + \mathbf{a}_0^{\mathrm{T}}\mathbb{P}_0$, it follows from $H_n = I$ that $\mathscr{L}(P^2) = \|\mathbf{a}_n\|^2 + \cdots + \|\mathbf{a}_0\|^2 > 0$. Since $C_{n+1} = A_n^{\mathrm{T}}$, equation (3.3.12) becomes $A_n H_{n+1} = \operatorname{diag}\{H_k, \ldots, H_k\} A_n$. Since $\mathbb{P}_0 = 1$ and $\mathscr{L}1 = 1$, $H_0 = 1$. Suppose that H_n is an identity matrix. Then $\operatorname{diag}\{H_k, \ldots, H_k\}$ is also an identity matrix. Thus, it follows from the rank condition and the above equation that H_{n+1} is an identity matrix. The proof is completed by induction. □

Both these theorems are extensions of Favard's theorem for orthogonal polynomials of one variable. Since the three-term relation and the rank condition characterize the orthogonality, we should be able to extract the essential information on orthogonal polynomials by studying the three-term relation; indeed, this has been carried out systematically for orthogonal polynomials in one variable. For several variables, the coefficients of (3.3.1) and (3.3.3) are matrices and they are unique only up to matrix similarity (for (3.3.3), unitary similarity) since an orthogonal basis is not unique and there is no natural order among orthogonal polynomials of the same degree. Hence, it is much harder to extract information from these coefficients.

3.3.3 Centrally symmetric integrals

In this subsection we show that a centrally symmetric linear functional can be described by a property of the coefficient matrices of the three-term relation.

Let $\mathscr{L}$ be a positive definite linear functional. Of special interest are examples of $\mathscr{L}$ expressible as integrals with respect to a nonnegative weight function with finite moments, that is, $\mathscr{L}f = \int f(x)W(x)\,\mathrm{d}x$; in such cases we shall work with the weight function W instead of the functional $\mathscr{L}$.

Definition 3.3.9 Let $\Omega \subset \mathbb{R}^d$ be the support set of the weight function W. Then W is *centrally symmetric* if

$$x \in \Omega \Rightarrow -x \in \Omega \quad \text{and} \quad W(x) = W(-x).$$

We also call the linear functional $\mathscr{L}$ centrally symmetric if it satisfies

$$\mathscr{L}(x^\alpha) = 0, \quad \alpha \in \mathbb{N}^d, \quad |\alpha| \text{ an odd integer.}$$

It is easily seen that when $\mathscr{L}$ is expressible as an integral with respect to W, the two definitions are equivalent. As examples we mention the rotation invariant weight function $(1 - \|x\|^2)^{\mu-1/2}$ on the unit ball B^d and the product weight function $\prod_{i=1}^{d}(1-x_i)^{a_i}(1+x_i)^{b_i}$ on the cube $[-1,1]^d$, which is centrally symmetric only if $a_i = b_i$. Weight functions on the simplex are not centrally symmetric since the simplex is not symmetric with respect to the origin.

A centrally symmetric $\mathscr{L}$ has a relatively simple structure which allows us to gain information about the structure of orthogonal polynomials. The following theorem connects the central symmetry of $\mathscr{L}$ to the coefficient matrices in the three-term relation.

Theorem 3.3.10 *Let $\mathscr{L}$ be a positive definite linear functional, and let $\{B_{n,i}\}$ be the coefficient matrices in* (3.3.3) *for the orthogonal polynomials with respect to $\mathscr{L}$. Then $\mathscr{L}$ is centrally symmetric if and only if $B_{n,i} = 0$ for all $n \in \mathbb{N}_0$ and $1 \le i \le d$.*

Proof First assume that $\mathscr{L}$ is centrally symmetric. From (3.3.2),

$$B_{0,i} = \mathscr{L}(x_i \mathbb{P}_0 \mathbb{P}_0^{\mathrm{T}}) = \mathscr{L}(x_i) = 0.$$

Therefore from (3.3.10) $\mathbb{P}_1 = \sum x_i D_{0,i}^{\mathrm{T}}$, which implies that the constant terms in the components of $\mathbb{P}_1$ vanish. Suppose we have proved that $B_{k,i} = 0$ for $1 \le k \le n-1$. Then, by (3.3.10), since $E_k = 0$ for $k = 0,1,\ldots,k-1$ we can show by induction that the components of $\mathbb{P}_n$ are sums of monomials of even degree if n is even and of odd degree if n is odd. By definition,

$$B_{n,i} = \mathscr{L}(x_i \mathbb{P}_n \mathbb{P}_n^{\mathrm{T}}) = \sum_{j=0}^{d}\sum_{k=0}^{d} \mathscr{L}\left[x_i D_{n-1,j}^{\mathrm{T}}(x_j \mathbb{P}_{n-1} + A_{n-2,j}^{\mathrm{T}} \mathbb{P}_{n-2}) \right.$$
$$\left. \times (x_k \mathbb{P}_{n-1} + A_{n-2,k}^{\mathrm{T}} \mathbb{P}_{n-2})^{\mathrm{T}} D_{n-1,k}\right].$$

Since the components of $(x_j \mathbb{P}_{n-1} + A_{n-2,j}^{\mathrm{T}} \mathbb{P}_{n-2})(x_k \mathbb{P}_{n-1} + A_{n-2,k}^{\mathrm{T}} \mathbb{P}_{n-2})^{\mathrm{T}}$ are polynomials that are sums of monomials of even degree, their multiples by x_i are sums of monomials of odd degree. Since $\mathscr{L}$ is centrally symmetric, $B_{n,i} = 0$.

However, assuming that $B_{n,i} = 0$ for all $n \in \mathbb{N}_0$, the above proof shows that the components of $\mathbb{P}_n$ are sums of monomials of even degree if n is even and sums of monomials of odd degree if n is odd. From $B_{0,i} = \mathscr{L}(x_i \mathbb{P}_0 \mathbb{P}_0^{\mathrm{T}}) = \mathscr{L}(x_i) = 0$ it follows that $\mathscr{L}(x_i) = 0$. We now use induction to prove that

$$\{B_{k,i} = 0 : 1 \le k \le n\} \Rightarrow \{\mathscr{L}(x_1^{\alpha_1} \cdots x_d^{\alpha}) = 0 : \alpha_1 + \cdots + \alpha_d = 2n+1\}.$$

Suppose that the claim has been proved for $k \le n-1$. By (3.3.3) and (3.3.2),

$$\begin{aligned} A_{n-1,j} B_{n,i} A_{n-1,k}^{\mathrm{T}} &= \mathscr{L}[x_i A_{n-1,j} \mathbb{P}_n (A_{n-1,k} \mathbb{P}_n)^{\mathrm{T}}] \\ &= \mathscr{L}[x_i (x_j \mathbb{P}_{n-1} - A_{n-2,j}^{\mathrm{T}} \mathbb{P}_{n-2})(x_k \mathbb{P}_{n-1} - A_{n-2,k}^{\mathrm{T}} \mathbb{P}_{n-2})^{\mathrm{T}}] \\ &= \mathscr{L}(x_i x_j x_k \mathbb{P}_{n-1} \mathbb{P}_{n-1}^{\mathrm{T}}) - \mathscr{L}(x_i x_j \mathbb{P}_{n-1} \mathbb{P}_{n-2}^{\mathrm{T}}) A_{n-2,k} \\ &\quad - A_{n-2,j}^{\mathrm{T}} \mathscr{L}(x_i x_k \mathbb{P}_{n-1} \mathbb{P}_{n-2}^{\mathrm{T}}) + A_{n-2,j}^{\mathrm{T}} B_{n-2,i} A_{n-2,k}. \end{aligned}$$

Then $\mathscr{L}(x_i x_j \mathbb{P}_{n-1}\mathbb{P}_{n-2}^{\mathrm{T}}) = 0$ and $\mathscr{L}(x_i x_k \mathbb{P}_{n-1}\mathbb{P}_{n-2}^{\mathrm{T}}) = 0$, since the elements of these matrices are polynomials that are sums of odd powers. Therefore, from $B_{n,i} = 0$ we conclude that $\mathscr{L}(x_i x_j x_k \mathbb{P}_{n-1}\mathbb{P}_{n-1}^{\mathrm{T}}) = 0$. Multiplying the above matrix from the left and the right by $A_{l,p}$ and $A_{l,p}^{\mathrm{T}}$, respectively, for $1 \le p \le d$ and $l = n-2, \ldots, 1$ and using the three-term relation, we can repeat the above argument and derive $\mathscr{L}(x_1^{\alpha_1} \cdots x_d^{\alpha_d}) = 0$, $\alpha_1 + \cdots + \alpha_d = 2n+1$, in finitely many steps. □

One important corollary of the above proof is the following result.

Theorem 3.3.11 *Let $\mathscr{L}$ be a centrally symmetric linear functional. Then an orthogonal polynomial of degree n with respect to $\mathscr{L}$ is a sum of monomials of even degree if n is even and a sum of monomials of odd degree if n is odd.*

For examples see various bases of the classical orthogonal polynomials on the unit disk in Section 2.3.

If a linear functional $\mathscr{L}$ or a weight function W is not centrally symmetric but becomes centrally symmetric under a nonsingular affine transformation then the orthogonal polynomials should have a structure similar to that of the centrally symmetric case. The coefficients $B_{n,i}$ in the three-term relations, however, will not be zero. It turns out that they satisfy a commutativity condition, which suggests the following definition.

Definition 3.3.12 If the coefficient matrices $B_{n,i}$ in the three-term relation satisfy

$$B_{n,i}B_{n,j} = B_{n,j}B_{n,i}, \quad 1 \le i, j \le d, \quad n \in \mathbb{N}_0, \tag{3.3.13}$$

then the linear functional $\mathscr{L}$ (or the weight function W) is called *quasi-centrally-symmetric*.

Since the condition $B_{n,i} = 0$ implies (3.3.13), central symmetry implies quasi-central symmetry.

Theorem 3.3.13 *Let W be a weight function defined on $\Omega \subset \mathbb{R}^d$. Suppose that W becomes centrally symmetric under the nonsingular linear transformation*

$$u \mapsto x, \quad x = Tu + a, \quad \det T > 0, \quad x, u, a \in \mathbb{R}^d.$$

Then, W is quasi-centrally-symmetric.

Proof Let $W^*(u) = W(Tu+a)$ and $\Omega^* = \{u : x = Tu + a, \quad x \in \Omega\}$. Denote by $\mathbb{P}_n$ the orthonormal polynomials associated to W; then, making the change of variables $x \mapsto Tx + a$ shows that the corresponding orthonormal polynomials for W^* are given by $\mathbb{P}_n^*(u) = \sqrt{\det T}\, \mathbb{P}_n(Tu+a)$. Let $B_{n,i}(W)$ denote the matrices associated with W. Then, by (3.3.2),

$$\begin{aligned} B_{n,i}(W) &= \int_\Omega x_i \mathbb{P}_n(x)\mathbb{P}_n^{\mathrm{T}}(x)W(x)\,\mathrm{d}x \\ &= \int_{\Omega^*}\left(\sum_{j=1}^d t_{ij}u_j + a_i\right)\mathbb{P}_n^*(u)\mathbb{P}_n^{*\mathrm{T}}(u)W^*(u)\,\mathrm{d}u \\ &= \sum_{j=1}^d t_{ij}B_{n,j}(W^*) + a_i I. \end{aligned}$$

By assumption W^* is centrally symmetric on Ω^*, which implies that $B_{n,i}(W^*) = 0$ by Theorem 3.3.10. Then $B_{n,i}(W) = a_i I$, which implies that $B_{n,i}(W)$ satisfies (3.3.13). Hence W is quasi-centrally-symmetric. □

How to characterize quasi-central symmetry through the properties of the linear functional $\mathscr{L}$ or the weight function is not known.

3.3.4 Examples

To get a sense of what the three-term relations are like, we give several examples before continuing our study. The examples chosen are the classical orthogonal polynomials of two variables on the square, the triangle and the disk. For a more concise notation, we will write orthonormal polynomials of two variables as P_k^n. The polynomials P_k^n, $0 \le k \le n$, constitute a basis of $\mathscr{V}_n^2$, and the polynomial vector $\mathbb{P}_n$ is given by $\mathbb{P}_n = (P_0^n \quad \cdots \quad P_n^n)^{\mathrm{T}}$. The three-term relations are

$$x_i \mathbb{P}_n = A_{n,i}\mathbb{P}_{n+1} + B_{n,i}\mathbb{P}_n + A_{n-1,i}^{\mathrm{T}}\mathbb{P}_{n-1}, \qquad i = 1,2,$$

where the sizes of the matrices are $A_{n,i} : (n+1)\times(n+2)$ and $B_{n,i} : (n+1)\times(n+1)$. Closed formulae for the coefficient matrices are derived using (3.3.2) and the explicit formulae for an orthonormal basis. We leave the verification to the reader.

Product polynomials on the square

In the simplest case, that of a square, we can work with general product polynomials. Let p_n be the orthonormal polynomials with respect to a weight function w on $[-1,1]$; they satisfy the three-term relation

$$xp_n = a_n p_{n+1} + b_n p_n + a_{n-1}p_{n-1}, \qquad n \ge 0.$$

Let W be the product weight function defined by $W(x,y) = w(x)w(y)$. One basis of orthonormal polynomials is given by

$$P_k^n(x,y) = p_{n-k}(x)p_k(y), \quad 0 \le k \le n.$$

The coefficient matrices of the three-term relation are then of the following form:

$$A_{n,1} = \begin{bmatrix} a_n & & \bigcirc & 0 \\ & \ddots & & \vdots \\ \bigcirc & & a_0 & 0 \end{bmatrix} \quad \text{and} \quad A_{n,2} = \begin{bmatrix} 0 & a_0 & & \bigcirc \\ \vdots & & \ddots & \\ 0 & \bigcirc & & a_n \end{bmatrix};$$

$$B_{n,1} = \begin{bmatrix} b_n & & \bigcirc \\ & \ddots & \\ \bigcirc & & b_0 \end{bmatrix} \quad \text{and} \quad B_{n,2} = \begin{bmatrix} b_0 & & \bigcirc \\ & \ddots & \\ \bigcirc & & b_n \end{bmatrix}.$$

If w is centrally symmetric then $b_k = 0$; hence $B_{n,i} = 0$.

Classical polynomials on the triangle

For the weight function

$$W(x,y) = x^{\alpha-1/2} y^{\beta-1/2} (1-x-y)^{\gamma-1/2},$$

with normalization constant given by (setting $|\kappa| = \alpha+\beta+\gamma$)

$$w_{\alpha,\beta,\gamma} = \left(\int_{T^2} W(x,y)\, \mathrm{d}x\, \mathrm{d}y \right)^{-1} = \frac{\Gamma(|\kappa|+\frac{3}{2})}{\Gamma(\alpha+\frac{1}{2})\Gamma(\beta+\frac{1}{2})\Gamma(\gamma+\frac{1}{2})},$$

we consider the basis of orthonormal polynomials, see (2.4.2),

$$P_k^n(x,y) = (h_{k,n})^{-1} P_{n-k}^{(2k+\beta+\gamma,\alpha-1/2)}(2x-1)(1-x)^k P_k^{(\gamma-1/2,\beta-1/2)}\left(\frac{2y}{1-x}-1\right),$$

where

$$(h_{k,n})^2 = \frac{w_{\alpha,\beta,\gamma}}{(2n+|\kappa|+\frac{1}{2})(2k+\beta+\gamma)} \times \frac{\Gamma(n+k+\beta+\gamma+1)\Gamma(n-k+\alpha+\frac{1}{2})\Gamma(k+\beta+\frac{1}{2})\Gamma(k+\gamma+\frac{1}{2})}{(n-k)!k!\Gamma(n+k+|\kappa|+\frac{1}{2})\Gamma(k+\beta+\gamma)}.$$

The coefficient matrices in the three-term relations are as follows:

$$A_{n,1} = \begin{bmatrix} a_{0,n} & & & \bigcirc & 0 \\ & a_{1,n} & & & 0 \\ & & \ddots & & \vdots \\ \bigcirc & & & a_{n,n} & 0 \end{bmatrix}, \quad B_{n,1} = \begin{bmatrix} b_{0,n} & & & \bigcirc \\ & b_{1,n} & & \\ & & \ddots & \\ \bigcirc & & & b_{n,n} \end{bmatrix},$$

where

$$a_{k,n} = \frac{h_{k,n+1}}{h_{k,n}} \frac{(n-k+1)(n+k+|\kappa|+\frac{1}{2})}{(2n+|\kappa|+\frac{1}{2})(2n+|\kappa|+\frac{3}{2})},$$

$$b_{k,n} = \frac{1}{2} - \frac{(\beta+\gamma+2k)^2 - (\alpha-\frac{1}{2})^2}{2(2n+|\kappa|-\frac{1}{2})(2n+|\kappa|+\frac{3}{2})};$$

$$A_{n,2} = \begin{bmatrix} e_{0,n} & d_{0,n} & & & \bigcirc & 0 \\ c_{1,n} & e_{1,n} & d_{1,n} & & & \vdots \\ & \ddots & \ddots & \ddots & & \vdots \\ & & c_{n-1,n} & & d_{n-1,n} & 0 \\ \bigcirc & & & c_{n,n} & e_{n,n} & d_{n,n} \end{bmatrix},$$

where

$$c_{k,n} = \frac{h_{k-1,n+1}(n-k+1)(n-k+2)(k+\beta-\frac{1}{2})(k+\gamma-\frac{1}{2})}{h_{k,n}(2n+|\kappa|+\frac{1}{2})(2n+|\kappa|+\frac{3}{2})(2k+\beta+\gamma)(2k+\beta+\gamma-1)},$$

$$e_{k,n} = -\left(1+\frac{(\beta-\frac{1}{2})^2-(\gamma-\frac{1}{2})^2}{(2k+\beta+\gamma+1)(2k+\beta+\gamma-1)}\right)\frac{a_{n,k}}{2},$$

$$d_{k,n} = \frac{h_{k+1,n+1}(n+k+|\kappa|+\frac{1}{2})(n+k+|\kappa|+\frac{3}{2})(k+1)(k+\beta+\gamma)}{h_{k,n}(2n+|\kappa|+\frac{1}{2})(2n+|\kappa|+\frac{3}{2})(2k+\beta+\gamma)(2k+\beta+\gamma+1)};$$

and finally

$$B_{n,2} = \begin{bmatrix} f_{0,n} & g_{0,n} & & & \bigcirc \\ g_{0,n} & f_{1,n} & g_{1,n} & & \\ & \ddots & \ddots & \ddots & \\ & & g_{n-2,n} & f_{n-1,n} & g_{n-1,n} \\ \bigcirc & & & g_{n-1,n} & f_{n,n} \end{bmatrix},$$

where

$$f_{k,n} = \left(1+\frac{(\beta-\frac{1}{2})^2-(\gamma-\frac{1}{2})^2}{(2k+\beta+\gamma+1)(2k+\beta+\gamma-1)}\right)\frac{1-b_{n,k}}{2},$$

$$g_{k,n} = -2\frac{h_{k+1,n}}{h_{k,n}} \times \frac{(n-k+\alpha-\frac{1}{2})(n+k+|\kappa|+\frac{1}{2})(k+1)(k+\beta+\gamma)}{(2n+|\kappa|-\frac{1}{2})(2n+|\kappa|+\frac{3}{2})(2k+\beta+\gamma)(2k+\beta+\gamma+1)}.$$

Classical polynomials on the disk

With respect to the weight function $W(x,y) = (1-x^2-y^2)^{\mu-1/2}$, we consider the orthonormal polynomials, see (2.3.1),

$$P_k^n(x,y) = (h_{k,n})^{-1} C_{n-k}^{k+\mu+1/2}(x)(1-x^2)^{k/2} C_k^{\mu}\left(\frac{y}{\sqrt{1-x^2}}\right),$$

where

$$(h_{k,n})^2 = \frac{2\pi(2\mu+1)}{2^{2k+4\mu}[\Gamma(\mu)]^2} \frac{\Gamma(n+k+2\mu+1)\Gamma(2\mu+k)}{(2n+2\mu+1)(k+\mu)[\Gamma(k+\mu+\frac{1}{2})]^2(n-k)!k!}.$$

If $\mu = 0$, these formulae hold under the limit relation $\lim_{\mu \to 0} \mu^{-1} C_n^{\mu}(x) = (2/n) T_n(x)$. In this case $B_{n,i} = 0$, since the weight function is centrally symmetric. The matrices $A_{n,1}$ and $A_{n,2}$ are of the same form as in the case of classical orthogonal polynomials on the triangle, with $e_{k,n} = 0$ and

$$a_{k,n} = \frac{h_{k,n+1}}{h_{k,n}} \frac{n-k+1}{2n+2\mu+1}, \qquad d_{k,n} = \frac{h_{k+1,n+1}}{h_{k,n}} \frac{(k+1)(2k+2\mu+1)}{2(k+\mu)(2n+2\mu+1)},$$
$$c_{k,n} = -\frac{h_{k-1,n+1}}{h_{k,n}} \frac{(k+2\mu-1)(n-k+1)(n-k+2)}{2(k+\mu)(2k+2\mu-1)(2n+2\mu+1)}.$$

3.4 Jacobi Matrices and Commuting Operators

For a family of orthogonal polynomials in one variable we can use the coefficients of the three-term relation to define a Jacobi matrix J, as in Subsection 1.3.2. This semi-infinite matrix defines an operator on the sequence space ℓ^2. It is well known that the spectral measure of this operator is closely related to the measure that defines the linear function $\mathscr{L}$ with respect to which the polynomials are orthogonal.

For orthogonal polynomials in several variables, we can use the coefficient matrices to define a family of tridiagonal matrices J_i, $1 \le i \le d$, as follows:

$$J_i = \begin{bmatrix} B_{0,i} & A_{0,i} & & \bigcirc \\ A_{0,i}^{\mathrm{T}} & B_{1,i} & A_{1,i} & \\ & A_{1,i}^{\mathrm{T}} & B_{2,i} & \ddots \\ \bigcirc & & \ddots & \ddots \end{bmatrix}, \qquad 1 \le i \le d. \tag{3.4.1}$$

We call the J_i *block Jacobi matrices*. It should be pointed out that their entries are coefficient matrices of the three-term relation, whose sizes increase to infinity going down the main diagonal. As part of the definition of the block Jacobi matrices, we require that the matrices $A_{n,i}$ and $B_{n,i}$ satisfy the rank conditions (3.3.7) and (3.3.8). This requirement, however, implies that the coefficient matrices have to satisfy several other conditions. Indeed, the three-term relation and the rank conditions imply that the $\mathbb{P}_n$ are orthonormal, for which we have

Theorem 3.4.1 *Let $\{\mathbb{P}_n\}$ be a sequence of orthonormal polynomials satisfying the three-term relation* (3.3.3). *Then the coefficient matrices satisfy the commutativity conditions*

$$\begin{aligned} A_{k,i}A_{k+1,j} &= A_{k,j}A_{k+1,i}, \\ A_{k,i}B_{k+1,j} + B_{k,i}A_{k,j} &= B_{k,j}A_{k,i} + A_{k,j}B_{k+1,i}, \\ A_{k-1,i}^{\mathrm{T}}A_{k-1,j} + B_{k,i}B_{k,j} + A_{k,i}A_{k,j}^{\mathrm{T}} &= A_{k-1,j}^{\mathrm{T}}A_{k-1,i} + B_{k,j}B_{k,i} + A_{k,j}A_{k,i}^{\mathrm{T}}, \end{aligned} \tag{3.4.2}$$

for $i \ne j$, $1 \le i, j \le d$, and $k \ge 0$, where $A_{-1,i} = 0$.

Proof By Theorem 3.3.8 there is a linear functional $\mathscr{L}$ that makes $\mathbb{P}_k$ orthonormal. Using the recurrence relation, there are two different ways of calculating the matrices $\mathscr{L}(x_i x_j \mathbb{P}_k \mathbb{P}_{k+2}^{\mathrm{T}})$, $\mathscr{L}(x_i x_j \mathbb{P}_k \mathbb{P}_k^{\mathrm{T}})$ and $\mathscr{L}(x_i x_j \mathbb{P}_k \mathbb{P}_{k+1}^{\mathrm{T}})$. These calculations lead to the desired matrix equations. For example, by the recurrence relation (3.3.3),

$$\begin{aligned}\mathscr{L}(x_i x_j \mathbb{P}_k \mathbb{P}_{k+2}^{\mathrm{T}}) &= \mathscr{L}(x_i \mathbb{P}_k x_j \mathbb{P}_{k+2}^{\mathrm{T}}) \\ &= \mathscr{L}\left[(A_{k,i}\mathbb{P}_{k+1} + \cdots)(\cdots + A_{k+1,j}^{\mathrm{T}}\mathbb{P}_{k+1})^{\mathrm{T}}\right] = A_{k,i}A_{k+1,j}\end{aligned}$$

and

$$\mathscr{L}(x_i x_j \mathbb{P}_k \mathbb{P}_{k+2}^{\mathrm{T}}) = \mathscr{L}(x_j \mathbb{P}_k x_i \mathbb{P}_{k+2}^{\mathrm{T}}) = A_{k,j}A_{k+1,i},$$

which leads to the first equation in (3.4.2). □

The reason that the relations in (3.4.2) are called commutativity conditions will become clear soon. We take these conditions as part of the definition of the block Jacobi matrices J_i.

The matrices J_i can be considered as a family of linear operators which act via matrix multiplication on ℓ^2. The domain of the J_i consists of all sequences in ℓ^2 for which matrix multiplication yields sequences in ℓ^2. As we shall see below, under proper conditions the matrices J_i form a family of commuting self-adjoint operators and the spectral theorem implies that the linear functional $\mathscr{L}$ in Theorem 3.3.7 has an integral representation. In order to proceed, we need some notation from the spectral theory of self-adjoint operators in a Hilbert space (see, for example, Riesz and Sz. Nagy [1955] or Rudin [1991]).

Let $\mathbb{H}$ be a separable Hilbert space and let $\langle\cdot,\cdot\rangle$ denote the inner product in $\mathbb{H}$. Each self-adjoint operator T in $\mathbb{H}$ is associated with a spectral measure E on $\mathbb{R}$ such that $T = \int x\,\mathrm{d}E(x)$; E is a projection-valued measure defined for the Borel sets of $\mathbb{R}$ such that $E(\mathbb{R})$ is the identity operator in $\mathbb{H}$ and $E(B \cap C) = E(B) \cap E(C)$ for Borel sets $B, C \subseteq \mathbb{R}$. For any $\phi \in \mathbb{H}$ the mapping $B \to \langle E(B)\phi, \phi\rangle$ is an ordinary measure defined for the Borel sets $B \subseteq \mathbb{R}$ and denoted $\langle E\phi, \phi\rangle$. The operators in the family $\{T_1, \ldots, T_d\}$ in $\mathbb{H}$ *commute*, by definition, if their spectral measures commute: that is, if $E_i(B)E_j(C) = E_i(C)E_j(B)$ for any $i, j = 1, \ldots, d$ and any two Borel sets $B, C \subseteq \mathbb{R}$. If $T_1, \ldots, T_d$ commute then $E = E_1 \otimes \cdots \otimes E_d$ is a spectral measure on $\mathbb{R}^d$ with values that are self-adjoint projections in $\mathbb{H}$. In particular, E is the unique measure such that

$$E(B_1 \times \cdots \times B_d) = E_1(B_1) \cdots E_d(B_d)$$

for any Borel sets $B_1, \ldots, B_d \subseteq \mathbb{R}$. The measure E is called the spectral measure of the commuting family $T_1, \ldots, T_d$. A vector $\Phi_0 \in \mathbb{H}$ is a *cyclic vector* in $\mathbb{H}$ with respect to the commuting family of self-adjoint operators $T_1, \ldots, T_d$ in $\mathbb{H}$ if the linear manifold $\{P(T_1, \ldots, T_d)\Phi_0\colon P \in \Pi^d\}$ is dense in $\mathbb{H}$. The spectral theorem for $T_1, \ldots, T_d$ states:

Theorem 3.4.2 *If $\{T_1,\dots,T_d\}$ is a commuting family of self-adjoint operators with cyclic vector Φ_0 then $T_1,\dots,T_d$ are unitarily equivalent to multiplication operators $X_1,\dots,X_d$,*

$$(X_i f)(x) = x_i f(x), \quad 1 \le i \le d,$$

defined on $L^2(\mu)$, where the measure μ is defined by $\mu(B) = \langle E(B)\Phi_0, \Phi_0\rangle$ for the Borel set $B \subset \mathbb{R}^d$.

The unitary equivalence means that there exists a unitary mapping $U : \mathbb{H} \to L^2(\mu)$ such that $UT_iU^{-1} = X_i$, $1 \le i \le d$. If the operators in Theorem 3.4.2 are bounded and their spectra are denoted by S_i then the measure μ has support $S \subset S_1 \times \cdots \times S_d$. Note that the measure μ satisfies

$$\langle \Phi_0, P(J_1,\dots,J_d)\Phi_0\rangle = \int_S P(x)\, d\mu(x).$$

We apply the spectral theorem to the block Jacobi operators $J_1,\dots,J_d$ on ℓ^2. Let $\{\phi_\alpha\}_{\alpha\in\mathbb{N}_0^d}$ denote the canonical orthonormal basis for $\mathbb{H} = \ell^2$. Introduce the notation $\Phi_k = (\phi_\alpha)_{|\alpha|=k}$ for the column vector whose elements are ϕ_α ordered according to lexicographical order. Then each $f \in \mathbb{H}$ can be written as $f = \sum \mathbf{a}_k^{\mathrm{T}}\Phi_k$. Let $\|\cdot\|_2$ be the matrix norm induced by the Euclidean norm for vectors. First we consider bounded operators.

Lemma 3.4.3 *The operator J_i defined via* (3.4.1) *is bounded if and only if* $\sup_{k\ge0}\|A_{k,i}\|_2 < \infty$ *and* $\sup_{k\ge0}\|B_{k,i}\|_2 < \infty$.

Proof For any $f \in \mathbb{H}$, $f = \sum \mathbf{a}_k^{\mathrm{T}}\Phi_k$, we have $\|f\|_{\mathbb{H}}^2 = \langle f, f\rangle = \sum \mathbf{a}_k^{\mathrm{T}}\mathbf{a}_k$. It follows from the definition of J_i that

$$\begin{aligned} J_i f &= \sum_{k=0}^{\infty} \mathbf{a}_k^{\mathrm{T}}(A_{k,i}\Phi_{k+1} + B_{k,i}\Phi_k + A_{k-1,i}^{\mathrm{T}}\Phi_{k-1}) \\ &= \sum_{k=0}^{\infty} (\mathbf{a}_{k-1}^{\mathrm{T}}A_{k-1,i} + \mathbf{a}_k^{\mathrm{T}}B_{k,i} + \mathbf{a}_{k+1}^{\mathrm{T}}A_{k,i}^{\mathrm{T}})\Phi_k, \end{aligned}$$

where we define $A_{-1,i} = 0$. Hence, if $\sup_{k\ge0}\|A_{k,i}\|_2$ and $\sup_{k\ge0}\|B_{k,i}\|_2$ are bounded then it follows from

$$\begin{aligned} \|J_i f\|_{\mathbb{H}}^2 &= \sum_{k=0}^{\infty} \|\mathbf{a}_{k-1}^{\mathrm{T}}A_{k-1,i} + \mathbf{a}_k^{\mathrm{T}}B_{k,i} + \mathbf{a}_{k+1}^{\mathrm{T}}A_{k,i}^{\mathrm{T}}\|_2^2 \\ &\le 3\Big(2\sup_{k\ge0}\|A_{k,i}\|_2^2 + \sup_{k\ge0}\|B_{k,i}\|_2^2\Big)\|f\|_{\mathbb{H}}^2 \end{aligned}$$

that J_i is a bounded operator. Conversely, suppose that $\|A_{k,i}\|_2$, say, approaches infinity for a subsequence of $\mathbb{N}_0$. Let $\mathbf{a}_k$ be vectors such that $\|\mathbf{a}_k\|_2 = 1$ and $\|\mathbf{a}_k^{\mathrm{T}}A_{k,i}\|_2 = \|A_{k,i}\|_2$. Then $\|\mathbf{a}_k\Phi_k\| = \|\mathbf{a}_k\|_2 = 1$. Therefore, it follows from

$$\|J_i\|_{\mathbb{H}}^2 \ge \|J_i\mathbf{a}_k^{\mathrm{T}}\Phi_k\|_{\mathbb{H}}^2 = \|\mathbf{a}_k^{\mathrm{T}}A_{k,i}\|_2^2 + \|\mathbf{a}_k^{\mathrm{T}}B_{k,i}\|_2^2 + \|\mathbf{a}_k^{\mathrm{T}}A_{k-1,i}^{\mathrm{T}}\|_2^2 \ge \|A_{k,i}\|_2^2$$

that J_i is unbounded. □

Lemma 3.4.4 *Suppose that J_i, $1 \le i \le d$, is bounded. Then $J_1,\ldots,J_d$ are self-adjoint operators and they pairwise commute.*

Proof Since J_i is bounded, it is self-adjoint if it is symmetric. That is, we need to show that $\langle J_i f, g\rangle = \langle f, J_i g\rangle$. Let $f = \sum \mathbf{a}_k^{\mathrm{T}}\Phi_k$ and $g = \sum \mathbf{b}_k^{\mathrm{T}}\Phi_k$. From the definition of J_i,

$$\begin{aligned}\langle J_i f, g\rangle &= \sum \mathbf{a}_{k-1}^{\mathrm{T}}(A_{k-1,i} + \mathbf{a}_k^{\mathrm{T}} B_{k,i} + \mathbf{a}_{k+1}^{\mathrm{T}} A_{k,i}^{\mathrm{T}})\mathbf{b}_k \\ &= \sum \mathbf{b}_k^{\mathrm{T}}(A_{k-1,i}^{\mathrm{T}}\mathbf{a}_{k-1} + B_{k,i}\mathbf{a}_k + A_{k,i}\mathbf{a}_{k+1}) = \langle f, J_i g\rangle.\end{aligned}$$

Also, since the J_i are bounded, to show that $J_1,\ldots,J_d$ commute we need only verify that

$$J_k J_j f = J_j J_k f \qquad \forall f \in \mathbb{H}.$$

A simple calculation shows that this is equivalent to the commutativity conditions (3.4.2). □

Lemma 3.4.5 *Suppose that J_i, $1 \le i \le d$, is bounded. Then $\Phi_0 = \phi_0 \in \mathbb{H}$ is a cyclic vector with respect to $J_1,\ldots,J_d$, and*

$$\Phi_n = \mathbb{P}_n(J_1 \quad \cdots \quad J_d)\Phi_0 \tag{3.4.3}$$

where $\mathbb{P}_n(x)$ is a polynomial vector as in (3.2.4).

Proof We need only prove (3.4.3), since it implies that Φ_0 is a cyclic vector. The proof uses induction. Clearly $\mathbb{P}_0 = 1$. From the definition of J_i,

$$J_i\Phi_0 = A_{0,i}\Phi_1 + B_{0,i}\Phi_0, \qquad 1 \le i \le d.$$

Multiplying by $D_{0,i}^{\mathrm{T}}$ from (3.3.9) and summing over $i = 1,\ldots,d$, we get

$$\Phi_1 = \sum_{i=1}^{d} D_{0,i}^{\mathrm{T}} J_i \Phi_0 - \sum_{i=1}^{d} D_{0,i}^{\mathrm{T}} B_{0,i}\Phi_0 = \left(\sum_{i=1}^{d} D_{0,i}^{\mathrm{T}} J_i - E_0\right)\Phi_0,$$

where $E_0 = \sum D_{0,i}B_{0,i}$. Therefore $\mathbb{P}_1(x) = \sum_{i=1}^{d} x_i D_{0,i}^{\mathrm{T}} - E_0$. Since $D_{0,i}^{\mathrm{T}}$ is of size $r_1^d \times r_0^d = d \times 1$, $\mathbb{P}_1$ is of the desired form. Likewise, for $k \ge 1$,

$$J_i\Phi_k = A_{k,i}\Phi_{k+1} + B_{k,i}\Phi_k + A_{k-1,i}^{\mathrm{T}}\Phi_{k-1}, \qquad 1 \le i \le d,$$

which implies, by induction, that

$$\begin{aligned}\Phi_{k+1} &= \sum_{i=1}^{d} D_{k,i}^{\mathrm{T}} J_i \Phi_k - E_k\Phi_k - F_k\Phi_{k-1} \\ &= \left(\sum_{i=1}^{d} D_{k,i}^{\mathrm{T}} J_i \mathbb{P}_k(J) - E_k\mathbb{P}_k(J) - F_k\mathbb{P}_{k-1}(J)\right)\Phi_0,\end{aligned}$$

where $J = (J_1 \quad \cdots \quad J_d)^{\mathrm{T}}$ and E_k and F_k are as in (3.3.10). Consequently

$$\mathbb{P}_{k+1}(x) = \sum_{i=1}^{d} x_i D_{k,i}^{\mathrm{T}} \mathbb{P}_k(x) - E_k \mathbb{P}_k(x) - F_k \mathbb{P}_{k-1}(x).$$

Clearly, every component of $\mathbb{P}_{k+1}$ is a polynomial in Π_{k+1}^d. □

The last two lemmas show that we can apply the spectral theorem to the bounded operators $J_1, \ldots, J_d$. Hence, we can state the following.

Lemma 3.4.6 *If J_i, $1 \le i \le d$, is bounded then there exists a measure $\mu \in \mathscr{M}$ with compact support such that $J_1, \ldots, J_d$ are unitarily equivalent to the multiplication operators $X_1, \ldots, X_d$ on $L^2(\mathrm{d}\mu)$. Moreover, the polynomials $\{\mathbb{P}_n\}_{n=0}^{\infty}$ in Lemma 3.4.5 are orthonormal with respect to μ.*

Proof Since $\mu(B) = \langle E(B)\Phi_0, \Phi_0 \rangle$ in Theorem 3.4.2,

$$\int \mathbb{P}_n(x)\mathbb{P}_m^{\mathrm{T}}(x)\,\mathrm{d}\mu(x) = \langle \mathbb{P}_n \Phi_0, \mathbb{P}_m^{\mathrm{T}} \Phi_0 \rangle = \langle \Phi_n, \Phi_m^{\mathrm{T}} \rangle.$$

This proves that the $\mathbb{P}_n$ are orthonormal. □

Unitary equivalence associates the cyclic vector Φ_0 with the function $P(x) = 1$ and the orthonormal basis $\{\Phi_n\}$ in $\mathbb{H}$ with the orthonormal basis $\{\mathbb{P}_n\}$ in $L^2(\mathrm{d}\mu)$.

Theorem 3.4.7 *Let $\{\mathbb{P}_n\}_{n=0}^{\infty}$, $\mathbb{P}_0 = 1$, be a sequence in Π^d. Then the following statements are equivalent.*

(i) *There exists a determinate measure $\mu \in \mathscr{M}$ with compact support in $\mathbb{R}^d$ such that $\{\mathbb{P}_n\}_{n=0}^{\infty}$ is orthonormal with respect to μ.*
(ii) *The statement (ii) in Theorem 3.3.8 holds, together with*

$$\sup_{k \ge 0} \|A_{k,i}\|_2 < \infty \quad \text{and} \quad \sup_{k \ge 0} \|B_{k,i}\|_2 < \infty, \qquad 1 \le i \le d. \tag{3.4.4}$$

Proof On the one hand, if μ has compact support then the multiplication operators $X_1, \ldots, X_d$ in $L^2(\mathrm{d}\mu)$ are bounded. Since these operators have the J_i as representation matrices with respect to the orthonormal basis $\mathbb{P}_n$, (3.4.4) follows from Lemma 3.4.3.

On the other hand, by Theorem 3.3.8 the three-term relation and the rank conditions imply that the $\mathbb{P}_n$ are orthonormal; hence, we can define Jacobi matrices J_i. The condition (3.4.4) shows that the operators J_i are bounded. The existence of a measure with compact support follows from Lemma 3.4.6. That the measure is determinate follows from Theorem 3.2.17. □

For block Jacobi matrices, the formal commutativity $J_i J_j = J_j J_i$ is equivalent to the conditions in (3.4.2), which is why we call them commutativity conditions. For bounded self-adjoint operators, the commutation of the spectral measures is equivalent to formal commutation. There are, however, examples of unbounded self-adjoint operators with a common dense domain such that they formally commute but their spectral measures do not commute (see Nelson [1959]). Theorem 3.4.7 shows that the measures with unbounded support set give rise to unbounded Jacobi matrices. There is a sufficient condition for the commutation of operators in Nelson [1959], which turns out to be applicable to a family of unbounded operators $J_1, \ldots, J_d$ that satisfy an additional condition. The result is as follows.

Theorem 3.4.8 *Let* $\{\mathbb{P}_n\}_{n=0}^{\infty}$, $\mathbb{P}_0 = 1$, *be a sequence in* Π^d *that satisfies the three-term relation* (3.3.3) *and the rank conditions* (3.3.7) *and* (3.3.8). *If*

$$\sum_{k=0}^{\infty} \frac{1}{\|A_{k,i}\|_2} = \infty, \qquad 1 \le i \le d, \tag{3.4.5}$$

then there exists a determinate measure $\mu \in \mathcal{M}$ *such that the* $\mathbb{P}_n$ *are orthonormal with respect to* μ.

The condition (3.4.5) is a well-known condition in one variable and implies the classical Carleman condition (3.2.16) for the determinacy of a moment problem; see p. 24 of Akhiezer [1965]. For the proof of this theorem and a discussion of the condition (3.4.5), see Xu [1993b].

3.5 Further Properties of the Three-Term Relation

In this section we discuss further properties of the three-term relation. To some extent, the properties discussed below indicate major differences between three-term relations in one variable and in several variables. They indicate that the three-term relation in several variables is not as strong as that in one variable; for example, there is no analogue of orthogonal polynomials of the second kind. These differences reflect the essential difficulties in higher dimensions.

3.5.1 Recurrence formula

For orthogonal polynomials in one variable, the three-term relation is equivalent to a recursion relation. For any given sequence of $\{a_n\}$ and $\{b_n\}$ we can use

$$p_{n+1} = \frac{1}{a_n}(x - b_n)p_n - \frac{a_{n-1}}{a_n}p_{n-1}, \qquad p_0 = 1, \quad p_{-1} = 0$$

to generate a sequence of polynomials, which by Favard's theorem is a sequence of orthogonal polynomials provided that $a_n > 0$.

In several variables, however, the recurrence formula (3.3.10) without additional conditions is not equivalent to the three-term relation. As a consequence, although Theorem 3.3.8 is an extension of the classical Favard theorem for one variable, it is not as strong as that theorem. In fact, one direction of the theorem says that if there is a sequence of polynomials $\mathbb{P}_n$ that satisfies the three-term relation and the rank condition then it is orthonormal. But it does not answer the question when and which $\mathbb{P}_n$ will satisfy such a relation. The conditions under which the recurrence relation (3.3.10) is equivalent to the three-term relation are discussed below.

Theorem 3.5.1 *Let* $\{\mathbb{P}_k\}_{k=0}^{\infty}$ *be defined by* (3.3.10). *Then there is a linear functional* $\mathscr{L}$ *that is positive definite and makes* $\{\mathbb{P}_k\}_{k=0}^{\infty}$ *an orthonormal basis for* Π^d *if and only if* $B_{k,i}$ *are symmetric and* $A_{k,i}$ *satisfy the rank condition* (3.3.7) *and together they satisfy the commutativity conditions* (3.4.2).

This theorem reveals one major difference between orthogonal polynomials of one variable and of several variables; namely, the three-term relation in the case of several variables is different from the recurrence relation. For the recurrence formula to generate a sequence of orthogonal polynomials it is necessary that its coefficients satisfy the commutativity conditions. We note that, even if the set $\{\mathbb{P}_n\}_{n=0}^{\infty}$ is only orthogonal rather than orthonormal, $S_n^{-1}\mathbb{P}_n$ will be orthonormal, where S_n is a nonsingular matrix that satisfies $S_n S_n^{\mathrm{T}} = \mathscr{L}(\mathbb{P}_n \mathbb{P}_n^{\mathrm{T}})$. Therefore, our theorem may be stated in terms of polynomials that are only orthogonal. However, it has to be appropriately modified since the three-term relation in this case is (3.3.1) and the commutativity conditions become more complicated.

For the proof of Theorem 3.5.1 we will need several lemmas. One reason that the recurrence relation (3.3.10) is not equivalent to the three-term relation (3.3.1) lies in the fact that D_n^{T} is only a left inverse of A_n; that is, $D_n^{\mathrm{T}} A_n = I$, but $A_n D_n^{\mathrm{T}}$ is not equal to the identity. We need to understand the kernel space of $I - A_n D_n^{\mathrm{T}}$, which is related to the kernel space of $\{A_{n-1,i} - A_{n-1,j} : 1 \le i, j \le n\}$. In order to describe the latter object, we introduce the following notation.

Definition 3.5.2 For any given sequence of matrices $C_1, \ldots, C_d$, where each C_i is of size $s \times t$, a joint matrix Ξ_C is defined as follows. Let $\Xi_{i,j}$, $1 \le i, j \le d$, $i \ne j$, be block matrices defined by

$$\Xi_{i,j} = (\cdots |0| C_j^{\mathrm{T}} | \cdots | -C_i^{\mathrm{T}} |0| \cdots), \qquad \Xi_{i,j} : t \times ds,$$

that is, the only two nonzero blocks are C_j^{T} at the ith block and $-C_i^{\mathrm{T}}$ at the jth block. The matrix Ξ_C is then defined by using the $\Xi_{i,j}$ as blocks in the lexicographical order of $\{(i,j) : 1 \le i < j \le d\}$,

$$\Xi_C = [\Xi_{1,2}^{\mathrm{T}} | \Xi_{1,3}^{\mathrm{T}} | \cdots | \Xi_{d-1,d}^{\mathrm{T}}].$$

The matrix Ξ_C is of size $ds \times \binom{d}{2} t$.

For example, for $d = 2$ and $d = 3$, respectively, Ξ_C is given by

$$\begin{bmatrix} C_2 \\ -C_1 \end{bmatrix} \quad \text{and} \quad \begin{bmatrix} C_2 & C_3 & 0 \\ -C_1 & 0 & C_3 \\ 0 & -C_1 & -C_2 \end{bmatrix}.$$

The following lemma follows readily from the definition of Ξ_C.

Lemma 3.5.3 *Let* $Y = (Y_1^{\mathrm{T}} \quad \cdots \quad Y_d^{\mathrm{T}})^{\mathrm{T}}$, $Y_i \in \mathbb{R}^s$. *Then*

$$\Xi_C^{\mathrm{T}} Y = 0 \qquad \text{if and only if} \qquad C_i^{\mathrm{T}} Y_j = C_j^{\mathrm{T}} Y_i, \quad 1 \le i < j \le d.$$

By definition, Ξ_{A_n} and $\Xi_{A_n^{\mathrm{T}}}$ are matrices of different sizes. Indeed,

$$\Xi_{A_n} : dr_n^d \times \binom{d}{2} r_{n+1}^d \qquad \text{and} \qquad \Xi_{A_n^{\mathrm{T}}} : dr_{n+1}^d \times \binom{d}{2} r_n^d.$$

We will need to find the ranks of these two matrices. Because of the relation (3.3.4), we first look at Ξ_{L_n} and $\Xi_{L_n^{\mathrm{T}}}$, where the matrices $L_{n,i}$ are defined by (3.3.5).

Lemma 3.5.4 *For* $n \in \mathbb{N}_0$ *and* $1 \le i < j \le d$, $L_{n,i}L_{n+1,j} = L_{n,j}L_{n+1,i}$.

Proof From (3.3.5) it follows that, for any $x \in \mathbb{R}^d$,

$$L_{n,i}L_{n+1,j}\mathbf{x}^{n+2} = x_i x_j \mathbf{x}^n = L_{n,j}L_{n+1,i}\mathbf{x}^{n+2}.$$

Taking partial derivatives, with respect to $\mathbf{x}$, of the above identity we conclude that $(L_{n,i}L_{n+1,j} - L_{n,j}L_{n+1,i})\mathbf{e}_k = 0$ for every element $\mathbf{e}_k$ of the standard basis of $\mathbb{R}^{r_{n+2}}$. Hence, the desired identity follows. □

Lemma 3.5.5 *For* $d \ge 2$ *and* $n \ge 1$, $\operatorname{rank} \Xi_{L_n^{\mathrm{T}}} = dr_{n+1}^d - r_{n+2}^d$.

Proof We shall prove that the dimension of the null space of $\Xi_{L_n^{\mathrm{T}}}^{\mathrm{T}}$ is r_{n+2}^d. Let $Y = (Y_1^{\mathrm{T}} \quad \cdots \quad Y_d^{\mathrm{T}})^{\mathrm{T}} \in \mathbb{R}^{dr_{n+1}}$, where $Y_i \in \mathbb{R}^{r_{n+1}}$. Consider the homogeneous equation in dr_{n+1}^d variables $\Xi_{L_n^{\mathrm{T}}}^{\mathrm{T}} Y = 0$, which by Lemma 3.5.3, is equivalent to

$$L_{n,i}Y_j = L_{n,j}Y_i, \qquad 1 \le i < j \le d. \tag{3.5.1}$$

There is exactly a single 1 in each row of $L_{n,i}$ and the rank of $L_{n,i}$ is r_n^d. Define $\mathcal{N}_n = \{\alpha \in \mathbb{N}_0^d : |\alpha| = n\}$ and, for each $1 \le i \le d$, $\mathcal{N}_{n,i} = \{\alpha \in \mathbb{N}_0^d : |\alpha| = n \text{ and } \alpha_i \ne 0\}$. Let $\mu(\mathcal{N})$ denote the number of elements in the set $\mathcal{N}$. Counting the number of integer solutions of $|\alpha| = n$ shows that $\mu(\mathcal{N}_n) = r_n^d$, $\mu(\mathcal{N}_{n,i}) = r_{n-1}^d$. We can consider the $L_{n,i}$ as transforms from $\mathcal{N}_n$ to $\mathcal{N}_{n,i}$. Fix a one-to-one correspondence between the elements of $\mathcal{N}_n$ and the elements of a vector in $\mathbb{R}^{r_n}$, and write $Y_i|_{\mathcal{N}_{n+1,j}} = L_{n,j}Y_i$. The linear systems of equations (3.5.1) can be written as

$$Y_i|_{\mathcal{N}_{n+1,j}} = Y_j|_{\mathcal{N}_{n+1,i}}, \qquad 1 \le i < j \le d. \tag{3.5.2}$$

This gives $\binom{d}{2} r_n^d$ equations in $d r_{n+1}^d$ variables of Y, not all of which are independent. For any distinct integers i, j and k,

$$Y_i|_{\mathcal{N}_{n+1,j}\cap\mathcal{N}_{n+1,k}} = Y_j|_{\mathcal{N}_{n+1,i}\cap\mathcal{N}_{n+1,k}}, \qquad Y_i|_{\mathcal{N}_{n+1,k}\cap\mathcal{N}_{n+1,j}} = Y_k|_{\mathcal{N}_{n+1,i}\cap\mathcal{N}_{n+1,j}},$$
$$Y_j|_{\mathcal{N}_{n+1,k}\cap\mathcal{N}_{n+1,i}} = Y_k|_{\mathcal{N}_{n+1,j}\cap\mathcal{N}_{n+1,i}}.$$

Thus, there are exactly $\mu(\mathcal{N}_{n+1,j}\cap\mathcal{N}_{n+1,k}) = r_{n-1}^d$ duplicated equations among these three systems of equations. Counting all combinations of the three systems generated by the equations in (3.5.2) gives $\binom{d}{3} r_{n-1}^d$ equations; however, among them some are counted more than once. Repeating the above argument for the four distinct systems generated by the equations in (3.5.1), we have that there are $\mu(\mathcal{N}_{n,i}\cap\mathcal{N}_{n,j}\cap\mathcal{N}_{n,k}) = r_{n-2}^d$ equations that are duplicated. There are $\binom{d}{4}$ combinations of the four different systems of equations. We then need to consider five systems of equations, and so on. In this way, it follows from the inclusion–exclusion formula that the number of duplicated equations in (3.5.2) is

$$\binom{d}{3} r_{n-1}^d - \binom{d}{4} r_{n-2}^d + \cdots + (-1)^{d+1} r_{n-d+2}^d = \sum_{k=3}^{d} (-1)^{k+1} \binom{d}{k} r_{n-k+2}^d.$$

Thus, among the $\binom{d}{2} r_n^d$ equations (3.5.1), the number that are independent is

$$\binom{d}{2} r_n^d - \sum_{k=3}^{d} (-1)^{k+1} \binom{d}{k} r_{n-k+2}^d = \sum_{k=2}^{d} (-1)^{k} \binom{d}{k} r_{n-k+2}^d.$$

Since the dimension of the null space is equal to the number of variables minus the number of independent equations, and there are $d r_{n+1}^d$ variables, we have

$$\dim\ker \Xi_{L_n^{\mathrm{T}}}^{\mathrm{T}} = d r_{n+1}^d - \sum_{k=2}^{d} (-1)^{k} \binom{d}{k} r_{n-k+2}^d = r_{n+2}^d,$$

where the last equality follows from a Chu–Vandermonde sum (see Proposition 1.2.3). The proof is complete. □

Lemma 3.5.6 *For $d \ge 2$ and $n \ge 1$,* $\operatorname{rank} \Xi_{L_n} = d r_n^d - r_{n-1}^d$.

Proof We shall prove that the dimension of the null space of $\Xi_{L_n}^{\mathrm{T}}$ is r_{n-1}^d. Suppose that $Y = (Y_1^{\mathrm{T}} \ \cdots \ Y_d^{\mathrm{T}})^{\mathrm{T}} \in \mathbb{R}^{d r_n}$, where $Y_i \in \mathbb{R}^{r_n}$. Consider the homogeneous equation $\Xi_{L_n}^{\mathrm{T}} Y = 0$. By Lemma 3.5.3 this equation is equivalent to the system of linear equations

$$L_{n,i}^{\mathrm{T}} Y_j = L_{n,j}^{\mathrm{T}} Y_i, \quad 1 \le i < j \le d. \tag{3.5.3}$$

We take $\mathcal{N}_n$ and $\mathcal{N}_{n,i}$ to be as in the proof of the previous lemma. Note that $L_{n,i}^{\mathrm{T}} Y_j$ is a vector in $\mathbb{R}^{r_{n+1}}$, whose coordinates corresponding to $\mathcal{N}_{n,i}$ are those of Y_j and whose other coordinates are zeros. Thus, equation (3.5.3) implies that the elements of Y_j are nonzero only when they correspond to $\mathcal{N}_{n,j}\cap\mathcal{N}_{n,i}$ being in $L_{n,i}^{\mathrm{T}} Y_j$, that is, $(L_{n,i}^{\mathrm{T}}Y_j)|_{\mathcal{N}_{n,i}\cap\mathcal{N}_{n,j}}$; the nonzero elements of Y_i are $(L_{n,j}^{\mathrm{T}}Y_i)|_{\mathcal{N}_{n,i}\cap\mathcal{N}_{n,j}}$. Moreover, these two vectors are equal. Since for any $X \in \mathbb{R}^{r_{n+1}}$ we have $L_{n-1,j}L_{n,i}X = X|_{\mathcal{N}_{n,i}\cap\mathcal{N}_{n,j}}$, which follows from $L_{n-1,i}L_{n,j}\mathbf{x}^{n+1} = x_i x_j \mathbf{x}^{n-1}$, from the fact that $L_{n,i}L_{n,i}^{\mathrm{T}} = I$ we have

$$(L_{n,i}^{\mathrm{T}}Y_j)|_{\mathcal{N}_{n,i}\cap\mathcal{N}_{n,j}} = L_{n-1,j}L_{n,i}(L_{n,i}^{\mathrm{T}}Y_j) = L_{n-1,j}Y_j = Y_j|_{\mathcal{N}_{n-1,j}}.$$

Therefore, the nonzero elements of Y_i and Y_j satisfy $Y_i|_{\mathcal{N}_{n-1,i}} = Y_j|_{\mathcal{N}_{n-1,j}}$, $1 \le i < j \le d$. Thus there are exactly $\mu(\mathcal{N}_{n-1,i}) = r_{n-1}^d$ independent variables in the solution of (3.5.3). That is, the dimension of the null space of $\Xi_{L_n}^{\mathrm{T}}$ is r_{n-1}^d. □

From the last two lemmas we can derive the ranks of Ξ_{A_n} and $\Xi_{A_n^{\mathrm{T}}}$.

Proposition 3.5.7 *Let $A_{n,i}$ be the matrices in the three-term relation. Then*

$$\operatorname{rank} \Xi_{A_n} = dr_n^d - r_{n-1}^d, \qquad \textit{and} \qquad \operatorname{rank} \Xi_{A_n^{\mathrm{T}}} = dr_{n+1}^d - r_{n+2}^d.$$

Proof By (3.3.4) and the definition of Ξ_C it readily follows that

$$\Xi_{A_n} \operatorname{diag}\{G_{n+1}, \dots, G_{n+1}\} = \operatorname{diag}\{G_n, \dots, G_n\}\, \Xi_{L_n},$$

where the block diagonal matrix on the left-hand side is of size $\binom{d}{2} r_{n+1}^d \times \binom{d}{2} r_{n+1}^d$ and that on the right-hand side is of size $dr_n^d \times dr_n^d$. Since G_n and G_{n+1} are both invertible, so are these two block matrices. Therefore $\operatorname{rank} \Xi_{A_n} = \operatorname{rank} \Xi_{L_n}$. Thus, the first equality follows from the corresponding equality in Lemma 3.5.6. The second is proved similarly. □

Our next lemma deals with the null space of $I - A_n D_n^{\mathrm{T}}$.

Lemma 3.5.8 *Let $\{A_{n,i}\}$ be the coefficient matrices in the three-term relation* (3.3.1), *and let D_n^{T} be the left inverse of A_n, defined in* (3.3.9). *Then $Y = (Y_1^{\mathrm{T}} \ \cdots \ Y_d^{\mathrm{T}})^{\mathrm{T}}$, $Y_i \in \mathbb{R}^{r_n}$, is in the null space of $I - A_n D_n^{\mathrm{T}}$ if and only if $A_{n-1,i}Y_j = A_{n-1,j}Y_i$, $1 \le i, j \le d$, or, equivalently, $\Xi_{A_{n-1}^{\mathrm{T}}}^{\mathrm{T}} Y = 0$.*

Proof Let the singular-value decompositions of A_n and D_n^{T} be of the form given in Subsection 3.3.1. Both A_n and D_n are of size $dr_n^d \times r_{n+1}^d$. It follows that

$$I - A_n D_n^{\mathrm{T}} = W^{\mathrm{T}} \begin{pmatrix} 0 & -\Sigma\Sigma_1 \\ 0 & I \end{pmatrix} W,$$

where W is a unitary matrix, Σ and Σ_1 are invertible diagonal matrices and I is the identity matrix of size $dr_n^d - r_{n+1}^d$, from which it follows that $\operatorname{rank}(I - A_n D_n^{\mathrm{T}}) = dr_n^d - r_{n+1}^d$. Hence

$$\dim \ker(I - A_n D_n^{\mathrm{T}}) = dr_n^d - \operatorname{rank}(I - A_n D_n^{\mathrm{T}}) = r_{n+1}^d.$$

Since $D_n^{\mathrm{T}} A_n = I$, it follows that $(I - A_n D_n^{\mathrm{T}}) A_n = A_n - A_n = 0$, which implies that the columns of A_n form a basis for the null space of $I - A_n D_n^{\mathrm{T}}$. Therefore, if $(I - A_n D_n^{\mathrm{T}}) Y = 0$ then there exists a vector $\mathbf{c}$ such that $Y = A_n \mathbf{c}$. But the first commutativity condition in (3.4.2) is equivalent to $\Xi^{\mathrm{T}}_{A^{\mathrm{T}}_{n-1}} A_n = 0$; it follows that $\Xi^{\mathrm{T}}_{A^{\mathrm{T}}_{n-1}} Y = 0$. This proves one direction. For the other direction, suppose that $\Xi^{\mathrm{T}}_{A^{\mathrm{T}}_{n-1}} Y = 0$. By Proposition 3.5.7,

$$\dim \ker \Xi^{\mathrm{T}}_{A^{\mathrm{T}}_{n-1}} = d r_n^d - \operatorname{rank} \Xi^{\mathrm{T}}_{A^{\mathrm{T}}_{n-1}} = r_{n+1}^d.$$

From $\Xi^{\mathrm{T}}_{A^{\mathrm{T}}_{n-1}} Y = 0$ it also follows that the columns of A_n form a basis for $\ker \Xi^{\mathrm{T}}_{A^{\mathrm{T}}_{n-1}}$. Therefore, there exists a vector $\mathbf{c}^*$ such that $Y = A_n \mathbf{c}^*$, which, by $D_n^{\mathrm{T}} A_n = I$, implies that $(I - A_n D_n^{\mathrm{T}}) Y = 0$. □

Finally, we are in a position to prove Theorem 3.5.1.

Proof of Theorem 3.5.1 Evidently, we need only prove the "if" part of the theorem. Suppose that $\{\mathbb{P}_n\}_{n=0}^{\infty}$ is defined recursively by (3.3.10) and that the matrices $A_{n,i}$ and $B_{n,i}$ satisfy the rank and commutativity conditions. We will use induction. For $n = 0$, $\mathbb{P}_0 = 1$ and $\mathbb{P}_{-1} = 0$; thus

$$\mathbb{P}_1 = \sum_{i=1}^{d} D_{0,i}^{\mathrm{T}} x_i - \sum_{i=1}^{d} D_{0,i}^{\mathrm{T}} B_{0,i}.$$

Since A_0 and D_0 are both $d \times d$ matrices and $D_0^{\mathrm{T}} A_0 = I$, it follows that $A_0 D_0^{\mathrm{T}} = I$, which implies that $A_{0,i} D_{0,j}^{\mathrm{T}} = \delta_{i,j}$. Therefore

$$A_{0,j} \mathbb{P}_1 = x_j - B_{0,j} = x_j \mathbb{P}_0 - B_{0,j} \mathbb{P}_0.$$

Assuming we have proved that

$$A_{k,j} \mathbb{P}_{k+1} = x_j \mathbb{P}_k - B_{k,j} \mathbb{P}_k - A_{k-1,j}^{\mathrm{T}} \mathbb{P}_{k-1}, \quad 0 \le k \le n-1, \tag{3.5.4}$$

we now show that this equation also holds for $k = n$. First we prove that

$$A_{n,j} \mathbb{P}_{n+1} = x_j \mathbb{P}_n - B_{n,j} \mathbb{P}_n - A_{n-1,j}^{\mathrm{T}} \mathbb{P}_{n-1} + \mathbb{Q}_{n-2,j}, \tag{3.5.5}$$

where the components of $\mathbb{Q}_{n-2,j}$ are elements of the polynomial space Π_{n-2}^d. Since each component of $\mathbb{P}_n$ is a polynomial of degree at most n, we can write

$$\mathbb{P}_n = G_{n,n} \mathbf{x}^n + G_{n,n-1} \mathbf{x}^{n-1} + G_{n,n-2} \mathbf{x}^n + \cdots,$$

for some matrices $G_{n,k}$ and also write $G_n = G_{n,n}$. Upon expanding both sides of (3.5.5) in powers of $\mathbf{x}^k$, it suffices to show that the highest three coefficients in the expansions are equal. Since $\mathbb{P}_{n+1}$ is defined by the recurrence relation (3.3.10), comparing the coefficients shows that we need to establish that

$A_{n,j}\sum_{i=1}^{d} D_{n,i}^{\mathrm{T}} X_{n,i} = X_{n,j}$, $1 \le j \le d$, where $X_{n,j}$ is one of $Q_{n,j}$, $R_{n,j}$ and $S_{n,j}$, which are as follows:

$$\begin{aligned} Q_{n,j} &= G_n L_{n,j}, \\ R_{n,j} &= G_{n,n-1} L_{n-1,j} - B_{n,j} G_n, \\ S_{n,j} &= G_{n,n-2} L_{n-2,j} - B_{n,j} G_{n,n-1} - A_{n-1,j}^{\mathrm{T}} G_{n-1}; \end{aligned} \tag{3.5.6}$$

see (3.3.5) for the $L_{n,j}$. Written in a more compact form, the equations that we need to show are

$$A_n D_n^{\mathrm{T}} Q_n = Q_n, \qquad A_n D_n^{\mathrm{T}} R_n = R_n, \qquad A_n D_n^{\mathrm{T}} S_n = S_n.$$

We proceed by proving that the columns of Q_n, R_n and S_n belong to the null space of $I - A_n D_n^{\mathrm{T}}$. But, by Lemmas 3.5.8 and 3.5.3, this reduces to showing that

$$A_{n-1,i} Q_{n,j} = A_{n-1,j} Q_{n,i}, \qquad A_{n-1,i} R_{n,j} = A_{n-1,j} R_{n,i},$$
$$A_{n-1,i} S_{n,j} = A_{n-1,j} S_{n,i}.$$

These equations can be seen to follow from Lemma 3.5.4, equations (3.4.2) and the following:

$$\begin{aligned} G_{n-1} L_{n-1,i} &= A_{n-1,i} G_n, \\ G_{n-1,n-2} L_{n-2,i} &= A_{n-1,i} G_{n,n-1} - B_{n-1,i} G_{n-1}, \\ G_{n-1,n-3} L_{n-3,i} &= A_{n-1,i} G_{n,n-2} - B_{n-1,i} G_{n-1,n-2} - A_{n-2,i}^{\mathrm{T}} G_{n-2}, \end{aligned}$$

which are obtained by comparing the coefficients of (3.5.4) for $k = n-1$. For example,

$$\begin{aligned} A_{n-1,i} R_{n,j} &= A_{n-1,i}(G_{n,n-1} L_{n-1,j} - B_{n,j} G_n) \\ &= G_{n-1,n-2} L_{n-2,i} L_{n-1,j} - B_{n-1,i} G_{n-1} L_{n-1,j} - A_{n-1,i} B_{n,j} G_n \\ &= G_{n-1,n-2} L_{n-2,i} L_{n-1,j} - (B_{n-1,i} A_{n-1,j} + A_{n-1,i} B_{n,j}) G_n \\ &= A_{n-1,j} R_{n,i}. \end{aligned}$$

The other two equations follow similarly. Therefore, we have proved (3.5.5). In particular, from (3.5.4) and (3.5.5) we obtain

$$A_{k,i} G_{k+1} = G_k L_{k,i}, \qquad 0 \le k \le n, \quad 1 \le i \le d, \tag{3.5.7}$$

which, since $A_{n,i}$ satisfies the rank condition, implies that G_{n+1} is invertible, as in the proof of Theorem 3.3.7.

To complete the proof, we now prove that $\mathbb{Q}_{n-2,i} = 0$. On the one hand, multiplying equation (3.5.5) by $D_{n,i}^{\mathrm{T}}$ and summing for $1 \le i \le d$, it follows from the recurrence formulae (3.3.10) and $D_n^{\mathrm{T}} A_n = I$ that

$$\sum_{j=1}^{d} D_{n,j}^{\mathrm{T}} \mathbb{Q}_{n-2,j} = \sum_{j=1}^{d} D_{n,j}^{\mathrm{T}} A_{n,j} \mathbb{P}_{n+1} - \mathbb{P}_{n+1} = 0. \tag{3.5.8}$$

On the other hand, it follows from (3.5.4) and (3.5.5) that

$$\begin{aligned} A_{n-1,i}\mathbb{Q}_{n-2,j} = & -x_i x_j \mathbb{P}_{n-1} + A_{n-1,i}A_{n,j}\mathbb{P}_{n+1} \\ & + (A_{n-1,i}B_{n,j} + B_{n-1,i}A_{n-1,j})\mathbb{P}_n \\ & + (A_{n-1,i}A_{n-1,j}^{\mathrm{T}} + B_{n-1,i}B_{n-1,j} + A_{n-2,i}^{\mathrm{T}}A_{n-2,j})\mathbb{P}_{n-1} \\ & + (A_{n-2,i}^{\mathrm{T}}B_{n-2,j} + B_{n-1,i}A_{n-2,j}^{\mathrm{T}})\mathbb{P}_{n-2} + A_{n-2,i}^{\mathrm{T}}A_{n-3,j}^{\mathrm{T}}\mathbb{P}_{n-3}, \end{aligned}$$

which implies by the commutativity conditions (3.4.2) that

$$A_{n-1,i}\mathbb{Q}_{n-2,j} = A_{n-1,j}\mathbb{Q}_{n-2,i}.$$

We claim that this equation implies that there is a vector $\mathbb{Q}_n$ such that $\mathbb{Q}_{n-2,i} = A_{n,i}\mathbb{Q}_n$, which, once established, completes the proof upon using (3.5.8) and the fact that $D_n^{\mathrm{T}}A_n = I$. Indeed, the above equation by (3.5.7), is equivalent to $L_{n-1,i}X_j = L_{n-1,j}X_i$. Define $\mathbb{Q}_n$ by $\mathbb{Q}_n = G_{n+1}^{-1}X$, where X is defined as a vector that satisfies $L_{n,i}X = X_i$, in which $X_i = G_{n,i}^{-1}\mathbb{Q}_{n-2,i}$. In other words, recalling the notations used in the proof of Lemma 3.5.5, X is defined as a vector that satisfies $X|_{\mathcal{N}_{n,i}} = X_i$. Since $\mathcal{N} = \bigcup \mathcal{N}_{n,i}$ and

$$X|_{\mathcal{N}_{n,i}\cap\mathcal{N}_{n,j}} = L_{n-1,i}L_{n,j}X = L_{n-1,i}X_j = L_{n-1,j}X_i = X|_{\mathcal{N}_{n,i}\cap\mathcal{N}_{n,j}},$$

we see that X, and hence $\mathbb{Q}_n$, is well defined. □

3.5.2 General solutions of the three-term relation

Next we turn our attention to another difference between orthogonal polynomials of one and several variables. From the three-term relation and Favard's theorem, orthogonal polynomials in one variable can be considered as solutions of the difference equation

$$a_k y_{k+1} + b_k y_k + a_{k-1} y_{k-1} = x y_k, \qquad k \geq 1,$$

where $a_k > 0$ and the initial conditions are $y_0 = 1$ and $y_1 = a_0^{-1}(x - b_0)$. It is well known that this difference equation has another solution that satisfies the initial conditions $y_0 = 0$ and $y_1 = a_0^{-1}$. The components of the solution, denoted by q_k, are customarily called associated polynomials, or polynomials of the second kind. Together, these two sets of solutions share many interesting properties, and $\{q_n\}$ plays an important role in areas such as the problem of moments, the spectral theory of the Jacobi matrix and continuous fractions.

By the three-term relation (3.3.3), we can consider orthogonal polynomials in several variables to be solutions of the multi-parameter finite difference equations

$$x_i Y_k = A_{k,i}Y_{k+1} + B_{k,i}Y_k + A_{k-1,i}^{\mathrm{T}}Y_{k-1}, \qquad 1 \leq i \leq d, k \geq 1, \tag{3.5.9}$$

where the $B_{n,i}$ are symmetric, the $A_{n,i}$ satisfy the rank conditions (3.3.8) and (3.3.7) and the initial values are given by

$$Y_0 = a, \quad Y_1 = \mathbf{b}, \qquad a \in \mathbb{R}, \quad \mathbf{b} \in \mathbb{R}^d. \tag{3.5.10}$$

One might expect that there would be other linearly independent solutions of (3.5.9) which could be defined as associated polynomials in several variables. However, it turns out rather surprisingly that (3.5.9) has no solutions other than the system of orthogonal polynomials. This result is formulated as follows.

Theorem 3.5.9 *If the multi-parameter difference equation* (3.5.9) *has a solution* $\mathbb{P} = \{\mathbb{P}_k\}_{k=0}^{\infty}$ *for the particular initial values*

$$Y_0^* = 1, \qquad Y_1^* = A_0^{-1}(x - B_0) \tag{3.5.11}$$

then all other solutions of (3.5.9) *and* (3.5.10) *are multiples of* $\mathbb{P}$ *with the possible exception of the first component. More precisely, if* $Y = \{Y_k\}_{k=0}^{\infty}$ *is a solution of* (3.5.9) *and* (3.5.10) *then* $Y_k = h\mathbb{P}_k$ *for all* $k \geq 1$, *where* h *is a function independent of* k.

Proof The assumption and the extension of Favard's theorem in Theorem 3.3.8 imply that $\mathbb{P} = \{\mathbb{P}_k\}_{k=0}^{\infty}$ forms a sequence of orthonormal polynomials. Therefore, the coefficient matrices of (3.5.9) satisfy the commutativity conditions (3.4.2).

Suppose that a sequence of vectors $\{Y_k\}$ satisfies (3.5.9) and the initial values (3.5.10). From equation (3.5.9),

$$A_{k,i}Y_{k+1} = x_iY_k - B_{k,i}Y_k - A_{k-1,i}^{\mathrm{T}}Y_{k-1}, \qquad 1 \leq i \leq d.$$

Multiplying the ith equation by $A_{k-1,j}$ and the jth equation by $A_{k-1,i}$, it follows from the first equation of (3.4.2) that

$$A_{k-1,i}(x_jY_k - B_{k,j}Y_k - A_{k-1,j}^{\mathrm{T}}Y_{k-1}) = A_{k-1,j}(x_iY_k - B_{k,i}Y_k - A_{k-1,i}^{\mathrm{T}}Y_{k-1}), \tag{3.5.12}$$

for $1 \leq i, j \leq d$ and $k \geq 1$. In particular, the case $k = 1$ gives a relation between Y_0 and Y_1, which, upon using the second equation of (3.4.2) for $k = 1$ and the fact that $A_{0,i}A_{0,j}^{\mathrm{T}}$ and $B_{0,i}$ are numbers, we can rewrite as

$$(x_i - B_{0,i})A_{0,j}Y_1 = (x_j - B_{0,j})A_{0,i}Y_1.$$

However, x_i and x_j are independent variables; Y_1 must be a function of x, and moreover, has to satisfy $A_{0,i}Y_1 = (x_i - b_i)h(x)$, where h is a function of x. Substituting the latter equation into the above displayed equation, we obtain $b_i = B_{0,i}$. The case $h(x) = 1$ corresponds to the orthogonal polynomial solution $\mathbb{P}$. Since $D_0^{\mathrm{T}} = A_0^{-1}$, from (3.5.11) we have

$$Y_1 = \sum_{i=1}^{d} D_{0,i}^{\mathrm{T}}(x_i - B_{0,i})h(x) = A_0^{-1}(x - B_0)h(x) = h(x)\mathbb{P}_1.$$

Therefore, upon using the left inverse D_k^{T} of A_k from (3.3.9) we see that every solution of (3.5.9) satisfies

$$Y_{k+1} = \sum_{i=1}^{d} D_{k,i}^{\mathrm{T}}x_iY_k - E_kY_k - F_kY_{k-1}, \tag{3.5.13}$$

where E_k and F_k are defined as in (3.3.10); consequently,

$$Y_2 = \sum_{i=1}^{d} D_{1,i}^{\mathrm{T}}(x_i I - B_{1,i})h(x)\mathbb{P}_1 - F_1 Y_0 = h(x)\mathbb{P}_2 + F_1(h(x) - Y_0).$$

Since $\mathbb{P}$ is a solution of equation (3.5.9) with initial conditions (3.5.11), it follows from (3.5.12) that

$$A_{1,i}(x_j\mathbb{P}_2 - B_{2,j}\mathbb{P}_2 - A_{1,j}^{\mathrm{T}}\mathbb{P}_1) = A_{1,j}(x_i\mathbb{P}_2 - B_{2,i}\mathbb{P}_2 - A_{1,i}^{\mathrm{T}}\mathbb{P}_1).$$

Multiplying this equation by h and subtracting it from (3.5.12) with $k = 2$, we conclude from the formulae for Y_1 and Y_2 that

$$A_{1,i}(x_j I - B_{2,j})F_1(h(x) - Y_0) = A_{1,j}(x_i I - B_{2,i})F_1(h(x) - Y_0).$$

If $Y_0 = h(x)$ then it follows from (3.5.13) and $Y_1 = h(x)\mathbb{P}_1$ that $Y_k = h(x)\mathbb{P}_k$ for all $k \geq 0$, which is the conclusion of the theorem. Now assume that $h(x) \neq Y_0$. Thus, $h(x) - Y_0$ is a nonzero number so, from the previous formula,

$$A_{1,i}(x_j I - B_{2,j})F_1 = A_{1,j}(x_i I - B_{2,i})F_1.$$

However, since x_i and x_j are independent variables, we conclude that for this equality to hold it is necessary that $A_{1,i}F_1 = 0$, which implies that $F_1 = 0$ because $D_n^{\mathrm{T}} A_n = 1$. We then obtain from the formula for Y_2 that $Y_2 = h(x)\mathbb{P}_2$. Thus, by (3.5.13), $Y_k = h(x)\mathbb{P}_k$ for all $k \geq 1$, which concludes the proof. □

3.6 Reproducing Kernels and Fourier Orthogonal Series

Let $\mathscr{L}(f) = \int f\,\mathrm{d}\mu$ be a positive definite linear functional, where $\mathrm{d}\mu$ is a nonnegative Borel measure with finite moments. Let $\{P_\alpha^n\}$ be a sequence of orthonormal polynomials with respect to $\mathscr{L}(f)$. For any function f in $L^2(\mathrm{d}\mu)$, we can consider its Fourier orthogonal series with respect to $\{P_\alpha^n\}$:

$$f \sim \sum_{n=0}^{\infty} \sum_{|\alpha|=n} a_\alpha^n(f) P_\alpha^n \qquad \text{with} \qquad a_\alpha^n(f) = \int f(x) P_\alpha^n(x)\,\mathrm{d}\mu.$$

Although the bases of orthonormal polynomials are not unique, the Fourier series in fact is independent of the particular basis. Indeed, using the vector notation $\mathbb{P}_n$, the above definition of the Fourier series may be written as

$$f \sim \sum_{n=0}^{\infty} \mathbf{a}_n^{\mathrm{T}}(f)\mathbb{P}_n \qquad \text{with} \qquad \mathbf{a}_n(f) = \int f(x)\mathbb{P}_n(x)\,\mathrm{d}\mu. \tag{3.6.1}$$

Furthermore, recalling that $\mathscr{V}_n^d$ denotes the space of orthogonal polynomials of degree exactly n, as in (3.1.2), it follows from the orthogonality that the expansion can be considered as

$$f \sim \sum_{n=0}^{\infty} \mathrm{proj}_{\mathscr{V}_n^d} f, \qquad \text{where} \qquad \mathrm{proj}_{\mathscr{V}_n^d} f = \mathbf{a}_n^{\mathrm{T}}(f)\mathbb{P}_n.$$

We shall denote the projection operator $\mathrm{proj}_{\mathcal{V}_n^d} f$ by $\mathbf{P}_n(f)$. In terms of an orthonormal basis, it can be written as

$$\mathbf{P}_n(f;x) = \int f(y)\mathbf{P}_n(x,y)\,\mathrm{d}\mu(y), \qquad \mathbf{P}_n(x,y) = \mathbb{P}_n^{\mathrm{T}}(x)\mathbb{P}_n(y). \tag{3.6.2}$$

The projection operator is independent of the particular basis. In fact, since two different orthonormal bases differ by a factor that is an orthogonal matrix (Theorem 3.2.14), it is evident from (3.6.2) that $\mathbf{P}_n(f)$ depends on $\mathcal{V}_n^d$ rather than a particular basis of $\mathcal{V}_n^d$. We define the nth partial sum of the Fourier orthogonal expansion (3.6.1) by

$$S_n(f) = \sum_{k=0}^{n} \mathbf{a}_k^{\mathrm{T}}(f)\mathbb{P}_k = \int \mathbf{K}_n(\cdot,y)f(y)\,\mathrm{d}\mu, \tag{3.6.3}$$

where the kernel $\mathbf{K}_n(\cdot,\cdot)$ is defined by

$$\mathbf{K}_n(x,y) = \sum_{k=0}^{n}\sum_{|\alpha|=k} P_\alpha^k(x)P_\alpha^k(y) = \sum_{k=0}^{n} \mathbf{P}_k(x,y) \tag{3.6.4}$$

and is often called the nth reproducing kernel, for reasons to be given below.

3.6.1 Reproducing kernels

For the definition of $\mathbf{K}_n$ it is not necessary to use an orthonormal basis; any orthogonal basis will do. Moreover, the definition makes sense for any moment functional.

Let $\mathcal{L}$ be a moment functional, and let $\mathbb{P}_n$ be a sequence of orthogonal polynomials with respect to $\mathcal{L}$. In terms of $\mathbb{P}_n$, the kernel function $\mathbf{K}_n$ takes the form

$$\mathbf{K}_n(x,y) = \sum_{k=0}^{n} \mathbb{P}_k^{\mathrm{T}}(x)H_k^{-1}\mathbb{P}_k(y), \qquad H_k = \mathcal{L}(\mathbb{P}_k\mathbb{P}_k^{\mathrm{T}}).$$

Theorem 3.6.1 *Let $\mathcal{V}_k^d$, $k \ge 0$, be defined by means of a moment functional $\mathcal{L}$. Then $\mathbf{K}_n(\cdot,\cdot)$ depends only on $\mathcal{V}_k^d$ rather than on a particular basis of $\mathcal{V}_k^d$.*

Proof Let $\mathbb{P}_k$ be a basis of $\mathcal{V}_k^d$. If $\mathbb{Q}_k$ is another basis then there exists an invertible matrix M_k, independent of x, such that $\mathbb{P}_k(x) = M_k\mathbb{Q}_k(x)$. Let $H_k(\mathbb{P}) = H_k = \mathcal{L}(\mathbb{P}_k\mathbb{P}_k^{\mathrm{T}})$ and $H_k(\mathbb{Q}) = \mathcal{L}(\mathbb{Q}_k\mathbb{Q}_k^{\mathrm{T}})$. Then $H_k^{-1}(\mathbb{P}) = (M_k^{\mathrm{T}})^{-1}H_k^{-1}(\mathbb{Q})M_k^{-1}$. Therefore $\mathbb{P}_k^{\mathrm{T}}H_k^{-1}(\mathbb{P})\mathbb{P}_k = \mathbb{Q}_k^{\mathrm{T}}H_k^{-1}(\mathbb{Q})\mathbb{Q}_k$, which proves the stated result. □

Since the definition of $\mathbf{K}_n$ does not depend on the particular basis in which it is expressed, it is often more convenient to work with an orthonormal basis when $\mathcal{L}$ is positive definite. The following theorem justifies the name *reproducing kernel*.

Theorem 3.6.2 *Let $\mathscr{L}$ be a positive definite linear functional on all P in the polynomial space Π^d. Then, for all $P \in \Pi_n^d$,*

$$P(x) = \mathscr{L}\big(\mathbf{K}_n(x,\cdot)P(\cdot)\big).$$

Proof Let $\mathbb{P}_n$ be a sequence of orthogonal polynomials with respect to $\mathscr{L}$ and let $\mathscr{L}(\mathbb{P}_n\mathbb{P}_n^{\mathrm{T}}) = H_n$. For $P \in \Pi_n^d$ we can expand it in terms of the basis $\{\mathbb{P}_0, \mathbb{P}_1, \ldots, \mathbb{P}_n\}$ by using the orthogonality property:

$$P(x) = \sum_{k=0}^{n} [\mathbf{a}_k(P)]^{\mathrm{T}} H_k^{-1}\mathbb{P}_k(x) \qquad \text{with} \qquad \mathbf{a}_k(P) = \mathscr{L}(P\mathbb{P}_k).$$

But this equation is equivalent to

$$P(x) = \sum_{k=0}^{n} \mathscr{L}(P\mathbb{P}_k)^{\mathrm{T}} H_k^{-1}\mathbb{P}_k(x) = \mathscr{L}\big[\mathbf{K}_n(x,\cdot)P(\cdot)\big],$$

which is the desired result. □

For orthogonal polynomials in one variable, the reproducing kernel enjoys a compact expression called the Christoffel–Darboux formula. The following theorem is an extension of this formula for several variables.

Theorem 3.6.3 *Let $\mathscr{L}$ be a positive definite linear functional, and let $\{\mathbb{P}_k\}_{k=0}^{\infty}$ be a sequence of orthogonal polynomials with respect to $\mathscr{L}$. Then, for any integer $n \ge 0$, $x, y \in \mathbb{R}^d$,*

$$\sum_{k=0}^{n} \mathbb{P}_k^{\mathrm{T}}(x) H_k^{-1}\mathbb{P}_k(y)$$
$$= \frac{\big[A_{n,i}\mathbb{P}_{n+1}(x)\big]^{\mathrm{T}} H_n^{-1}\mathbb{P}_n(y) - \mathbb{P}_n^{\mathrm{T}}(x) H_n^{-1}\big[A_{n,i}\mathbb{P}_{n+1}(y)\big]}{x_i - y_i}, \tag{3.6.5}$$

for $1 \le i \le d$, where $x = (x_1 \quad \cdots \quad x_d)$ and $y = (y_1 \quad \cdots \quad y_d)$.

Proof By Theorem 3.3.1, the orthogonal polynomials $\mathbb{P}_n$ satisfy the three-term relation (3.3.1). Let us write

$$\Sigma_k = \big[A_{k,i}\mathbb{P}_{k+1}(x)\big]^{\mathrm{T}} H_k^{-1}\mathbb{P}_k(y) - \mathbb{P}_k(x)^{\mathrm{T}} H_k^{-1}\big[A_{k,i}\mathbb{P}_{k+1}(y)\big].$$

From the three-term relation (3.3.1),

$$\begin{aligned}
\Sigma_k &= \big[x_i\mathbb{P}_k(x) - B_{k,i}\mathbb{P}_k(x) - C_{k,i}\mathbb{P}_{k-1}(x)\big]^{\mathrm{T}} H_k^{-1}\mathbb{P}_k(y) \\
&\quad - \mathbb{P}_k^{\mathrm{T}}(x) H_k^{-1}\big[y_i\mathbb{P}_k(y) - B_{k,i}\mathbb{P}_k(y) - C_{k,i}\mathbb{P}_{k-1}(y)\big] \\
&= (x_i - y_i)\mathbb{P}_k^{\mathrm{T}}(x) H_k^{-1}\mathbb{P}_k(y) - \mathbb{P}_k^{\mathrm{T}}(x)\left(B_{k,i}^{\mathrm{T}} H_k^{-1} - H_k^{-1} B_{k,i}\right)\mathbb{P}_k(y) \\
&\quad - \big[\mathbb{P}_{k-1}^{\mathrm{T}}(x) C_{k,i}^{\mathrm{T}} H_k^{-1}\mathbb{P}_k(y) - \mathbb{P}_k^{\mathrm{T}}(x) H_k^{-1} C_{k,i}\mathbb{P}_{k-1}(y)\big].
\end{aligned}$$

By the definition of H_k and (3.3.2), both H_k and $B_{k,i}H_k$ are symmetric matrices; hence $B_{k,i}H_k = (B_{k,i}H_k)^{\mathrm{T}} = H_k B_{k,i}^{\mathrm{T}}$, which implies that $B_{k,i}^{\mathrm{T}} H_k^{-1} = H_k^{-1} B_{k,i}$. Therefore, the second term on the right-hand side of the above expression for Σ_k is zero. From the third equation of (3.3.2), $C_{k,i}^{\mathrm{T}} H_k^{-1} = H_{k-1}^{-1} A_{k-1,i}$. Hence, the third term on the right-hand side of the expression for Σ_k is

$$-\mathbb{P}_{k-1}^{\mathrm{T}}(x) H_{k-1}^{-1} A_{k-1,i} \mathbb{P}_k(y) + \mathbb{P}_k^{\mathrm{T}}(x) A_{k-1,i}^{\mathrm{T}} H_{k-1}^{-1} \mathbb{P}_{k-1}(y) = -\Sigma_{k-1}.$$

Consequently, the expression for Σ_k can be rewritten as

$$(x_i - y_i)\mathbb{P}_k^{\mathrm{T}}(x) H_k^{-1} \mathbb{P}_k(y) = \Sigma_k - \Sigma_{k-1}.$$

Summing this identity from 0 to n and noting that $\Sigma_{-1} = 0$, we obtain the stated equation. □

Corollary 3.6.4 *Let $\{\mathbb{P}_k\}_{k=0}^{\infty}$ be as in Theorem 3.6.3. Then*

$$\sum_{k=0}^{n} \mathbb{P}_k^{\mathrm{T}}(x) H_k^{-1} \mathbb{P}_k(x) = \mathbb{P}_n^{\mathrm{T}}(x) H_n^{-1} \left[A_{n,i} \partial_i \mathbb{P}_{n+1}(x)\right] - \left[A_{n,i} \mathbb{P}_{n+1}(x)\right]^{\mathrm{T}} H_n^{-1} \partial_i \mathbb{P}_n(x),$$

where $\partial_i = \partial/\partial x_i$ denotes the partial derivative with respect to x_i.

Proof Since $\mathbb{P}_n(x)^{\mathrm{T}} H_n^{-1} [A_{n,i} \mathbb{P}_{n+1}(x)]$ is a scalar function, it is equal to its own transpose. Thus

$$\mathbb{P}_n(x)^{\mathrm{T}} H_n^{-1} \left[A_{n,i} \mathbb{P}_{n+1}(x)\right] = \left[A_{n,i} \mathbb{P}_{n+1}(x)\right]^{\mathrm{T}} H_n^{-1} \mathbb{P}_n(x).$$

Therefore the numerator of the right-hand side of the Christoffel–Darboux formula (3.6.5) can be written as

$$\left[A_{n,i} \mathbb{P}_{n+1}(x)\right]^{\mathrm{T}} H_n^{-1} \left[\mathbb{P}_n(y) - \mathbb{P}_n(x)\right] - \mathbb{P}_n(x)^{\mathrm{T}} H_n^{-1} A_{n,i} \left[\mathbb{P}_{n+1}(y) - \mathbb{P}_{n+1}(x)\right].$$

Thus the desired result follows from (3.6.5) on letting $y_i \to x_i$. □

If $\mathbb{P}_n$ are orthonormal polynomials then the formulae in Theorem 3.6.3 and Corollary 3.6.4 hold with $H_k = I$. We state this case as follows, for easy reference.

Theorem 3.6.5 *Let $\mathbb{P}_n$ be a sequence of orthonormal polynomials. Then*

$$\mathbf{K}_n(x,y) = \frac{\left[A_{n,i} \mathbb{P}_{n+1}(x)\right]^{\mathrm{T}} \mathbb{P}_n(y) - \mathbb{P}_n^{\mathrm{T}}(x) \left[A_{n,i} \mathbb{P}_{n+1}(y)\right]}{x_i - y_i} \tag{3.6.6}$$

and

$$\mathbf{K}_n(x,x) = \mathbb{P}_n^{\mathrm{T}}(x) \left[A_{n,i} \partial_i \mathbb{P}_{n+1}(x)\right] - \left[A_{n,i} \mathbb{P}_{n+1}(x)\right]^{\mathrm{T}} \partial_i \mathbb{P}_n(x). \tag{3.6.7}$$

It is interesting to note that, for each of the above formulae, although the right-hand side seems to depend on i, the left-hand side shows that it does not. By (3.6.3), the function $\mathbf{K}_n$ is the kernel for the Fourier partial sum; thus it is important to derive a compact formula for $\mathbf{K}_n$, that is, one that contains no summation. In the case of one variable, the right-hand side of the Christoffel–Darboux formula gives the desired compact formula. For several variables, however, the numerator of the right-hand side of (3.6.6) is still a sum, since $\mathbb{P}_n^{\mathrm{T}}(x)A_{n,i}\mathbb{P}_{n+1}(y)$ is a linear combination of $P_\alpha^n(x)P_\beta^{n+1}(y)$. For some special weight functions, including the classical orthogonal polynomials on the ball and on the simplex, we are able to find a compact formula; this will be discussed in Chapter 8.

For one variable the function $1/\mathbf{K}_n(x,x)$ is often called the *Christoffel function*. We will retain the name in the case of several variables, and we define

$$\Lambda_n(x) = \left[\mathbf{K}_n(x,x)\right]^{-1}. \tag{3.6.8}$$

Evidently this function is positive; moreover, it satisfies the following property.

Theorem 3.6.6 *Let $\mathscr{L}$ be a positive definite linear functional. For an arbitrary point $x \in \mathbb{R}^d$,*

$$\Lambda_n(x) = \min\{\mathscr{L}[P^2] : \ P(x) = 1, \ P \in \Pi_n^d\};$$

that is, the minimum is taken over all polynomials $P \in \Pi_n^d$ subject to the condition $P(x) = 1$.

Proof Since $\mathscr{L}$ is positive definite, there exists a corresponding sequence $\{\mathbb{P}_k\}$ of orthonormal polynomials. If P is a polynomial of degree at most n, then P can be written in terms of the orthonormal basis as follows:

$$P(x) = \sum_{k=0}^{n} \mathbf{a}_k^{\mathrm{T}}(P)\mathbb{P}_k(x) \quad \text{where} \quad \mathbf{a}_k(P) = \mathscr{L}(P\mathbb{P}_k).$$

From the orthonormal property of $\mathbb{P}_k$ it follows that

$$\mathscr{L}(P^2) = \sum_{k=0}^{n} \mathbf{a}_k^{\mathrm{T}}(P)\mathbf{a}_k(P) = \sum_{k=0}^{n} \|\mathbf{a}_k(P)\|^2.$$

If $P(x) = 1$ then, by Cauchy's inequality,

$$\begin{aligned} 1 = [P(x)]^2 &= \left(\sum_{k=0}^{n} \mathbf{a}_k^{\mathrm{T}}(P)\mathbb{P}_k(x)\right)^2 \le \left(\sum_{k=0}^{n} \|\mathbf{a}_k(P)\|\,\|\mathbb{P}_k(x)\|\right)^2 \\ &\le \sum_{k=0}^{n} \|\mathbf{a}_k(P)\|^2 \sum_{k=0}^{n} \|\mathbb{P}_k(x)\|^2 = \mathscr{L}(P^2)\mathbf{K}_n(x,x), \end{aligned}$$

where equality holds if and only if $\mathbf{a}_k(P) = \left[\mathbf{K}_n(x,x)\right]^{-1}\mathbb{P}_k(x)$. The proof is complete. □

Theorem 3.6.6 states that the function Λ_n is the solution of an extremum problem. One can also take the formula in the theorem as the definition of the Christoffel function.

3.6.2 Fourier orthogonal series

Let $\mathscr{L}$ be a positive definite linear functional, and let $\mathbb{H}_{\mathscr{L}}$ be the Hilbert space of real-valued functions defined on $\mathbb{R}^d$ with inner product $\langle f,g\rangle = \mathscr{L}(fg)$. Then the standard Hilbert space theory applies to the Fourier orthogonal series defined in (3.6.1).

Theorem 3.6.7 *Let $\mathscr{L}$ be a positive definite linear functional, and $f \in \mathbb{H}_{\mathscr{L}}$. Then among all the polynomials P in Π_n^d, the value of*

$$\mathscr{L}\left(|f(x) - P(x)|^2\right)$$

becomes minimal if and only if $P = S_n(f)$.

Proof Let $\{\mathbb{P}_k\}$ be an orthonormal basis of $\mathscr{V}_k$. For any $P \in \Pi_n^d$ there exist $\mathbf{b}_k$ such that

$$P(x) = \sum_{k=0}^{n} \mathbf{b}_k^{\mathrm{T}} \mathbb{P}_k(x),$$

and $S_n(f)$ satisfies equation (3.6.3). Following a standard argument,

$$0 \le \mathscr{L}\left(|f-P|^2\right) = \mathscr{L}\left(f^2\right) - 2\sum \mathbf{b}_k^{\mathrm{T}} \mathscr{L}\left(f\mathbb{P}_k\right) + \sum_k \sum_j \mathbf{b}_k^{\mathrm{T}} \mathscr{L}\left(\mathbb{P}_k \mathbb{P}_j^{\mathrm{T}}\right) \mathbf{b}_j$$

$$= \mathscr{L}\left(f^2\right) - 2\sum_{k=0}^{n} \mathbf{b}_k^{\mathrm{T}} \mathbf{a}_k(f) + \sum_{k=0}^{n} \mathbf{b}_k^{\mathrm{T}} \mathbf{b}_k$$

$$= \mathscr{L}\left(f^2\right) - \sum_{k=0}^{n} \mathbf{a}_k^{\mathrm{T}}(f)\mathbf{a}_k(f) + \sum_{k=0}^{n} [\mathbf{a}_k^{\mathrm{T}}(f)\mathbf{a}_k(f) + \mathbf{b}_k^{\mathrm{T}}\mathbf{b}_k - 2\mathbf{b}_k^{\mathrm{T}}\mathbf{a}_k(f)].$$

By Cauchy's inequality the third term on the right-hand side is nonnegative; moreover, the value of $\mathscr{L}(|f-P|^2)$ is minimal if and only if $\mathbf{b}_k = \mathbf{a}_k(f)$, or $P = S_n(f)$. □

In the case when the minimum is attained, we have Bessel's inequality

$$\sum_{k=0}^{\infty} \mathbf{a}_k^{\mathrm{T}}(f)\mathbf{a}_k(f) \le \mathscr{L}\left(|f|^2\right).$$

Moreover, the following result holds as in standard Hilbert space theory.

Theorem 3.6.8 *If Π^d is dense in $\mathbb{H}_{\mathscr{L}}$ then $S_n(f)$ converges to f in $\mathbb{H}_{\mathscr{L}}$, and we have Parseval's identity*

$$\sum_{k=0}^{\infty} \mathbf{a}_k^{\mathrm{T}}(f)\mathbf{a}_k(f) = \mathscr{L}\left(|f|^2\right).$$

The definition of the Fourier orthogonal series does not depend on the special basis of $\mathscr{V}_n^d$. If $\mathbb{P}_n$ is a sequence of polynomials that are orthogonal rather than orthonormal then the Fourier series will take the form

$$f \sim \sum_{n=0}^{\infty} \mathbf{a}_n^{\mathrm{T}}(f) H_n^{-1} \mathbb{P}_n, \qquad \mathbf{a}_n(f) = \mathscr{L}(f\mathbb{P}_n), \qquad H_n = \mathscr{L}\left(\mathbb{P}_n \mathbb{P}_n^{\mathrm{T}}\right). \quad (3.6.9)$$

The presence of the inverse matrix H_k^{-1} makes it difficult to find the Fourier orthogonal series using a basis that is orthogonal but not orthonormal.

An alternative method is to use biorthogonal polynomials to find the Fourier orthogonal series. Let $\mathbb{P}_n = \{P_\alpha^n\}$ and $\mathbb{Q}_n = \{Q_\alpha^n\}$ be two sequences of orthogonal polynomials with respect to $\mathscr{L}$. They are biorthogonal if $\mathscr{L}\left(P_\alpha^n Q_\beta^n\right) = d_\alpha \delta_{\alpha,\beta}$, where d_α is nonzero. If all $d_\alpha = 1$, the two systems are called biorthonormal. If $\mathbb{P}_n$ and $\mathbb{Q}_n$ are biorthonormal then the coefficients of the Fourier orthogonal series in terms of $\mathbb{P}_n$ can be found as follows:

$$f \sim \sum_{n=0}^{\infty} \mathbf{a}_n^{\mathrm{T}}(f) \mathbb{P}_n, \qquad \mathbf{a}_n(f) = \mathscr{L}(f\mathbb{Q}_n);$$

that is, the Fourier coefficient associated with P_α^n is given by $\mathscr{L}(fQ_\alpha^n)$. The idea of using biorthogonal polynomials to study Fourier orthogonal expansions was first applied to the case of classical orthogonal polynomials on the unit ball (see Section 5.2), and can be traced back to the work of Hermite; see Chapter XII in Vol. II of Erdélyi *et al.* [1953].

We finish this section with an extremal problem for polynomials in several variables. Let G_n be the leading-coefficient matrix corresponding to an orthonormal polynomial vector $\mathbb{P}_n$ as before. Denote by $\|G_n\|$ the spectral norm of G_n; $\|G_n\|^2$ is equal to the largest eigenvalue of $G_n G_n^{\mathrm{T}}$.

Theorem 3.6.9 *Let $\mathscr{L}$ be a positive definite linear functional. Then*

$$\min\{\mathscr{L}(P^2) : P = \mathbf{a}^{\mathrm{T}}\mathbf{x}^n + \cdots \in \Pi_n^d, \|\mathbf{a}\| = 1\} = \|G_n\|^{-2}.$$

Proof Since $\{\mathbb{P}_0, \ldots, \mathbb{P}_n\}$ forms a basis of Π_n^d, we can rewrite P as

$$P(x) = \mathbf{a}^{\mathrm{T}} G_n^{-1} \mathbb{P}_n + \mathbf{a}_{n-1}^{\mathrm{T}} \mathbb{P}_{n-1} + \cdots + \mathbf{a}_0^{\mathrm{T}} \mathbb{P}_0.$$

For $\|\mathbf{a}\|_2 = 1$, it follows from Bessel's inequality that

$$\begin{aligned}\mathscr{L}(P^2) &\geq \mathbf{a}^{\mathrm{T}} G_n^{-1} (G_n^{-1})^{\mathrm{T}} \mathbf{a} + \|\mathbf{a}_{n-1}\|^2 + \cdots + \|\mathbf{a}_0\|^2 \\ &\geq \mathbf{a}^{\mathrm{T}} G_n^{-1} (G_n^{-1})^{\mathrm{T}} \mathbf{a} \geq \lambda_{\min}\end{aligned}$$

where $\lambda_{\min}$ is the smallest eigenvalue of $(G_n^{\mathrm{T}} G_n)^{-1}$, which is equal to the reciprocal of the largest eigenvalue of $G_n G_n^{\mathrm{T}}$. Thus, we have proved that

$$\mathscr{L}(P^2) \geq \|G_n\|^{-2}, \qquad P(x) = \mathbf{a}^{\mathrm{T}}\mathbf{x}^n + \cdots, \quad \|\mathbf{a}\| = 1.$$

Choosing $\mathbf{a}$ such that $\|\mathbf{a}\| = 1$ and $\mathbf{a}^{\mathrm{T}} G_n^{-1} (G_n^{-1})^{\mathrm{T}} \mathbf{a} = \lambda_{\min}$, we see that the polynomial $P = \mathbf{a}^{\mathrm{T}} G_n^{-1} \mathbb{P}_n$ attains the lower bound. □

In Theorem 3.2.14, $\det G_n$ can be viewed as an extension of the leading coefficient k_n (see Proposition 1.3.7) of orthogonal polynomials in one variable in the context there. The above theorem shows that $\|G_n\|$ is also a natural extension of γ_n. In view of the formula (3.3.4), it is likely that $\|G_n\|$ is more important.

3.7 Common Zeros of Orthogonal Polynomials in Several Variables

For polynomials in one variable the nth orthonormal polynomial p_n has exactly n distinct zeros, and these zeros are the eigenvalues of the truncated Jacobi matrix J_n. These facts have important applications in quadrature formulae, as we saw in Subsection 1.3.2.

A zero of a polynomial in several variables is an example of an algebraic variety. In general, a zero of a polynomial in two variables can be either a single point or an algebraic curve on the plane. The structures of zeros are far more complicated in several variables. However, if common zeros of orthogonal polynomials are used then at least part of the theory in one variable can be extended to several variables.

A *common zero* of a set of polynomials is a zero for every polynomial in the set. Let $\mathscr{L}$ be a positive definite linear functional. Let $\mathbb{P}_n = \{P_\alpha^n\}$ be a sequence of orthonormal polynomials associated with $\mathscr{L}$. A common zero of $\mathbb{P}_n$ is a zero of every P_α^n. Clearly, we can consider zeros of $\mathbb{P}_n$ as zeros of the polynomial subspace $\mathscr{V}_n^d$.

We start with two simple properties of common zeros. First we need a definition: if x is a zero of $\mathbb{P}_n$ and at least one partial derivative of $\mathbb{P}_n$ at x is not zero then x is called a *simple zero* of $\mathbb{P}_n$.

Theorem 3.7.1 *All zeros of $\mathbb{P}_n$ are distinct and simple. Two consecutive polynomials $\mathbb{P}_n$ and $\mathbb{P}_{n-1}$ do not have common zeros.*

Proof Recall the Christoffel–Darboux formula in Theorem 3.6.3, which gives

$$\sum_{k=0}^{n-1} \mathbb{P}_k^{\mathrm{T}}(x)\mathbb{P}_k(x) = \mathbb{P}_{n-1}^{\mathrm{T}}(x) A_{n-1,i} \partial_i \mathbb{P}_n(x) - \mathbb{P}_n^{\mathrm{T}}(x) A_{n-1,i}^{\mathrm{T}} \partial_i \mathbb{P}_{n-1}(x),$$

where $\partial_i = \partial/\partial x_i$ denotes the partial derivative with respect to x_i. If x is a zero of $\mathbb{P}_n$ then

$$\sum_{k=0}^{n-1} \mathbb{P}_k^{\mathrm{T}}(x)\mathbb{P}_k(x) = \mathbb{P}_{n-1}^{\mathrm{T}}(x) A_{n-1,i} \partial_i \mathbb{P}_n(x).$$

Since the left-hand side is positive, neither $\mathbb{P}_{n-1}(x)$ nor $\partial_i \mathbb{P}_n(x)$ can be zero. □

Using the coefficient matrices of the three-term relation (3.3.3), we define the *truncated block Jacobi matrices* $J_{n,i}$ as follows:

$$J_{n,i} = \begin{bmatrix} B_{0,i} & A_{0,i} & & & \bigcirc \\ A_{0,i}^{\mathrm{T}} & B_{1,i} & A_{1,i} & & \\ & \ddots & \ddots & \ddots & \\ & & A_{n-3,i}^{\mathrm{T}} & B_{n-2,i} & A_{n-2,i} \\ \bigcirc & & & A_{n-2,i}^{\mathrm{T}} & B_{n-1,i} \end{bmatrix}, \qquad 1 \le i \le d.$$

Note that $J_{n,i}$ is a square matrix of order $N = \dim \Pi_{n-1}^d$. We say that $\Lambda = (\lambda_1 \quad \cdots \quad \lambda_d)^{\mathrm{T}} \in \mathbb{R}^d$ is a *joint eigenvalue* of $J_{n,1}, \ldots, J_{n,d}$ if there is a $\xi \neq 0$, $\xi \in \mathbb{R}^N$, such that $J_{n,i}\xi = \lambda_i \xi$ for $i = 1, \ldots, d$; the vector ξ is called a *joint eigenvector* associated with Λ.

The common zeros of $\mathbb{P}_n$ are characterized in the following theorem.

Theorem 3.7.2 *A point $\Lambda = (\lambda_1 \quad \cdots \quad \lambda_d)^{\mathrm{T}} \in \mathbb{R}^d$ is a common zero of $\mathbb{P}_n$ if and only if it is a joint eigenvalue of $J_{n,1}, \ldots, J_{n,d}$; moreover, the joint eigenvectors of Λ are constant multiples of $(\mathbb{P}_0^{\mathrm{T}}(\Lambda) \quad \cdots \quad \mathbb{P}_{n-1}^{\mathrm{T}}(\Lambda))^{\mathrm{T}}$.*

Proof If $\mathbb{P}_n(\Lambda) = 0$ then it follows from the three-term relation that

$$\begin{aligned} B_{0,i}\mathbb{P}_0(\Lambda) + A_{0,i}\mathbb{P}_1(\Lambda) &= \lambda_i \mathbb{P}_0(\Lambda), \\ A_{k-1,i}^{\mathrm{T}}\mathbb{P}_{k-1}(\Lambda) + B_{k,i}\mathbb{P}_k(\Lambda) + A_{k,i}\mathbb{P}_{k+1}(\Lambda) &= \lambda_i \mathbb{P}_k(\Lambda), \qquad 1 \le k \le n-2, \\ A_{n-2,i}^{\mathrm{T}}\mathbb{P}_{n-2}(\Lambda) + B_{n-1,i}\mathbb{P}_{n-1}(\Lambda) &= \lambda_i \mathbb{P}_{n-1}(\Lambda), \end{aligned}$$

for $1 \le i \le d$. From the definition of $J_{n,i}$ it follows that, on the one hand,

$$J_{n,i}\xi = \lambda_i \xi, \qquad \xi = \left(\mathbb{P}_0^{\mathrm{T}}(\Lambda) \quad \cdots \quad \mathbb{P}_{n-1}^{\mathrm{T}}(\Lambda)\right)^{\mathrm{T}}.$$

Thus, Λ is the eigenvalue of J_n with joint eigenvector ξ.

On the other hand, suppose that $\Lambda = (\lambda_1 \quad \cdots \quad \lambda_d)$ is an eigenvalue of J_n and that J_n has a joint eigenvector ξ for Λ. Write $\xi = (\mathbf{x}_0^{\mathrm{T}} \quad \cdots \quad \mathbf{x}_{n-1}^{\mathrm{T}})^{\mathrm{T}}$, $\mathbf{x}_j \in \mathbb{R}^{r_j}$. Since $J_{n,i}\xi = \lambda_i \xi$, it follows that $\{\mathbf{x}_j\}$ satisfies a three-term relation:

$$\begin{aligned} B_{0,i}\mathbf{x}_0 + A_{0,i}\mathbf{x}_1 &= \lambda_i \mathbf{x}_0, \\ A_{k-1,i}^{\mathrm{T}}\mathbf{x}_{k-1} + B_{k,i}\mathbf{x}_k + A_{k,i}\mathbf{x}_{k+1} &= \lambda_i \mathbf{x}_k, \qquad 1 \le k \le n-2, \\ A_{n-2,i}^{\mathrm{T}}\mathbf{x}_{n-2} + B_{n-1,i}\mathbf{x}_{n-1} &= \lambda_i \mathbf{x}_{n-1}, \end{aligned}$$

for $1 \le i \le d$. First we show that $\mathbf{x}_0$, and thus ξ, is nonzero. Indeed, if $\mathbf{x}_0 = 0$ then it follows from the first equation in the three-term relation that $A_{0,i}\mathbf{x}_1 = 0$, which implies that $A_0\mathbf{x}_1 = 0$. Since A_0 is a $d \times d$ matrix and it has full rank, it follows that $\mathbf{x}_1 = 0$. With $\mathbf{x}_0 = 0$ and $\mathbf{x}_1 = 0$, it then follows from the three-term relation that $A_{1,i}\mathbf{x}_2 = 0$, which leads to $A_1\mathbf{x}_2 = 0$. Since A_1 has full rank, $\mathbf{x}_2 = 0$. Continuing this process leads to $\mathbf{x}_i = 0$ for $i \ge 3$. Thus, we end up with $\xi = 0$, which contradicts the assumption that ξ is an eigenvector. Let us assume that $\mathbf{x}_0 = 1 = \mathbb{P}_0$ and define $\mathbf{x}_n \in \mathbb{R}^{r_n}$ as $\mathbf{x}_n = 0$. We will prove that $\mathbf{x}_j = \mathbb{P}_j(\Lambda)$ for all $1 \le j \le n$. Since the last equation in the three-term relation of $\mathbf{x}_j$ can be written as

$$A_{n-2,i}^{\mathrm{T}} \mathbf{x}_{n-2} + B_{n-1,i} \mathbf{x}_{n-1} + A_{n-1,i} \mathbf{x}_n = \lambda_i \mathbf{x}_{n-1},$$

it follows that $\{\mathbf{x}_k\}_{k=0}^n$ and $\{\mathbb{P}_k(\Lambda)\}_{k=0}^n$ satisfy the same three-term relation. Thus so does $\{\mathbf{y}_k\} = \{\mathbb{P}_k(\Lambda) - \mathbf{x}_k\}$. But since $\mathbf{y}_0 = 0$, it follows from the previous argument that $\mathbf{y}_k = 0$ for all $1 \le k \le n$. In particular, $\mathbf{y}_n = \mathbb{P}_n(\Lambda) = 0$. The proof is complete. □

From this theorem follow several interesting corollaries.

Corollary 3.7.3 *All the common zeros of $\mathbb{P}_n$ are real and are points in $\mathbb{R}^d$.*

Proof This follows because the $J_{n,i}$ are symmetric matrices and the eigenvalues of a symmetric matrix are real and hence are points in $\mathbb{R}^d$. □

Corollary 3.7.4 *The polynomials in $\mathbb{P}_n$ have at most* $\dim \Pi_{n-1}^d$ *common zeros.*

Proof Since $J_{n,i}$ is a square matrix of size $\dim \Pi_{n-1}^d$, it can have at most that many eigenvectors. □

In one variable, the nth orthogonal polynomial p_n has $n = \dim \Pi_{n-1}^1$ distinct real zeros. The situation becomes far more complicated in several variables. The following theorem characterizes when $\mathbb{P}_n$ has the maximum number of common zeros.

Theorem 3.7.5 *The orthogonal polynomial $\mathbb{P}_n$ has $N = \dim \Pi_{n-1}^d$ distinct real common zeros if and only if*

$$A_{n-1,i} A_{n-1,j}^{\mathrm{T}} = A_{n-1,j} A_{n-1,i}^{\mathrm{T}}, \qquad 1 \le i, j \le d. \tag{3.7.1}$$

Proof According to Theorem 3.7.2 a zero of $\mathbb{P}_n$ is a joint eigenvalue of $J_{n,1}, \ldots, J_{n,d}$ whose eigenvectors forms a one-dimensional space. This implies that $\mathbb{P}_n$ has $N = \dim \Pi_{n-1}^d$ distinct zeros if and only if $J_{n,1}, \ldots, J_{n,d}$ have N distinct eigenvalues, which is equivalent to stating that $J_{n,1}, \ldots, J_{n,d}$ can be simultaneously diagonalized by an invertible matrix. Since a family of matrices is simultaneously diagonalizable if and only if it is a commuting family,

$$J_{n,i} J_{n,j} = J_{n,j} J_{n,i}, \qquad 1 \le i, j \le d.$$

From the definition of $J_{n,i}$ and the commutativity conditions (3.4.2), the above equation is equivalent to the condition

$$A_{n-2,i}^{\mathrm{T}} A_{n-2,j} + B_{n-1,i} B_{n-1,j} = A_{n-2,j}^{\mathrm{T}} A_{n-2,i} + B_{n-1,j} B_{n-1,i}.$$

The third equation in (3.4.2) then leads to the desired result. □

The condition (3.7.1) is trivial for orthogonal polynomials in one variable, as then $A_i = a_i$ are scalars. The condition is usually not satisfied, since matrix multiplication is not commutative; it follows that in general $\mathbb{P}_n$ does not have $\dim \Pi_{n-1}^d$ common zeros. Further, the following theorem gives a necessary condition for $\mathbb{P}_n$ to have the maximum number of common zeros. Recall the notation Ξ_C for joint matrices in Definition 3.5.2.

Theorem 3.7.6 *A necessary condition for the orthogonal polynomial $\mathbb{P}_n$ to have $\dim \Pi_{n-1}^d$ distinct real common zeros is*

$$\operatorname{rank}\left(\Xi_{B_n}^{\mathsf{T}} B_n\right) = \operatorname{rank}\left(\Xi_{A_{n-1}}^{\mathsf{T}} A_{n-1}\right) = \binom{n+d-2}{n} + \binom{n+d-3}{n-1}.$$

In particular, for $d = 2$, the condition becomes

$$\operatorname{rank}(B_{n,1}B_{n,2} - B_{n,2}B_{n,1}) = \operatorname{rank}(A_{n-1,1}^{\mathsf{T}} A_{n-1,2} - A_{n-1,2}^{\mathsf{T}} A_{n-1,1}) = 2.$$

Proof If $\mathbb{P}_n$ has $\dim \Pi_{n-1}^d$ distinct real common zeros then, by Theorem 3.7.5, the matrices $A_{n-1,i}$ satisfy equation (3.7.1). It follows that the third commutativity condition in (3.4.2) is reduced to

$$B_{n,i}B_{n,j} - B_{n,j}B_{n,i} = A_{n-1,j}^{\mathsf{T}} A_{n-1,i} - A_{n-1,i}^{\mathsf{T}} A_{n-1,j},$$

which, using the fact that the $B_{n,i}$ are symmetric matrices, can be written as $\Xi_{B_n}^{\mathsf{T}} B_n = -\Xi_{A_{n-1}}^{\mathsf{T}} A_{n-1}$. Using the rank condition (3.3.8), Proposition 3.5.7 and a rank inequality, we have

$$\begin{aligned} \operatorname{rank} \Xi_{B_n}^{\mathsf{T}} B_n &= \operatorname{rank} \Xi_{A_{n-1}}^{\mathsf{T}} A_{n-1} \\ &\geq \operatorname{rank} A_{n-1} + \operatorname{rank} \Xi_{A_{n-1}} - d r_{n-1}^d \\ &\geq r_n^d + d r_{n-1}^d - r_{n-2}^d - d r_{n-1}^d = r_n^d - r_{n-2}^d. \end{aligned}$$

To prove the desired result, we now prove the reverse inequality. On the one hand, using the notation Ξ_C the first equation in (3.4.2) can be written as $\Xi_{A_{n-1}^{\mathsf{T}}}^{\mathsf{T}} A_n = 0$. By Proposition 3.5.7 and the rank condition (3.3.8), it follows that the columns of A_n form a basis for the null space of $\Xi_{A_{n-1}^{\mathsf{T}}}^{\mathsf{T}}$. On the other hand, the condition (3.7.1) can be written as $\Xi_{A_{n-1}^{\mathsf{T}}}^{\mathsf{T}} (A_{n-1,1} \quad \cdots \quad A_{n-1,d})^{\mathsf{T}} = 0$, which shows that the columns of the matrix $(A_{n-1,1} \quad \cdots \quad A_{n-1,d})^{\mathsf{T}}$ belong to the null space of $\Xi_{A_{n-1}^{\mathsf{T}}}^{\mathsf{T}}$. So, there is a matrix $S_n : r_{n+1}^d \times r_{n-1}^d$ such that $(A_{n-1,1} \quad \cdots \quad A_{n-1,d})^{\mathsf{T}} = A_n S_n$. It follows from

$$r_{n-1}^d = \operatorname{rank}(A_{n-1,1} \quad \cdots \quad A_{n-1,d})^{\mathsf{T}} = \operatorname{rank} A_n S_n \leq \operatorname{rank} S_n \leq r_{n-1}^d$$

that $\operatorname{rank} S_n = r_{n-1}^d$. Using the condition (3.7.1) with n in place of $n-1$, we have

$$\Xi_{B_{n+1}}^{\mathsf{T}} B_{n+1} S_n = -\Xi_{A_n}^{\mathsf{T}} A_n S_n = -\Xi_{A_n}^{\mathsf{T}} (A_{n-1,1} \quad \cdots \quad A_{n-1,d})^{\mathsf{T}} = 0,$$

where the last equality follows from the first equation in (3.4.2). Hence

$$\operatorname{rank}\left(\Xi_{B_{n+1}}^{\mathrm{T}} B_{n+1}\right) = r_{n+1}^d - \dim\ker \Xi_{B_{n+1}}^{\mathrm{T}} B_{n+1}$$
$$\le r_{n+1}^d - \operatorname{rank} S_n = r_{n+1}^d - r_{n-1}^d,$$

which completes the proof. □

If $\mathscr{L}$ is a quasi-centrally-symmetric linear functional then $\operatorname{rank}[\Xi_{B_n}^{\mathrm{T}} B_n] = 0$, since it follows from Lemma 3.5.3 that

$$\Xi_{B_n}^{\mathrm{T}} B_n = 0 \quad \Longleftrightarrow \quad B_{n,i}B_{n,j} = B_{n,j}B_{n,i},$$

where we have used the fact that the $B_{n,i}$ are symmetric. Therefore, a immediate corollary of the above result is the following.

Corollary 3.7.7 *If $\mathscr{L}$ is a quasi-centrally-symmetric linear functional then $\mathbb{P}_n$ does not have $\dim \Pi_{n-1}^d$ distinct common zeros.*

The implications of these results are discussed in the following section.

3.8 Gaussian Cubature Formulae

In the theory of orthogonal polynomials in one variable, Gaussian quadrature formulae played an important role. Let w be a weight function defined on $\mathbb{R}$. A quadrature formula with respect to w is a weighted sum of a finite number of function values that gives an approximation to the integral $\int f w\,dx$; it gives the exact value of the integral for polynomials up to a certain degree, which is called the degree of precision of the formula. Among all quadrature formulae a Gaussian quadrature formula has the maximum degree of precision, which is $2n-1$ for a formula that uses n nodes, as discussed in Theorem 1.3.13. Furthermore, a Gaussian quadrature formula exists if and only if the nodes are zeros of an orthogonal polynomial of degree n with respect to w. Since such a polynomial always has n distinct real zeros, the Gaussian quadrature formula exists for every suitable weight function.

Below we study the analogues of the Gaussian quadrature formulae for several variables, called the *Gaussian cubature formulae*. The existence of such formulae depends on the existence of common zeros of orthogonal polynomials in several variables.

Let $\mathscr{L}$ be a positive definite linear functional. A cubature formula of degree $2n-1$ with respect to $\mathscr{L}$ is a linear functional

$$\mathscr{I}_n(f) = \sum_{k=1}^{N} \lambda_k f(x_k), \qquad \lambda_k \in \mathbb{R}, \quad x_k \in \mathbb{R}^d,$$

such that $\mathscr{L}(f) = \mathscr{I}_n(f)$ for all $f \in \Pi^d_{2n-1}$ and $\mathscr{L}(f^*) \neq \mathscr{L}_{2n-1}(f^*)$ for at least one $f^* \in \Pi^d_{2n}$. The points x_k are called the *nodes*, and the numbers λ_k the *weights*, of the cubature formula. Such a formula is called *minimal* if N, the number of nodes, is minimal. If a cubature formula has all positive weights, it is called a positive cubature.

We are interested in the minimal cubature formulae. To identify a cubature formula as minimal, it is necessary to know the minimal number of nodes in advance. The following theorem gives a lower bound for the numbers of nodes of cubature formulae; those formulae that attain the lower bound are minimal.

Theorem 3.8.1 *The number of nodes of a cubature formula of degree* $2n-1$ *satisfies* $N \geq \dim \Pi^d_{n-1}$.

Proof If a cubature formula $\mathscr{L}_n$ of degree $2n-1$ has less than $\dim \Pi^d_{n-1}$ nodes then the linear system of equations $P(x_k) = 0$, where the x_k are nodes of the cubature formula and $P \in \Pi^d_{n-1}$, has $\dim \Pi^d_{n-1}$ unknowns that are coefficients of P but fewer than $\dim \Pi^d_{n-1}$ equations. Therefore, there will be at least one nonzero polynomial $P \in \Pi^d_{n-1}$ which vanishes on all nodes. Since the formula is of degree $2n-1$, it follows that $\mathscr{L}(P^2) = \mathscr{L}_n(P^2) = 0$, which contradicts the fact that $\mathscr{L}$ is positive definite. □

For $d = 1$, the lower bound in Theorem 3.8.1 is attained by the Gaussian quadrature formulae. As an analogy of this, we make the following definition.

Definition 3.8.2 A cubature formula of degree $2n-1$ with $\dim \Pi^d_{n-1}$ nodes is called a Gaussian cubature.

The main result in this section is the characterization of Gaussian cubature formulae. We start with a basic lemma of Mysovskikh [1976], which holds the key to the proof of the main theorem in this section and which is of considerable interest in itself. We need a definition: an algebraic hypersurface of degree m in $\mathbb{R}^d$ is a zero set of a polynomial of degree m in d variables.

Lemma 3.8.3 *The pairwise distinct points* x_k, $1 \leq k \leq N = \dim \Pi^d_{n-1}$, *on* $\mathbb{R}^d$ *are not on an algebraic hypersurface of degree* $n-1$ *if and only if there exist* r^d_n *polynomials of degree n such that their nth degree terms are linearly independent, for which these points are the common zeros.*

Proof That the x_k are not on a hypersurface of degree $n-1$ means that the $N \times N$ matrix (x_k^β), where the columns are arranged according to the lexicographical order in $\{\beta : 0 \leq |\beta| \leq n-1\}$ and the rows correspond to $1 \leq k \leq N$, is nonsingular. Therefore, we can construct r^d_n polynomials

$$Q^n_\alpha = x^\alpha - R_\alpha, \qquad |\alpha| = n, \quad R_\alpha(x) = \sum_{|\beta|\le n-1} C_{\alpha,\beta} x^\beta \in \Pi^d_{n-1} \tag{3.8.1}$$

by solving the linear system of equations

$$R^n_\alpha(x_k) = x_k^\alpha, \qquad 1 \le k \le N,$$

for each fixed α. Clearly, the set of polynomials Q^n_α is linearly independent and the conditions $Q^n_\alpha(x_k) = 0$ are satisfied.

However, suppose that there are r^d_n polynomials Q^n_α of degree n which have x_k as common zeros and that their nth terms are linearly independent. Because of their linear independence, we can assume that these polynomials are of the form (3.8.1). If the points x_k are on an algebraic hypersurface of degree $n-1$ then the matrix $X_n = (x_k^\beta)$, where $1 \le k \le N$ and $|\beta| \le n-1$, is singular. Suppose that $\operatorname{rank} X_n = M-1$, $M \le N$. Assume that the first $M-1$ rows of this matrix are linearly independent. For each k define a vector $U_k = (x_k^\beta)_{|\beta|\le n-1}$ in $\mathbb{R}^N$; then $U_1,\ldots,U_{M-1}$ are linearly independent and there exist scalars $c_1,\ldots,c_M$, not all zero, such that $\sum_{k=1}^M c_k U_k = 0$. Moreover, by considering the first component of this vector equation it follows that at least two c_j are nonzero. Assume that $c_1 \ne 0$ and $c_M \ne 0$. By (3.8.1) the equations $Q^n_\alpha(x_k) = 0$ can be written as

$$x_k^\alpha - \sum_{|\beta|\le n-1} C_{\alpha,\beta} x_k^\beta = 0, \qquad |\alpha| = n, \quad k = 1,\ldots,N.$$

Since the summation term can be written as $C_\alpha^{\mathrm{T}} U_k$, with C_α a vector in $\mathbb{R}^N$, it follows from $\sum_{k=1}^M c_k U_k = 0$ that $\sum_{k=1}^M c_k x_k^\alpha = 0$, $|\alpha| = n$. Using the notation for the vectors U_k again, it follows from the above equation that $\sum_{k=1}^M c_k x_{k,i} U_k = 0$, $1 \le i \le d$, where we have used the notation $x_k = (x_{k,1},\ldots,x_{k,d})$. Multiplying the equation $\sum_{k=1}^M c_k U_k = 0$ by $x_{M,i}$ and then subtracting it from the equation $\sum_{k=1}^M c_k x_{k,i} U_k = 0$, we obtain

$$\sum_{k=1}^{M-1} c_k (x_{k,i} - x_{M,i}) U_k = 0, \qquad 1 \le i \le d.$$

Since $U_1,\ldots,U_{M-1}$ are linearly independent, it follows that $c_k(x_{k,i} - x_{M,i}) = 0$, $1 \le k \le M-1$, $1 \le i \le d$. Since $c_1 \ne 0$, $x_{1,i} = x_{M,i}$ for $1 \le i \le M-1$, which is to say that $x_1 = x_M$, a contradiction to the assumption that the points are pairwise distinct. □

The existence of a Gaussian cubature of degree $2n-1$ is characterized by the common zeros of orthogonal polynomials. The following theorem was first proved by Mysovskikh [1970]; our proof is somewhat different.

Theorem 3.8.4 *Let $\mathscr{L}$ be a positive definite linear functional, and let $\mathbb{P}_n$ be the corresponding orthogonal polynomials. Then $\mathscr{L}$ admits a Gaussian cubature formula of degree $2n-1$ if and only if $\mathbb{P}_n$ has $N = \dim \Pi^d_{n-1}$ common zeros.*

Proof Suppose that $\mathbb{P}_n$ has N common zeros. By Theorem 3.7.1 and Corollary 3.7.3, all these zeros are real and distinct. Let us denote these common zeros by $x_1, x_2, \ldots, x_N$. From the Christoffel–Darboux formula (3.6.5) and the formula for $\mathbf{K}_n(x,x)$, it follows that the polynomial $\ell_k(x) = \mathbf{K}_n(x,x_k)/\mathbf{K}_n(x_k,x_k)$ satisfies the condition $\ell_k(x_j) = \delta_{k,j}$, $1 \le k, j \le N$. Since the ℓ_k are of degree $n-1$ they serve as fundamental interpolation polynomials, and it follows that

$$L_n(f,x) = \sum_{k=1}^{N} f(x_k) \frac{\mathbf{K}(x,x_k)}{\mathbf{K}(x_k,x_k)}, \qquad L_n(f) \in \Pi_{n-1}^d, \tag{3.8.2}$$

is a Lagrange interpolation based on x_k, that is, $L_n(f,x_k) = f(x_k)$, $1 \le k \le N$. By Lemma 3.8.3 the points x_k are not on an algebraic hypersurface of degree $n-1$. Therefore, it readily follows that $L_n(f)$ is the unique Lagrange interpolation formula. Thus the formula

$$\mathscr{L}[L_n(f)] = \sum_{k=1}^{N} \Lambda_k f(x_k), \qquad \Lambda_k = [\mathbf{K}_n(x_k,x_k)]^{-1}, \tag{3.8.3}$$

is a cubature formula for $\mathscr{L}$ which is exact for all polynomials in Π_{n-1}^d. However, since $\mathbb{P}_n$ is orthogonal to Π_{n-1}^d and the nodes are zeros of $\mathbb{P}_n$, it follows that, for any vector $\mathbf{a} \in \mathbb{R}^{r_n}$,

$$\mathscr{L}(\mathbf{a}^{\mathrm{T}}\mathbb{P}_n) = 0 = \sum_{k=1}^{N} \Lambda_k \mathbf{a}^{\mathrm{T}}\mathbb{P}_n(x_k).$$

Since any polynomial $P \in \Pi_n^d$ can be written as $P = \mathbf{a}^{\mathrm{T}}\mathbb{P}_n + R$ for some $\mathbf{a}$ with $R \in \Pi_{n-1}^d$,

$$\mathscr{L}(P) = \mathscr{L}(R) = \sum_{k=1}^{N} \Lambda_k R(x_k) = \sum_{k=1}^{N} \Lambda_k P(x_k) \qquad \forall P \in \Pi_n^d. \tag{3.8.4}$$

Thus, the cubature formula (3.8.3) is exact for all polynomials in Π_n^d. Clearly, we can repeat this process. By orthogonality and the fact that $\mathbb{P}_n(x_k) = 0$,

$$\mathscr{L}(x_i \mathbf{a}^{\mathrm{T}}\mathbb{P}_n) = 0 = \sum_{k=1}^{N} \Lambda_k x_{k,i} \mathbf{a}^{\mathrm{T}}\mathbb{P}_n(x_k), \qquad 1 \le i \le d,$$

which, by the recurrence relation (3.3.10) and the relation (3.8.4), implies that $\mathscr{L}(\mathbf{a}^{\mathrm{T}}\mathbb{P}_{n+1}) = \sum_{k=1}^{N} \Lambda_k \mathbf{a}^{\mathrm{T}}\mathbb{P}_{n+1}(x_k)$. Therefore, it readily follows that the cubature formula (3.8.3) is exact for all polynomials in Π_{n+1}^d. Because $\mathbb{P}_n$ is orthogonal to polynomials of degree less than n, we can apply this process on x^α, $|\alpha| \le n-2$, and conclude that the cubature formula (3.8.3) is indeed exact for Π_{2n-1}^d.

Now suppose that a Gaussian cubature formula exists and it has x_k, $1 \le k \le N = \dim \Pi_{n-1}^d$, as its nodes. Since $\mathscr{L}$ is positive definite, these nodes cannot all lie on an algebraic hypersurface of degree $n-1$. Otherwise there would be a $Q \in \Pi_{n-1}$, not identically zero, such that Q vanished on all nodes, which would imply that $\mathscr{L}(Q^2) = 0$. By Lemma 3.8.3 there exist linearly independent polynomials P_j^n, $1 \le j \le r_n^d$, which have the x_k as common zeros. By the Gaussian cubature formula, for any $R \in \Pi_{n-1}$,

$$\mathscr{L}(RP_j^n) = \sum_{k=1}^{N} \Lambda_k (RP_j^n)(x_k) = 0.$$

Thus, the polynomials P_j^n are orthogonal to Π_{n-1}^d and we can write them in the form (3.8.1). □

It is worthwhile mentioning that the above proof is similar to the proof for the Gaussian quadrature formulae. The use of interpolation polynomials leads to the following corollary.

Corollary 3.8.5 *A Gaussian cubature formula takes the form* (3.8.3)*; in particular, it is a positive cubature.*

In the proof of Theorem 3.8.4, the Gaussian cubature formula (3.8.3) was derived by integrating the Lagrange interpolation polynomial $L_n f$ in (3.8.2). This is similar to the case of one variable. For polynomial interpolation in one variable, it is known that the Lagrange interpolation polynomials based on the distinct nodes always exist and are unique. The same, however, is not true for polynomial interpolation in several variables. A typical example is that of the interpolation of six points on the plane by a polynomial of degree 2. Assume that the six points lie on the unit circle $x^2 + y^2 = 1$; then the polynomial $p(x,y) = x^2 + y^2 - 1$ will vanish on the nodes. If P is an interpolation polynomial of degree 2 on these nodes then so is $P + ap$ for any given number a. This shows that the Lagrange interpolation polynomial, if it exists, is not unique. If the nodes are zeros of a Gaussian cubature formula then they admit a unique Lagrange interpolation polynomial.

Theorem 3.8.6 *If x_k, $1 \le k \le \dim \Pi_{n-1}^d$, are nodes of a Gaussian cubature formula then for any given $\{y_i\}$ there is a unique polynomial $P \in \Pi_{n-1}^d$ such that $P(x_i) = y_i$, $1 \le i \le \dim \Pi_{n-1}^d$.*

Proof The interpolating polynomial takes the form (3.8.2) with $y_k = f(x_k)$. If all $f(x_i) = 0$ then $P(x_i) = 0$ and the Gaussian cubature formula shows that $\mathscr{L}(P^2) = 0$, which implies $P = 0$. □

Theorem 3.8.4 characterizes the cubature formulae through the common zeros of orthogonal polynomials. Together with Theorem 3.7.5 we have the following.

Theorem 3.8.7 *Let $\mathscr{L}$ be a positive definite linear functional, and let $\mathbb{P}_n$ be the corresponding orthonormal polynomials, which satisfy the three-term relation* (3.3.3)*. Then a Gaussian cubature formula exists if and only if* (3.7.1) *holds.*

Since Corollary 3.7.7 states that (3.7.1) does not hold for a quasi-centrally-symmetric linear functional, it follows that:

Corollary 3.8.8 *A quasi-centrally-symmetric linear functional does not admit a Gaussian cubature formula.*

In particular, there is no Gaussian cubature formula for symmetric classical weight functions on the square or for the classical weight function on the unit disk, or for their higher-dimensional counterparts. An immediate question is whether there are any weight functions which admit a Gaussian cubature formula. It turns out that there are such examples, as will be shown in Subsection 5.4.2.

3.9 Notes

Section 3.2 The general properties of orthogonal polynomials in several variables were studied in Jackson [1936], Krall and Sheffer [1967], Bertran [1975], Kowalski [1982a, b], Suetin [1999], and a series of papers of Xu [1993a, b, 1994a–e] in which the vector–matrix notion is emphasized.

The classical references for the moment problems in one variable are Shohat and Tamarkin [1943] and Akhiezer [1965]. For results on the moment problem in several variables, we refer to the surveys Fuglede [1983] and Berg [1987], as well as Putinar and Vasilescu [1999] and the references therein. Although in general positive polynomials in several variables cannot be written as sums of squares of polynomials, they can be written as sums of squares of rational functions (Hilbert's 17th problem); see the recent survey Reznick [2000] and the references therein.

Section 3.3 The three-term relation and Favard's theorem in several variables were first studied by Kowalski [1982a, b], in which the vector notation was used in the form $x\mathbb{P}_n = (x_1\mathbb{P}_n^{\mathrm{T}} \quad \cdots \quad x_d\mathbb{P}_n^{\mathrm{T}})$ and Favard's theorem was stated under an additional condition that takes a rather complicated form. The additional condition was shown to be equivalent to rank $C_k = r_k^d$, where $C_k = (C_{k,1} \quad \cdots \quad C_{k,d})$ in Xu [1993a]. The relation for orthonormal polynomials is stated and proved in Xu [1994a]. The three-term relations were used by Barrio, Peña and Sauer [2010] for evaluating orthogonal polynomials. The matrices in the three-term relations were computed for further orthogonal polynomials of two variables in Area, Godoy, Ronveaux and Zarzo [2012]. The structure of orthogonal polynomials in lexicographic order, rather than graded lexicographic order, was studied in Delgado, Geronimo, Iliev and Marcellán [2006].

Section 3.4 The relation between orthogonal polynomials of several variables and self-adjoint commuting operators was studied in Xu [1993b, 1994a]; see also Gekhtman and Kalyuzhny [1994]. An extensive study from the operator theory perspective was carried out in Cichoń, Stochel and Szafraniec [2005].

Sections 3.7 and 3.8 The study of Gaussian cubature formulae started with the classical paper of Radon [1948]. Mysovskikh and his school continued the study and made considerable progress; Mysovskikh obtained, for example, the basic properties of the common zeros of $\mathbb{P}_n$ and most of the results in Section 3.8; see Mysovskikh [1981] and the references therein. The fact that the common zeros are joint eigenvalues of block Jacobi matrices and Theorem 3.7.5 were proved in Xu [1993a, 1994b]. Mysovskikh [1976] proved a version of Theorem 3.7.5 in two variables without writing it in terms of coefficient matrices of the three-term relation. A different necessary and sufficient condition for the existence of the Gaussian cubature formula was given recently in Lasserre [2012]. There are examples of weight functions that admit Gaussian cubature formulae for all n (Schmid and Xu [1994], Berens, Schmid and Xu [1995b]).

Möller [1973] proved the following important results: the number of nodes, N, of a cubature formula of degree $2n-1$ in two variables must satisfy

$$N \geq \dim \Pi^2_{n-1} + \tfrac{1}{2}\operatorname{rank}(A_{n-1,1}A^{\mathrm{T}}_{n-1,2} - A_{n-1,2}A^{\mathrm{T}}_{n-1,1}), \tag{3.9.1}$$

which we have stated in matrix notation; moreover, for the quasi-centrally-symmetric weight function, $\operatorname{rank}(A_{n-1,1}A^{\mathrm{T}}_{n-1,2} - A_{n-1,2}A^{\mathrm{T}}_{n-1,1}) = 2[n/2]$. Möller also showed that if a cubature formula attains the lower bound (3.9.1) then its nodes must be common zeros of a subset of orthogonal polynomials of degree n. Similar results hold in the higher dimensional setting. Much of the later study in this direction has been influenced by Möller's work. There are, however, only a few examples in which the lower bound (3.9.1) is attained; see, for example, Möller [1976], Morrow and Patterson [1978], Schmid [1978, 1995], and Xu [2013]. The bound in (3.9.1) is still not sharp, even for radial weight functions on the disk; see Verlinden and Cools [1992]. For cubature formulae of even degree, see Schmid [1978] and Xu [1994d]. In general, all positive cubature formulae are generated by common zeros of quasi-orthogonal polynomials (Möller [1973], Xu [1994f, 1997a]), which are orthogonal to polynomials of a few degrees lower, a notion that can also be stated in terms of the m-orthogonality introduced by Möller. A polynomial p is said to be m-orthogonal if $\int pq\,\mathrm{d}\mu = 0$ for all $q \in \Pi^d$ such that $pq \in \Pi^d_m$. The problem can be stated in the language of polynomial ideals and varieties. The essential result says that if the variety of an ideal of m-orthogonal polynomials consists of only real points and the size of the variety is more or less equal to the codimension of the ideal then there is a cubature formula of degree m; see Xu [1999c]. Concerning the basic question of finding a set of orthogonal polynomials or m-orthogonal polynomials with a large number of common zeros, there is no general method yet. For further results in this direction we refer to the papers cited above and to the references therein. Putinar [1997, 2000] studied cubature formulae using an operator theory approach. The books by Stroud [1971], Engels [1980] and Mysovskikh [1981] deal with cubature formulae in general.

4

Orthogonal Polynomials on the Unit Sphere

In this chapter we consider orthogonal polynomials with respect to a weight function defined on the unit sphere, the structure of which is not covered by the discussion in the previous chapter. Indeed, if $\mathrm{d}\mu$ is a measure supported on the unit sphere of $\mathbb{R}^d$ then the linear functional $\mathscr{L}(f) = \int f \, \mathrm{d}\mu$ is not positive definite in the space of polynomials, as $\int (1 - \|x\|^2)^2 \, \mathrm{d}\mu = 0$. It is positive definite in the space of polynomials restricted to the unit sphere, which is the space in which these orthogonal polynomials are defined.

We consider orthogonal polynomials with respect to the surface measure on the sphere first; these are the spherical harmonics. Our treatment will be brief, since most results and proofs will be given in a more general setting in Chapter 7. The general structure of orthogonal polynomials on the sphere will be derived from the close connection between the orthogonal structures on the sphere and on the unit ball. This connection goes both ways and can be used to study classical orthogonal polynomials on the unit ball. We will also discuss a connection between the orthogonal structures on the unit sphere and on the simplex.

4.1 Spherical Harmonics

The Fourier analysis of continuous functions on the unit sphere $S^{d-1} := \{x : \|x\| = 1\}$ in $\mathbb{R}^d$ is performed by means of spherical harmonics, which are the restrictions of homogeneous harmonic polynomials to the sphere. In this section we present a concise overview of the theory and a construction of an orthogonal basis by means of Gegenbauer polynomials. Further results can be deduced as special cases of theorems in Chapter 7, by taking the weight function there as 1.

We assume $d \geq 3$ unless stated otherwise. Let $\Delta = \partial_1^2 + \cdots + \partial_d^2$ be the Laplace operator, where $\partial_i = \partial / \partial x_i$.

Definition 4.1.1 For $n = 0, 1, 2, \ldots$ let $\mathcal{H}_n^d$ be the linear space of harmonic polynomials, homogeneous of degree n, on $\mathbb{R}^d$, that is,

$$\mathcal{H}_n^d = \left\{ P \in \mathcal{P}_n^d : \Delta P = 0 \right\}.$$

Theorem 4.1.2 *For* $n = 0, 1, 2, \ldots$

$$\dim \mathcal{H}_n^d = \dim \mathcal{P}_n^d - \dim \mathcal{P}_{n-2}^d = \binom{n+d-1}{d-1} - \binom{n+d-3}{d-1}.$$

Proof Briefly, one shows that Δ maps $\mathcal{P}_n^d$ onto $\mathcal{P}_{n-2}^d$; this uses the fact that $\Delta\left[\|x\|^2 P(x)\right] \neq 0$ whenever $P(x) \neq 0$. □

There is a "Pascal triangle" for the dimensions. Let $a_{d,n} = \dim \mathcal{H}_n^d$; then $a_{d,0} = 1, a_{2,n} = 2$ for $n \geq 1$ and $a_{d+1,n} = a_{d+1,n-1} + a_{d,n}$ for $n \geq 1$. Note that $a_{3,n} = 2n+1$ and $a_{4,n} = (n+1)^2$. The basic device for constructing orthogonal bases is to set up the differential equation $\Delta\big(g(x_{m+1}, \ldots, x_d) f(\sum_{j=m}^d x_j^2, x_m)\big) = 0$, where g is harmonic and homogeneous and f is a polynomial.

Proposition 4.1.3 *For $1 \leq m < d$, suppose that $g(x_{m+1}, \ldots, x_d)$ is harmonic and homogeneous of degree s in its variables, and that $f(\sum_{j=m}^d x_j^2, x_m)$ is homogeneous of degree n in x; then $\Delta(gf) = 0$ implies that (see Definition 1.4.10)*

$$f = c\left(\sum_{j=m}^d x_j^2\right)^{n/2} C_n^\lambda\left(x_m \left(\sum_{j=m}^d x_j^2\right)^{-1/2}\right)$$

for some constant c and $\lambda = s + \frac{1}{2}(d-m-1)$.

Proof Let $\rho^2 = \sum_{j=m}^d x_j^2$ and apply the operator Δ to $g x_m^{n-2j} \rho^{2j}$. The product rule gives

$$\begin{aligned}
\Delta\left(g x_m^{n-2j} \rho^{2j}\right) &= x_m^{n-2j} \rho^{2j} \Delta g + g \Delta\big(x_m^{n-2j} \rho^{2j}\big) + 2 \sum_{i=m+1}^d \frac{\partial g}{\partial x_i} \frac{\partial \rho^{2j}}{\partial x_i} x_m^{n-2j} \\
&= 4gsj x_m^{n-2j} \rho^{2j-2} + g\big[2j(d-m-1+2n-2j) x_m^{n-2j} \rho^{2j-2} \\
&\qquad + (n-2j)(n-2j-1) x_m^{n-2j-2} \rho^{2j}\big].
\end{aligned}$$

Set $\Delta\big(g \sum_{j \leq n/2} c_j x_m^{n-2j} \rho^{2j}\big) = 0$ to obtain the recurrence equation

$$2(j+1)(d-m+2n+2s-3-2j) c_{j+1} + (n-2j)(n-2j-1) c_j = 0,$$

with solution

$$c_j = \frac{\left(-\frac{n}{2}\right)_j \left(\frac{1-n}{2}\right)_j}{j! \left(1-n-\frac{d-m-1}{2}-s\right)_j} c_0.$$

Now let $f(x) = \sum_{j\le n/2} c_j x_m^{n-2j}\rho^{2j} = c'\rho^n C_n^\lambda(\frac{x_m}{\rho})$, with $\lambda = s + (d-m-1)/2$, for some constant c'; then $\Delta(gf) = 0$. This uses the formula in Proposition 1.4.11 for Gegenbauer polynomials. □

The easiest way to start this process is with the real and imaginary parts of $(x_{d-1} + \sqrt{-1}x_d)^s$, which can be written in terms of Chebyshev polynomials of first and second kind, that is, $g_{s,0} = \rho^s T_s(\frac{x_{d-1}}{\rho})$ and $g_{s-1,1} = x_d\rho^{s-1}U_{s-1}(\frac{x_{d-1}}{\rho})$, where $\rho^2 = x_{d-1}^2 + x_d^2$, for $s \ge 1$ (and $g_{0,0} = 1$). Then Proposition 4.1.3, used inductively from $m = d-2$ to $m = 1$, produces an orthogonal basis for $\mathcal{H}_n^d$: fix $\mathbf{n} = (n_1, n_2, \ldots, n_{d-1}, 0)$ or $(n_1, n_2, \ldots, n_{d-1}, 1)$ and let

$$Y_{\mathbf{n}}(x) = g_{n_{d-1},n_d}(x)\prod_{j=1}^{d-2}\left[(x_j^2 + \cdots + x_d^2)^{n_j/2} C_{n_j}^{\lambda_j}\left(x_j(x_j^2 + \cdots + x_d^2)^{-1/2}\right)\right],$$

with $\lambda_j = \sum_{i=j+1}^d n_i + \frac{1}{2}(d-j-1)$. Note that $Y_{\mathbf{n}}(x)$ is homogeneous of degree $|\mathbf{n}|$ and that the number of such d-tuples is $\binom{n+d-2}{d-2} + \binom{n+d-3}{d-2} = \binom{n+d-1}{d-1} - \binom{n+d-3}{d-1} = \dim\mathcal{H}_n^d$. The fact that these polynomials are pairwise orthogonal follows easily from their expression in spherical polar coordinates. Indeed, let

$$\begin{aligned} x_1 &= r\cos\theta_{d-1},\\ x_2 &= r\sin\theta_{d-1}\cos\theta_{d-2},\\ &\vdots \\ x_{d-1} &= r\sin\theta_{d-1}\cdots\sin\theta_2\cos\theta_1,\\ x_d &= r\sin\theta_{d-1}\cdots\sin\theta_2\sin\theta_1, \end{aligned} \tag{4.1.1}$$

with $r \ge 0$, $0 \le \theta_1 < 2\pi$ and $0 \le \theta_i \le \pi$ for $i \ge 2$. In these coordinates the surface measure on S^{d-1} and the normalizing constant are

$$\mathrm{d}\omega = \prod_{j=1}^{d-2}(\sin\theta_{d-j})^{d-j-1}\,\mathrm{d}\theta_1\mathrm{d}\theta_2\cdots\mathrm{d}\theta_{d-1}, \tag{4.1.2}$$

$$\sigma_{d-1} = \int_{S^{d-1}}\mathrm{d}\omega = \frac{2\pi^{d/2}}{\Gamma(\frac{d}{2})}, \tag{4.1.3}$$

where σ_{d-1} is the surface area of S^{d-1}.

Theorem 4.1.4 *In the spherical polar coordinate system, let*

$$Y_{\mathbf{n}}(x) = r^{|\mathbf{n}|}g'(x)\prod_{j=1}^{d-2}\left[(\sin\theta_{d-j})^{\beta_j}C_{n_j}^{\lambda_j}(\cos\theta_{d-j})\right],$$

where $\beta_j = \sum_{i=j+1}^d n_i$, $g'(x) = \cos n_{d-1}\theta_1$ *for* $n_d = 0$ *and* $\sin(n_{d-1}+1)\theta_1$ *for* $n_d = 1$. *Then* $\{Y_{\mathbf{n}}\}$ *is a mutually orthogonal basis of* $\mathcal{H}_n^d$ *and the* L^2 *norms of the polynomials are*

$$\sigma_{d-1}^{-1}\int_{S^{d-1}} Y_{\mathbf{n}}^2\,\mathrm{d}\omega = a_{\mathbf{n}}\prod_{j=1}^{d-2}\frac{\lambda_j(2\lambda_j)_{n_j}\left(\frac{d-j}{2}\right)_{\beta_j}}{n_j!\left(\frac{d-j+1}{2}\right)_{\beta_j}(n_j+\lambda_j)},$$

where $a_{\mathbf{n}} = \frac{1}{2}$ *if* $n_{d-1}+n_d > 0$, *else* $a_{\mathbf{n}} = 1$.

The calculation of the norm uses the formulae in Subsection 1.4.3. We also note that in the expansion of $Y_{\mathbf{n}}(x)$ as a sum of monomials the highest term in lexicographic order is (a multiple of) $x^{\mathbf{n}}$.

It is an elementary consequence of Green's theorem that homogeneous harmonic polynomials of differing degrees are orthogonal with respect to the surface measure. In the following $\partial/\partial n$ denotes the normal derivative.

Proposition 4.1.5 *Suppose that* f, g *are polynomials on* $\mathbb{R}^d$; *then*

$$\int_{S^{d-1}}\left(\frac{\partial f}{\partial n}g - \frac{\partial g}{\partial n}f\right)\mathrm{d}\omega = \int_{B^d}(g\Delta f - f\Delta g)\,\mathrm{d}x$$

and if f, g *are homogeneous then*

$$(\deg f - \deg g)\int_{S^{d-1}} fg\,\mathrm{d}\omega = \int_{B^d}(g\Delta f - f\Delta g)\,\mathrm{d}x.$$

Proof The first statement is Green's theorem specialized to the ball B^d and its boundary S^{d-1}. The normal derivative on the sphere is $\partial f/\partial n = \sum_{i=1}^d x_i\partial f/\partial x_i = (\deg f)f$. □

Clearly, if f, g are spherical harmonics of different degrees then they are orthogonal in $L^2\left(S^{d-1}, \mathrm{d}\omega\right)$. There is another connection to the Laplace operator. For $x \in \mathbb{R}^d$, consider the spherical polar coordinates (r, ξ) with $r > 0$ and $\xi \in \mathscr{S}^{d-1}$ defined by $x = r\xi$.

Proposition 4.1.6 *In spherical coordinates* $(r, \xi_1, \ldots, \xi_{d-1})$, *the Laplace operator can be written as*

$$\Delta = \frac{\partial^2}{\partial r^2} + \frac{d-1}{r}\frac{\partial}{\partial r} + \frac{1}{r^2}\Delta_0, \tag{4.1.4}$$

where the spherical part Δ_0 *is given by*

$$\Delta_0 = \Delta^{(\xi)} - \langle\xi, \nabla^{(x)}\rangle^2 - (d-2)\langle\xi, \nabla^{(\xi)}\rangle;$$

here $\Delta^{(\xi)}$ *and* $\nabla^{(\xi)}$ *denote the Laplacian and the gradient with respect to* $(\xi_1, \ldots, \xi_{d-1})$, *respectively.*

Proof Write r and ξ_i in terms of x by the change of variables $x \mapsto (r, \xi_1, \ldots, \xi_d)$ and compute the partial derivatives $\partial \xi_i / \partial y_i$; the chain rule implies that

$$\frac{\partial}{\partial x_i} = \xi_i \frac{\partial}{\partial r} + \frac{1}{r}\Big(\frac{\partial}{\partial \xi_i} - \xi_i \langle \xi, \nabla^{(\xi)} \rangle\Big), \qquad 1 \le i \le d,$$

where for $i = d$ we use the convention that $\xi_d^2 = 1 - \xi_1^2 - \cdots - \xi_{x-1}^2$ and define $\partial / \partial \xi_d = 0$. Repeating the above operation and simplifying, we obtain the second-order derivatives after some tedious computation:

$$\frac{\partial^2}{\partial \xi_i^2} = \xi_i^2 \frac{\partial^2}{\partial r^2} + \frac{1-\xi_i^2}{r}\frac{\partial}{\partial r} + \frac{1}{r^2}\Big(\frac{\partial^2}{\partial \xi_i^2} - (1-\xi_i^2)\langle \xi, \nabla^{(\xi)}\rangle - \xi_i \langle \xi, \nabla^{(\xi)}\rangle \frac{\partial}{\partial \xi_i}\Big)$$
$$+ \Big(\xi_i \frac{\partial}{\partial \xi_i} - \xi_i^2 \langle \xi, \nabla^{(\xi)}\rangle\Big)\Big(\frac{1}{r}\frac{\partial}{\partial r} - \frac{1}{r^2}\Big) - \Big(\xi_i \frac{\partial}{\partial \xi_i} - \xi_i^2 \langle \xi, \nabla^{(\xi)}\rangle\Big)\langle x, \nabla^{(x)}\rangle.$$

Adding these equations for $1 \le i \le d$, we see that the last two terms become zero, since $\xi_1^2 + \cdots + \xi_d^2 = 1$, and we end up with the equation

$$\Delta = \frac{\partial^2}{\partial r^2} + \frac{d-1}{r}\frac{\partial}{\partial r} + \frac{1}{r^2}\left(\Delta^{(\xi)} - (d-1)\langle \xi, \nabla^{(\xi)}\rangle - \sum_{i=1}^{d-1} \xi_i \langle \xi, \nabla^{(\xi)}\rangle \frac{\partial}{\partial \xi_i}\right).$$

The term with the summation sign can be verified to be $\langle \xi, \nabla^{(\xi)}\rangle^2 - \langle \xi, \nabla^{(\xi)}\rangle$, from which the stated formula follows. □

The differential operator Δ_0 is called the Laplace–Beltrami operator.

Theorem 4.1.7 *For* $n = 0, 1, 2, \ldots,$

$$\Delta_0 Y = -n(n+d-2)Y \qquad \forall Y \in \mathscr{H}_n^d. \tag{4.1.5}$$

Proof Let $Y \in \mathscr{H}_n^d$. Since Y is homogeneous, $Y_n(x) = r^n Y_n(x')$. As Y is harmonic, (4.1.4) shows that

$$0 = \Delta Y(x) = n(n-1)r^{n-2}Y_n(x') + (d-1)nr^{n-2}Y_n(x') + r^{n-2}\Delta_0 Y_n(x'),$$

which is, when restricted to the sphere, equation (4.1.5). □

For a given n, let $P_n(x,y)$ be the reproducing kernel for $\mathscr{H}_n^d$ (as functions on the sphere); then

$$P_n(x,y) = \sum_{\mathbf{n}} \|Y_{\mathbf{n}}\|^{-2} Y_{\mathbf{n}}(x) Y_{\mathbf{n}}(y), \quad |\mathbf{n}| = n,\ n_d = 0 \text{ or } 1.$$

However the linear space $\mathscr{H}_n^d$ is invariant under the action of the orthogonal group $O(d)$ (that is, $f(x) \to f(xw)$ where $w \in O(d)$); the surface measure is also invariant, hence $P_n(xw, yw) = P_n(x,y)$ for all $w \in O(d)$. This implies that $P_n(x,y)$ depends only on the distance between x and y (both points being on the sphere) or, equivalently, on $\langle x, y\rangle = \sum_{i=1}^d x_i y_i$, the inner product. Fix $y = (1, 0, \ldots, 0)$;

the fact that $P_n(x,y) = f(\langle x,y\rangle) = f(x_1)$ must be the restriction of a harmonic homogeneous polynomial to the sphere shows that

$$P_n(x,y) = a_n \|x\|^n \|y\|^n C_n^{(d-2)/2}(\langle x,y\rangle/\|x\|\,\|y\|)$$

for some constant a_n. With respect to normalized surface measure on S^{d-1} we have $\sigma_{d-1}^{-1}\int_{S^{d-1}} P_n(x,x)\mathrm{d}\omega(x) = \dim \mathscr{H}_n^d$, but $P_n(x,x)$ is constant and thus $a_n C_n^{(d-2)/2}(1) = \dim \mathscr{H}_n^d$. Hence $a_n = (2n+d-2)/(d-2)$. Thus, we have proved the following:

Theorem 4.1.8 *The reproducing kernel $P_n(x,y)$ of $\mathscr{H}_n^d$ satisfies*

$$P_n(x,y) = \frac{n+\lambda}{\lambda}\|x\|^n\|y\|^n C_n^\lambda\left(\frac{\langle x,y\rangle}{\|x\|\,\|y\|}\right), \qquad \lambda = \frac{d-2}{2}. \tag{4.1.6}$$

The role of the generic *zonal harmonic* P_n in the Poisson summation kernel will be addressed in later chapters. Applying the reproducing kernel to the harmonic polynomial $x \mapsto \|x\|^n C_n^{(d-2)/2}(\langle x,z\rangle/\|x\|)$ for a fixed $z \in S^{d-1}$ implies the so-called Funk–Hecke formula

$$\sigma_{d-1}^{-1}\frac{2n+d-2}{d-2}\int_{S^{d-1}} C_n^{(d-2)/2}(\langle x,y\rangle) f(x)\,\mathrm{d}\omega(x) = f(y),$$

for any $f \in \mathscr{H}_n^d$ and $y \in S^{d-1}$.

4.2 Orthogonal Structures on S^d and on B^d

Our main results on the orthogonal structure on the unit sphere will be deduced by lifting those on the unit ball B^d to the unit sphere S^d, where $B^d := \{x \in \mathbb{R}^d : \|x\| \le 1\}$ and $S^d \subset \mathbb{R}^{d+1}$. We emphasize that we shall work with the unit sphere S^d instead of S^{d-1} in this section, since otherwise we would be working with B^{d-1}.

Throughout this section we fix the following notation: for $y \in \mathbb{R}^{d+1}$, write $y = (y_1, \ldots, y_d, y_{d+1}) = (y', y_{d+1})$ and use polar coordinates

$$y = r(x, x_{d+1}), \qquad \text{where} \quad r = \|y\|, \quad (x, x_{d+1}) \in S^d. \tag{4.2.1}$$

Note that $(x, x_{d+1}) \in S^d$ implies immediately that $x \in B^d$. Here are the weight functions on S^d that we shall consider:

Definition 4.2.1 A weight function H defined on $\mathbb{R}^{d+1}$ is called S-symmetric if it is even with respect to y_{d+1} and centrally symmetric with respect to the variables y', that is, if

$$H(y', y_{d+1}) = H(y', -y_{d+1}) \qquad \text{and} \qquad H(y', y_{d+1}) = H(-y', y_{d+1}).$$

We further assume that the integral of H over the sphere S^d is nonzero.

Examples of S-symmetric functions include $H(y) = W(y_1^2,\dots,y_{d+1}^2)$, which is even in each of its variables, and $H(y) = W(y_1,\dots,y_d)h(y_{d+1})$, where W is centrally symmetric on $\mathbb{R}^d$ and h is even on $\mathbb{R}$. Note that the weight function $\prod_{i<j}|y_i - y_j|^\alpha$ is not an S-symmetric function, since it is not even with respect to y_{d+1}. Nevertheless, this function is centrally symmetric. In fact, it is easy to see that:

Lemma 4.2.2 *If H is an S-symmetric weight function on $\mathbb{R}^{d+1}$ then it is centrally symmetric, that is, $H(y) = H(-y)$ for $y \in \mathbb{R}^{d+1}$.*

In association with a weight function H defined on $\mathbb{R}^{d+1}$, define a weight function W_H^B on B^d by

$$W_H^B(x) = H(x, \sqrt{1-\|x\|^2}), \qquad x \in B^d.$$

If H is S-symmetric then the assumption that H is centrally symmetric with respect to the first d variables implies that W_H^B is centrally symmetric on B^d. Further, define a pair of weight functions on B^d,

$$W_1^B(x) = 2W_H^B(x)\Big/\sqrt{1-\|x\|^2} \qquad \text{and} \qquad W_2^B(x) = 2W_H^B(x)\sqrt{1-\|x\|^2},$$

respectively. We shall show that orthogonal polynomials with respect to H on S^d are closely related to those with respect to W_1^B and W_2^B on B^d. We need the following elementary lemma.

Lemma 4.2.3 *For any integrable function f defined on S^d,*

$$\int_{S^d} f(y)\,\mathrm{d}\omega_d = \int_{B^d} \left[f\left(x, \sqrt{1-\|x\|^2}\right) + f\left(x, -\sqrt{1-\|x\|^2}\right)\right] \frac{\mathrm{d}x}{\sqrt{1-\|x\|^2}}.$$

Proof For $y \in S^d$ write $y = (\sqrt{1-t^2}x, t)$, where $x \in S^{d-1}$ and $-1 \le t \le 1$. It follows that

$$\mathrm{d}\omega_d(y) = (1-t^2)^{(d-2)/2}\,\mathrm{d}t\,\mathrm{d}\omega_{d-1}(x).$$

Making the change of variables $y \mapsto (\sqrt{1-t^2}x, t)$ gives

$$\begin{aligned}
\int_{S^d} f(y)\,\mathrm{d}\omega_d &= \int_{-1}^{1}\int_{S^{d-1}} f\left(\sqrt{1-t^2}x, t\right)\mathrm{d}\omega_{d-1}(1-t^2)^{(d-2)/2}\mathrm{d}t \\
&= \int_0^1\int_{S^{d-1}} \left[f\left(\sqrt{1-t^2}x, t\right) + f\left(\sqrt{1-t^2}x, -t\right)\right]\mathrm{d}\omega_{d-1}(1-t^2)^{(d-2)/2}\mathrm{d}t \\
&= \int_0^1\int_{S^{d-1}} \left[f(rx, \sqrt{1-r^2}) + f\left(rx, -\sqrt{1-r^2}\right)\right]\mathrm{d}\omega_{d-1}r^{d-1}\frac{\mathrm{d}r}{\sqrt{1-r^2}},
\end{aligned}$$

from which the stated formula follows from the standard parameterization of the integral over B^d in spherical polar coordinates $y = rx$ for $y \in B^d$, $0 < r \le 1$ and $x \in S^{d-1}$. □

Recall the notation $\mathscr{V}_n^d(W)$ for the space of orthonormal polynomials of degree n with respect to the weight function W. Denote by $\{P_\alpha\}_{|\alpha|=n}$ and $\{Q_\alpha\}_{|\alpha|=n}$ systems of orthonormal polynomials that form bases for $\mathscr{V}_n^d(W_1^B)$ and $\mathscr{V}_n^d(W_2^B)$, respectively. Keeping in mind the notation (4.2.1), define

$$Y_\alpha^{(1)}(y) = r^n P_\alpha(x) \quad \text{and} \quad Y_\beta^{(2)}(y) = r^n x_{d+1} Q_\beta(x), \tag{4.2.2}$$

where $|\alpha| = n$, $|\beta| = n-1$ and $\alpha, \beta \in \mathbb{N}_0^d$. These functions are, in fact, homogeneous orthogonal polynomials on S^d.

Theorem 4.2.4 *Let H be an S-symmetric weight function defined on $\mathbb{R}^{d+1}$. Then $Y_\alpha^{(1)}$ and $Y_\beta^{(2)}$ in* (4.2.2) *are homogeneous polynomials of degree $|\alpha|$ on $\mathbb{R}^{d+1}$, and they satisfy*

$$\int_{S^d} Y_\alpha^{(i)}(y) Y_\beta^{(j)}(y) H(y)\, d\omega_d = \delta_{\alpha,\beta}\delta_{i,j}, \qquad i,j = 1,2,$$

where if $i = j = 1$ then $|\alpha| = |\beta| = n$ and if $i = j = 2$ then $|\alpha| = |\beta| = n-1$.

Proof Since both W_1^B and W_2^B are centrally symmetric, it follows from Theorem 3.3.11 that P_α and Q_α are sums of monomials of even degree if $|\alpha|$ is even and sums of monomials of odd degree if $|\alpha|$ is odd. This allows us to write, for example,

$$P_\alpha(x) = \sum_{k=0}^{n} \sum_{|\gamma|=n-2k} a_\gamma x^\gamma, \quad a_\gamma \in \mathbb{R}, \quad x \in B^d,$$

where $|\alpha| = n$, which implies that

$$Y_\alpha^{(1)}(y) = r^n P_\alpha(x) = \sum_{k=0}^{n} r^{2k} \sum_{|\gamma|=n-2k} a_\gamma y'^\gamma.$$

Since $r^2 = y_1^2 + \cdots + y_{d+1}^2$ and $y' = (y_1, \ldots, y_d)$, this shows that $Y_\alpha^{(1)}(y)$ is a homogeneous polynomial of degree n in y. Upon using $rx_{d+1} = y_{d+1}$, a similar proof shows that $Y_\alpha^{(2)}$ is homogeneous of degree n.

Since $Y_\alpha^{(1)}$, when restricted to S^d, is independent of x_{d+1} and since $Y_\alpha^{(2)}$ contains a single factor x_{d+1}, it follows that $Y_\alpha^{(1)}$ and $Y_\beta^{(2)}$ are orthogonal with respect to $H(y)\, d\omega_d$ on S^d for any α and β. Since H is even with respect to its last variable, it follows from Lemma 4.2.3 that

$$\begin{aligned}\int_{S^d} Y_\alpha^{(1)}(x) Y_\beta^{(1)}(x) H(x)\, d\omega_d &= 2\int_{B^d} P_\alpha(x) P_\beta(x) \frac{H(x, \sqrt{1-\|x\|^2})}{\sqrt{1-\|x\|^2}}\, dx \\ &= \int_{B^d} P_\alpha(x) P_\beta(x) W_1^B(x)\, dx = \delta_{\alpha,\beta}\end{aligned}$$

and, similarly, using the fact that $x_{d+1}^2 = 1 - \|x\|^2$, we see that the polynomials $Y_\alpha^{(2)}$ are orthonormal. □

In particular, ordinary spherical harmonics (the case $H(y)=1$) are related to the orthogonal polynomials with respect to the radial weight functions $W_0(x) = 1/\sqrt{1-\|x\|^2}$ and $W_1(x) = \sqrt{1-\|x\|^2}$ on B^d, both of which are special cases of the classical weight functions $W_\mu(x) = w_\mu(1-\|x\|^2)^{\mu-1/2}$; see Subsection 5.2. For $d=2$, the spherical harmonics on S^1 are given in polar coordinates $(y_1,y_2) = r(x_1,x_2) = r(\cos\theta,\sin\theta)$ by the formulae $r^n\cos n\theta$ and $r^n\sin n\theta$, which can be written as

$$Y_n^{(1)}(y_1,y_2) = r^n T_n(x_1) \qquad \text{and} \qquad Y_n^{(2)}(y_1,y_2) = r^n x_2 U_{n-1}(x_1),$$

where, with $t=\cos\theta$, $T_n(t)=\cos n\theta$ and $U_n(t) = \sin(n+1)\theta/\sin\theta$ are Chebyshev polynomials of the first and the second kind and are orthogonal with respect to $1/\sqrt{1-t^2}$ and $\sqrt{1-t^2}$, respectively; see Subsection 1.4.3.

Theorem 4.2.4 shows that the orthogonal polynomials with respect to an S-symmetric weight function are homogeneous. This suggests the following definition.

Definition 4.2.5 For a given weight function H, denote by $\mathscr{H}_n^{d+1}(H)$ the space of homogeneous orthogonal polynomials of degree n with respect to $H\,d\omega$ on S^d.

Using this definition, Theorem 4.2.4 can be restated as follows: the relation (4.2.2) defines a one-to-one correspondence between an orthonormal basis of $\mathscr{H}_n^{d+1}(H)$ and an orthonormal basis of $\mathscr{V}_n^d(W_1^B) \oplus x_{d+1}\mathscr{V}_{n-1}^d(W_2^B)$. Consequently, the orthogonal structure on S^d can be studied using the orthogonality structure on B^d.

Theorem 4.2.6 *Let H be an S-symmetric function on $\mathbb{R}^{d+1}$. For each $n\in\mathbb{N}_0$,*

$$\dim \mathscr{H}_n^{d+1}(H) = \binom{n+d}{d} - \binom{n+d-2}{d} = \dim \mathscr{P}_n^{d+1} - \dim \mathscr{P}_{n-2}^{d+1}.$$

Proof From the orthogonality in Theorem 4.2.4, the polynomials in $\{Y_\alpha^{(1)}, Y_\beta^{(2)}\}$ are linearly independent. Hence, it follows that

$$\dim \mathscr{H}_n^{d+1}(H) = r_n^d + r_{n-1}^d = \binom{n+d-1}{n} + \binom{n+d-2}{n-1},$$

where we use the convention that $\binom{k}{j}=0$ if $j<0$. Using the identity $\binom{n+m}{n} - \binom{n+m-1}{n} = \binom{n+m-1}{n-1}$, it is easy to verify that $\dim\mathscr{H}_n^{d+1}(H)$ is given by the formula stated in Proposition 4.1.2. □

Theorem 4.2.7 *Let H be an S-symmetric function on $\mathbb{R}^{d+1}$. For each $n\in\mathbb{N}_0$,*

$$\mathscr{P}_n^{d+1} = \bigoplus_{k=0}^{[n/2]} \|y\|^{2k}\mathscr{H}_{n-2k}^{d+1}(H);$$

that is, if $P \in \mathscr{P}_n^{d+1}$ then there is a unique decomposition

$$P(y) = \sum_{k=0}^{[n/2]} \|y\|^{2k} P_{n-2k}(y), \qquad P_{n-2k} \in \mathscr{H}_{n-2k}^{d+1}(H).$$

Proof Since P is homogeneous of degree n, we can write $P(y) = r^n P(x, x_{d+1})$, using the notation in (4.2.1). Using $x_{d+1}^2 = 1 - \|x\|^2$ whenever possible, we further write

$$P(y) = r^n P(x) = r^n [p(x) + x_{d+1} q(x)], \qquad x \in B^d,$$

where p and q are polynomials of degree at most n and $n-1$, respectively. Moreover, if n is even then p is a sum of monomials of even degree and q is a sum of monomials of odd degree; and if n is odd, then p is a sum of monomials of odd degree and q is a sum of monomials of even degree. Since both $\{P_\alpha\}$ and $\{Q_\alpha\}$ form bases for Π_n^d and since the weight functions W_1^B and W_2^B are centrally symmetric, we have the unique expansions

$$p(x) = \sum_{k=0}^{[n/2]} \sum_{|\alpha|=n-2k} a_\alpha P_\alpha(x) \qquad \text{and} \qquad q(x) = \sum_{k=0}^{[(n-1)/2]} \sum_{|\beta|=n-2k-1} b_\beta Q_\beta(x).$$

Therefore, by the definitions of $Y_\alpha^{(1)}$ and $Y_\beta^{(2)}$,

$$P(y) = \sum_{k=0}^{[n/2]} r^{2k} \sum_{|\alpha|=n-2k} a_\alpha Y_\alpha^{(1)}(y) + y_{d+1} \sum_{k=0}^{[(n-1)/2]} r^{2k} \sum_{|\beta|=n-2k-1} b_\beta Y_\beta^{(2)}(y),$$

which is the desired decomposition. The uniqueness of the decomposition of $P(y)$ follows from the orthogonality in Theorem 4.2.4. □

Let H be an S-symmetric weight function and let $Y_\alpha^{(1)}$ and $Y_\alpha^{(2)}$ be orthonormal polynomials with respect to H such that they form an orthonormal basis of $\mathscr{H}_n^{d+1}(H)$. For $f \in L^2(H, S^d)$, its Fourier orthogonal expansion is defined by

$$f \sim \sum_\alpha \left[a_\alpha^{(1)}(f) Y_\alpha^{(1)} + a_\alpha^{(2)}(f) Y_\alpha^{(2)} \right],$$

where $a_\alpha^{(i)}(f) = \int_{S^d} f(x) Y_\alpha^{(i)}(x) H(x)\, d\omega$. We define the nth component by

$$P_n(f;x) = \sum_{|\alpha|=n} \left[a_\alpha^{(1)}(f) Y_\alpha^{(1)}(x) + a_\alpha^{(2)}(f) Y_\alpha^{(2)}(x) \right]. \tag{4.2.3}$$

The expansion $f \sim \sum_{n=0}^\infty P_n(f)$ is also called the Laplace series. As in the case of orthogonal expansions for a weight function supported on a solid domain, the component $P_n(f;x)$ can be viewed as the orthogonal projection of f on the space

$\mathscr{H}_n^{d+1}(H)$ and, being so, it is independent of the choice of basis of $\mathscr{H}_n^{d+1}(H)$. Define

$$P_n(H;x,y) = \sum_{|\alpha|=n} \left[Y_\alpha^{(1)}(x) Y_\alpha^{(1)}(y) + Y_\alpha^{(2)}(x) Y_\alpha^{(2)}(y) \right]; \tag{4.2.4}$$

this is the reproducing kernel of $\mathscr{H}_n^{d+1}(H)$ since it evidently satisfies

$$\int_{S^d} P_n(H;x,y) Y(y) H(y)\, d\omega(y) = Y(x), \qquad Y \in \mathscr{H}_n^{d+1}(H),$$

which can be taken as the definition of $P_n(H;x,y)$. For ordinary spherical harmonic polynomials, P_n is a so-called zonal polynomial; see Section 4.1. The function $P_n(H)$ does not depend on the choice of a particular basis. It serves as an integral kernel for the component

$$S_n(f;x) = \int_{S^d} f(y) P_n(H;x,y) H(y) d\omega(y). \tag{4.2.5}$$

As a consequence of Theorem 4.2.4, there is also a relation between the reproducing kernels of $\mathscr{H}_n^{d+1}(H)$ and $\mathscr{V}_n^d(W_1^B)$. Let us denote the reproducing kernel of $\mathscr{V}_n^d(W)$ by $\mathbf{P}_n(W;x,y)$; see (3.6.2) for its definition.

Theorem 4.2.8 *Let H be an S-symmetric function and let W_1^B be associated with H. Recall the notation* (4.2.1). *Then*

$$\begin{aligned} \mathbf{P}_n(W_1^B;x,y) = \tfrac{1}{2} \Big[& P_n\Big(H;(x,x_{d+1}),\Big(y,\sqrt{1-\|y\|^2}\Big)\Big) \\ & + P_n\Big(H;(x,x_{d+1}),\Big(y,-\sqrt{1-\|y\|^2}\Big)\Big)\Big], \end{aligned} \tag{4.2.6}$$

where $x_{d+1} = \sqrt{1-\|x\|}$ and $\|x\|$ is the Euclidean norm of x.

Proof From Theorem 4.2.4 and (4.2.4), we see that for $z = r(x,x_{d+1}) \in \mathbb{R}^{d+1}$ and $y \in B^d$ with $y_{d+1} = \sqrt{1-\|x\|^2}$,

$$P_n\big(H;z,(y,y_{d+1})\big) = r^n \mathbf{P}_n(W_1^B;x,y) + z_{d+1} y_{d+1} r^{n-1} \mathbf{P}_{n-1}(W_2^B;x,y),$$

from which the stated equation follows on using

$$r^n \mathbf{P}_n(W_1^B;x,y) = \frac{1}{2}\Big[P_n\Big(H;z,\Big(y,\sqrt{1-\|y\|^2}\Big)\Big) + P_n\Big(H;z,\Big(y,-\sqrt{1-\|y\|^2}\Big)\Big)\Big]$$

with $r=1$. □

In particular, as a consequence of the explicit formula of the zonal harmonic (4.1.6), we have the following corollary.

Corollary 4.2.9 *Let $W_0(x) := 1/\sqrt{1-\|x\|^2}$, $x \in B^d$. The reproducing kernel $\mathbf{P}_n(W_0;\cdot,\cdot)$ of $\mathscr{V}_n^d(W_0)$ satisfies*

$$\mathbf{P}_n(W_0;x,y) = \frac{n+\frac{d-1}{2}}{d-1}\left[C_n^{d/2}\left(\langle x,y\rangle + \sqrt{1-\|x\|^2}\sqrt{1-\|y\|^2}\right)\right.$$
$$\left. + C_n^{d/2}\left(\langle x,y\rangle - \sqrt{1-\|x\|^2}\sqrt{1-\|y\|^2}\right)\right].$$

Moreover, by Theorem 4.2.4 and (4.2.2), the spherical harmonics in $\mathscr{H}_n^{d+1}$ that are even in x_{d+1} form an orthogonal basis of $\mathscr{V}_n^d(W_0)$. In particular, an orthonormal basis of $\mathscr{V}_n^d(W_0)$ can be deduced from the explicit basis given in Theorem 4.1.4. The weight function W_0, however, is a special case of the classical weight functions on the unit ball to be discussed in the next chapter.

4.3 Orthogonal Structures on B^d and on S^{d+m-1}

We consider a relation between the orthogonal polynomials on the ball B^d and those on the sphere S^{d+m-1}, where $m \ge 1$.

A function f defined on $\mathbb{R}^m$ is called positively homogeneous of order σ if $f(tx) = t^\sigma f(x)$ for $t > 0$.

Definition 4.3.1 The weight function H defined on $\mathbb{R}^{d+m}$ is called *admissible* if

$$H(x) = H_1(\mathbf{x}_1)H_2(\mathbf{x}_2), \qquad \mathbf{x} = (\mathbf{x}_1,\mathbf{x}_2) \in \mathbb{R}^{d+m}, \quad \mathbf{x}_1 \in \mathbb{R}^d, \quad \mathbf{x}_2 \in \mathbb{R}^m,$$

where we assume that H_1 is a centrally symmetric function, and that H_2 is positively homogeneous of order 2τ and even in each of its variables.

Examples of admissible weight functions include $H(x) = c\prod_{i=1}^{d+m}|x_i|^{\kappa_i}$ for $\kappa_i \ge 0$, in which H_1 and H_2 are of the same form but with fewer variables, and $H(x) = cH_1(\mathbf{x}_1)\prod|x_i^2 - x_j^2|^{\beta_{i,j}}$, where $d+1 \le i < j \le d+m$ and $2\tau = \sum_{i,j}\beta_{ij}$.

Associated with H, we define a weight function W_H^m on B^d by

$$W_H^m(x) = H_1(x)(1-\|x\|^2)^{\tau+(m-2)/2}, \qquad x \in B^d. \tag{4.3.1}$$

For convenience, we assume that H has unit integral on S^{d+m-1} and W_H^m has unit integral on B^d. We show that the orthogonal polynomials with respect to H and W_H^m are related, using the following elementary lemma.

Lemma 4.3.2 *Let d and m be positive integers and $m \ge 2$. Then*

$$\int_{S^{d+m-1}} f(y)\,\mathrm{d}\omega = \int_{B^d}(1-\|x\|^2)^{(m-2)/2}\int_{S^{m-1}} f(x, \sqrt{1-\|x\|^2}\eta)\,\mathrm{d}\omega(\eta)\,\mathrm{d}x.$$

Proof Making the change of variables $y \mapsto (x, \sqrt{1-\|x\|^2}\eta)$, $x \in B^d$ and $\eta \in S^{m-1}$, in the integral over S^{d+m-1} shows that $d\omega_{d+m}(y) = (1-\|x\|^2)^{(m-2)/2}\,dx\,d\omega_m(\eta)$. □

When $m = 1$, the lemma becomes, in a limiting sense, Lemma 4.2.3.

In the following we denote by $\{P_\alpha(W_H^m)\}$ a sequence of orthonormal polynomials with respect to W_H^m on B^d. Since H_1, and thus W_H^m, is centrally symmetric, it follows from Theorem 3.3.11 that $P_\alpha(W_H^m)$ is a sum of monomials of even degree if $|\alpha|$ is even and a sum of monomials of odd degree if $|\alpha|$ is odd. These polynomials are related to homogeneous orthogonal polynomials with respect to H on S^{d+m-1}.

Theorem 4.3.3 *Let H be an admissible weight function. Then the functions*

$$Y_\alpha(y) = \|y\|^{|\alpha|} P_\alpha(W_H^m; \mathbf{x}_1), \qquad y = r(\mathbf{x}_1, \mathbf{x}_2) \in \mathbb{R}^{d+m}, \quad \mathbf{x}_1 \in B^d,$$

are homogeneous polynomials in y and are orthonormal with respect to $H(y)d\omega$ on S^{d+m-1}.

Proof We first prove that Y_α is orthogonal to polynomials of lower degree. It is sufficient to prove that Y_α is orthogonal to $g_\gamma(y) = y^\gamma$ for $\gamma \in \mathbb{N}_0^{d+m}$ and $|\gamma| \le |\alpha| - 1$. From Lemma 4.3.2,

$$\begin{aligned}&\int_{S^{d+m-1}} Y_\alpha(y) g_\gamma(y) H(y)\, d\omega \\ &\quad = \int_{B^d} P_\alpha(W_H^m; \mathbf{x}_1) \left[\int_{S^{m-1}} g_\gamma\left(\mathbf{x}_1, \sqrt{1-|\mathbf{x}_1|^2}\,\eta \right) H_2(\eta)\, d\omega(\eta) \right] W_H^m(\mathbf{x}_1)\, d\mathbf{x}_1.\end{aligned}$$

If g_γ is odd with respect to at least one of its variables $y_{d+1}, \ldots, y_{d+m+1}$ then we can conclude, using the fact that H_2 is even in each of its variables, that the integral inside the square bracket is zero. Hence, Y_α is orthogonal to g_γ in this case. If g_γ is even in every variable of $y_{d+1}, \ldots, y_{d+m}$ then the function inside the square bracket will be a polynomial in $\mathbf{x}_1$ of degree at most $|\alpha| - 1$, from which we conclude that Y_α is orthogonal to g_γ by the orthogonality of P_α with respect to W_H^m on B^d. Moreover,

$$\int_{S^{d+m-1}} Y_\alpha(y) Y_\beta(y) H(y)\, d\omega = \int_{B^d} P_\alpha(W_H^m; \mathbf{x}_1) P_\beta(W_H^m; \mathbf{x}_1) W_H^m(\mathbf{x}_1)\, d\mathbf{x}_1,$$

which shows that $\{Y_\alpha\}$ is an orthonormal set. The fact that Y_α is homogeneous of degree n in y follows as in the proof of Theorem 4.2.4. □

If in the limiting case $m = 1$, the integral over S^{m-1} is taken as point evaluations at 1 and -1 with weights $\frac{1}{2}$ for each point, then we recover a special case of Theorem 4.2.4. In this case $W_H^m(x) = H(x, \sqrt{1-\|x\|^2})/\sqrt{1-\|x\|^2}$.

In the following we show that an orthogonal basis of $\mathscr{H}_k^{m+d}(H)$ can be derived from orthogonal polynomials on B^d and on S^m. Let H_2 be defined as in Definition 4.3.1. Then, by Theorem 4.2.4, the space $\mathscr{H}_k^m(H_2)$ has an orthonormal basis $\{S_\beta^k\}$ such that each S_β^k is homogeneous, and S_β^k is even in each of its variables if k is even and odd in each of its variables if k is odd.

For $x \in \mathbb{R}^d$, we write $|x| := x_1 + \cdots + x_d$ in the following.

Theorem 4.3.4 *Let $\{S_\beta^k\}$ be an orthonormal basis of $\mathscr{H}_k^m(H_2)$ as above, and let $\{P_\alpha^{n-k}(W_H^{m+2k})\}$ be an orthonormal basis for $\mathscr{V}_{n-k}^d(W_H^{m+2k})$ with respect to the weight function W_H^{m+2k} on B^d, where $0 \le k \le n$. Then the polynomials*

$$Y_{\alpha,\beta,k}^n(\mathbf{y}) = r^n A_{m,k} P_\alpha^{n-k}(W_H^{m+2k}; \mathbf{y}_1') S_\beta^k(\mathbf{y}_2'), \qquad \mathbf{y} = r(\mathbf{y}_1', \mathbf{y}_2'),\ r = \|\mathbf{y}\|,$$

where $\mathbf{y}_1' \in B^d$, $\mathbf{y}_2' \in B^m$ and $[A_{m,k}]^{-2} = \int_{B^d} W_H^m(\mathbf{x})(1-\|\mathbf{x}\|^2)^k d\mathbf{x}$, are homogeneous of degree n in $\mathbf{y}$ and form an orthonormal basis for $\mathscr{H}_n^{d+m}(H)$.

Proof Since the last property of S_β^k in Theorem 4.2.4 implies that it has the same parity as n, and $P_\alpha^{n-k}(W_H^{m+2k})$ has the same parity as $n-k$, the fact that $Y_{\alpha,\beta,k}^n$ is homogeneous of degree n in $\mathbf{y}$ follows as in the proof of Theorem 4.3.3.

We prove that $Y_{\alpha,\beta,k}^n$ is orthogonal to all polynomials of lower degree. Again, it is sufficient to show that $Y_{\alpha,\beta,k}^n$ is orthogonal to $g_\gamma(\mathbf{y}) = \mathbf{y}^\gamma$, $|\gamma|_1 \le n-1$. Using the notation $\mathbf{y} = r(\mathbf{y}_1', \mathbf{y}_2')$, we write g_γ as

$$g_\gamma(\mathbf{y}) = r^{|\gamma|_1} \mathbf{y}_1'^{\gamma_1} \mathbf{y}_2'^{\gamma_2}, \qquad |\gamma_1|_1 + |\gamma_2|_1 = |\gamma|_1 \le n-1.$$

Using the fact $\|\mathbf{y}_2'\|^2 = 1 - \|\mathbf{y}_1'\|^2$ and the integral formula in Lemma 4.3.2, we conclude that

$$\begin{aligned}
&\int_{S^{d+m-1}} Y_{\alpha,\beta,k}^n(\mathbf{y}) g_\gamma(\mathbf{y}) H(\mathbf{y})\, d\omega_{d+m} \\
&\quad = [A_{m,k}]^{-1} \int_{B^d} P_\alpha^{n-k}(W_H^{m+2k}; \mathbf{y}_1') \mathbf{y}_1'^{\gamma_1} (1 - \|\mathbf{y}_1'\|^2)^{(|\gamma_2|_1 - k)/2} \\
&\qquad \times W_H^{m+2k}(\mathbf{y}_1')\, d\mathbf{y}_1' \left(\int_{S^{m-1}} S_\beta^k(\eta) \eta^{\gamma_2} H_2(\eta)\, d\omega_m(\eta) \right),
\end{aligned}$$

where we have used the fact that $W_H^{m+2k}(\mathbf{x}) = W_H^m(\mathbf{x})(1 - \|\mathbf{x}\|^2)^k$. We can show that this integral is zero by considering the following cases. If $|\gamma_2|_1 < k$ then the integral in the square brackets is zero because of the orthogonality of S_β^k. If $|\gamma_2|_1 \ge k$ and $|\gamma_2|_1 - k$ is an odd integer then $|\gamma_2|_1$ and k have different parity; hence, since S_β^k is homogeneous, a change of variable $\eta \mapsto -\eta$ leads to the conclusion that the integral in the square brackets is zero. If $|\gamma_2|_1 \ge k$ and $|\gamma_2|_1 - k$ is an even integer then $\mathbf{y}_1'^{\gamma_1}(1 - \|\mathbf{y}_1'\|^2)^{(|\gamma_2|_1 - k)/2}$ is a polynomial of degree $|\gamma_1|_1 + |\gamma_2|_1 - k \le n - 1 - k$; hence, the integral is zero by the orthogonality of $P_\alpha^{n-k}(W_H^{m+2k})$. The same consideration also shows that the polynomial $Y_{\alpha,\beta,k}^n$ has norm 1 in $L^2(H, S^{d+m})$, since $P_\alpha^{n-k}(W_H^{m+2k})$ and S_β^n are normalized.

Finally, we show that $\{Y^n_{\alpha,\beta,k}\}$ forms a basis of $\mathscr{H}^{m+d}_n(H)$. By orthogonality, the elements of $\{Y^n_{\alpha,\beta,k}\}$ are linearly independent; since $\dim \mathscr{H}^m_k(H) = r^{m-1}_k + r^{m-1}_{k-1}$, their cardinality is equal to

$$\sum_{k=0}^{n} r^d_{n-k}(r^{m-1}_k + r^{m-1}_{k-1}) = \sum_{k=0}^{n} r^d_{n-k} r^{m-1}_k + \sum_{k=0}^{n-1} r^d_{n-1-k} r^{m-1}_k,$$

which is the same as the dimension of $\mathscr{H}^{m+d-1}_n(H)$, as can be verified by the combinatorial identity

$$\sum_{k=0}^{n} r^d_{n-k} r^{m-1}_k = \sum_{k=0}^{n} \binom{n-k+d-1}{n-k}\binom{k+m-2}{k} = \binom{n+d+m-2}{n} = r^{d+m-1}_n.$$

This completes the proof. □

There is also a relation between the reproducing kernels of $\mathscr{H}^{d+m+1}_n(H)$ and $\mathscr{V}^d_n(W^m_H)$, which we now establish.

Theorem 4.3.5 *Let H be an admissible weight function defined on $\mathbb{R}^{d+m+1}$ and W^m_H be the associated weight function on B^d. Then*

$$\mathbf{P}_n(W^m_H;\mathbf{x}_1,\mathbf{y}_1) = \int_{S^m} P_n\left(H;x,\left(\mathbf{y}_1,\sqrt{1-\|\mathbf{y}_1\|^2}\eta\right)\right) H_2(\eta)\,\mathrm{d}\omega_m(\eta)$$

where $x = (\mathbf{x}_1,\mathbf{x}_2) \in S^{d+m}$ with $\mathbf{x}_1 \in B^d$, $y = (\mathbf{y}_1,\mathbf{y}_2) \in S^{d+m}$ with $\mathbf{y}_1 \in B^d$ and $\mathbf{y}_2 = \|y\|\eta \in B^{m+1}$ and $\eta \in S^m$.

Proof Since $P_n(H)$ is the reproducing kernel of $\mathscr{H}^{d+m+1}_n(H)$, we can write it in terms of the orthonormal basis $\{Y^n_{\alpha,\beta,k}\}$ in Theorem 4.3.4:

$$P_n(H;x,y) = \sum_k \sum_\alpha \sum_\beta Y^n_{\alpha,\beta,k}(x) Y^n_{\alpha,\beta,k}(y).$$

Integrating $P_n(H;x,y)$ with respect to $H_2(\mathbf{y}_2)$ over S^m, we write $\|\mathbf{y}_2\|^2 = 1 - \|\mathbf{y}_1\|^2$ and use the fact that S_β is homogeneous and orthogonal with respect to $H_2(\mathbf{y}_2)\,\mathrm{d}\omega_m$ to conclude that the integral of $Y^n_{\alpha,\beta,k}$ on S^m results in zero for all $k \neq 0$, while for $k = 0$ we have that $\beta = 0$ and $Y^n_{\alpha,0,0}(y)$ are the polynomials $P^n_\alpha(W^m_H;\mathbf{y}_1)$. Hence, we conclude that

$$\int_{S^m} P_n\left(H;x,\left(\mathbf{y}_1,\sqrt{1-\|y\|^2}\eta\right)\right) H_2(\eta)\,\mathrm{d}\omega_m(\eta) = \sum_\alpha P^n_\alpha(W^m_H;\mathbf{x}_1) P^n_\alpha(W^m_H;\mathbf{y}_1) = \mathbf{P}_n(W^m_H;\mathbf{x}_1,\mathbf{y}_1),$$

by the definition of the reproducing kernel. □

In the case that H, H_1 and H_2 are all constants, $W_H^m(x)$ becomes the case $\mu = \frac{m-1}{2}$ of the classical weight function W_μ on the ball B^d,

$$W_\mu(x) := (1 - \|x\|^2)^{\mu - 1/2}, \qquad x \in B^d. \tag{4.3.2}$$

As a consequence of the general result in Theorem 4.3.3, we see that, for $m = 1, 2, \ldots$, the orthogonal polynomials in $\mathscr{V}_n^d(W_{(m-1)/2})$ can be identified with the spherical harmonics in $\mathscr{H}_n^{d+m}$. In particular, it is possible to deduce an orthogonal basis of $V_n^d(W_{(m-1)/2})$ from the basis of $\mathscr{H}_n^{d+m}$, which can be found from Theorem 4.1.4. Since the orthogonal polynomials for the classical weight function W_μ will be studied in the next chapter, we shall not derive this basis here. One consequence of this connection is the following corollary of Theorem 4.3.5 and the explicit formula of the zonal harmonic (4.1.6):

Corollary 4.3.6 *For $m = 1, 2, 3, \ldots$, the reproducing kernel of $\mathscr{V}_n^d(W_\mu)$ with $\mu = \frac{m}{2}$ satisfies*

$$\begin{aligned}\mathbf{P}(W_\mu; x, y) = & \frac{n+\lambda}{\lambda} c_\mu \int_{-1}^{1} C_n^\lambda\left(\langle x, y\rangle + \sqrt{1-\|x\|^2}\sqrt{1-\|y\|^2}\,t\right) \\ & \times (1-t^2)^{\mu-1}\,\mathrm{d}t,\end{aligned}$$

where $\lambda = \mu + \frac{d-2}{2}$ and $c_\mu = \Gamma(\mu + \frac{1}{2})/(\sqrt{\pi}\Gamma(\mu))$.

This formula holds for all $\mu > 0$, as will be proved in Chapter 8. When $\mu = 0$, the reproducing kernel for W_0 is given in Corollary 4.2.9, which can be derived from the above corollary by taking the limit $\mu \to 0$.

4.4 Orthogonal Structures on the Simplex

In this section we consider the orthogonal polynomials on the simplex

$$T^d := \{x \in \mathbb{R}^d : x_1 \geq 0, \ldots, x_d \geq 0, 1 - |x|_1 \geq 0\},$$

where $|x|_1 = x_1 + \cdots + x_d$. The structure of the orthogonal polynomials can be derived from that on the unit sphere or from that on the unit ball. The connection is based on the following lemma.

Lemma 4.4.1 *Let $f(x_1^2, \ldots, x_d^2)$ be an integrable function defined on B^d. Then*

$$\int_{B^d} f(y_1^2, \ldots, y_d^2)\,\mathrm{d}y = \int_{T^d} f(x_1, \ldots, x_d) \frac{\mathrm{d}x}{\sqrt{x_1 \cdots x_d}}.$$

Proof Since $f(x_1^2, \ldots, x_d^2)$ is an even function in each of its variables, the left-hand side of the stated formula can be written as 2^d times the integral over $B_+^d = \{x \in B^d : x_1 \geq 0, \ldots, x_d \geq 0\}$. It can be seen to be equal to the right-hand side upon making the change of variables $y_1 = x_1^2, \ldots, y_d = x_d^2$ in the integral over B_+^d, for which the Jacobian is $2^{-d}(y_1, \ldots, y_d)^{-1/2}$. □

Definition 4.4.2 Let $W^B(x) = W(x_1^2, \ldots, x_d^2)$ be a weight function defined on B^d. Associated with W^B define a weight function on T^d by

$$W^T(y) = W(y_1, \ldots, y_d)/\sqrt{y_1 \cdots y_d}, \qquad y = (y_1, \ldots, y_d) \in T^d.$$

Evidently, the weight function W^B is invariant under the group $\mathbb{Z}_2^d$, that is, invariant under sign changes. If a polynomial is invariant under $\mathbb{Z}_2^d$ then it is necessarily of even degree since it is even in each of its variables.

Definition 4.4.3 Define $\mathscr{V}_{2n}^d(W^B, \mathbb{Z}_2^d) = \{P \in \mathscr{V}_{2n}^d(W^B) : P(x) = P(xw),\ w \in \mathbb{Z}_2^d\}$. Let P_α^{2n} be the elements of $\mathscr{V}_{2n}^d(W^B, \mathbb{Z}_2^d)$; define polynomials R_α^n by

$$P_\alpha^{2n}(x) = R_\alpha^n(x_1^2, \ldots, x_d^2). \tag{4.4.1}$$

Theorem 4.4.4 *Let W^B and W^T be defined as above. The relation* (4.4.1) *defines a one-to-one correspondence between an orthonormal basis of $\mathscr{V}_n^d(W^T)$ and an orthonormal basis of $\mathscr{V}_{2n}(W^B, \mathbb{Z}_2^d)$.*

Proof Assume that $\{R_\alpha\}_{|\alpha|=n}$ is a set of orthonormal polynomials in $\mathscr{V}_n^d(W^T)$. If $\beta \in \mathbb{N}_0^d$ has one odd component then the integral of $P_\alpha(x)x^\beta$ with respect to W^B over B^d is zero. If all components of β are even and $|\beta| < 2n$ then it can be written as $\beta = 2\tau$ with $\tau \in \mathbb{N}_0^d$ and $|\tau| \le n-1$. Using Lemma 4.4.1, we obtain

$$\int_{B^d} P_\alpha(x) x^\beta W^B(x)\,dx = \int_{T^d} R_\alpha(y) y^\tau W^T(y)\,dy = 0,$$

by the orthogonality of R_α. This relation also shows that R_α is orthogonal to all polynomials of degree at most $n-1$, if P_α is an orthogonal polynomial in $\mathscr{V}_n^d(W^B, \mathbb{Z}_2^d)$. Moreover, Lemma 4.4.1 also shows that the $L^2(W^B; B^d)$ norm of P_α is the same as the $L^2(W^T; T^d)$ norm of R_α. □

The connection also extends to the reproducing kernels. Let us denote by $\mathbf{P}_n(W^\Omega; \cdot, \cdot)$ the reproducing kernel of $\mathscr{V}_n^d(W^\Omega, \Omega^d)$ for $\Omega = B$ or T.

Theorem 4.4.5 *For $n = 0, 1, 2, \ldots$ and $x, y \in T^d$,*

$$\mathbf{P}_n(W^T; x, y) = 2^{-d} \sum_{\varepsilon \in \mathbb{Z}_2^d} \mathbf{P}_{2n}(W^B; \sqrt{x}, \varepsilon\sqrt{y}), \tag{4.4.2}$$

where $\sqrt{x} := (\sqrt{x_1}, \ldots, \sqrt{x_d})$ and $\varepsilon u := (\varepsilon_1 u_1, \ldots, \varepsilon_d u_d)$.

Proof Directly from the definition of the reproducing kernel, it is not difficult to see that

$$\mathbf{P}(x,y) := 2^{-d} \sum_{\varepsilon \in \mathbb{Z}_2^d} \mathbf{P}_{2n}(W^B; x, \varepsilon y), \qquad x, y \in B^d,$$

is the the reproducing kernel of $\mathcal{V}_{2n}^d(W^B, \mathbb{Z}_2^d)$ and, as such, can be written as

$$\mathbf{P}(x,y) = \sum_{|\alpha|=2n} P_\alpha^{2n}(W^B; x) P_\alpha^{2n}(W^B; y)$$

in terms of an orthonormal basis of $\mathcal{V}_{2n}^d(W^B, \mathbb{Z}_2^d)$. The polynomial $P_\alpha^{2n}(W^B; x)$ is necessarily even in each of its variables, so that (4.4.2) follows from Theorem 4.4.4 and the definition of the reproducing kernel with respect to W^T. □

As a consequence of this theorem and Theorem 4.2.4, we also obtain a relation between the orthogonal polynomials on T^d and the homogeneous orthogonal polynomials on S^d. Let $H(y) = W(y_1^2, \ldots, y_{d+1}^2)$ be a weight function defined on $\mathbb{R}^{d+1}$ and assume its restriction to S^d is not zero. Then H is an S-symmetric weight function; see Definition 4.2.1. Associated with H define a weight function W^T on the simplex T^d by

$$W_H^T(x) = 2W(x_1, \ldots, x_d, 1 - |x|_1)\sqrt{x_1 \cdots x_d(1 - |x|_1)}, \qquad x \in T^d.$$

Let R_α be an element of $\mathcal{V}_n^d(W_H^T)$. Using the coordinates $y = r(x, x_{d+1})$ for $y \in \mathbb{R}^{d+1}$, define

$$S_\alpha^{2n}(y) = r^{2n} R_\alpha^n(x_1^2, \ldots, x_d^2). \tag{4.4.3}$$

By Theorems 4.2.4 and 4.4.4, we see that S_α^{2n} is a homogeneous orthogonal polynomial in $\mathcal{H}_{2n}^d(H)$ and is invariant under $\mathbb{Z}_2^{d+1}$. Denote the subspace of $\mathbb{Z}_2^{d+1}$ invariant elements of $\mathcal{H}_{2n}^d(H)$ by $\mathcal{H}_{2n}^d(H, \mathbb{Z}_2^{d+1})$. Then S_α^{2n} is an element of $\mathcal{H}_{2n}^d(H, \mathbb{Z}_2^{d+1})$, and we conclude:

Theorem 4.4.6 *Let H and W_H^T be defined as above. The relation* (4.4.3) *defines a one-to-one correspondence between an orthonormal basis of $\mathcal{V}_{2n}(W^B, \mathbb{Z}_2^d)$ and an orthonormal basis of $\mathcal{V}_n^d(W^T)$.*

As an immediate consequence of the above two theorems, we have

Corollary 4.4.7 *Let H and W^B be defined as above. Then*

$$\dim \mathcal{V}_{2n}^d(W^B, \mathbb{Z}_2^d) = \binom{n+d}{n} \quad \text{and} \quad \dim \mathcal{H}_{2n}^d(H, \mathbb{Z}_2^{d+1}) = \binom{n+d-1}{n}.$$

Working on the simplex T^d, it is often convenient to use homogeneous coordinates $x = (x_1, \ldots, x_{d+1})$ with $|x|_1 = 1$; that is, we identify T^d with

$$T_{\text{hom}}^d := \{x \in \mathbb{R}^{d+1} : x_1 \geq 0, \ldots, x_{d+1} \geq 0, |x|_1 = 1\}$$

in $\mathbb{R}^{d+1}$. This allows us to derive a homogeneous basis of orthogonal polynomials. Indeed, since H is even in each of its variables, the space $\mathscr{H}_{2n}^d(H,\mathbb{Z}_2^{d+1})$ has a basis that is even in each of its variables and whose elements can be written in the form $S_\alpha^n(x_1^2,\ldots,x_{d+1}^2)$.

Proposition 4.4.8 *Let H be defined as above and $\{S_\alpha^n(x_1^2,\ldots,x_{d+1}^2) : |\alpha| = n,\ \alpha \in \mathbb{N}_0^d\}$ be an orthonormal homogeneous basis of $\mathscr{H}_{2n}^d(H,\mathbb{Z}_2^{d+1})$. Then $\{S_\alpha^n(x_1,\ldots,x_{d+1}) : |\alpha| = n,\ \alpha \in \mathbb{N}_0^d\}$ forms an orthonormal homogeneous basis of $\mathscr{V}_n^d(W_H^T)$.*

Proof It follows from Lemmas 4.2.3 and 4.4.1 that

$$\int_{S^d} f(y_1^2,\ldots,y_{d+1}^2)\,\mathrm{d}\omega = 2\int_{|x|_1=1} f(x_1,\ldots,x_{d+1})\frac{\mathrm{d}x}{\sqrt{x_1\cdots x_{d+1}}}, \tag{4.4.4}$$

where we use the homogeneous coordinates in the integral over T^d and write the domain of integration as $\{|x|_1 = 1\}$ to make this clear. This formula can also be obtained by writing the integral over S^d as 2^{d+1} times the integral over S_+^d and then making the change of variables $x_i \mapsto y_i^2$, $1 \le i \le d+1$. The change of variables gives the stated result. □

For reproducing kernels, this leads to the following relation:

Theorem 4.4.9 *For $n = 0,1,2,\ldots$ and $x,y \in T_{\mathrm{hom}}^d$,*

$$\mathbf{P}_n(W^T;x,y) = 2^{-d-1}\sum_{\varepsilon\in\mathbb{Z}_2^{d+1}} P_{2n}(H;\sqrt{x},\varepsilon\sqrt{y}), \tag{4.4.5}$$

where $\sqrt{x} := (\sqrt{x_1},\ldots,\sqrt{x_{d+1}})$ and $\varepsilon u := (\varepsilon_1 u_1,\ldots,\varepsilon_d u_{d+1})$.

In particular, for $H(x) = 1$ the weight function W_H^T becomes

$$W_0(x) = \frac{1}{\sqrt{x_1\ldots x_d(1-|x|_1)}}, \qquad x \in T^d.$$

Moreover, by Definition 4.4.2 this weight is also closely related to $W_0^B(x) = 1/\sqrt{1-\|x\|^2}$ on the unit ball. In particular, it follows that the orthogonal polynomials with respect to W_0 on T^d are equivalent to those spherical harmonics on S^d that are even in every variable and, in turn, are related to the orthogonal polynomials with respect to W_0^B that are even in every variable. Furthermore, from the explicit formula for the zonal harmonic (4.1.6) we can deduce the following corollary.

Corollary 4.4.10 *For W_0 on T^d, the reproducing kernel $\mathbf{P}_n(W_0;\cdot,\cdot)$ of $\mathscr{V}_n^d(W_0)$ satisfies, with $\lambda = (d-1)/2$,*

$$\mathbf{P}_n(W_0;x,y) = \frac{(2n+\lambda)(\lambda)_n}{\lambda(\frac{1}{2})_n} \frac{1}{2^{d+1}} \sum_{\varepsilon \in \mathbb{Z}_2^{d+1}} P_n^{\lambda-1/2,-1/2}\left(2z(x,y,\varepsilon)^2 - 1\right)$$

for $x,y \in T^d$, *where* $z(x,y,\varepsilon) = \sqrt{x_1 y_1}\varepsilon_1 + \cdots + \sqrt{x_{d+1} y_{d+1}}\varepsilon_{d+1}$ *with* $x_{d+1} = 1 - |x|_1$ *and* $y_{d+1} = 1 - |y|$.

Proof Taking $W^T = W_0$ in (4.4.5) and using (4.1.6), we conclude that

$$\mathbf{P}_n(W_0;x,y) = \frac{2n+\lambda}{\lambda} \frac{1}{2^{d+1}} \sum_{\varepsilon \in \mathbb{Z}_2^{d+1}} C_{2n}^{\lambda}\left(z(x,y,\varepsilon)\right),$$

which gives, by the relation $C_{2n}^{\lambda}(t) = \frac{(\lambda)_n}{(\frac{1}{2})_n} P_n^{(\lambda-1/2,-1/2)}(2t^2-1)$, the stated formula. □

4.5 Van der Corput–Schaake Inequality

Just as there are Bernstein inequalities relating polynomials of one variable and their derivatives, there is an inequality relating homogeneous polynomials and their gradients. The inequality will be needed in Chapter 6. For $n = 1, 2, \ldots$ and $p \in \mathscr{P}_n^d$, we define a sup-norm on $\mathscr{P}_n^d$ by

$$\|p\|_S := \sup\{|p(x)| : \|x\| = 1\},$$

and a similar norm related to the gradient:

$$\|p\|_\partial = \frac{1}{n} \sup\{|\langle u, \nabla p(x)\rangle| : \|x\| = 1 = \|u\|\}.$$

Because $\sum_{i=1}^d x_i \partial p(x)/\partial x_i = np(x)$ it is clear that $\|p\|_S \le \|p\|_\partial$. The remarkable fact, proven by van der Corput and Schaake [1935], is that equality always holds for real polynomials. Their method was to first prove it in $\mathbb{R}^2$ and then extend it to any $\mathbb{R}^d$. The $\mathbb{R}^2$ case depends on polar coordinates and simple properties of trigonometric polynomials. *In this section all polynomials are real valued.* The maximum number of sign changes of a homogeneous polynomial on S^1 provides a key device.

Lemma 4.5.1 *Suppose that* $p \in \mathscr{P}_n^2$; *then* p *can be written as*

$$c \prod_{j=1}^{n-2m} \langle x, u^j \rangle \prod_{i=1}^{m} \left(\|x\|^2 + \langle x, v^i\rangle^2\right),$$

for some m, $c \in \mathbb{R}$ *and* $u^1, \ldots, u^{n-2m} \in S^1$, $v^1, \ldots, v^m \in \mathbb{R}^2$. *The trigonometric polynomial* $\widetilde{p}(\theta) = p(\cos\theta, \sin\theta)$ *has no more than* $2n$ *sign changes in any interval* $\theta_0 < \theta < \theta_0 + 2\pi$ *when* $\widetilde{p}(\theta_0) \neq 0$.

Proof By rotation of coordinates if necessary, we can assume that $p(1,0) \neq 0$; thus $\widetilde{p}(\theta) = (\sin\theta)^n g(\cot\theta)$ where g is a polynomial of degree n. Hence $g(t) = c_0 \prod_{j=1}^{n-2m}(t - r_j) \prod_{i=1}^{m}(t^2 + b_i t + c_i)$ with real numbers c_0, r_j, b_i, c_i and $b_i^2 - 4c_i < 0$ for each i (and some m). This factorization implies the claimed expression for p and also exhibits the possible sign changes of $\widetilde{p}$. □

Theorem 4.5.2 *Let* $p(x_1,x_2) \in \mathscr{P}_n^2$ *and* $M = \sup|p(x)|$ *for* $x_1^2 + x_2^2 = 1$*; then* $|\langle u, \nabla p(x)\rangle| \leq Mn\|u\|(x_1^2 + x_2^2)^{(n-1)/2}$ *for any* $u \in R^2$.

Proof Since ∇p is homogeneous of degree $n-1$, it suffices to prove the claim for $\|x\| = 1$. Accordingly, let $x_0 = (\cos\theta_0, \sin\theta_0)$ for an arbitrary $\theta_0 \in (-\pi, \pi]$ and make the change of variables

$$\begin{aligned} y &= (r\cos(\theta - \theta_0), r\sin(\theta - \theta_0)) \\ &= (x_1\cos\theta_0 + x_2\sin\theta_0, -x_1\sin\theta_0 + x_2\cos\theta_0), \end{aligned}$$

where $x = (r\cos\theta, r\sin\theta)$. Clearly $(\frac{\partial p}{\partial y_1})^2 + (\frac{\partial p}{\partial y_2})^2 = (\frac{\partial p}{\partial x_1})^2 + (\frac{\partial p}{\partial x_2})^2 = \|\nabla p\|^2$, all terms being evaluated at the same point. Express p in terms of y_1, y_2 as $\sum_{j=0}^{n} a_j y_1^{n-j} y_2^j$; then $(\partial p/\partial y_1)^2 + (\partial p/\partial y_2)^2$ all terms being evaluated at $y = (1,0)$, corresponding to $x = x_0$, equals $n^2 a_0^2 + a_1^2$. If $a_1 = 0$ then $\|\nabla p(x_0)\|^2 = n^2 a_0^2 = n^2 p(x_0)^2 \leq n^2 M^2$, as required. Assume now that $a_1 \neq 0$ and, by way of contradiction, assume that $n^2 a_0^2 + a_1^2 > n^2 M^2$. There is an angle $\phi \neq 0, \pi$ such that

$$a_0 \cos n\phi + \frac{a_1}{n}\sin n\phi = \sqrt{a_0^2 + \frac{a_1^2}{n^2}}.$$

Let $f(\theta) = a_0\cos n\theta + (a_1/n)\sin n\theta - \sum_{j=0}^{n} a_j (\cos\theta)^{n-j}(\sin\theta)^j$. Denote by $[x]$ the integer part of $x \in \mathbb{R}$. Because

$$\cos n\theta = \cos^n\theta + \sum_{1}^{[n/2]} \binom{n}{2j} (-1)^j (\cos\theta)^{n-2j}(\sin\theta)^{2j},$$

$$\sin n\theta = n\cos^{n-1}\theta\sin\theta + \sum_{1}^{[(n-1)/2]} \binom{n}{2j+1} (-1)^j (\cos\theta)^{n-1-2j}(\sin\theta)^{2j+1},$$

it follows that $f(\theta) = (\sin\theta)^2 g(\cos\theta, \sin\theta)$, where g is a homogeneous polynomial of degree $n-2$. Consider the values

$$f\left(\phi + \frac{j\pi}{n}\right) = (-1)^j\sqrt{a_0^2 + \frac{a_1^2}{n^2}} - p(y^j),$$

where $y^j = (\cos\phi_j, \sin\phi_j)$ and $\phi_j = \phi + j\pi/n - \theta_0, 0 \leq j \leq 2n-1$. Since $|p(y^j)| \leq M$, we have $(-1)^j f(\phi + j\pi/n) \geq \sqrt{a_0^2 + a_1^2/n^2} - M > 0$. This implies that $g(\cos\theta, \sin\theta)$ changes sign $2n$ times in the interval $\phi < \theta < \phi + 2\pi$. This contradicts the fact that $g \in \mathscr{P}^2_{n-2}$ and can change its sign at most $2(n-2)$ times by the lemma. Thus $\sqrt{a_0^2 + a_1^2/n^2} \le M$. □

Van der Corput and Schaake proved a stronger result: if $\sqrt{a_0^2 + a_1^2/n^2} = M$ for some particular x_0 then $|a_0| = |p(x_0)| = M$ (so that $\|\nabla p\|$ is maximized at the same point as $|p|$) or $\|\nabla p(x)\|$ is constant on the circle, and $p(\cos\theta, \sin\theta) = M \cos n(\theta - \theta_0)$ for some θ_0.

Theorem 4.5.3 *Let $p \in \mathscr{P}^d_n$; then $|\langle u, \nabla p(x)\rangle| \le \|p\|_S \|u\| \|x\|^{n-1}$ for any $u \in \mathbb{R}^d$.*

Proof For any $x_0 \in \mathbb{R}^d$ with $\|x_0\| = 1$ there is a plane E through the origin containing both x_0 and $\nabla p(x_0)$ (it is unique, in general). By rotation of coordinates transform x_0 to $(1, 0, \dots)$ and E to $\{(x_1, x_2, 0, \dots)\}$. In this coordinate system $\nabla p(x_0) = (v_1, v_2, 0, \dots)$ and, by Theorem 4.5.2, $(v_1^2 + v_2^2)^{1/2} \le n \sup\{|p(x)| : \|x\| = 1 \text{ and } x \in E\} \le n\|p\|_S$. □

Corollary 4.5.4 *For $p \in \mathscr{P}^d_n$ and $n \ge 1$, $\|p\|_S = \|p\|_\partial$.*

There is another way of defining $||p||_\partial$.

Proposition 4.5.5 *For $p \in \mathscr{P}^d_n$,*

$$\|p\|_\partial = \frac{1}{n!} \sup \left\{ \prod_{j=1}^n \left| \langle y^{(j)}, \nabla \rangle p(x) \right| : x \in S^{d-1}, y^{(j)} \in S^{d-1} \text{ for all } j \right\}.$$

Proof The claim is true for $n = 1$. Assume that it is true for some n, and let $p \in \mathscr{P}^d_{n+1}$; for any $y^{(n+1)} \in S^{d-1}$ the function $g(x) = \langle y^{(n+1)}, \nabla \rangle p(x)$ is homogeneous of degree n. Thus

$$\begin{aligned} &\sup\{|\langle y^{(n+1)}, \nabla \rangle p(x)| : x \in S^{d-1}\} = \|g\|_\partial \\ &= \frac{1}{n!} \sup \left\{ \prod_{j=1}^n \langle y^{(j)}, \nabla \rangle g(x) : x \in S^{d-1}, y^{(j)} \in S^{d-1} \text{ for all } j \right\}, \end{aligned}$$

by the inductive hypothesis. But

$$\|p\|_\partial = \frac{1}{n+1} \sup\{|\langle y^{(n+1)}, \nabla \rangle p(x)| : x, y^{(n+1)} \in S^{d-1}\},$$

and this proves the claim for $n+1$. □

4.6 Notes

There are several books that have sections on classical spherical harmonics. See, for example, Dunkl and Ramirez [1971], Stein and Weiss [1971], Helgason [1984], Axler, Bourdon and Ramey [1992], Groemer [1996] and Müller [1997]. Two recent books that contain the development of spherical harmonics and their applications are Atkinson and Han [2012], which includes application in the numerical analysis of partial differential equations, and Dai and Xu [2013], which addresses approximation theory and harmonic analysis on the sphere. The Jacobi polynomials appear as spherical harmonics with an invariant property in Braaksma and Meulenbeld [1968] and Dijksma and Koornwinder [1971].

A study of orthogonal structures on the sphere and on the ball was carried out in Xu [1998a, 2001a]. The relation between the reproducing kernels of spherical harmonics and of orthogonal polynomials on the unit ball can be used to relate the behavior of the Fourier orthogonal expansions on the sphere and those on the ball, as will be discussed in Chapter 9. The relation between orthogonal polynomials with respect to $1/\sqrt{1-\|x\|^2}$ on B^d and ordinary spherical harmonics on S^d, and more generally W_μ with $\mu = (m-1)/2$ on B^d and spherical harmonics on S^{d+m}, is observed in the explicit formulae for the biorthogonal polynomials (see Section 5.2) in the work of Hermite, Didon, Appell and Kampé de Fériet; see Chapter XII, Vol. II, of Erdélyi *et al.* [1953].

The relation between the orthogonal structures on the sphere, ball and simplex was studied in Xu [1998c]. The relation between the reproducing kernels will be used in Chapter 8 to derive a compact formula for the Jacobi weight on the ball and on the simplex. These relations have implications for other topics, such as cubature formulae and approximations on these domains; see Xu [2006c].

5

Examples of Orthogonal Polynomials in Several Variables

In this chapter we present a number of examples of orthogonal polynomials in several variables. Most of these polynomials, but not all, are separable in the sense that they are given in terms of the classical orthogonal polynomials of one variable. Many of our examples are extensions of orthogonal polynomials in two variables to more than two variables. We concentrate mostly on basic results in this chapter and discuss, for example, only classical results on regular domains, leaving further results to later chapters when they can be developed and often extended in more general settings.

One essential difference between orthogonal polynomials in one variable and in several variables is that the bases of $\mathcal{V}_n^d$ are not unique for $d \geq 2$. For $d = 1$, there is essentially one element in $\mathcal{V}_n^1$ since it is one dimensional. For $d \geq 2$, there can be infinitely many different bases. Besides orthonormal bases, several other types of base are worth mentioning. One is the *monic orthogonal* basis defined by

$$Q_\alpha(x) = x^\alpha + R_\alpha(x), \qquad |\alpha| = n, \quad R_\alpha \in \Pi_{n-1}^d; \tag{5.0.1}$$

that is, each Q_α of degree n contains exactly one monomial of degree n. Because the elements of a basis are not necessarily orthogonal, it is sometimes useful to find a *biorthogonal basis*, that is, another basis P_α of $\mathcal{V}_n^d$ such that $\langle Q_\alpha, P_\beta \rangle = 0$ whenever $\alpha \neq \beta$. The relations between various orthogonal bases were discussed in Section 3.2.

5.1 Orthogonal Polynomials for Simple Weight Functions

In this section we consider weight functions in several variables that arise from weight functions of one variable and whose orthogonal polynomials can thus be easily constructed. We consider two cases: one is the product weight function and the other is the radial weight function.

5.1.1 Product weight functions

For $1 \le j \le d$, let w_j be a weight function on a subset I_j of $\mathbb{R}$ and let $\{p_{j,n}\}_{n=0}^{\infty}$ be a sequence of polynomials that are orthogonal with respect to w_j. Let W be the product weight function

$$W(x) := w_1(x_1) \cdots w_d(x_d), \qquad x \in I_1 \times \cdots \times I_d.$$

The product structure implies immediately that

$$P_\alpha(x) := p_{1,\alpha_1}(x_1) \cdots p_{d,\alpha_d}(x_d), \qquad \alpha \in \mathbb{N}_0^d,$$

is an orthogonal polynomial of degree $|\alpha|$ with respect to W on $I_1 \times \cdots \times I_d$. Furthermore, denote by $\mathcal{V}_n^d(W)$ the space of orthogonal polynomials with respect to W. We immediately have the following:

Theorem 5.1.1 *For $n = 0,1,2,\ldots$, the set $\{P_\alpha : |\alpha| = n\}$ is a mutually orthogonal basis of $\mathcal{V}_n^d(W)$.*

If the polynomials $p_{j,n}$ are orthonormal with respect to w_j for each j then the polynomials P_α are also orthonormal with respect to W. Furthermore, the polynomial P_α is also a constant multiple of monic orthogonal polynomials. In fact, a monic orthogonal basis, up to a constant multiple, is also an orthonormal basis only when the weight function is of product type. As an example we consider the multiple Jacobi weight function

$$W_{\mathbf{a},\mathbf{b}}(x) = \prod_{i=1}^{d} (1-x_i)^{a_i}(1+x_i)^{b_i} \tag{5.1.1}$$

on the domain $[-1,1]^d$. An orthogonal basis P_α for the space $\mathcal{V}_n^d(W_{\mathbf{a},\mathbf{b}})$ is given in terms of the *Jacobi polynomials* $P_{\alpha_j}^{(a_j,b_j)}$ by

$$P_\alpha(W_{\mathbf{a},\mathbf{b}};x) = P_{\alpha_1}^{(a_1,b_1)}(x_1) \cdots P_{\alpha_d}^{(a_d,b_d)}(x_d), \qquad |\alpha| = n, \tag{5.1.2}$$

where, to emphasize the dependence of the polynomials P_α on the weight function, we include $W_{\mathbf{a},\mathbf{b}}$ in the notation.

The product Hermite and Laguerre polynomials are two other examples. They are given in later subsections.

5.1.2 Rotation-invariant weight functions

A rotation-invariant weight function W on $\mathbb{R}^d$ is a function of $\|x\|$, also called a radial function. Let w be a weight function defined on $[0,\infty)$ such that it is determined by its moments. We consider the rotation-invariant weight function

$$W_{\mathrm{rad}}(x) = w(\|x\|), \qquad x \in \mathbb{R}^d.$$

An orthogonal basis for $\mathscr{V}_n^d(W)$ can be constructed from the spherical harmonics and orthogonal polynomials of one variable.

For $0 \le j \le n/2$, let $\{Y_\nu^{n-2j} : 1 \le \nu \le a_{n-2j}^d\}$ denote an orthonormal basis for the space $\mathscr{H}_{n-2j}^d$ of spherical harmonics, where $a_n^d = \dim \mathscr{H}_n^d$. Fix j and n with $0 \le j \le n/2$, and let

$$w_{j,n}(t) := |t|^{2n-4j+d-1} w(|t|), \qquad t \in \mathbb{R}.$$

Denote by $p_m^{(2n-4j+d-1)}$ the polynomials orthogonal with respect to $w_{j,n}$.

Theorem 5.1.2 *For $0 \le j \le n/2$ and $1 \le \nu \le a_{n-2j}^d$, define*

$$P_{j,\nu}^n(x) := p_{2j}^{(2n-4j+d-1)}(\|x\|) Y_\nu^{n-2j}(x). \tag{5.1.3}$$

Then the $P_{j,\nu}^n$ form an orthonormal basis of $\mathscr{V}_n^d(W_{\mathrm{rad}})$.

Proof Since the weight function $|t|^{2n-4j+d-1}w(|t|)$ is an even function, the polynomial $p_{2j}^{(2n-4j+d-1)}(t)$ is also even, so that $P_{j,\nu}(x)$ is indeed a polynomial of degree n in x. To prove the orthogonality, we use an integral with polar coordinates,

$$\int_{\mathbb{R}^d} f(x)\, \mathrm{d}x = \int_0^\infty r^{d-1} \int_{S^{d-1}} f(rx')\, \mathrm{d}\omega(x') \mathrm{d}r. \tag{5.1.4}$$

Since $Y_\nu^{n-2j}(x) = r^{n-2j} Y_\nu^{n-2j}(x')$, it follows from the orthonormality of the Y_ν^{n-2j} that

$$\begin{aligned} \int_{\mathbb{R}^d} P_{j,\nu}^n(x) P_{j',\nu'}^m(x) W_{\mathrm{rad}}(x)\, \mathrm{d}x &= \delta_{n-2j,m-2j'} \delta_{\nu,\nu'} \sigma_{d-1} \\ &\times \int_0^\infty r^{2n-4j+d-1} p_{2j}^{(2n-4j+d-1)}(r) p_{2j'}^{(2n-4j+d-1)}(r) w(r)\, \mathrm{d}r. \end{aligned}$$

The last integral is 0 if $j \ne j'$ by the orthogonality with respect to $w_{j,n}$. This establishes the orthogonality of the $P_{j,\nu}^n$. To show that they form a basis of $\mathscr{V}_n^d(W_{\mathrm{rad}})$, we need to show that

$$\sum_{0 \le j \le n/2} \dim \mathscr{H}_{n-2j}^d = \dim \mathscr{V}_n^d(W_{\mathrm{rad}}) = \dim \mathscr{P}_n^d,$$

which, however, follows directly from Theorem 4.2.7. □

5.1.3 Multiple Hermite polynomials on $\mathbb{R}^d$

The multiple Hermite polynomials on $\mathbb{R}^d$ are polynomials orthogonal with respect to the classical weight function

$$W^H(x) = \mathrm{e}^{-\|x\|^2}, \qquad x \in \mathbb{R}^d.$$

The normalization constant of W^H is $\pi^{-d/2}$, determined by using (5.1.4):

$$\int_{\mathbb{R}^d} \mathrm{e}^{-\|x\|^2} \mathrm{d}x = \prod_{i=1}^{d} \int_{\mathbb{R}} \mathrm{e}^{-x_i^2} \mathrm{d}x_i = \pi^{d/2}.$$

The Hermite weight function is both a product function and a rotationally invariant function. The space $\mathcal{V}_n^d(W^H)$ of orthogonal polynomials has, therefore, two explicit mutually orthogonal bases.

First orthonormal basis Since W^H is a product of Hermite weight functions of one variable, the multiple Hermite polynomials defined by

$$P_\alpha(W^H;x) = H_{\alpha_1}(x_1)\cdots H_{\alpha_d}(x_d), \qquad |\alpha| = n, \tag{5.1.5}$$

form a mutually orthogonal basis for $\mathcal{V}_n^d$ with respect to W^H, as shown in Theorem 5.1.1. If we replace H_{α_i} by the orthonormal Hermite polynomials, then the basis becomes orthonormal.

Second orthonormal basis Since W^H is rotationally invariant, it has a basis given by

$$P_{j,\nu}(W^H;x) = [h_{j,n}^H]^{-1} L_j^{n-2j+(d-2)/2}(\|x\|^2) Y_\nu^{n-2j}(x), \tag{5.1.6}$$

where L_j^a is a Laguerre polynomial and the constant $h_{j,n}^H$ is given by the formula

$$[h_{j,n}^H]^2 = \left(\frac{d}{2}\right)_{n-2j} \Big/ (2j)!.$$

That the $P_{j,\nu}$ are mutually orthogonal follows from Theorem 5.1.2, since, by the orthogonality of the L_m^α, the polynomials $L_m^{n-2j+(d-2)/2}(t^2)$ can be seen to be orthogonal with respect to $|t|^{2n-2j+d-1}\mathrm{e}^{-t^2}$ upon making the change of variable $t \mapsto \sqrt{t}$. Since the Y_ν^{n-2j} are orthonormal with respect to $\mathrm{d}\omega/\sigma_{d-1}$, the normalization constant is derived from

$$\left(h_{j,n}^H\right)^2 = \frac{\sigma_{d-1}}{\pi^{d/2}} \int_0^\infty \left(L_j^{n-2j+(d-2)/2}(t)\right)^2 \mathrm{e}^{-t} t^{n-2j+(d-2)/2}\, \mathrm{d}t$$

and the L^2 norm of the Laguerre polynomials.

The orthogonal polynomials in $\mathcal{V}_n^d(W^H)$ are eigenfunctions of a second-order differential operator: for $n = 0, 1, 2\ldots$,

$$\left(\Delta - 2\sum_{i=1}^{d} x_i \frac{\partial}{\partial x_i}\right) P = -2nP \qquad \forall P \in \mathcal{V}_n^d(W^H). \tag{5.1.7}$$

This fact follows easily from the product structure of the basis (5.1.5) and the differential equation satisfied by the Hermite polynomials (Subsection 1.4.1).

5.1.4 Multiple Laguerre polynomials on $\mathbb{R}_+^d$

These are orthogonal polynomials with respect to the weight function

$$W_\kappa^L(x) = x_1^{\kappa_1} \cdots x_d^{\kappa_d} \mathrm{e}^{-|x|}, \qquad \kappa_i > -1, \quad x \in \mathbb{R}_+^d,$$

which is a product of Laguerre weight functions in one variable. The normalization constant is $\int_{\mathbb{R}_+^d} W_\kappa^L(x)\,\mathrm{d}x = \prod_{i=1}^d \Gamma(\kappa_i + 1)$.

The multiple Laguerre polynomials defined by

$$P_\alpha(W_\kappa^L; x) = L_{\alpha_1}^{\kappa_1}(x_1) \cdots L_{\alpha_d}^{\kappa_d}(x_d), \qquad |\alpha| = n, \tag{5.1.8}$$

are polynomials that are orthogonal with respect to W_κ^L and, as shown in Theorem 5.1.1, the set $\{P_\alpha(W_\kappa^L) : |\alpha| = n\}$ is a mutually orthogonal basis for $\mathscr{V}_n^d(W_\kappa^L)$, and it becomes an orthonormal basis if we replace L_{α_i} by the orthonormal Laguerre polynomials.

The orthogonal polynomials in $\mathscr{V}_n^d(W_\kappa^L)$ are eigenfunctions of a second-order differential operator. For $n = 0, 1, 2, \ldots$,

$$\sum_{i=1}^d x_i \frac{\partial^2 P}{\partial x_i^2} + \sum_{i=1}^d (\kappa_i + 1 - x_i) \frac{\partial P}{\partial x_i} = -nP. \tag{5.1.9}$$

This fact follows from the product nature of (5.1.8) and the differential equation satisfied by the Laguerre polynomials (Subsection 1.4.2).

5.2 Classical Orthogonal Polynomials on the Unit Ball

The classical orthogonal polynomials on the unit ball B^d of $\mathbb{R}^d$ are defined with respect to the weight function

$$W_\mu^B(x) = (1 - \|x\|^2)^{\mu - 1/2}, \qquad \mu > -\tfrac{1}{2}, \quad x \in B^d. \tag{5.2.1}$$

The normalization constant of W_μ^B is given by

$$w_\mu^B = \left(\int_{B^d} W_\mu^B(x)\,\mathrm{d}x \right)^{-1} = \frac{2}{\sigma_{d-1}} \frac{\Gamma(\mu + \frac{d+1}{2})}{\Gamma(\mu + \frac{1}{2}) \Gamma(\frac{d}{2})} = \frac{\Gamma(\mu + \frac{d+1}{2})}{\pi^{d/2} \Gamma(\mu + \frac{1}{2})}.$$

The value of w_μ^B can be verified by the use of the polar coordinates $x = rx'$, $x' \in S^{d-1}$, which leads to the formula

$$\int_{B^d} f(x)\,\mathrm{d}x = \int_0^1 r^{d-1} \int_{S^{d-1}} f(rx')\,\mathrm{d}\omega(x')\,\mathrm{d}r \tag{5.2.2}$$

and a beta integral. The polynomials P in $\mathscr{V}_n^d$ with respect to W_μ^B are eigenfunctions of the second-order differential operator $\mathscr{D}_\mu$, that is,

$$\mathscr{D}_\mu P = -(n+d)(n + 2\mu - 1)P, \tag{5.2.3}$$

where

$$\mathscr{D}_\mu := \Delta - \sum_{j=1}^{d} \frac{\partial}{\partial x_j} x_j \left((2\mu - 1) + \sum_{i=1}^{d} x_i \frac{\partial}{\partial x_i} \right).$$

This fact will be proved in Subsection 8.1.1. In the following we will give several explicit bases of $\mathscr{V}_n^d$ with respect to W_μ^B. They all satisfy the differential equation (5.2.3).

5.2.1 Orthonormal bases

We give two explicit orthonormal bases in this subsection.

First orthonormal basis Since the weight function W_μ^B is rotationally invariant, it has a basis given explicitly in terms of spherical polar coordinates, as in (5.1.3).

Proposition 5.2.1 *For* $0 \le j \le \frac{n}{2}$ *and* $\{Y_\nu^{n-2j} : 1 \le \nu \le a_{n-2j}^d\}$ *an orthonormal basis of* $\mathscr{H}_{n-2j}^d$*, the polynomials*

$$P_{j,\nu}^n(W_\mu^B; x) = (h_{j,n})^{-1} P_j^{(\mu-1/2, n-2j+(d-2)/2)}(2\|x\|^2 - 1) Y_\nu^{n-2j}(x), \qquad (5.2.4)$$

(cf. (5.1.2)) form an orthonormal basis of $\mathscr{V}_n^d(W_\mu^B)$*; the constant is given by*

$$(h_{j,n})^2 = \frac{(\mu + \frac{1}{2})_j (\frac{d}{2})_{n-j} (n - j + \mu + \frac{d-1}{2})}{j!(\mu + \frac{d+1}{2})_{n-j}(n + \mu + \frac{d-1}{2})}.$$

Proof The orthogonality of the $P_{j,i}^n$ follows easily from Theorem 5.1.2 and the orthogonality of the Jacobi polynomials (Subsection 1.4.4). The constant, using (5.2.2) and the orthonormality of the Y_ν^{n-2j}, comes from

$$\left(h_{j,n}^B\right)^2 = w_\mu^B \sigma_{d-1} \int_0^1 \left[P_j^{(\mu-1/2, n-2j+(d-2)/2)}(2r^2 - 1) \right]^2 \times r^{2n-4j+d-1}(1 - r^2)^{\mu-1/2}\, dr,$$

which can be evaluated, upon changing the variable $2r^2 - 1$ to t, from the L^2 norm of the Jacobi polynomial $P_n^{(a,b)}(t)$ and the properties of the Pochhammer symbol. □

Second orthonormal basis We need the following notation. Associated with $x = (x_1, \ldots, x_d) \in \mathbb{R}^d$, for each j define by $\mathbf{x}_j$ a truncation of x, namely

$$\mathbf{x}_0 = 0, \qquad \mathbf{x}_j = (x_1, \ldots, x_j), \quad 1 \le j \le d.$$

Note that $\mathbf{x}_d = x$. Associated with $\alpha = (\alpha_1, \ldots, \alpha_d)$, define

$$\alpha^j := (\alpha_j, \ldots, \alpha_d), \quad 1 \le j \le d, \qquad \text{and} \qquad \alpha^{d+1} := 0.$$

Recall that C_k^λ denotes the Gegenbauer polynomial associated with the weight function $(1-x^2)^{\lambda-1/2}$ on $[-1,1]$; see Subsection 1.4.3.

Proposition 5.2.2 *Let $\mu > 0$. For $\alpha \in \mathbb{N}_0^d$, the polynomials P_α given by*

$$P_\alpha(W_\mu^B;x) = (h_\alpha)^{-1}\prod_{j=1}^d (1-\|\mathbf{x}_{j-1}\|^2)^{\alpha_j/2} C_{\alpha_j}^{\lambda_j}\left(\frac{x_j}{\sqrt{1-\|\mathbf{x}_{j-1}\|^2}}\right),$$

where $\lambda_j = \mu + |\alpha^{j+1}| + \frac{d-j}{2}$, form an orthonormal basis of $\mathcal{V}_n^d(W_\mu^B)$; the constants h_α are defined by

$$(h_\alpha)^2 = \frac{(\mu+\frac{d}{2})_{|\alpha|}}{(\mu+\frac{d+1}{2})_{|\alpha|}} \prod_{j=1}^d \frac{(\mu+\frac{d-j}{2})_{|\alpha^j|}(2\mu+2|\alpha^{j+1}|+d-j)_{\alpha_j}}{(\mu+\frac{d-j+1}{2})_{|\alpha^j|}\alpha_j!}.$$

If $\mu = 0$, the basis comes from $\lim_{\mu\to 0}\mu^{-1}C_n^\mu(x) = \frac{2}{n}T_n(x)$ and the constant needs to be modified accordingly.

Proof The orthonormality and the normalization constants can both be verified by the use of the formula

$$\int_{B^d} f(x)\,\mathrm{d}x = \int_{B^{d-1}}\int_{-1}^1 f\left(\mathbf{x}_{d-1}, \sqrt{1-\|\mathbf{x}_{d-1}\|^2}y\right)\mathrm{d}y\sqrt{1-\|\mathbf{x}_{d-1}\|^2}\,\mathrm{d}\mathbf{x}_{d-1},$$

which follows from the simple change of variable $x_d \mapsto y\sqrt{1-\|\mathbf{x}_{d-1}\|^2}$. Using this formula repeatedly to reduce the number of iterations of the integral, that is, by reducing the dimension d by 1 at each step, we end up with

$$\begin{aligned} &w_\mu^B \int_{B^d} P_\alpha(W_\mu^B;x)P_\beta(W_\mu^B;x)W_\mu^B(x)\,\mathrm{d}x \\ &\quad = (h_\alpha)^{-2} w_\mu^B \prod_{j=1}^d \int_{-1}^1 \left(C_{\alpha_j}^{\mu+|\alpha^{j+1}|+(d-j)/2}(t)\right)^2 \\ &\qquad \times (1-t^2)^{\mu+|\alpha^{j+1}|+(d-j-1)/2}\,\mathrm{d}t\ \delta_{\alpha,\beta}, \end{aligned}$$

where the last integral can be evaluated by using the normalization constant of the Gegenbauer polynomials. □

5.2.2 Appell's monic orthogonal and biorthogonal polynomials

The definition of these polynomials can be traced back to the work of Hermite, Didon, Appell, and Kampé de Fériet; see Appell and de Fériet [1926] and Chapter XII, Vol. II, of Erdélyi *et al.* [1953]. There are two families of these polynomials.

Monic orthogonal Appell polynomials V_α The first family of polynomials, V_α, is defined by the generating function

$$(1-2\langle a,x\rangle+\|a\|^2)^{-\mu-(d-1)/2}=\sum_{\alpha\in\mathbb{N}_0^d} a^\alpha V_\alpha(x), \tag{5.2.5}$$

where $a\in B^d$. We prove two properties of V_α in the following.

Proposition 5.2.3 *The polynomials V_α are orthogonal with respect to W_μ^B on B^d.*

Proof From the generating function of the Gegenbauer polynomials and (5.2.5),

$$(1-2\langle a,x\rangle+\|a\|^2)^{-\mu-(d-1)/2}=\sum_{n=0}^{\infty}\|a\|^n C_n^{\mu+(d-1)/2}(\langle a/\|a\|,x\rangle).$$

Replacing a by at and comparing the coefficients of t^n gives

$$\|a\|^n C_n^{\mu+(d-1)/2}(\langle a/\|a\|,x\rangle)=\sum_{|\alpha|=n} a^\alpha V_\alpha(x).$$

Hence, in order to prove that V_α is orthogonal to polynomials of lower degree, it is sufficient to prove that

$$\int_{B^d}\|a\|^n C_n^{\mu+(d-1)/2}(\langle a/\|a\|,x\rangle)x^\beta W_\mu^B(x)\,\mathrm{d}x=0$$

for all monomials x^β with $|\beta|\le n-1$. Since W_μ^B is rotation invariant, we can assume that $a=(0,\ldots,0,c)$ without loss of generality, which leads to $C_n^{\mu+(d-1)/2}(\langle a/\|a\|,x\rangle)=C_n^{\mu+(d-1)/2}(x_d)$; making the change of variables $x_i=y_i\sqrt{1-x_d^2}$ for $i=1,\ldots,d-1$, the above integral is equal to

$$\int_{-1}^{1}C_n^{\mu+(d-1)/2}(x_d)x_d^{\beta_d}(1-x_d^2)^{\mu+(|\beta|-\beta_d)/2+(d-2)/2}\,\mathrm{d}x_d$$
$$\times\int_{B^{d-1}}y_1^{\beta_1}\cdots y_{d-1}^{\beta_{d-1}}(1-\|y\|^2)^{\mu-1/2}\,\mathrm{d}y.$$

If any β_i for $1\le i\le d-1$ is odd then the second integral is zero. If all β_i for $1\le i\le d-1$ are even then $x_d^{\beta_d}(1-x_d^2)^{|\beta|/2-\beta_d/2}$ is a polynomial of degree $|\beta|\le n-1$; hence, the first integral is equal to zero by the orthogonality of the Gegenbauer polynomials. □

Proposition 5.2.4 *The polynomials V_α are constant multiples of monic orthogonal polynomials; more precisely,*

$$V_\alpha(x)=\frac{2^{|\alpha|}}{\alpha!}\left(\mu+\frac{d-1}{2}\right)_{|\alpha|}x^\alpha+R_\alpha(x),\qquad R_\alpha\in\Pi_{n-1}^d.$$

Proof Let $\lambda = \mu + (d-1)/2$ in this proof. We use the multinomial and negative binomial formulae to expand the generating function as follows:

$$\begin{aligned}(1-2\langle a,x\rangle + \|a\|^2)^{-\lambda} &= [1-a_1(2x_1-a_1)-\cdots-a_d(2x_d-a_d)]^{-\lambda}\\ &= \sum_\beta \frac{(\lambda)_{|\beta|}}{\beta!} a^\beta (2x_1-a_1)^{\beta_1}\cdots(2x_d-a_d)^{\beta_d}\\ &= \sum_\beta \frac{(\lambda)_{|\beta|}}{\beta!} \sum_\gamma \frac{(-\beta_1)_{\gamma_1}\cdots(-\beta_d)_{\gamma_d}}{\gamma!} 2^{|\beta|-|\gamma|} x^{\beta-\gamma} a^{\gamma+\beta}.\end{aligned}$$

Therefore, upon making the change in summation indices $\beta_i + \gamma_i = \alpha_i$ and comparing the coefficients of a^α on both sides, we obtain the following explicit formula for $V_\alpha(x)$:

$$V_\alpha(x) = 2^{|\alpha|} x^\alpha \sum_\gamma \frac{(\lambda)_{|\alpha|-|\gamma|}(-\alpha_1+\gamma_1)_{\gamma_1}\cdots(-\alpha_d+\gamma_d)_{\gamma_d}}{(\alpha-\gamma)!\gamma!} 2^{-2|\gamma|} x^{-2\gamma},$$

from which the fact that V_α is a constant multiple of a monic orthogonal polynomial is evident. □

Moreover, using the formulae

$$(\lambda)_{m-k} = \frac{(-1)^k(\lambda)_m}{(1-\lambda-m)_k} \quad \text{and} \quad \frac{(-m+k)_k}{(m-k)!} = \frac{(-1)^k(-m)_{2k}}{m!}$$

as well as

$$2^{-2k}(-m)_{2k} = \left(-\frac{m}{2}\right)_k \left(\frac{1-m}{2}\right)_k,$$

which can all be verified directly from the expression for the Pochhammer symbol $(a)_b$, given in Definition 1.1.3, we can rewrite the formula as

$$\begin{aligned}V_\alpha(x) =& \frac{2^{|\alpha|}x^\alpha}{\alpha!}\left(\mu+\frac{d-1}{2}\right)_{|\alpha|}\\ &\times F_B\left(-\frac{\alpha}{2}, \frac{\mathbf{1}-\alpha}{2}; -|\alpha|-\mu-\frac{d-3}{2}; \frac{1}{x_1^2},\ldots,\frac{1}{x_d^2}\right),\end{aligned}$$

where $\mathbf{1}-\alpha = (1-\alpha_1,\ldots,1-\alpha_d)$ and F_B is the Lauricella hypergeometric series of d variables defined in Section 1.2.

Biorthogonal Appell polynomials U_α The second family of polynomials, U_α, can be defined by the generating function

$$\left[(1-\langle a,x\rangle)^2 + \|a\|^2(1-\|x\|^2)\right]^{-\mu} = \sum_{\alpha\in\mathbb{N}_0^d} a^\alpha U_\alpha(x). \tag{5.2.6}$$

An alternative definition is given by the following Rodrigues-type formula.

Proposition 5.2.5 *The polynomials U_α satisfy*

$$U_\alpha(x) = \frac{(-1)^{|\alpha|}(2\mu)_{|\alpha|}}{2^{|\alpha|}(\mu+\frac{1}{2})_{|\alpha|}\alpha!} \times (1-\|x\|^2)^{-\mu+1/2} \frac{\partial^{|\alpha|}}{\partial x_1^{\alpha_1}\cdots\partial x_d^{\alpha_d}}(1-\|x\|^2)^{|\alpha|+\mu-1/2}.$$

Proof Let $\langle a,\nabla\rangle = \sum_{i=1}^d a_i\partial/\partial x_i$, where ∇ denotes the gradient operator. The following formula can be easily verified by induction:

$$\langle a,\nabla\rangle^n(1-\|x\|^2)^b = \sum_{0\le j\le n/2} \frac{n!2^{n-2j}(-b)_{n-j}}{j!(n-2j)!}\langle a,x\rangle^{n-2j}(1-\|x\|^2)^{b-n+j}\|a\|^{2j}.$$

Putting $b = n+\mu-\frac{1}{2}$ and multiplying the above formula by $(1-\|x\|^2)^{-\mu+1/2}$ gives

$$(1-\|x\|^2)^{-\mu+1/2}\langle a,\nabla\rangle^n(1-\|x\|^2)^{n+\mu-1/2} = \sum_{0\le j\le n/2}(-1)^{n-j}\frac{n!2^{n-2j}(j+\mu+\frac{1}{2})_{n-j}}{j!(n-2j)!}\langle a,x\rangle^{n-2j}\|a\|^{2j}(1-\|x\|^2)^j,$$

where we have used $(-n-\mu+\frac{1}{2})_{n-j} = (-1)^{n-j}(j+\mu+\frac{1}{2})_{n-j}$. However, expanding the generating function shows that it is equal to

$$(1-\langle a,x\rangle)^{-2\mu}\left(1+\frac{\|a\|^2(1-\|x\|^2)}{(1-\langle a,x\rangle)^2}\right)^{-\mu} = \sum_{i=0}^\infty\sum_{j=0}^\infty(-1)^j\frac{(\mu)_j(2\mu+2j)_i}{i!j!}\|a\|^{2j}\langle a,x\rangle^i(1-\|x\|^2)^j.$$

Making use of the formula

$$(\mu)_j(2\mu+2j)_i = \frac{(\mu)_j(2\mu)_{i+2j}}{(2\mu)_{2j}} = \frac{(2\mu)_{i+2j}}{2^{2j}(\mu+\frac{1}{2})_j},$$

and setting $i+2j=n$, we obtain

$$\left[(1-\langle a,x\rangle)^2+\|a\|^2(1-\|x\|^2)\right]^{-\mu} = \sum_{n=0}^\infty\left(\frac{(2\mu)_n(-1)^n}{n!(\mu+\frac{1}{2})_n2^n} \times \sum_{0\le j\le n/2}(-1)^{n-j}\frac{n!2^{n-2j}(j+\mu+\frac{1}{2})_{n-j}}{j!(n-2j)!}\langle a,x\rangle^{n-2j}\|a\|^{2j}(1-\|x\|^2)^j\right).$$

Comparing these formulae gives

$$[(1-\langle a,x\rangle)^2+\|a\|^2(1-\|x\|^2)]^{-\mu} = \sum_{n=0}^\infty\frac{(2\mu)_n(-1)^n}{n!(\mu+\frac{1}{2})_n2^n}(1-\|x\|^2)^{-\mu+1/2}\langle a,\nabla\rangle^n(1-\|x\|^2)^{n+\mu-1/2}.$$

Consequently, comparing the coefficients of a^α in the above formula and in (5.2.6) finishes the proof. □

The most important property of the polynomials U_α is that they are biorthogonal to V_α. More precisely, the following result holds.

Proposition 5.2.6 *The polynomials $\{U_\alpha\}$ and $\{V_\alpha\}$ are biorthogonal:*

$$w_\mu^B \int_{B^d} V_\alpha(x) U_\beta(x) W_\mu^B(x)\,\mathrm{d}x = \frac{\mu + \frac{d-1}{2}}{|\alpha| + \mu + \frac{d-1}{2}} \frac{(2\mu)_{|\alpha|}}{\alpha!} \delta_{\alpha,\beta}.$$

Proof Since the set $\{V_\alpha\}$ forms an orthogonal basis, we only need to consider the case $|\beta| \ge |\alpha|$, which will also show that $\{U_\beta\}$ is an orthogonal basis. Using the Rodrigues formula and integration by parts,

$$\begin{aligned} & w_\mu^B \int_{B^d} V_\alpha(x) U_\beta(x) W_\mu^B(x)\,\mathrm{d}x \\ & = w_\mu^B \frac{(2\mu)_{|\alpha|}}{2^{|\alpha|}(\mu+\frac{1}{2})_{|\alpha|}\alpha!} \int_{B^d} \left[\frac{\partial^{|\beta|}}{\partial x_1^{\beta_1} \cdots \partial x_d^{\beta_d}} V_\alpha(x) \right] (1-\|x\|^2)^{|\alpha|+\mu-1/2}\,\mathrm{d}x. \end{aligned}$$

However, since V_α is a constant multiple of a monic orthogonal polynomial and $|\beta| \ge |\alpha|$,

$$\frac{\partial^{|\beta|}}{\partial x_1^{\beta_1} \cdots \partial x_d^{\beta_d}} V_\alpha(x) = 2^{|\alpha|} \left(\mu + \tfrac{d-1}{2}\right)_{|\alpha|} \delta_{\alpha,\beta},$$

from which we can invoke the formula for the normalization constant w_μ^B of W_μ^B to conclude that

$$\int_{B^d} V_\alpha(x) U_\beta(x) W_\mu^B(x)\,\mathrm{d}x = \frac{w_\mu^B}{w_{|\alpha|+\mu}} \frac{(2\mu)_{|\alpha|}(\mu+\frac{d-1}{2})_{|\alpha|}}{(\mu+\frac{1}{2})_{|\alpha|}\alpha!} \delta_{\alpha,\beta}.$$

The constant can be simplifed using the formula for w_μ^B. □

We note that the biorthogonality also shows that U_α is an orthogonal polynomial with respect to W_μ^B. However, U_α is not a monic orthogonal polynomial. To see what U_α is, we derive an explicit formula for it.

Proposition 5.2.7 *The polynomial U_α is given by*

$$U_\alpha(x) = \frac{(2\mu)_{|\alpha|}}{\alpha!} \sum_\beta \frac{(-1)^{|\beta|}(-\alpha_1)_{2\beta_1} \cdots (-\alpha_d)_{2\beta_d}}{2^{2|\beta|}\beta!(\mu+\frac{1}{2})_{|\beta|}} x^{\alpha-2\beta}(1-\|x\|^2)^{|\beta|}.$$

Proof Let $\phi : \mathbb{R} \mapsto \mathbb{R}$ be a real differentiable function. Using the chain rule and induction gives

$$\frac{d^n}{dx^n}\phi(x^2) = \sum_{j=0}^{[n/2]} (2x)^{n-2j} \frac{(-n)_{2j}}{j!} \phi^{(n-j)}(x^2),$$

where $\phi^{(m)}$ denotes the mth derivative of ϕ. Hence, for a function $\Phi : \mathbb{R}^d \to \mathbb{R}$, we have

$$\begin{aligned} &\frac{\partial^{|\alpha|}}{\partial x_1^{\alpha_1} \cdots \partial x_d^{\alpha_d}} \Phi(x_1^2, \ldots, x_d^2) \\ &\quad = 2^{|\alpha|} \sum_{\beta} \frac{(-\alpha_1)_{2\beta_1} \cdots (-\alpha_d)_{2\beta_d}}{2^{2|\beta|} \beta!} x^{\alpha - 2\beta} (\partial^{\alpha-\beta}\Phi)(x_1^2, \ldots, x_d^2), \end{aligned}$$

where $\partial^{\alpha-\beta}\Phi$ denotes the partial derivative of $(|\alpha| - |\beta|)$th order of Φ. In particular, applying the formula to $\Phi(x) = (1 - x_1 - \cdots - x_d)^{|\alpha|+\mu-1/2}$ gives

$$\begin{aligned} &\frac{\partial^{|\alpha|}}{\partial x_1^{\alpha_1} \cdots \partial x_d^{\alpha_d}} (1 - \|x\|^2)^{|\alpha|+\mu-1/2} \\ &\qquad = (-1)^{|\alpha|} 2^{|\alpha|} \left(\mu + \tfrac{1}{2}\right)_{|\alpha|} (1 - \|x\|^2)^{\mu-1/2} \\ &\qquad\quad \times \sum_{\beta} \frac{(-1)^{|\beta|} (-\alpha_1)_{2\beta_1} \cdots (-\alpha_d)_{2\beta_d}}{2^{2|\beta|} \beta! (\mu + \frac{1}{2})_{|\beta|}} x^{\alpha-2\beta} (1 - \|x\|^2)^{|\beta|}. \end{aligned}$$

The stated formula follows from the Rodrigues–type formula for U_α. □

Using the definition of the Lauricella hypergeometric series F_B (see Subsection 1.2.1), we can rewrite the above formula as

$$U_\alpha(x) = \frac{(2\mu)_{|\alpha|} x^\alpha}{\alpha!} F_B\Big(-\frac{\alpha}{2}, \frac{\mathbf{1} - \alpha}{2}; \mu + \frac{1}{2}; \frac{1 - \|x\|^2}{x_1^2}, \ldots, \frac{1 - \|x\|^2}{x_d^2}\Big).$$

There are a great number of formulae for both Appell polynomials, V_α and U_α; we refer to Appell and de Fériet [1926] and Erdélyi *et al.* [1953].

5.2.3 Reproducing kernel with respect to W_μ^B on B^d

Let $\mathbf{P}_n(W_\mu^B; \cdot, \cdot)$ denote the reproducing kernel of $\mathcal{V}_n^d(W_\mu^B)$ as defined in (3.6.2). This kernel satisfies a remarkable and concise formula.

Theorem 5.2.8 *For $\mu > 0$, $x, y \in B^d$,*

$$\begin{aligned} \mathbf{P}_n(W_\mu^B; x, y) = c_\mu \frac{n+\lambda}{\lambda} \int_{-1}^{1} C_n^\lambda \left(\langle x, y \rangle + t\sqrt{1 - \|x\|^2}\sqrt{1 - \|y\|^2} \right) \\ \times (1 - t^2)^{\mu-1} \, dt, \end{aligned} \tag{5.2.7}$$

where $\lambda = \mu + \frac{d-1}{2}$ and $c_\mu = \Gamma(\mu + \frac{1}{2})/[\sqrt{\pi}\Gamma(\mu)]$.

In the case $\mu = m/2$ and $m = 1, 2, 3, \ldots$ this expression was established in Corollary 4.3.6. For $\mu > 0$, the proof will appear as a special case of a more general result in Theorem 8.1.16. When $\mu = 0$, the formula for $\mathbf{P}(W_0;x,y)$ is found in Corollary 4.2.9.

It is worth mentioning that the kernel $\mathbf{P}_n(W_\mu^B;x,y)$ is rotation invariant, in the sense that $\mathbf{P}_n(W_\mu^B;x,y) = \mathbf{P}_n(W_\mu^B;x\sigma,y\sigma)$ for each rotation σ; this is obviously a consequence of the rotation invariance of W_μ^B.

For the constant weight $W_{1/2}(x) = 1$, there is another formula for the reproducing kernel. Let $\mathbf{P}(x,y) = \mathbf{P}_n(W_{1/2};x,y)$.

Theorem 5.2.9 *For $n = 1,2,3,\ldots$,*

$$\mathbf{P}_n(x,y) = \frac{n+\frac{d}{2}}{\frac{d}{2}} \int_{S^{d-1}} C_n^{d/2}(\langle x,\xi\rangle) C_n^{d/2}(\langle \xi,y\rangle)\, d\omega(\xi). \tag{5.2.8}$$

Proof Using the explicit orthonormal basis of $P_j^n := P_{j,\beta}^n(W_{1/2})$ in Proposition 5.2.1, with $\mu = \frac{1}{2}$, we obtain on the one hand

$$\mathbf{P}_n(x,y) = \sum_{0\le 2j\le n} \sum_{\beta} P_{j,\beta}^n(x) P_{j,\beta}^n(y). \tag{5.2.9}$$

On the other hand, by (5.2.7) we have for $\xi \in S^{d-1}$

$$\mathbf{P}_n(x,\xi) = \frac{n+\frac{d}{2}}{\frac{d}{2}} C_n^{d/2}(\langle x,\xi\rangle), \qquad \xi \in S^{d-1}, \quad x \in B^d. \tag{5.2.10}$$

Furthermore, when restricted to S^{d-1}, $P_{j,\beta}^n$ becomes a spherical harmonic,

$$P_{j,\beta}^n(\xi) = H_n Y_{j,n-2k}(\xi), \qquad \xi \in S^{d-1},$$

where, using the fact that $P_k^{(0,b)}(1) = 1$, we see that the formula

$$H_n = (h_{j,n})^{-1} P_j^{(0,n-2j+(d-2)/2)}(1) = \sqrt{\frac{n+\frac{d}{2}}{\frac{d}{2}}},$$

is independent of j. Consequently, integrating over S^{d-1} we obtain

$$\sigma_d^{-1} \int_{S^{d-1}} P_{j,\beta}^n(\xi) P_{j',\beta'}^n(\xi)\, d\omega(\xi) = H_n^2 \delta_{j,j'} \delta_{\beta,\beta'} = \frac{n+\frac{d}{2}}{\frac{d}{2}} \delta_{j,j'} \delta_{\beta,\beta'}.$$

Multiplying the above equation by $P_{j,\beta}^n(x)$ and $P_{j',\beta}^n(y)$ and summing over all j, j', β, β', the right-hand side becomes $\mathbf{P}_n(x,y)$, by (5.2.9), whereas the left-hand side becomes, by (5.2.9) and (5.2.10), the right-hand side of (5.2.8). This completes the proof. □

5.3 Classical Orthogonal Polynomials on the Simplex

Let T^d denote the simplex in the Euclidean space $\mathbb{R}^d$,

$$T^d = \{x \in \mathbb{R}^d : x_1 \ge 0, \ldots, x_d \ge 0, 1 - x_1 - \cdots - x_d \ge 0\}.$$

For $d = 2$, the region T^2 is the triangle with vertices at $(0,0)$, $(0,1)$ and $(1,0)$. The classical orthogonal polynomials on T^d are orthogonal with respect to the weight function

$$W_\kappa^T(x) = x_1^{\kappa_1 - 1/2} \cdots x_d^{\kappa_d - 1/2}(1 - |x|)^{\kappa_{d+1} - 1/2}, \quad \kappa_i > -\tfrac{1}{2}, \tag{5.3.1}$$

where $x \in T^d$ and $|x| = x_1 + \cdots + x_d$ is the usual ℓ^1 norm for $x \in T^d$. Upon using the formula

$$\int_{T^d} f(x)\,\mathrm{d}x = \int_{T^{d-1}} \int_0^1 f(\mathbf{x}_{d-1}, (1 - |\mathbf{x}_{d-1}|)y)\,\mathrm{d}y(1 - |\mathbf{x}_{d-1}|)\,\mathrm{d}\mathbf{x}_{d-1}, \tag{5.3.2}$$

which follows from the change of variables $x_d \mapsto y(1 - |\mathbf{x}_{d-1}|)$, we see that the normalization constant w_κ^T of W_κ^T is given by the Dirichlet integral

$$\begin{aligned} (w_\kappa^T)^{-1} &= \int_{T^d} x_1^{\kappa_1 - 1/2} \cdots x_d^{\kappa_d - 1/2}(1 - |x|)^{\kappa_{d+1} - 1/2}\,\mathrm{d}x \\ &= \frac{\Gamma(\kappa_1 + \frac{1}{2}) \cdots \Gamma(\kappa_{d+1} + \frac{1}{2})}{\Gamma(|\kappa| + \frac{d+1}{2})}, \end{aligned} \tag{5.3.3}$$

where $|\kappa| = \kappa_1 + \cdots + \kappa_{d+1}$. The orthogonal polynomials in $\mathcal{V}_n^d(W_\kappa^T)$ are eigenfunctions of a second-order differential operator. For $n = 0, 1, 2, \ldots$,

$$\begin{aligned} &\sum_{i=1}^d x_i(1 - x_i)\frac{\partial^2 P}{\partial x_i^2} - 2\sum_{1 \le i < j \le d} x_i x_j \frac{\partial^2 P}{\partial x_i \partial x_j} \\ &\quad + \sum_{i=1}^d \left[\left(\kappa_i + \tfrac{1}{2}\right) - \left(|\kappa| + \tfrac{d+1}{2}\right)x_i\right]\frac{\partial P}{\partial x_i} = \lambda_n P, \end{aligned} \tag{5.3.4}$$

where $\lambda_n = -n(n + |\kappa| + \frac{d-1}{2})$ and $\kappa_i > -\frac{1}{2}$ for $1 \le i \le d + 1$. The proof of equation (5.3.4) will be given in Subsection 8.2.1. For this weight function, we give three explicit orthogonal bases.

An orthonormal basis To state this basis we use the notation of $\mathbf{x}_j$ and α^j as in the second orthonormal basis on B^d (Subsection 5.2.1). For $\kappa = (\kappa_1, \ldots, \kappa_{d+1})$ and $0 \le j \le d + 1$, let $\kappa^j := (\kappa_j, \ldots, \kappa_{d+1})$.

Proposition 5.3.1 *For $\alpha \in \mathbb{N}_0^d$ and $|\alpha| = n$, the polynomials*

$$P_\alpha(W_\kappa^T; x) = (h_\alpha)^{-1} \prod_{j=1}^d (1 - |\mathbf{x}_{j-1}|)^{\alpha_j} P_{\alpha_j}^{(a_j, b_j)}\left(\frac{2x_j}{1 - |\mathbf{x}_{j-1}|} - 1\right),$$

where $a_j = 2|\alpha^{j+1}| + |\kappa^{j+1}| + \frac{d-j-1}{2}$ *and* $b_j = \kappa_j - \frac{1}{2}$, *form an orthonormal basis of* $\mathcal{V}_n^d(W_\kappa^T)$, *in which the constant is given by*

$$(h_\alpha)^2 = \frac{1}{\left(|\kappa| + \frac{d+1}{2}\right)_{2|\alpha|}} \prod_{j=1}^{d} \frac{(a_j+b_j+1)_{2\alpha_j}(a_j+1)_{\alpha_j}(b_j+1)_{\alpha_j}}{(a_j+b_j+1)_{\alpha_j}\alpha_j!}.$$

Proof Using the second expression of the Jacobi polynomial as a hypergeometric function in Proposition 1.4.14, we see that $P_\alpha(W_\kappa^T;x)$ is indeed a polynomial of degree $|\alpha|$ in x. To verify the orthogonality and the value of the constant we use the formula (5.3.2), which allows us to reduce the integral over T^d to a multiple of d integrals of one variable, so that the orthogonality will follow from that of the Jacobi polynomials. Indeed, using the formula (5.3.2) repeatedly gives

$$\begin{aligned} w_\kappa^T &\int_{T^d} P_\alpha(W_\kappa^T;x) P_\beta(W_\kappa^T;x) W_\kappa^T(x)\, dx \\ &= w_\kappa^T (h_\alpha)^{-2} \prod_{j=1}^{d} \int_0^1 [P_{\alpha_j}^{(a_j,b_j)}(2t-1)]^2 (1-t)^{a_j} t^{b_j}\, dt\, \delta_{\alpha,\beta}. \end{aligned}$$

Hence, using the norm for the Jacobi polynomials, we obtain that

$$[h_\alpha]^2 = w_\kappa^T \prod_{j=1}^{d} \frac{\Gamma(a_j+1)\Gamma(b_j+1)}{\Gamma(a_j+b_j+2)} \frac{(a_j+1)_{\alpha_j}(b_j+1)_{\alpha_j}(a_j+b_j+\alpha_j+1)}{(a_j+b_j+2)_{\alpha_j}(a_j+b_j+2\alpha_j+1)\alpha_j!},$$

which simplifies, using $a_d = \kappa_{d+1}$, $a_{j-1}+1 = a_j+b_j+2\alpha_j+2$, and $a_1+b_1+2\alpha_1+2 = |\kappa| + 2|\alpha| + \frac{d+1}{2}$, to the stated formula. □

Next we give two bases that are biorthogonal, each of which has its own distinguished feature. The first is the monic orthogonal basis and the second is given by the Rodrigues formulas.

Monic orthogonal basis For $\alpha \in \mathbb{N}_0^d$ and $|\alpha| = n$, we define

$$V_\alpha^T(x) = \sum_{0 \le \beta \le \alpha} (-1)^{n+|\beta|} \prod_{i=1}^{d} \binom{\alpha_i}{\beta_i} \frac{(\kappa_i+\frac{1}{2})_{\alpha_i}}{(\kappa_i+\frac{1}{2})_{\beta_i}} \frac{(|\kappa| + \frac{d-1}{2})_{n+|\beta|}}{(|\kappa| + \frac{d-1}{2})_{n+|\alpha|}} x^\beta,$$

where $\beta \le \alpha$ means $\beta_1 \le \alpha_1, \ldots, \beta_d \le \alpha_d$.

Proposition 5.3.2 *For* V_α^T *as defined above, the set* $\{V_\alpha^T : |\alpha| = n\}$ *forms a monic orthogonal basis of* $\mathcal{V}_n^d$ *with respect to* W_κ^T.

Proof The fact that $V_\alpha^T(x)$ equals x^α plus a polynomial of lower degree is evident. We will prove that V_α^T is orthogonal to polynomials of lower degree with respect to W_κ^T, for which a different basis for the space Π_n^d is used. Let

$$X_\gamma(x) = x_1^{\gamma_1} \cdots x_d^{\gamma_d} (1-|x|)^{\gamma_{d+1}},$$

where $\gamma \in \mathbb{N}_0^{d+1}$. Then the X_γ with $|\gamma| = n$ form a basis of Π_n^d; that is,

$$\Pi_n^d = \operatorname{span}\{x^\alpha : \alpha \in \mathbb{N}_0^d, |\alpha| \le n\} = \operatorname{span}\{X_\gamma : \gamma \in \mathbb{N}_0^{d+1}, |\gamma| = n\}.$$

This can be seen easily from applying the binomial theorem to $(1-|x|)^{\gamma_{d+1}}$. Thus, to prove the orthogonality of V_α^T it is sufficient to show that V_α^T is orthogonal to X_γ for all $\gamma \in \mathbb{N}_0^{d+1}$ satisfying $|\gamma| = n-1$. Using (5.3.3) and the formula for V_α^T, it follows for $|\gamma| = n-1$ and $|\alpha| = n$ that

$$\begin{aligned}
w_\kappa^T &\int_{T^d} V_\alpha^T(x) X_\gamma(x) W_\kappa^T(x)\,\mathrm{d}x \\
&= \sum_{\beta \le \alpha} \Bigg((-1)^{n+|\beta|} \prod_{i=1}^d \binom{\alpha_i}{\beta_i} \frac{(\kappa_i+\frac12)_{\alpha_i}}{(\kappa_i+\frac12)_{\beta_i}} \frac{(|\kappa|+\frac{d-1}{2})_{n+|\beta|}}{(|\kappa|+\frac{d-1}{2})_{n+|\alpha|}} \\
&\qquad \times \frac{\prod_{i=1}^d \Gamma(\beta_i+\gamma_i+\kappa_i+\frac12)\Gamma(\gamma_{d+1}+\kappa_{d+1}+\frac12)}{\Gamma(|\gamma|+|\beta|+|\kappa|+\frac{d+1}{2})} \Bigg) \\
&= (-1)^n \frac{\prod_{i=1}^d \Gamma(\alpha_i+\kappa_i+\frac12)\Gamma(\gamma_i+1)\Gamma(\gamma_{d+1}+\kappa_{d+1}+\frac12)}{\Gamma(2n+|\kappa|+\frac{d+1}{2})} \\
&\qquad \times \prod_{i=1}^d \sum_{\beta_i=0}^{\alpha_i} (-1)^{\beta_i} \binom{\alpha_i}{\beta_i} \binom{\beta_i+\gamma_i+\kappa_i-\frac12}{\gamma_i}.
\end{aligned}$$

A Chu–Vandermonde sum shows that, for $m, n \in \mathbb{N}$ and $a \in \mathbb{R}$,

$$\begin{aligned}
\sum_{k=0}^n (-1)^k \binom{n}{k} \binom{a+k}{m} &= \frac{(-1)^m(-a)_m}{m!}\, {}_2F_1\left(\begin{matrix} -n, 1+a \\ 1+a-m \end{matrix}; 1\right) \\
&= \frac{(-1)^m(-a)_m(-m)_n}{m!(1+a-m)_n} = \frac{(-m)_n}{m!(1+a)_{n-m}}.
\end{aligned}$$

In particular, the sum becomes zero, because of the factor $(-m)_n$, if $m < n$. Since $|\gamma| = n-1$, there is at least one i for which $\gamma_i < \alpha_i$; consequently, with $n = \alpha_i$, $m = \gamma_i$ and $a = \gamma_i + \kappa_i - \frac12$, we conclude that V_α^T is orthogonal to X_γ. □

Further results on V_α^T, including a generating function and its L^2 norm, will be given in Subsection 8.2.3.

Rodrigues formula and biorthogonal basis It is easy to state another basis, U_α^T, that is biorthogonal to the basis V_α^T. Indeed, we can define U_α^T by

$$\begin{aligned}
U_\alpha^T(x) &= x_1^{-\kappa_1+1/2} \cdots x_d^{-\kappa_d+1/2} (1-|x|)^{-\kappa_{d+1}+1/2} \\
&\quad \times \frac{\partial^{|\alpha|}}{\partial x_1^{\alpha_1} \cdots \partial x_d^{\alpha_d}} x_1^{\alpha_1+\kappa_1-1/2} \cdots x_d^{\alpha_d+\kappa_d-1/2} (1-|x|)^{|\alpha|+\kappa_{d+1}-1/2}.
\end{aligned}$$

These are analogues of the Rodrigues formula for orthogonal polynomials on the simplex. For $d = 2$ and up to a constant multiple, the U_α are also called Appell polynomials; see Appell and de Fériet [1926] and Chapter XII, Vol. II, of Erdélyi *et al.* [1953].

Proposition 5.3.3 *The polynomials U_α^T are orthogonal polynomials and they are biorthogonal to the polynomials V_α^T:*

$$\int_{T^d} V_\beta^T(x) U_\alpha^T(x) W_\kappa^T(x)\,\mathrm{d}x = \frac{\prod_{i=1}^d (\kappa_i + \frac{1}{2})_{\alpha_i} (\kappa_{d+1} + \frac{1}{2})_{|\alpha|}}{(|\kappa| + \frac{d+1}{2})_{|\alpha|}} \alpha!\, \delta_{\alpha,\beta}.$$

Proof It is evident from the definition that U_α^T is a polynomial of degree n. Integrating by parts leads to

$$\begin{aligned} w_\kappa^T &\int_{T^d} V_\beta^T(x) U_\alpha^T(x) W_\kappa^T(x)\,\mathrm{d}x \\ &= w_\kappa^T \int_{T^d} \left(\frac{\partial^{|\alpha|}}{\partial x_1^{\alpha_1} \ldots \partial x_d^{\alpha_d}} V_\beta(x) \right) \prod_{i=1}^d x_i^{\alpha_i + \kappa_i - 1/2} (1 - |x|)^{|\alpha| + \kappa_{d+1} - 1/2}\,\mathrm{d}x. \end{aligned}$$

Since V_β^T is an orthogonal polynomial with respect to W_κ^T, the integral on the left-hand side is zero for $|\beta| > |\alpha|$. For $|\beta| \le |\alpha|$ the fact that V_β^T is a monic orthogonal polynomial gives

$$\frac{\partial^{|\alpha|}}{\partial x_1^{\alpha_1} \cdots \partial x_d^{\alpha_d}} V_\beta^T(x) = \alpha!\, \delta_{\alpha,\beta},$$

from which the stated formula follows. The fact that the U_α^T are biorthogonal to V_β^T shows that they are orthogonal polynomials with respect to W_κ^T. □

Reproducing kernel with respect to W_κ on T^d Let $\mathbf{P}_n(W_\kappa^T;\cdot,\cdot)$ denote the reproducing kernel of $\mathscr{V}_n^d(W_\kappa)$ as defined in (3.6.2). This kernel satisfies a remarkable and concise formula.

Theorem 5.3.4 *For $\kappa_i > 0$, $1 \le i \le d$, $x, y \in T^d$, and $\lambda_k := |\kappa| + \frac{d-1}{2}$,*

$$\begin{aligned} \mathbf{P}_n(W_\kappa^T; x, y) = & \frac{(2n + \lambda_\kappa)(\lambda_\kappa)_n}{\lambda_\kappa (\frac{1}{2})_n} c_\kappa \qquad (5.3.5) \\ & \times \int_{[-1,1]^{d+1}} P_n^{(\lambda_\kappa - 1/2, -1/2)} \left(2z(x,y,t)^2 - 1\right) \prod_{i=1}^{d+1} (1 - t_i^2)^{\kappa_i - 1}\,\mathrm{d}t, \end{aligned}$$

where $z(x,y,t) = \sqrt{x_1 y_1}\, t_1 + \cdots + \sqrt{x_d y_d}\, t_d + \sqrt{1-|x|}\sqrt{1-|y|}\, t_{d+1}$ and $[c_\kappa]^{-1} = \int_{[-1,1]^{d+1}} \prod_{i=1}^{d+1} (1 - t_i^2)^{\kappa_i - 1} \mathrm{d}t$. If some $\kappa_i = 0$ then the formula holds under the limit relation (1.5.1).

This theorem will be proved in Subsection 8.2.2. We note that when $\kappa = 0$ we have the limiting case $\mathbf{P}(W_0; x, y)$, which was given in Corollary 4.4.10.

5.4 Orthogonal Polynomials via Symmetric Functions

We now present a non-classical example, two families of orthogonal polynomials derived from symmetric polynomials.

5.4.1 Two general families of orthogonal polynomials

These orthogonal polynomials are derived from symmetric and antisymmetric polynomials. Although they are orthogonal with respect to unusual measures, they share an interesting property, that of having the largest number of distinct common zeros, which will be discussed in the next subsection.

We recall some basic facts about symmetric polynomials. A polynomial $f \in \Pi^d$ is called symmetric if f is invariant under any permutation of its variables. The elementary symmetric polynomials in Π_k^d are given by

$$e_k = e_k(x_1,\ldots,x_d) = \sum_{1\le i_1<\ldots<i_k\le d} x_{i_1}\ldots x_{i_k}, \qquad k = 1,2,\ldots,d,$$

and they satisfy the generating function

$$\prod_{i=1}^{d}(1+rx_i) = \sum_{k=0}^{d} e_k(x_1,\ldots,x_d)r^k.$$

The polynomials e_k form a linear basis for the subspace of symmetric polynomials. Hence, any symmetric polynomial f can be uniquely represented in terms of elementary symmetric polynomials. Consider the mapping

$$u:\ x = (x_1,\ldots,x_d) \mapsto (e_1(x),\ldots,e_d(x)), \tag{5.4.1}$$

where the e_i are elementary symmetric polynomials in x, which is one-to-one on the region $S = \{x \in \mathbb{R}^d :\ x_1 < x_2 < \cdots < x_d\}$; denote the image of S by R, that is,

$$R = \{u \in \mathbb{R}^d :\ u = u(x),\ x_1 < x_2 < \cdots < x_d,\ x \in \mathbb{R}^d\}. \tag{5.4.2}$$

Let $D(x)$ denote the Jacobian of this mapping. We claim that

$$D(x) = \det\left[\left(\frac{\partial e_i}{\partial x_j}\right)_{i,j}\right] = \prod_{1\le i<j\le d}(x_i - x_j). \tag{5.4.3}$$

To prove this equation we use the following identities, which follow from the definition of the elementary symmetric functions:

$$\frac{\partial e_k}{\partial x_j} = e_{k-1}(x_1,\ldots,x_{j-1},x_{j+1},\ldots,x_d)$$

and

$$\begin{aligned} &e_{k-1}(x_1,\ldots,x_{j-1},x_{j+1},\ldots,x_d) - e_{k-1}(x_2,\ldots,x_d) \\ &\quad = (x_1 - x_j)e_{k-2}(x_2,\ldots,x_{j-1},x_{j+1},\ldots,x_d). \end{aligned}$$

Subtracting the first column from the other columns in the determinant, using the above two identities and taking out the factors, we obtain

$$D(x) = \prod_{i=2}^{d}(x_1 - x_i)D(x_2,\ldots,x_{d-1}),$$

from which the formula (5.4.3) follows.

Since the square of the Jacobian is a symmetric polynomial in x, under the mapping (5.4.1), it becomes a polynomial in u, which we denote by $\Delta(u)$, that is,

$$\Delta(u) = \prod_{1\le i<j\le d}(x_i - x_j)^2, \qquad u \in R.$$

Let $\mathrm{d}\mu$ be a nonnegative measure on $\mathbb{R}$. Define a measure ν on R as the image of the product measure $\mathrm{d}\mu(x) = \mathrm{d}\mu(x_1)\cdots\mathrm{d}\mu(x_d)$ under the mapping (5.4.1), that is,

$$\mathrm{d}\nu(u) = \mathrm{d}\mu(x_1)\cdots\mathrm{d}\mu(x_d), \qquad u \in R.$$

Our examples are orthogonal polynomials with respect to the measures

$$[\Delta(u)]^{1/2}\,\mathrm{d}\nu \qquad \text{and} \qquad [\Delta(u)]^{-1/2}\,\mathrm{d}\nu,$$

denoted by $P_\alpha^{n,1/2}$ and $P_\alpha^{n,-1/2}$, respectively.

The explicit formulae for these polynomials are given below. Let $\{p_n\}$ be orthonormal polynomials with respect to $\mathrm{d}\mu$ on $\mathbb{R}$. For $\alpha = (\alpha_1,\ldots,\alpha_d)$, $n \ge \alpha_1 \ge \cdots \ge \alpha_d \ge 0$ and $n \in \mathbb{N}_0$, set

$$P_\alpha^{n,-1/2}(u) = \sum_{\beta\in S_d} p_{\alpha_1}(x_{\beta_1})\cdots p_{\alpha_d}(x_{\beta_d}), \quad u \in R, \tag{5.4.4}$$

where S_d denotes the symmetric group on d objects, so that the summation is performed over all permutations β of $\{1,2,\ldots,d\}$. By the fundamental theorem on symmetric functions, $P_\alpha^{n,-1/2}$ is a polynomial in u of degree n.

We now define the second set of orthogonal polynomials. For $\alpha = (\alpha_1,\ldots,\alpha_d)$, $0 \le \alpha_1 \le \cdots \le \alpha_d = n$ and $n \in \mathbb{N}_0$, set

$$P_\alpha^{n,1/2}(u) = \frac{E_\alpha^n(x)}{D(x)}, \qquad \text{where} \qquad E_\alpha^n(x) = \det\left(p_{\alpha_i+d-i}(x_j)\right)_{i,j=1}^{d}. \tag{5.4.5}$$

Since $E_\alpha^n(x)$ vanishes for $x_i = x_j$, $i \ne j$, and since $D(x) = \prod_{1\le i<j\le d}(x_i - x_j)$ is a factor of $E_\alpha^n(x)$, permutation of the variables in E_α^n and D can only change the signs. This happens to both E_α^n and D, so E_α^n/D is a symmetric polynomial. Since the polynomial D is of degree $d-1$ in each of its variables x_i and the polynomial E_α^n is of degree $n+d-1$ in x_i, it follows that E_α^n/D is a symmetric polynomial of degree n in x_i or, equivalently, a polynomial in u of degree n.

Let us now verify that the systems $\{P_\alpha^{n,-1/2}\}$ and $\{P_\alpha^{n,1/2}\}$ are orthogonal on R with respect to $\Delta^{-1/2}\,\mathrm{d}\nu$ and $\Delta^{1/2}\,\mathrm{d}\nu$, respectively. For any polynomial f in u, that is, for any symmetric polynomial in x, changing variables gives

$$\int_R f(u)[\Delta(u)]^{-1/2}\,\mathrm{d}\nu(u) = \int_S f(u(x))\,\mathrm{d}\mu(x) = \frac{1}{d!}\int_{\mathbb{R}^d} f(u(x))\,\mathrm{d}\mu(x),$$

and, similarly,

$$\int_R f(u)[\Delta(u)]^{1/2}\,\mathrm{d}\nu(u) = \int_S f(u(x))\Delta^2(x)\,\mathrm{d}\mu(x).$$

In particular, it follows that

$$\begin{aligned}\int_R P_\alpha^{n,-1/2}(u)P_\beta^{m,-1/2}(u)[\Delta(u)]^{-1/2}\,\mathrm{d}\nu(u)\\ = \frac{1}{d!}\int_{\mathbb{R}^d} P_\alpha^{n,-1/2}(u(x))P_\beta^{m,-1/2}(u(x))\,\mathrm{d}\mu(x).\end{aligned}$$

From the last expression and (5.4.4) we see that the polynomials of the system $\{P_\alpha^{n,-1/2}\}$ are mutually orthogonal: the above integral is nonzero only when $n=m$ and $\alpha=\beta$. Moreover, if $\alpha'_1,\ldots,\alpha'_{d'}$, $1\le d'\le d$, are the distinct elements in α, and m_i is the number of occurrences of α'_i, then

$$\int_R \left[P_\alpha^{n,-1/2}(u)\right]^2[\Delta(u)]^{-1/2}\,\mathrm{d}\nu(u) = m_1!\cdots m_{d'}!$$

Hence, $\left\{1/\sqrt{m_1!\cdots m_{d'}!}P_\alpha^{n,-1/2}\right\}$ forms an orthonormal family on R.

Similarly, for the system $\{P_\alpha^{n,1/2}\}$,

$$\begin{aligned}\int_R P_\alpha^{n,1/2}(u)P_\beta^{m,1/2}(u)[\Delta(u)]^{1/2}\,\mathrm{d}\nu(u) &= \int_S E_\alpha^n(x)E_\beta^m(x)\,\mathrm{d}\mu(x)\\ &= \frac{1}{d!}\int_{\mathbb{R}^d} E_\alpha^n(x)E_\beta^m(x)\,\mathrm{d}\mu(x),\end{aligned}$$

from which we can proceed as above. The system $\{P_\alpha^{n,1/2}\}$ is orthonormal.

In the case $d=2$, the mapping (5.4.1) takes the form

$$u=(u_1,u_2):\quad u_1=x+y,\quad u_2=xy,$$

and the function Δ takes the form $\Delta(u_1,u_2)=(x-y)^2=u_1^2-4u_2$. The orthogonal polynomials (5.4.4) are exactly the Koornwinder polynomials, discussed in Section 2.7, on the domain bounded by two lines and a parabola.

5.4.2 Common zeros and Gaussian cubature formulae

The orthogonal polynomials with respect to $\Delta^{\pm 1/2}$ have the maximal number of common zeros. Let $\mathbb{P}_n^{(\pm 1/2)}$ denote orthogonal bases of degree n with respect to $[\Delta(u)]^{\pm 1/2}\,\mathrm{d}\nu$, respectively.

Theorem 5.4.1 *The orthogonal polynomials in $\mathbb{P}_n^{(1/2)}$ or $\mathbb{P}_n^{(-1/2)}$ with respect to the measures $[\Delta(u)]^{1/2}\,\mathrm{d}\nu$ or $[\Delta(u)]^{-1/2}\,\mathrm{d}\nu$, respectively, have $\dim\Pi_n^d$ common zeros.*

Proof Let $p_n(x)$ be orthonormal polynomials with respect to $d\mu(x)$ on $\mathbb{R}$, as in Subsection 5.4.1. Let $\{x_{k,n}\}_{k=1}^n$ be the zeros of p_n, ordered by $x_{1,n} < \cdots < x_{n,n}$. From the formula (5.4.4) it follows that $P_\alpha^{n,-1/2}$ vanishes on all $u_{\alpha,n}$, which is the image of $(x_{\alpha_1,n},\ldots,x_{\alpha_d,n})$ under the map (5.4.1), where the $x_{\alpha_i,n}$ are the zeros of p_n. Using the symmetric mapping (5.4.1) and taking into account the restriction $x_1 \le x_2 \le \cdots \le x_d$ in the definition of R, we see that the number of distinct common zeros of $\mathbb{P}_n^{(-1/2)}$ is equal to the cardinality of the set $\{\gamma \in \mathbb{N}^d : 0 \le \gamma_d \le \cdots \le \gamma_1 \le n-1\}$. Denote the cardinality by $N_{n-1,d}$. It follows by setting $\gamma_d = k$ that

$$N_{n-1,d} = \sum_{k=0}^{n-1} \#\{\gamma \in \mathbb{N}^d : \gamma_1 \le \cdots \le \gamma_{d-1} \le k\} = \sum_{k=0}^{n-1} N_{k,d-1},$$

which is the same as the equation $\dim \Pi_{n-1}^d = \sum r_k^d$, and induction shows that $N_{n,d} = \dim \Pi_{n-1}^d$. Therefore, $\mathbb{P}_n^{(-1/2)}$ has $\dim \Pi_{n-1}^d$ distinct common zeros.

Similarly, let $t_{k,n+d-1}$ be the zeros of the orthogonal polynomial p_{n+d-1}. From the formula (5.4.5) it follows that $P_\alpha^{n,1/2}$ vanishes on all $u_{\gamma,n}$, which are the images of $(t_{\gamma_1,n+d-1},\ldots,t_{\gamma_d,n+d-1})$ under the mappings (5.4.1). The number of distinct common zeros of $\mathbb{P}_n^{1/2}$ is equal to the cardinality of the set $\{\gamma \in \mathbb{N}^d : 0 \le \gamma_d < \cdots < \gamma_1 \le n+d-2\}$, which is easily seen to be equal to the cardinality of the set defined in the previous paragraph. Hence, $\mathbb{P}_n^{(1/2)}$ has $\dim \Pi_{n-1}^d$ distinct common zeros. □

Together with Theorem 3.8.4, this shows that Gaussian cubature formulae exist for all weight functions $\Delta(u)^{\pm 1/2}\,d\nu$. Moreover, we can write down the cubature formulae explicitly. Let us recall from Subsection 5.4.1 that p_n is the nth orthogonal polynomial with respect to $d\mu$ on $\mathbb{R}$ and that the set $\{x_{k,n}\}_{k=1}^n$ contains the zeros of p_n. The Gaussian quadrature formula with respect to $d\mu$ takes the form

$$\int_{\mathbb{R}} f(x)\,d\mu = \sum_{k=1}^n f(x_{k,n})\lambda_{k,n}, \qquad f \in \Pi_{2n-1}^d. \tag{5.4.6}$$

Let $u_{\gamma,n}$ denote the image of $x_{\gamma,n} = (x_{\gamma_1,n},\ldots,x_{\gamma_d,n})$ under the map (5.4.1).

Theorem 5.4.2 *Using the notation in Subsection 5.4.1, both the measures $[\Delta(u)]^{1/2}d\nu$ and $[\Delta(u)]^{-1/2}d\nu$ admit Gaussian cubature formulae. Moreover, the formulae of degree $2n-1$ take the form*

$$\int_R f(u)[\Delta(u)]^{-1/2}d\nu(u) = \sum_{\gamma_1=1}^n \sum_{\gamma_2=1}^{\gamma_1} \cdots \sum_{\gamma_d=1}^{\gamma_{d-1}} \Lambda_{\gamma,n}^{-1/2} f(u_{\gamma,n})$$

and

$$\int_R f(u)[\Delta(u)]^{1/2}d\nu(u) = \sum_{\gamma_1=1}^{n+d-1} \sum_{\gamma_2=1}^{\gamma_1-1} \cdots \sum_{\gamma_d=1}^{\gamma_{d-1}-1} \Lambda_{\gamma,n}^{1/2} f(u_{\gamma,n+d-1}),$$

where

$$\Lambda_{\gamma,n}^{-1/2} = \frac{\lambda_{\gamma_1,n}^{m_1}\cdots\lambda_{\gamma_{d'},n}^{m_{d'}}}{m_1!\cdots m_{d'}!} \qquad \text{and} \qquad \Lambda_{\gamma,n}^{1/2} = J(x_{\gamma,n})^2\lambda_{\gamma_1,n}\cdots\lambda_{\gamma_d,n}$$

and where, for a given multi-index $\gamma \in \mathbb{N}_0^d$, $\gamma'_1,\ldots,\gamma'_{d'}$ *are its distinct elements and* $m_1,\ldots,m_{d'}$ *their respective multiplicities.*

Proof We need only determine the weights $\{\Lambda_{\alpha,n}^{1/2}\}$ and $\{\Lambda_{\alpha,n}^{-1/2}\}$ in the two formulae.

First notice that by changing variables the first cubature formula can be rewritten as

$$\frac{1}{d!}\int_{\mathbb{R}^d} f(x)\,\mathrm{d}\mu(x) = \sum_{\gamma_1=1}^{n}\sum_{\gamma_2=1}^{\gamma_1}\cdots\sum_{\gamma_d=1}^{\gamma_{d-1}} \Lambda_{\gamma,n}^{-1/2} f(x_{\gamma,n}),$$

which is exact for symmetric polynomials f whose degree in each variable is at most $2n-1$. Let $\ell_{k,n}$ be the polynomial of degree $n-1$ uniquely determined by $\ell_{k,n}(x_{j,n}) = \delta_{k,j}$, $1 \le j \le n$. Let

$$f(x) = \sum_{\beta\in S_d} \ell_{\alpha_1,n}(x_{\beta_1})\cdots\ell_{\alpha_d,n}(x_{\beta_d}),$$

which is symmetric with degree $n-1$ in each of its variables. There are $d!$ summands in the definition of f. On the one hand we can write the integration on the left-hand side of the first cubature formula as a sum of products of one-variable integrals. Using (5.4.6) the integral is equal to $\lambda_{\alpha_1,n}\cdots\lambda_{\alpha_d,n}$. On the other hand, the right-hand side of the first cubature formula equals $m_1!\cdots m_{d'}!\Lambda_{\alpha,n}^{-1/2}$, since only those terms with $\gamma = \alpha$ count.

In the same way as above, the second cubature formula can be rewritten as

$$\frac{1}{d!}\int_{\mathbb{R}^d} f(x)J(x)^2\,\mathrm{d}\mu(x) = \sum_{\gamma_1=1}^{n+d-1}\sum_{\gamma_2=1}^{\gamma_1-1}\cdots\sum_{\gamma_d=1}^{\gamma_{d-1}-1} \Lambda_{\gamma,n}^{1/2} f(x_{\gamma,n}),$$

which is exact for symmetric polynomials f whose degree in each variable is at most $n-1$. Now let

$$f(x) = \left[\det(\ell_{\alpha_i}(x_j))_{i,j=1}^d\right]^2/D(x)^2$$

which is a symmetric polynomial with degree $2n-1$ in each of its variables. Clearly, the right-hand side of the second cubature formula is simply $\Lambda_{\alpha,n}^{1/2}/D(x_{\alpha,n})^2$, while, using the Lagrange expansion of a determinant,

$$\det\left(\ell_{\alpha_i}(x_j)\right)_{i,j=1}^d = \sum_{\beta\in S_d} \operatorname{sign}(\beta)\,\ell_{\alpha_1}(x_{\beta_1})\cdots\ell_{\alpha_d}(x_{\beta_d}),$$

we see that the left-hand side of the formula is equal to $\lambda_{\alpha_1,n}^2\ldots\lambda_{\alpha_d,n}^2$, which proves the theorem. □

5.5 Chebyshev Polynomials of Type $\mathscr{A}_d$

The Chebyshev polynomials of type $\mathscr{A}_d$ are orthogonal polynomials derived from orthogonal exponential functions that are invariant under the reflection group $\mathscr{A}_d$, just as the Chebyshev polynomials are derived from $\mathrm{e}^{i\theta} \pm \mathrm{e}^{-i\theta}$, which is invariant or skew-invariant under $\mathbb{Z}_2$. They are the d-variable extension of the second family of Koornwinder polynomials, discussed in Section 2.9.

The space $\mathbb{R}^d$ can be identified with the hyperplane

$$\mathbb{R}^{d+1}_H := \{\mathbf{t} \in \mathbb{R}^{d+1} : t_1 + \cdots + t_{d+1} = 0\} \subset \mathbb{R}^{d+1}.$$

In this section we shall adopt the convention of using a bold letter, such as $\mathbf{t} = (t_1, \ldots, t_{d+1})$, to denote the homogeneous coordinates in $\mathbb{R}^{d+1}_H$ that satisfy $t_1 + \cdots + t_{d+1} = 0$. The root lattice of the group $\mathscr{A}_d$ is generated by reflections σ_{ij}, defined by, in homogeneous coordinates,

$$\mathbf{t}\sigma_{ij} := \mathbf{t} - 2\frac{\langle \mathbf{t}, \mathbf{e}_{i,j}\rangle}{\langle \mathbf{e}_{i,j}, \mathbf{e}_{i,j}\rangle}\mathbf{e}_{i,j} = \mathbf{t} - (t_i - t_j)\mathbf{e}_{i,j}, \qquad \text{where } \mathbf{e}_{i,j} := e_i - e_j.$$

This group can be identified with the symmetric group of $d+1$ elements. The fundamental domain Ω of $\mathscr{A}_d$ is defined by

$$\Omega = \left\{\mathbf{t} \in \mathbb{R}^{d+1}_H : -1 \le t_i - t_j \le 1,\ 1 \le i < j \le d+1\right\},$$

which is the domain that tiles $\mathbb{R}^{d+1}_H$ under A_d in the sense that

$$\sum_{\mathbf{k} \in \mathbb{Z}^{d+1}_H} \chi_\Omega(\mathbf{t} + \mathbf{k}) = 1, \qquad \text{for almost all } \mathbf{t} \in \mathbb{R}^{d+1}_H. \tag{5.5.1}$$

The dual lattice of $\mathbb{Z}^{d+1}_H$ is given by $\{\mathbf{k}/(d+1) : \mathbf{k} \in \mathbb{H}\}$, where

$$\mathbb{H} := \{\mathbf{k} \in \mathbb{Z}^{d+1} \cap \mathbb{R}^{d+1}_H : k_1 \equiv \cdots \equiv k_{d+1} \mod d+1\}.$$

There is a close relation between tiling and exponential functions. We define the exponential functions

$$\phi_{\mathbf{k}}(\mathbf{t}) := \exp\left(\frac{2\pi \mathrm{i}}{d+1}\mathbf{k}^{\mathrm{T}}\mathbf{t}\right), \qquad \mathbf{k} \in \mathbb{H}.$$

It is easy to see that these functions are periodic in the sense that $\phi_{\mathbf{k}}(\mathbf{t}) = \phi_{\mathbf{k}}(\mathbf{t} + \mathbf{j})$ $\forall \mathbf{j} \in \mathbb{Z}^{d+1} \cap \mathbb{R}^{d+1}_H$. Furthermore, they are orthogonal:

$$\frac{1}{\sqrt{d+1}} \int_\Omega \phi_{\mathbf{k}}(\mathbf{t})\overline{\phi_{\mathbf{j}}(\mathbf{t})}\,\mathrm{d}\mathbf{t} = \delta_{\mathbf{k},\mathbf{j}}. \tag{5.5.2}$$

This orthogonality can be deduced by taking the Fourier transform of (5.5.1) and using the Poisson summation formula. For $d = 2$ the orthogonality is verified directly in Proposition 2.9.1. However, the choice of Ω is different from that of Section 2.9, which is the reason for the different normalization of the integral over Ω.

Define the invariant operator $\mathscr{P}^+$ and the skew-invariant operator $\mathscr{P}^-$ by

$$\mathscr{P}^{\pm} f(\mathbf{t}) := \frac{1}{(d+1)!}\left[\sum_{\sigma\in\mathscr{A}^+} f(\mathbf{t}\sigma) \pm \sum_{\sigma\in\mathscr{A}^-} f(\mathbf{t}\sigma)\right],$$

where $\mathscr{A}^+$ ($\mathscr{A}^-$) contains even (odd) elements in $\mathscr{A}_d$. The invariant and skew-invariant functions defined by, respectively,

$$\mathsf{TC}_{\mathbf{k}}(\mathbf{t}) := \mathscr{P}^+\phi_{\mathbf{k}}(\mathbf{t}), \quad \mathbf{k}\in\Lambda, \qquad \text{and} \qquad \mathsf{TS}_{\mathbf{k}}(\mathbf{t}) := \frac{1}{\mathrm{i}}\mathscr{P}^-\phi_{\mathbf{k}}(\mathbf{t}), \quad \mathbf{k}\in\Lambda^0,$$

are analogues of cosine and sine functions; here

$$\begin{aligned}\Lambda &:= \{\mathbf{k}\in\mathbb{H} : k_1 \ge k_2 \ge \cdots \ge k_{d+1}\},\\ \Lambda^0 &:= \{\mathbf{k}\in\mathbb{H} : k_1 > k_2 > \cdots > k_{d+1}\}.\end{aligned}$$

The fundamental domain Ω of the $\mathscr{A}_d$ lattice is the union of congruent simplexes under the action of the group $\mathscr{A}_d$. We fix one simplex as

$$\triangle := \left\{\mathbf{t}\in\mathbb{R}_H^{d+1} : 0 \le t_i - t_j \le 1, 1 \le i \le j \le d+1\right\}$$

and define an inner product on $\triangle$ by

$$\langle f,g\rangle_\triangle := \frac{1}{|\triangle|}\int_\triangle f(\mathbf{t})\overline{g(\mathbf{t})}\,\mathrm{d}\mathbf{t} = \frac{(d+1)!}{\sqrt{d+1}}\int_\triangle f(\mathbf{t})\overline{g(\mathbf{t})}\,\mathrm{d}\mathbf{t}.$$

Proposition 5.5.1 *For $\mathbf{k},\mathbf{j}\in\Lambda$, we have*

$$\langle \mathsf{TC}_{\mathbf{k}}, \mathsf{TC}_{\mathbf{j}}\rangle_\triangle = \frac{\delta_{\mathbf{k},\mathbf{j}}}{|\mathbf{k}\mathscr{A}_d|} \qquad \textit{and} \qquad \langle \mathsf{TS}_{\mathbf{k}}, \mathsf{TS}_{\mathbf{j}}\rangle_\triangle = \frac{\delta_{\mathbf{k},\mathbf{j}}}{(d+1)!}, \qquad \mathbf{k},\mathbf{j}\in\Lambda^0,$$

where $|\mathbf{k}\mathscr{A}_d|$ denotes the cardinality of the orbit $\mathbf{k}\mathscr{A}_d := \{\mathbf{k}\sigma : \sigma\in A_d\}$.

Proof Both these relations follow from the fact that if $f(\mathbf{t})\overline{g(\mathbf{t})}$ is invariant under $\mathscr{A}_d$ then

$$\frac{1}{\sqrt{d+1}}\int_\Omega f(\mathbf{t})\overline{g(\mathbf{t})}\,\mathrm{d}\mathbf{t} = \langle f,g\rangle_\triangle$$

and from the orthogonality of $\phi_{\mathbf{k}}(\mathbf{t})$ over Ω. □

The Chebyshev polynomials of type $\mathscr{A}_d$ are images of the generalized cosine and sine functions under the change of variables $\mathbf{t}\mapsto z$, where $z_1,\ldots,z_d$ denote the first d elementary symmetric functions of $\mathrm{e}^{2\pi\mathrm{i}t_1},\ldots,\mathrm{e}^{2\pi\mathrm{i}t_{d+1}}$ and are defined by

$$z_k := \binom{d+1}{k}^{-1} \sum_{\substack{J\subset\mathbb{N}_{d+1}\\ |J|=k}} \mathrm{e}^{2\pi\mathrm{i}\sum_{j\in J} t_j}, \tag{5.5.3}$$

where $\mathbb{N}_{d+1} = \{1,2,\ldots,d+1\}$. Let $\mathbf{v}^k := \left(\{d+1-k\}^k, \{-k\}^{d+1-k}\right)$ with $\{t\}^k$ means that t is repeated k times. Then $\mathbf{v}^k/(d+1)$, $k = 1,2,\ldots,d$, are vertices of the simplex $\triangle$ and $z_{\mathbf{k}}$ equals $\mathsf{TC}_{\mathbf{v}^k}(\mathbf{t})$, as can be seen from $(\mathbf{v}_k)^{\mathsf{T}}\mathbf{t} = (d+1)(t_1+\cdots+t_k)$, by the homogeneity of $\mathbf{t}$. Thus $\mathsf{TC}_{\mathbf{k}}$ is a symmetric polynomial in $\mathrm{e}^{2\pi\mathrm{i}t_1},\ldots,\mathrm{e}^{2\pi\mathrm{i}t_{d+1}}$ and hence a polynomial in $z_1,\ldots,z_d$. For $\mathsf{TS}_{\mathbf{k}}$ we need the following lemma:

Lemma 5.5.2 *Let*

$$\mathbf{v}^\circ := \left(\frac{d(d+1)}{2}, \frac{(d-2)(d+1)}{2}, \ldots, \frac{(2-d)(d+1)}{2}, \frac{-d(d+1)}{2}\right).$$

Then

$$\mathsf{TS}_{\mathbf{v}^\circ}(\mathbf{t}) = \frac{(2\mathrm{i})^{d(d+1)/2}}{(d+1)!} \prod_{1 \le \mu < \nu \le d+1} \sin \pi(t_\mu - t_\nu). \tag{5.5.4}$$

Proof Let $\beta := (d, d-1, \ldots, 1, 0) \in \mathbb{N}_0^{d+1}$. Then

$$\exp\left(\frac{2\pi \mathrm{i}}{d+1}\mathbf{v}^\circ \cdot \mathbf{t}\right) = \exp[-\pi \mathrm{i} d(t_1 + \cdots + t_{d+1})]\exp(2\pi \mathrm{i}\beta \cdot \mathbf{t}) = \exp(2\pi \mathrm{i}\beta \cdot \mathbf{t}).$$

Using the Vandermonde determinant (see, for example, p. 40 of Macdonald [1992]),

$$\sum_{\sigma \in S_{d+1}} \rho(\sigma)\sigma(x^\beta) = \det\left(x_i^{d+1-j}\right)_{1 \le i,j \le d+1} = \prod_{1 \le i < j \le d+1} (x_i - x_j)$$

and setting $x_j = \mathrm{e}^{2\pi \mathrm{i} t_j}$, we obtain

$$\begin{aligned} \mathsf{TS}_{\mathbf{v}^\circ}(\mathbf{t}) = \frac{1}{|\mathscr{G}|}\sum_{\sigma \in \mathscr{G}} \rho(\sigma)\exp\left(\frac{2\pi \mathrm{i}}{d+1}\mathbf{v}^\circ \cdot \mathbf{t}\right) &= \frac{1}{(d+1)!}\prod_{1 \le \mu < \nu \le d+1}\left(\mathrm{e}^{2\mathrm{i}\pi t_\mu} - \mathrm{e}^{2\mathrm{i}\pi t_\nu}\right) \\ &= \frac{1}{(d+1)!}\prod_{1 \le \mu < \nu \le d+1}\left(\mathrm{e}^{\mathrm{i}\pi t_{\mu-\nu}} - \mathrm{e}^{\mathrm{i}\pi t_{\nu-\mu}}\right), \end{aligned}$$

where the second equality follows from $t_1 + \cdots + t_{d+1} = 0$; from this we have (5.5.4). □

The same argument that proves (5.5.4) also shows that

$$\mathsf{TS}_{\mathbf{k}+\mathbf{v}^\circ}(\mathbf{t}) = \det\left(x_i^{\lambda_j + \beta}\right)_{1 \le i,j \le d+1} \qquad \forall \mathbf{k} \in \Lambda,$$

where $x_i = \mathrm{e}^{2\pi \mathrm{i} t_i}$ and $\lambda := (k_1 - k_{d+1}, k_2 - k_{d+1}, \ldots, k_d - k_{d+1}, 0)$. Since $\lambda \in \Lambda$ is a partition, $\mathsf{TS}_{\mathbf{k}+\mathbf{v}^\circ}(\mathbf{t})$ is divisible by $\mathsf{TS}_{\mathbf{v}^\circ}$ in the ring $\mathbb{Z}[x_1, \ldots, x_d]$ (cf. p. 40 of Macdonald [1995]) and the quotient

$$s_\lambda(x_1, \ldots, x_d) = \mathsf{TS}_{\mathbf{k}+\mathbf{v}^\circ}(\mathbf{t})/\mathsf{TS}_{\mathbf{v}^\circ}(\mathbf{t})$$

is a symmetric polynomial in $x_1, \ldots, x_d$, which is the Schur function in the variables $x_1, \ldots, x_d$ corresponding to the partition λ. In particular, s_λ is a polynomial in the elementary symmetric polynomials $z_1, \ldots, z_d$; more precisely, it is a polynomial in z of degree $(k_1 - k_{d+1})/(d+1)$, as shown by formula (3.5) in Macdonald [1995]. This is our analogue of the Chebyshev polynomials of the second kind.

Definition 5.5.3 Define the index mapping $\alpha : \Lambda \to \mathbb{N}_0^d$ by

$$\alpha_i = \alpha_i(\mathbf{k}) := \frac{k_i - k_{i+1}}{d+1}, \qquad 1 \le i \le d. \tag{5.5.5}$$

Under the change of variables $\mathbf{t} \mapsto z = (z_1, \ldots, z_k)$ in (5.5.3), define

$$T_\alpha(z) := \mathsf{TC}_\mathbf{k}(\mathbf{t}) \quad \text{and} \quad U_\alpha(z) := \frac{\mathsf{TS}_{\mathbf{k}+\mathbf{v}^\circ}(\mathbf{t})}{\mathsf{TS}_{\mathbf{v}^\circ}(\mathbf{t})}, \qquad \alpha \in \mathbb{N}_0^d.$$

The polynomials $T_\alpha(z)$ and $U_\alpha(z)$ are called the Chebyshev polynomials type $\mathscr{A}_d$, of the first and the second kind, respectively.

The polynomials T_α and U_α are of degree $|\alpha|$ in z. They are analogues of the classical Chebyshev polynomials of the first kind and the second kind, respectively. In particular, we shall show that they are respectively orthogonal on the domain $\triangle^*$, the image of $\triangle$ under $\mathbf{t} \mapsto z$. We need the Jacobian of the change of variables. Let us define

$$w(x) = w(x(\mathbf{t})) := \prod_{1 \le \mu < \nu \le d+1} \sin^2 \pi(t_\mu - t_\nu)$$

under this change of variables.

Lemma 5.5.4 *The Jacobian of the change of variables* $\mathbf{t} \mapsto x$ *is given by*

$$\det \frac{\partial(z_1, z_2, \ldots, z_d)}{\partial(t_1, t_2, \ldots, t_d)} = \prod_{k=1}^{d} \frac{2\pi \mathrm{i}}{\binom{d+1}{k}} \prod_{1 \le \mu < \nu \le d+1} |\sin \pi(t_\mu - t_\nu)|.$$

Proof Regarding $t_1, t_2, \ldots, t_{d+1}$ as independent variables, one sees that

$$\frac{\partial z_k}{\partial t_j} = \frac{2\pi \mathrm{i}}{\binom{d+1}{k}} \sum_{j \in I \subseteq \mathbb{N}_{d+1}, |I| = k} \mathrm{e}^{2\pi \mathrm{i} \sum_{\nu \in I} t_\nu}.$$

For each fixed j, let $\mathbb{N}_m^{(j)}$ denote the set of $\mathbb{N}_m \subset \{j\}$ and split $I \subset \mathbb{N}_{d+1}$ into two parts, one containing $\{j, d+1\}$ and the other not. We then obtain, after canceling the common factor,

$$\begin{aligned}
\frac{\partial z_k}{\partial t_j} - \frac{\partial z_k}{\partial t_{d+1}} &= \frac{2\pi \mathrm{i}}{\binom{d+1}{k}} \left(\sum_{j \in I \subseteq \mathbb{N}_{d+1}^{\{d+1\}}, |I| = k} - \sum_{d+1 \in I \subseteq \mathbb{N}_{d+1}^{\{j\}}, |I| = k} \right) \mathrm{e}^{2\pi \mathrm{i} \sum_{\nu \in I} t_\nu} \\
&= \frac{2\pi \mathrm{i}}{\binom{d+1}{k}} \left(\mathrm{e}^{2\pi \mathrm{i} t_j} - \mathrm{e}^{2\pi \mathrm{i} t_{d+1}} \right) \sum_{I \subseteq \mathbb{N}_d^{\{j\}}, |I| = k-1} \mathrm{e}^{2\pi \mathrm{i} \sum_{\nu \in I} t_\nu}.
\end{aligned}$$

Hence, setting $f_{j,k}^d := \sum_{I \subseteq \mathbb{N}_d^{\{j\}}, |I| = k-1} \mathrm{e}^{2\pi \mathrm{i} \sum_{\nu \in I} t_\nu}$ and defining the matrix $F_d := \left(f_{j,k}^d\right)_{1 \le j,k \le d}$, we have

$$\det \frac{\partial(z_1,z_2,\ldots,z_d)}{\partial(t_1,t_2,\ldots,t_d)} = \prod_{k=1}^{d} \frac{2\pi \mathrm{i}}{\binom{d+1}{k}} \prod_{j=1}^{d} \left(\mathrm{e}^{2\pi \mathrm{i} t_j} - \mathrm{e}^{2\pi \mathrm{i} t_{d+1}}\right) \det F_d.$$

By definition $f_{j,1}^d = 1$; splitting $\mathbb{N}^{\{j\}}$ into a part containing d and a part not containing d, it follows that

$$\begin{aligned} f_{j,k}^d - f_{d,k}^d &= \left(\mathrm{e}^{2\pi \mathrm{i} t_d} - \mathrm{e}^{2\pi \mathrm{i} t_j}\right) \sum_{I \subseteq \mathbb{N}_d^{\{j,d\}}, |I|=k-2} \mathrm{e}^{2\pi \mathrm{i} \sum_{\nu \in I} t_\nu} \\ &= \left(\mathrm{e}^{2\pi \mathrm{i} t_d} - \mathrm{e}^{2\pi \mathrm{i} t_j}\right) f_{j,k-1}^{d-1}, \end{aligned}$$

which allows us to use induction and show that

$$\begin{aligned} \det F_d &= \prod_{1 \le \mu < d-1} \left(\mathrm{e}^{2\pi \mathrm{i} t_\mu} - \mathrm{e}^{2\pi \mathrm{i} t_d}\right) \det F_{d-1} = \cdots \\ &= \prod_{1 \le \mu < \nu \le d} \left(\mathrm{e}^{2\pi \mathrm{i} t_\mu} - \mathrm{e}^{2\pi \mathrm{i} t_\nu}\right). \end{aligned}$$

The use of (5.5.4) now proves the lemma. □

From the homogeneity of $\mathbf{t}$, the z_k in (5.5.3) satisfy $z_k = \overline{z_{d+1-k}}$. Hence, we can define real coordinates $x_1, \ldots, x_d$ by

$$x_k = \frac{z_k + z_{d+1-k}}{2}, \quad x_{d+1-k} = \frac{z_k - z_{d+1-k}}{2\mathrm{i}} \qquad \text{for} \quad 1 \le k \le \lfloor \frac{d}{2} \rfloor,$$

and we define $x_{(d+1)/2} = \frac{1}{\sqrt{2}} z_{(d+1)/2}$ when d is odd. Instead of making the change of variables $\mathbf{t} \mapsto z$, we consider $\mathbf{t} \mapsto x$ of $\mathbb{R}_H^{d+1} \mapsto \mathbb{R}^d$. The image of the simplex $\triangle$ under this mapping is

$$\triangle^* := \left\{ x = x(\mathbf{t}) \in \mathbb{R}^d : \mathbf{t} \in \mathbb{R}_H^{d+1}, \prod_{1 \le i < j \le d+1} \sin \pi(t_i - t_j) \ge 0 \right\}.$$

For $d = 2$, this is the second family of Koornwinder polynomials, presented in Section 2.9, and $\triangle^*$ is the region bounded by the hypercycloid.

The Chebyshev polynomials are orthogonal with respect to the weight functions $W_{-1/2}$ and $W_{1/2}$, respectively, on $\triangle^*$, where

$$W_\alpha(z) := \prod_{1 \le \mu < \nu \le d+1} |\sin \pi(t_\mu - t_\nu)|^{2\alpha}.$$

We will need the cases $\alpha = -\frac{1}{2}$ and $\alpha = \frac{1}{2}$ of the weighted inner product

$$\langle f, g \rangle_{W_\alpha} := c_\alpha \int_{\Delta^*} f(z) \overline{g(z)} W_\alpha(z) \, \mathrm{d}z,$$

where c_α is a normalization constant, $c_\alpha := 1/\int_{\Delta^*} W_\alpha(z) \, \mathrm{d}z$.

The orthogonality of T_κ and U_κ, respectively, then follows from the orthogonality of $\mathsf{TC}_\mathbf{k}$ and $\mathsf{TS}_\mathbf{k}$ with the change of variables. More precisely, Proposition 5.5.1 leads to the following theorem:

Theorem 5.5.5 *The polynomials T_α and U_α are orthogonal polynomials with respect to $W_{-1/2}$ and $W_{1/2}$, respectively, and*

$$\langle T_\alpha, T_\beta \rangle_{W_{-1/2}} = \frac{\delta_{\alpha,\beta}}{|\mathbf{k}\mathscr{A}_d|} \quad \text{and} \quad \langle U_\alpha, U_\beta \rangle_{W_{1/2}} = \frac{\delta_{\alpha,\beta}}{(d+1)!}$$

for $\alpha, \beta \in \mathbb{N}_0^d$, where $\mathbf{k}$ is defined by $\alpha = \alpha(\mathbf{k})$.

Both $T_\alpha(z)$ and $U_\alpha(z)$ are polynomials of degree $|\alpha| = \alpha_1 + \cdots + \alpha_d$ in z. Moreover, they both satisfy a simple recursion relation.

Theorem 5.5.6 *Let P_α denote either T_α or U_α. Then*

$$\overline{P_\alpha(z)} = P_{\alpha_d, \alpha_{d-1}, \ldots, \alpha_1}(z), \qquad \alpha \in \mathbb{N}_0^d, \tag{5.5.6}$$

and P_α satisfies the recursion relation

$$\binom{d+1}{i} z_i P_\alpha(z) = \sum_{\mathbf{j} \in \mathbf{v}^i \mathscr{A}_d} P_{\alpha + \alpha(\mathbf{j})}(z), \qquad \alpha \in \mathbb{N}_0^d, \tag{5.5.7}$$

in which the components of $\alpha(\mathbf{j})$, $\mathbf{j} \in \mathbf{v}^i \mathscr{A}_d$, have values in $\{-1, 0, 1\}$, $U_\alpha(z) = 0$ whenever α has a component $\alpha_i = -1$ and

$$T_0(z) = 1, \quad T_{\varepsilon_k}(z) = z_k, \qquad 1 \le k \le d,$$

$$U_0(z) = 1, \quad U_{\varepsilon_k}(z) = \binom{d+1}{k} z_k, \qquad 1 \le k \le d.$$

Proof The relation (5.5.6) follows readily from the fact that $-(k_i - k_j) = k_j - k_i$ and (5.5.5). For a proof of the recursion relations we need

$$\mathsf{TC}_{\mathbf{j}}(\mathbf{t}) \mathsf{TC}_{\mathbf{k}}(\mathbf{t}) = \frac{1}{|\mathscr{A}_d|} \sum_{\sigma \in \mathscr{A}_d} \mathsf{TC}_{\mathbf{k} + \mathbf{j}\sigma}(\mathbf{t}), \qquad \mathbf{j}, \mathbf{k} \in \Lambda, \tag{5.5.8}$$

$$\mathsf{TC}_{\mathbf{j}}(\mathbf{t}) \mathsf{TS}_{\mathbf{k}}(\mathbf{t}) = \frac{1}{|\mathscr{A}_d|} \sum_{\sigma \in \mathscr{A}_d} \mathsf{TS}_{\mathbf{k} + \mathbf{j}\sigma}(\mathbf{t}), \qquad \mathbf{j}, \mathbf{k} \in \Lambda, \tag{5.5.9}$$

which follow, using simple computation, from the definitions of $\mathsf{TC}_{\mathbf{k}}$ and $\mathsf{TS}_{\mathbf{k}}$. The relation (5.5.7) follows immediately from (5.5.8) and (5.5.9). Further, using (5.5.9) it is not difficult to verify that

$$z_k \mathsf{TS}_{\mathbf{v}^\circ}(\mathbf{t}) = \frac{k!(d+1-k)!}{(d+1)!} \mathsf{TS}_{\mathbf{v}^\circ + \mathbf{v}^k}(\mathbf{t}), \qquad 1 \le k \le d,$$

from which the values of P_0 and P_{ε_k} are obtained readily. If α has a component $\alpha_i = -1$ then, by (5.5.5), $k_i(\alpha) = k_{i+1}(\alpha) - (d+1)$. A quick computation shows that then $k_i(\alpha) + v_i^\circ = k_{i+1}(\alpha) + v_{i+1}^\circ$, which implies that $\mathbf{k} + \mathbf{v}^\circ \in \partial\Lambda$, so that $\mathsf{TS}_{\mathbf{k}+\mathbf{v}^\circ}(\mathbf{t}) = 0$ and $U_\alpha(z) = 0$. □

By (5.5.6) together with $z_k = z_{d-k+1}$ we can derive a sequence of real orthogonal polynomials from either $\{T_\alpha\}$ or $\{U_\alpha\}$.

5.6 Sobolev Orthogonal Polynomials on the Unit Ball

Sobolev orthogonal polynomials are orthogonal with respect to an inner product that involves derivatives. In this section we consider two types of Sobolev orthogonal polynomials on the unit ball B^d.

5.6.1 Sobolev orthogonal polynomials defined via the gradient operator

In this subsection we consider the inner product defined by

$$\langle f,g\rangle_\nabla := \frac{\lambda}{\sigma_d}\int_{B^d} \nabla f(x)\cdot\nabla g(x)\,\mathrm{d}x + \frac{1}{\sigma_d}\int_{S^{d-1}} f(x)g(x)\,\mathrm{d}\omega(x), \tag{5.6.1}$$

where λ is a fixed positive number and $x\cdot y$ stands for the inner product of vectors x and y. Since the positive definiteness of the bilinear form (5.6.1) can be easily verified, it is an inner product and consequently polynomials orthogonal with respect to $\langle f,g\rangle_\nabla$ exist. Let $\mathscr{V}_n^d(\nabla)$ denote the space of orthogonal polynomials of degree n with respect to (5.6.1).

In analogy to (5.2.4), a basis of $\mathscr{V}_n^d(\nabla)$ can be given in terms of polynomials of the form

$$Q_{j,\nu}^n(x) := q_j(2\|x\|^2-1)Y_\nu^{n-2j}(x), \qquad 0\le 2j\le n, \tag{5.6.2}$$

where $\{Y_\nu^{n-2j}: 0\le \nu\le a_{n-2j}^d\}$, with $a_m^d := \dim\mathscr{H}_m^d$, is an orthonormal basis of $\mathscr{H}_{n-2j}^d$ and q_j is a polynomial of degree j in one variable. We need the following lemma.

Lemma 5.6.1 *Let $Q_{j,\nu}^n$ be defined as above. Then*

$$\Delta\left[(1-\|x\|^2)Q_{j,\nu}^n(x)\right] = 4\left(\mathscr{J}_\beta q_j\right)(2r^2-1)Y_\nu^{n-2j}(x),$$

where Δ is the Laplacian operator, $\beta = n-2j+\frac{d-2}{2}$ and

$$(\mathscr{J}_\beta q_j)(s) = (1-s^2)q_j''(s) + [\beta-1-(\beta+3)s]q_j'(s) - (\beta+1)q_j(s).$$

Proof Using spherical polar coordinates, it follows from (4.1.4) and (4.1.5) that

$$\begin{aligned}\Delta\left[(1-\|x\|^2)Q_{j,\nu}^n(x)\right] &= \Delta\left[(1-r^2)q_j(2r^2-1)r^{n-2j}Y_\nu^{n-2j}(x')\right]\\ &= 4r^{n-2j}\left[4r^2(1-r^2)q_j''(2r^2-1)\right.\\ &\quad +2((\beta+1)-(\beta+3)r^2)q_j'(2r^2-1)\\ &\quad \left.-(\beta+1)q_j(2r^2-1)\right]Y_\nu^{n-2j}(x').\end{aligned}$$

Setting $s\mapsto 2r^2-1$ gives the stated result. □

It is interesting that λ does not appear in the $Q_{j,\nu}^n$.

Theorem 5.6.2 *For $0 \le j \le n/2$, let $\{Y_\nu^{n-2j} : 1 \le \nu \le a_{n-2j}^d\}$ be an orthonormal basis of $\mathscr{H}_{n-2j}^d$. Then a mutually orthogonal basis $\{Q_{j,\nu}^n : 0 \le j \le \frac{n}{2}, 1 \le \nu \le a_{n-2j}^d\}$ for $\mathscr{V}_n^d(\nabla)$ is given by*

$$
\begin{aligned}
Q_{0,\nu}^n(x) &= Y_\nu^n(x), \\
Q_{j,\nu}^n(x) &= (1-\|x\|^2) P_{j-1}^{(1,n-2j+(d-2)/2)}(2\|x\|^2-1) Y_\nu^{n-2j}(x)
\end{aligned}
\tag{5.6.3}
$$

for $0 < j \le n/2$. Furthermore, the norms of these polynomials are

$$
\langle Q_{0,\nu}^n, Q_{0,\nu}^n \rangle_\nabla = n\lambda + 1, \qquad \langle Q_{j,\nu}^n, Q_{j,\nu}^n \rangle_\nabla = \frac{2j^2}{n+\frac{d-2}{2}} \lambda. \tag{5.6.4}
$$

Proof As in Proposition 5.2.1, we only need to prove the orthogonality. We start with Green's identity

$$
\int_{B^d} \nabla f(x) \cdot \nabla g(x) \, \mathrm{d}x = \int_{S^{d-1}} f(x) \frac{\mathrm{d}}{\mathrm{d}r} g(x) \, \mathrm{d}\omega - \int_{B^d} f(x) \Delta g(x) \, \mathrm{d}x,
$$

where $\mathrm{d}/\mathrm{d}r$ is the normal derivative, which coincides with the derivative in the radial direction. The above identity can be used to rewrite the inner product $\langle \cdot, \cdot \rangle_\nabla$ as

$$
\langle f, g \rangle_\nabla = \frac{1}{\sigma_d} \int_{S^{d-1}} f(x) \left[\lambda \frac{\mathrm{d}}{\mathrm{d}r} g(x) + g(x) \right] \mathrm{d}\omega - \frac{\lambda}{\sigma_d} \int_{B^d} f(x) \Delta g(x) \, \mathrm{d}x.
$$

First we consider the case $j = 0$, that is, the orthogonality of $Q_{0,\nu}^n = Y_\nu^n$. Setting $j = 0$ in (5.2.4) shows that Y_ν^n is an orthogonal polynomial in $\mathscr{V}_n^d(W_\mu^B)$. Since $Q_{j,\nu}^n(x)|_{r=1} = 0$,

$$
\frac{\mathrm{d}}{\mathrm{d}r} Q_{j,\nu}^n(x)|_{r=1} = -2 P_{j-1}^{(1,n-2j+(d-2)/2)}(1) Y_\nu^{n-2j}(x')
$$

and $\Delta Q_{j,\nu}^m \in \Pi_{m-2}^d$, it follows from the above expression for $\langle f, g \rangle_\nabla$ that, for $m < n$, $j > 0$ and $0 \le \mu \le \sigma_{m-2j}$,

$$
\langle Q_{0,\nu}^n, Q_{j,\mu}^m \rangle_\nabla = -2 P_{j-1}^{(1,n-2j+(d-2)/2)}(1) \frac{\lambda}{\sigma_d} \int_{S^{d-1}} Y_\nu^n(x') Y_\mu^m(x') \, \mathrm{d}\omega(x') = 0.
$$

Furthermore, using the fact that $(\mathrm{d}/\mathrm{d}r) Y_\nu^n(x)|_{r=1} = n Y_\nu^n(x')$, the same consideration shows that

$$
\langle Q_{0,\nu}^n, Q_{0,\nu}^m \rangle_\nabla = (\lambda n + 1) \frac{1}{\sigma_d} \int_{S^{d-1}} \left[Y_\nu^n(x') \right]^2 \mathrm{d}\omega(x') \, \delta_{n,m} = (\lambda n + 1) \delta_{n,m}.
$$

Next we consider $Q_{j,\nu}^n$ for $j \ge 1$. In this case $Q_{j,\nu}^n(x)|_{r=1} = 0$ since it contains the factor $1 - \|x\|^2$, which is zero on S^{d-1}. Consequently, the first term in

$$\langle Q^n_{j,\nu}, Q^m_{l,\mu}\rangle_\nabla = \frac{1}{\sigma_d}\int_{\mathbb{S}^{d-1}} Q^n_{j,\nu}(x')\left(\lambda \frac{\mathrm{d}}{\mathrm{d}r} Q^m_{l,\mu} + Q^m_{l,\mu}\right)(x')\,\mathrm{d}\omega(x')$$
$$- \frac{\lambda}{\omega_d}\int_{\mathbb{B}^d} Q^n_{j,\nu}(x)\Delta Q^m_{l,\mu}(x)\,\mathrm{d}x$$

is zero. For the second term, we use Lemma 5.6.1 to derive a formula for $\Delta Q^n_{j,\nu}$. The formula in the lemma gives

$$\Delta Q^n_{j,\nu}(x) = 4\left(\mathscr{J}_\beta P^{(1,\beta)}_{j-1}\right)(2r^2-1)Y^{n-2j}_\nu(x), \qquad \beta = n-2j+\tfrac{d-2}{2}.$$

On the other hand, the differential equation satisfied by the Jacobi polynomial becomes, for $P^{(1,\beta)}_{j-1}$,

$$(1-s^2)y'' - [1-\beta+(3+\beta)s]y' + (j-1)(j+\beta+1)y = 0,$$

which implies that $(\mathscr{J}_\beta P^{(1,\beta)}_{j-1})(s) = -j(j+\beta)P^{(1,\beta)}_{j-1}(s)$. Consequently, denoting by $P^n_{j,\nu}(W^B_\mu;x)$ the polynomials in (5.2.4) without their normalization constant, we obtain

$$\begin{aligned}\Delta Q^n_{j,\nu}(x) &= -4j(j+\beta)P^{(1,\beta)}_{j-1}(2r^2-1)Y^{n-2j}_\nu(x)\\ &= -4j(n-j+\tfrac{d-2}{2})P^{n-2}_{j-1,\nu}(W_1;x).\end{aligned} \tag{5.6.5}$$

Hence, using the fact that $Q^m_{l,\mu}(x) = (1-\|x\|^2)P^{m-2}_{l-1,\mu}(W_{3/2};x)$, we derive from (5.6.5) that

$$\begin{aligned}&\int_{\mathbb{B}^d} Q^m_{l,\mu}(x)\Delta Q^n_{j,\nu}(x)\,\mathrm{d}x\\ &= -4j(n-j+\tfrac{d-2}{2})\int_{\mathbb{B}^d} P^{m-2}_{l-1,\mu}(W_{3/2};x)P^{n-2}_{j-1,\nu}(W_{3/2};x)(1-\|x\|^2)\,\mathrm{d}x\\ &= -4j(n-j+\tfrac{d-2}{2})\int_{\mathbb{B}^d}\left[P^{n-2}_{j-1,\nu}(W_{3/2};x)\right]^2(1-\|x\|^2)\,\mathrm{d}x\,\delta_{n,m}\delta_{j,l}\delta_{\nu,\mu}.\end{aligned}$$

Using the norm of $P^n_{j,\nu}(W^B_\mu;x)$ computed in the proof of Proposition 5.2.1, we can then verify (5.6.4) for $j \ge 1$. □

From the explicit form of the basis (5.6.3) it follows that $Q^n_{j,\nu}$ is related to polynomials orthogonal with respect to $W_{3/2}(x) = 1-\|x\|^2$. In fact, we have

$$Q^n_{j,\nu}(x) = (1-\|x\|^2)P^{n-2}_{j-1,\nu}(W_{3/2};x), \qquad j \ge 1,$$

which has already been used in the above proof. An immediate consequence is the following corollary.

Corollary 5.6.3 *For $n \ge 1$,*

$$\mathscr{V}^d_n(\nabla) = \mathscr{H}^d_n \oplus (1-\|x\|^2)\mathscr{V}^d_{n-2}(W_{3/2}).$$

Recall that the classical orthogonal polynomials in $\mathscr{V}_n^d(W_\mu^B)$ are eigenfunctions of a second-order differential operator $\mathscr{D}_\mu$ for $\mu > -\frac{1}{2}$; see (5.2.3). Usng the relation (5.6.5), we can deduce the following result.

Corollary 5.6.4 *The polynomials P in $\mathscr{V}_n^d(\nabla)$ are eigenfunctions of $D_{-1/2}$, and*

$$\mathscr{D}_{-1/2}P = -(n+d)(n-2)\Delta P \qquad \forall P \in \mathscr{V}_n^d(\nabla).$$

5.6.2 Sobolev orthogonal polynomials defined via the Laplacian operator

We consider the inner product defined by

$$\langle f, g\rangle_\Delta := a_d \int_{B^d} \Delta[(1-\|x\|^2)f(x)]\,\Delta[(1-\|x\|^2)g(x)]\,\mathrm{d}x,$$

where $a_d = (4d\sigma_{d-1})^{-1}$ so that $\langle 1,1\rangle_\Delta = 1$. To see that it is an inner product, we need to show only that $\langle f,f\rangle_\Delta > 0$ if $f \neq 0$. However, if $\Delta(1-\|x\|^2)f(x) = 0$ then $(1-\|x\|^2)f(x)$ is a solution of the Dirichlet problem with zero boundary condition for the Laplace operator on the unit ball, so that by the uniqueness of the Dirichlet problem f must be the zero function. Let $\mathscr{V}_n^d(\Delta)$ denote the space of orthogonal polynomials of degree n with respect to (5.6.1).

As in the case of $\langle \cdot,\cdot\rangle$, a basis of $\mathscr{V}_n^d(\Delta)$ can be given in terms of polynomials of the form

$$R_{j,\nu}^n(x) := q_j(2\|x\|^2 - 1)Y_\nu^{n-2j}(x), \qquad 0 \le 2j \le n, \tag{5.6.6}$$

where $\{Y_\nu^{n-2j} : 0 \le \nu \le a_{n-2j}^d\}$, with $a_m^d := \dim \mathscr{H}_m^d$, is an orthonormal basis of $\mathscr{H}_{n-2j}^d$ and q_j is a polynomial of degree j in one variable. We need the following lemma, in which $\mathscr{J}_\beta$ is defined as in Lemma 5.6.1.

Lemma 5.6.5 *Let $\beta > -1$. The polynomials p_j^β defined by*

$$p_0^\beta(s) = 1, \qquad p_j^\beta(s) = (1-s)P_{j-1}^{(2,\beta)}(s), \quad j \ge 1,$$

are orthogonal with respect to the inner product $(f,g)_\beta$ defined by

$$(f,g)_\beta := \int_{-1}^1 (\mathscr{J}_\beta f)(s)(\mathscr{J}_\beta g)(s)(1+s)^\beta\,\mathrm{d}s, \qquad \beta > -1.$$

Proof The three-term relation of the Jacobi polynomials gives

$$(1-s)P_{j-1}^{(2,\beta)}(s) = \frac{2}{2j+\beta+1}\left[(j+1)P_{j-1}^{(1,\beta)}(s) - jP_j^{(1,\beta)}(s)\right]$$

and the differential equation satisfied by $P_{j-1}^{(1,\beta)}$ is

$$(1-s^2)y'' + [-1+\beta-(3+\beta)s]y' + (j-1)(j+\beta+1)y = 0.$$

Using these two facts, we easily deduce that

$$\begin{aligned}
\frac{1}{2}(2j+\beta+1)\mathscr{I}_\beta\left[(1-s)P_{j-1}^{(2,\beta)}(s)\right] &= (j+1)\mathscr{I}_\beta P_{j-1}^{(1,\beta)}(s) - j\mathscr{I}_\beta P_{j-1}^{(1,\beta)}(s)\\
&= (j+1)\Big([-(j-1)(j+\beta+1)-(\beta+1)]P_{j-1}^{(1,\beta)}(s)\\
&\qquad -j[-j(j+\beta+2)-(\beta+1)]P_j^{(1,\beta)}(s)\Big)\\
&= -j(j+1)\left[(j+\beta)P_{j-1}^{(1,\beta)}(s)-(j+\beta+1)P_j^{(1,\beta)}(s)\right].
\end{aligned}$$

We need one more formula for the Jacobi polynomials (p. 782, formula (22.7.8) in Abramowitz and Stegun [1970]):

$$(2j+\beta+1)P_j^{(0,\beta)}(s) = (j+\beta+1)P_j^{(1,\beta)}(s) - (j+\beta)P_{j-1}^{(1,\beta)}(s),$$

which implies immediately that

$$\mathscr{I}_\beta\left[(1-s)P_{j-1}^{(2,\beta)}(s)\right] = 2j(j+1)P_j^{(0,\beta)}(s).$$

Hence, for $j, j' \ge 1$, we conclude that

$$\begin{aligned}
(p_j^\beta, p_{j'}^\beta)_\beta &= \int_{-1}^1 \mathscr{I}_\beta\left[(1-s)P_{j-1}^{(2,\beta)}(s)\right]\mathscr{I}_\beta\left[(1-s)P_{j'-1}^{(2,\beta)}(s)\right](1+s)^\beta \mathrm{d}s\\
&= 4j(j+1)j'(j'+1)\int_{-1}^1 P_j^{(0,\beta)}(s)P_{j'}^{(0,\beta)}(s)(1+s)^\beta \mathrm{d}s = 0
\end{aligned}$$

whenever $j \neq j'$. Furthermore, for $j \ge 1$ we have

$$(p_0^\beta, p_j^\beta)_\beta = -2j(j+1)(\beta+1)\int_{-1}^1 P_j^{(0,\beta)}(s)(1+s)^\beta \mathrm{d}s = 0,$$

since $(\mathscr{I}_\beta p_0^\beta)(s) = (\mathscr{I}_\beta 1)(s) = -(\beta+1)$. □

Theorem 5.6.6 *A mutually orthogonal basis for $\mathscr{V}_n^d(\Delta)$ is given by*

$$\begin{aligned}
R_{0,\nu}^n(x) &= Y_\nu^n(x),\\
R_{j,\nu}^n(x) &= (1-\|x\|^2)P_{j-1}^{(2,n-2j+(d-2)/2)}Y_\nu^{n-2j}(x), \qquad 1 \le j \le \tfrac{n}{2},
\end{aligned} \tag{5.6.7}$$

where $\{Y_\nu^{n-2j} : 1 \le \nu \le \sigma_{n-2j}\}$ is an orthonormal basis of $\mathscr{H}_{n-2j}^d$. Furthermore

$$\langle R_{0,\nu}^n, R_{0,\nu}^n\rangle_\Delta = \frac{2n+d}{d}, \qquad \langle R_{j,\nu}^n, R_{j,\nu}^n\rangle_\Delta = \frac{8j^2(j+1)^2}{d(n+\frac{d}{2})}. \tag{5.6.8}$$

Proof First we show that $R_{j,\nu}^n \in \mathscr{V}_n^d(\Delta)$. Let p_j^β be defined as in the previous lemma. Set $q_j := p_j^{\beta_{n-2j}}$ with $\beta_k = k+(d-2)/2$. Using (5.2.2), Lemma 5.6.1 and the orthonormality of Y_ν^{n-2j}, we obtain

$$\langle R^n_{j,\nu}, R^{n'}_{j',\nu'}\rangle_\Delta = \delta_{\nu,\nu'}\delta_{n-2j,n'-2j'}\frac{1}{4d}\int_0^1 r^{d+2(n-2j)-1}4^2 \\ \times (\mathscr{I}_{\beta_{n-2j}}q_j)(2r^2-1)]^2(\mathscr{I}_{\beta_{n'-2j'}}q_{j'})(2r^2-1)\,\mathrm{d}r.$$

The change of variable $r \mapsto \sqrt{\frac{1}{2}(1+s)}$ shows then that

$$\langle R^n_{j,\nu}, R^{n'}_{j',\nu'}\rangle_\Delta = \delta_{\nu,\nu'}\delta_{n-2j,n'-2j'}\frac{1}{2^{\beta_{n-2j}}d}(q_j,q_{j'})_{\beta_{n-2j}}, \tag{5.6.9}$$

which proves the orthogonality by Lemma 5.6.5. To compute the norm of $Q^n_{0,\nu}$ we use the fact that

$$\Delta[(1-\|x\|^2)Y^{n-2j}_\nu(x)] = -2dY^n_\nu(x) - 4\langle x,\nabla\rangle Y^n_\nu = -2(d+2n)Y^n_\nu(x),$$

by Euler's formula on homogeneous polynomials, which shows that

$$\langle Q^n_{0,\nu}, Q^n_{0,\nu}\rangle_\Delta = \frac{(2n+d)^2}{\omega_{d-1}d}\int_0^1 r^{d-1+2n}\,\mathrm{d}r\int_{S^{d-1}}[Y^n_\nu(x)]^2\,\mathrm{d}x = \frac{2n+d}{d}.$$

Furthermore, using equation (5.6.9), the proof of Lemma 5.6.5 shows that

$$\begin{aligned}\langle Q^n_{j,\nu}, Q^n_{j,\nu}\rangle_\Delta &= \frac{1}{2^{\beta_j}d}(p_j,p_{j'})_{\beta_j} \\ &= \frac{4j^2(j+1)^2}{2^{\beta_j}d}\int_{-1}^1\left[P_j^{(0,\beta_j)}(s)\right]^2(1+s)^{\beta_j}\,\mathrm{d}s \\ &= \frac{8j^2(j+1)^2}{(\beta_j+2j+1)d} = \frac{8j^2(j+1)^2}{(n+\frac{d}{2})d}\end{aligned}$$

using the expression for the norm of the Jacobi polynomial. □

From the explicit formula for the basis (5.6.7) it follows that $R^n_{j,\nu}$ is related to the polynomials orthogonal with respect to $W_{5/2}(x) = (1-\|x\|^2)^2$ on B^d. In fact we have

$$R^n_{j,\nu}(x) = (1-\|x\|^2)P^{n-2}_{j-1,\nu}(W_{5/2};x), \qquad j \geq 1,$$

which leads to the following corollary.

Corollary 5.6.7 *For $n \geq 1$,*

$$\mathscr{V}^d_n(\Delta) = \mathscr{H}^d_n \oplus (1-\|x\|^2)\mathscr{V}^d_{n-2}(W_{5/2}).$$

This corollary should be compared with Corollary 5.6.3. The polynomials in $\mathscr{V}^d_n(\Delta)$, however, are not all eigenfunctions of a second-order differential operator.

5.7 Notes

By the classical orthogonal polynomials we mean the polynomials that are orthogonal with respect to the Jacobi weight functions on the cube, ball or simplex in $\mathbb{R}^d$, the Laguerre weight function on $\mathbb{R}^d_+$, and the Hermite weight on $\mathbb{R}^d$. These polynomials are among the first that were studied.

The study of the Hermite polynomials of several variables was begun by Hermite. He was followed by many other authors; see Appell and de Fériet [1926] and Chapter XII, Vol. II, of Erdélyi *et al.* [1953]. Analogues of the Hermite polynomials W_H can be defined more generally for the weight function

$$W(x) = (\det A)^{1/2}\pi^{-d/2}\exp\left(-x^{\mathrm{T}}Ax\right), \qquad (5.7.1)$$

where A is a positive definite matrix. Since A is positive definite it can be written as $A = B^{\mathrm{T}}B$. Thus the orthogonal polynomials for W in (5.7.1) can be derived from the Hermite polynomials for W_H by a change of variables. One interesting result for such generalized Hermite polynomials is that two families of generalized Hermite polynomials, defined with respect to the matrices A and A^{-1}, can be biorthogonal.

For the history of the orthogonal polynomials V_α and U_α on B^d, we refer to Chapter XII, Vol. II of Erdélyi *et al.* [1953]. These polynomials were studied in detail in Appell and de Fériet [1926]. They are sometimes used to derive explicit formulae for Fourier expansions. Rosier [1995] used them to study Radon transforms. Orthogonal bases for ridge polynomials were discussed in Xu [2000b], together with a Funk–Hecke type formula for orthogonal polynomials. Compact formulae (5.2.7) for the reproducing kernels were proved in Xu [1999a] and used to study expansion problems. The formula (5.2.8) was proved in Petrushev [1999] in the context of approximation by ridge functions and in Xu [2007] in connection with Radon transforms. The orthogonal polynomials for the unit weight function on B^d were used in Maiorov [1999] to study questions related to neural networks.

The orthogonal polynomials U_α on T^d were defined in Appell and de Fériet [1926] for $d = 2$. The polynomials V_α on T^d are extensions of those for the unit weight function on T^d defined in Grundmann and Möller [1978]. The formula (5.3.5) for the reproducing kernel appeared in Xu [1998d]. A product formula for orthogonal polynomials on the simplex was established in Koornwinder and Schwartz [1997]. The orthogonal polynomials on the simplex also appeared in the work of Griffiths [1979], Griffiths and Spanò [2011, 2013] and Rosengren [1998, 1999]. A basis for the orthogonal polynomials on the simplex can be given in terms of Bernstein polynomials, as shown in Farouki, Goodman and Sauer [2003] and Waldron [2006]. Further properties of the Rodrigues formulae on the simplex were given in Aktaş and Xu [2013]. A probability interpretation of these and other related polynomials was given in Griffiths and Spanò [2011].

The orthogonal polynomials defined via symmetric functions were studied in Berens, Schmid and Xu [1995b]. The family of Gaussian cubature formulae based on their common zeros is the first family for all n and in all dimensions to be discovered. The polynomials $E_\alpha^n(x)$ were studied in Karlin and McGregor [1962, 1975].

The Chebyshev polynomials of type $\mathscr{A}_d$ were studied systematically by Beerends [1991]. They are generalizations of the Koornwinder polynomials for $d = 2$ and of the partial results in Eier and Lidl [1974, 1982], Dunn and Lidl [1982], Ricci [1978] and Barcry [1984]. We have followed the presentation in Li and Xu [2010], who studied these polynomials from the point of view of tiling and discrete Fourier analysis and also studied their common zeros. It turned out that the set of orthogonal polynomials $\{U_\alpha : |\alpha| = n\}$ of degree n has $\dim \Pi_{n-1}^d$ distinct real common zeros in $\triangle^*$, so that the Gaussian cubature formulae exist for $W_{1/2}$ on $\triangle^*$ by Theorem 3.8.4. Gaussian cubature, however, does not exist for $W_{-1/2}$. For further results and references, we refer to Beerends [1991] and Li and Xu [2010].

The symmetric orthogonal polynomials associated with $\mathscr{A}_d$ are related to the BC_n-type orthogonal polynomials in several variables; see, for example, Vretare [1984], Beerends and Opdam [1993] and van Diejen [1999]. They are special cases of the Jacobi polynomials for root systems studied in Opdam [1988]. The Chebyshev polynomials in the form of symmetric trigonometric orthogonal polynomials have also been studied recently for compact simple Lie groups; see Nesterenko, Patera, Szajewska and Tereszkiewicz [2010] and Moody and Patera [2011], and the references therein, but only a root system of the $\mathscr{A}_d$ type leads to a full basis of algebraic orthogonal polynomials.

For the Sobolev orthogonal polynomials on the unit ball, the orthogonal basis for the inner product $\langle \cdot, \cdot \rangle_\triangle$ was constructed in Xu [2006b], in response to a problem in the numerical solution of a Poisson equation on the disk raised by Atkinson and Hansen [2005]. The orthogonal basis for $\langle \cdot, \cdot \rangle_\nabla$ was constructed in Xu [2008], in which an orthogonal basis was also constructed for a second inner product defined via the gradient,

$$\langle f, g \rangle_{\nabla, \mathrm{II}} := \frac{\lambda}{\sigma_d} \int_{B^d} \nabla f(x) \cdot \nabla g(x)\, \mathrm{d}x + f(0)g(0).$$

As shown in Corollary 5.6.4, orthogonal polynomials with respect to $\langle \cdot, \cdot \rangle_\nabla$ are eigenfunctions of the differential operator $\mathscr{D}_\mu$ in the limiting case $\mu = -\frac{1}{2}$. Eigenfunctions of $\mathscr{D}_\mu$ for further singular cases, $\mu = -\frac{3}{2}, -\frac{5}{2}, \ldots$, were studied in Piñar and Xu [2009].

Despite extensive research into the Sobolev orthogonal polynomials in one variable, study of the Sobolev polynomials in several variables started only recently. We refer to Lee and Littlejohn [2005], Bracciali, Delgado, Fernández, Pérez and Piñar [2010], Aktaş and Xu [2013], and Pérez, Piñar and Xu [2013].

Besides the unit ball, the only other case that has been studied carefully is that of the simplex. Let $\mathscr{D}_\kappa$ denote the differential operator on the right-hand side of (5.3.4). Then the orthogonal polynomials in $\mathscr{V}_n^d(W_\kappa)$ on the simplex are eigenfunctions of $\mathscr{D}_\kappa$ for $\kappa_i - \frac{1}{2}$, $1 \le i \le d$. The limiting cases where some, or all, $\kappa_i = -1$ are studied in Aktaş and Xu [2013]. In each case a complete basis was found for the differential equation and an inner product of the Sobolev type was constructed such that the eigenfunctions are orthogonal with respect to the inner product.

6

Root Systems and Coxeter Groups

There is a far reaching extension of classical orthogonal polynomials that uses finite reflection groups. This chapter presents the part of the theory needed for our analysis. We refer to the books by Coxeter [1973], Grove and Benson [1985] and Humphreys [1990] for the algebraic structure theorems. We will begin with the orthogonal groups, definitions of reflections and root systems and descriptions of the infinite families of finite reflection groups. A key part of the chapter is the definition and fundamental theorems for the differential–difference (*Dunkl*) operators.

6.1 Introduction and Overview

For $x,y \in \mathbb{R}^d$ the inner product is $\langle x,y\rangle = \sum_{j=1}^{d} x_j y_j$ and the norm is $\|x\| = \langle x,x\rangle^{1/2}$. A matrix $w = (w_{ij})_{i,j=1}^{d}$ is called orthogonal if $ww^{\mathrm{T}} = I_d$, where w^{T} denotes the transpose of w and I_d is the $d \times d$ identity matrix. Equivalent conditions for orthogonality are the following:

1. w is invertible and $w^{-1} = w^{\mathrm{T}}$;
2. for each $x \in \mathbb{R}^d$, $\|x\| = \|xw\|$;
3. for each $x,y \in \mathbb{R}^d$, $\langle x,y\rangle = \langle xw, yw\rangle$;
4. the rows of w form an orthonormal basis for $\mathbb{R}^d$.

The set of orthogonal matrices is closed under multiplication and inverses (by condition (2), for example) and forms the *orthogonal group*, denoted $O(d)$. Condition (4) shows that $O(d)$ is a closed bounded subset of all $d \times d$ matrices and hence is a compact group. If $w \in O(d)$ then $\det w = \pm 1$. The subgroup $SO(d) = \{w \in O(d) : \det w = 1\}$ is called the *special orthogonal group*.

Definition 6.1.1 The right regular representation of $O(d)$ is the homomorphism $w \mapsto R(w)$ of linear maps of Π^d to itself (endomorphisms), given by $R(w)p(x) = p(xw)$ for all $x \in \mathbb{R}^d, p \in \Pi^d$.

Further, $R(w)$ is an automorphism of $\mathscr{P}_n^d$ for each $n, w \in O(d)$. Note that $R(w_1w_2) = R(w_1)R(w_2)$ (by the homomorphism property of R). The Laplacian Δ commutes with each $w \in O(d)$, that is, $\Delta(R(w)p)(x) = (\Delta p)(xw)$ for any $x \in \mathbb{R}^d$. In the present context the basic tool for constructing orthogonal transformations is the reflection.

Definition 6.1.2 For a nonzero $u \in \mathbb{R}^d$, the reflection along u, denoted by σ_u, is defined by

$$x\sigma_u = x - 2\frac{\langle x,u\rangle}{\|u\|^2}u.$$

Writing $\sigma_u = I_d - 2\left(uu^{\mathrm{T}}\right)^{-1}u^{\mathrm{T}}u$ shows that $\sigma_u = \sigma_u^{\mathrm{T}}$ and $\sigma_u\sigma_u = I_d$ (note that $uu^{\mathrm{T}} = \|u\|^2$ while $u^{\mathrm{T}}u$ is a matrix). The matrix entries of σ_u are $(\sigma_u)_{ij} = \delta_{ij} - 2u_iu_j/\|u\|^2$. It is clear that $x\sigma_u = x$ exactly when $\langle x,u\rangle = 0$, that is, the invariant set for σ_u is the hyperplane $u^\perp = \{x : \langle x,u\rangle = 0\}$. Also, $u\sigma_u = -u$ and any nonzero multiple of u determines the same reflection. Since σ_u has one eigenvector for the eigenvalue -1 and $d-1$ independent eigenvectors for the eigenvalue $+1$ (any basis for $u^\perp$), it follows that $\det\sigma_u = -1$.

Proposition 6.1.3 *Suppose that* $x^{(i)}, y^{(i)} \in \mathbb{R}^d$, $\|x^{(i)}\|^2 = \|y^{(i)}\|^2 = 1$ *for* $i = 1,2$, *and* $\langle x^{(1)},y^{(1)}\rangle = \langle x^{(2)},y^{(2)}\rangle$; *then there is a product* w *of reflections such that* $x^{(1)}w = x^{(2)}$ *and* $y^{(1)}w = y^{(2)}$.

Proof If $x^{(1)} = x^{(2)}$ then put $y^{(3)} = y^{(1)}$, else let $u = x^{(1)} - x^{(2)} \neq 0$, so that $x^{(1)}\sigma_u = x^{(2)}$, and let $y^{(3)} = y^{(1)}\sigma_u$. If $y^{(3)} = y^{(2)}$ then the construction is finished. Otherwise, let $v = y^{(3)} - y^{(2)}$; then $y^{(3)}\sigma_v = y^{(2)}$ and $x^{(2)}\sigma_v = x^{(2)}$ since $\langle x^{(2)},v\rangle = \langle x^{(2)},y^{(3)}\rangle - \langle x^{(2)},y^{(2)}\rangle = \langle x^{(1)},y^{(1)}\rangle - \langle x^{(2)},y^{(2)}\rangle = 0$. One of σ_u, σ_v and $\sigma_u\sigma_v$ is the desired product. □

The following is crucial for analyzing groups generated by reflections.

Proposition 6.1.4 *Suppose that* u,v *are linearly independent in* $\mathbb{R}^d$, *and set* $\cos\theta = \langle u,v\rangle/(\|u\|\,\|v\|)$; *then* $\sigma_u\sigma_v$ *is a plane rotation in* $\operatorname{span}\{u,v\}$ *through an angle* 2θ.

Proof Assume that $\|u\| = \|v\| = 1$; thus $\cos\theta = \langle u,v\rangle$ and $\|v - \langle u,v\rangle u\| = \sin\theta$, where $0 < \theta < \pi$. Let $v' = (\sin\theta)^{-1}(v - \langle u,v\rangle u)$, so that $\{u,v'\}$ is an orthonormal basis for $\operatorname{span}\{u,v\}$. With respect to this basis σ_u, σ_v, and $\sigma_u\sigma_v$ have the matrix representations

$$\sigma_u = \begin{bmatrix} -1 & 0 \\ 0 & 1 \end{bmatrix},$$
$$\sigma_v = \begin{bmatrix} -\cos 2\theta & -\sin 2\theta \\ -\sin 2\theta & \cos 2\theta \end{bmatrix},$$
$$\sigma_u \sigma_v = \begin{bmatrix} \cos 2\theta & \sin 2\theta \\ -\sin 2\theta & \cos 2\theta \end{bmatrix},$$

and $\sigma_u \sigma_v$ is a rotation. □

For two nonzero vectors u, v, denote $\cos\measuredangle(u,v) = \langle u,v\rangle / (\|u\|\,\|v\|)$. Consequently, for a given $m = 1,2,3,\ldots, (\sigma_u\sigma_v)^m = I_d$ if and only if $\cos\measuredangle(u,v) = \cos(\pi j/m)$ for some integer j. Since $(\sigma_u\sigma_v)^{-1} = \sigma_v\sigma_u$ for any two reflections σ_u and σ_v we see that σ_u and σ_v commute if and only if $\langle u,v\rangle = 0$. The conjugate of a reflection is also a reflection:

Lemma 6.1.5 *Let $u \in \mathbb{R}^d, u \neq 0$, and let $w \in O(d)$; then $w^{-1}\sigma_u w = \sigma_{uw}$.*

Proof For $x \in \mathbb{R}^d$,

$$xw^{-1}\sigma_u w = x - \frac{2\langle xw^{-1}, u\rangle}{\|u\|^2} uw = x - \frac{2\langle x, uw\rangle}{\|u\|^2} uw.$$ □

6.2 Root Systems

The idea of a root system seems very simple, yet in fact it is a remarkably deep concept, with many ramifications in algebra and analysis.

Definition 6.2.1 A root system is a finite set R of nonzero vectors in $\mathbb{R}^d$ such that $u, v \in R$ implies that $u\sigma_v \in R$. If, additionally, $u, v \in R$ and $v = cu$ for some scalar $c \in \mathbb{R}$ implies that $c = \pm 1$ then R is said to be reduced.

Clearly $u \in R$ implies that $-u = u\sigma_u \in R$ for any root system. The set $\{u^\perp : u \in R\}$ is a finite set of hyperplanes; thus there exists $u_0 \in \mathbb{R}^d$ such that $\langle u, u_0\rangle \neq 0$ for all $u \in R$. With respect to u_0 define the set of positive roots

$$R_+ = \{u \in R : \langle u, u_0\rangle > 0\},$$

so that $R = R_+ \cup (-R_+)$.

Definition 6.2.2 The Coxeter group $W = W(R)$ generated by the root system R is the subgroup of $O(d)$ generated by $\{\sigma_u : u \in R\}$.

Note that for the purpose of studying the group W one can replace R by a reduced root system (for example $\{u/\|u\| : u \in R\}$). There is a very useful polynomial associated with R.

Definition 6.2.3 For any reduced root system R, the *discriminant*, or alternating polynomial, is given by

$$a_R(x) = \prod_{u \in R_+} \langle x, u \rangle .$$

Theorem 6.2.4 *For $v \in R$ and $w \in W(R)$,*

$$a_R(x\sigma_v) = -a_R(x) \qquad \textit{and} \qquad a_R(xw) = \det w\, a_R(x).$$

Proof Assume that $v \in R_+$; by definition $R_+ \backslash \{v\}$ is a disjoint union $E_1 \cup E_2$, where $E_1 = \{u \in R_+ : u\sigma_v = u\}$ and $u \in E_2$ implies that $u\sigma_v = c_u u'$ for some $u' \in E_2$ and $u' \neq u$ and $c_u = \pm 1$. Thus $u'\sigma_v = c_u u$ and so $\langle x, u\rangle \langle x, u'\rangle$ is invariant under σ_v (note that $\langle x\sigma_v, u\rangle = \langle x, u\sigma_v\rangle$ because $\sigma_v^{\mathrm{T}} = \sigma_v$). Then $a_R(x\sigma_v) = \langle x, -v\rangle \pi_1 \pi_2$, where $\pi_1 = \prod_{u \in E_1} \langle x, u\rangle$ and $\pi_2 = \prod_{u \in E_2} \langle x, u\sigma_v\rangle$. The second product is a permutation having an even number of sign changes of $\prod_{u \in E_2} \langle x, u\rangle$. This shows that $a_R(x\sigma_v) = -a_R(x)$. Let $w = \sigma_{u_1}\sigma_{u_2} \cdots \sigma_{u_n}$ be a product of n reflections, where $u_1, \ldots, u_n \in R_+$; then $a_R(xw) = (-1)^n a_R(x)$. Obviously $\det w = (-1)^n$. □

There is an amusing connection with harmonic polynomials: a product $\prod_{v \in E} \langle x, v\rangle$ is harmonic exactly when the elements of E are nonzero multiples of some system R_+ of positive roots.

Lemma 6.2.5 *Suppose that u is a nonzero vector in $\mathbb{R}^d$ and $p(x)$ is a polynomial such that $p(x\sigma_u) = -p(x)$ for all $x \in \mathbb{R}^d$; then $p(x)$ is divisible by $\langle x, u\rangle$.*

Proof The divisibility property is invariant under $O(d)$, so one can assume that $u = (1, 0, \ldots, 0)$. Any polynomial p can be written as $\sum_{j=0}^n x_1^j p_j(x_2, \ldots, x_d)$; then $p(x\sigma_u) = \sum_{j=0}^n (-x_1)^j p_j(x_2, \ldots, x_d)$. Thus $p(x\sigma_u) = -p(x)$ implies that $p_j = 0$ unless j is odd. □

Theorem 6.2.6 *Suppose E is a finite set of nonzero vectors in $\mathbb{R}^d$; then $\Delta \prod_{v \in E} \langle x, v\rangle = 0$ if and only if there are scalars $c_v, v \in E$ such that $\{c_v v : v \in E\} = R_+$ for some reduced root system and such that no vector in E is a scalar multiple of another vector in E.*

Proof First we show $\Delta a_R = 0$ for any reduced root system. The polynomial $p = \Delta a_R$ satisfies $p(x\sigma_u) = -p(x)$ for any $u \in R_+$ because Δ commutes with $R(\sigma_u)$. By Lemma 6.2.5 $p(x)$ is divisible by $\langle x, u\rangle$, but $\deg p \leq \deg a_R - 2$ and thus $p = 0$.

Now suppose that $\Delta p = 0$ for $p(x) = \prod_{v \in E} \langle x, v\rangle$, for a set E. Assume that $\|v\| = 1$ for every $v \in E$. Let u be an arbitrary element of E. Without loss of generality assume that $u = (1, 0, \ldots, 0)$ and expand $p(x)$ as $\sum_{j=0}^n x_1^j p_j(x_2, \ldots, x_d)$ for some n. Then $\Delta p(x) = \sum_{j=0}^n x_1^j [\Delta p_j + (j+2)(j+1) p_{j+2}]$. Since p is a multiple of $\langle x, u\rangle$, $p_0 = 0$. This implies that $p_{2j} = 0$ for each $j = 0, 1, 2, \ldots$ and thus that

$p(x)/\langle x,u\rangle$ is even in x_1. Further, $p_1 \neq 0$ or else $p = 0$; thus $x_1 = \langle x,u\rangle$ is not a repeated factor of p. The product $\prod\{\langle x,v\rangle : v \in E, v \neq u\}$ is invariant under σ_u. This shows that the set $\{v\sigma_u : v \in E\}$ has the same elements as $\{\pm v : v \in E\}$ up to multiplication of each by ± 1. Hence $E \cup (-E)$ satisfies the definition of a root system. □

Theorem 6.2.7 *For any root system R, the group $W(R)$ is finite; if also R is reduced then the set of reflections contained in $W(R)$ is exactly $\{\sigma_u : u \in R_+\}$.*

Proof Let $E = \text{span}(R)$ have dimension r, so that W can be considered as a subgroup of $O(r)$ and every element of $E^\perp$ is invariant under W. The definition of a root system shows that $uw \in R$ for every $u \in R$ and $w \in W$. Since R contains a basis for E, this shows that any basis element v of E has a finite orbit $\{vw : w \in W\}$, hence W is finite. If $w = \sigma_v \in W$ is a reflection then $\det w = -1$ and so $a_R(xw) = -a_R(x)$. By Lemma 6.2.5, $a_R(x)$ is divisible by $\langle x,v\rangle$. But linear factors are irreducible and the unique factorization theorem shows that some multiple of v is an element of R. □

Groups of the type $W(R)$ are called *finite reflection* groups or *Coxeter* groups. The dimension of span(R) is called the *rank* of R. If R can be expressed as a disjoint union of non-empty sets $R_1 \cup R_2$ with $\langle u,v\rangle = 0$ for every $u \in R_1, v \in R_2$ then each R_i $(i = 1,2)$ is itself a root system and $W(R) = W(R_1) \times W(R_2)$, a direct product. Further, $W(R_1)$ and $W(R_2)$ act on the orthogonal subspaces span(R_1) and span(R_2), respectively. In this case the root system R and the reflection group $W(R)$ are called *decomposable*. Otherwise the system and group are *indecomposable* on *irreducible*. There is a complete classification of indecomposable finite reflection groups. Some more concepts are now needed for the discussion.

Assume that the rank of R is d, that is, span$(R) = \mathbb{R}^d$. The set of hyperplanes $H = \{v^\perp : v \in R_+\}$ divides $\mathbb{R}^d$ into connected open components called *chambers*. A theorem states that the order of the group equals the number of chambers. Recall that the positive roots are defined in terms of some vector $u_0 \in \mathbb{R}^d$. The connected component of the complement of H which contains u_0 is called the fundamental chamber. The roots corresponding to the bounding hyperplanes of this chamber are called *simple roots*, and they form a basis of $\mathbb{R}^d$. The ("simple") reflections corresponding to the simple roots are denoted $s_i, i = 1,\ldots,d$. Let m_{ij} be the order of s_is_j (clearly $m_{ii} = 1$ and $m_{ij} = 2$ if and only if $s_is_j = s_is_j$, for $i \neq j$). The following is the fundamental theorem in this topic, proved by H. S. M. Coxeter; see Coxeter and Moser [1965, p. 122].

Theorem 6.2.8 *The group $W(R)$ is isomorphic to the abstract group generated by $\{s_i : 1 \leq i \leq d\}$ subject to the relations $(s_is_j)^{m_{ij}} = 1$.*

The Coxeter diagram is a graphical way of displaying the above relations: it is a graph with d nodes corresponding to simple reflections; nodes i and j are joined with an edge when $m_{ij} > 2$; an edge is labeled by m_{ij} when $m_{ij} > 3$. The root system is indecomposable if and only if the Coxeter diagram is connected. We proceed to the description of the infinite families of indecomposable root systems and two exceptional cases. The systems are named by a letter and a subscript indicating the rank. It is convenient to introduce the standard basis vectors $\varepsilon_i = \left(0,\ldots,0,\overset{i}{1},0,\ldots,0\right)$ for $\mathbb{R}^d$, $1 \le i \le d$; the superscript i above the element '1' indicates that it occurs in the ith position.

6.2.1 Type A_{d-1}

For a type A_{d-1} system, the roots are given by $R = \left\{v_{(i,j)} = \varepsilon_i - \varepsilon_j : i \ne j\right\} \subset \mathbb{R}^d$. The span is exactly $(1,1,\ldots,1)^\perp$ and thus the rank is $d-1$. The reflection $\sigma_{ij} = \sigma_{v_{(i,j)}}$ interchanges the components x_i and x_j of each $x \in \mathbb{R}^d$ and is called a transposition, often denoted by (i,j). Thus $W(R)$ is exactly the symmetric or permutation group S_d on d objects. Choose $u_0 = (d, d-1, \ldots, 1)$; then $R_+ = \{v_{(i,j)} = \varepsilon_i - \varepsilon_j : i > j\}$ and the simple roots are $\{\varepsilon_i - \varepsilon_{i+1} : 1 \le i \le d-1\}$. The corresponding reflections are the adjacent transpositions $(i, i+1)$. The general fact about reflection groups, that simple reflections generate $W(R)$, specializes to the well-known statement that adjacent transpositions generate the symmetric group. Since $(i,i+1)(i+1,i+2)$ is a permutation of period 3, the structure constants satisfy $m_{i,i+1} = 3$ and $m_{ij} < 3$ otherwise. The Coxeter diagram is

It is clear that for any two roots u, v there is a permutation $w \in W(R)$ such that $uw = v$; hence any two reflections are conjugate in the group, $w^{-1}\sigma_u w = \sigma_{uw}$. The alternating polynomial is

$$a_R(x) = \prod\nolimits_{u \in R_+} \langle x, u \rangle = \prod_{1 \le i < j \le d} (x_i - x_j).$$

The fundamental chamber is $\{x : x_1 > x_2 > \cdots > x_d\}$; if desired, the chambers can be restricted to $\operatorname{span}(R) = \left\{x : \sum_{i=1}^d x_i = 0\right\}$.

6.2.2 Type B_d

The root system for a type B_d system is $R = \{v_{(i,j)} = \varepsilon_i - \varepsilon_j : i \ne j\} \cup \{u_{(i,j)} = \operatorname{sign}(j-i)(\varepsilon_i + \varepsilon_j) : i \ne j\} \cup \{\pm\varepsilon_i\}$. The reflections corresponding to the three different sets of roots are

$$x\sigma_{ij} = \left(x_1, \dots, \overset{i}{x_j}, \dots, \overset{j}{x_i}, \dots\right),$$

$$x\tau_{ij} = \left(x_1, \dots, -\overset{i}{x_j}, \dots, -\overset{j}{x_i}, \dots\right),$$

$$x\sigma_i = \left(x_1, \dots, -\overset{i}{x_i}, \dots\right),$$

for $v_{(i,j)}, u_{(i,j)}$ and ε_i respectively. The group $W(R)$ is denoted W_d; it is the full symmetry group of the hyperoctahedron $\{\pm\varepsilon_1, \pm\varepsilon_2, \dots, \pm\varepsilon_d\} \subset \mathbb{R}^d$ (also of the hypercube) and is thus called the *hyperoctahedral group*. Its elements are exactly the $d \times d$ permutation matrices with entries ± 1 (that is, each row and each column has exactly one nonzero element ± 1). With the same $u_0 = (d, d-1, \dots, 1)$ as in the previous subsection, the positive root system is $R_+ = \{\varepsilon_i - \varepsilon_j, \varepsilon_i + \varepsilon_j : i < j\} \cup \{\varepsilon_i : 1 \le i \le d\}$ and the simple roots are to be found in $\{\varepsilon_i - \varepsilon_{i+1} : i < d\} \cup \{\varepsilon_d\}$. The order of $\sigma_{d-1,d}\sigma_d$ is 4. The Coxeter diagram is

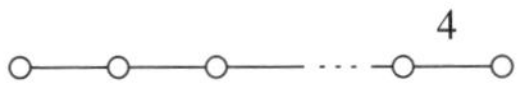

This group has two conjugacy classes of reflections, $\{\sigma_{ij}, \tau_{ij} : i < j\}$ and $\{\sigma_i : 1 \le i \le d\}$. The alternating polynomial is

$$a_R(x) = \prod_{i=1}^{d} x_i \prod_{1 \le i < j \le d} \left(x_i^2 - x_j^2\right).$$

The fundamental chamber is $\{x : x_1 > x_2 > \cdots > x_d > 0\}$.

6.2.3 Type $I_2(m)$

The type $I_2(m)$ systems are dihedral and correspond to symmetry groups of regular m-gons in $\mathbb{R}^2$ for $m \ge 3$. Using a complex coordinate system $z = x_1 + \mathrm{i}x_2$ and $\bar{z} = x_1 - \mathrm{i}x_2$, a rotation through the angle θ can be expressed as $z \mapsto z e^{\mathrm{i}\theta}$, and the reflection along $(\sin\phi, -\cos\phi)$ is $z \mapsto \bar{z} e^{2\mathrm{i}\phi}$. Now let $\omega = e^{2\pi\mathrm{i}/m}$; then the reflection along

$$v_{(j)} = \left(\sin\frac{\pi j}{m}, -\cos\frac{\pi j}{m}\right)$$

corresponds to $\sigma_j : z \mapsto \bar{z}\omega^j$ for $1 \le j \le 2m$; note that $v_{(m+j)} = -v_{(j)}$. Choose

$$u_0 = \left(\cos\frac{\pi}{2m}, \sin\frac{\pi}{2m}\right);$$

then the positive roots are $\{v_{(j)} : 1 \le j \le m\}$ and the simple roots are $v_{(1)}, v_{(m)}$. Then $\sigma_m\sigma_1$ maps z to $z\omega$ and has period m. The Coxeter diagram is 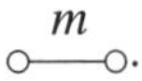. A simple calculation shows that $\sigma_j\sigma_n\sigma_j = \sigma_{2j-n}$ for any n, j; thus there are two conjugacy classes of reflections $\{\sigma_{2i}\}, \{\sigma_{2i+1}\}$ when m is even but only one class when m is odd. There are three special cases for m: $I_2(3)$ is isomorphic to A_2;

$I_2(4) = B_2$; and $I_2(6)$ is called G_2 in the context of Weyl groups. To compute the alternating polynomial, note that

$$x_1 \sin\frac{\pi j}{m} - x_2 \cos\frac{\pi j}{m} = \frac{1}{2}\left[\exp\left(\mathrm{i}\pi\left(\frac{1}{2} - \frac{j}{m}\right)\right)\right]\left(z - \bar{z}\omega^j\right)$$

for each $j = 1, \ldots, m$; thus

$$a_R(x) = 2^{-m}\mathrm{i}^{-1}\left(z^m - \bar{z}^m\right).$$

The fundamental chamber is $\{(r\cos\theta, r\sin\theta) : r > 0, 0 < \theta < \pi/m\}$.

6.2.4 Type D_d

For our purposes, the type D_d system can be considered as a special case of the system for type $B_d, d \geq 4$. The root system is a subset of that for B_d, namely $R = \{\varepsilon_i - \varepsilon_j : i \neq j\} \cup \{\mathrm{sign}(j-i)(\varepsilon_i + \varepsilon_j) : i \neq j\}$. The associated group is the subgroup of W_d of elements with an even number of sign changes (equivalently, the subgroup fixing the polynomial $\prod_{j=1}^{d} x_j$). The simple roots are $\{\varepsilon_i - \varepsilon_{i+1} : i < d\} \cup \{\varepsilon_{d-1} + \varepsilon_d\}$ and the Coxeter diagram is

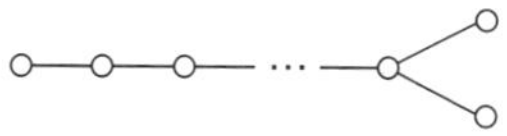

The alternating polynomial is

$$a_R(x) = \prod_{1 \leq i < j \leq d} \left(x_i^2 - x_j^2\right)$$

and the fundamental chamber is $\{x : x_1 > x_2 > \cdots > |x_d| > 0\}$.

6.2.5 Type H_3

The type H_3 system generates the symmetry group of the regular dodecahedron and of the regular icosahedron. The algebraic number (the "golden ratio") $\tau = \frac{1}{2}(1 + \sqrt{5})$, which satisfies $\tau^2 = \tau + 1$, is crucial here. Note that $\tau^{-1} = \tau - 1 = \frac{1}{2}(\sqrt{5} - 1)$. For the choice $u_0 = (3, 2\tau, \tau)$ the positive root system is $R_+ = \{(2,0,0), (0,2,0), (0,0,2), (\tau, \pm\tau^{-1}, \pm 1), (\pm 1, \tau, \pm\tau^{-1}), (\tau^{-1}, \pm 1, \tau), (-\tau^{-1}, 1, \tau), (\tau^{-1}, 1, -\tau)\}$, where the choices of signs in $\pm$ are independent of each other. The full root system $R = R_+ \cup (-R_+)$ is called the icosidodecahedron as a configuration in $\mathbb{R}^3$. Thus there are 15 reflections in the group $W(R)$. It is the symmetry group of the icosahedron $Q_{12} = \{(0, \pm\tau, \pm 1), (\pm 1, 0, \pm\tau), (\pm\tau, \pm 1, 0)\}$ (which has 12 vertices and 20 triangular faces) and of the dodecahedron $Q_{20} = \{(0, \pm\tau^{-1}, \pm\tau), (\pm\tau, 0, \pm\tau^{-1}), (\pm\tau^{-1}, \pm\tau, 0), (\pm 1, \pm 1, \pm 1)\}$ (which has 20 vertices and 12 pentagonal faces); see Coxeter [1973] for the details.

To understand the geometry of this group, consider the spherical Coxeter complex of R, namely the great circles on the unit sphere which are intersections with the planes $\{v^\perp : v \in R_+\}$. There are fivefold intersections at the vertices of the icosahedron (so the subgroup of $W(R)$ fixing such a vertex, the so-called stabilizer, is of the type $I_2(5)$), and there are threefold intersections at the vertices of the dodecahedron (with stabilizer group $I_2(3)$). The fundamental chamber meets the sphere in a spherical triangle with vertices at $(\tau+2)^{-1/2}(\tau,1,0)$, $3^{-1/2}(1,1,1)$, $2^{-1}(1,\tau,\tau^{-1})$ and the simple roots are $v_1 = (\tau,-\tau^{-1},-1)$, $v_2 = (-1,\tau,-\tau^{-1})$ and $v_3 = (\tau^{-1},-1,\tau)$. The angles between the simple roots are calculated as $\cos\measuredangle(v_1,v_2) = -\frac{1}{2} = \cos\frac{2}{3}\pi, \cos\measuredangle(v_1,v_3) = 0, \cos\measuredangle(v_2,v_3) = -\frac{1}{2}\tau = \cos\frac{4}{5}\pi$. Thus the Coxeter diagram is 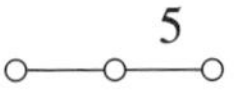. The alternating polynomial can be expanded, by computer algebra for instance, but we do not need to write it down here.

6.2.6 Type F_4

For a type F_4 system the Coxeter diagram is 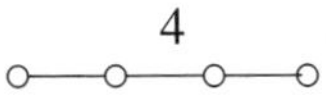. The group contains W_4 as a subgroup (of index 3). One conjugacy class of roots consists of $R_1 = \{\varepsilon_i - \varepsilon_j : i \neq j\} \cup \{\operatorname{sign}(j-i)(\varepsilon_i + \varepsilon_j) : i \neq j\}$ and the other class is $R_2 = \{\pm 2\varepsilon_i : 1 \le i \le 4\} \cup \{\pm\varepsilon_1 \pm \varepsilon_2 \pm \varepsilon_3 \pm \varepsilon_4\}$. Then $R = R_1 \cup R_2$ and there are 24 positive roots. With the orthogonal coordinates $y_1 = (-x_1+x_2)/\sqrt{2}, y_2 = (x_1+x_2)/\sqrt{2}$, $y_3 = (-x_3+x_4)/\sqrt{2}, y_4 = (x_3+x_4)/\sqrt{2}$, the alternating polynomial is

$$a_R(x) = 2^6 \prod_{1 \le i < j \le 4} (x_i^2 - x_j^2)(y_i^2 - y_j^2).$$

6.2.7 Other types

There is a four-dimensional version of the icosahedron corresponding to the root system H_4, with diagram 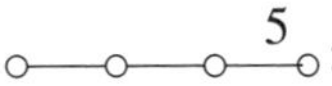; it is the symmetry group of the 600 cell (Coxeter [1973]) and was called the hecatonicosahedroidal group by Grove [1974]. In addition there are the exceptional Weyl groups E_6, E_7, E_8, but they will not be studied in this book.

6.2.8 Miscellaneous results

For any root system, the subgroup generated by a subset of simple reflections (the result of deleting one or more nodes from the Coxeter diagram) is called a *parabolic subgroup* of $W(R)$. For example, the parabolic subgroups of the type A_{d-1} diagram are Young subgroups, with $S_m \times S_{d-m}$ maximal for each d with $1 \le m \le d$. Removing one interior node from the B_d diagram results in a subgroup of the form $S_m \times W_{d-m}$.

As mentioned before, the number of chambers of the Coxeter complex (the complement of $\{v^\perp : v \in R_+\}$) equals the order of the group $W(R)$. For each chamber C there is a unique $w \in W(R)$ such that $C_0 w = C$, where C_0 is the fundamental chamber. Thus for each $v \in R$ there exists $w \in W(R)$ such that vw is a simple root, assuming that R is reduced, and so each σ_v is conjugate to a simple reflection. (Any chamber could be chosen as the fundamental chamber and its walls would then determine a set of simple roots.) It is also true that the positive roots when expressed as linear combinations of the simple roots (recall that they form a basis for span(R)) have all coefficients nonnegative.

The number of conjugacy classes of reflections equals the number c of connected components of the Coxeter diagram after all edges with an even label have been removed. Also, the quotient $G = W/W'$ of W by the commutator subgroup W', that is, the maximal abelian image, is $\mathbb{Z}_2^c$. The argument is easy: G is the group subject to the same relations but now with commuting generators. Denote these generators by $s_1', s_2', \ldots$, so that $(s_i')^2 = 1$ and G is a direct product of $\mathbb{Z}_2$ factors. Consider two reflections s_1, s_2 linked by an edge with an odd label $2m+1$ (recall that the label "3" is to be understood when no label is indicated); this means that $(s_1 s_2)^{2m+1} = 1$ and thus $s_2 = w^{-1} s_1 w$ with $w = (s_2 s_1)^m$; also $(s_1')^{2m+1}(s_2')^{2m+1} = s_1' s_2' = 1$. This implies that $s_1' = s_2'$ and thus that s_1, s_2 are conjugate in W. Relations of the form $(s_2 s_3)^{2m} = 1$ have no effect on s_2', s_3'. Thus simple reflections in different parts of the modified diagram have different images in G and cannot be conjugate in W. By the above remarks it suffices to consider the conjugacy classes of simple reflections.

There is a *length function* on $W(R)$: for any w in the group it equals both the number of factors in the shortest product $w = s_{i_1} s_{i_2} \cdots s_{i_m}$ in terms of simple reflections and the number of positive roots made negative by w, that is, $|R_+ w \cap (-R_+)|$. For the group S_d it is the number of adjacent transpositions required to express a permutation; it is also the number of inversions in a permutation considered as a listing of the numbers $1, 2, \ldots, d$. For example, the permutation $(4,1,3,2)$ has length $3+0+1+0 = 4$ (the first entry is larger than the three following entries, and so on), and $(1,2,3,4), (4,3,2,1)$ have lengths 0 and 6 respectively.

6.3 Invariant Polynomials

Any subgroup of $O(d)$ has representations on the spaces of homogeneous polynomials $\mathscr{P}_n^d$ of any given degree n. These representations are defined as $w \mapsto R(w)$ where $R(w)p(x) = p(xw)$; sometimes this equation is written as $wp(x)$, when there is no danger of confusion. The effect on the gradient is as follows: let $\nabla p(x)$ denote the row vector $(\partial p/\partial x_1 \quad \cdots \quad \partial p/\partial x_d)$; then

$$\nabla [R(w)p](x) = \left(\frac{\partial p(xw)}{\partial x_1} \quad \cdots \quad \frac{\partial p(xw)}{\partial x_d} \right) w^{\mathrm{T}} = R(w) \nabla p(x) w^{\mathrm{T}}.$$

Definition 6.3.1 For a finite subgroup W of $O(d)$, let Π^W denote the space of W-invariant polynomials $\{p \in \Pi^d : R(w)p = p \text{ for all } w \in W\}$ and set $\mathscr{P}_n^W = \mathscr{P}_n^d \cap \Pi^W$, the space of invariant homogeneous polynomials of degree $n = 0, 1, 2, \ldots$

There is an obvious projection onto the space of invariants:

$$\pi_W p = \frac{1}{|W|} \sum_{w \in W} R(w) p.$$

When W is a finite reflection group, Π^W has a uniquely elegant structure; there is a set of algebraically independent homogeneous generators whose degrees are fundamental constants associated with W. The following statement is known as Chevalley's theorem.

Theorem 6.3.2 *Suppose that R is a root system in $\mathbb{R}^d$ and that $W = W(R)$; then there exist d algebraically independent W-invariant homogeneous polynomials $\{q_j : 1 \le j \le d\}$ of degrees n_j, such that Π^W is the ring of polynomials generated by $\{q_j\}$, $|W| = n_1 n_2 \cdots n_d$ and the number of reflections in W is $\sum_{j=1}^d (n_j - 1)$.*

Corollary 6.3.3 *The Poincaré series for W has the factorization*

$$\sum_{n=0}^{\infty} \left(\dim \mathscr{P}_n^W\right) t^n = \prod_{j=1}^{d} (1 - t^{n_j})^{-1}.$$

Proof The algebraic independence means that the set of homogeneous polynomials $\{q_1^{m_1} q_2^{m_2} \cdots q_d^{m_d} : \sum_{j=1}^d n_j m_j = k\}$ is linearly independent for any $k = 0, 1, 2, \ldots$; thus it is a basis for $\mathscr{P}_k^W$. The cardinality of the set equals the coefficient of t^k in the product. □

It suffices to study the invariants of indecomposable root systems; note that if $\dim \operatorname{span}(R) = m < d$ then the orthogonal complement of the span provides $d - m$ invariants (the coordinate functions) of degree 1. When the rank of R is d and R is indecomposable, the group $W(R)$ is irreducibly represented on $\mathbb{R}^d$; this is called the reflection representation. The numbers $\{n_j : 1 \le j \le d\}$ are called the *fundamental degrees* of $W(R)$ in this situation. The coefficient of t^k in the product $\prod_{j=1}^d [1 + (n_j - 1)t]$ is the number of elements of $W(R)$ whose fixed-point set is of codimension k (according to a theorem of Shephard and Todd [1954]).

Lemma 6.3.4 *If the rank of a root system R is d then the lowest degree of a non-constant invariant polynomial is 2, and if R is indecomposable then there are no proper invariant subspaces of $\mathbb{R}^d$.*

Proof Identify $\mathbb{R}^d$ with $\mathscr{P}_1^d$, and suppose that E is an invariant subspace. Any polynomial in E has the form $p(x) = \langle x, u \rangle$ for some $u \in \mathbb{R}^d$. By assumption

$R(\sigma_v)p(x) = \langle x\sigma_v, u\rangle = \langle x, u\sigma_v\rangle \in E$ for any $v \in R$; thus $p(x) - R(\sigma_v)p(x) = \langle x, u - u\sigma_v\rangle \in E$ and $\langle x, u - u\sigma_v\rangle = 2\langle x, v\rangle\langle u, v\rangle/\|v\|^2$. If p is W-invariant then this implies that $\langle u, v\rangle = 0$ for all $v \in R$, but span $(R) = \mathbb{R}^d$ and so $u = 0$.

However, if R is indecomposable and $u \neq 0$ then there exists $v \in R$ with $\langle u, v\rangle \neq 0$ and the equation for $\langle x, u - u\sigma_v\rangle$ shows that $\langle x, v\rangle \in E$. Let

$$R_1 = \{v_1 \in R : \langle x, v_1\rangle \in E\}.$$

Any root $v_2 \in R$ satisfies either $v_2 \in R_1$ or $\langle v_2, v_1\rangle = 0$ for any $v_1 \in R_1$. By definition $R_1 = R$ and $E = \mathscr{P}_1^d$. □

Clearly $q(x) = \sum_{j=1}^d x_j^2$ is invariant for any subgroup of $O(d)$. Here is another general theorem about the fundamental invariants.

Theorem 6.3.5 *Let $\{q_j : 1 \le j \le d\}$ be fundamental invariants for the root system R, and let $J(x)$ be the Jacobian matrix $\left(\partial q_i(x)/\partial x_j\right)_{i,j=1}^d$; then* $\det J(x) = c a_R(x)$ *(see Definition 6.2.3) for some scalar multiple c.*

Proof The algebraic independence implies that $\det J \neq 0$. For any invariant q and $v \in R$, $\nabla q(x\sigma_v) = \nabla q(x)\sigma_v$; thus any row in $J(x\sigma_v)$ equals the corresponding row of $J(x)$ multiplied by σ_v. Now $\det J(x\sigma_v) = \det J(x)\det\sigma_v = -\det J(x)$, and so $\det J(x)$ has the alternating property. By Lemma 6.2.5, $\det J$ is a multiple of a_R. Further, $\det J$ is a homogeneous polynomial of degree $\sum_{j=1}^d (n_j - 1)$, which equals the number of reflections in $W(R)$, the degree of a_R. □

6.3.1 Type A_{d-1} invariants

The type A_{d-1} invariants are of course the classical symmetric polynomials. The fundamental invariants are usually taken to be the elementary symmetric polynomials $e_1, e_2, \ldots, e_d$, defined by $\sum_{j=0}^d e_j(x)t^j = \prod_{i=1}^d (1 + tx_i)$. Restricted to the span $(1, 1, \ldots, 1)^\perp$ of the root system $e_1 = 0$, the fundamental degrees of the group are $2, 3, \ldots, d$; by way of verification note that the product of the degrees is $d!$ and the sum $\sum_{j=2}^d (j-1) = \frac{1}{2}d(d-1)$, the number of transpositions. In this case $\det J(x) = \prod_{i<j}(x_i - x_j)$, the Vandermonde determinant. To see this, argue by induction on d; (this is trivial for $d = 1$). Let e_i' denote the elementary symmetric function of degree i in $(x_1 \quad x_2 \quad \cdots \quad x_{d-1})$; then $e_i = e_i' + x_d e_{i-1}'$. The ith row of $J(x)$ is

$$\left(\cdots \quad \frac{\partial}{\partial x_j}e_i' + x_d\frac{\partial}{\partial x_j}e_{i-1}' \quad \cdots \quad e_{i-1}'\right).$$

The first row is $(1\ 1\ \cdots\ 1)$ and the terms e_d' in the last row are 0. Now subtract x_d times the first row from the second row, subtract x_d times the resulting second row from the third row and so on. The resulting matrix has ith row (for $1 \le i \le d-1$ with $e'_0 = 1$) equal to

$$\left(\cdots \quad \frac{\partial}{\partial x_j} e'_i \quad \cdots \quad \sum_{j=0}^{i-1} e'_{i-1-j} (-x_d)^j \right),$$

and last row equal to $\left(0 \quad \cdots \quad 0 \quad \sum_{j=0}^{d-1} e'_{d-1-j} (-x_d)^j \right)$. The last entry is $\prod_{j=1}^{d-1} (x_j - x_d)$, and the proof is complete by induction.

6.3.2 Type B_d invariants

The type B_d invariants are generated by the elementary symmetric functions q_j in the squares of x_i, that is, $\sum_{j=0}^{d} q_j(x) t^j = \prod_{i=1}^{d} \left(1 + t x_i^2\right)$. The fundamental degrees are $2, 4, 6, \ldots, 2d$, the order of the group is $2^d d!$ and the number of reflections is $\sum_{j=1}^{d} (2j-1) = d^2$. Also, $\det J(x) = 2^d \prod_{j=1}^{d} x_j \prod_{i<j} \left(x_i^2 - x_j^2\right)$. To see this, note that $\partial / \partial x_i = 2x_i \partial / \partial \left(x_i^2\right)$ for each i, and use the result for A_{d-1}.

6.3.3 Type D_d invariants

The system with type D_d invariants has all but one of the fundamental invariants of the type B_d system, namely $q_1, q_2, \ldots, q_{d-1}$, and it also has $e_d = x_1 \cdots x_d$. The fundamental degrees are $2, 4, 6, \ldots, 2d-2, d$, the order of the group is $2^{d-1} d!$ and the number of reflections is $\sum_{j=1}^{d-1} (2j-1) + (d-1) = d^2 - d$.

6.3.4 Type $I_2(m)$ invariants

For a system with type $I_2(m)$ invariants, in complex coordinates Z and $\bar{z}$ for $\mathbb{R}^2$, where $z = x_1 + \mathrm{i} x_2$, the fundamental invariants are $z\bar{z}$ and $z^m + \bar{z}^m$. The fundamental degrees are 2 and m, the order of the group is $2m$ and the number of reflections is m.

6.3.5 Type H_3 invariants

The fundamental invariants for a type H_3 system are

$$q_1(x) = \sum_{j=1}^{3} x_j^2,$$

$$q_2(x) = \prod \{ \langle x, u \rangle : u \in Q_{12}, \langle (3, 2\tau, \tau), u \rangle > 0 \},$$

$$q_3(x) = \prod \{ \langle x, u \rangle : u \in Q_{20}, \langle (3, 2\tau, \tau), u \rangle > 0 \}.$$

That is, the invariants q_2, q_3 are products of inner products with half the vertices of the icosahedron Q_{12} or of the dodecahedron Q_{20}, respectively. The same argument used for the alternating polynomial a_R shows that the products of $q_2(x\sigma_v)$ and $q_3(x\sigma_v)$ for any $v \in R$ have the same factors as $q_2(x)$, $q_3(x)$ (respectively) with an even number of sign changes. By direct verification, q_2 is not a multiple of q_1 and

q_3 is not a polynomial in q_1 and q_2. Thus the fundamental degrees are $2,6,10$, the order of the group is 120 and there are $2+6+10-3=15$ reflections.

6.3.6 Type F_4 invariants

For a type F_4 system, using the notation from type B_4 (in which $q_i(x)$ denotes the elementary symmetric function of degree i in $x_1^2,\ldots,x_4^2$), a set of fundamental invariants is given by

$$\begin{aligned} g_1 &= q_1, \\ g_2 &= 6q_3 - q_1q_2, \\ g_3 &= 24q_4 + 2q_2^2 - q_1^2q_2, \\ g_4 &= 288q_2q_4 - 8q_2^3 - 27q_1^3q_3 + 12q_1^2q_2^2. \end{aligned}$$

These formulae are based on the paper by Ignatenko [1984]. The fundamental degrees are $2,6,8,12$ and the order of the group is 1152.

6.4 Differential–Difference Operators

Keeping in mind the goal of studying orthogonal polynomials for weights which are invariant under some finite reflection group, we now consider invariant differential operators. This rather quickly leads to difficulties: it may be hard to construct a sufficient number of such operators for an arbitrary group (although this will be remedied), but the biggest problem is that they do not map polynomials to polynomials unless the polynomials are themselves invariant. So, to deal with arbitrary polynomials a new class of operators is needed. There are two main concepts: first, the reflection operator $p \mapsto [p(x) - p(x\sigma_v)]/\langle x, v\rangle$ for a given nonzero $v \in \mathbb{R}^d$, the numerator of which is divisible by $\langle x, v\rangle$ because it has the alternating property for σ_v (see Lemma 6.2.5); second, the reflection operators corresponding to the positive roots of some root system need to be assembled in a way which incorporates parameters and mimics the properties of derivatives. It turns out that this is possible to a large extent, except for the customary product rule, and the number of independent parameters equals the number of conjugacy classes of reflections.

Fix a reduced root system R, with positive roots R_+ and associated reflection group $W = W(R)$. The parameters are specified in the form of a multiplicity function (the terminology comes from analysis on homogeneous spaces formed from compact Lie groups).

Definition 6.4.1 A multiplicity function is a function $v \mapsto \kappa_v$ defined on a root system with the property that $\kappa_u = \kappa_v$ whenever σ_u is conjugate to σ_v in W, that is, when there exists $w \in W$ such that $uw = v$. The values can be numbers or formal parameters (“transcendental extensions” of $\mathbb{Q}$).

Definition 6.4.2 The first-order differential–difference operators $\mathscr{D}_i$ are defined, in coordinate form for $1 \le i \le d$ and in coordinate-free form for $u \in \mathbb{R}^d$ by, for $p \in \Pi^d$,

$$\mathscr{D}_i p(x) = \frac{\partial p(x)}{\partial x_i} + \sum_{v \in R_+} \kappa_v \frac{p(x) - p(x\sigma_v)}{\langle x, v\rangle} v_i,$$

$$\mathscr{D}_u p(x) = \langle u, \nabla p(x)\rangle + \sum_{v \in R_+} \kappa_v \frac{p(x) - p(x\sigma_v)}{\langle x, v\rangle} \langle u, v\rangle .$$

Clearly, $\mathscr{D}_i = \mathscr{D}_{\varepsilon_i}$ and $\mathscr{D}_u$ maps $\mathscr{P}_n^d$ into $\mathscr{P}_{n-1}^d$ for $n \ge 1$; that is, $\mathscr{D}_u$ is a homogeneous operator of degree -1. Under the action of W, $\mathscr{D}_u$ transforms like the directional derivative.

Proposition 6.4.3 *For $u \in \mathbb{R}^d$ and $w \in W$, $\mathscr{D}_u R(w)p = R(w)\mathscr{D}_{uw} p$ for $p \in \Pi^d$.*

Proof It is convenient to express $\mathscr{D}_u$ as a sum over R and to divide by 2; thus

$$\begin{aligned}
\mathscr{D}_u R(w)p(x) &= \langle uw, \nabla p(xw)\rangle + \tfrac{1}{2}\sum_{v \in R} \kappa_v \frac{p(xw) - p(x\sigma_v w)}{\langle x, v\rangle} \langle u, v\rangle \\
&= \langle uw, \nabla p(xw)\rangle + \tfrac{1}{2}\sum_{v \in R} \kappa_v \frac{p(xw) - p(xw\sigma_{vw})}{\langle xw, vw\rangle} \langle uw, vw\rangle \\
&= \langle uw, \nabla p(xw)\rangle + \tfrac{1}{2}\sum_{z \in R} \kappa_{zw^{-1}} \frac{p(xw) - p(xw\sigma_z)}{\langle xw, z\rangle} \langle uw, z\rangle \\
&= R(w)\mathscr{D}_{uw} p(x).
\end{aligned}$$

In the third line we changed the summation variable from v to z and then used the property of κ from Definition 6.4.1 giving $\kappa_{zw^{-1}} = \kappa_z$. □

The important aspect of differential–difference operators is their commutativity. Some lemmas will be established before the main proof. An auxiliary operator is convenient.

Definition 6.4.4 For $v \in \mathbb{R}^d$ and $v \ne 0$ define the operator ρ_v by

$$\rho_v f(x) = \frac{f(x) - f(x\sigma_v)}{\langle x, v\rangle}.$$

Lemma 6.4.5 *For $u, v \in \mathbb{R}^d$ and $v \ne 0$,*

$$\langle u, \nabla\rangle \rho_v f(x) - \rho_v \langle u, \nabla\rangle f(x) = \frac{\langle u, v\rangle}{\langle x, v\rangle}\left(\frac{2\langle v, \nabla f(x\sigma_v)\rangle}{\langle v, v\rangle} - \frac{f(x) - f(x\sigma_v)}{\langle x, v\rangle}\right).$$

Proof Recall that $\langle u, \nabla R(\sigma_v) f\rangle = \langle u\sigma_v, R(\sigma_v)\nabla f\rangle$; thus

$$\langle u, \nabla \rangle \rho_v f(x) = \frac{\langle u, \nabla f(x) \rangle - \langle u\sigma_v, \nabla f(x\sigma_v) \rangle}{\langle x, v \rangle} - \frac{\langle u, v \rangle [f(x) - f(x\sigma_v)]}{\langle x, v \rangle^2}$$

and

$$\rho_v \langle u, \nabla \rangle f(x) = \frac{\langle u, \nabla f(x) \rangle - \langle u, \nabla f(x\sigma_v) \rangle}{\langle x, v \rangle}.$$

Subtracting the second quantity from the first gives the claimed equation since $u - u\sigma_v = (2\langle u, v \rangle / \langle v, v \rangle)v$. □

In the proof below that $\mathscr{D}_u \mathscr{D}_z = \mathscr{D}_z \mathscr{D}_u$, Lemma 6.4.5 takes care of the interaction between differentiation and differencing. The next lemma handles the difference parts. Note that the product of two reflections is either the identity or a plane rotation (the plane being the span of the corresponding roots).

Lemma 6.4.6 *Suppose that $B(x,y)$ is a bilinear form on $\mathbb{R}^d$ such that $B(x\sigma_v, y\sigma_v) = B(y,x)$ whenever $v \in \mathrm{span}(x,y)$, and let w be a plane rotation in W; then*

$$\sum \left\{ \frac{\kappa_u \kappa_v B(u,v)}{\langle x, u \rangle \langle x, v \rangle} : u, v \in R_+, \sigma_u \sigma_v = w \right\} = 0, \qquad x \in \mathbb{R}^d,$$

$$\sum \{ \kappa_u \kappa_v B(u,v) \rho_u \rho_v : u, v \in R_+, \sigma_u \sigma_v = w \} = 0,$$

where the two equations refer to rational functions and operators, respectively.

Proof Let E be the plane of w (the orthogonal complement of E is pointwise fixed under w); thus $\sigma_u \sigma_v = w$ implies that $u, v \in E$. Let $R_0 = R_+ \cap E$ and let $W_0 = W(R_0)$, the reflection group generated by R_0. Because W_0 is a dihedral group, $\sigma_v w \sigma_v = w^{-1}$ for any $v \in R_0$. Denote the first sum above by $s(x)$. We will show that s is W_0-invariant. Fix $z \in R_0$; it suffices to show that $s(x\sigma_z) = s(x)$. Consider the effect of σ_z on R_0: the map $\sigma_v \to \sigma_z \sigma_v \sigma_z$ is an involution and so define $\tau v = \varepsilon(v) v\sigma_z$ for $v \in R_0$, where $\tau v \in R_0$ and $\varepsilon(v) = \pm 1$. Then $\sigma_z \sigma_v \sigma_z = \sigma_{\tau v}$, also $\tau^2 v = v$ and $\varepsilon(\tau v) = \varepsilon(v)$. Then

$$\begin{aligned} s(x\sigma_z) &= \sum \left\{ \frac{\kappa_u \kappa_v B(u,v)}{\langle x\sigma_z, u \rangle \langle x\sigma_z, v \rangle} : u, v \in R_0, \sigma_u \sigma_v = w \right\} \\ &= \sum \left\{ \frac{\kappa_{\tau u} \kappa_{\tau v} B(\tau u, \tau v)}{\langle x\sigma_z, \tau u \rangle \langle x\sigma_z, \tau v \rangle} : u, v \in R_0, \sigma_{\tau u} \sigma_{\tau v} = w \right\} \\ &= \sum \left\{ \frac{\kappa_u \kappa_v B(\varepsilon(u) u\sigma_z, \varepsilon(v) v\sigma_z)}{\langle x, \varepsilon(u) u \rangle \langle x, \varepsilon(v) v \rangle} : u, v \in R_0, \sigma_z \sigma_u \sigma_v \sigma_z = w \right\} \\ &= \sum \left\{ \frac{\kappa_u \kappa_v B(v,u)}{\langle x, u \rangle \langle x, v \rangle} : u, v \in R_0, \sigma_u \sigma_v = \sigma_z w \sigma_z = w^{-1} \right\} \\ &= s(x). \end{aligned}$$

The second line follows by a change in summation variables and the third line uses the multiplicity function property that $\kappa_{\tau u} = \kappa_u$ because $\sigma_{\tau u} = \sigma_z \sigma_u \sigma_z$ and also $\langle x\sigma_z, \tau u\rangle = \langle x, (\tau u)\sigma_z\rangle$; the fourth line uses the assumption about the form B and also $(\sigma_u \sigma_v)^{-1} = \sigma_v \sigma_u$. Now let $q(x) = s(x)\prod_{v\in R_0}\langle x, v\rangle$; then $q(x)$ is a polynomial of degree $|R_0| - 2$ and it has the alternating property for W_0. Thus $q(x) = 0$.

Let $f(x)$ be an arbitrary polynomial and, summing over $u, v \in R_0, \sigma_u\sigma_v = w$, consider

$$\begin{aligned}&\sum \kappa_u \kappa_v B(u,v)\,\rho_u \rho_v f(x)\\ &\quad = \sum \kappa_u \kappa_v B(u,v)\left(\frac{f(x) - f(x\sigma_v)}{\langle x,u\rangle\langle x,v\rangle} - \frac{f(x\sigma_u) - f(xw)}{\langle x,u\rangle\langle x, v\sigma_u\rangle}\right).\end{aligned}$$

The coefficient of $f(x)$ in the sum is 0 by the first part of the lemma. For a fixed $z \in R_0$ the term $f(x\sigma_z)$ appears twice, for the case $\sigma_u\sigma_z = w$ with coefficient $\kappa_u\kappa_z B(u,z)/(\langle x,u\rangle\langle x,z\rangle)$ and for the case $\sigma_z\sigma_v = w$ with coefficient $\kappa_z\kappa_v B(z,v)/(\langle x,z\rangle\langle x, v\sigma_z\rangle)$. But $\sigma_v = \sigma_z\sigma_u\sigma_z$ and thus $v = \tau u$ (using the notation of the previous paragraph), and so the second coefficient equals

$$\frac{\kappa_z\kappa_u B(z,\tau u)}{\langle x,z\rangle\langle x,\varepsilon(u)u\rangle} = \frac{\kappa_z\kappa_u B((\tau u)\sigma_z, z\sigma_z)}{\langle x,z\rangle\langle x,\varepsilon(u)u\rangle} = \frac{\kappa_z\kappa_u B(\varepsilon(u)u, -z)}{\langle x,z\rangle\langle x,\varepsilon(u)u\rangle},$$

which cancels out the first coefficient. To calculate the coefficient of $f(xw)$ we note that, for any $z \in R_0$, $\sigma_z\sigma_v = w$ if and only if $\sigma_{\tau v}\sigma_z = w$; thus

$$\frac{\kappa_z\kappa_v B(z,v)}{\langle x,z\rangle\langle x,v\sigma_z\rangle} = \frac{\kappa_z\kappa_{\tau v} B(z,(\tau v)\sigma_z)}{\langle x,z\rangle\langle x,\tau v\rangle} = \frac{\kappa_z\kappa_{\tau v} B(\tau v, -z)}{\langle x,z\rangle\langle x,\tau v\rangle},$$

which shows that the coefficient is $-s(x)$ and hence 0. □

Corollary 6.4.7 *Under the same hypotheses,*

$$\sum\{\kappa_u\kappa_v B(u,v)\,\rho_u\rho_v : u,v\in R_+\} = 0.$$

Proof Decompose the sum into parts $\sigma_u\sigma_v = w$ for rotations $w \in W$ and a part $\sigma_u\sigma_v = 1$ (that is, $u = v$). Each rotation contributes 0 by the lemma, and $\rho_u^2 = 0$ for each u because $\rho_u f(x\sigma_u) = \rho_u f(x)$ for any polynomial f. □

Theorem 6.4.8 *For $t, u \in \mathbb{R}^d$, $\mathscr{D}_t\mathscr{D}_u = \mathscr{D}_u\mathscr{D}_t$.*

Proof For any polynomial f, the action of the commutator can be expressed as follows:

$$
\begin{aligned}
&(\mathscr{D}_t\mathscr{D}_u - \mathscr{D}_u\mathscr{D}_t)f \\
&\quad = (\langle t,\nabla\rangle\langle u,\nabla\rangle - \langle u,\nabla\rangle\langle t,\nabla\rangle)f \\
&\qquad + \sum_{v\in R_+}\kappa_v\left[\langle u,v\rangle(\langle t,\nabla\rangle\rho_v - \rho_v\langle t,\nabla\rangle) - \langle t,v\rangle(\langle u,\nabla\rangle\rho_v - \rho_v\langle u,\nabla\rangle)\right]f \\
&\qquad + \sum_{v,z\in R_+}\kappa_v\kappa_z(\langle t,v\rangle\langle u,z\rangle - \langle u,v\rangle\langle t,z\rangle)\rho_v\rho_z f.
\end{aligned}
$$

The first line of the right-hand side is trivially zero, the second line vanishes by Lemma 6.4.5 and the third line vanishes by Lemma 6.4.6 applied to the bilinear form $B(x,y) = \langle t,x\rangle\langle u,y\rangle - \langle u,x\rangle\langle t,y\rangle$. To see that B satisfies the hypothesis, let $v = ax+by$ with $a,b\in\mathbb{R}$ and $v\neq 0$; then $B(v,y) = aB(x,y), B(x,v) = bB(x,y)$ and

$$
\begin{aligned}
B(x\sigma_v, y\sigma_v) &= B(x,y)\left(1 - \frac{2a\langle x,v\rangle}{\langle v,v\rangle} - \frac{2b\langle y,v\rangle}{\langle v,v\rangle}\right) \\
&= B(x,y)(1-2) = B(y,x).
\end{aligned}
$$

□

The norm $\|x\|^2$ is invariant for every reflection group. The corresponding invariant operator is called the *h-Laplacian* (the prefix h refers to the weight function $\prod_{v\in R_+}|\langle x,v\rangle|^{\kappa_v}$, discussed later on in this book) and is defined as $\Delta_h = \sum_{i=1}^d \mathscr{D}_i^2$. The definition is independent of the choice of orthogonal basis; this is a consequence of the following formulation (recall that ε_i denotes the ith unit basis vector).

Theorem 6.4.9 *For any polynomial $f(x)$,*

$$
\sum_{i=1}^d \mathscr{D}_i^2 f(x) = \Delta f(x) + \sum_{v\in R_+}\kappa_v\left(\frac{2\langle v,\nabla f(x)\rangle}{\langle v,x\rangle} - \|v\|^2\frac{f(x)-f(x\sigma_v)}{\langle v,x\rangle^2}\right).
$$

Proof Write ∂_i for $\partial/\partial x_i$; then

$$
\mathscr{D}_i^2 f(x) = \partial_i^2 f(x) + \sum_{v\in R_+}\kappa_v v_i(\partial_i\rho_v + \rho_v\partial_i)f(x) + \sum_{u,v\in R_+}\kappa_u\kappa_v u_i v_i\rho_u\rho_v f(x).
$$

The middle term equals

$$
\sum_{v\in R_+}\kappa_v v_i\left(\frac{2\partial_i f(x)}{\langle v,x\rangle} - v_i\frac{f(x)-f(x\sigma_v)}{\langle v,x\rangle^2} - \frac{\langle \varepsilon_i + \varepsilon_i\sigma_v, \nabla f(x\sigma_v)\rangle}{\langle v,x\rangle}\right).
$$

Summing the first two parts of this term over $1\le i\le d$ clearly yields the reflection part (the sum over R_+) of the claim. Since $v\sigma_v = -v$, the third part sums to $\sum_{v\in R_+}\kappa_v\langle v+v\sigma_v, \nabla f(x\sigma_v)\rangle/\langle v,x\rangle = 0$.

Finally, the double sum that is the last term in $\mathscr{D}_i^2 f(x)$ produces $\sum_{u,v\in R_+}\kappa_u\kappa_v\langle u,v\rangle\rho_u\rho_v f(x)$, which vanishes by Lemma 6.4.6 applied to the bilinear form $B(x,y) = \langle x,y\rangle$. □

The product rule for $\mathscr{D}_u$ is more complicated than that for differentiation. We begin with a case involving a first-degree factor.

Proposition 6.4.10 *Let $t,u \in \mathbb{R}^d$; then*

$$\mathscr{D}_u\left(\langle\cdot,t\rangle f\right)(x) - \langle x,t\rangle\,\mathscr{D}_u f(x) = \langle u,t\rangle\, f(x) + 2\sum_{v\in R_+} \kappa_v \frac{\langle u,v\rangle\,\langle t,v\rangle}{\|v\|^2} f(x\sigma_v).$$

Proof This is a routine calculation which uses the relation $\langle x,t\rangle - \langle x\sigma_v,t\rangle = 2\langle x,v\rangle\,\langle v,t\rangle/\|v\|^2$. □

Proposition 6.4.11 *If W acts irreducibly on $\mathbb{R}^d$ (span$(R) = \mathbb{R}^d$ and R is indecomposable) then $\mathscr{D}_u\langle x,t\rangle = \left(1+\frac{2}{d}\sum_{v\in R_+}\kappa_v\right)\langle u,t\rangle$ for any $t,u\in\mathbb{R}^d$.*

Proof From the product rule it suffices to consider the bilinear form $B(t,u) = \sum_{v\in R_+}\kappa_v\langle u,v\rangle\,\langle t,v\rangle/\|v\|^2$. This is clearly symmetric and invariant under W, since $B(t\sigma_v,u\sigma_v) = B(t,u)$ for each $v\in R_+$. Suppose that positive values are chosen for κ_v; then the form can be diagonalized so that $B(t,u) = \sum_{i=1}^d \lambda_i\langle t,\varepsilon_i'\rangle\,\langle u,\varepsilon_i'\rangle$ for some orthonormal basis $\{\varepsilon_i'\}$ of $\mathbb{R}^d$ and numbers λ_i. But the subspace $\{x : B(x,x) = \lambda_1\langle x,x\rangle\}$ is W-invariant since it includes ε_1'; by the irreducibility assumption it must be the whole of $\mathbb{R}^d$. Thus $B(t,u) = \lambda_1\langle t,u\rangle$ for any $t,u\in\mathbb{R}^d$. To evaluate the constant, consider $\sum_{i=1}^d B(\varepsilon_i,\varepsilon_i) = d\lambda_1 = \sum_{i=1}^d\sum_{v\in R_+}\kappa_v v_i^2/\|v\|^2 = \sum_{v\in R_+}\kappa_v$. □

Here is a more general product rule, which is especially useful when one of the polynomials is W-invariant.

Proposition 6.4.12 *For polynomials f,g and $u\in\mathbb{R}^d$,*

$$\begin{aligned}\mathscr{D}_u(fg)(x) &= g(x)\mathscr{D}_u f(x) + f(x)\langle u,\nabla g(x)\rangle\\ &\quad + \sum_{v\in R_+}\kappa_v f(x\sigma_v)\,\langle u,v\rangle\,\frac{g(x)-g(x\sigma_v)}{\langle v,x\rangle}.\end{aligned}$$

For conciseness in certain proofs, the gradient corresponding to $\mathscr{D}_u$ is useful: let $\nabla_\kappa f(x) = (\mathscr{D}_i f(x))_{i=1}^d$, considered as a row vector. Then $\nabla_\kappa R(w)f(x) = \nabla_\kappa f(xw)\,w^{-1}$ for $w\in W$, and $\mathscr{D}_u f = \langle u,\nabla_\kappa f\rangle$.

6.5 The Intertwining Operator

There are classical fractional integral transforms which map one family of orthogonal polynomials onto another. Some of these transforms can be interpreted as special cases of the intertwining operator for differential–difference operators. Such an object should be a linear map V on polynomials with at least the property that $\mathscr{D}_i V p(x) = V(\partial/\partial x_i)p(x)$ for each i; it turns out that it is uniquely defined by the additional assumptions $V1 = 1$ and $V\mathscr{P}_n^d \subset \mathscr{P}_n^d$, that is, homogeneous polynomials are preserved. It is easy to see that the formal inverse always exists, that is, an operator T such that $T\mathscr{D}_i = (\partial/\partial x_i)T$.

Heuristically, let $Tp(x) = \exp\left(\sum_{i=1}^d x_i \mathscr{D}_i^{(y)}\right) p(y)|_{y=0}$. Now apply the formal series to a polynomial, where $\mathscr{D}_i^{(y)}$ acts on y; then set $y = 0$, obtaining a polynomial in x. Here is the precise definition.

Definition 6.5.1 For $n = 0, 1, 2, \ldots$ define T on $\mathscr{P}_n^d$ by

$$Tp(x) = \frac{1}{n!}\left(\sum_{i=1}^d x_i \mathscr{D}_i^{(y)}\right)^n p(y)$$

for $p \in \mathscr{P}_n^d$. Then T is extended to Π^d by linearity.

Of course, the commutativity of $\{\mathscr{D}_i\}$ allows us to write down the nth power and to prove the intertwining property.

Proposition 6.5.2 *For any polynomial p and $1 \le i \le d$,*

$$\frac{\partial}{\partial x_i} Tp(x) = T\mathscr{D}_i p(x).$$

Proof It suffices to let $p \in \mathscr{P}_n^d$. Then

$$\frac{\partial}{\partial x_i} Tp(x) = \frac{n}{n!}\mathscr{D}_i^{(y)}\left(\sum_{j=1}^d x_j \mathscr{D}_j^{(y)}\right)^{n-1} p(y) = T\mathscr{D}_i p(x),$$

because $\mathscr{D}_i^{(y)}$ commutes with multiplication by x_j. □

Note that the existence of T has been proven for any multiplicity function, but nothing has been shown about the existence of an inverse. The situation where T has a nontrivial kernel was studied by Dunkl, de Jeu and Opdam [1994]. The κ_v values which occur are called singular values; they are related to the structure of the group W and involve certain negative rational numbers. We will show that the inverse exists whenever $\kappa_v \ge 0$ (in fact, for values corresponding to integrable weight functions). The construction depends on a detailed analysis of the operator $\sum_{j=1}^d x_j \mathscr{D}_j = \langle x, \nabla_\kappa \rangle$. A certain amount of algebra and group theory will be required, but nothing really more complicated than the concepts of group algebra and some matrix theory. The following is a routine calculation.

Proposition 6.5.3 *For any smooth function f on $\mathbb{R}^d$,*

$$\sum_{j=1}^d x_j \mathscr{D}_j f(x) = \sum_{j=1}^d x_j \frac{\partial}{\partial x_j} f(x) + \sum_{v \in R_+} \kappa_v [f(x) - f(x\sigma_v)].$$

Note that if $f \in \mathscr{P}_n^d$ then $\sum_{j=1}^d x_j \mathscr{D}_j f(x) = (n + \sum_v \kappa_v) f(x) - \sum_v \kappa_v f(x\sigma_v)$. We will construct functions $c_n(w)$ on W with the property that

$$\sum_{w \in W} c_n(w) \langle xw, \nabla_\kappa f(xw) \rangle = f(x),$$

for $n \geq 1$. This requires the use of (matrix) exponentials in the group algebra.

Let $\mathbb{K}$ be some extension field of the rational numbers $\mathbb{Q}$ containing at least $\{\kappa_v\}$; then $\mathbb{K}W$ is the algebra $\{\sum_w a(w)w : a(w) \in \mathbb{K}\}$ with the multiplication rule $\sum_w a(w)w \sum_{w'} b(w')w' = \sum_w \left(\sum_z a(z) b(z^{-1}w)\right) w$. The center of $\mathbb{K}W$ is the span of the conjugacy classes ξ (a conjugacy class is a subset of W with the following property: $w, w' \in \xi$ if and only if there exists $z \in W$ such that $w' = z^{-1}wz$). We use ξ as the name of both the subset of W and the element of $\mathbb{K}W$. Since $z^{-1}\sigma_v z = \sigma_{vz}$ for $v \in R_+$, the reflections are in conjugacy classes by themselves. Label them $\xi_1, \xi_2, \ldots, \xi_r$ (recall that r is the number of connected components of the Coxeter diagram after all edges with an even label have been removed); further, let $\kappa_i' = \kappa_v$ for any v with $\sigma_v \in \xi_i, 1 \leq i \leq r$.

The group algebra $\mathbb{K}W$ can be interpreted as an algebra of $|W| \times |W|$ matrices $\sum_w a(w) M(w)$, where $M(w)$ is a permutation matrix with $M(w)_{y,z} = 1$ if $y^{-1}z = w$, else $M(w)_{y,z} = 0$, for $y, z \in W$. On the one hand, a class ξ_i corresponds to $\sum\{M(\sigma_v) : \sigma_v \in \xi_i\}$, which has integer entries; thus all eigenvalues of ξ_i are algebraic integers (that is, the zeros of monic polynomials with integer coefficients). On the other hand, ξ_i is a (real) symmetric matrix and can be diagonalized. Thus there is an orthogonal change of basis of the underlying vector space, and $\mathbb{R}^{|W|} = \bigoplus_\lambda E_\lambda$ where ξ_i acts as $\lambda 1$ on each eigenspace E_λ. Since each σ_v commutes with ξ_i, the transformed matrices for σ_v have a corresponding block structure (and we have $\sigma_v E_\lambda = E_\lambda$, equality as spaces). Further, since all the group elements commute with ξ_i, the projections of σ_v onto E_λ for $\sigma_v \in \xi_i$ are all conjugate to each other. Hence each projection has the same multiplicities of its eigenvalues, say 1 with multiplicity n_0 and -1 with multiplicity n_1 (the projections are certainly involutions); thus $\lambda(n_0 + n_1) = |\xi_i|(n_0 - n_1)$, taking the trace of the projection of ξ_i to E_λ. This shows that λ is rational, satisfies $|\lambda| \leq |\xi_i|$ and is an algebraic integer, and so all the eigenvalues of ξ_i are integers in the interval $[-|\xi_i|, |\xi_i|]$. Note that this is equivalent to the eigenvalues of $\sum\{1 - \sigma : \sigma \in \xi_i\} = |\xi_i| 1 - \xi_i$ being integers in $[0, 2|\xi_i|]$.

Definition 6.5.4 For $s \in \mathbb{R}$ and $1 \leq i \leq r$, the functions $q_i(w;s)$ are defined by

$$\exp[s(|\xi_i| 1 - \xi_i)] = \sum_{w \in W} q_i(w;s) w.$$

Proposition 6.5.5 *For $1 \leq i \leq r$ and $w \in W$:*

(i) *$q_i(1;0) = 1$ and $q_i(w;0) = 0$ for $w \neq 1$;*
(ii) *$s < 0$ implies $q_i(w;s) \geq 0$;*

(iii) $q_i(w;s)$ *is a linear combination of the members of* $\{e^{s\lambda} : \lambda$ is an eigenvalue of $|\xi_i| 1 - \xi_i\}$;
(iv) $\partial q_i(w;s)/\partial s = \sum_{\sigma \in \xi_i} [q_i(w;s) - q_i(w\sigma;s)]$;
(v) $\sum_w q_i(w;s) = 1$.

Proof The values $q_i(w;0)$ are determined by $\exp(0) = 1 \in \mathbb{Q}W$. For $s < 0$, $\sum_w q_i(w;s)w = \exp(s|\xi_i|)\exp(-s\xi_i)$. The matrix for $-s\xi_i$ has all entries nonnegative, and the product of any two conjugacy classes is a nonnegative integer linear combination of other classes; thus $q_i(w;s) \geq 0$. Part (iii) is a standard linear algebra result (and recall that ξ_i corresponds to a self-adjoint matrix). For part (iv),

$$\begin{aligned} \frac{\partial}{\partial s}\sum_w q_i(w;s)w &= (\exp[s(|\xi_i| 1 - \xi_i)])\,(|\xi_i| 1 - \xi_i) \\ &= \sum_w \sum_{\sigma\in\xi_i} [q_i(w;s)w - q_i(w;s)w\sigma] \\ &= \sum_{\sigma\in\xi_i}\sum_w [q_i(w;s)w - q_i(w\sigma;s)w], \end{aligned}$$

replacing w by $w\sigma$ in the last sum. Finally, the trivial homomorphism $w \mapsto 1$ extended to $\mathbb{K}W$ maps $|\xi_i| 1 - \xi_i$ to 0, and $\exp(0) = 1$. □

The classes of reflections are combined in the following.

Definition 6.5.6 The functions $q_\kappa(w;s)$ are given by

$$\sum_{w\in W} q_\kappa(w;s)w = \prod_{i=1}^{r}\left(\sum_w q_i(w;s\kappa_i')\,w\right).$$

Since the classes ξ_i are in the center of $\mathbb{K}W$, this is a product of commuting factors. Proposition 6.5.5 shows that $q_\kappa(w;s)$ is a linear combination of products of terms like $\exp(\kappa_i'\lambda s)$, where λ is an integer and $0 \leq \lambda \leq 2|\xi_i|$. Also,

$$\sum_{w\in W} q_\kappa(w;s)w = \exp\left(s\sum_{v\in R_+}\kappa_v(1-\sigma_v)\right)$$

and

$$\frac{\partial}{\partial s} q_\kappa(w;s) = \sum_{v\in R_+}\kappa_v[q_\kappa(w;s) - q_\kappa(w\sigma_v;s)].$$

Definition 6.5.7 For $\kappa_v \geq 0$, $n \geq 1$ and $w \in W$, set

$$c_n(w) = \int_{-\infty}^{0} q_\kappa(w;s)e^{ns}\,\mathrm{d}s.$$

Proposition 6.5.8 *With the hypotheses of Definition 6.5.7 we have*

(i) $c_n(w) \geq 0$;
(ii) $nc_n(w) + \sum_{v\in R_+}\kappa_v[c_n(w) - c_n(w\sigma_v)] = \delta_{1,w}$ (1 *if* $w = 1$, *else* 0);
(iii) $\sum_{w\in W} c_n(w) = 1/n$.

Proof The positivity follows from that of $q_\kappa(w;s)$. Multiply both sides of the differential relation just before Definition 6.5.7 by e^{ns} and integrate by parts over $-\infty < s \le 0$, obtaining

$$q_\kappa(w;0) - \lim_{s\longrightarrow -\infty} e^{ns} q_\kappa(w;s) - nc_n(w) = \sum_{v\in R_+} \kappa_v \left[c_n(w) - c_n(w\sigma_v)\right].$$

Since $q_\kappa(w;s)$ is a sum of terms $e^{\lambda s}$ with $\lambda \ge 0$ and $n \ge 1$, the integral for c_n is absolutely convergent and the above limit is zero. From Proposition 6.5.5 $q_\kappa(w;0) = 1$ if $w = 1$, else $q_\kappa(W;0)$ is 0. Finally, $\sum_w c_n(w) = \int_{-\infty}^0 \sum_w q_\kappa(w;s) e^{ns} ds = \int_{-\infty}^0 e^{ns} ds = 1/n$ (Proposition 6.5.5, part (v)). □

Proposition 6.5.9 *For $n \ge 1$ let $f \in \mathscr{P}_n^d$; then*

$$\sum_{w\in W} c_n(w) \langle xw, \nabla_\kappa f(xw)\rangle = f(x).$$

Proof Indeed,

$$\begin{aligned}
&\sum_{w\in W} c_n(w) \langle xw, \nabla_\kappa f(xw)\rangle \\
&\quad= \sum_{w\in W} c_n(w) \left(nf(xw) + \sum_{v\in R_+} \kappa_v \left[f(xw) - f(xw\sigma_v)\right]\right) \\
&\quad= \sum_{w\in W} c_n(w) nf(xw) + \sum_{v\in R_+} \kappa_v \sum_{w\in W} f(xw)\left[c_n(w) - c_n(w\sigma_v)\right] \\
&\quad= \sum_{w\in W} q_\kappa(w;0) f(xw) = f(x).
\end{aligned}$$

The last line uses (ii) in Proposition 6.5.8. □

Corollary 6.5.10 *For any polynomial $f \in \Pi^d$,*

$$\sum_{w\in W} \int_{-\infty}^0 q_\kappa(w;s) \langle xw, \nabla_\kappa f(e^s xw)\rangle e^s ds = f(x) - f(0).$$

Proof Express f as a sum of homogeneous components; if g is homogeneous of degree n then $\nabla_\kappa g(e^s xw) = e^{(n-1)s} \nabla_\kappa g(xw)$. □

The above results can be phrased in the language of forms.

Definition 6.5.11 An $\mathbb{R}^d$-valued polynomial $\mathbf{f}(x) = (f_i(x))_{i=1}^d$ (with each $f_i \in \Pi^d$) is a κ-closed 1-form if $\mathscr{D}_i f_j = \mathscr{D}_j f_i$ for each i, j or a κ-exact 1-form if there exists $g \in \Pi^d$ such that $\mathbf{f} = \nabla_\kappa g$.

The commutativity of $\{\mathscr{D}_i\}$ shows that each κ-exact 1-form is κ-closed. The corollary shows that g is determined up to a constant by $\nabla_\kappa g$. The construction of the intertwining operator depends on the fact that κ-closed 1-forms are κ-exact. This will be proven using the functions $c_n(w)$.

Lemma 6.5.12 *If* $\mathbf{f}$ *is a* κ*-closed* 1*-form and* $u \in \mathbb{R}^d$ *then*

$$\mathscr{D}_u \langle x, \mathbf{f}(x) \rangle = \left(1 + \sum_{i=1}^{d} x_i \frac{\partial}{\partial x_i}\right) \langle u, \mathbf{f}(x) \rangle + \sum_{v \in R_+} \kappa_v \langle u, \mathbf{f}(x) - \mathbf{f}(x\sigma_v)\sigma_v \rangle .$$

Proof Apply the product rule (Proposition 6.4.10) to $x_i f_i(x)$ to obtain

$$\begin{aligned}
\mathscr{D}_u \langle x, \mathbf{f}(x) \rangle &= \sum_{i=1}^{d} \Big(x_i \mathscr{D}_u f_i(x) + u_i f_i(x) \Big) + 2 \sum_{v \in R_+} \kappa_v \frac{\langle u, v \rangle \langle v, \mathbf{f}(x\sigma_v) \rangle}{\|v\|^2} \\
&= \langle u, \mathbf{f}(x) \rangle + 2 \sum_{v \in R_+} \kappa_v \frac{\langle u, v \rangle \langle v, \mathbf{f}(x\sigma_v) \rangle}{\|v\|^2} + \sum_{i,j=1}^{d} x_i u_j \mathscr{D}_i f_j(x) \\
&= \left(1 + \sum_{i=1}^{d} x_i \frac{\partial}{\partial x_i}\right) \langle u, \mathbf{f}(x) \rangle + 2 \sum_{v \in R_+} \kappa_v \frac{\langle u, v \rangle \langle v, \mathbf{f}(x\sigma_v) \rangle}{\|v\|^2} \\
&\quad + \sum_{v \in R_+} \kappa_v \langle u, \mathbf{f}(x) - \mathbf{f}(x\sigma_v) \rangle .
\end{aligned}$$

In the third line we replaced $\mathscr{D}_j f_i$ by $\mathscr{D}_i f_j$ (employing the κ-closed hypothesis). The use of Proposition 6.5.3 produces the fourth line. The terms involving $\mathbf{f}(x\sigma_v)$ add up to $\sum_{v \in R_+} \kappa_v \langle u, -\mathbf{f}(x\sigma_v)\sigma_v \rangle$. □

Theorem 6.5.13 *Suppose that* $\mathbf{f}$ *is a homogeneous* κ*-closed* 1*-form with each* $f_i \in \mathscr{P}_n^d$ *for* $n \geq 0$*; then the homogeneous polynomial defined by*

$$F(x) = \sum_{w \in W} c_{n+1}(w) \langle xw, \mathbf{f}(xw) \rangle$$

satisfies $\nabla_\kappa F = \mathbf{f}$*, and* $\mathbf{f}$ *is* κ*-exact.*

Proof We will show that $\mathscr{D}_u F(x) = \langle u, \mathbf{f}(x) \rangle$ for any $u \in \mathbb{R}^d$. Note that

$$\mathscr{D}_u [R(w) \langle x, \mathbf{f}(x) \rangle] = R(w) [\mathscr{D}_{uw} \langle x, \mathbf{f}(x) \rangle] ;$$

thus

$$\begin{aligned}
&\sum_{w \in W} c_{n+1}(w) \mathscr{D}_u [\langle xw, \mathbf{f}(xw) \rangle] \\
&\quad = \sum_{w \in W} c_{n+1}(w) \left((n+1) \langle uw, \mathbf{f}(xw) \rangle + \sum_{v \in R_+} \kappa_v \langle uw, \mathbf{f}(xw) - \mathbf{f}(xw\sigma_v)\sigma_v \rangle \right) \\
&\quad = \sum_{w \in W} \langle uw, \mathbf{f}(xw) \rangle \left(c_{n+1}(w)(n+1) + \sum_{v \in R_+} \kappa_v [c_{n+1}(w) - c_{n+1}(w\sigma_v)] \right) \\
&\quad = \sum_{w \in W} \langle uw, \mathbf{f}(xw) \rangle q_\kappa(w;0) = \langle u, \mathbf{f}(x) \rangle .
\end{aligned}$$

The second line comes from the lemma, then in the sum involving $\mathbf{f}(xw\sigma_v)$ the summation variable w is replaced by $w\sigma_v$. □

Corollary 6.5.14 *For any κ-closed 1-form $\mathbf{f}$ the polynomial F defined by*

$$F(x) = \sum_{w \in W} \int_{-\infty}^{0} q_\kappa(w;s) \langle xw, \mathbf{f}(e^s xw) \rangle e^s \, ds$$

satisfies $\nabla_\kappa F = \mathbf{f}$, and $\mathbf{f}$ is κ-exact.

In the following we recall the notation $\partial_i f(x) = \partial f(x)/\partial x_i$; $\partial_i f(xw)$ denotes the ith derivative evaluated at xw.

Definition 6.5.15 The operators $V_n : \mathscr{P}_n^d \to \mathscr{P}_n^d$ are defined inductively by $V_0(a) = a$ for any constant $a \in \mathbb{R}$ and

$$V_n f(x) = \sum_{w \in W} c_n(w) \left(\sum_{i=1}^{d} (xw)_i V_{n-1} [\partial_i f(xw)] \right), \qquad n \geq 1, \quad f \in \mathscr{P}_n^d.$$

Theorem 6.5.16 *For $f \in \mathscr{P}_n^d$, $n \geq 1$ and $1 \leq i \leq d$, $\mathscr{D}_i V_n f = V_{n-1} \partial_i f$; the operators V_n are uniquely defined by these conditions and $V_0 1 = 1$.*

Proof Clearly $\mathscr{D}_i V_0 a = 0$. Suppose that the statement is true for $n-1$; then $\mathbf{g} = (V_{n-1}\partial_i f)_{i=1}^d$ is a κ-closed 1-form, homogeneous of degree $n-1$ because $\mathscr{D}_j g_i = V_{n-2}\partial_j \partial_i f = \mathscr{D}_i g_j$. Hence the polynomial

$$F(x) = \sum_{w \in W} c_n(w) \langle xw, \mathbf{g}(xw) \rangle$$

is homogeneous of degree n and satisfies $\mathscr{D}_i F = g_i = V_{n-1}\partial_i f$ for each i, by Theorem 6.5.13. The uniqueness property is also proved inductively: suppose that V_{n-1} is uniquely determined; then by Proposition 6.5.9 there is a unique homogeneous polynomial F such that $\mathscr{D}_i F = V_{n-1}\partial_i f$ for each i. □

Definition 6.5.17 The *intertwining operator* V is defined on Π^d as the linear extension of the formal sum $\bigoplus_{n=0}^{\infty} V_n$; that is, if $f = \sum_{n=0}^{m} f_n$ with $f_n \in \mathscr{P}_n^d$ then $Vf = \sum_{n=0}^{m} V_n f_n$.

The operator V, as expected, commutes with each $R(w)$.

Proposition 6.5.18 *Let $f \in \Pi^d$ and $w \in W$; then $VR(w)f = R(w)Vf$.*

Proof For any $u \in \mathbb{R}^d$, $\mathscr{D}_u V f(x) = V \langle u, \nabla f(x) \rangle$. Substitute uw for u and xw for x to obtain on the one hand

$$\begin{aligned}R(w)\left(\mathscr{D}_{uw}Vf\right)(x) &= V\langle uw, \nabla f(xw)\rangle \\ &= V\langle uw, \nabla\left[R(w)f\right](x)w\rangle \\ &= V\langle u, \nabla\left[R(w)f\right](x)\rangle \\ &= \mathscr{D}_u V\left[R(w)f\right](x);\end{aligned}$$

on the other hand, $\mathscr{D}_u\left[R(w)Vf\right](x) = R(w)\left(\mathscr{D}_{uw}Vf\right)(x)$. Therefore

$$\nabla_\kappa\left[R(w)Vf\right] = \nabla_\kappa\left[VR(w)f\right],$$

which shows that $VR(w)f - R(w)Vf$ is a constant by Proposition 6.5.9. In the homogeneous decomposition of f, the constant term is clearly invariant under V and $R(w)$; thus $VR(w)f = R(w)Vf$. □

The inductive type of definition used for V suggests that there are boundedness properties which involve similarly defined norms. The following can be considered as an analogue of power series with summable coefficients.

Definition 6.5.19 For $f \in \Pi^d$ let $\|f\|_A = \sum_{n=0}^{\infty}\|f_n\|_S$, where $f = \sum_{n=0}^{\infty} f_n$ with each $f_n \in \mathscr{P}_n^d$ and $\|g\|_S = \sup_{\|x\|=1}|g(x)|$. Let $A(B^d)$ be the closure of Π^d in the A-norm.

Clearly $A\left(B^d\right)$ is a commutative Banach algebra under pointwise operations and is contained in $C\left(B^d\right)\cap C^\infty(\{x : \|x\| < 1\})$. Also, $\|f\|_S \le \|f\|_A$. We will show that V is bounded in the A-norm. The van der Corput–Schaake inequality (see Theorem 4.5.3) motivates the definition of a gradient-type norm associated with ∇_κ.

Definition 6.5.20 Suppose that $f \in \mathscr{P}_n^d$; for $n = 0$ let $\|f\|_\kappa = f$, and for $n \ge 1$ let

$$\|f\|_\kappa = \frac{1}{n!}\sup\left\{\left|\prod_{i=1}^{n}\langle y^{(i)}, \nabla_\kappa\rangle f(x)\right| : y^{(1)},\ldots,y^{(n)} \in S^{d-1}\right\}.$$

Proposition 6.5.21 *For $n \ge 0$ and $f \in \mathscr{P}_n^d$, $\|f\|_S \le \|f\|_\kappa$.*

Proof Proceeding inductively, suppose that the statement is true for $\mathscr{P}_{n-1}^d$ and $n \ge 1$. Let $x \in S^{d-1}$ and, by Proposition 6.5.9, let

$$f(x) = \sum_{w\in W} c_n(w)\,\langle xw, \nabla_\kappa f(xw)\rangle.$$

Thus $|f(x)| \le \sum_w c_n(w)\left|\langle xw, \nabla_\kappa f(xw)\rangle\right|$; $c_n(w) \ge 0$ by Proposition 6.5.8. For each w, $\left|\langle xw, \nabla_\kappa f(xw)\rangle\right| \le \sup_{\|u\|=1}\left|\langle u, \nabla_\kappa f(xw)\rangle\right|$. But, for given u and w, let $g(x) = \langle u, \nabla_\kappa f(xw)\rangle = \left\langle uw^{-1}, \nabla_\kappa R(w)f(x)\right\rangle \in \mathscr{P}_{n-1}^d$, and the inductive hypothesis implies that

$$
\begin{aligned}
|g(x)| &\le \frac{1}{(n-1)!}\sup\left\{\left|\prod_{i=1}^{n-1}\langle y^{(i)},\nabla_\kappa\rangle g(x)\right| : y^{(1)},\ldots,y^{(n-1)}\in S^{d-1}\right\}\\
&= \frac{1}{(n-1)!}\sup\left\{\left|\prod_{i=1}^{n-1}\langle y^{(i)},\nabla_\kappa\rangle\langle uw^{-1},\nabla_\kappa\rangle R(w)f(x)\right| :\right.\\
&\qquad\qquad\left. y^{(1)},\ldots,y^{(n-1)}\in S^{d-1}\right\}\\
&\le n\|R(w)f\|_\kappa = n\|f\|_\kappa .
\end{aligned}
$$

Note that $\|R(w)f\|_\kappa = \|f\|_\kappa$ for any $w\in W$, because

$$\nabla_\kappa R(w)f(x) = R(w)\,\nabla_\kappa f(x)w^{-1}.$$

Thus $|f(x)| \le \sum_w c_n(w)n\|f\|_\kappa$ and $\sum_w c_n(w) = 1/n$ by Proposition 6.5.8. □

The norm $\|f\|_\partial$ was defined in van der Corput and Schaake [1935]; it is the same as $\|f\|_\kappa$ but with all parameters 0.

Proposition 6.5.22 *For $n\ge 0$ and $f\in\mathscr{P}_n^d$, $\|Vf\|_\kappa = \|f\|_\partial$.*

Proof For any $y^{(1)},\ldots,y^{(n)}\in S^{d-1}$,

$$\prod_{i=1}^{n}\langle y^{(i)},\nabla_\kappa\rangle Vf(x) = V\prod_{i=1}^{n}\langle y^{(i)},\nabla\rangle f(x) = \prod_{i=1}^{n}\langle y^{(i)},\nabla\rangle f(x);$$

the last equation follows from $V1 = 1$. Taking the supremum over $\{y^{(i)}\}$ shows that $\|Vf\|_\kappa = \|f\|_\partial$. □

Corollary 6.5.23 *For $f\in\Pi^d$ and $x\in B^d$, we have $|Vf(x)|\le\|f\|_A$ and $\|Vf\|_A\le\|f\|_A$.*

Proof Write $f=\sum_{n=0}^{\infty} f_n$ with $f_n\in\mathscr{P}_n^d$; then $\|Vf_n\|_\kappa = \|f_n\|_\partial = \|f_n\|_S$ by Theorem 4.5.3 and $|Vf_n(x)|\le\|Vf_n\|_\kappa$ for $\|x\|\le 1$ and each $n\ge 0$. Thus $\|Vf_n\|_S\le\|f_n\|_S$ and $\|Vf\|_A\le\|f\|_A$. □

It was originally conjectured that a stronger bound held: namely, $|Vf(x)|\le\sup\{|f(y)| : y\in\mathrm{co}(xW)\}$, where $\mathrm{co}(xW)$ is the convex hull of $\{xw : w\in W\}$. This was later proven by Rösler [1999].

6.6 The κ-Analogue of the Exponential

For $x,y\in\mathbb{C}^d$ the function $\exp(\langle x,y\rangle)$ has important applications, such as its use in Fourier transforms and as a reproducing kernel for polynomials (Taylor's theorem in disguise): for $f \in \Pi^d$, the formal series $\exp\left(\sum_{i=1}^d x_i\partial/\partial y_i\right) f(y)|_{y=0} = f(x)$. Also $(\partial/\partial x_i)\exp(\langle x,y\rangle) = y_i \exp(\langle x,y\rangle)$ for each i. There is an analogous function for the differential–difference operators. In this section $\kappa_v \geq 0$ is assumed throughout. A superscript such as (x) on an operator refers to the $\mathbb{R}^d$ variable on which it acts.

Definition 6.6.1 For $x,y \in \mathbb{R}^d$ let $K(x,y) = V^{(x)}\exp(\langle x,y\rangle)$ and $K_n(x,y) = (1/n!)V^{(x)}(\langle x,y\rangle^n)$ for $n \geq 0$.

It is clear that for each $y \in \mathbb{R}^d$ the function $f_y(x) = \exp(\langle x,y\rangle)$ is in $A\left(B^d\right)$ and that $\|f_y\|_A = \exp(\|y\|)$; thus the sum $K(x,y) = \sum_{n=0}^\infty K_n(x,y)$ converges absolutely, and uniformly on closed bounded sets. Here are the important properties of $K_n(x,y)$. Of course, $K_0(x,y) = 1$.

Proposition 6.6.2 *For $n \geq 1$ and $x,y \in \mathbb{R}^d$:*

(i) $K_n(x,y) = \sum_{w\in W} c_n(w)\langle xw,y\rangle K_{n-1}(xw,y)$;
(ii) $|K_n(x,y)| \leq \frac{1}{n!}\max_{w\in W}|\langle xw,y\rangle|^n$;
(iii) $K_n(xw,yw) = K_n(x,y)$ *for any* $w \in W$;
(iv) $K_n(y,x) = K_n(x,y)$;
(v) $\mathscr{D}_u^{(x)} K_n(x,y) = \langle u,y\rangle K_{n-1}(x,y)$ *for any* $u \in \mathbb{R}^d$.

Proof Fix $y \in \mathbb{R}^d$ and let $g_y(x) = \langle x,y\rangle^n$. By the construction of V we have

$$K_n(x,y) = \frac{1}{n!}Vg_y(x) = \frac{1}{n!}\sum_{w\in W} c_n(w)\sum_{i=1}^d (xw)_i V\partial_i g_y(xw)$$

and $\partial_i g_y(x) = ny_i\langle x,y\rangle^{n-1}$; thus $V\partial_i g_y(xw) = n(n-1)!y_i K_{n-1}(xw,y)$, implying part (i). Part (ii) is proved inductively using part (i) and the facts that $c_n(w) \geq 0$ and $\sum_w c_n(w) = 1/n$.

For part (iii),

$$K_n(xw,yw) = \frac{1}{n!}R(w)Vg_{yw}(x) = \frac{1}{n!}VR(w)g_{yw}(x)$$

and $R(w)g_{yw}(x) = \langle xw,yw\rangle^n = \langle x,y\rangle^n$, hence $K_n(xw,yw) = K_n(x,y)$.

Suppose that $K_{n-1}(y,x) = K_{n-1}(x,y)$ for all x,y; then, by (i) and (iii),

$$\begin{aligned}K_n(y,x) &= \sum_{w\in W} c_n(w)\langle yw,x\rangle K_{n-1}(yw,x)\\ &= \sum_{w\in W} c_n(w)\langle xw^{-1},y\rangle K_{n-1}(xw^{-1},y)\\ &= \sum_{w\in W} c_n(w^{-1})\langle xw,y\rangle K_{n-1}(xw,y)\\ &= K_n(x,y).\end{aligned}$$

Note that $\sum_w q_\kappa(w;s)w = \exp\left(s\sum_{v\in R_+}(1-\sigma_v)\right)$ is in the center of the group algebra $\mathbb{K}W$ and is mapped onto itself under the transformation $w \to w^{-1}$ (because each $\sigma_v^{-1} = \sigma_v$); hence $q_\kappa\left(w^{-1};s\right) = q_\kappa(w;s)$ for each w,s, which implies that $c_n\left(w^{-1}\right) = c_n(w)$.

Let $u \in \mathbb{R}^d$; then

$$\mathscr{D}_u^{(x)} K_n(x,y) = \frac{1}{n!}\mathscr{D}_u V g_y(x) = \frac{1}{n!} V\langle u,\nabla\rangle g_y(x)$$

by the intertwining of V, so that

$$\mathscr{D}_u^{(x)} K_n(x,y) = \frac{n}{n!}\langle u,y\rangle V^{(x)}\left(\langle x,y\rangle^{n-1}\right) = \langle u,y\rangle K_{n-1}(x,y).$$

□

Corollary 6.6.3 *For $x,y,u \in \mathbb{R}^d$ the following hold:*

(i) $|K(x,y)| \le \exp\left(\max_{w\in W}|\langle xw,y\rangle|\right)$;
(ii) $K(x,0) = K(0,y) = 1$;
(iii) $\mathscr{D}_u^{(x)} K(x,y) = \langle u,y\rangle K(x,y)$.

Observe that part (iii) of the corollary can be given an inductive proof:

$$\begin{aligned} K_n(xw,yw) &= \sum_{z\in W} c_n(z)\langle xwz,yw\rangle K_{n-1}(xwz,yw) \\ &= \sum_{z\in W} c_n(z)\left\langle xwzw^{-1},y\right\rangle K_{n-1}\left(xwzw^{-1},y\right) \\ &= \sum_{z\in W} c_n\left(w^{-1}zw\right)\langle xz,y\rangle K_{n-1}(xz,y), \end{aligned}$$

and $c_n\left(w^{-1}zw\right) = c_n(z)$ because $\sum_w q_\kappa(w;s)w$ is in the center of $\mathbb{K}W$.

The projection of $K(x,y)$ onto invariants is called the κ-Bessel function for the particular root system and multiplicity function.

Definition 6.6.4 For $x,y \in \mathbb{R}^d$ let

$$K_W(x,y) = \frac{1}{|W|}\sum_{w\in W} K(x,yw).$$

Thus $K_W(x,0) = 1$ and $K_W(xw,y) = K_W(x,yw) = K_W(x,y)$ for each $w \in W$. The name "Bessel" stems from the particular case $W = \mathbb{Z}_2$, where K_W is expressed as a classical Bessel function.

6.7 Invariant Differential Operators

This section deals with the large class of differential–difference operators of arbitrary degree as well as with the method for finding certain commutative sets of differential operators which are related to the multiplicity function κ_v. Since the operators $\mathscr{D}_u$ commute, they generate a commutative algebra of polynomials in $\{\mathscr{D}_i : 1 \leq i \leq d\}$ (see Definition 6.4.2 for $\mathscr{D}_u$ and $\mathscr{D}_i$).

Definition 6.7.1 For $p \in \Pi^d$ set $p(\mathscr{D}) = p(\mathscr{D}_1, \mathscr{D}_2, \ldots, \mathscr{D}_d)$; that is, each x_i is replaced by $\mathscr{D}_i$.

For example, the polynomial $\|x\|^2$ corresponds to Δ_h. The action of W on these operators is dual to its action on polynomials.

Proposition 6.7.2 *Let $p \in \Pi^d$ and $w \in W$; then*

$$p(\mathscr{D})R(w) = R(w)\left[R(w^{-1})p(\mathscr{D})\right].$$

If $p \in \Pi^W$, the algebra of invariant polynomials, then $p(\mathscr{D})R(w) = R(w)p(\mathscr{D})$.

Proof Each polynomial is a sum of terms such as $q(x) = \prod_{i=1}^n \langle u^{(i)}, x\rangle$, with each $u^{(i)} \in \mathbb{R}^d$. By Proposition 6.4.3, $q(\mathscr{D})R(w) = R(w)\prod_{i=1}^n \langle u^{(i)}w, \nabla_\kappa\rangle$. Also, $R(w^{-1})q(x) = \prod_{i=1}^n \langle u^{(i)}, xw^{-1}\rangle = \prod_{i=1}^n \langle u^{(i)}w, x\rangle$. This shows that $q(\mathscr{D})R(w) = R(w)\left[R(w^{-1})q(\mathscr{D})\right]$. If a polynomial p is invariant then $R(w^{-1})p = p$, which implies that $p(\mathscr{D})R(w) = R(w)p(\mathscr{D})$. □

The effect of $p(\mathscr{D})$ on $K(x,y)$ is easy to calculate:

$$p(\mathscr{D}^{(x)})K(x,y) = p(y)K(x,y), \qquad x, y \in \mathbb{R}^d.$$

The invariant polynomials for several reflection groups were described in Subsections 4.3.1–4.3.6. The motivation for the rest of this section comes from the observation that Δ_h restricted to Π^W acts as a differential operator:

$$\Delta_h p(x) = \Delta p(x) + \sum_{v \in R_+} \kappa_v \frac{2\langle v, \nabla p(x)\rangle}{\langle v, x\rangle}, \qquad \text{for } p \in \Pi^W.$$

It turns out that this holds for any operator $q(\mathscr{D})$ with $q \in \Pi^W$. However, explicit forms such as that for Δ_h are not easily obtainable (one reason is that there are many different invariants for different groups; only $\|x\|^2$ is invariant for each group). Heckman [1991a] proved this differential operator result. Some background is needed to present the theorem. The basic problem is that of characterizing differential operators among all linear operators on polynomials.

Definition 6.7.3 Let $\mathscr{L}(\Pi^d)$ be the set of linear transformations of $\Pi^d \to \Pi^d$ ("endomorphisms"). For $n \geq 0$ let $\mathscr{L}_{\partial,n}(\Pi^d) = \text{span}\{a(x)\partial^\alpha : a \in \Pi^d, \alpha \in \mathbb{N}_0^d, |\alpha| \leq n\}$, where $\partial^\alpha = \partial_1^{\alpha_1} \cdots \partial_d^{\alpha_d}$, the space of differential operators of degree $\leq n$ with polynomial coefficients. Further, let $\mathscr{L}_\partial(\Pi^d) = \bigcup_{n=0}^\infty \mathscr{L}_{\partial,n}(\Pi^d)$.

Definition 6.7.4 For $S, T \in \mathscr{L}(\Pi^d)$ define an operator $ad(S)T \in \mathscr{L}(\Pi^d)$ by

$$ad(S)Tf = (ST - TS)f \qquad \text{for } f \in \Pi^d.$$

The name *ad* refers to the adjoint map from Lie algebra theory. The elements of $\mathscr{L}_{\partial,0}$ are just multiplier operators on $\Pi^d : f \mapsto pf$ for a polynomial $p \in \Pi^d$; we use the same symbol (p) for this operator. The characterization of $\mathscr{L}_\partial(\Pi^d)$ involves the action of $ad(p)$ for arbitrary $p \in \Pi^d$.

Proposition 6.7.5 *If $T \in \mathscr{L}(\Pi^d)$ and $ad(x_i)T = 0$ for each i then $T \in \mathscr{L}_{\partial,0}(\Pi^d)$ and $Tf = (T1)f$ for each $f \in \Pi^d$.*

Proof By the hypothesis, $x_i Tf(x) = T(x_i f)(x)$ for each i and $f \in \Pi^d$. The set $E = \{f \in \Pi^d : Tf = (T1)f\}$ is a linear space, contains 1 and is closed under multiplication by each x_i; hence $E = \Pi^d$. □

The usual product rule holds for *ad*: for $S, T_1, T_2 \in \mathscr{L}(\Pi^d)$, a trivial verification shows that

$$ad(S)T_1T_2 = [ad(S)T_1]\,T_2 + T_1\,[ad(S)\,T_2].$$

There is a commutation fact:

$$S_1S_2 = S_2S_1 \quad \text{implies that} \quad ad(S_1)ad(S_2) = ad(S_2)ad(S_1).$$

This holds because both sides of the second equation applied to T equal $S_1S_2T - S_1TS_2 - S_2TS_1 + TS_2S_1$. In particular, $ad(p)\,ad(q) = ad(q)\,ad(p)$ for multipliers $p, q \in \Pi^d$. For any $p \in \Pi^d$, $ad(p)\partial_i = -\partial_i p \in \mathscr{L}_{\partial,0}(\Pi^d)$, a multiplier.

It is not as yet proven that $\mathscr{L}_\partial(\Pi^d)$ is an algebra; this is a consequence of the following.

Proposition 6.7.6 *Suppose that $m, n \geq 0$ and $\alpha, \beta \in \mathbb{N}_0^d$, with $|\alpha| = m, |\beta| = n$ and $p, q \in \Pi^d$; then $p\partial^\alpha q\partial^\beta = pq\partial^{\alpha+\beta} + T$, with $T \in \mathscr{L}_{\partial,m+n-1}(\Pi^d)$.*

Proof Proceed inductively with the degree-1 factors of ∂^α; a typical step uses $\partial_i qS = q\partial_i S + (\partial_i q)S$ for some operator S. □

Corollary 6.7.7 *Suppose that $n \geq 1$ and $T \in \mathscr{L}_{\partial,n}(\Pi^d)$; then $ad(p)T \in \mathscr{L}_{\partial,n-1}(\Pi^d)$ and $ad(p)^{n+1}T = 0$, for any $p \in \Pi^d$.*

In fact Corollary 6.7.7 contains the characterization announced previously. The effect of $ad(x_i)$ on operators of the form $q\partial^\alpha$ can be calculated explicitly.

Lemma 6.7.8 *Suppose that $\alpha \in \mathbb{N}_0^d, q \in \Pi^d$ and $1 \leq i \leq d$; then*

$$ad(x_i)\,q\partial^\alpha = (-\alpha_i)\,q\partial_1^{\alpha_1} \cdots \partial_i^{\alpha_i - 1} \cdots \partial_d^{\alpha_d}.$$

Proof Since x_i commutes with q and $\partial_j^{\alpha_j}$ for $j \neq i$, it suffices to find $ad(x_i)\,\partial_i^{\alpha_i}$. Indeed, by the Leibniz rule, $x_i\partial_i^{\alpha_i} f - \partial_i^{\alpha_i}(x_i f) = -\sum_{j=1}^{\alpha_i} \binom{\alpha_i}{j} \partial_i^j(x_i)\,\partial_i^{\alpha_i - j}(f) = -\alpha_i \partial_i^{\alpha_i - 1}(f)$ for any $f \in \Pi^d$. □

Theorem 6.7.9 *For $n \geq 1, T \in \mathscr{L}(\Pi^d)$ the following are equivalent:*

(i) $T \in \mathscr{L}_{\partial,n}(\Pi^d)$;
(ii) $ad(p)^{n+1} T = 0$ *for each* $p \in \Pi^d$;
(iii) $ad(x_1)^{\alpha_1} ad(x_2)^{\alpha_2} \cdots ad(x_d)^{\alpha_d}\, T = 0$ *for each* $\alpha \in \mathbb{N}_0^d$ *with* $|\alpha| = n+1$.

Proof We will show that (i) ⇒ (ii) ⇒ (iii) ⇒ (i). The part (i) ⇒ (ii) is already proven. Suppose now that (ii) holds; for any $y \in \mathbb{R}^d$ let $p = \sum_{i=1}^d y_i x_i$. The multinomial theorem applies to the expansion of $\left(\sum_{i=1}^d y_i\, ad\,(x_i)\right)^{n+1}$ because the terms commute with each other. Thus

$$0 = ad(p)^{n+1} T = \sum_{|\alpha|=n+1} \binom{|\alpha|}{\alpha} y^\alpha\, ad(x_1)^{\alpha_1} ad\,(x_2)^{\alpha_2} \cdots ad(x_d)^{\alpha_d} T.$$

Considering this as a polynomial identity in the variable y, we deduce that each coefficient is zero. Thus (ii) ⇒ (iii).

Suppose that (iii) holds. Proposition 6.7.5 shows that (i) is true for $n = 0$. Proceed by induction and suppose that (iii) implies (i) for $n-1$. Note that Lemma 6.7.8 shows that there is a biorthogonality relation

$$ad(x_1)^{\alpha_1}\, ad(x_2)^{\alpha_2} \cdots ad(x_d)^{\alpha_d} \left(q\partial^\beta\right) = \prod_{i=1}^d (-\beta_i)_{\alpha_i}\, q\partial^{\beta-\alpha},$$

where $\partial^{\beta-\alpha} = \prod_{i=1}^d \partial_i^{\beta_i - \alpha_i}$ and negative powers of ∂_i are considered to be zero (of course, the Pochhammer symbol $(-\beta_i)_{\alpha_i} = 0$ for $\alpha_i > \beta_i$). For each $\alpha \in \mathbb{N}_0^d$ with $|\alpha| = n$, let

$$S_\alpha = (-1)^n \prod_{i=1}^d (\alpha_i!)^{-1} ad(x_1)^{\alpha_1}\, ad(x_2)^{\alpha_2} \cdots ad(x_d)^{\alpha_d}\, T.$$

By hypothesis, $ad(x_i)\, S_\alpha = 0$ for each i and, by Proposition 6.7.5, S_α is a multiplier, say, $q_\alpha \in \Pi^d$. Now let $T_n = \sum_{|\alpha|=n} q_\alpha \partial^\alpha$. By construction and the biorthogonality relation, we have

$$ad(x_1)^{\alpha_1}\, ad(x_2)^{\alpha_2} \cdots ad(x_d)^{\alpha_d}\, (T - T_n) = 0$$

for any $\alpha \in \mathbb{N}_0^d$ with $|\alpha| = n$. By the inductive hypothesis, $T - T_n \in \mathscr{L}_{\partial,n-1}(\Pi^d)$; thus $T \in \mathscr{L}_{\partial,n}(\Pi^d)$ and (i) is true. □

We now apply this theorem to polynomials in $\{q_j : 1 \leq j \leq d\}$, the fundamental invariant polynomials of W (see Chevalley's theorem 6.3.2). Let $\mathscr{L}_\partial(\Pi^W)$ denote the algebra generated by $\left\{\partial/\partial q_i : 1 \leq i \leq d\right\} \cup \Pi^W$, where the latter are considered as multipliers.

Theorem 6.7.10 *Let $p \in \Pi^W$; then the restriction of $p(\mathscr{D})$ to Π^W coincides with an operator $D_p \in \mathscr{L}_\partial(\Pi^W)$. The correspondence $p \to D_p$ is an algebra homomorphism.*

Proof By the previous theorem we need to show that there is a number n such that $ad(g)^{n+1} p(\mathscr{D}) f = 0$ for each $f, g \in \Pi^W$. Let n be the degree of p as a polynomial in x; then p is a sum of terms such as $\prod_{i=1}^m \langle u^{(i)}, x\rangle$ with $u^{(i)} \in \mathbb{R}^d$ and $m \le n$. By the multinomial Leibniz rule,

$$ad(g)^{n+1} \prod_{i=1}^{m} \langle u^{(i)}, \nabla_\kappa \rangle$$
$$= \sum \left\{ \binom{n+1}{\beta} \prod_{i=1}^{m} \left(ad(g)^{\beta_i} \langle u^{(i)}, \nabla_\kappa \rangle \right) : \beta \in \mathbb{N}_0^m, |\beta| = n+1 \right\}.$$

For any β in the sum there must be at least one index i with $\beta_i \ge 2$, since $n+1 > m$. But if $g \in \Pi^W$ and $q \in \Pi^d$ then $\langle u, \nabla_\kappa \rangle (gq) = \mathscr{D}_u(gq) = g\mathscr{D}_u q + q\langle u, \nabla\rangle g$; that is, $ad(g)\langle u, \nabla_\kappa\rangle$ is the same as multiplication by $\langle u, \nabla\rangle g$ and $ad(g)^2 \langle u, \nabla_\kappa \rangle = 0$. Thus $D_p \in \mathscr{L}_\partial(\Pi^W)$. Furthermore, if $p_1, p_2 \in \Pi^W$ then the restriction of $p_1(\mathscr{D})p_2(\mathscr{D})$ to Π^W is clearly the product of the respective restrictions, so that $D_{p_1 p_2} = D_{p_1} D_{p_2}$. □

This shows that D_p is a sum of terms such as

$$f(q_1, q_2, \ldots, q_d) \prod_{i=1}^{d} \left(\frac{\partial}{\partial q_i} \right)^{\alpha_i}$$

with $\alpha \in \mathbb{N}_0^d$ and $|\alpha| \le n$. This can be expressed in terms of x. Recall the Jacobian matrix $J(x) = \left(\partial q_i(x)/\partial x_j \right)_{i,j=1}^d$ with $\det J(x) = c a_R(x)$, a scalar multiple of the alternating polynomial $a_R(x)$ (see Theorem 6.3.5). The inverse of $J(x)$ exists for each x such that $\langle x, v\rangle \ne 0$ for each $v \in R_+$, and the entries are rational functions whose denominators are products of factors $\langle x, v\rangle$. Since $\partial/\partial q_j = \sum_{i=1}^d \left(J(x)^{-1} \right)_{i,j} \partial/\partial x_i$, we see that each D_p can be expressed as a differential operator on $x \in \mathbb{R}^d$ with rational coefficients (and singularities on $\{x : \prod_{v \in R_+} \langle x, v\rangle = 0\}$, the union of the reflecting hyperplanes). This is obvious for Δ_h.

The κ-Bessel function is an eigenfunction of each $D_p, p \in \Pi^W$; this implies that the correspondence $p \to D_p$ is one to one.

Proposition 6.7.11 *For $x, y \in \mathbb{R}^d$, if $p \in \Pi^W$ then*

$$D_p^{(x)} K_W(x, y) = p(y) K_W(x, y).$$

Proof It is clear that K_W is the absolutely convergent sum over $n \ge 0$ of the sequence $|W|^{-1} \sum_{w \in W} K_n(x, yw)$ and that $p\left(\mathscr{D}^{(x)}\right)$ can be applied term by term. Indeed,

$$D_p^{(x)} K_W(x,y) = p(\mathscr{D}^{(x)}) K_W(x,y) = \frac{1}{|W|} \sum_{w \in W} p(\mathscr{D}^{(x)}) K(x, yw)$$
$$= \frac{1}{|W|} \sum_{w \in W} p(yw) K(x, yw) = \frac{1}{|W|} p(y) \sum_{w \in W} K(x, yw)$$
$$= p(y) K_W(x,y).$$

The third equals sign uses (iii) in Corollary 6.6.3. □

Opdam [1991] established the existence of K_W as the unique real-entire joint eigenfunction of all invariant differential operators commuting with $D_p\, p(x) = \|x\|^2$, that is, the differential part of Δ_h, the eigenvalues specified in Proposition 6.7.11.

6.8 Notes

The fundamental characterization of reflection groups in terms of generators and relations was established by Coxeter [1935]. Chevalley [1955] proved that the ring of invariants of a finite reflection group is a polynomial ring (that is, it is isomorphic to the ordinary algebra of polynomials in d variables). It is an "heuristic feeling" of the authors that this is the underlying reason why reflection groups and orthogonal polynomials go together. For background information, geometric applications and proofs of theorems about reflection groups omitted in the present book, the reader is referred to Coxeter and Moser [1965], Grove and Benson [1985] and Humphreys [1990].

Historically, one of the first results that involved reflection-invariant operators and spherical harmonics appeared in Laporte [1948] (our theorem 6.2.6 for the case $\mathbb{R}^3$).

Differential–difference operators, intertwining operators and κ-exponential functions were developed in a series of papers by Dunkl [1988, 1989a, 1990, 1991]. The construction of the differential–difference operators came as a result of several years of research aimed at understanding the structure of non-invariant orthogonal polynomials with respect to invariant weight functions. Differential operators do not suffice, owing to their singularities on the reflecting hyperplanes.

Subsequently to Dunkl's 1989 paper, Heckman [1991a] constructed operators of a trigonometric type, associated with Weyl groups. There are similar operators in which elliptic functions replace the linear denominators (Buchstaber, Felder and Veselov [1994]). Rösler [1999] showed that the intertwining operator can be expressed as an integral transform with positive kernel; this was an existence proof, so the finding of explicit forms is, as of the time of writing, an open problem. In some simple cases, however, an explicit form is known; see the next chapter and its notes.

7

Spherical Harmonics Associated with Reflection Groups

In this chapter we study orthogonal polynomials on spheres associated with weight functions that are invariant under reflection groups. A theory of homogeneous orthogonal polynomials in this setting, called h-harmonics, can be developed in almost complete analogy to that for the ordinary spherical harmonics. The results include several inner products under which orthogonality holds, explicit expressions for the reproducing kernels given in terms of the intertwining operator and an integration which shows that the average of the intertwining operator over the sphere removes the action of the operator. As examples, we discuss the h-harmonics associated with $\mathbb{Z}_2^d$ and those in two variables associated with dihedral groups. Finally, we discuss the analogues of the Fourier transform associated with reflection groups.

7.1 h-Harmonic Polynomials

We use the notation of the previous chapter and start with definitions.

Definition 7.1.1 Let R_+ be the system of positive roots of a reflection group W acting in $\mathbb{R}^d$. Let κ be a multiplicity function as in Definition 6.4.1. Associated with R_+ and κ we define invariant weight functions

$$h_\kappa(x) := \prod_{v \in R_+} |\langle v, x\rangle|^{\kappa_v}, \qquad x \in \mathbb{R}^d.$$

Throughout this chapter we use the notation γ_κ and λ_κ as follows:

$$\gamma_\kappa := \sum_{v \in R_+} \kappa_v \qquad \text{and} \qquad \lambda_\kappa := \gamma_\kappa + \frac{d-2}{2}. \tag{7.1.1}$$

Sometimes we will write γ for γ_κ and λ for λ_κ, and particularly in the proof sections when no confusion is likely. Note that the weight function h_κ is positively homogeneous of degree γ_κ.

Recall that the operators $\mathscr{D}_i$, $1 \le i \le d$, from Definition 6.4.2 commute and that the h-Laplacian is defined by $\Delta_h = \sum_{i=1}^{d} \mathscr{D}_i^2$.

Definition 7.1.2 A polynomial P is h-harmonic if $\Delta_h P = 0$.

We split the differential and difference parts of Δ_h, denoting them by L_h and D_h, respectively; that is, $\Delta_h = L_h + D_h$ with

$$\begin{aligned} L_h f &= \frac{\Delta(f h_\kappa) - f\Delta h_\kappa}{h_\kappa}, \\ D_h f(x) &= -2\sum_{v\in R_+} \kappa_v \frac{f(x) - f(x\sigma_v)}{\langle v, x\rangle^2} \|v\|^2. \end{aligned} \tag{7.1.2}$$

Lemma 7.1.3 *Both L_h and D_h commute with the action of the reflection group W.*

Proof We have $L_h f = \Delta f + 2\langle \nabla h_\kappa, \nabla f\rangle / h_\kappa$. We need to prove that $R(w)L_h f = L_h R(w) f$ for every $w \in W$. The Laplacian commutes with $R(w)$ since W is a subgroup of $O(d)$. Every term in $2\langle \nabla h_\kappa, \nabla f\rangle / h_\kappa = \frac{1}{2}[\Delta(h_\kappa^2 f) - f\Delta(h_\kappa^2) - h_\kappa^2 \Delta f]/h_\kappa^2$ commutes with $R(w)$ since h_κ^2 is W-invariant. Since Δ_h commutes with $R(w)$ and $D_h = \Delta_h - L_h$, D_h commutes with the action of W. □

In the proof of the following lemmas and theorem assume that $\kappa_v \ge 1$. Analytic continuation can be used to extend the range of validity to $\kappa_v \ge 0$. Moreover, we will write $\mathrm{d}\omega$ for $\mathrm{d}\omega_{d-1}$ whenever no confusion can result from this abbreviation.

Lemma 7.1.4 *The operator D_h is symmetric on polynomials in $L^2(h_\kappa^2\, \mathrm{d}\omega)$.*

Proof For any polynomial f, $D_h f \in L^2(h_\kappa^2 \mathrm{d}\omega)$ since each factor $\langle x, v\rangle^2$ in the denominators of the terms of $D_h f$ is canceled by $h_\kappa^2(x) = \prod_{v\in R_+} |\langle x, v\rangle|^{2\kappa_v}$ for each $\kappa_v \ge 1$. For polynomials f and g,

$$\begin{aligned} &\int_{S^{d-1}} f D_h \bar{g} h_\kappa^2 \,\mathrm{d}\omega \\ &\quad = -\sum_{v\in R_+} \kappa_v \left(\int_{S^{d-1}} f(x)\bar{g}(x)\langle x, v\rangle^{-2} h_\kappa^2(x)\,\mathrm{d}\omega \right. \\ &\qquad \left. - \int_{S^{d-1}} f(x)\bar{g}(x\sigma_v)\langle x, v\rangle^{-2} h_\kappa(x)^2\,\mathrm{d}\omega \right) \|v\|^2 = \int_{S^{d-1}} (D_h f)\bar{g} h_\kappa^2\,\mathrm{d}\omega, \end{aligned}$$

where in the second integral in the sum we have replaced x by $x\sigma_v$; this step is valid by the W-invariance of h_κ^2 under each σ_v and the relation $\langle x\sigma_v, v\rangle = \langle x, v\sigma_v\rangle = -\langle x, v\rangle$. □

Lemma 7.1.5 *For $f, g \in C^2(B^d)$,*

$$\int_{S^{d-1}} \frac{\partial f}{\partial n} g h_\kappa^2 \, d\omega = \int_{B^d} (g L_h f + \langle \nabla f, \nabla g \rangle) h_\kappa^2 \, dx,$$

where $\partial f / \partial n$ is the normal derivative of f.

Proof Green's first identity states that

$$\int_{S^{d-1}} \frac{\partial f_1}{\partial n} f_2 \, d\omega = \int_{B^d} (f_2 \Delta f_1 + \langle \nabla f_1, \nabla f_2 \rangle) \, dx$$

for $f_1, f_2 \in C^2(B^d)$. In this identity, first put $f_1 = f h_\kappa$ and $f_2 = g h_\kappa$ to get

$$\int_{S^{d-1}} \left(f g \frac{\partial h_\kappa}{\partial n} + g h_\kappa \frac{\partial f}{\partial n} \right) h_\kappa \, d\omega = \int_{B^d} \left[g h_\kappa \Delta (f h_\kappa) + \langle \nabla (f h_\kappa), \nabla (g h_\kappa) \rangle \right] dx$$

and then put $f_1 = h_\kappa$ and $f_2 = f g h_\kappa$ to get

$$\int_{S^{d-1}} f g \frac{\partial h_\kappa}{\partial n} \, d\omega = \int_{B^d} \left[f g h_\kappa \Delta h_\kappa + \langle \nabla (f g h_\kappa), \nabla h_\kappa \rangle \right] dx.$$

Subtracting the second equation from the first, we obtain

$$\int_{S^{d-1}} g \frac{\partial f}{\partial n} h_\kappa^2 \, d\omega = \int_{B^d} \left(g h_\kappa [\Delta (f h_\kappa) - f \Delta h_\kappa] + h_\kappa^2 \langle \nabla f, \nabla g \rangle \right) dx$$

by using the product rule for ∇ repeatedly. □

Theorem 7.1.6 *Suppose that f and g are h-harmonic homogeneous polynomials of different degrees; then $\int_{S^{d-1}} f(x) g(x) h_\kappa^2(x) \, d\omega = 0$.*

Proof Since f is homogeneous, by Euler's formula, $\partial f / \partial n = (\deg f) f$. Hence, by Green's identity in Lemma 7.1.5,

$$\begin{aligned} (\deg f - \deg g) \int_{S^{d-1}} f g h_\kappa^2 \, d\omega &= \int_{B^d} (g L_h f - f L_h g) h_\kappa^2 \, dx \\ &= \int_{B^d} (g D_h f - f D_h g) h_\kappa^2 \, dx = 0, \end{aligned}$$

using the symmetry of D_h from Lemma 7.1.4. □

Hence, the h-harmonics are homogeneous orthogonal polynomials with respect to $h_\kappa^2 \, d\omega$. Denote the space of h-harmonic homogeneous polynomials of degree n by

$$\mathscr{H}_n^d(h_\kappa^2) = \mathscr{P}_n^d \cap \ker \Delta_h,$$

using notation from Section 4.2. If h_κ^2 is S-symmetric then it follows from Theorem 4.2.7 that $\mathscr{P}_n^d$ admits a unique decomposition in terms of $\mathscr{H}_j^d(h_\kappa^2)$. The same result also holds for the general h-harmonic polynomial h_κ from Definition 7.1.1.

Theorem 7.1.7 *For each $n \in \mathbb{N}_0$, $\mathscr{P}_n^d = \bigoplus_{j=0}^{[n/2]} \|x\|^{2j} \mathscr{H}_{n-2j}^d(h_\kappa^2)$; that is, there is a unique decomposition $P(x) = \sum_{j=0}^{[n/2]} \|x\|^{2j} P_{n-2j}(x)$ for $P \in \mathscr{P}_n^d$ with $P_{n-2j} \in \mathscr{H}_{n-2j}^d(h_\kappa^2)$. In particular,*

$$\dim \mathscr{H}_n^d(h_\kappa^2) = \binom{n+d-1}{n} - \binom{n+d-3}{n-2}.$$

Proof We proceed by induction. From the fact that $\Delta_h \mathscr{P}_n^d \subset \mathscr{P}_{n-2}^d$ we obtain $\mathscr{P}_0^d = \mathscr{H}_0^d$, $\mathscr{P}_1^d = \mathscr{H}_1^d$ and $\dim \mathscr{H}_n^d(h_\kappa^2) \geq \dim \mathscr{P}_n^d - \dim \mathscr{P}_{n-2}^d$. Suppose that the statement is true for $m = 0, 1, \ldots, n-1$ for some n. Then $\|x\|^2 \mathscr{P}_{n-2}^d$ is a subspace of $\mathscr{P}_n^d$ and is isomorphic to $\mathscr{P}_{n-2}^d$, since homogeneous polynomials are determined by their values on S^{d-1}. By the induction hypothesis $\|x\|^2 \mathscr{P}_{n-2}^d = \bigoplus_{j=0}^{[n/2]-1} \|x\|^{2j+2} \mathscr{H}_{n-2-2j}^d(h_\kappa^2)$, and by the previous theorem $\mathscr{H}_n^d(h_\kappa^2) \perp \mathscr{H}_{n-2-2j}^d(h_\kappa^2)$ for each $j = 0, 1, \ldots, [n/2]-1$. Hence $\mathscr{H}_n^d(h_\kappa^2) \perp \|x\|^2 \mathscr{P}_{n-2}^d$ in $L^2(h_\kappa^2 \, d\omega)$. Thus $\dim \mathscr{H}_n^d(h_\kappa^2) + \dim \mathscr{P}_{n-2}^d \leq \dim \mathscr{P}_n^d$, and so $\dim \mathscr{H}_n^d(h_\kappa^2) + \dim(\|x\|^2 \mathscr{P}_{n-2}^d) = \dim \mathscr{P}_n^d$. □

Corollary 7.1.8 *$\mathscr{P}_n^d \cap (\mathscr{P}_{n-2}^d)^\perp = \mathscr{H}_n^d(h_\kappa^2)$; that is, if $p \in \mathscr{P}_n^d$ and $p \perp P_{n-2}^d$ then p is h-harmonic.*

A basis of $\mathscr{H}_n^d(h_\kappa^2)$ can be generated by a simple procedure. First, we state

Lemma 7.1.9 *Let s be a real number and $g \in \mathscr{P}_n^d$. Then*

$$\begin{aligned} \mathscr{D}_i(\|x\|^s g) &= s x_i \|x\|^{s-2} g + \|x\|^s \mathscr{D}_i g, \\ \Delta_h(\|x\|^s g) &= 2s \left(\frac{d}{2} + \frac{s}{2} - 1 + n + \gamma_\kappa \right) \|x\|^{s-2} g + \|x\|^s \Delta_h g, \end{aligned} \tag{7.1.3}$$

where, if $s < 2$, both these identities are restricted to $\mathbb{R}^d \setminus \{0\}$.

Proof Since $\|x\|^s$ is invariant under the action of the reflection group, it follows from the product formula in Proposition 6.4.12 that

$$\mathscr{D}_i(\|x\|^s g) = \partial_i(\|x\|^s) g + \|x\|^s \mathscr{D}_i g,$$

from which the first identity follows. To prove the second identity, use the split $\Delta_h = L_h + D_h$. It is easy to see that $D_h(\|x\|^s g) = \|x\|^s D_h g$. Hence, using the fact that $L_h f = \Delta f + 2\langle \nabla h_\kappa, \nabla f \rangle / h_\kappa$, the second identity follows from

$$\Delta(\|x\|^s g) = 2s \left(\frac{d}{2} + \frac{s}{2} - 1 + n \right) \|x\|^{s-2} g + \|x\|^s \Delta g$$

and

$$\langle \nabla(\|x\|^s g), \nabla h_\kappa \rangle = s \|x\|^{s-2} \langle g, \nabla h_\kappa \rangle + \|x\|^s \langle \nabla g, \nabla h_\kappa \rangle,$$

where we have used the relation $\sum (x_i \partial h_\kappa / \partial x_i) = \gamma_\kappa h_\kappa$. □

Lemma 7.1.10 *For* $1 \le i \le d$, $\Delta_h[x_i f(x)] = x_i \Delta_h f(x) + 2\mathscr{D}_i f(x)$.

Proof By the product rules for Δ and for ∇,

$$\Delta_h[x_i f(x)] = x_i \Delta f(x) + 2\frac{\partial f}{\partial x_i} + \sum_{v \in R_+} \kappa_v \left[\frac{2\langle v, \nabla f\rangle x_i}{\langle v, x\rangle} + \frac{2f(x)v_i}{\langle v, x\rangle} - \|v\|^2 \frac{x_i f(x) - (x\sigma_v)_i f(x\sigma_v)}{\langle v, x\rangle^2} \right],$$

which, upon using the identities

$$x_i f(x) - (x\sigma_v)_i f(x\sigma_v) = x_i[f(x) - f(x\sigma_v)] + (x_i - (x\sigma_v)_i) f(x\sigma_v)$$

and

$$x_i - (x\sigma_v)_i = \frac{2\langle x, v\rangle v_i}{\|v\|^2},$$

is seen to be equal to $x_i \Delta_h f(x) + 2\mathscr{D}_i f(x)$. □

Definition 7.1.11 For any $\alpha \in \mathbb{N}_0^d$, define homogeneous polynomials H_α by

$$H_\alpha(x) := \|x\|^{2\lambda_\kappa + 2|\alpha|} \mathscr{D}^\alpha \|x\|^{-2\lambda_\kappa}, \tag{7.1.4}$$

where $\mathscr{D}^\alpha = \mathscr{D}_1^{\alpha_1} \cdots \mathscr{D}_d^{\alpha_d}$ and λ_κ is as in (7.1.1).

Recall that $\varepsilon_i = (0, \ldots, 1, \ldots, 0)$, $1 \le i \le d$, denotes the standard basis elements of $\mathbb{R}^d$. In the following proof we write λ for λ_κ.

Theorem 7.1.12 *For each* $\alpha \in \mathbb{N}^d$, H_α *is an h-harmonic polynomial of degree* $|\alpha|$, *that is,* $H_\alpha \in \mathscr{H}_{|\alpha|}^d(h_\kappa^2)$. *Moreover, the* H_α *satisfy the recursive relation*

$$H_{\alpha+\varepsilon_i}(x) = -2(\lambda_\kappa + |\alpha|) x_i H_\alpha(x) + \|x\|^2 \mathscr{D}_i H_\alpha(x). \tag{7.1.5}$$

Proof First we prove that H_α is a homogeneous polynomial of degree $|\alpha|$, using induction on $n = |\alpha|$. Clearly $H_0(x) = 1$. Assume that H_α has been proved to be a homogeneous polynomial of degree n for $|\alpha| = n$. Using the first identity in (7.1.3) it follows that

$$\begin{aligned} \mathscr{D}_i H_\alpha =& (2\lambda + 2|\alpha|) x_i \|x\|^{2\lambda - 2 + 2|\alpha|} \mathscr{D}^\alpha (\|x\|^{-2\lambda}) \\ &+ \|x\|^{2\lambda + 2|\alpha|} \mathscr{D}_i \mathscr{D}^\alpha (\|x\|^{-2\lambda}), \end{aligned}$$

from which the recursive formula (7.1.5) follows from the definition of H_α. Since $\mathscr{D}_i : \mathscr{P}_n^d \mapsto \mathscr{P}_{n-1}^d$, it follows from the recursive formula that $H_{\alpha+\varepsilon_i}$ is a homogeneous polynomial of degree $n + 1 = |\alpha| + 1$.

Next we prove that H_α is an h-harmonic polynomial, that is, that $\Delta_h H_\alpha = 0$. Setting $a = -2n - 2\lambda$ in the second identity of (7.1.3), we conclude that

$$\Delta_h(\|x\|^{-2n-2\lambda} g) = \|x\|^{-2n-2\lambda} \Delta_h g,$$

for $g \in \mathscr{P}_n^d$. In particular, for $g = 1$ and $n = 0$, $\Delta_h(\|x\|^{-2\lambda}) = 0$ for $x \in \mathbb{R}^d \setminus \{0\}$. Hence, setting $g = H_\alpha$ and $|\alpha| = n$ and using the fact that $\mathscr{D}_1, \ldots, \mathscr{D}_d$ commute, it follows that

$$\Delta_h H_\alpha = \|x\|^{2n+2\lambda} \mathscr{D}^\alpha \Delta_h(\|x\|^{-2\lambda}) = 0,$$

which holds for all $x \in \mathbb{R}^d$ since H_α is a polynomial. □

The formula (7.1.4) defines a one-to-one correspondence between x^α and H_α. Since every h-harmonic in $\mathscr{H}_n^d(h_\kappa^2)$ can be written as a linear combination of x^α with $|\alpha| = n$, the set $\{H_\alpha : |\alpha| = n\}$ contains a basis of $\mathscr{H}_n^d(h_\kappa^2)$. However, the h-harmonics in this set are not linearly independent since there are $\dim \mathscr{P}_n^d$ of them, which is more than $\dim \mathscr{H}_n^d(h_\kappa^2)$. Nevertheless, it is not hard to derive a basis from them.

Corollary 7.1.13 *The set $\{H_\alpha : |\alpha| = n\}$ contains bases of $\mathscr{H}_n^d(h_\kappa^2)$; one particular basis can be taken as $\{H_\alpha : |\alpha| = n, \alpha_d = 0, 1\}$.*

Proof The linear dependence relations among the members of $\{H_\alpha : |\alpha| = n\}$ are given by

$$H_{\beta+2\varepsilon_1} + \cdots + H_{\beta+2\varepsilon_d} = \|x\|^{2n+2\lambda} \mathscr{D}^\beta \Delta_h(\|x\|^{-2\lambda}) = 0$$

for every $\beta \in \mathbb{N}^d$ such that $|\beta| = n - 2$. These relations number exactly $\dim \mathscr{P}_{n-2}^d = \#\{\beta \in \mathbb{N}^d : |\beta| = n - 2\}$. For each relation we can exclude one polynomial from the set $\{H_\alpha : |\alpha| = n\}$. The remaining $\dim \mathscr{H}_n^d(h_\kappa^2)$ $(= \dim \mathscr{P}_n^d - \dim \mathscr{P}_{n-2}^d)$ harmonics still span $\mathscr{H}_n^d(h_\kappa^2)$; hence they form a basis for $\mathscr{H}_n^d(h_\kappa^2)$. The basis is not unique, since we can exclude any of the polynomials $H_{\beta+2\varepsilon_1}, \ldots, H_{\beta+2\varepsilon_d}$ for each dependent relation. Excluding $H_{\beta+2\varepsilon_d}$ for all $|\beta| = n - 2$ from $\{H_\alpha : |\alpha| = n\}$ is one way to obtain a basis, and it generalizes the spherical harmonics from Section 4.1. □

The polynomials in the above basis are not mutually orthogonal, however. In fact, constructing orthonormal bases with explicit formulae is a difficult problem for most reflection groups. At the present such bases are known only in the case of abelian groups $\mathbb{Z}_2^d$ (see Section 7.5) and the dihedral groups acting on polynomials of two variables (see Section 7.6).

Since (7.1.4) defines a map which takes x^α to H_α, this map must be closely related to the projection operator from $\mathscr{P}_n^d$ to $\mathscr{H}_n^d(h_\kappa^2)$.

Definition 7.1.14 An operator $\operatorname{proj}_{n,h}$ is defined on the space of homogeneous polynomials $\mathscr{P}_n^d$ by

$$\operatorname{proj}_{n,h} P(x) = \frac{(-1)^n}{2^n (\lambda_\kappa)_n} \|x\|^{2\lambda_\kappa + 2|\alpha|} P(\mathscr{D})\{\|x\|^{-2\lambda_\kappa}\},$$

where $(\lambda_\kappa)_n$ is a Pochhammer symbol.

Theorem 7.1.15 *The operator* $\mathrm{proj}_{n,h}$ *is the projection operator from* $\mathscr{P}_n^d$ *to* $\mathscr{H}_n^d(h_\kappa^2)$*, and it is given by*

$$\mathrm{proj}_{n,h} P(x) = \sum_{j=0}^{[n/2]} \frac{1}{4^j j!(-\lambda_\kappa - n + 1)_j} \|x\|^{2j} \Delta_h^j P; \tag{7.1.6}$$

moreover, every $P \in \mathscr{P}_n^d$ *has the expansion*

$$P(x) = \sum_{j=0}^{[n/2]} \frac{1}{4^j j!(\lambda_\kappa + 1 + n - 2j)_j} \|x\|^{2j} \mathrm{proj}_{n-2j,h} \Delta_h^j P(x). \tag{7.1.7}$$

Proof Use induction on $n = |\alpha|$. The case $n = 0$ is evident. Suppose that the equation has been proved for all α such that $|\alpha| = n$. Then, with $P(x) = x^\alpha$ and $|\alpha| = n$,

$$\mathscr{D}^\alpha \|x\|^{-2\lambda} = (-1)^n 2^n (\lambda)_n \|x\|^{-2\lambda-2n} \times \sum_{j=0}^{[n/2]} \frac{1}{4^j j!(-\lambda - n + 1)_j} \|x\|^{2j} \Delta_h^j (x^\alpha).$$

Applying $\mathscr{D}_i$ to this equation and using the first identity in (7.1.3) with $g = \Delta_h^j(x^\alpha)$, we conclude, after carefully computing the coefficients, that

$$\mathscr{D}_i \mathscr{D}^\alpha \|x\|^{-2\lambda} = (-1)^n 2^n (\lambda)_n (-2\lambda - 2n) \|x\|^{-2\lambda-2n-2} \times \sum_{j=0}^{[(n+1)/2]} \frac{1}{4^j j!(-\lambda - n)_j} \|x\|^{2j} \left[x_i \Delta_h^j (x^\alpha) + 2j \Delta_h^{j-1} \mathscr{D}_i (x^\alpha) \right],$$

where we have also used the fact that $\mathscr{D}_i$ commutes with Δ_h. Now, using the identity

$$\Delta_h^j [x_i f(x)] = x_i \Delta_h^j f(x) + 2j \mathscr{D}_i \Delta_h^{j-1} f(x), \qquad j = 1, 2, 3, \ldots,$$

which follows from Lemma 7.1.10, and the fact that $(-1)^n 2^n (\lambda)_n (-2\lambda - 2n) = (-1)^{n+1} 2^{n+1} (\lambda)_{n+1}$, we conclude that equation (7.1.6) holds for $P(x) = x_i x^\alpha$, which completes the induction.

To prove (7.1.7), recall that every $P \in \mathscr{P}_n^d$ has a unique decomposition $P(x) = \sum_{j=0}^{[n/2]} \|x\|^{2j} P_{n-2j}$ by Theorem 7.1.7, where $P_{n-2j} \in \mathscr{H}_{n-2j}^d(h_\kappa^2)$. It follows from the second identity of (7.1.3) that

$$\Delta_h^j P = \sum_{i=j}^{[n/2]} 4^j (-i)_j (-\lambda + n + i)_j \|x\|^{2i-2j} P_{n-2i}, \tag{7.1.8}$$

from which we obtain that

$$\begin{aligned}\operatorname{proj}_{n,h} P &= \sum_{j=0}^{[n/2]} \|x\|^{2j} \sum_{i=j}^{[n/2]} \frac{4^j(-i)_j(-\lambda-n+i)_j}{4^j j!(-\lambda-n+1)_j} \|x\|^{2i-2j} P_{n-2i} \\ &= \sum_{i=0}^{[n/2]} \|x\|^{2i} P_{n-2i} \sum_{j=0}^{i} \frac{(-i)_j(-\lambda-n+i)_j}{j!(-\lambda-n+1)_j} \\ &= \sum_{i=0}^{[n/2]} \|x\|^{2i} P_{n-2i}\, {}_2F_1\left(\begin{matrix}-i, -\lambda-n+i \\ -\lambda-n+1\end{matrix}; 1\right).\end{aligned}$$

By the Chu–Vandermonde identity the hypergeometric function ${}_2F_1$ is zero except when $i=0$, which yields $\operatorname{proj}_{n,h} P = P_n$. Finally, using $\Delta_h^j P$ in place of P and taking into account (7.1.8), we conclude that

$$\operatorname{proj}_{n-2j,h} \Delta_h^j P = (-j)_j(-\lambda-n+j)_j P_{n-2j};$$

a simple conversion of Pochhammer symbols completes the proof. □

An immediate consequence of this theorem is the following corollary.

Corollary 7.1.16 *For each* $\alpha \in \mathbb{N}_0^d$,

$$H_\alpha(x) = (-1)^n 2^n (\lambda_\kappa)_n \operatorname{proj}_{n,h}(x^\alpha).$$

The projection operator turns out to be related to the adjoint operator $\mathscr{D}_i^*$ of $\mathscr{D}_i$ in $L^2(h_\kappa^2\, d\omega)$. The adjoint $\mathscr{D}_i^*$ on $\mathscr{H}_n^d(h_\kappa^2)$ is defined by

$$\int_{S^{d-1}} p(\mathscr{D}_i q) h_\kappa^2\, d\omega = \int_{S^{d-1}} (\mathscr{D}_i^* p) q h_\kappa^2\, d\omega, \qquad p \in \mathscr{H}_n^d(h_\kappa^2), \quad q \in \mathscr{H}_{n+1}^d(h_\kappa^2).$$

It follows that $\mathscr{D}_i^*$ is a linear operator that maps $\mathscr{H}_n^d(h_\kappa^2)$ into $\mathscr{H}_{n+1}^d(h_\kappa^2)$.

Theorem 7.1.17 *For* $p \in \mathscr{H}_n^d(h_\kappa^2)$,

$$\mathscr{D}_i^* p = 2(n+\lambda_\kappa+1)\left[x_i p - (2n+2\lambda_\kappa)^{-1} \|x\|^2 \mathscr{D}_i p\right].$$

Proof First assume that $|\alpha| = |\beta| = n$. It follows from (7.1.5) that

$$\begin{aligned}&(2\lambda+2n) \int_{S^{d-1}} x_i H_\alpha H_\beta h_\kappa^2\, d\omega \\ &\quad = \int_{S^{d-1}} (\mathscr{D}_i H_\alpha) H_\beta h_\kappa^2\, d\omega - \int_{S^{d-1}} H_{\alpha+\varepsilon_i} H_\beta h_\kappa^2\, d\omega = 0.\end{aligned}$$

Since $\{H_\alpha : |\alpha| = n\}$ contains a basis for $\mathscr{H}_n^d(h_\kappa^2)$, it follows that

$$\int_{S^{d-1}} x_i p q h_\kappa^2\, d\omega = 0, \qquad 1 \le i \le d, \quad p, q \in \mathscr{H}_n^d(h_\kappa^2). \tag{7.1.9}$$

Using this equation and the recursive relation (7.1.5) twice we conclude that, for $|\alpha| = n$ and $|\beta| = n+1$,

$$\int_{S^{d-1}} H_\alpha(\mathscr{D}_i H_\beta) h_\kappa^2 \, d\omega = (2\lambda + 2n + 2) \int_{S^{d-1}} x_i H_\alpha H_\beta h_\kappa^2 \, d\omega$$
$$= -\frac{2\lambda + 2n + 2}{2\lambda + 2n} \int_{S^{d-1}} H_{\alpha+\varepsilon_i} H_\beta h_\kappa^2 \, d\omega,$$

which gives, using the definition of $\mathscr{D}_i^*$,

$$\mathscr{D}_i^* H_\alpha = -\frac{n + \lambda + 1}{n + \lambda} H_{\alpha+\varepsilon_i}.$$

Using the recurrence relation (7.1.4) again, we obtain the desired result for $p = H_\alpha$, which completes the proof by Corollary 7.1.13. □

We can also obtain a formula for the adjoint operator defined on $L^2(h^2 d\mu, \mathbb{R}^d)$, where we define the measure μ by

$$d\mu = (2\pi)^{-d/2} e^{-\|x\|^2/2} \, dx.$$

Let $\mathscr{V}_n^d(h_\kappa^2 \, d\mu)$ denote the space of orthogonal polynomials of degree n with respect to $h_\kappa^2 \, d\mu$. An orthogonal basis of $\mathscr{V}_n^d(h_\kappa^2 \, d\mu)$ is given by

$$P_{\nu,j}^n(x) = L_j^{n-2j+\lambda_\kappa}\left(\frac{1}{2}\|x\|^2\right) Y_{n-2j,\nu}^h(x), \tag{7.1.10}$$

where $Y_{n-2j,\nu}^h \in \mathscr{H}_{n-2j}^d(h_\kappa^2)$ and $0 \le 2j \le n$. Its verification follows as in (5.1.6). We introduce the following normalization constants for convenience:

$$\begin{aligned} c_h = c_{h,d} &= \left(\int_{\mathbb{R}^d} h_\kappa^2(x) \, d\mu\right)^{-1}, \\ c_h' = c_{h,d}' &= \sigma_{d-1}\left(\int_{S^{d-1}} h_\kappa^2 \, d\omega\right)^{-1}. \end{aligned} \tag{7.1.11}$$

Using polar coordinates we have for $p \in \mathscr{P}_{2m}$ that

$$\int_{\mathbb{R}^d} p(x) h_\kappa^2(x) \, d\mu = \frac{1}{(2\pi)^{d/2}} \int_0^\infty r^{2\gamma_\kappa + 2m + d - 1} e^{-r^2/2} \, dr \int_{S^{d-1}} p h_\kappa^2 \, d\omega$$
$$= 2^{m+\gamma_\kappa} \frac{\Gamma(\lambda_\kappa + m + 1)}{\Gamma(\frac{d}{2})} \frac{1}{\sigma_{d-1}} \int_{S^{d-1}} p h_\kappa^2 \, d\omega,$$

which implies, in particular, that

$$c_{h,d}' = 2^{\gamma_\kappa} \frac{\Gamma(\lambda_\kappa + 1)}{\Gamma(\frac{d}{2})} c_{h,d}.$$

We denote the adjoint operator of $\mathscr{D}_i$ on $L^2(h_\kappa^2 \, d\mu, \mathbb{R}^d)$ by $\widetilde{\mathscr{D}}_i^*$, to distinguish it from the $L^2(h_\kappa^2 \, d\omega, S^{d-1})$ adjoint $\mathscr{D}_i^*$.

Theorem 7.1.18 *The adjoint $\widetilde{\mathscr{D}}_i^*$ acting on $L^2(h_\kappa^2\,\mathrm{d}\mu, \mathbb{R}^d)$ is given by*

$$\widetilde{\mathscr{D}}_i^* p(x) = x_i p(x) - \mathscr{D}_i p(x), \qquad p \in \Pi^d.$$

Proof Assume that $\kappa_v \geq 1$. Analytic continuation can be used to extend the range of validity to $\kappa_v \geq 0$. Let p and q be two polynomials. Integrating by parts, we obtain

$$\begin{aligned}\int_{\mathbb{R}^d} \big[\partial_i p(x)\big] q(x) h_\kappa^2(x)\,\mathrm{d}\mu = &-\int_{\mathbb{R}^d} p(x)\big[\partial_i q(x)\big] h_\kappa^2(x)\,\mathrm{d}\mu \\ &+ \int_{\mathbb{R}^d} p(x)q(x)\big[-2h_\kappa(x)\partial_i h_\kappa(x) + h_\kappa^2(x)x_i\big]\,\mathrm{d}\mu.\end{aligned}$$

For a fixed root v,

$$\begin{aligned}&\int_{\mathbb{R}^d} \frac{p(x) - p(x\sigma_v)}{\langle x, v\rangle} q(x) h_\kappa^2(x)\,\mathrm{d}\mu \\ &\quad = \int_{\mathbb{R}^d} \frac{p(x)q(x)}{\langle x, v\rangle} h_\kappa^2(x)\,\mathrm{d}\mu - \int_{\mathbb{R}^d} \frac{p(x\sigma_v)q(x)}{\langle x, v\rangle} h_\kappa^2(x)\,\mathrm{d}\mu \\ &\quad = \int_{\mathbb{R}^d} \frac{p(x)q(x)}{\langle x, v\rangle} h_\kappa^2(x)\,\mathrm{d}\mu + \int_{\mathbb{R}^d} \frac{p(x)q(x\sigma_v)}{\langle x, v\rangle} h_\kappa^2(x)\,\mathrm{d}\mu,\end{aligned}$$

where in the second integral we have replaced x by $x\sigma_v$, which changes $\langle x, v\rangle$ to $\langle x\sigma_v, v\rangle = -\langle x, v\rangle$ and leaves h_κ^2 invariant. Note also that

$$h_\kappa(x)\partial_i h_\kappa(x) = \sum_{v\in R_+} \kappa_v \frac{v_i}{\langle x, v\rangle} h_\kappa^2(x).$$

Combining these ingredients, we obtain

$$\begin{aligned}&\int_{\mathbb{R}^d} \mathscr{D}_i p(x) q(x) h_\kappa^2(x)\,\mathrm{d}\mu \\ &\quad = \int_{\mathbb{R}^d} \Big[p(x)\big[x_i q(x) - \partial_i q(x)\big] \\ &\qquad + \sum_{v\in R_+} \big(\kappa_v v_i\, p(x)\big[-2q(x) + q(x) + q(x\sigma_v)\big]/\langle x, v\rangle\big)\Big] h_\kappa^2(x)\,\mathrm{d}\mu;\end{aligned}$$

the term inside the large square brackets is exactly $p(x)\big[x_i q(x) - \mathscr{D}_i q(x)\big]$. □

7.2 Inner Products on Polynomials

For polynomials in Π^d, a natural inner product is $\langle p, q\rangle_\partial = p(\partial)q(0)$, $p, q \in \mathscr{P}_n^d$, where $p(\partial)$ means that x_i has been replaced by ∂_i in $p(x)$. The reproducing kernel of this inner product is $\langle x, y\rangle^n/n!$, since $(\langle x, \partial_y\rangle^n/n!)q(y) = q(x)$ for $q \in \mathscr{P}_n^d$ and $x \in \mathbb{R}^d$. With the goal of constructing the Poisson kernel for h-harmonics, we consider the action of the intertwining operator V (Section 6.5) on this inner product. As in the previous chapter, a superscript such as (x) on an operator refers to the $\mathbb{R}^d$ variable on which it acts.

Proposition 7.2.1 *If $p \in \mathscr{P}_n^d$ then $K_n(x, \mathscr{D}^{(y)})p(y) = p(x)$ for all $x \in \mathbb{R}^d$, where $K_n(x, \mathscr{D}^{(y)})$ is the operator formed by replacing y_i by $\mathscr{D}_i$ in $K_n(x,y)$.*

Proof If $p \in \mathscr{P}_n^d$ then $p(x) = (\langle x, \partial^{(y)}\rangle^n / n!)p(y)$. Applying $V^{(x)}$ leads to $V^{(x)}p(x) = K_n(x, \partial^{(y)})p(y)$. The left-hand side is independent of y, so applying $V^{(y)}$ to both sides gives $V^{(x)}p(x) = K_n(x, \mathscr{D}^{(y)})V^{(y)}p(y)$. Thus the desired identity holds for all Vp with $p \in \mathscr{P}_n^d$, which completes the proof since V is one to one. □

Definition 7.2.2 The bilinear form $\langle p, q\rangle_h$ (known as an *h inner product*) equals $p(\mathscr{D}^{(x)})q(x)$ for $p, q \in \mathscr{P}_n^d$ and $n \in \mathbb{N}_0^d$ and, since homogenous polynomials of different degrees are orthogonal to each other, this extends by linearity to all polynomials.

We remark that if $\kappa = 0$, that is, $h_\kappa(x) = 1$, then $\langle p, q\rangle_h$ is the same as $\langle p, q\rangle_\partial$.

Theorem 7.2.3 *For $p, q \in \mathscr{P}_n^d$,*

$$\langle p, q\rangle_h = K_n(\mathscr{D}^{(x)}, \mathscr{D}^{(y)})p(x)q(y) = \langle q, p\rangle_h.$$

Proof By Proposition 7.2.1, $p(x) = K_n(x, \mathscr{D}^{(y)})p(y)$. The operators $\mathscr{D}^{(x)}$ and $\mathscr{D}^{(y)}$ commute and thus

$$\langle p, q\rangle_h = K_n(\mathscr{D}^{(x)}, \mathscr{D}^{(y)})p(y)q(x) = K_n(\mathscr{D}^{(y)}, \mathscr{D}^{(x)})p(y)q(x)$$

by part (iv) of Lemma 6.6.2. The latter expression equals $\langle q, p\rangle_h$. □

Theorem 7.2.4 *If $p \in \mathscr{P}_n^d$ and $q \in \mathscr{H}_n^d(h_\kappa^2)$ then*

$$\langle p, q\rangle_h = c_h \int_{\mathbb{R}^d} pqh_\kappa^2 \, \mathrm{d}\mu = 2^n \left(\gamma_\kappa + \tfrac{d}{2}\right)_n c_h' \int_{S^{d-1}} pqh_\kappa^2 \, \mathrm{d}\omega.$$

Proof Since $p(\mathscr{D})q(x)$ is a constant, it follows from Theorem 7.1.18 that

$$\begin{aligned} \langle p, q\rangle_h &= c_h \int_{\mathbb{R}^d} p(\mathscr{D})q(x)h_\kappa^2(x) \, \mathrm{d}\mu \\ &= c_h \int_{\mathbb{R}^d} q(x) \left[p(\widetilde{\mathscr{D}}^*)1\right] h_\kappa^2(x) \, \mathrm{d}\mu \\ &= c_h \int_{\mathbb{R}^d} q(x) \left[p(x) + s(x)\right] h_\kappa^2(x) \, \mathrm{d}\mu \end{aligned}$$

for a polynomial s of degree less than n, where using the relation $\widetilde{\mathscr{D}}_i^* g(x) = x_i g(x) - \mathscr{D}_i g(x)$ repeatedly and the fact that the degree of $\mathscr{D}_i g$ is lower than that of g shows that $p(\widetilde{\mathscr{D}}^*)1 = p(x) + s(x)$. But since $q \in \mathscr{H}_n^d(h_\kappa^2)$, it follows from the polar integral that $\int_{\mathbb{R}^d} q(x)s(x)h_\kappa^2(x) \mathrm{d}\mu = 0$. This gives the first equality in the statement. The second follows from the use of polar coordinates as in the formula below equations (7.1.11). □

Thus $\langle p,q\rangle_h$ is positive definite. We can also give an integral representation of $\langle p,q\rangle_h$ for all $p,q\in\Pi^d$. We need two lemmas.

Lemma 7.2.5 *Let $p,q\in\mathscr{P}_n^d$ and write*

$$p(x)=\sum_{j=0}^{\lfloor n/2\rfloor}\|x\|^{2j}p_{n-2j}(x)\qquad \textit{and}\qquad q(x)=\sum_{j=0}^{\lfloor n/2\rfloor}\|x\|^{2j}q_{n-2j}(x),$$

with $p_{n-2j},q_{n-2j}\in\mathscr{H}_{n-2j}^d(h_\kappa^2)$. Then

$$\langle p,q\rangle_h=\sum_{j=0}^{\lfloor n/2\rfloor}4^j j!(n-2j+\lambda_\kappa+1)_j\langle p_{n-2j},q_{n-2j}\rangle_h.$$

Proof The expansions of p and q in h-harmonics are unique by Theorem 7.1.7. Using the definition of Δ_h (see the text before Definition 7.1.2) and $\langle p_{n-2j},q_{n-2j}\rangle_h$,

$$\langle p,q\rangle_h=\sum_{j=0}^{\lfloor n/2\rfloor}\sum_{i=0}^{\lfloor n/2\rfloor}\Delta_h^i p_{n-2i}(\mathscr{D})\left[\|x\|^{2j}q_{n-2j}(x)\right].$$

By the second identity of (7.1.3),

$$\Delta_h^i\left(\|x\|^{2j}q_{n-2j}(x)\right)=4^i(-j)_i(-n-\lambda+j)_i\|x\|^{2j-2i}q_{n-2j}(x),$$

which is zero if $i>j$. If $i<j$ then $\langle\|x\|^{2j}q_{n-2j}(x),\|x\|^{2i}p_{n-2i}(x)\rangle_h=0$ by the same argument and the fact that the pairing is symmetric (Theorem 7.2.3). Hence, the only remaining terms are those with $j=i$, which are given by $4^j j!(-n-\lambda+j)_j p_{n-2j}(\mathscr{D})q_{n-2j}(x)$. □

Lemma 7.2.6 *Let $p\in\mathscr{H}_n^d(h_\kappa^2)$, $n,m\in\mathbb{N}_0^d$; then*

$$\mathrm{e}^{-\Delta_h/2}\|x\|^{2m}p(x)=(-1)^m m!2^m L_m^{n+\lambda_\kappa}\left(\frac{1}{2}\|x\|^2\right)p(x).$$

Proof We observe that, for $p\in\mathscr{P}_n^d$, $\mathrm{e}^{-\Delta_h/2}p$ is a finite sum. From (7.1.3),

$$\mathrm{e}^{-\Delta_h/2}\|x\|^{2m}p(x)=\sum_{j=0}^{m}\frac{(-1)^j2^j}{j!}(-m)_j(-n-m-\lambda)_j\|x\|^{2m-2j}p(x).$$

Using the fact that $(-1)^j(1-m-a)_j=(a)_m/(a)_{m-j}$ with $a=n+\lambda+1$, the stated identity follows from the expansion of the Laguerre polynomial in Subsection 1.4.2. □

Theorem 7.2.7 *For polynomials p,q*

$$\langle p,q\rangle_h=c_h\int_{\mathbb{R}^d}\left(\mathrm{e}^{-\Delta_h/2}p\right)\left(\mathrm{e}^{-\Delta_h/2}q\right)h_\kappa^2\,\mathrm{d}\mu.$$

Proof By Lemma 7.2.5, it suffices to establish this identity for p,q of the form

$$p(x) = \|x\|^{2j} p_m(x), \quad q(x) = \|x\|^{2j} q_m(x), \quad \text{with} \quad p_m, q_m \in \mathscr{H}_m^d(h_\kappa^2).$$

By the second identity of (7.1.3),

$$\begin{aligned}\langle p,q\rangle_h &= 4^j j!\left(m+\gamma+\tfrac{d}{2}\right)_j \langle p_m, q_m\rangle_h \\ &= 4^j j!(m+\lambda+1)_j 2^m(\lambda+1)_j c_h' \int_{S^{d-1}} p_m q_m h_\kappa^2 \,\mathrm{d}\omega \\ &= 2^{m+2j} j!(\lambda+1)_{m+j} c_h' \int_{S^{d-1}} p_m q_m h_\kappa^2 \,\mathrm{d}\omega.\end{aligned}$$

The right-hand side of the stated formula, by Lemma 7.2.6, is seen to equal

$$\begin{aligned}& c_h \int_{\mathbb{R}^d} (j!2^j)^2 [L_j^{m+\lambda}\left(\tfrac{1}{2}\|x\|^2\right)]^2 p_m(x) q_m(x) h_\kappa^2(x)\,\mathrm{d}\mu \\ &= 2^{-\gamma} \frac{\Gamma(\frac{d}{2})}{\Gamma(\lambda+1)} c_h'(j!)^2 2^{m+\gamma+2j} \frac{\Gamma(\lambda+1+m+j)}{j!\Gamma(\frac{d}{2})} \int_{S^{d-1}} p_m q_m h_\kappa^2 \,\mathrm{d}\omega\end{aligned}$$

upon using polar coordinates in the integral and the relation between c_h and c_h'. These two expressions are clearly the same. □

As an application of Theorem 7.2.4, we state the following result.

Proposition 7.2.8 *For $p \in \mathscr{P}_n^d$ and $q \in \mathscr{H}_n^d$, where $\mathscr{H}_n^d$ is the space of ordinary harmonics of degree n, we have*

$$c_h' \int_{S^{d-1}} pVqh_\kappa^2 \,\mathrm{d}\omega = \frac{\left(\frac{d}{2}\right)_n}{\left(\gamma_\kappa + \frac{d}{2}\right)_n \sigma_{d-1}} \int_{S^{d-1}} pq \,\mathrm{d}\omega.$$

Proof Since $\Delta_h V = V\Delta$, $q \in \mathscr{H}_n^d$ implies that $Vq \in \mathscr{H}_n^d(h_\kappa^2)$. Apply Theorem 7.2.4 with Vq in place of q; then

$$\begin{aligned}2^n\left(\gamma+\tfrac{d}{2}\right)_n c_h' \int_{S^{d-1}} pVqh_\kappa^2 \,\mathrm{d}\omega &= p(\mathscr{D})Vq(x) \\ &= Vp(\partial)q(x) = p(\partial)q(x) \\ &= 2^n\left(\tfrac{d}{2}\right)_n \sigma_{d-1} \int_{S^{d-1}} pq\,\mathrm{d}\omega,\end{aligned}$$

where the first equality follows from Theorem 7.2.4, the second follows from the intertwining property of V, the third follows from the fact that $p(\partial)q(x)$ is a constant and $V1 = 1$ and the fourth follows from Theorem 7.2.4 with $\kappa = 0$. □

An immediate consequence of this theorem is the following biorthogonal relation.

Corollary 7.2.9 *Let $\{S_\alpha\}$ be an orthonormal basis of $\mathscr{H}_n^d$. Then $\{VS_\alpha\}$ and $\{S_\alpha\}$ are biorthogonal; more precisely,*

$$\int_{S^{d-1}} (VS_\alpha) S_\beta h_\kappa^2 \, d\omega = \frac{\left(\frac{d}{2}\right)_n}{\left(\gamma_\kappa + \frac{d}{2}\right)_n c_h' \sigma_{d-1}} \delta_{\alpha,\beta}.$$

In particular, $\{VS_\alpha\}$ *is a basis of* $\mathscr{H}_n^d(h_\kappa^2)$.

The corollary follows from setting $p = S_\alpha$ and $q = S_\beta$ in the proposition. It should be pointed out that the biorthogonality is restricted to elements of $\mathscr{H}_n^d$ and $\mathscr{H}_n^d(h_\kappa^2)$ for the same n. In general S_α is not orthogonal to VS_β with respect to $h_\kappa^2 \, d\omega$ if $|\beta| < |\alpha|$.

7.3 Reproducing Kernels and the Poisson Kernel

If $f \in L^2(h_\kappa^2 d\omega)$ then we can expand f as an h-harmonic series. Its nth component is the orthogonal projection defined by

$$S_n(h_\kappa^2; f, x) := c_h' \int_{S^{d-1}} P_n(h_\kappa^2; x, y) f(y) h_\kappa^2(y) \, d\omega(y), \tag{7.3.1}$$

where $P_n(h_\kappa^2; x, y)$ denotes the reproducing kernel of $\mathscr{H}_n^d(h_\kappa^2)$ and is consistent with the notation (4.2.4). Recall $K_n(x,y)$ defined in Definition 6.6.1.

Theorem 7.3.1 *For* $n \in \mathbb{N}_0$ *and for* $x, y \in \mathbb{R}^d$,

$$P_n(h_\kappa^2; x, y) = \sum_{0 \le j \le n/2} \frac{(\lambda_\kappa + 1)_n 2^{n-2j}}{(1 - n - \lambda_\kappa)_j j!} \|x\|^{2j} \|y\|^{2j} K_{n-2j}(x, y).$$

Proof The kernel $P_n(h_\kappa^2; x, y)$ is uniquely defined by the fact that it reproduces the space $\mathscr{H}_n^d(h_\kappa^2)$. Let $f \in \mathscr{H}_n^d(h_\kappa^2)$. Then by Proposition 7.2.1 $f(y) = K_n(\mathscr{D}^{(x)}, y) f(x)$. Fix y and let $p(x) = K_n(x, y)$, so that $f(y) = \langle p, f \rangle_h$. Expand $p(x)$ as $p(x) = \sum_{0 \le j \le n/2} \|x\|^{2j} p_{n-2j}(x)$ with $p_{n-2j} \in \mathscr{H}_{n-2j}^d(h_\kappa^2)$; it then follows from Theorem 7.2.4 that

$$f(y) = \langle p, f \rangle_h = \langle p_n, f \rangle_h = 2^n (\lambda + 1)_n \, c_h' \int_{S^{d-1}} p_n f h_\kappa^2 \, d\omega.$$

Thus, by the reproducing property, $P_n(h_\kappa^2; x, y) = 2^n (\lambda + 1)_n p_n(x)$ with $p_n = \operatorname{proj}_{n,h} p$. Hence, the stated identity follows from Theorem 7.1.15 and the fact that $\Delta_h^j p(x) = \Delta_h^j K_n(x, y) = \|y\|^{2j} K_{n-2j}(x, y)$; see (v) in Proposition 6.6.2. □

Corollary 7.3.2 *For* $n \in \mathbb{N}_0$ *and* $\|y\| \le \|x\| = 1$,

$$P_n(h_\kappa^2; x, y) = \frac{n + \lambda_\kappa}{\lambda_\kappa} V\left[C_n^{\lambda_\kappa}\left(\left\langle \cdot, \frac{y}{\|y\|} \right\rangle\right)\right](x) \|y\|^n.$$

Proof Since the intertwining operator V is linear and $\langle \cdot, y \rangle^m$ is homogeneous of degree m in y, for $\|x\| = 1$ write

$$P_n(h_\kappa^2, x, y) = V\left\{\sum_{j=0}^{[n/2]} \frac{(\lambda)_n 2^{n-2j}}{(1-n-\lambda)_j j!(n-2j)!}\left\langle \cdot, \frac{y}{\|y\|}\right\rangle^{n-2j}\right\}(x)\|y\|^n.$$

Set $t = \langle \cdot, y/\|y\|\rangle$ and denote the expression inside the braces by $L_n(t)$. Since $(1-n-\lambda)_j = (-1)^j(\lambda)_n/(\lambda)_{n-j}$, the proof of Proposition 1.4.11 shows that $L_n(t) = [(\lambda+1)_n/(\lambda)_n]C_n^\lambda(t)$, which gives the stated formula. □

The Poisson or reproducing kernel $P(h_\kappa^2; x, y)$ is defined by the property

$$f(y) = c_h' \int_{S^{d-1}} f(x)P(h_\kappa^2; x, y)h_\kappa^2(x)\,\mathrm{d}\omega(x)$$

for each $f \in \mathscr{H}_n^d(h_\kappa^2)$, $n \in \mathbb{N}_0$ and $\|y\| < 1$.

Theorem 7.3.3 *Fix $y \in B^d$; then*

$$P(h_\kappa^2; x, y) = V\left(\frac{1-\|y\|^2}{(1-2\langle \cdot, y\rangle + \|y\|^2)^{\lambda_\kappa+1}}\right)(x)$$

for $\|y\| < 1 = \|x\|$.

Proof Denote by f_y the function which is the argument of V in the statement. We claim that $f_y \in A(B^d)$ (see Definition 6.5.19) with $\|f\|_A = (1-\|y\|^2)(1-\|y\|)^{-2-2\lambda}$. Indeed,

$$\begin{aligned} f_y(x) &= (1-\|y\|^2)(1+\|y\|^2)^{-1-\lambda}\left(1-\frac{2\langle x, y\rangle}{1+\|y\|^2}\right)^{-d/2-\gamma} \\ &= (1-\|y\|^2)(1+\|y\|^2)^{-1-\lambda}\sum_{n=0}^{\infty}\frac{(1+\lambda)_n 2^n}{n!(1+\|y\|^2)^n}\langle x, y\rangle^n. \end{aligned}$$

But $\|\langle x, y\rangle^n\|_\infty = \|y\|^n$ and $0 \le 2\|y\| < 1 + \|y\|^2$ for $\|y\| < 1$, so that

$$\begin{aligned} \|f_y\|_A &= (1-\|y\|^2)(1+\|y\|^2)^{-1-\lambda}\left(1-\frac{2\|y\|}{1+\|y\|^2}\right)^{-1-\lambda} \\ &= (1-\|y\|^2)(1-\|y\|)^{-2-2\lambda}. \end{aligned}$$

Thus, Vf_y is defined and continuous for $\|x\| \le 1$. Since $\|y\| < 1$, the stated identity follows from $P(h_\kappa^2; x, y) = \sum_{n=0}^\infty P_n(h_\kappa^2; x, y)$, Corollary 7.3.2 and the Poisson kernel of the Gegenbauer polynomials. □

The Poisson kernel $P(h_\kappa^2; x, y)$ satisfies the following two properties:

$$0 \le P(h_\kappa^2; x, y), \qquad \|y\| \le \|x\| = 1,$$
$$c_h' \int_{S^{d-1}} P(h_\kappa^2; x, y)h_\kappa^2(y)\,\mathrm{d}\omega(y) = 1.$$

The first follows from the fact that V is a positive operator (Rösler [1998]), and the second is a consequence of the reproducing property.

The formula in Corollary 7.3.2 also allows us to prove a formula that is the analogue of the classical Funk–Hecke formula for harmonics. Denote by w_λ the normalized weight function

$$w_\lambda(t) = B(\lambda + \tfrac{1}{2}, \tfrac{1}{2})^{-1}(1-t^2)^{\lambda-1/2}, \qquad t \in [-1,1],$$

whose orthogonal polynomials are the Gegenbauer polynomials. Then the Funk–Hecke formula for h-harmonics is as follows.

Theorem 7.3.4 *Let f be a continuous function on $[-1,1]$. Let $Y_n^h \in \mathscr{H}_n^d(h_\kappa^2)$. Then*

$$c_h' \int_{S^{d-1}} Vf(\langle x, \cdot \rangle)(y) Y_n^h(y) h_\kappa^2(y)\, d\omega = \tau_n(f) Y_n^h(x), \qquad x \in S^{d-1},$$

where $\tau_n(f)$ is a constant defined by

$$\tau_n(f) = \frac{1}{C_n^{\lambda_\kappa}(1)} \int_{-1}^{1} f(t) C_n^{\lambda_\kappa}(t) w_{\lambda_\kappa}(t)\, dt.$$

Proof First assume that f is a polynomial of degree m. Then we can write f in terms of the Gegenbauer polynomials as

$$f(t) = \sum_{k=0}^{m} a_k \frac{\lambda + k}{\lambda} C_k^\lambda(t) = \sum_{k=0}^{m} a_k \widetilde{C}_k^\lambda(1) \widetilde{C}_k^\lambda(t),$$

where $\widetilde{C}_k^\lambda$ stands for a Gegenbauer polynomial that is orthonormal with respect to w_λ. Using this orthonormality of the $\widetilde{C}_k^\lambda$, the coefficients a_k are given by the formula

$$a_k = \frac{1}{C_k^\lambda(1)} \int_{-1}^{1} f(t) C_k^\lambda(t) w_\lambda(t)\, dt,$$

where we have used the fact that $[\widetilde{C}_k^\lambda(1)]^2 = [(k+\lambda)/\lambda] C_k^\lambda(1)$. For $x, y \in S^{d-1}$, it follows from the formula for $P_n(h_\kappa^2)$ in Corollary 7.3.2 that

$$Vf(\langle x, \cdot \rangle)(y) = \sum_{k=0}^{m} a_k \frac{\lambda + k}{\lambda} V C_k^\lambda(\langle x, \cdot \rangle)(y) = \sum_{k=0}^{m} a_k P_k^h(x,y).$$

Since $\mathscr{P}_n^h$ is the reproducing kernel of the space $\mathscr{H}_n^d(h_\kappa^2)$ we have, for any $Y_n^h \in \mathscr{H}_n^d(h_\kappa^2)$,

$$\int_{S^{d-1}} Vf(\langle x, \cdot \rangle)(y) Y_n^h(y) h_\kappa^2(y)\, d\omega = \lambda_n Y_n^h(x), \qquad x \in S^{d-1},$$

where if $m < n$ then $\lambda_n = 0$, so that both sides of the equation become zero. Together with the formula for $\tau_n(f)$ this gives the stated formula for f a polynomial.

If f is a continuous function on $[-1,1]$ then choose P_m to be a sequence of polynomials such that P_m converges to f uniformly on $[-1,1]$. The intertwining operator V is positive and $|V(g)(x)| \le \sup\{g(y) : |y| \le 1\}$ for $|x| \le 1$. Let

$g_x(y) = f(\langle x,y\rangle) - P_m(\langle x,y\rangle)$. It follows that, for m sufficiently large, $|g_x(y)| \le \varepsilon$ for $|y| \le 1$ and $|V(g_x)(y)| \le \varepsilon$ for all $x \in S^{d-1}$, from which we deduce from the dominated convergence theorem that the stated result holds for f. □

If $\kappa = 0$ then $V = id$ and the theorem reduces to the classical Funk-Hecke formula.

7.4 Integration of the Intertwining Operator

Although a compact formula for the intertwining operator is not known in the general case, its integral over S^{d-1} can be computed. This will have several applications in the later development. We start with

Lemma 7.4.1 *Let $m \ge 0$ and $S_m \in \mathscr{H}_m^d$ be an ordinary harmonic. Then*

$$c_h' \int_{S^{d-1}} V(\|\cdot\|^{2j} S_m)(x)\, h_\kappa^2(x)\, d\omega = \frac{(\frac{d}{2})_j}{(\lambda_\kappa+1)_j}\delta_{m,0}.$$

Proof Since $\|x\|^{2j} S_m$ is a homogeneous polynomial of degree $m+2j$, so is $V(\|\cdot\|^{2j} S_m)$. Using the projection operator $\operatorname{proj}_{k,h} : \mathscr{P}_k^d \mapsto \mathscr{H}_k^h(h_\kappa^2)$ and (7.1.7), we obtain

$$V(\|\cdot\|^{2j} S_m) = \sum_{i=0}^{[m/2]+j} \Bigg(\|x\|^{2i} \frac{1}{4^i i!} \frac{1}{(\lambda+m+2j-2i+1)_i} \times \operatorname{proj}_{m+2j-2i,h}[\Delta_h^i V(\|\cdot\|^{2j} S_m)] \Bigg).$$

Since $\mathscr{H}_k^d(h_\kappa^2) \perp 1$ for $k > 0$ with respect to $h_\kappa^2\, d\omega$, it follows that the integral of the function $\operatorname{proj}_{m+2j-2i,h} \Delta_h^i V(\|\cdot\|^{2j} S_m)$ is zero unless $m+2j-2i = 0$, which shows that m is even. In the case where $2i = m+2j$, we have

$$\int_{S^{d-1}} V(\|\cdot\|^{2j} S_m)(x)\, h_\kappa^2(x)\, d\omega = \frac{1}{4^{j+m/2}(j+\frac{m}{2})!}(\lambda+1)_{j+m/2} \int_{S^{d-1}} \Delta_h^{j+m/2} V(\|\cdot\|^{2j} S_m)\, h_\kappa^2(x)\, d\omega,$$

since now $\operatorname{proj}_{0,h} = id$. By the intertwining property of V, it follows that

$$\Delta_h^{j+m/2} V(\|\cdot\|^{2j} S_m) = V \Delta^{j+m/2}(\|\cdot\|^{2j} S_m).$$

Using the identity

$$\Delta(\|x\|^{2j} g_m) = 4j(m+j-1+\tfrac{d}{2})\|x\|^{2j-2} g_m + \|x\|^{2j} \Delta g_m$$

for $g_m \in \mathscr{P}_m^d$, $m = 0,1,2,\ldots$ and the fact that $\Delta S_m = 0$, we see that

$$\Delta^{j+m/2}(\|\cdot\|^{2j} S_m)(x) = 4^j j! \left(\tfrac{d}{2}\right)_j \Delta^{m/2} S_m(x),$$

which is zero if $m > 0$. For $m = 0$, use the fact that $S_0(x) = 1$ and put the constants together to finish the proof. □

Recall from Section 5.2 the normalized weight function W_μ on the unit ball B^d and the notation $\mathcal{V}_n^d(W_\mu)$ for the space of polynomials orthogonal with respect to W_μ on B^d.

Lemma 7.4.2 *For $P_n \in \mathcal{V}_n^d(W_{\gamma-1/2})$, $\gamma = \gamma_\kappa$ and $n \in \mathbb{N}_0$ and $P_0(x) = 1$,*

$$c_h' \int_{S^{d-1}} V(P_n)(x)\, h_\kappa^2(x)\, \mathrm{d}\omega = \delta_{0,n}.$$

Proof Recall that an orthonormal basis for the space $\mathcal{V}_n^d(W_{\gamma-1/2})$ was given in Proposition 5.2.1; it suffices to show that

$$c_h' \int_{S^{d-1}} V\big(P_j^{(\gamma-1,n-2j+(d-2)/2)}(2\|\cdot\|^2-1) Y_{n-2j,\nu}\big)(x)\, h_\kappa^2(x)\, \mathrm{d}\omega = \delta_{0,n}$$

for $0 \le 2j \le n$ and $Y_{n-2j,\nu} \in \mathcal{H}_{n-2j}^d$. If $2j < n$, this follows from Lemma 7.4.1. For the remaining case, $2j = n$, use the fact that $P_j^{(a,b)}(-t) = (-1)^j P_j^{(b,a)}(t)$, together with

$$(-1)^j P_j^{((d-2)/2,\gamma-1)}(1-2t^2) = (-1)^j \frac{(\frac{d}{2})_j}{j!}\, {}_2F_1\left(\begin{matrix} -j, j+\lambda \\ \frac{d}{2} \end{matrix}; t^2\right),$$

and the expansion of ${}_2F_1$ to derive that

$$\begin{aligned}
& c_h' \int_{S^{d-1}} V\left(P_j^{(\gamma-1,n-2j+(d-2)/2)}(2\|\cdot\|^2-1) S_{n-2j,\beta}\right)(x)\, h_\kappa^2(x)\, \mathrm{d}\omega \\
&= (-1)^j \frac{(\frac{d}{2})_j}{j!} \sum_{i=0}^{j} \frac{(-j)_i (j+\lambda)_i}{(\frac{d}{2})_i} \int_{S^{d-1}} V(\|\cdot\|^{2j})\, h_\kappa^2\, \mathrm{d}\omega \\
&= (-1)^j \frac{(\frac{d}{2})_j}{j!}\, {}_2F_1\left(\begin{matrix} -j, j+\lambda \\ \lambda+1 \end{matrix}; 1\right) \\
&= (-1)^j \frac{(\frac{d}{2})_j (1-j)_j}{j! (\lambda+1)_j},
\end{aligned}$$

by the Chu–Vandermonde identity. This is zero if $j > 0$ or $n = 2j > 0$. For $n = 0$ use the facts that $P_0^{(a,b)}(x) = 1$ and $V1 = 1$. □

Theorem 7.4.3 *Let V be the intertwining operator. Then*

$$\int_{S^{d-1}} Vf(x) h_\kappa^2(x)\, \mathrm{d}\omega = A_\kappa \int_{B^d} f(x)(1-\|x\|^2)^{\gamma_\kappa - 1}\, \mathrm{d}x$$

for $f \in L^2(h_\kappa; S^{d-1})$ such that both integrals are finite; here $A_\kappa = w_{\gamma_\kappa}/c_h'$ and w_γ is the normalization constant of $W_{\gamma-1}$.

Proof Expand f into an orthogonal series with respect to the orthonormal basis $P^n_{j,\beta}(W_{\gamma-1/2})$ in Proposition 5.2.1. Since V is a linear operator,

$$\int_{S^{d-1}} Vf(x)\,h^2_\kappa(x)\,\mathrm{d}\omega = \sum_{n=0}^{\infty}\sum_{\beta,j} a^n_{j,\beta}(f)\int_{S^{d-1}} V\left[P^n_{j,\beta}(W_{\gamma-1/2})\right](x)\,h^2_\kappa(x)\,\mathrm{d}\omega.$$

By Lemma 7.4.2, only the constant term is nonzero; hence

$$c'_h\int_{S^{d-1}} Vf(x)\,h^2_\kappa(x)\,\mathrm{d}\omega = a^0_{0,0}(f) = w_\gamma\int_{B^d} f(x)(1-\|x\|^2)^{\gamma-1}\,\mathrm{d}x,$$

which is the stated result. □

Lemma 7.4.4 *For an integrable function* $f:\mathbb{R}\mapsto\mathbb{R}$,

$$\int_{S^{d-1}} f(\langle x,y\rangle)\,\mathrm{d}\omega(y) = \sigma_{d-2}\int_{-1}^{1} f(s\|x\|)(1-s^2)^{(d-3)/2}\,\mathrm{d}s, \qquad x\in\mathbb{R}^d.$$

Proof The left-hand side is evidently invariant under rotations, which means that it is a radial function, that is, a function that depends only on $r=\|x\|$. Hence, we may assume that $x=(r,0,\ldots,0)$, so that

$$\int_{S^{d-1}} f(\langle x,y\rangle)\,\mathrm{d}\omega(y) = \sigma_{d-2}\int_0^\pi f(r\cos\theta)(\sin\theta)^{d-2}\,\mathrm{d}\theta$$

where $y=(\cos\theta,\sin\theta\,y')$, $y'\in S^{d-2}$, which gives the stated formula. □

The following special case of Theorem 7.4.3 is of interest in itself.

Corollary 7.4.5 *Let* $g:\mathbb{R}\mapsto\mathbb{R}$ *be a function such that all the integrals below are defined. Then*

$$\int_{S^{d-1}} Vg(\langle x,\cdot\rangle)(y)\,h^2_\kappa(y)\,\mathrm{d}\omega(y) = B_\kappa\int_{-1}^{1} g(t\|x\|)(1-t^2)^{\lambda_\kappa-1}\,\mathrm{d}t,$$

where $B_\kappa=[B(\lambda_\kappa+\frac12,\frac12)c'_h]^{-1}$.

Proof Using the formulae in Lemma 7.4.4 and Theorem 7.4.3,

$$\begin{aligned} I(x) &= \int_{S^{d-1}} Vg(\langle x,\cdot\rangle)(y)h^2_\kappa(y)\,\mathrm{d}\omega = A_\kappa\int_{B^d} g(\langle x,y\rangle)(1-\|y\|^2)^{\gamma-1}\mathrm{d}y\\ &= A_\kappa\sigma_{d-2}\int_0^1 r^{d-1}\int_{S^{d-1}} g(r\langle x,y'\rangle)\,\mathrm{d}\omega(y')(1-r^2)^{\gamma-1}\mathrm{d}r\\ &= A_\kappa\sigma_{d-2}\int_0^1 r^{d-1}\int_{-1}^{1} g(sr\|x\|)(1-s^2)^{(d-3)/2}\,\mathrm{d}s(1-r^2)^{\gamma-1}\mathrm{d}r. \end{aligned}$$

Making the change of variable $s\mapsto t/r$ in the last formula and interchanging the order of the integrations, we have

$$I(x) = A_\kappa \sigma_{d-2} \int_{-1}^{1} g(t\|x\|) \int_{|t|}^{1} (r^2 - t^2)^{(d-3)/2} r (1 - r^2)^{\gamma - 1} \mathrm{d}r\, \mathrm{d}t$$
$$= A_\kappa \sigma_{d-2} \times \tfrac{1}{2} \int_0^1 u^{(d-3)/2} (1-u)^{\gamma-1} \mathrm{d}u \int_{-1}^{1} g(t\|x\|)(1-t^2)^{\gamma+(d-3)/2} \mathrm{d}t,$$

where the last step follows from setting $u = (1-r^2)/(1-t^2)$. This is the stated formula. Instead of keeping track of the constant, we can determine it by setting $g = 1$ and using the fact that $V1 = 1$. □

These formulae will be useful for studying the convergence of the h-harmonic series, as will be seen in Chapter 9. The integration formula for V in Theorem 7.4.3 also implies the following formula.

Theorem 7.4.6 *Suppose that f is a polynomial and ϕ is a function such that both the integrals below are finite. Then*

$$\int_{\mathbb{R}^d} V f(x) \phi(\|x\|) h_\kappa^2(x)\, \mathrm{d}x = A_\kappa \int_{\mathbb{R}^d} f(x) \psi(\|x\|)\, \mathrm{d}x,$$

where $\psi(t) = \int_t^\infty r(r^2 - t^2)^{\gamma_\kappa - 1} \phi(r)\, \mathrm{d}r$ and A_κ is as in Theorem 7.4.3.

Proof Write f in terms of its homogeneous components, $f = \sum_{n=0}^\infty f_n$, where $f_n \in \mathscr{P}_n^d$. Then, using polar coordinates $x = rx'$,

$$\int_{\mathbb{R}^d} V f(x) \phi(\|x\|) h_\kappa^2(x)\, \mathrm{d}x$$
$$= \sum_{n=0}^\infty \int_0^\infty r^{d-1+2\gamma+n} \phi(r)\, \mathrm{d}r \int_{S^{d-1}} V f_n(x') h_\kappa^2(x')\, \mathrm{d}\omega$$
$$= A_\kappa \sum_{n=0}^\infty \int_0^\infty r^{d-1+2\gamma+n} \phi(r)\, \mathrm{d}r \int_{B^d} f_n(x)(1 - \|x\|^2)^{\gamma-1} \mathrm{d}x$$
$$= A_\kappa \int_0^\infty r^{d-1+2\gamma} \phi(r) \int_{B^d} f(rx)(1 - \|x\|^2)^{\gamma-1} \mathrm{d}x\, \mathrm{d}r,$$

where we have used the integral formula in Theorem 7.4.3. Using the polar coordinates $x = \rho x'$ in the integral over B^d and setting $t = r\rho$, we see that the last integral is equal to

$$A_\kappa \int_0^\infty r\phi(r) \int_0^r t^{d-1} (r^2 - t^2)^{\gamma-1} \int_{S^{d-1}} f(tx')\, \mathrm{d}\omega\, \mathrm{d}t\, \mathrm{d}r$$
$$= A_\kappa \int_0^\infty t^{d-1} \int_{S^{d-1}} f(tx')\, \mathrm{d}\omega \left(\int_t^\infty r(r^2 - t^2)^{\gamma-1} \phi(r)\, \mathrm{d}r \right) \mathrm{d}t,$$

which can be seen to be the stated formula upon using the polar coordinates again. □

Corollary 7.4.7 *Let f be a polynomial. Then*

$$\int_{\mathbb{R}^d} V f(x) h_\kappa^2(x) \mathrm{e}^{-\|x\|^2/2}\, \mathrm{d}x = b_\kappa \int_{\mathbb{R}^d} f(x) \mathrm{e}^{-\|x\|^2/2}\, \mathrm{d}x,$$

where $b_\kappa = 2^{\gamma_\kappa - 1} \Gamma(\gamma_\kappa) A_\kappa$.

Proof Taking $\phi(t) = \mathrm{e}^{-t^2/2}$ in Theorem 7.4.3 we have

$$\psi(t) = \int_{t^2/2}^{\infty} (u - \tfrac{1}{2}t^2)^{\gamma-1} \mathrm{e}^{-u}\, \mathrm{d}u = 2^{\gamma-1} \mathrm{e}^{-t^2/2} \int_0^\infty v^{\gamma-1} \mathrm{e}^{-v}\, \mathrm{d}v,$$

from which the stated formula follows. □

7.5 Example: Abelian Group $\mathbb{Z}_2^d$

Here we consider h-harmonics for the abelian group $\mathbb{Z}_2^d$, the group of sign changes. The weight function is defined by

$$h_\kappa(x) = \prod_{i=1}^d |x_i|^{\kappa_i}, \qquad \kappa_i \ge 0, \quad x \in \mathbb{R}^d, \tag{7.5.1}$$

and it is invariant under the sign changes of the group $\mathbb{Z}_2^d$. This case serves as an example of the general results. Moreover, its simple structure allows us to make many results more specific than those available for a generic reflection group. This is especially true in connection with an orthogonal basis and the intertwining operator.

For the weight function h_κ defined in (7.5.1), we have

$$\gamma_\kappa = |\kappa| := \kappa_1 + \cdots + \kappa_d \qquad \text{and} \qquad \lambda_\kappa = |\kappa| + \tfrac{d-2}{2}.$$

The normalization constant $c'_{h,d}$ defined in (7.1.11) is given by

$$c'_h = \frac{\pi^{d/2}}{\Gamma(\frac{d}{2})} \frac{\Gamma(|\kappa| + \frac{d}{2})}{\Gamma(\kappa_1 + \frac{1}{2}) \cdots \Gamma(\kappa_d + \frac{1}{2})}$$

which follows from

$$\int_{S^{d-1}} |x_1|^{2\kappa_1} \cdots |x_d|^{2\kappa_d}\, \mathrm{d}\omega = 2 \frac{\Gamma(\kappa_1 + \frac{1}{2}) \cdots \Gamma(\kappa_d + \frac{1}{2})}{\Gamma(|\kappa| + \frac{d}{2})}.$$

7.5.1 Orthogonal basis for h-harmonics

Associated with h_κ^2 in (7.5.1) are the Dunkl operators, given by

$$\mathscr{D}_j f(x) = \partial_j f(x) + \kappa_j \frac{f(x) - f(x_1, \ldots, -x_j, \ldots, x_d)}{x_j} \tag{7.5.2}$$

for $1 \le j \le d$ and the h-Laplacian, given by

$$\Delta_h f(x) = \Delta f(x) + \sum_{j=1}^d \kappa_j \left(\frac{2}{x_j} \frac{\partial f}{\partial x_j} - \frac{f(x) - f(x_1, \ldots, -x_j, \ldots, x_d)}{x_j^2} \right). \tag{7.5.3}$$

We first state an orthonormal basis for $\mathscr{H}_n^d(h_\kappa^2)$ in terms of the generalized Gegenbauer polynomials $C_n^{(\lambda,\mu)}$ defined in Subsection 1.5.2. For $d = 2$, the orthonormal basis is given as follows.

Theorem 7.5.1 *Let $d = 2$ and $h_\kappa(x_1,x_2) = c'_{2,h}|x_1|^{\kappa_1}|x_2|^{\kappa_2}$. A mutually orthogonal basis for $\mathscr{H}_n^2(h_\kappa^2)$ is given by*

$$\begin{aligned} Y_n^1(x) &= r^n C_n^{(\kappa_2,\kappa_1)}(\cos\theta), \\ Y_n^2(x) &= r^n \sin\theta\, C_{n-1}^{(\kappa_2+1,\kappa_1)}(\cos\theta), \end{aligned} \tag{7.5.4}$$

where we use the polar coordinates $x = r(\cos\theta, \sin\theta)$ and set $Y_0^2(x) = 0$.

Proof Since Y_n^1 is even in x_2 and Y_m^2 is odd in x_2, we see that Y_n^1 and Y_n^2 are orthogonal. The integral of a function $f(x_1,x_2)$ can be written as

$$\int_{S^1} f(x_1,x_2)\,\mathrm{d}\omega = \int_{-\pi}^{\pi} f(\cos\theta,\sin\theta)\,\mathrm{d}\theta,$$

which can be converted to an integral over $[-1,1]$ if $f(x_1,x_2)$ is even in x_2. Clearly, $Y_n^1Y_m^1$ is even in x_2 and so is $Y_n^2Y_m^2$. The orthogonality of Y_n^1 and Y_m^2 follows from that of $C_n^{(\kappa_2,\kappa_1)}(t)$ and $C_n^{(\kappa_2+1,\kappa_1)}(t)$, respectively. □

To state a basis for $d \ge 3$, we use the spherical coordinates in (4.1.1) and the notation in Section 5.2. Associated with $\kappa = (\kappa_1,\ldots,\kappa_d)$, define

$$\kappa^j = (\kappa_j,\ldots,\kappa_d), \qquad 1 \le j \le d.$$

Since κ^d consists of only the last element of κ, write $\kappa^d = \kappa_d$. Define α^j for $\alpha \in \mathbb{N}_0^{d-1}$ similarly.

Theorem 7.5.2 *For $d > 2$ and $\alpha \in \mathbb{N}_0^d$, define*

$$Y_\alpha(x) := (h_\alpha)^{-1} r^{|\alpha|} g_\alpha(\theta_1) \prod_{j=1}^{d-2} (\sin\theta_{d-j})^{|\alpha^{j+1}|} C_{\alpha_j}^{(\lambda_j,\kappa_j)}(\cos\theta_{d-j}); \tag{7.5.5}$$

$g_\alpha(\theta)$ equals $C_{\alpha_{d-1}}^{(\kappa_d,\kappa_{d-1})}(\cos\theta)$ for $\alpha_d = 0$ and $\sin\theta\, C_{\alpha_{d-1}-1}^{(\kappa_d+1,\kappa_{d-1})}(\cos\theta)$ for $\alpha_d = 1$; $|\alpha^j| = \alpha_j + \cdots + \alpha_d$, $|\kappa^j| = \kappa_j + \cdots + \kappa_d$, $\lambda_j = |\alpha^{j+1}| + |\kappa^{j+1}| + \frac{d-j-1}{2}$ and

$$[h_\alpha^n]^2 = \frac{a_\alpha}{(|\kappa| + \frac{d}{2})_n} \prod_{j=1}^{d-1} h_{\alpha_j}^{(\lambda_j,\kappa_j)} (\kappa_j + \lambda_j)_{\alpha_j}, \qquad a_\alpha = \begin{cases} 1 & \text{if } \alpha_d = 0, \\ \kappa_d + \frac{1}{2} & \text{if } \alpha_d = 1. \end{cases}$$

Here $h_n^{(\lambda,\mu)}$ denotes the normalization constant of $C_n^{(\lambda,\mu)}$. Then $\{Y_\alpha : |\alpha| = n, \alpha_d = 0,1\}$ is an orthonormal basis of $\mathscr{H}_n^d(h_\kappa^2)$.

Proof These formulae could be verified directly by the use of the spherical coordinates. We choose, however, to give a different proof, making use of Δ_h. We start with the following decomposition of $\mathscr{P}_n^d$:

$$\mathscr{P}_n^d = \sum_{l=0}^{n} x_1^{n-l} \mathscr{H}_l^{d-1}(h_{\kappa'}^2) + r^2 \mathscr{P}_{n-2}^d,$$

where $\kappa' = (\kappa_2, \ldots, \kappa_d)$. This follows from the fact that $f \in \mathscr{P}_n^d$ can be written as

$$f(x) = \|x\|^2 F(x) + x_1\phi_1(x') + \phi_2(x'), \qquad x = (x_1, x'),$$

where $F \in \mathscr{P}_{n-2}^d$, $\phi_1 \in \mathscr{P}_{n-1}^d$ and $\phi_2 \in \mathscr{P}_n^d$. We then apply the canonical decomposition in Theorem 7.1.7 to ϕ_i and collect terms according to the power of x_1, after replacing $\|x'\|^2$ by $\|x\|^2 - x_1^2$. For any $p \in \mathscr{H}_n^d(h_\kappa^2)$, the decomposition allows us to write

$$p(x) = \sum_{l=0}^{n} x_1^{n-l} p_l(x') + \|x\|^2 q(x), \qquad q \in \mathscr{P}_{n-2}^d.$$

Therefore, using the formula for the harmonic projection operator $\operatorname{proj}_{n,h}$ in (7.1.6), it follows that

$$p(x) = \sum_{l=0}^{n} \operatorname{proj}_{n,h}[x_1^{n-l} p_l(x')] = \sum_{l=0}^{n} \sum_{j=0}^{[n/2]} \frac{\|x\|^{2j} \Delta_h^j [x_1^{n-l} p_l(x')]}{4^j j! (-\lambda_\kappa - n + 1)_j}.$$

Write $\Delta_{h,d}$ for Δ_h to indicate the dependence on d. Since our h_κ is a product of mutually orthogonal factors, it follows from the definition of Δ_h that $\Delta_{h,d} = \mathscr{D}_1^2 + \Delta_{h,d-1}$, where $\Delta_{h,d-1}$ acts with respect to $x' = (x_2, \ldots, x_d)$. Therefore, since $p_l \in \mathscr{H}_l^{d-1}(h_{\kappa'}^2)$, we have that

$$\Delta_{h,d}[x_1^{n-l} p_l(x')] = \mathscr{D}_1^2(x_1^{n-l}) p_l(x').$$

By the definition of $\mathscr{D}_1$, see (7.5.2), it follows easily that

$$\mathscr{D}_1^2(x_1^m) = \big(m + [1-(-1)^m]\kappa_1\big)\big(m - 1 + [1-(-1)^{m-1}]\kappa_1\big) x_1^{m-2}.$$

Using this formula repeatedly, for $m = n - l$ even we obtain

$$\Delta_{h,d}^j (x_1^{n-l} p_l(x')) = 2^{2j} \left(-\tfrac{n-l}{2}\right)_j \left(-\tfrac{n-l-1}{2} - \kappa_1\right)_j x_1^{n-l-2j} p_l(x').$$

Therefore, for $n - l$ even,

$$\begin{aligned}
&\operatorname{proj}_{n,h}(x_1^{n-l} p_l(x')) \\
&\quad = \sum_{j=0}^{[n-l/2]} \frac{(-\frac{n-l}{2})_j (-\frac{n-l-1}{2} - \kappa_1)_j}{(-n - \lambda_\kappa + 1)_j j!} \|x\|^{2j} x_1^{n-l-2j} p_l(x') \\
&\quad = p_l(x') x_1^{n-l} {}_2F_1\left(\begin{matrix} -\frac{n-l}{2}, -\frac{n-l-1}{2} - \kappa_1 \\ -n - \lambda_\kappa + 1 \end{matrix}; \frac{\|x\|^2}{x_1^2}\right) \\
&\quad = \binom{n + |\kappa| - 2 + \frac{d}{2}}{\frac{n-l}{2}}^{-1} p_l(x') \|x\|^{n-l} P_{(n-l)/2}^{(l+\lambda_\kappa - \kappa_1 - 1/2, \kappa_1 - 1/2)}\left(2\frac{x_1^2}{\|x\|^2} - 1\right).
\end{aligned}$$

A similar equation holds for $n - l$ odd. By the definition of the generalized Gegenbauer polynomials,

$$\operatorname{proj}_{n,h}[x_1^{n-l} p_l(x')] = c p_l(x') \|x\|^{n-l} C_{n-l}^{(l+|\kappa| - \kappa_1 + (d-2)/2, \kappa_1)}(\cos\theta_1),$$

where c is a constant; since $p_l \in \mathscr{H}_l^d(h_{\kappa'}^2)$ it admits a similar decomposition to F_ℓ. This process can be continued until we reach the case of h-harmonic polynomials of two variables, x_1 and x_2, which can be written as linear combinations of the spherical h-harmonics in (7.5.4). Therefore, taking into account that in spherical coordinates $x_j/r_j = \cos\theta_{d-j}$ and $r_{j+1}/r_j = \sin\theta_{j-1}$ where $r_j^2 = x_j^2 + \cdots + x_d^2$, we conclude that any polynomial in $\mathscr{H}_n^d(h_\kappa^2)$ can be uniquely presented as a linear combination of functions of the form $\|x\|^n Y_\alpha$. The value of h_α is determined by

$$h_\alpha^2 = c_h' \int_{S^{d-1}} \left(g_\alpha(\theta_1) \prod_{j=1}^{d-2} (\sin\theta_{d-j})^{|\alpha^{j+1}|} C_{\alpha_j}^{(\lambda_j,\kappa_j)}(\cos\theta_{d-j}) \right)^2 h_\kappa^2(x)\,\mathrm{d}\omega,$$

where the integrand is given in spherical coordinates; this allows us to convert into a product of integrals, which can be evaluated by the L^2 norm of $C_n^{(\lambda,\mu)}$ in Subsection 1.5.2 and gives a formula for the constant h_α. □

Let us look at the case $d = 2$ again. According to Theorem 7.5.1, both the functions $r^n C_n^{(\kappa_1,\kappa_2)}(\cos\theta)$ and $r^n \sin\theta\, C_{n-1}^{(\kappa_1+1,\kappa_2)}(\cos\theta)$ are h-harmonic polynomials. Hence, they satisfy the same second order differential–difference equation, $\Delta_h f(x) = 0$. Furthermore, since $r^n C_{2n}^{(\kappa_1,\kappa_2)}(\cos\theta)$ is even in both x_1 and x_2 it is invariant under $\mathbb{Z}_2^2$; it follows that the equation when applied to this function becomes a differential equation,

$$\Delta f + \frac{2\kappa_1}{x_1}\frac{\partial f}{\partial x_1} + \frac{2\kappa_2}{x_2}\frac{\partial f}{\partial x_2} = 0.$$

Changing variables to polar coordinates by setting $x_1 = r\cos\theta$ and $x_2 = r\sin\theta$ leads to

$$\frac{\partial f}{\partial x_1} = \cos\theta\frac{\partial f}{\partial r} - \frac{\sin\theta}{r}\frac{\partial f}{\partial \theta}, \qquad \frac{\partial f}{\partial x_1} = \sin\theta\frac{\partial f}{\partial r} + \frac{\cos\theta}{r}\frac{\partial f}{\partial \theta},$$

$$\Delta f = \frac{1}{r^2}\frac{\partial^2 f}{\partial \theta^2} + \frac{1}{r}\frac{\partial}{\partial r}\left(r\frac{\partial f}{\partial r}\right),$$

from which we conclude that the differential equation takes the form

$$\frac{1}{r^2}\frac{\partial^2 f}{\partial \theta^2} + \frac{1}{r}\frac{\partial}{\partial r}\left(r\frac{\partial f}{\partial r}\right) + \frac{2(\kappa_1+\kappa_2)}{r}\frac{\partial f}{\partial r} + \frac{2}{r^2}\left(\kappa_2\frac{\cos\theta}{\sin\theta} - \kappa_1\frac{\sin\theta}{\cos\theta}\right)\frac{\partial f}{\partial \theta} = 0.$$

Consequently, applying this equation to $f = r^{2n} g(\cos\theta)$ we end up with a differential equation satisfied by the polynomial $C_{2n}^{(\kappa_1,\kappa_2)}(\cos\theta)$, which is a Jacobi polynomial by definition. We summarize the result as the following.

Proposition 7.5.3 *The Jacobi polynomial* $P_n^{(\kappa_1-1/2,\kappa_2-1/2)}(\cos 2\theta)$ *satisfies the differential equation*

$$\frac{\mathrm{d}^2 g}{\mathrm{d}\theta^2} + 2\left(\kappa_2\frac{\cos\theta}{\sin\theta} - \kappa_1\frac{\sin\theta}{\cos\theta}\right)\frac{\mathrm{d}g}{\mathrm{d}\theta} - 4n(n+\kappa_1+\kappa_2)g = 0.$$

The idea of deriving a differential equation in this manner will be used in the case of several variables in Section 8.1, where the differential equations satisfied by the orthogonal polynomials on the ball will be derived from the eigenequations of the h-Laplace–Beltrami operator.

7.5.2 Intertwining and projection operators

For the weight function h_κ invariant under $\mathbb{Z}_2^d$ in (7.5.1), the intertwining operator V enjoys an explicit formula, given below:

Theorem 7.5.4 *For the weight function h_κ associated with $\mathbb{Z}_2^d$,*

$$Vf(x) = \int_{[-1,1]^d} f(x_1t_1,\dots,x_dt_d)\prod_{i=1}^d c_{\kappa_i}(1+t_i)(1-t_i^2)^{\kappa_i-1}\mathrm{d}t,$$

where $c_\lambda = [B(\frac{1}{2},\lambda)]^{-1}$. If any $\kappa_i = 0$, the formula holds under the limit (1.5.1).

Proof The facts that $V1 = 1$ and that $V : \mathscr{P}_n^d \mapsto \mathscr{P}_n^d$ are evident from the definition of V in Section 6.5. Thus, we need only verify that $\mathscr{D}_iV = V\partial_i$. From the definition of $\mathscr{D}_i$, write

$$\mathscr{D}_if = \partial_if + \widetilde{\mathscr{D}}_if, \qquad \widetilde{\mathscr{D}}_if(x) = \kappa_i\frac{f(x)-f(x-2x_i\varepsilon_i)}{x_i}.$$

Taking $i=1$, for example, we consider

$$\widetilde{\mathscr{D}}_1Vf(x) = \kappa_1\int_{[-1,1]^d}\frac{f(x_1t_1,\dots,x_dt_d)-f(-x_1t_1,x_2t_2,\dots,x_dt_d)}{x_1} \times\prod_{i=1}^d(1+t_i)\prod_{i=1}^d c_{\kappa_i}(1-t_i^2)^{\kappa_i-1}\mathrm{d}t.$$

Since the difference in the integral is an odd function of t_1, it follows from integration by parts that

$$\widetilde{\mathscr{D}}_1Vf(x) = \frac{2\kappa_1}{x_1}\int_{[-1,1]^d} f(x_1t_1,\dots,x_dt_d)t_1\prod_{i=2}^d(1+t_i)\prod_{i=1}^d c_{\kappa_i}(1-t_i^2)^{\kappa_i-1}\mathrm{d}t$$
$$= \int_{[-1,1]^d}\partial_1f(x_1t_1,\dots,x_dt_d)(1-t_1)\prod_{i=1}^d(1+t_i)\prod_{i=1}^d c_{\kappa_i}(1-t_i^2)^{\kappa_i-1}\mathrm{d}t.$$

Furthermore, directly from the definition of V, we have

$$\partial_1Vf(x) = \int_{[-1,1]^d}\partial_1f(x_1t_1,\dots,x_dt_d)t_1\prod_{i=1}^d(1+t_i)\prod_{i=1}^d c_{\kappa_i}(1-t_i^2)^{\kappa_i-1}\mathrm{d}t.$$

The stated result follows from adding the last two equations together. □

As an immediate consequence of the explicit formula for V in Theorem 7.5.4 and of Corollary 7.3.2, we obtain an integral formula for the reproducing kernel of $\mathscr{H}_n^d(h_\kappa^2)$ when h_κ is invariant under $\mathbb{Z}_2^d$.

Theorem 7.5.5 *For $h_\kappa^2 \mathrm{d}\omega$ on S^{d-1} and $\lambda_\kappa = |\kappa| + \frac{d-2}{2}$,*

$$P_n(h_\kappa^2; x, y) = \frac{n+\lambda_\kappa}{\lambda_\kappa} \int_{[-1,1]^d} C_n^{\lambda_\kappa}(x_1 y_1 t_1 + \cdots + x_d y_d t_d) \times \prod_{i=1}^d (1+t_i) \prod_{i=1}^d c_{\kappa_i}(1-t_i^2)^{\kappa_i - 1} \mathrm{d}t.$$

In the case $d = 2$, the dimension of $\mathscr{H}_n^2(h_\kappa^2)$ is 2 and the orthonormal basis is that given in (7.5.4). The theorem when restricted to polynomials that are even in x_2 gives

Corollary 7.5.6 *For $\lambda, \mu \geq 0$,*

$$C_n^{(\lambda,\mu)}(\cos\theta) C_n^{(\lambda,\mu)}(\cos\phi) = \frac{n+\lambda+\mu}{\lambda+\mu} C_n^{(\lambda,\mu)}(1) c_\lambda c_\mu \int_{-1}^1 \int_{-1}^1 C_n^{\lambda+\mu}(t\cos\theta\cos\phi + s\sin\theta\sin\phi)(1+t) \times (1-t^2)^{\mu-1}(1-s^2)^{\lambda-1} \mathrm{d}t\, \mathrm{d}s.$$

In particular, as $\mu \to 0$ the above equation becomes the classical product formula for the Gegenbauer polynomials,

$$\frac{C_n^\lambda(x) C_n^\lambda(y)}{C_n^\lambda(1)} = c_\lambda \int_{-1}^1 C_n^\lambda(xy + s\sqrt{1-x^2}\sqrt{1-y^2})(1-s^2)^{\lambda-1} \mathrm{d}s.$$

As another consequence of the corollary, set $\theta = 0$ in the formula to conclude that

$$C_n^{(\lambda,\mu)}(x) = \frac{n+\lambda+\mu}{\lambda+\mu} c_\mu \int_{-1}^1 C_n^{\lambda+\mu}(xt)(1+t)(1-t^2)^{\mu-1} \mathrm{d}t.$$

Comparing with the formulae in Subsection 1.5.2, we see that the above expression is the same as $VC_n^{\lambda+\mu}(x) = C_n^{(\lambda,\mu)}(x)$, and this explains why we denoted the integral transform in Subsection 1.5.2 by V.

Many identities and transforms for Jacobi and Gegenbauer polynomials can be derived from these examples. Let us mention just one more. By Proposition 7.2.8 and Corollary 7.2.9,

$$r^n C_n^{(\kappa_1,\kappa_2)}(\cos\theta) = \text{const} \times V\left[(x_1^2 + x_2^2)^n T_n\left(x_1 / \sqrt{x_1^2 + x_2^2}\right)\right],$$

since V maps ordinary harmonics to h-harmonics; here T_n is a Chebyshev polynomial of the first kind. Using the formula for V, the above equation becomes

$$C_n^{(\kappa_1,\kappa_2)}(\cos\theta)$$
$$= \text{const} \times \int_{-1}^{1}\int_{-1}^{1}(t_1^2\sin^2\theta + t_2^2\cos^2\theta)^{n/2}$$
$$\times T_n\left(\frac{t_2\cos\theta}{(t_2^2\cos^2\theta + t_1^2\sin^2\theta)^{1/2}}\right)(1+t_2)(1-t_1^2)^{\kappa_1-1}(1-t_2^2)^{\kappa_2-1}dt_1\,dt_2,$$

where the constant can be determined by setting $\theta = 0$. In particular, for $\kappa_2 = 0$ this formula is a special case of an integral due to Feldheim and Vilenkin (see, for example, Askey [1975, p. 24]).

The integral formula for V in Theorem 7.5.4 also gives an explicit formula for the Poisson kernel, which can be used to study the Abel means defined by

$$S(f,rx') = c_h' \int_{S^{d-1}} f(y)P(h_\kappa^2; rx', y)h_\kappa^2(y)\,d\omega(y), \qquad x = rx', \quad x' \in S^{d-1},$$

where $r < 1$ for integrable functions f on S^{d-1}. We have the following theorem.

Theorem 7.5.7 *If f is continuous on S^{d-1} then*

$$\lim_{r\to 1-} S(f,rx') = f(x'), \qquad x = rx', \quad x' \in S^{d-1}.$$

Proof Let $A_\delta = \{y \in S^{d-1} : \|x'-y\| < \delta\}$ in this proof. Since $P(h_\kappa^2) \ge 0$ and its integral over S^{d-1} is 1,

$$|S(f,rx') - f(x')| = c_h' \int_{S^{d-1}} [f(x) - f(y)]P(h_\kappa^2; rx', y)h_\kappa^2(y)\,d\omega(y)$$
$$\le c_h'\left(\int_{A_\delta} + \int_{S^{d-1}\setminus A_\delta}\right)|f(x) - f(y)|P(h_\kappa^2; rx', y)h_\kappa^2(y)\,d\omega(y)$$
$$\le \sup_{\|x'-y\|\le\delta} |f(rx) - f(y)| + 2\|f\|_\infty c_h' \int_{S^{d-1}\setminus A_\delta} P^h(h_\kappa^2; rx', y)h_\kappa^2(y)\,d\omega(y),$$

where $\|f\|_\infty$ is the maximum of f over S^{d-1}. Since f is continuous over S^{d-1}, we need to prove only that the last integral converges to zero when $r \to 1$. By the explicit formula for $P(h_\kappa^2)$ in Theorem 7.3.3, it suffices to show that

$$\limsup_{r\to 1} \int_{S^{d-1}} \int_{[0,1]^d} \frac{1}{[1 - 2r(x_1'y_1t_1 + \cdots + x_d'y_dt_d) + r^2]^{|\kappa|+d/2}}$$
$$\times \prod_{i=1}^{d} c_{\kappa_i}(1-t_i^2)^{\kappa_i-1}\,dt\,h_\kappa^2(y)\,dy$$

is finite for every $x' \in S^{d-1}$. Let $B_\sigma = \{t = (t_1,\ldots,t_d) : t_i > 1-\sigma, 1 \le i \le d\}$ with $\sigma = \frac{1}{4}\delta^2$. For $y \in S^{d-1} \setminus A_\delta$ and $t \in B_\sigma$, use the fact that $\langle x', y\rangle = 1 - \frac{1}{2}\|x'-y\|^2$ to get

$$|x_1'y_1t_1 + \cdots + x_d'y_dt_d| = |\langle x', y\rangle - x_1'y_1(1-t_1) - \cdots - x_d'y_d(1-t_d)|$$
$$\le 1 - \tfrac{1}{2}\delta^2 + \max_i(1-t_i) \le 1 - \tfrac{1}{4}\delta^2,$$

from which it follows readily that

$$1-2r(x_1'y_1t_1+\cdots+x_d'y_dt_d)+r^2 \geq 1+r^2-2r\left(1-\tfrac{1}{4}\delta^2\right)=(1-r)^2+\tfrac{1}{2}r\delta^2 \geq \tfrac{1}{4}\delta^2.$$

For $t \in [0,1]^d \setminus B_\sigma$, we have that $t_i \leq 1-\sigma$ for at least one i. Assume that $t_1 \leq 1-\sigma$. We can assume also that $x_1' \neq 0$, since otherwise t_1 does not appear in the integral and we can repeat the above argument for $(t_2,\ldots,t_d) \in [0,1]^{d-1}$. It follows that

$$\begin{aligned}1-2r(x_1'y_1t_1+\cdots+x_d'y_dt_d)+r^2 &= r^2x_1'^2(1-t_1^2)+\sum_{i=1}^d (y_i-rx_i't_i)^2\\ &\quad + r^2\left(1-x_1'^2-\sum_{i=2}^d x_i'^2t_i^2\right)\\ &\geq r^2x_1'^2(1-t_1^2) \geq r^2\sigma x_1'^2 > 0.\end{aligned}$$

Therefore, for each $x = rx'$ the denominator of the integrand is nonzero and the expression is finite as $r \to 1$. □

7.5.3 Monic orthogonal basis

Our definition of the monomial orthogonal polynomials is an analogue of that for the generating function of the Gegenbauer polynomials, and it uses the intertwining operator V.

Definition 7.5.8 Define polynomials $\widetilde{R}_\alpha(x)$ by

$$V\left(\frac{1}{(1-2\langle b,\cdot\rangle+\|b\|^2\|x\|^2)^{\lambda_\kappa}}\right)(x) = \sum_{\alpha\in\mathbb{N}_0^d} b^\alpha \widetilde{R}_\alpha(x), \qquad x \in \mathbb{R}^d.$$

Let F_B be the Lauricella hypergeometric series of type B defined in Section 1.2. Write $\mathbf{1} = (1,\ldots,1)$. We derive the properties of $\widetilde{R}_\alpha$ in what follows.

Proposition 7.5.9 *The polynomials $\widetilde{R}_\alpha$ satisfy the following properties:*

(i) $\widetilde{R}_\alpha \in \mathscr{P}_n^d$ *and*

$$\widetilde{R}_\alpha(x) = \frac{2^{|\alpha|}(\lambda_\kappa)_{|\alpha|}}{\alpha!}\sum_\gamma \frac{(-\frac{\alpha}{2})_\gamma(-\frac{\alpha+\mathbf{1}}{2})_\gamma}{(-|\alpha|-\lambda_\kappa+1)_{|\gamma|}\gamma!}\|x\|^{2|\gamma|}V_\kappa(x^{\alpha-2\gamma}),$$

where the series terminates, as the summation is over all γ such that $\gamma \leq \frac{\alpha}{2}$;

(ii) $\widetilde{R}_\alpha \in \mathscr{H}_n^d(h_\kappa^2)$ *and*

$$\widetilde{R}_\alpha(x) = \frac{2^{|\alpha|}(\lambda_\kappa)_{|\alpha|}}{\alpha!}V_\kappa[S_\alpha(\cdot)](x) \qquad \textit{for } \|x\| = 1,$$

where

$$S_\alpha(y) = y^\alpha F_B\left(-\tfrac{\alpha}{2}, \tfrac{-\alpha+\mathbf{1}}{2}; -|\alpha| - \lambda_\kappa + 1; \frac{1}{y_1^2}, \ldots, \frac{1}{y_d^2}\right).$$

Furthermore,

$$\sum_{|\alpha|=n} b^\alpha \widetilde{R}_\alpha(x) = \frac{\lambda_\kappa}{n+\lambda_\kappa} P_n(h_\kappa^2; b, x), \qquad \|x\| = 1,$$

where $P_n(h_\kappa^2; y, x)$ is the reproducing kernel of $\mathcal{H}_n^d(h_\kappa^2)$.

Proof Using the multinomial and binomial formulae, we write

$$\begin{aligned}
(1 - 2\langle a, y\rangle + \|a\|^2)^{-\lambda} &= [1 - a_1(2y_1 - a_1) - \cdots - a_d(2y_d - a_d)]^{-\lambda} \\
&= \sum_\beta \frac{(\lambda)_{|\beta|}}{\beta!} a^\beta (2y_1 - a_1)^{\beta_1} \cdots (2y_d - a_d)^{\beta_d} \\
&= \sum_\beta \frac{(\lambda)_{|\beta|}}{\beta!} \sum_\gamma \frac{(-\beta_1)_{\gamma_1} \cdots (-\beta_d)_{\gamma_d}}{\gamma!} \times 2^{|\beta| - |\gamma|} y^{\beta-\gamma} a^{\gamma+\beta}.
\end{aligned}$$

Changing summation indices according to $\beta_i + \gamma_i = \alpha_i$ and using the expressions

$$(\lambda)_{m-k} = \frac{(-1)^k (\lambda)_m}{(1-\lambda-m)_k} \qquad \text{and} \qquad \frac{(-m+k)_k}{(m-k)!} = \frac{(-1)^k(-m)_{2k}}{m!}$$

as well as $2^{-2k}(-m)_{2k} = (-\frac{m}{2})_k(\frac{1}{2}(1-m))_k$, we can rewrite the formula as

$$\begin{aligned}
&(1 - 2\langle a, y\rangle + \|a\|^2)^{-\lambda} \\
&\quad = \sum_\alpha a^\alpha \frac{2^{|\alpha|}(\lambda)_{|\alpha|}}{\alpha!} \sum_\gamma \frac{(-\frac{\alpha}{2})_\gamma (\frac{-\alpha+\mathbf{1}}{2})_\gamma}{(-|\alpha| - \lambda + 1)_{|\gamma|} \gamma!} y^{\alpha - 2\gamma} \\
&\quad = \sum_\alpha a^\alpha \frac{2^{|\alpha|}(\lambda)_{|\alpha|}}{\alpha!} y^\alpha F_B\left(-\tfrac{\alpha}{2}, \tfrac{\mathbf{1}-\alpha}{2}; -|\alpha| - \lambda + 1; \frac{1}{y_1^2}, \ldots, \frac{1}{y_d^2}\right).
\end{aligned}$$

Using the left-hand side of the first equation of this proof with the function

$$(1 - 2\langle b, y\rangle + \|x\|^2\|b\|^2)^{-\lambda} = (1 - 2\langle \|x\|b, y/\|x\|\rangle + \big\|\|x\|b\big\|^2)^{-\lambda}$$

and applying V with respect to y gives the expression for $\widetilde{R}_\alpha$ in (i). If $\|x\| = 1$ then the second equation gives the expression for $\widetilde{R}_\alpha$ in (ii). We still need to show that $\widetilde{R}_\alpha \in \mathcal{H}_n^d(h_\kappa^2)$. Let $\|x\| = 1$. For $\|y\| \le 1$ the generating function of the Gegenbauer polynomials gives

$$(1 - 2\langle b, y\rangle + \|b\|^2)^{-\lambda} = \sum_{n=0}^\infty \|b\|^n C_n^\lambda(\langle b/\|b\|, y\rangle).$$

Applying V_κ to y in the above equation gives

$$\sum_{|\alpha|=n} b^\alpha \widetilde{R}_\alpha(x) = \|b\|^n V_\kappa[C_n^\lambda(\langle b/\|b\|, \cdot\rangle)](x), \qquad \|x\| = 1.$$

Using Corollary 7.3.2 we can see that $\sum_{|\alpha|=n} b^\alpha \widetilde{R}_\alpha(x)$ is a constant multiple of $P_n(h_\kappa^2;x,b)$. Consequently, for any b, $\sum b^\alpha \widetilde{R}_\alpha(x)$ is an element in $\mathscr{H}_n^d(h_\kappa^2)$; therefore, so is $\widetilde{R}_\alpha$. □

In the following let $\lfloor \frac{\alpha}{2} \rfloor$ denote $(\lfloor \frac{\alpha_1}{2} \rfloor, \dots, \lfloor \frac{\alpha_d}{2} \rfloor)$ for $\alpha \in \mathbb{N}_0^d$.

Proposition 7.5.10 *For $\alpha \in \mathbb{N}_0^d$, let $\beta := \alpha - \lfloor \frac{\alpha+1}{2} \rfloor$. Then*

$$\widetilde{R}_\alpha(x) = \frac{2^{|\alpha|}(\lambda_\kappa)_{|\alpha|}}{\alpha!} \frac{(\frac{1}{2})_{\alpha-\beta}}{(\kappa+\frac{1}{2})_{\alpha-\beta}} R_\alpha(x),$$

where R_α is given by

$$R_\alpha(x) = x^\alpha F_B\left(-\beta, -\alpha+\beta-\kappa+\tfrac{1}{2}; -|\alpha|-\lambda_\kappa+1; \frac{\|x\|^2}{x_1^2}, \dots, \frac{\|x\|^2}{x_d^2}\right).$$

In particular, $R_\alpha(x) = x^\alpha + \|x\|^2 Q_\alpha(x)$ for $Q_\alpha \in \mathscr{P}_{n-2}^d$.

Proof By considering m even and m odd separately, it is easy to verify that

$$c_\kappa \int_{-1}^1 t^{m-2k}(1+t)(1-t^2)^{\kappa-1}\,\mathrm{d}t = \frac{(\frac{1}{2})_{\lfloor (m+1)/2 \rfloor}}{(\kappa+\frac{1}{2})_{\lfloor (m+1)/2 \rfloor}} \frac{(-\lfloor \frac{m+1}{2} \rfloor - \kappa + \frac{1}{2})_k}{(-\lfloor \frac{m+1}{2} \rfloor + \frac{1}{2})_k}$$

for $\kappa \ge 0$. Hence, again using the explicit formula for V, the formula for $\widetilde{R}_\alpha$ in part (i) of Proposition 7.5.9 becomes

$$\widetilde{R}_\alpha(x) = \frac{2^{|\alpha|}(\lambda)_{|\alpha|}}{\alpha!} \frac{(\frac{1}{2})_{\lfloor(\alpha+1)/2\rfloor}}{(\kappa+\frac{1}{2})_{\lfloor(\alpha+1)/2\rfloor}} \times \sum_\gamma \frac{(-\frac{\alpha}{2})_\gamma (\frac{-\alpha+1}{2})_\gamma}{(-|\alpha|-\lambda+1)_{|\gamma|}\gamma!} \frac{(-\lfloor \frac{\alpha+1}{2} \rfloor - \kappa + \frac{1}{2})_\gamma}{(-\lfloor \frac{\alpha+1}{2} \rfloor + \frac{1}{2})_\gamma} \|x\|^{2|\gamma|} x^{\alpha-2\gamma}.$$

Using the fact that

$$\left(-\tfrac{\alpha}{2}\right)_\gamma \left(\tfrac{-\alpha+1}{2}\right)_\gamma = \left(-\alpha + \left\lfloor \tfrac{\alpha+1}{2} \right\rfloor\right)_\gamma \left(-\left\lfloor \tfrac{\alpha+1}{2} \right\rfloor + \tfrac{1}{2}\right)_\gamma,$$

the above expression for $\widetilde{R}_\alpha$ can be written in terms of F_B as stated in the proposition. Note that the function F_B is a finite series, since $(-n)_m = 0$ if $m > n$, from which the last assertion of the proposition follows. □

If all components of α are even then we can write R_α using a Lauricella function of type A, denoted by $F_A(c, \alpha; \beta; x)$ in Section 1.2.

Proposition 7.5.11 *Let $\beta \in \mathbb{N}_0^d$. Then*

$$R_{2\beta}(x) = (-1)^{|\beta|} \frac{(\kappa+\frac{1}{2})_\beta}{(n+\lambda_\kappa)_{|\beta|}} \|x\|^{2|\beta|}$$
$$\times F_A\left(-\beta, |\beta|+\lambda_\kappa; \kappa+\tfrac{1}{2}; \frac{x_1^2}{\|x\|^2}, \ldots, \frac{x_d^2}{\|x\|^2}\right).$$

Proof For $\alpha = 2\beta$ the formula in terms of F_B becomes

$$R_{2\beta}(x) = \sum_{\gamma \le \beta} \frac{(-\beta)_\gamma(-\beta-\kappa+\frac{1}{2})_\gamma}{(-2|\beta|-\rho+1)_{|\gamma|}\gamma!} \|x\|^{2|\gamma|} x^{2\beta-2\gamma},$$

where $\gamma \le \beta$ means that $\gamma_1 < \beta_1, \ldots, \gamma_{d+1} < \beta_{d+1}$; note that $(-\beta)_\gamma = 0$ if $\gamma > \beta$. Changing the summation index according to $\gamma_i \mapsto \beta_i - \gamma_i$ and using the formula $(a)_{n-m} = (-1)^m (a)_n/(1-n-a)_m$ to rewrite the Pochhammer symbols, so that we have, for example,

$$(\kappa+\tfrac{1}{2})_{\beta-\gamma} = \frac{(-1)^\gamma(\kappa+\frac{1}{2})_\beta}{(-\beta-\kappa+\frac{1}{2})_\gamma}, \qquad (\beta-\gamma)! = (\mathbf{1})_{\beta-\gamma} = \frac{(-1)^{|\gamma|}\beta!}{(-\beta)_\gamma},$$

we can rewrite the summation as the stated formula in F_A. □

The restriction of R_α on the sphere is monic: $R_\alpha(x) = x^\alpha + Q_\alpha(x)$. Recall the h-harmonics H_α defined in Definition 7.1.11.

Proposition 7.5.12 *The polynomials R_α satisfy the following:*

(i) $R_\alpha(x) = \mathrm{proj}_{n,h} x^\alpha$, $n = |\alpha|$ *and* $R_\alpha(x) = \dfrac{(-1)^n}{2^n(\lambda_\kappa)_n} H_\alpha(x)$;

(ii) $\|x\|^2 \mathscr{D}_i R_\alpha(x) = -2(n+\lambda_\kappa)\,[R_{\alpha+e_i}(x) - x_i R_\alpha(x)]$;

(iii) $\{R_\alpha : |\alpha| = n, \alpha_d = 0, 1\}$ *is a basis of* $\mathscr{H}_n^d(h_\kappa^2)$.

Proof Since $R_\alpha \in \mathscr{H}_n^d(h_\kappa^2)$ and $R_\alpha(x) = x^\alpha - \|x\|^2 Q(x)$, where $Q \in \mathscr{P}_{n-2}^d$, it follows that $R_\alpha(x) = \mathrm{proj}_n x^\alpha$. The relation to H_α follows from Definition 7.1.14. The second and third assertions are reformulations of the properties of H_α. □

In the following we compute the L^2 norm of R_α. Let

$$\|f\|_{2,\kappa} := \left(c_h' \int_{S^{d-1}} |R_\alpha(x)|^2 \, d\omega\right)^{1/2}.$$

Theorem 7.5.13 *For $\alpha \in \mathbb{N}_0^d$, let $\beta = \alpha - \lfloor \alpha + \frac{1}{2} \rfloor$. Then*

$$\|R_\alpha\|_{2,\kappa}^2 = \frac{\rho\,(\kappa+\frac{1}{2})_\alpha}{(\lambda_\kappa)_{|\alpha|}} \sum_\gamma \frac{(-\beta)_\gamma\,(-\alpha+\beta-\kappa+\frac{1}{2})_\gamma}{(-\alpha-\kappa+\frac{1}{2})_\gamma \gamma!(|\alpha|-|\gamma|+\lambda_\kappa)}$$
$$= 2\lambda_\kappa \frac{\beta!\,(\kappa+\frac{1}{2})_{\alpha-\beta}}{(\lambda_\kappa)_{|\alpha|}} \int_0^1 \prod_{i=1}^d C_{\alpha_i}^{(1/2,\kappa_i)}(t)\, t^{|\alpha|+2\lambda_\kappa-1}\, dt. \qquad (7.5.6)$$

Proof Using a beta-type integral,

$$c_h' \int_{S^d} x^{2\sigma} h_\kappa^2(x)\, d\omega = \frac{\Gamma(\lambda_\kappa+1}{\Gamma(|\sigma|+\lambda_\kappa+1)} \prod_{i=1}^{d+1} \frac{\Gamma(\sigma_i+\kappa_i+\frac{1}{2})}{\Gamma(\kappa_i+\frac{1}{2})} = \frac{(\kappa+\frac{1}{2})_\sigma}{(\lambda+1)_{|\sigma|}},$$

it follows from the explicit formula for R_α in Proposition 7.5.10 that

$$\begin{aligned} c_h' \int_{S^d} |R_\alpha(x)|^2 h_\kappa^2(x)\, d\omega &= c_h' \int_{S^d} R_\alpha(x) x^\alpha h_\kappa^2(x)\, d\omega \\ &= \sum_\gamma \frac{(-\beta)_\gamma(-\alpha+\beta-\kappa+\frac{1}{2})_\gamma(\kappa+\frac{1}{2})_{\alpha-\gamma}}{(-|\alpha|-\lambda+1)_{|\gamma|}\gamma!(\lambda+1)_{|\alpha|-|\gamma|}}. \end{aligned}$$

Now using $(a)_{n-m} = (-1)^m (a)_n/(1-n-a)_m$ and $(-a)_n/(-a+1)_n = a/(a-n)$ to rewrite the sum gives the first equation in (7.5.6). To derive the second equation, we show that the sum in the first equation can be written as an integral. We define a function

$$F(r) = \sum_{\gamma\le\beta} \frac{(-\beta)_\gamma(-\alpha+\beta-\kappa+\frac{1}{2})_\gamma}{(-\alpha-\kappa+\frac{1}{2})_\gamma\gamma!(|\alpha|-|\gamma|+\lambda)} r^{|\alpha|-|\gamma|+\lambda}.$$

Evidently, $F(1)$ is the sum in the first equation in (7.5.6). Moreover, the latter equation is a finite sum over $\gamma\le\beta$ as $(-\beta)_\gamma = 0$ for $\gamma>\beta$; it follows that $F(0)=0$. Hence the sum $F(1)$ is given by $F(1) = \int_0^1 F'(r)\, dr$. The derivative of F can be written as

$$\begin{aligned} F'(r) &= \sum_\gamma \frac{(-\beta)_\gamma(-\alpha+\beta-\kappa+\frac{1}{2})_\gamma}{(-\alpha-\kappa+\frac{1}{2})_\gamma\gamma!} r^{|\alpha|-|\gamma|+\lambda-1} \\ &= r^{|\alpha|+\lambda-1} \prod_{i=1}^d \sum_{\gamma_i} \frac{(-\beta_i)_{\gamma_i}(-\alpha_i+\beta_i-\kappa_i+\frac{1}{2})_{\gamma_i}}{(-\alpha_i-\kappa_i+\frac{1}{2})_{\gamma_i}\gamma_i!} r^{-\gamma_i} \\ &= r^{|\alpha|+\lambda-1} \prod_{i=1}^d {}_2F_1\left(\begin{matrix} -\beta_i, -\alpha_i+\beta_i-\kappa_i+\frac{1}{2} \\ -\alpha_i-\kappa_i+\frac{1}{2} \end{matrix}; \frac{1}{r}\right). \end{aligned}$$

The Jacobi polynomial $P_n^{(a,b)}$ can be written in terms of the hypergeometric series ${}_2F_1$, (4.22.1) of Szegő [1975]:

$$P_n^{(a,b)}(t) = \binom{2n+a+b}{n}\left(\frac{t-1}{2}\right)^n {}_2F_1\left(\begin{matrix} -n, -n-a \\ -2n-a-b \end{matrix}; \frac{2}{1-t}\right).$$

Using this formula with $n=\beta_i$, $a=\alpha_i-2\beta_i+\kappa_i-\frac{1}{2}$, $b=0$ and $r=\frac{1}{2}(1-t)$, and then using $P_n^{(a,b)}(t) = (-1)^n P_n^{(b,a)}(-t)$, we conclude that

$$F'(r) = \frac{(\kappa+\frac{1}{2})_{\alpha-\beta}\beta!}{(\kappa+\frac{1}{2})_\alpha} r^{|\alpha|-|\beta|+\lambda-1} \prod_{i=1}^d P_{\beta_i}^{(0,\alpha_i-2\beta_i+\kappa_i-1/2)}(2r-1).$$

Consequently, it follows that

$$F(1) = \frac{(\kappa+\frac{1}{2})_{\alpha-\beta}\beta!}{(\kappa+\frac{1}{2})_{\alpha}} \int_0^1 \prod_{i=1}^{d} P_{\beta_i}^{(0,\alpha_i-2\beta_i+\kappa_i-1/2)}(2r-1) r^{|\alpha|-|\beta|+\lambda-1}\, \mathrm{d}r.$$

By the definition of $C_n^{(\lambda,\mu)}$ (see Definition 1.5.5), $P_{\beta_i}^{(0,\alpha_i-2\beta_i+\kappa_1-1/2)}(2t^2-1) = C_\alpha^{(1/2,\kappa_i)}(t)$ if α_i is even and $tP_{\beta_i}^{(0,\alpha_i-2\beta_i+\kappa_1-1/2)}(2t^2-1) = C_\alpha^{(1/2,\kappa_i)}(t)$ if α_i is odd. Hence, making the change of variables $r \mapsto t^2$ in the above integral leads to the second equation in (7.5.6). □

Since R_α is monic, $x^\alpha - R_\alpha(x) = Q_\alpha$ is a polynomial of lower degree when restricted to the sphere, and R_α is orthogonal to lower-degree polynomials, it follows from standard Hilbert space theory that Q_α is the best approximation to x^α among all polynomials of lower degree. In other words R_α has the smallest L^2 norm among all polynomials of the form $x^\alpha - P(x)$, $P \in \Pi_{n-1}^d$ on S^{d-1}:

$$\|R_\alpha\|_2 = \min_{P\in\Pi_{n-1}^{d+1}} \|x^\alpha - P\|_2, \qquad |\alpha| = n.$$

Corollary 7.5.14 *Let $\alpha \in \mathbb{N}_0^d$ and $n = |\alpha|$. Then*

$$\inf_{P\in\Pi_{n-1}^d} \|x^\alpha - P(x)\|_2^2 = \frac{2\rho\,(\kappa+\frac{1}{2})_\alpha}{(\rho)_{|\alpha|}} \int_0^1 \prod_{i=1}^{d} \frac{C_{\alpha_i}^{(1/2,\kappa_i)}(t)}{k_{\alpha_i}^{(1/2,\kappa_i)}} t^{|\alpha|+2\rho-1}\, \mathrm{d}t,$$

where $k_n^{(\lambda,\mu)}$ is the leading coefficient of $C_n^{(\lambda,\mu)}(t)$.

7.6 Example: Dihedral Groups

We now consider the h-harmonics associated with the dihedral group, denoted by I_k, $k \ge 2$, which is the group of symmetries of the regular k-gon with root system $I_2(k)$ given in Subsection 6.3.4. We use complex coordinates $z = x_1 + \mathrm{i}x_2$ and identify $\mathbb{R}^2$ with $\mathbb{C}$. For a fixed k, define $\omega = \mathrm{e}^{\mathrm{i}\pi/k}$. Then the rotations in I_k consist of $z \mapsto z\omega^{2j}$ and the reflections in I_k consist of $z \mapsto \bar{z}\omega^{2j}$, $0 \le j \le k-1$. The associated weight function is a product of powers of the linear functions whose zero sets are the mirrors of the reflections in I_k. For parameters $\alpha, \beta \ge 0$, or $\alpha \ge 0$ and $\beta = 0$ when k is odd, define

$$h_{\alpha,\beta}(z) = \left|\frac{z^k - \bar{z}^k}{2\mathrm{i}}\right|^\alpha \left|\frac{z^k + \bar{z}^k}{2}\right|^\beta.$$

As a function in (x_1, x_2), this is positively homogeneous of degree $\gamma = k(\alpha+\beta)$. Since $z^k - \bar{z}^k = \prod_{j=0}^{k-1}(z - \bar{z}\omega^{2j})$ and $z^k + \bar{z}^k = \prod_{j=0}^{k-1}(z - \bar{z}\omega^{2j+1})$, the reflection in the line $z - \bar{z}\omega^j = 0$ is given by $z \mapsto \bar{z}\omega^j$, $0 \le j \le 2k-1$. The corresponding group is I_{2k} when $\beta > 0$ and I_k when $\beta = 0$.

The normalization constant of $h^2_{\alpha,\beta}$ is determined by

$$c_{\alpha,\beta}\int_{-\pi}^{\pi} h^2_{\alpha,\beta}(\mathrm{e}^{\mathrm{i}\theta})\,\mathrm{d}\theta = 1 \qquad \text{with} \qquad c_{\alpha,\beta} = \left[2B\left(\alpha+\tfrac{1}{2},\beta+\tfrac{1}{2}\right)\right]^{-1}.$$

From $z = x_1 + \mathrm{i}x_2$, it follows that

$$\frac{\partial}{\partial z} = \frac{1}{2}\left(\frac{\partial}{\partial x_1} - \mathrm{i}\frac{\partial}{\partial x_2}\right), \qquad \frac{\partial}{\partial \bar{z}} = \frac{1}{2}\left(\frac{\partial}{\partial x_1} + \mathrm{i}\frac{\partial}{\partial x_2}\right).$$

Definition 7.6.1 For a dihedral group W on $\mathbb{R}^2 \cong \mathbb{C}$ and associated weight function $h_{\alpha,\beta}$, let $\mathscr{D} = \frac{1}{2}(\mathscr{D}_1 - \mathrm{i}\mathscr{D}_2)$ and $\overline{\mathscr{D}} = \frac{1}{2}(\mathscr{D}_1 + \mathrm{i}\mathscr{D}_2)$. Also, let $\mathscr{K}_h$ denote the kernel of $\overline{\mathscr{D}}$.

Note that $\Delta_h = \mathscr{D}_1^2 + \mathscr{D}_2^2 = (\mathscr{D}_1 + \mathrm{i}\mathscr{D}_2)(\mathscr{D}_1 - \mathrm{i}\overline{\mathscr{D}}_2) = 4\mathscr{D}\overline{\mathscr{D}}$.

Lemma 7.6.2 *For $\omega \in \mathbb{C}$, $|\omega| = 1$, let $v = (-\operatorname{Im}\omega, \operatorname{Re}\omega) \in \mathbb{R}^2$; then, for a polynomial in x or z,*

$$\frac{1}{2}\frac{f(x) - f(x\sigma_v)}{\langle x, v\rangle}(-\operatorname{Im}\omega + \mathrm{i}\varepsilon\operatorname{Re}\omega) = \frac{f(z) - f(\bar{z}\omega^2)}{z - \bar{z}\omega^2}\xi,$$

where $\xi = 1$ for $\varepsilon = 1$ and $\xi = -\omega^2$ for $\varepsilon = -1$.

Proof An elementary calculation shows that $2\langle x, v\rangle(-\operatorname{Im}\omega + \mathrm{i}\operatorname{Re}\omega) = z - \bar{z}\omega^2$ and $x\sigma_v = x - 2\langle x, v\rangle v = \bar{z}\omega^2$, which yields the stated formula. □

As a consequence of the factorization of $z^k \pm \bar{z}^k$ and Lemma 7.6.2, we have

$$\mathscr{D}f(z) = \frac{\partial f}{\partial z} + \alpha\sum_{j=0}^{k-1}\frac{f(z) - f(\bar{z}\omega^{2j})}{z - \bar{z}\omega^{2j}} + \beta\sum_{j=0}^{k-1}\frac{f(z) - f(\bar{z}\omega^{2j+1})}{z - \bar{z}\omega^{2j+1}},$$

$$\overline{\mathscr{D}}f(z) = \frac{\partial f}{\partial \bar{z}} - \alpha\sum_{j=0}^{k-1}\frac{f(z) - f(\bar{z}\omega^{2j})}{z - \bar{z}\omega^{2j}}\omega^{2j} - \beta\sum_{j=0}^{k-1}\frac{f(z) - f(\bar{z}\omega^{2j+1})}{z - \bar{z}\omega^{2j+1}}\omega^{2j+1}.$$

We denote the space of h-harmonics associated with $h^2_{\alpha,\beta}$ by $\mathscr{H}_n(h^2_{\alpha,\beta})$.

7.6.1 An orthonormal basis of $\mathscr{H}_n(h^2_{\alpha,\beta})$

Let $\mathscr{Q}_h$ denote the closed span of of $\ker\overline{\mathscr{D}} \cap \Pi^2$ in $L^2(h^2_{\alpha,\beta}(\mathrm{e}^{\mathrm{i}\theta})\,\mathrm{d}\theta)$. For $\alpha = \beta = 0$, the usual orthogonal basis for $\mathscr{H}_n(h^2_{\alpha,\beta})$, $n \geq 1$, is $\{z^n, \bar{z}^n\}$; the members of the basis are annihilated by $\partial/\partial\bar{z}$ and $\partial/\partial z$, respectively. This is not in general possible for $\alpha > 0$. Instead, we have

Proposition 7.6.3 *If $\phi_n \in \mathscr{H}_n(h^2_{\alpha,\beta}) \cap \mathscr{Q}_h$, $n \geq 0$, then $\bar{z}\bar{\phi}_n(z) \in \mathscr{H}_{n+1}(h^2_{\alpha,\beta})$ and $\bar{z}\bar{\phi}_n(z)$ is orthogonal to $\ker\overline{\mathscr{D}}$ in $L^2(h^2_{\alpha,\beta}(\mathrm{e}^{\mathrm{i}\theta})\,\mathrm{d}\theta)$.*

Proof By the above definition of $\overline{\mathscr{D}}$ and Theorem 7.1.17, the adjoint $\overline{\mathscr{D}}^*$ satisfies

$$\overline{\mathscr{D}}^* f(z) = (1+n+\gamma)\left[\bar{z}f(z) - (n+\gamma)^{-1}|z|^2\mathscr{D}f(z)\right]$$

if $f \in \mathscr{H}_n(h^2_{\alpha,\beta})$ and $\overline{\mathscr{D}}^* f \in \mathscr{H}_{n+1}(h^2_{\alpha,\beta})$. Since $\mathscr{D}\bar{\phi}_n(z) = \left(\overline{\mathscr{D}}\phi_n(z)\right)^- = 0$, this shows that $\overline{\mathscr{D}}^*\bar{\phi}_n(z) = (1+n+\gamma)\bar{z}\bar{\phi}_n(z)$. By the definition of an adjoint, the latter function is orthogonal to $\ker\overline{\mathscr{D}}$. □

Corollary 7.6.4 *If $\phi_n \in \mathscr{H}_n(h^2_{\alpha,\beta}) \cap \mathscr{Q}_h$ then $\{\phi_n(z), \bar{z}\bar{\phi}_n(z) : n = 0,1,2\ldots\}$ is an orthogonal basis for $L^2(h^2_{\alpha,\beta}(\mathrm{e}^{\mathrm{i}\theta})\,\mathrm{d}\theta)$.*

Proof This follows from the facts that $\dim\mathscr{H}_0(h^2_{\alpha,\beta}) = 1$, $\dim\mathscr{H}_n(h^2_{\alpha,\beta}) = 2$, $n \geq 1$ and $L^2(h^2_{\alpha,\beta}(\mathrm{e}^{\mathrm{i}\theta})\,\mathrm{d}\theta) = \bigoplus_{n=0}^{\infty}\mathscr{H}_n(h^2_{\alpha,\beta})$. □

Since $\mathscr{D}p = 0$ implies that $\Delta_h p = 0$, the corollary shows that we only need to find homogeneous polynomials in $\mathscr{K}_h = \ker\overline{\mathscr{D}}$. Most of the work in describing $\mathscr{K}_h$ reduces to the case $k = 1$; indeed, we have

Proposition 7.6.5 *Let $f(z) = g(z^k)$ and $\xi = z^k$. Then*

$$\mathscr{D}f(z) = kz^{k-1}\left(\frac{\partial g}{\partial \xi} + \alpha\frac{g(\xi)-g(\bar{\xi})}{\xi-\bar{\xi}} + \beta\frac{g(\xi)-g(-\bar{\xi})}{\xi+\bar{\xi}}\right),$$

$$\overline{\mathscr{D}}f(z) = k\bar{z}^{k-1}\left(\frac{\partial g}{\partial \bar{\xi}} - \alpha\frac{g(\xi)-g(\bar{\xi})}{\xi-\bar{\xi}} + \beta\frac{g(\xi)-g(-\bar{\xi})}{\xi+\bar{\xi}}\right).$$

Proof Substituting g into the formulae for $\mathscr{D}f(z)$ and $\overline{\mathscr{D}}f(z)$ and making use of

$$(\bar{z}\omega^{2j})^k = \bar{z}^k = \bar{\xi}, \qquad (\bar{z}\omega^{2j+1})^k = -\bar{\xi},$$

we use the following two formulae to evaluate the sum of the resulting equations:

$$\sum_{j=0}^{k-1}\frac{1}{t-\omega^{2j}} = \frac{kt^{k-1}}{t^k-1}, \qquad \sum_{j=0}^{k-1}\frac{\omega^{2j}}{t-\omega^{2j}} = \frac{k}{t^k-1}, \qquad t \in \mathbb{C}.$$

These formulae can be verified as follows. Multiplying the first formula by $t^k - 1$, the sum on the left-hand side becomes the Lagrange interpolation polynomial of kt^{k-1} based on the points ω^{2j}, $0 \leq j \leq k-1$, which is equal to the expression kt^{k-1} on the right-hand side by the uniqueness of the interpolation problem. The second formula follows from the first on using $\omega^{2j} = t - (t - \omega^{2j})$. We then use these formulae with $t = z/\bar{z}$ and $t = z/(\bar{z}\omega)$, respectively, to finish the proof. □

Moreover, the following proposition shows that if $g(z^k) \in \mathscr{K}_h$ then $z^j g(z^k) \in \mathscr{K}_h$ for $0 \leq j < k$, which allows the complete listing of the homogeneous polynomials in $\mathscr{K}_h$.

Proposition 7.6.6 *If $\overline{\mathscr{D}}g(z^k) = 0$ then $\overline{\mathscr{D}}\left[z^j g(z^k)\right] = 0$ for $0 \le j \le k-1$.*

Proof We use induction, based on the formula

$$\overline{\mathscr{D}}[zf(z)] = z\overline{\mathscr{D}}f(z) - \alpha \sum_{j=0}^{k-1} \omega^{2j} f(\bar{z}\omega^{2j}) - \beta\omega \sum_{j=0}^{k-1} \omega^{2j} f(\bar{z}\omega^{2j+1}),$$

which is verified by noting that a typical term in $\overline{\mathscr{D}}$ is of the form

$$\omega \frac{zf(z) - \bar{z}f(\bar{z}\omega^j)}{z - \bar{z}\omega^j} = z\omega^j \frac{f(z) - f(\bar{z}\omega^j)}{z - \bar{z}\omega^j} + \omega^j f(\bar{z}\omega^j).$$

Let $f(z) = z^j g(z^k)$ with $0 \le j \le k-2$ and assume that $\overline{\mathscr{D}}f = 0$. Then the above formula becomes

$$\overline{\mathscr{D}}[z^{j+1}g(z^k)] = z\overline{\mathscr{D}}f(z) - \left(\alpha z^j g(z^k) + \beta\omega z^j g(-z^k)\right) \sum_{l=0}^{k-1} \omega^{2l(j+1)}.$$

The latter sum is zero if $0 \le j+1 \le k-1$ since ω^2 is a kth root of unity. Thus, $\overline{\mathscr{D}}(z^{j+1}g(z^k)) = 0$ by the induction hypothesis. □

We are now ready to state bases for $\mathscr{K}_h$, and thus bases for $\mathscr{H}(h^2_{\alpha,\beta})$.

The groups $I_1 = \mathbb{Z}_2$ and $I_2 = \mathbb{Z}_2^2$

Set $k = 1$ and $\alpha, \beta \ge 0$. In polar coordinates the corresponding measure on the circle is $(\sin^2\theta)^\alpha(\cos^2\theta)^\beta \mathrm{d}\theta$, associated with the group I_2. If $\beta = 0$ then the measure is associated with the group I_1.

The basis of $\mathscr{H}_n(h^2_{\alpha,\beta})$ is given in Theorem 7.5.1 in terms of the generalized Gegenbauer polynomials defined in Subsection 1.5.2. For each n, some linear combination of the two elements of the basis is in $\mathscr{K}_h$. Define

$$f_n(z) = r^n \left(\frac{n + 2\alpha + \delta_n}{2\alpha + 2\beta} C_n^{(\alpha,\beta)}(\cos\theta) + \mathrm{i}\sin\theta\, C_{n-1}^{(\alpha+1,\beta)}(\cos\theta)\right), \qquad (7.6.1)$$

where $\delta_n = 2\beta$ if n is even and $\delta_n = 0$ if n is odd, and $z = re^{\mathrm{i}\theta}$. We will show that f_n is in $\mathscr{K}_h$. For this purpose it is convenient to express f_n in terms of products of powers of $\frac{1}{2}(z+\bar{z})$ and $\frac{1}{2}(z-\bar{z})$.

Proposition 7.6.7 *Let $\xi = \frac{1}{2}(z+\bar{z})$, $\eta = \frac{1}{2}(z-\bar{z})$ and let*

$$g_{2n} = \frac{(\beta+\frac{1}{2})_n}{(\alpha+\beta+1)_n} f_{2n} \qquad \text{and} \qquad g_{2n+1} = \frac{(\beta+\frac{1}{2})_{n+1}}{(\alpha+\beta+1)_n} f_{2n+1}.$$

Then, for $n \ge 0$,

$$\begin{aligned} g_{2n}(z) =& (-1)^n \sum_{j=0}^{n} \frac{\left(-n+\frac{1}{2}-\alpha\right)_{n-j}\left(-n+\frac{1}{2}-\beta\right)_j}{(n-j)!\,j!} \xi^{2n-2j}\eta^{2j} \\ &+ (-1)^{n-1} \sum_{j=0}^{n-1} \frac{\left(-n+\frac{1}{2}-\alpha\right)_{n-1-j}\left(-n+\frac{1}{2}-\beta\right)_j}{(n-1-j)!\,j!} \xi^{2n-1-2j}\eta^{2j+1} \end{aligned}$$

and

$$g_{2n+1}(z) = (-1)^{n+1} \sum_{j=0}^{n} \frac{\left(-n-\frac{1}{2}-\alpha\right)_{n+1-j} \left(-n-\frac{1}{2}-\beta\right)_j}{(n-j)!j!} \xi^{2n+1-2j} \eta^{2j} + (-1)^{n+1} \sum_{j=0}^{n} \frac{\left(-n-\frac{1}{2}-\alpha\right)_{n-j} \left(-n-\frac{1}{2}-\beta\right)_{j+1}}{(n-j)!j!} \xi^{2n-2j} \eta^{2j+1}.$$

Proof By the definition of generalized Gegenbauer polynomials, write g_n in terms of Jacobi polynomials. Recall from the proof of Proposition 1.4.14 that

$$P_n^{(a,b)}(t) = (-1)^n \sum_{j=0}^{n} \frac{(-n-a)_{n-j}(-n-b)_j}{(n-j)!j!} \left(\frac{1+t}{2}\right)^{n-j} \left(\frac{t-1}{2}\right)^j.$$

Set $t = (z^2+\bar{z}^2)/(2z\bar{z})$; then

$$\frac{1+t}{2} = \frac{1}{z\bar{z}} \left(\frac{z+\bar{z}}{2}\right)^2 \quad \text{and} \quad \frac{t-1}{2} = \frac{1}{z\bar{z}} \left(\frac{z-\bar{z}}{2}\right)^2.$$

Hence, for parameters a, b and $z = re^{i\theta}$,

$$r^{2n} P_n^{(a,b)}(\cos 2\theta) = (-1)^n \sum_{j=0}^{n} \frac{(-n-a)_{n-j}(-n-b)_j}{(n-j)!j!} \left(\frac{z+\bar{z}}{2}\right)^{2n-2j} \left(\frac{z-\bar{z}}{2}\right)^{2j}.$$

Writing g_{2n} and g_{2n+1} in terms of Jacobi polynomials and applying the above formula gives the stated results. □

From Proposition 7.6.5 with $k=1$ we have

$$\begin{aligned} \mathscr{D}[(z+\bar{z})^n(z-\bar{z})^m] &= [n+\beta(1-(-1)^n)]\,(z+\bar{z})^{n-1}(z-\bar{z})^m \\ &\quad + [m+\alpha(1-(-1)^m)]\,(z+\bar{z})^n(z-\bar{z})^{m-1}; \\ \overline{\mathscr{D}}[(z+\bar{z})^n(z-\bar{z})^m] &= [n+\beta(1-(-1)^n)]\,(z+\bar{z})^{n-1}(z-\bar{z})^m \\ &\quad - [m+\alpha(1-(-1)^m)]\,(z+\bar{z})^n(z-\bar{z})^{m-1}. \end{aligned}$$

Thus $\mathscr{D}+\overline{\mathscr{D}}$ and $\mathscr{D}-\overline{\mathscr{D}}$ have simple expressions on these basis elements (indicating the abelian nature of the group I_2). We can now prove some properties of f_n, (7.6.1).

Proposition 7.6.8 *Let $\alpha, \beta \ge 0$. Then we have the following.*

(i) $\overline{\mathscr{D}} f_n(z) = 0$, $n = 0, 1, \ldots$.

(ii) $\mathscr{D} f_{2n}(z) = 2\dfrac{(\alpha+\beta+1)_n}{(\beta+\frac{1}{2})_n} f_{2n-1}(z)$ *and*

$$\mathscr{D} f_{2n+1}(z) = 2(n+\alpha+\tfrac{1}{2})\frac{(\alpha+\beta+1)_n}{(\beta+\frac{1}{2})_n} f_{2n}(z).$$

(iii) *Let* $h_n = \left[2B(\alpha+\frac{1}{2},\beta+\frac{1}{2})\right]^{-1}\int_{-\pi}^{\pi}|f_n(e^{i\theta})|^2(\sin^2\theta)^{\alpha}(\cos^2\theta)^{\beta}d\theta$*; then*

$$h_{2n} = \frac{[\alpha+\beta+1]_n(\alpha+\frac{1}{2})_n}{n!(\beta+\frac{1}{2})_n} \quad \textit{and} \quad h_{2n+1} = \frac{(\alpha+\beta+1)_n(\alpha+\frac{1}{2})_{n+1}}{n!(\beta+\frac{1}{2})_{n+1}}.$$

Proof Observe that the symbolic involution given by $\alpha \leftrightarrow \beta, z+\bar{z} \leftrightarrow z-\bar{z}$ leaves each f_n invariant and interchanges $\mathscr{D}+\overline{\mathscr{D}}$ with $\mathscr{D}-\overline{\mathscr{D}}$. We will show that $\left(\mathscr{D}+\overline{\mathscr{D}}\right)g_{2n} = 2g_{2n-1}$, apply the involution to show that $\left(\mathscr{D}-\overline{\mathscr{D}}\right)g_{2n} = 2g_{2n-1}$ also and then use the same method for g_{2n+1}. This will prove (i) and (ii). Thus we write

$$\begin{aligned}
&\left(\mathscr{D}+\overline{\mathscr{D}}\right)g_{2n}(z) \\
&= (-1)^n\sum_{j=0}^{n}\left[\frac{\left(-n+\frac{1}{2}-\alpha\right)_{n-j}\left(-n+\frac{1}{2}-\beta\right)_j}{(n-j)!j!}(2n-2j)\right. \\
&\qquad\qquad \left.\times\left(\frac{z+\bar{z}}{2}\right)^{2n-1-2j}\left(\frac{z-\bar{z}}{2}\right)^{2j}\right] \\
&\quad +(-1)^{n-1}\sum_{j=0}^{n-1}\left[\frac{\left(-n+\frac{1}{2}-\alpha\right)_{n-1-j}\left(-n+\frac{1}{2}-\beta\right)_j}{(n-1-j)!j!}(2n-1-2j+2\beta)\right. \\
&\qquad\qquad \left.\times\left(\frac{z+\bar{z}}{2}\right)^{2n-2-2j}\left(\frac{z-\bar{z}}{2}\right)^{2j+1}\right] \\
&= 2g_{2n-1}(z).
\end{aligned}$$

The first sum loses the $j = n$ term; in the second sum we use

$$\left(-n+\tfrac{1}{2}-\beta\right)_j(2n-1-2j+2\beta) = -2\left(-n+\tfrac{1}{2}-\beta\right)_{j+1}.$$

Now we have

$$\begin{aligned}
&\left(\mathscr{D}+\overline{\mathscr{D}}\right)g_{2n+1}(z) \\
&= (-1)^{n+1}\sum_{j=0}^{n}\left[\frac{\left(-n-\frac{1}{2}-\alpha\right)_{n+1-j}\left(-n-\frac{1}{2}-\beta\right)_j}{(n-j)!j!}\right. \\
&\qquad\qquad \left.\times(2n+1-2j+2\beta)\left(\frac{z+\bar{z}}{2}\right)^{2n-2j}\left(\frac{z-\bar{z}}{2}\right)^{2j}\right] \\
&\quad +(-1)^{n+1}\sum_{j=0}^{n}\left[\frac{\left(-n-\frac{1}{2}-\alpha\right)_{n-j}\left(-n-\frac{1}{2}-\beta\right)_{j+1}}{(n-j)!j!}\right. \\
&\qquad\qquad \left.\times(2n-2j)\left(\frac{z+\bar{z}}{2}\right)^{2n-1-2j}\left(\frac{z-\bar{z}}{2}\right)^{2j+1}\right] \\
&= 2\left(n+\alpha+\tfrac{1}{2}\right)\left(n+\beta+\tfrac{1}{2}\right)g_{2n}.
\end{aligned}$$

Here the second sum loses the $j = n$ term, and we used

$$\left(-n-\tfrac{1}{2}-\alpha\right)_{n-j}\left(-n-\tfrac{1}{2}-\beta\right)_{j+1}$$
$$= \left(n+\alpha+\tfrac{1}{2}\right)\left(n+\beta+\tfrac{1}{2}\right)\left(-n+\tfrac{1}{2}-\alpha\right)_{n-1-j}\left(-n+\tfrac{1}{2}-\beta\right)_{j}.$$

In the first sum we used $\left(-n-\frac{1}{2}-\beta\right)_j(2n+1-2j+2\beta) = -2\left(-n-\frac{1}{2}-\beta\right)_{j+1}$. Like the second sum, the first sum equals $2\left(n+\alpha+\frac{1}{2}\right)\left(n+\beta+\frac{1}{2}\right)$ times the corresponding sum in the formula for g_{2n}.

The proof of (iii) follows from the formula

$$\int_{-\pi}^{\pi} g(\cos\theta)(\sin^2\theta)^{\alpha}(\cos^2\theta)^{\beta}\,\mathrm{d}\theta = 2\int_{-1}^{1} g(t)|t|^{2\alpha}(1-t^2)^{\beta-1/2}\,\mathrm{d}t,$$

and the norm of the generalized Gegenbauer polynomials. □

Note that if $\alpha = \beta = 0$ then $f_n(z) = z^n$, which follows from using the formula for the generalized Gegenbauer polynomials and taking limits after setting $(\alpha+\beta+1)_n/(\alpha+\beta+n) = (\alpha+\beta)_n/(\alpha+\beta)$.

The real and imaginary parts of f_n comprise an orthogonal basis for $\mathscr{H}_n(h^2_{\alpha,\beta})$. That they are orthogonal to each other follows from their parity in x_2. However, unlike the case of $\alpha = \beta = 0$, $f_n(z)$ and $\bar{f}_n(z)$ are not orthogonal. Corollary 7.6.4 shows that in fact $f_n(z)$ and $\bar{z}\bar{f}_n$ are orthogonal.

Proposition 7.6.9 *For $n \ge 0$,*

$$\bar{z}\bar{f}_{2n-1}(z) = -\frac{1}{2}\frac{\alpha+\beta}{n+\alpha+\beta}f_{2n}(z) + \frac{1}{2}\frac{2n+\alpha+\beta}{n+\alpha+\beta}\bar{f}_{2n}(z),$$
$$\bar{z}\bar{f}_{2n}(z) = \frac{1}{2n+2\beta+1}\left[(\beta-\alpha)f_{2n+1}(z) + (2n+\alpha+\beta+1)\bar{f}_{2n+1}(z)\right].$$

Proof By Corollary 7.6.4 we can write $\bar{z}\bar{f}_{n-1}(z)$ as a linear combination of $f_n(z)$ and $\bar{f}_n(z)$. From the formula in Proposition 7.6.7, it follows that the leading coefficients of z^n and $\bar{z}^n$ are given by

$$f_{2n}(z) = \frac{(\alpha+\beta+1)_{2n-1}}{2^{2n}(\beta+\frac{1}{2})_n n!}\left[(2n+\alpha+\beta)z^{2n} - (\alpha+\beta)\bar{z}^{2n}\right] + \cdots,$$
$$f_{2n+1}(z) = \frac{(\alpha+\beta+1)_{2n}}{2^{2n+1}(\beta+\frac{1}{2})_{n+1}n!}\left[(2n+\alpha+\beta+1)z^{2n+1} + (\alpha-\beta)\bar{z}^{2n+1}\right] + \cdots,$$

where the rest of the terms are linear combinations of powers of z and $\bar{z}$, each homogeneous of degree n in z and $\bar{z}$. Using these formulae to compare the coefficients of z^n and $\bar{z}^n$ in the expression $\bar{z}\bar{f}_{n-1}(z) = af_n(z) + b\bar{f}_n(z)$ leads to the stated formulae. □

These relations allow us to derive some interesting formulae for classical orthogonal polynomials. For example, the coefficient f_n satisfies the following integral expression.

Proposition 7.6.10 *For $n \geq 0$,*

$$c_\alpha c_\beta \int_{-1}^{1}\int_{-1}^{1} (s\cos\theta + \mathrm{i}t\sin\theta)^n (1+s)(1+t)(1-s^2)^{\beta-1}(1-t^2)^{\alpha-1}\,\mathrm{d}s\,\mathrm{d}t$$
$$= a_n f_n(\mathrm{e}^{\mathrm{i}\theta}) = a_n\left[\frac{n+2\alpha+\delta_n}{2\alpha+2\beta} C_n^{(\alpha,\beta)}(\cos\theta) + \mathrm{i}\sin\theta\, C_{n-1}^{(\alpha+1,\beta)}(\cos\theta)\right],$$

where

$$a_{2n} = \frac{(2n)!}{2^{2n}(\alpha+\beta+1)_n(\alpha+\frac{1}{2})_n},$$
$$a_{2n+1} = \frac{(2n+1)!}{2^{2n+1}(\alpha+\beta+1)_n(\alpha+\frac{1}{2})_{n+1}}.$$

Proof Let V denote the intertwining operator associated with $h^2_{\alpha,\beta}$. Since the real and imaginary parts of z^n are ordinary harmonics, the real and imaginary parts of $V(z^n)$ are elements of $\mathscr{H}_n(h^2_{\alpha,\beta})$. It follows that $V(z^n) = a_n f_n(z) + b_n \bar{z}\bar{f}_{n-1}(z)$. Since $f_n(z)$ and $\bar{z}\bar{f}_n(z)$ are orthogonal,

$$a_n h_n = \int_{-\pi}^{\pi} (Vz^n)(\mathrm{e}^{\mathrm{i}\theta})\bar{f}_n(\mathrm{e}^{\mathrm{i}\theta})\,\mathrm{d}\mu_{\alpha,\beta} = \frac{n!}{(\alpha+\beta+1)_n} 2\pi \int_{S^1} f_n(z)\bar{z}^n\,\mathrm{d}\omega,$$

where h_n is as in part (iii) of Proposition 7.6.8 and we have used Proposition 7.2.8 in the second equation. Using Theorem 7.2.4 and the expression for $\langle p,q\rangle_h$ in Definition 7.2.2, we conclude that

$$a_n h_n = \frac{1}{2^n(\alpha+\beta+1)_n}\left(\frac{\partial}{\partial x_1} - \mathrm{i}\frac{\partial}{\partial x_2}\right)^n f_n(z)$$
$$= \frac{1}{2^n(\alpha+\beta+1)_n}\left(2\frac{\partial}{\partial z}\right)^n f_n(z).$$

The value of $(\partial/\partial z)^n f_n(z)[n!]^{-1}$ is equal to the coefficient of z^n in f_n, which is given in the proof of Proposition 7.6.9. This gives the value of a_n. In a similar way we can compute b_n: it is given by a constant multiple of $(\partial/\partial z)^n \bar{z}\bar{f}_n(z)$. Hence, we conclude that $Vz^n = a_n f_n(z)$. The stated formula then follows from the closed formula for V in Theorem 7.5.4. □

In particular, if $\beta \to 0$ then the formula in Proposition 7.6.10 becomes a classical formula of the Dirichlet type:

$$c_\alpha \int_{-1}^{1} (\cos\theta + \mathrm{i}t\sin\theta)^n (1+t)(1-t^2)^{\alpha-1}\,\mathrm{d}t$$
$$= \frac{n!}{(2\alpha)_n} C_n^\alpha(\cos\theta) + \mathrm{i}\frac{n!}{(2\alpha+1)_n}\sin\theta\, C_{n-1}^{\alpha+1}(\cos\theta)$$

for $n = 1, 2, \ldots$ This is a result of Erdélyi [1965] for functions that are even in θ.

The group I_k

If k is an odd integer then the weight function h_α associated with the group I_k is the same as the weight function $h_{\alpha,\beta}$ associated with the group I_{2k} with $\beta = 0$. Thus, we only need to consider the case of I_k for even k.

Set $k = 2m$. By Proposition 7.6.6, the homogeneous polynomials of degree $mn + j$ in $\mathscr{K}_h$ are given by $z^j f_n(z^m)$, $0 \le j \le m-1$. Their real and imaginary parts comprise an orthogonal basis of $\mathscr{H}_{mn+j}(h^2_{\alpha,\beta})$; that is, if we define

$$p_{mn+j}(\theta) = \frac{n+2\alpha+\delta_n}{2\alpha+2\beta}\cos j\theta\, C_n^{(\alpha,\beta)}(\cos m\theta) - \sin\theta \sin j\theta\, C_{n-1}^{(\alpha+1,\beta)}(\cos m\theta),$$

$$q_{mn+j}(\theta) = \frac{n+2\alpha+\delta_n}{2\alpha+2\beta}\sin j\theta\, C_n^{(\alpha,\beta)}(\cos m\theta) + \sin\theta \cos j\theta\, C_{n-1}^{(\alpha+1,\beta)}(\cos m\theta),$$

where $\delta_n = 2\beta$ if n is even and $\delta = 0$ if n is odd, then

$$Y_{n,1}(x_1,x_2) = r^{mn+j} p_{mn+j}(\theta) \qquad \text{and} \qquad Y_{n,2}(x_1,x_2) = r^{mn+j} q_{mn+j}(\theta),$$

with polar coordinates $x_1 = r\cos\theta$ and $x_2 = r\sin\theta$, form an orthogonal basis for the space $\mathscr{H}_{mn+j}(h^2_{\alpha,\beta})$, $0 \le j \le m-1$. The two polynomials are indeed orthogonal since $Y_{n,1}$ is even in x_2 and $Y_{n,2}$ is odd in x_2.

Let us mention that in terms of the variable $t = \cos\theta$, the restriction of $h^2_{\alpha,\beta}$ associated with the group I_{2m} on the circle can be transformed into a weight function

$$w_{\alpha,\beta}(t) = \left|\sin m\theta\right|^{2\alpha} |\cos m\theta|^{2\beta} = \left|U_{m-1}(t)\sqrt{1-t^2}\right|^{2\alpha} |T_m(t)|^{2\beta} \tag{7.6.2}$$

defined on $[-1,1]$, where T_m and U_m are Chebyshev polynomials of the first and the second kind, respectively. In particular, under the map $\cos\theta \mapsto t$, p_{mn+j} is the orthogonal polynomial of degree $mn + j$ with respect to $w_{\alpha,\beta}$.

7.6.2 Cauchy and Poisson kernels

Let $\{\phi_n, \bar{z}\bar{\phi}_n\}_{n\ge 0}$ denote an orthonormal basis for $L^2(h^2_{\alpha,\beta}(\mathrm{e}^{\mathrm{i}\theta})\,\mathrm{d}\theta)$ associated with the group I_k; see Corollary 7.6.4. The Cauchy kernel is related to the projection onto a closed span of $\{\phi_n : n \ge 0\}$. It is defined by

$$C(I_k; z, w) = \sum_{n=0}^{\infty} \phi_n(z)\overline{\phi_n(w)}, \qquad |zw| < 1.$$

For any polynomial $f \in \mathscr{K}_h = \ker \overline{\mathscr{D}}$,

$$f(z) = c_{\alpha,\beta}\int_{-\pi}^{\pi} f(\mathrm{e}^{\mathrm{i}\theta}) C(I_k; z, \mathrm{e}^{\mathrm{i}\theta}) h^2_{\alpha,\beta}(\mathrm{e}^{\mathrm{i}\theta})\,\mathrm{d}\theta, \qquad |z| < 1.$$

The Poisson kernel which reproduces any h-harmonics in the disk was defined in Section 7.3. In terms of ϕ_n it is given by

$$\begin{aligned}P(I_k;z,w) &= \sum_{n=0}^{\infty} \phi_n(z)\overline{\phi_n(z)} + \sum_{n=0}^{\infty} \bar{z}\overline{\phi_n(z)}w\phi_n(w) \\ &= C(I_k;z,w) + \bar{z}wC(I_k;w,z).\end{aligned}$$

The following theorem shows that finding closed formulae for these kernels reduces to the cases $k=1$ and $k=2$.

Theorem 7.6.11 *For each weight function $h_{\alpha,\beta}$ associated with the group I_{2k},*

$$P(I_{2k};z,w) = \frac{1-|z|^2|w|^2}{1-\bar{z}w}C(I_{2k};z,w).$$

Moreover

$$P(I_{2k};z,w) = \frac{1-|z|^2|w|^2}{1-|z^k|^2|\bar{w}^k|^2}\frac{|1-z^k\bar{w}^k|^2}{|1-\bar{z}w|^2}P(I_2;z^k,w^k), \tag{7.6.3}$$

where the Poisson kernel $P(I_2;z,w)$ associated with $h(x+\mathrm{i}y) = |x|^\alpha|y|^\beta$ is given by

$$\begin{aligned}P(I_2;z,w) &= (1-|zw|^2)c_\alpha c_\beta \\ &\times \int_{-1}^{1}\int_{-1}^{1}\left[1+2(\operatorname{Im} z)(\operatorname{Im} w)s+2(\operatorname{Re} z)(\operatorname{Re} w)t+|zw|^2\right]^{-\alpha-\beta-1} \\ &\times (1+s)(1+t)(1-s^2)^{\beta-1}(1-t^2)^{\alpha-1}\mathrm{d}s\,\mathrm{d}t.\end{aligned}$$

Proof Let $C(I_k;z,w) = C_{\mathrm{Re}}(z,w) + \mathrm{i}C_{\mathrm{Im}}(z,w)$ with C_{Re} and C_{Im} real and symmetric. Then

$$\begin{aligned}\operatorname{Re} P(I_k;z,w) &= (1+\operatorname{Re} z\bar{w})C_{\mathrm{Re}}(z,w) - \operatorname{Im}\bar{z}w\,C_{\mathrm{Im}}(z,w), \\ \operatorname{Im} P(I_k;z,w) &= \operatorname{Im}\bar{z}w\,C_{\mathrm{Re}}(z,w) - (1-\operatorname{Re}\bar{z}w)C_{\mathrm{Im}}(z,w).\end{aligned}$$

Since $\operatorname{Im} P(z,w) = 0$, we have $C_{\mathrm{Im}}(z,w) = [(\operatorname{Im}\bar{z}w)/(1-\operatorname{Re}\bar{z}w)] \times C_{\mathrm{Re}}(z,w)$ and thus $P(I_k;z,w) = [(1-|z|^2|w|^2)/(1-\bar{z}w)]C(I_k;z,w)$.

Now let $\{\psi_n : n \ge 0\}$ be an orthonormal basis of $\ker\overline{\mathscr{D}}$ associated with $h(x+\mathrm{i}y) = |x|^\alpha|y|^\beta$. By Proposition 7.6.6 we conclude that the polynomials $z^j\psi_n(z^k)$ are in $\mathscr{K}_h$ and, in fact, are orthonormal h-harmonics in $\mathscr{H}_n(h^2_{\alpha,\beta})$. Thus

$$C(I_k;z,w) = \sum_{n=0}^{\infty}\psi_n(z^k)\bar{\psi}_n(w^k)\sum_{j=0}^{k-1}z^j\bar{w}^j = \frac{1-z^k\bar{w}^k}{1-z\bar{w}}C(I_2;z^k,w^k).$$

Hence, using $P(I_2;z,w) = [(1-|z|^2|w|^2)/(1-z\bar{w})]C(I_2;z,w)$ and these formulae, the relation (7.6.3) between $P(I_{2k};z,w)$ and $P(I_2;z,w)$ follows. The integral representation of $P(I_2;z,w)$ follows from Theorem 7.3.3 and the explicit formula for the intertwining operator V for $I_2 = \mathbb{Z}_2^2$ in Theorem 7.5.4. □

In particular, if $\beta = 0$ then the kernel $P(I_2; z, w)$ becomes the kernel $P(I_1; z, w)$ for the weight function $|x|^\alpha$ associated with $I_1 = \mathbb{Z}_2$:

$$P(I_1; z, w) = c_\alpha \int_{-1}^{1} \frac{1 - |zw|^2}{(1 + 2(\operatorname{Im} z)(\operatorname{Im} w) + 2(\operatorname{Re} z)(\operatorname{Re} w)t + |zw|^2)^{\alpha+1}} (1+t)(1-t^2)^{\alpha-1} \mathrm{d}t.$$

For the general odd-k dihedral group, we can use the formula

$$P(I_k; z, w) = \frac{1 - |z|^2|w|^2}{1 - |z^k|^2|\bar{w}^k|^2} \frac{|1 - z^k\bar{w}^k|^2}{|1 - z\bar{w}|^2} P(I_1; z^k, w^k)$$

to write down an explicit formula for $P(I_k; z, w)$.

In terms of the Lauricella function F_A of two variables (also called Appell's function),

$$F_A\binom{a_1, a_2}{b_1, b_2}; c; x, y\Big) = \sum_{m=0}^{\infty} \sum_{n=0}^{\infty} \frac{(c)_{m+n}(a_1)_m(a_2)_n}{(b_1)_m(b_2)_n m!n!} x^m y^n$$

(see Section 1.2), the Poisson kernel $P(I_2; w, z)$ can be written as

$$P(I_2; w, z) = \frac{1}{|1 + z\bar{w}|^{2(\alpha+\beta+1)}} \times F_A\left(\begin{matrix}\alpha+1, \beta+1\\ 2\alpha+1, 2\beta+1\end{matrix}; \alpha+\beta+1; \frac{4(\operatorname{Im} z)(\operatorname{Im} w)}{|1 + z\bar{w}|^2}, \frac{4(\operatorname{Re} z)(\operatorname{Re} w)}{|1 + z\bar{w}|^2}\right).$$

This follows from a formula in Bailey [1935, p. 77].

On the one hand the identity (7.6.3) gives an explicit formula for the Poisson kernel associated with the dihedral groups. On the other hand, the Poisson kernel is given in terms of the intertwining operator V in Theorem 7.3.3; the two formulae taken together suggest that the intertwining operator V might constitute an integral transform for any dihedral group.

7.7 The Dunkl Transform

The Fourier transform plays an important role in analysis. It is an isometry of $L^2(\mathbb{R}^d, \mathrm{d}x)$ onto itself. It is a remarkable fact that there is a Fourier transform for reflection-invariant measures which is an isometry of $L^2(\mathbb{R}^d, h_\kappa^2 \mathrm{d}x)$ onto itself.

Recall that $K(x, y) = V^{(x)} \mathrm{e}^{\langle x, y \rangle}$. Since V is a positive operator, $|K(x, \mathrm{i}y)| \le 1$. It plays a role similar to that of $\mathrm{e}^{\mathrm{i}\langle x, y \rangle}$ in classical Fourier analysis. Also recall that $\mathrm{d}\mu(x) = (2\pi)^{-d/2} \mathrm{e}^{-\|x\|^2/2} \mathrm{d}x$.

Definition 7.7.1 For $f \in L^1(h_\kappa^2 \mathrm{d}x)$, $y \in \mathbb{R}^d$, let

$$\widehat{f}(y) = (2\pi)^{-d/2} c_h \int_{\mathbb{R}^d} f(x) K(x, -\mathrm{i}y) h_\kappa^2(x) \,\mathrm{d}x$$

denote the Dunkl transform of f at y.

By the dominated convergence theorem $\widehat{f}$ is continuous on $\mathbb{R}^d$. Definition 7.7.1 was motivated by the following proposition.

Proposition 7.7.2 *Let p be a polynomial on $\mathbb{R}^d$ and $\nu(y) = \sum_{j=1}^d y_j^2$ for $y \in \mathbb{C}^d$; then*

$$c_h \int_{\mathbb{R}^d} \left[\mathrm{e}^{-\Delta_h/2} p(x)\right] K(x,y) h_\kappa^2(x)\,\mathrm{d}\mu(x) = \mathrm{e}^{\nu(y)/2} p(y).$$

Proof Let m be an integer larger than the degree of p, fix $y \in \mathbb{C}^d$ and let $q_m(x) = \sum_{j=0}^m K_j(x,y)$. Breaking p up into homogeneous components, it follows that $\langle q_m, p\rangle_h = p(y)$. By the formula in Theorem 7.2.7,

$$\langle q_m, p\rangle_h = c_h \int_{\mathbb{R}^d} \left(\mathrm{e}^{-\Delta_h/2} p\right)\left(\mathrm{e}^{-\Delta_h/2} q_m\right) h_\kappa^2\,\mathrm{d}\mu.$$

However, $\Delta_h^{(x)} K_n(x,y) = \nu(y) K_{n-2}(x,y)$ and so

$$\begin{aligned}\mathrm{e}^{-\Delta_h/2} q_m(x) &= \sum_{j=0}^{m} \sum_{l \le j/2} \frac{1}{l!}\left(\frac{-\nu(y)}{2}\right)^l K_{j-2l}(x,y)\\ &= \sum_{l \le m/2} \frac{1}{l!}\left(\frac{-\nu(y)}{2}\right)^l \sum_{s=0}^{m-2l} K_s(x,y).\end{aligned}$$

Now let $m \to \infty$. The double sum converges to $\mathrm{e}^{-\nu(y)/2} K(x,y)$ since it is dominated termwise by

$$\sum_{l=0}^{\infty} \frac{\|y\|^{2l}}{l!2^l} \sum_{s=0}^{\infty} \frac{\|x\|^s \|y\|^s}{s!} = \exp\left(\frac{\|y\|^2}{2} + \|x\|\,\|y\|\right),$$

which is integrable with respect to $\mathrm{d}\mu(x)$. By the dominated convergence theorem,

$$p(y) = \mathrm{e}^{\nu(y)/2} c_h \int_{\mathbb{R}^d} [\mathrm{e}^{-\Delta_h/2} p(x)] K(x,y) h_\kappa(x)^2\,\mathrm{d}\mu(x).$$

Multiplying both sides by $\mathrm{e}^{-\nu(y)}$ completes the proof. □

An orthogonal basis for $L^2(\mathbb{R}^d, h_\kappa^2\,\mathrm{d}x)$ is given as follows (compare with (7.1.10)).

Definition 7.7.3 For $m, n = 0, 1, 2, \ldots$ and $p \in \mathscr{H}_n^d(h_\kappa^2)$, let $\lambda_\kappa = \gamma_\kappa + \frac{d-2}{2}$. Then

$$\phi_m(p;x) = p(x) L_m^{n+\lambda_\kappa}(\|x\|^2) \mathrm{e}^{-\|x\|^2/2}, \qquad x \in \mathbb{R}^d.$$

Proposition 7.7.4 *For $k, l, m, n \in \mathbb{N}_0$, $p \in \mathscr{H}_n^d(h_\kappa^2)$ and $q \in \mathscr{H}_l^d(h_\kappa^2)$,*

$$\begin{aligned}&(2\pi)^{-d/2} c_h \int_{\mathbb{R}^d} \phi_m(p;x)\phi_k(q;x) h_\kappa^2(x)\,\mathrm{d}x\\ &\quad = \delta_{mk}\delta_{nl} 2^{-\lambda_\kappa - 1} \frac{(\lambda_\kappa + 1)_{n+m}}{m} c_h' \int_{S^{d-1}} p q h_\kappa^2\,\mathrm{d}\omega.\end{aligned}$$

Proof Using spherical polar coordinates and Definition 7.7.3, the integral on the left-hand side equals

$$c_h \int_0^\infty L_m^{n+\lambda_\kappa}(r^2) L_k^{l+\lambda_\kappa}(r^2) \mathrm{e}^{-r^2} r^{n+l+2\gamma_\kappa+d-1}\,\mathrm{d}r \,\frac{2^{1-d/2}}{\Gamma(\frac{d}{2})} \int_{S^{d-1}} pqh^2\,\mathrm{d}\omega.$$

The integral on the right-hand side is zero if $n \neq l$. Assume $n = l$ and make the change of variable $r^2 = t$. The first integral then equals $\frac{1}{2}\delta_{mk}\Gamma(n+\lambda_\kappa+1+m)/m!$. □

By Theorem 3.2.18 the linear span of $\{\phi_m(p) : m,n \geq 0, p \in \mathscr{H}_n^d(h_\kappa^2)\}$ is dense in $L^2(\mathbb{R}^d, h_\kappa^2\,\mathrm{d}x)$. Moreover, $\{\phi_m(p)\}$ is a basis of eigenfunctions of the Dunkl transform.

Theorem 7.7.5 *For* $m,n = 0,1,2,\ldots$, $p \in \mathscr{H}_n^d(h_\kappa^2)$ *and* $y \in \mathbb{R}^d$,

$$\left[\widehat{\phi_m(p)}\right](y) = (-\mathrm{i})^{n+2m}\phi_m(p;y).$$

Proof For brevity, let $A = n+\lambda_\kappa$. By Proposition 7.7.2 and Lemma 7.2.6,

$$\begin{aligned}(2\pi)^{-d/2} c_h &\int_{\mathbb{R}^d} L_j^A(\tfrac{1}{2}\|x\|^2) p(x) K(x,y) h_\kappa^2(x) \mathrm{e}^{-\|x\|^2/2}\,\mathrm{d}x \\ &= (-1)^j (j!2^j)^{-1} \mathrm{e}^{\nu(y)/2} \nu(y)^j p(y), \qquad y \in \mathbb{C}.\end{aligned}$$

We can change the argument in the Laguerre polynomial by using the identity

$$L_m^A(t) = \sum_{j=0}^m 2^j \frac{(A+1)_m}{(A+1)_j} \frac{(-1)^{m-j}}{(m-j)!} L_j^A(\tfrac{t}{2}), \qquad t \in \mathbb{R},$$

which can be proved by writing the generating function for the Laguerre polynomials as

$$(1-r)^{-A-1} \exp\left(\frac{-xr}{1-r}\right) = (1-r)^{-A-1} \exp\left(\frac{-\frac{1}{2}xt}{1-t}\right)$$

with $t = 2r/(1+r)$. Use this expression together with the above integral to obtain

$$\begin{aligned}(2\pi)^{-d/2} c_h &\int_{\mathbb{R}^d} L_m^A(\|x\|^2) p(x) \mathrm{e}^{-\|x\|^2/2} K(x,y) h_\kappa^2(x)\,\mathrm{d}x \\ &= \mathrm{e}^{\nu(y)/2} p(y) (-1)^m \frac{(A+1)_m}{m!} \sum_{j=0}^m \frac{(-m)_j}{(A+1)_j} \frac{[-\nu(y)]^j}{j!}.\end{aligned}$$

Now replace y by $-\mathrm{i}y$ with $y \in \mathbb{R}^d$; then $\nu(y)$ becomes $-\|y\|^2$ and $p(y)$ becomes $(-1)^m(-\mathrm{i})^n p(y)$, and the sum yields a Laguerre polynomial. The integral equals $(-1)^m(-\mathrm{i})^n p(y) L_m^A(\|y\|^2)\mathrm{e}^{-\|y\|^2/2}$. □

Since the eigenvalues of the Dunkl transform are powers of i, this proves its isometry properties.

Corollary 7.7.6 *The Dunkl transform extends to an isometry of $L^2(\mathbb{R}^d, h_\kappa^2\,\mathrm{d}x)$ onto itself. The square of the transform is the central involution; that is, if $f \in L^2(\mathbb{R}^d, h_\kappa^2\,\mathrm{d}x)$, $\widehat{f} = g$, then $\widehat{g}(x) = f(-x)$ for almost all $x \in \mathbb{R}^d$.*

The following proposition gives another interesting integral formula, as well as a method for finding the eigenfunctions of Δ_h. First, recall the definition of the Bessel function of index $A > -1/2$:

$$\begin{aligned} J_A(t) &= \frac{(\frac{t}{2})^A}{\sqrt{\pi}\Gamma(A+\frac{1}{2})} \int_{-1}^{1} \mathrm{e}^{\mathrm{i}ts}(1-s^2)^{A-1/2}\,\mathrm{d}s \\ &= \frac{(\frac{t}{2})^A}{\Gamma(A+1)} \sum_{m=0}^{\infty} \frac{1}{m!(A+1)_m}(-\tfrac{1}{4}t^2)^m, \qquad t > 0. \end{aligned}$$

Proposition 7.7.7 *Let $f \in \mathscr{H}_n^d(h_\kappa^2)$, $n = 0, 1, 2, \ldots$ and $y \in \mathbb{C}^d$; then*

$$c_h' \int_{S^{d-1}} f(x)K(x,y)h_\kappa^2(x)\,\mathrm{d}\omega = f(y) \sum_{m=0}^{\infty} \frac{\nu(y)^m}{2^{m+n}m!(\lambda_\kappa+1)_{m+n}}.$$

Moreover, if $y \in \mathbb{R}^d$, $\rho > 0$, then the function

$$g(y) = c_h' \int_{S^{d-1}} f(x)K(x,-\mathrm{i}y\rho)h_\kappa^2(x)\,\mathrm{d}\omega(x)$$

satisfies $\Delta_h g = -\rho^2 g$ and

$$g(y) = (-\mathrm{i})^n \Gamma(\lambda_\kappa+1)\Big(\frac{\rho\|y\|}{2}\Big)^{-\lambda_\kappa} f\Big(\frac{y}{\|y\|}\Big) J_{n+\lambda_\kappa}(\rho\|y\|).$$

Proof Since f is h-harmonic, $\mathrm{e}^{-\Delta_h/2} f = f$. In the formula of Proposition 7.7.2,

$$c_h \int_{\mathbb{R}^d} f(x)K(x,y)h_\kappa^2(x)\mathrm{d}\mu(x) = \mathrm{e}^{\nu(y)/2} f(y),$$

the part homogeneous of degree $n+2m$ in y, $m = 0, 1, 2, \ldots$, yields the equation

$$c_h \int_{\mathbb{R}^d} f(x)K_{n+2m}(x,y)h_\kappa^2(x)\,\mathrm{d}\mu(x) = \frac{\nu(y)^m}{2^m m!} f(y).$$

Then, using the integral formula in Theorem 7.2.4 and the fact that

$$\int_{S^{d-1}} f(x)K_j(x,y)h_\kappa^2(x)\,\mathrm{d}\omega(x) = 0$$

if $j < n$ or $j \not\equiv n$ modulo 2, we conclude that

$$\begin{aligned} &c_h' \int_{S^{d-1}} f(x)K(x,y)h_\kappa^2(x)\,\mathrm{d}\omega(x) \\ &\quad = \sum_{m=0}^{\infty} \frac{1}{2^{n+m}(\frac{d}{2}+\gamma)_{n+m}} c_h \int_{\mathbb{R}^d} f(x)K_{n+2m}(x,y)h_\kappa^2(x)\,\mathrm{d}\mu(x), \end{aligned}$$

which gives the stated formula.

Now replace y by $-\mathrm{i}\rho y$ for $\rho > 0$, $y \in \mathbb{R}^d$. Let $A = n + \lambda_\kappa$. This leads to an expression for g in terms of the Bessel function J_A. To find $\Delta_h g$ we can interchange the integral and $\Delta_h^{(y)}$, because the resulting integral of the series $\sum_{n=0}^{\infty} \Delta_h^{(y)} K_n(x, -\mathrm{i}y)$ converges absolutely. Indeed,

$$\begin{aligned}\Delta_h^{(y)} K(x, -\mathrm{i}\rho y) &= \Delta_h^{(y)} K(-\mathrm{i}\rho x, y) \\ &= \sum_{j=1}^{N} (-\mathrm{i}\rho x_j)^2 K(-\mathrm{i}\rho x, y) = -\rho^2 \|x\|^2 K(x, -\mathrm{i}\rho y);\end{aligned}$$

but $\|x\|^2 = 1$ on S^{d-1} and so $\Delta_h g = -\rho^2 g$. □

Proposition 7.7.8 *Suppose that f is a radial function in $L^1(\mathbb{R}^d, h_\kappa^2\,\mathrm{d}x)$; $f(x) = f_0(\|x\|)$ for almost all $x \in \mathbb{R}^d$. The Dunkl transform $\widehat{f}$ is also radial and has the form $\widehat{f}(x) = F_0(\|x\|)$ for all $x \in \mathbb{R}^d$, with*

$$F_0(\|x\|) = F_0(r) = \frac{1}{\sigma_{d-1} r^{\lambda_\kappa}} \int_0^\infty f_0(s) J_{\lambda_\kappa}(rs) s^{\lambda_\kappa + 1}\,\mathrm{d}s.$$

Proof Using polar coordinates and Corollary 7.4.5 we get

$$\begin{aligned}\widehat{f}(y) &= \frac{c_h}{(2\pi)^{d/2}} \int_0^\infty f_0(r) r^{d-1+2\gamma} \int_{S^{d-1}} K(rx', y) h_\kappa^2(x')\,\mathrm{d}\omega(x')\,\mathrm{d}r \\ &= \frac{c_h}{(2\pi)^{d/2}} B_\kappa \int_0^\infty f_0(r) r^{2\lambda_\kappa+1} \int_{-1}^{1} \mathrm{e}^{\mathrm{i}r\|y\|t} (1-t^2)^{\lambda_\kappa - 1/2}\,\mathrm{d}t\,\mathrm{d}r,\end{aligned}$$

from which the stated result follows from the definition of $J_A(r)$ and putting the constants together. □

The Dunkl transform diagonalizes each $\mathscr{D}_i$ (just as the Fourier–Plancherel transform diagonalizes $\partial/\partial x_i$). First we prove a symmetry relation with some technical restrictions.

Lemma 7.7.9 *Let $f\mathscr{D}_i g$ and $g\mathscr{D}_i f$ be in $L^1(\mathbb{R}^d, h_\kappa^2)$, and suppose that fg tends to zero at infinity. Then*

$$\int_{\mathbb{R}^d} (\mathscr{D}_j f) g h_\kappa^2\,\mathrm{d}x = -\int_{\mathbb{R}^d} f(\mathscr{D}_j g) h_\kappa^2\,\mathrm{d}x, \qquad j = 1, \ldots, d.$$

Proof The following integration by parts is justified by the assumption on fg at infinity. We also require $\kappa_v \geq 1$ for each $v \in R_+$ so that $1/\langle x, v\rangle$ is integrable for $h_\kappa^2\,\mathrm{d}x$. After the formula is established, it can be extended to $\kappa_i \geq 0$ by analytic continuation. Now

$$\begin{aligned}\int_{\mathbb{R}^d} (\mathscr{D}_j f) g h_\kappa^2 \, dx &= -\int_{\mathbb{R}^d} f(x) \frac{\partial}{\partial x_j} [g(x) h_\kappa(x)^2] \, dx \\ &\quad + \sum_{v \in R_+} \kappa_j(v) \int_{\mathbb{R}^d} \frac{f(x) - f(x\sigma_v)}{\langle x, v \rangle} g(x) h_\kappa^2(x) \, dx \\ &= -\int_{\mathbb{R}^d} \left[f(x) \frac{\partial}{\partial x_j} g(x) + 2f(x) \sum_{v \in R_+} \kappa_v \frac{v_j}{\langle x, v \rangle} g(x) \right] h_\kappa^2(x) \, dx \\ &\quad + \sum_{v \in R_+} \kappa_v v_j \int_{\mathbb{R}^d} f(x) \frac{g(x) + g(x\sigma_v)}{\langle x, v \rangle} h_\kappa^2(x) \, dx \\ &= -\int_{\mathbb{R}^d} f(\mathscr{D}_j g) h_\kappa^2 \, dx.\end{aligned}$$

In the above the substitution $x \mapsto x\sigma_v$, for which $\langle x, v \rangle$ becomes $\langle x\sigma_v, v \rangle = \langle x, v\sigma_v \rangle = -\langle x, v \rangle$, was used to show that

$$\int_{\mathbb{R}^d} \frac{f(x\sigma_v) g(x)}{\langle x, v \rangle} h_\kappa^2(x) \, dx = -\int_{\mathbb{R}^d} \frac{f(x) g(x\sigma_v)}{\langle x, v \rangle} h_\kappa^2(x) \, dx$$

since $h_\kappa^2 \, dx$ is W-invariant. □

Theorem 7.7.10 *For $f, \mathscr{D}_j f \in L^1(\mathbb{R}^d, h_\kappa^2 dx)$ and $y \in \mathbb{R}^d$, we have $\widehat{\mathscr{D}_j f}(y) = i y_j \widehat{f}(y)$ for $j = 1, \ldots, d$. The operator $-i\mathscr{D}_j$ is densely defined on $L^2(\mathbb{R}^d, h_\kappa^2 dx)$ and is self-adjoint.*

Proof For fixed $y \in \mathbb{R}^d$ put $g(x) = K(x, -iy)$ in Lemma 7.7.9. Then $\mathscr{D}_j g(x) = -i y_j K(x, -iy)$ and $\widehat{\mathscr{D}_j f}(y) = (-1)(-i y_j) \widehat{f}(y)$. The multiplication operator defined by $M_j f(y) = y_j f(y)$, $j = 1, \ldots, d$, is densely defined and self-adjoint on $L^2(\mathbb{R}^d, h_\kappa^2 dx)$. Further, $-i\mathscr{D}_j$ is the inverse image of M_j under the Dunkl transform, an isometric isomorphism. □

Corollary 7.7.11 *For $f \in L^1(\mathbb{R}^d)$, $j = 1, \ldots, d$ and $g_j(x) = x_j f(x)$, the transform $\widehat{g}_j(y) = i\mathscr{D}_j \widehat{f}(y)$, $y \in \mathbb{R}^d$.*

Let us consider the example of $\mathbb{Z}_2^2$ on $\mathbb{R}^2$, for which the weight function is $h_{\alpha,\beta} = |x|^\alpha |y|^\beta$. Recall the formula for the intertwining operator V in Theorem 7.5.4. Let us define

$$E_\alpha(x) = c_\alpha \int_{-1}^{1} e^{-isx} (1+s)(1-s^2)^{\alpha-1} ds$$

for $\alpha > 0$. Integrating by parts,

$$\begin{aligned}E_\alpha(x) &= c_\alpha \int_{-1}^{1} e^{-isx} (1-s^2)^{\alpha-1} ds - \frac{ix}{2\alpha} c_\alpha \int_{-1}^{1} e^{-isx} (1-s^2)^\alpha ds \\ &= \Gamma(\alpha + \tfrac{1}{2}) \left(\tfrac{1}{2}|x|\right)^{-\alpha+1/2} \left[J_{\alpha-1/2}(|x|) - i \operatorname{sign}(x) J_{\alpha+1/2}(|x|) \right].\end{aligned}$$

Using the definition of K and the formula for V, we then obtain

$$K(x, -\mathrm{i}y) = V\mathrm{e}^{\mathrm{i}\langle x,y\rangle} = E_\alpha(x_1y_1)E_\beta(x_2y_2).$$

If $\beta = 0$, so that we are dealing with the weight function $h_\alpha(x) = |x|^\alpha$ associated with $\mathbb{Z}_2$, then $K(x,-\mathrm{i}y) = E_\alpha(x_1y_1)$. In particular, recall the κ-Bessel function defined by

$$K_W(x,y) = \frac{1}{|W|}\sum_{w\in W} K(x, yw).$$

Then, in the case of $\mathbb{Z}_2$, we obtain

$$\begin{aligned} K_W(x,-\mathrm{i}y) &= \tfrac{1}{2}\big[K(x,-\mathrm{i}y) + K(x,\mathrm{i}y)\big] \\ &= \Gamma(\alpha+\tfrac{1}{2})\big(\tfrac{1}{2}|x_1y_1|\big)^{-\alpha+1/2} J_{\alpha-1/2}(|x_1y_1|), \end{aligned}$$

so that the transform $\widehat{f}$ specializes to the classical Hankel transform on $\mathbb{R}_+$.

Let us mention one interesting formula that follows from this set-up. By Corollary 7.4.5,

$$c_h' \int_{S^1} K(x,-\mathrm{i}y) h_{\alpha,\beta}(y)\, \mathrm{d}\omega(y) = c_{\alpha+\beta+1/2}\int_{-1}^{1} \mathrm{e}^{-\mathrm{i}\|x\|s}(1-s^2)^{\alpha+\beta-1/2}\,\mathrm{d}s.$$

Using the definition of the Bessel function and verifying the constants, we end up with the following formula:

$$\left(\frac{\|x\|}{2}\right)^{-(\alpha+\beta+1/2)} J_{\alpha+\beta+1/2}(\|x\|) = \frac{1}{2}\int_{S^1} E_\alpha(x_1y_1)E_\beta(x_2y_2)h_{\alpha,\beta}(y_1,y_2)\,\mathrm{d}\omega,$$

which gives an integral transform between Bessel functions.

7.8 Notes

If $\gamma_\kappa = \sum \kappa_v = 0$, h-harmonics reduce to ordinary harmonics. For ordinary harmonics, the definition of H_α in (7.1.4) is called Maxwell's representation (see, for example, Müller [1997, p. 69]). For h-harmonics, (7.1.4) is studied in Xu [2000a]. The proofs of Theorems 7.1.15 and 7.1.17 are different from the original proof in Dunkl [1988].

A maximum principle for h-harmonic polynomials was given in Dunkl [1988]: if f is a real non-constant h-harmonic polynomial then it cannot have an absolute minimum in B^d; if, in addition, f is W-invariant then f cannot have a local minimum in B^d.

The inner products were studied in Dunkl [1991]. The idea of using $\mathrm{e}^{\Delta_h/2}$ given in Theorem 7.2.7 first appeared in Macdonald [1982] in connection with $\langle p,q\rangle_\partial$.

The integration of the intertwining operator in Section 7.4 was proved and used in Xu [1997a] to study the summability of orthogonal expansions; see Chapter 9.

The $\mathbb{Z}_2^d$ case was studied in Xu [1997b]. The explicit expression for the intertwining operator V in Theorem 7.5.4 is essential for deeper analysis on the sphere; see Dai and Xu [2013] and its references. It remains essentially the only case for which a useful integral expression is known. The monic h-harmonics were studied in Xu [2005c].

An interesting connection with the orthogonal polynomials of the $\mathbb{Z}_2^d$ case appears in Volkmer [1999]. See also Kalnins, Miller and Tratnik [1991] for related families of orthogonal polynomials on the sphere.

The dihedral case was studied in Dunkl [1989a, b]. When $\beta = 0$, the polynomials f_n in Section 7.6 are a special case ($\beta = \alpha + 1$) of the Heisenberg polynomials $E_n^{(\alpha,\beta)}(z)$, defined in Dunkl [1982] through the generating function

$$\sum_{n=0}^{\infty} E_n^{(\alpha,\beta)}(z)t^n = (1-t\bar{z})^{-\alpha}(1-tz)^{-\beta}.$$

Many properties of the polynomials f_n can be derived using this generating function, for example, the properties in Proposition 7.6.8.

The weight functions in (7.6.2) are special cases of the so-called generalized Jacobi polynomials; see Badkov [1974] as well as Nevai [1979]. The dihedral symmetry allows us to write down explicit formulae for the orthogonal polynomials, which are available only for some generalized Jacobi weight functions.

The Dunkl transform for reflection-invariant measures was defined and studied in Dunkl [1992]. The positivity of V is not necessary for the proof. When the Dunkl transform is restricted to even functions on the real line, it becomes the Hankel transform. For the classical result on Fourier transforms and Hankel transforms, see Stein and Weiss [1971] and Helgason [1984]. One can define a convolution structure on the weighted space $L^2(h^2, \mathbb{R}^d)$ and study harmonic analysis in the Dunkl setting; most deeper results in analysis, however, have been established only the case of $\mathbb{Z}_2^d$ using the explicit integral formula of V; see Thangavelu and Xu [2005] and Dai and Wang [2010].

The only other case for which an explicit formula for the intertwining operator V is known is that for the symmetric group S_3, discovered in Dunkl [1995]. It is natural to speculate that V is an integral transform in all cases. However, even in the case of dihedral groups we do not know the complete answer. A partial result in the case of the dihedral group I_4 was given in Xu [2000e]. For the case B_2, partial results can be found in Dunkl [2007].

8

Generalized Classical Orthogonal Polynomials

In this chapter we study orthogonal polynomials that generalize the classical orthogonal polynomials. These polynomials are orthogonal with respect to weight functions that contain a number of parameters, and they reduce to the classical orthogonal polynomials when some parameters go to zero. Further, they satisfy a second-order differential–difference equation which reduces to the differential equation for the classical orthogonal polynomials. The generalized orthogonal polynomials that we consider are those on the unit ball and the simplex and the generalized Hermite and Laguerre polynomials.

8.1 Generalized Classical Orthogonal Polynomials on the Ball

The classical orthogonal polynomials on the ball B^d were discussed in Section 5.2. We study their generalizations by multiplying the weight function by a reflection-invariant function.

8.1.1 Definition and differential–difference equations

Definition 8.1.1 Let h_κ be a reflection-invariant weight function as in Definition 7.1.1. Define a weight function on B^d by

$$W_{\kappa,\mu}^B(x) = h_\kappa^2(x)(1-\|x\|^2)^{\mu-1/2}, \qquad \mu > -\tfrac{1}{2}.$$

The polynomials that are orthogonal to $W_{\kappa,\mu}^B$ are called generalized classical orthogonal polynomials on B^d.

Recall that c_h' is the normalization constant for h_κ^2 over S^{d-1}; it can be seen by using polar coordinates that the normalization constant of $W_{\kappa,\mu}^B$ is given by

$$w_{\kappa,\mu}^B = \left(\int_{B^d} W_{\kappa,\mu}^B(x)\,\mathrm{d}x\right)^{-1} = \frac{2c_h'}{\sigma_{d-1}B(\gamma_\kappa+\frac{d}{2},\mu+\frac{1}{2})},$$

where σ_{d-1} is the surface area of S^{d-1} and $\gamma_\kappa = \sum_{v\in R_+} \kappa_v$ is the homogeneous degree of h_κ.

If $\kappa = 0$, i.e., W is the orthogonal group then $h_\kappa(x) = 1$ and the weight function $W^B_{\kappa,\mu}(x)$ is reduced to $W^B_\mu(x) = (1-\|x\|^2)^{\mu-1/2}$, the weight function of the classical orthogonal polynomials on the ball.

We will study these polynomials by relating them to h-harmonics. Recall Theorem 4.2.4, which gives a correspondence between orthogonal polynomials on B^d and those on S^d; using the same notation for polar coordinates,

$$y = r(x, x_{d+1}) \qquad \text{where} \quad r = \|y\|, \quad (x, x_{d+1}) \in S^d,$$

we associate with $W^B_{\kappa,\mu}$ a weight function

$$h_{\kappa,\mu}(x, x_{d+1}) = h_\kappa(x)|x_{d+1}|^\mu = \prod_{v\in R_+} |\langle x, v\rangle|^{\kappa_v} |x_{d+1}|^\mu \tag{8.1.1}$$

defined on S^d, which is invariant under the reflection group $W \times \mathbb{Z}_2$. Throughout this section we shall set

$$\lambda_{\kappa,\mu} = \gamma_k + \mu + \frac{d-1}{2} \tag{8.1.2}$$

and will write γ and λ for γ_κ and $\lambda_{\kappa,\mu}$ in the proof sections.

The weight function $h_{\kappa,\mu}$ is homogeneous and is S-symmetric on $\mathbb{R}^{d+1}$ (Definition 4.2.1). By Theorem 4.2.4 the orthogonal polynomials

$$Y_\alpha(y) = r^n P_\alpha(x), \qquad P_\alpha \in \mathscr{V}^d_n(W^B_{\kappa,\mu})$$

(denoted $Y^{(1)}_\alpha$ in Section 4.2) are h-harmonics associated with $h_{\kappa,\mu}$. Consequently they satisfy the equation $\Delta_h^{W\times\mathbb{Z}_2} Y_\alpha = 0$, where we denote the h-Laplacian associated with $h_{\kappa,\mu}$ by $\Delta_h^{W\times\mathbb{Z}_2}$ to distinguish it from the usual h-Laplacian Δ_h.

For $v \in R_+$ (the set of positive roots of W), write $\tilde{v} = (v, 0) \in \mathbb{R}^{d+1}$, which is a vector in $\mathbb{R}^{d+1}$. By Theorem 6.4.9, for $y \in \mathbb{R}^{d+1}$,

$$\begin{aligned} \Delta_h^{W\times\mathbb{Z}_2} f(y) &= \Delta f(y) + \sum_{v\in R_+} k_v \Big(\frac{2\langle \nabla f(y), \tilde{v}\rangle}{\langle y, \tilde{v}\rangle} - \frac{f(y) - f(y\sigma_v)}{\langle y, \tilde{v}\rangle^2} \|\tilde{v}\|^2 \Big) \\ &\quad + \mu \Big(\frac{2}{y_{d+1}} \frac{\partial f(x)}{\partial y_{d+1}} - \frac{f(y) - f(y_1, \ldots, y_d, -y_{d+1})}{y^2_{d+1}} \Big) \end{aligned} \tag{8.1.3}$$

for $h_{\kappa,\mu}$, where $\nabla = (\partial_1 \quad \cdots \quad \partial_{d+1})^{\mathrm{T}}$ is the gradient operator and the reflection σ_v acts on the first d variables of y. We use $\Delta^{W\times\mathbb{Z}_2}_{\kappa,\mu}$ to derive a second-order differential–difference equation for polynomials orthogonal with respect to $W_{\kappa,\mu}$, by a change of variables.

Since the polynomials Y_α defined above are even in y_{d+1}, we need to deal only with the upper half space $\{y \in \mathbb{R}^{d+1} : y_{d+1} \geq 0\}$. In order to write the operator for $P^n_\alpha(W^B_{\kappa,\mu})$ in terms of $x \in B^d$, we choose the following mapping:

$$y \mapsto (r, x): \quad y_1 = rx_1, \ldots, y_d = rx_d, \quad y_{d+1} = r\sqrt{1 - x_1^2 - \cdots - x_d^2},$$

which is one-to-one from $\{y \in \mathbb{R}^{d+1} : y_{d+1} \ge 0\}$ to itself. We rewrite the h-Laplacian Δ_h in terms of the new coordinates (r,x).

Proposition 8.1.2 *Acting on functions on $\mathbb{R}^{d+1}$ that are even in y_{d+1}, the operator $\Delta_h^{W\times\mathbb{Z}_2}$ takes the form*

$$\Delta_h^{W\times\mathbb{Z}_2} = \frac{\partial^2}{\partial r^2} + \frac{2\lambda_{\kappa,\mu}+1}{r}\frac{\partial}{\partial r} + \frac{1}{r^2}\Delta_{h,0}^{W\times\mathbb{Z}_2}$$

in the coordinates (r,x) in $\{y\in\mathbb{R}^{d+1} : y_{d+1}\ge 0\}$, where the spherical part $\Delta_{h,0}^{W\times\mathbb{Z}_2}$, acting on functions in the variables x, is given by

$$\Delta_{h,0}^{W\times\mathbb{Z}_2} = \Delta_h^{(x)} - \langle x, \nabla^{(x)}\rangle^2 - 2\lambda_{\kappa,\mu}\langle x, \nabla^{(x)}\rangle,$$

in which Δ_h denotes the h-Laplacian for h_κ in $W_{\kappa,\mu}^B$.

Proof We make the change of variables $y \mapsto (r,x)$ on $\{y \in \mathbb{R}^{d+1} : y_{d+1} \ge 0\}$. Then, by Proposition 4.1.6, we conclude that

$$\Delta = \frac{\partial^2}{\partial r^2} + \frac{d}{r}\frac{\partial}{\partial r} + \frac{1}{r^2}\Delta_0,$$

where the spherical part Δ_0 is given by

$$\Delta_0 = \Delta^{(x)} - \langle x, \nabla^{(x)}\rangle^2 - (d-1)\langle x, \nabla^{(x)}\rangle; \tag{8.1.4}$$

here $\Delta^{(x)}$ and $\nabla^{(x)}$ denote the Laplacian and the gradient with respect to x, respectively. As shown in the proof of Proposition 4.1.6, the change of variables implies that

$$\frac{\partial}{\partial y_i} = x_i\frac{\partial}{\partial r} + \frac{1}{r}\left(\frac{\partial}{\partial x_i} - x_i\langle x, \nabla^{(x)}\rangle\right), \qquad 1 \le i \le d+1,$$

with $\partial/\partial_{d+1} = 0$. Since we are assuming that the operator acts on functions that are even in y_{d+1}, it follows from (8.1.1) that the last term in $\Delta_h^{W\times\mathbb{Z}_2}$, see (8.1.3), becomes

$$2\mu\frac{1}{y_{d+1}}\frac{\partial}{\partial y_{d+1}} = 2\mu\left(\frac{1}{r}\frac{\partial}{\partial r} - \frac{1}{r^2}\langle x, \nabla^x\rangle\right),$$

where we have used the fact that $y_{d+1} = rx_{d+1}$ and $x_{d+1}^2 = 1 - x_1^2 - \cdots - x_d^2$. Moreover, under the change of variables $y \mapsto (r,x)$ we can use (8.1.1) to verify that

$$\frac{\langle \tilde{v}, \nabla\rangle}{\langle y, \tilde{v}\rangle} = \frac{1}{r}\frac{\partial}{\partial r} + \frac{1}{r^2}\frac{\langle v, \nabla^{(x)}\rangle}{\langle x, v\rangle} - \frac{1}{r^2}\langle x, \nabla^{(x)}\rangle.$$

Since σ_v acts on the first d variables of y_{d+1}, the difference part in the sum in equation (8.1.3) becomes $r^{-2}(1-\sigma_v)/\langle x, v\rangle$. Upon summing over $v \in R_+$, we conclude from (8.1.3) that $\Delta_h^{W\times\mathbb{Z}_2}$ takes the stated form, with $\Delta_{0,h}^{W\times\mathbb{Z}_2}$ given by

$$\Delta_{h,0}^{W\times\mathbb{Z}_2} = \Delta_0 - 2(\gamma+\mu)\langle x, \nabla^{(x)}\rangle + \sum_{v\in R_+} \kappa_v \Big(\frac{2\langle v, \nabla^{(x)}\rangle}{\langle x, v\rangle} - \frac{1-\sigma_v}{\langle x, v\rangle^2}\|v\|^2\Big).$$

The desired formula for $\Delta_{0,h}^{W\times\mathbb{Z}_2}$ then follows from the formula for Δ_0 in (8.1.4) and from the formula for the h-Laplacian in Theorem 6.4.9. □

Recall that $\mathscr{V}_n^d(W_{\kappa,\mu}^B)$ stands for the space of orthogonal polynomials of degree n with respect to the weight function $W_{\kappa,\mu}^B$ on B^d.

Theorem 8.1.3 *The orthogonal polynomials in the space $\mathscr{V}_n^d(W_{\kappa,\mu}^B)$ satisfy the differential–difference equation*

$$\Big(\Delta_h - \langle x, \nabla\rangle^2 - 2\lambda_{\kappa,\mu}\langle x, \nabla\rangle\Big)P = -n(n+2\lambda_{\kappa,\mu})P,$$

where ∇ is the ordinary gradient operator acting on $x\in\mathbb{R}^d$.

Proof Let P_α be an orthogonal polynomial in $\mathscr{V}_n^d(W_{\kappa,\mu}^B)$. The homogeneous polynomial $Y_\alpha(y) = r^n P_\alpha(x)$ satisfies the equation $\Delta_h^{W\times\mathbb{Z}_2} Y_\alpha = 0$. Since Y_α is homogeneous of degree n, using Proposition 8.1.2 we conclude that

$$0 = \Delta_h^{W\times\mathbb{Z}_2} Y_\alpha(y) = r^{n-2}\big[n(n+d+2\gamma+2\mu-1)P_\alpha(x) + \Delta_{h,0}^{W\times\mathbb{Z}_2} P_\alpha(x)\big]$$

and the stated result follows from the formula for $\Delta_{h,0}^{W\times\mathbb{Z}_2}$. □

Written explicitly, the equation in Theorem 8.1.3 takes the form

$$\Delta_h P - \sum_{j=1}^d \frac{\partial}{\partial x_j} x_j\Big((2\gamma_\kappa+2\mu-1)P + \sum_{i=1}^d x_i\frac{\partial P}{\partial x_i}\Big) = -(n+d)(n+2\gamma_\kappa+2\mu-1)P.$$

In particular, if W is the rotation group $O(d)$ or $h_\kappa(x) = 1$ then $\Delta_h = \Delta$ and the difference part in the equation disappears. In that case the weight function $W_{\kappa,\mu}^B$ becomes the classical weight function $W_\mu(x) = (1-\|x\|^2)^{\mu-1/2}$, and the differential–difference equation becomes the second-order differential equation (5.2.3) satisfied by the classical orthogonal polynomials in Section 5.2.

Corollary 8.1.4 *Under the correspondence $Y(y) = r^n P(x)$, the partial differential equation satisfied by the classical orthogonal polynomials on B^d is equivalent to $\Delta_h Y = 0$ for $h_\kappa(y) = |y_{d+1}|^\mu$.*

The weight function given below is invariant under an abelian group and is more general than W_μ.

The abelian group $\mathbb{Z}_2^d$ The weight function is

$$W_{\kappa,\mu}^B(x) = \prod_{i=1}^d |x_i|^{2\kappa_i}(1-\|x\|^2)^{\mu-1/2}. \qquad (8.1.5)$$

The normalization constant of this weight function is given by

$$\int_{B^d} W^B_{\kappa,\mu}(x)\,\mathrm{d}x = \frac{\Gamma(\kappa_1+\frac12)\cdots\Gamma(\kappa_d+\frac12)\Gamma(\mu+\frac12)}{\Gamma(|\kappa|+\mu+\frac{d+1}{2})}.$$

The h-Laplacian associated with $h_\kappa^2(x) := \prod_{i=1}^d |x_i|^{\kappa_i}$ is given by

$$\Delta_h = \Delta + \sum_{i=1}^d \kappa_i \left(\frac{2}{x_i}\partial_i - \frac{1-\sigma_i}{x_i^2}\right), \tag{8.1.6}$$

where σ_i is defined by $x\sigma_i = x - 2x_i\varepsilon_i$ and ε_i is the ith coordinate vector. We just give Δ_h here, since the differential–difference equation can be easily written down using Δ_h.

Two other interesting cases are the symmetric group S_d and the hyperoctahedral group W_d. We state the corresponding operators and the weight functions below.

The symmetric group S_d The weight function is

$$W^B_{\kappa,\mu}(x) = \prod_{1\le i<j\le d} |x_i - x_j|^{2\kappa}(1-\|x\|^2)^{\mu-1/2} \tag{8.1.7}$$

and the corresponding h-Laplacian is given by

$$\Delta_h = \Delta + 2\kappa \sum_{1\le i<j\le d} \frac{1}{x_i - x_j}\left((\partial_i - \partial_j) - \frac{1-(i,j)}{x_i - x_j}\right), \tag{8.1.8}$$

where $(i,j)f(x)$ means that x_i and x_j in $f(x)$ are exchanged.

The hyperoctahedral group W_d The weight function is

$$W^B_{\kappa,\mu}(x) = \prod_{i=1}^d |x_i|^{2\kappa'} \prod_{1\le i\le j} |x_i^2 - x_j^2|^{2\kappa}(1-\|x\|^2)^{\mu-1/2}, \tag{8.1.9}$$

where κ and κ' are two nonnegative real numbers.

Since the group W_d has two conjugacy classes of reflections, see Subsection 6.2.2, there are two parameters for h_κ. The corresponding h-Laplacian is given by

$$\begin{aligned}\Delta_h = \Delta + \kappa' \sum_{i=1}^d \left(\frac{2}{x_i}\partial_i - \frac{1-\sigma_i}{x_i^2}\right) + 2\kappa \sum_{1\le i<j\le d}\left(\frac{\partial_i - \partial_j}{x_i - x_j} - \frac{1-\sigma_{ij}}{(x_i - x_j)^2}\right)\\ + 2\kappa \sum_{1\le i<j\le d}\left(\frac{\partial_i + \partial_j}{x_i + x_j} - \frac{1-\tau_{ij}}{(x_i + x_j)^2}\right),\end{aligned} \tag{8.1.10}$$

where σ_i, σ_{ij} and τ_{ij} denote the reflections in the hyperoctahedral group W_d defined by $x\sigma_i = x - 2x_i\varepsilon_i$, $x\sigma_{ij} = x(i,j)$ and $\tau_{ij} = \sigma_i\sigma_{ij}\sigma_i$, and the same notation is used to denote the operators defined by the action of these reflections.

Note that the equations on B^d are derived from the corresponding h-Laplacian in a similar way to the deduction of Proposition 7.5.3 for Jacobi polynomials.

8.1.2 Orthogonal basis and reproducing kernel

Just as in Section 5.2 for the classical orthogonal polynomials, an orthonormal basis of $\mathscr{V}_n^d(W_{\kappa,\mu}^B)$ can be given in terms of the h-harmonics associated with h_κ on S^{d-1} and the generalized Gegenbauer polynomials (Subsection 1.5.2) using polar coordinates. Let

$$\lambda_\kappa = \gamma_\kappa + \tfrac{d-1}{2}, \qquad \text{so that} \qquad \lambda_{\kappa,\mu} = \lambda_k + \mu.$$

Proposition 8.1.5 *For $0 \le j \le \frac{n}{2}$ let $\{Y_{\nu,n-2j}^h\}$ denote an orthonormal basis of $\mathscr{H}_{n-2j}^d(h_\kappa^2)$; then the polynomials*

$$P_{\beta,j}^n(W_{\kappa,\mu}^B;x) = (c_{j,n}^B)^{-1} P_j^{(\mu-1/2,n-2j+\lambda_\kappa-1/2)}(2\|x\|^2-1) Y_{\nu,n-2j}^h(x)$$

form an orthonormal basis of $\mathscr{V}_n^d(W_{\kappa,\mu}^B)$; the constant is given by

$$[c_{j,n}^B]^2 = \frac{(\mu+\frac{1}{2})_j(\lambda_k+\frac{1}{2})_{n-j}(n-j+\lambda_{\kappa,\mu})}{j!(\lambda_{\kappa,\mu}+1)_{n-j}(n+\lambda_{\kappa,\mu})}.$$

The proof of this proposition is almost identical to that of Proposition 5.2.1.

Using the correspondence between orthogonal polynomials on B^d and on S^d in (4.2.2), an orthogonal basis of $\mathscr{V}_n^d(W_{\kappa,\mu}^B)$ can also be derived from the h-harmonics associated with $h_{\kappa,\mu}$ on S^{d+1}. For a general reflection group, however, it is not easy to write down an explicit orthogonal basis. We give one explicit basis in a Rodrigues-type formula where the partial derivatives are replaced by Dunkl operators.

Definition 8.1.6 For $\alpha \in \mathbb{N}_0^d$, define

$$U_\alpha^\mu(x) := (1-\|x\|^2)^{-\mu+1/2}\mathscr{D}^\alpha(1-\|x\|^2)^{|\alpha|+\mu-1/2}.$$

When all the variables in κ are 0, $\mathscr{D}^\alpha = \partial^\alpha$ and we are back to orthogonal polynomials of degree n in d variables, as seen in Proposition 5.2.5.

Proposition 8.1.7 *Let $\alpha \in \mathbb{N}_0^d$ and $|\alpha| = n$. Then U_α^μ is a polynomial of degree n, and we have the recursive formula*

$$U_{\alpha+e_i}^\mu(x) = -(2\mu+1)x_i U_\alpha^{\mu+1}(x) + (1-\|x\|^2)\mathscr{D}_i U_\alpha^{\mu+1}(x).$$

Proof Since $(1-\|x\|^2)^a$ is invariant under the reflection group, a simple computation shows that

$$\begin{aligned}\mathscr{D}_i U_\alpha^{\mu+1}(x) &= \partial_i \left[(1-\|x\|^2)^{-\mu-1/2}\right] \mathscr{D}^\alpha (1-\|x\|^2)^{|\alpha|+\mu+1/2} \\ &\quad + (1-\|x\|^2)^{-\mu-1/2} \mathscr{D}^{\alpha+e_i}(1-\|x\|^2)^{|\alpha|+\mu+1/2} \\ &= (2\mu+1)x_i(1-\|x\|^2)^{-1}U_\alpha^{\mu+1}(x) + (1-\|x\|^2)^{-1}U_{\alpha+e_i}^{\mu}(x),\end{aligned}$$

which proves the recurrence relation. That U_α^μ is a polynomial of degree $|\alpha|$ is a consequence of this relation, which can be proved by induction if necessary. □

The following proposition plays a key role in proving that U_α is in fact an orthogonal polynomial with respect to $W_{\kappa,\mu}$ on B^d.

Proposition 8.1.8 *Assume that p and q are continuous functions and that p vanishes on the boundary of B^d. Then*

$$\int_{B^d} \mathscr{D}_i p(x) q(x) h_\kappa^2(x)\, dx = -\int_{B^d} p(x) \mathscr{D}_i q(x) h_\kappa^2(x)\, dx.$$

Proof The proof is similar to the proof of Theorem 7.1.18. We shall be brief. Assume that $\kappa_v \ge 1$. Analytic continuation can be used to extend the range of validity to $\kappa_v \ge 0$. Integration by parts shows that

$$\begin{aligned}\int_{B^d} \partial_i p(x) q(x) h_\kappa^2(x)\, dx &= -\int_{B^d} p(x)\partial_i [q(x)h_\kappa^2(x)]\, dx \\ &= -\int_{B^d} p(x)\partial_i q(x) h_\kappa^2(x)\, dx \\ &\quad -2\int_{B^d} p(x)q(x)h_\kappa(x)\partial_i h_\kappa(x)\, dx.\end{aligned}$$

Let D_i denote the difference part of $\mathscr{D}_i$: $\mathscr{D}_i = \partial_i + D_i$. For a fixed root v, a simple computation shows that

$$\int_{B^d} D_i p(x) q(x) h_\kappa^2(x)\, dx = \int_{B^d} p(x) \sum_{v \in R_+} \kappa_i v_i \frac{q(x)+q(x\sigma_v)}{\langle x, v\rangle} h_\kappa^2(x)\, dx.$$

Adding these two equations and using the fact that

$$h_\kappa(x)\partial_i h_\kappa(x) = \sum_{v\in R_+} \kappa_v v_i \frac{1}{\langle v, x\rangle} h_\kappa^2(x)$$

completes the proof. □

Theorem 8.1.9 *For $\alpha \in \mathbb{N}_0^d$ and $|\alpha| = n$, the polynomials U_α^μ are elements of $\mathscr{V}_n^d(W_{\kappa,\mu})$.*

Proof For each $\beta \in \mathbb{N}_0^d$ and $|\beta| < n$, we claim that

$$\mathscr{D}^\beta (1-\|x\|^2)^{n+\mu-1/2} = (1-\|x\|^2)^{n-|\beta|+\mu-1/2} Q_\beta(x) \tag{8.1.11}$$

for some $Q_\beta \in \Pi_n^d$. This follows from induction. The case $\beta = 0$ is trivial, with $Q_0(x) = 1$. Assume that (8.1.11) holds for $|\beta| < n-1$. Then

$$\mathscr{D}^{\beta+e_i}(1-\|x\|^2)^{n+\mu-1/2} = \mathscr{D}_i\left[(1-\|x\|^2)^{n-|\beta|+\mu-1/2}Q_\beta(x)\right]$$
$$= \partial_i\left[(1-\|x\|^2)^{n-|\beta|+\mu-1/2}\right]Q_\beta(x) + (1-\|x\|^2)^{n-|\beta|+\mu-1/2}\mathscr{D}_iQ_\beta(x)$$
$$= (1-\|x\|^2)^{n-|\beta|+\mu-3/2}\left[-2(n-|\beta|+\mu-\tfrac{1}{2})x_iQ_\beta(x) + (1-\|x\|^2)\mathscr{D}_iQ_\beta(x)\right]$$
$$= (1-\|x\|^2)^{n-|\beta+e_i|+\mu-1/2}Q_{\beta+e_i}(x),$$

where $Q_{\beta+e_i}$, as defined above, is clearly a polynomial of degree $|\beta|+1$, which completes the inductive proof. The identity (8.1.11) shows, in particular, that $\mathscr{D}^\beta(1-\|x\|^2)^{n+\mu-1/2}$ is a function that vanishes on the boundary of B^d if $|\beta|<n$. For any polynomial $p\in\Pi^d_{n-1}$, using Proposition 8.1.8 repeatedly gives

$$\int_{B^d} U^\mu_\alpha(x)p(x)h^2_\kappa(x)(1-\|x\|^2)^{\mu-1/2}\,dx$$
$$= \int_{B^d}\mathscr{D}^\alpha\left[(1-\|x\|^2)^{n+\mu-1/2}\right]p(x)h^2_\kappa(x)\,dx$$
$$= (-1)^n\int_{B^d}\mathscr{D}^\alpha p(x)(1-\|x\|^2)^{n+\mu-1/2}h^2_\kappa(x)\,dx = 0,$$

since $\mathscr{D}_i : \mathscr{P}^d_n \mapsto \mathscr{P}^d_{n-1}$, which implies that $\mathscr{D}^\alpha p(x)=0$ since $p\in\Pi^d_{n-1}$. □

There are other orthogonal bases of $\mathscr{V}^d_n(W_{\kappa,\mu})$, but few can be given explicitly for a general reflection group. For $d\ge 2$, the only exceptional case is $\mathbb{Z}_2^d$, which will be discussed in the next subsection.

The correspondence in (4.2.2) allows us to state a closed formula for the reproducing kernel $\mathbf{P}_n(W^B_{\kappa,\mu})$ of $\mathscr{V}^d_n(W^B_{\kappa,\mu})$ in terms of the intertwining operator V_κ associated with h_κ.

Theorem 8.1.10 *Let V_κ be the intertwining operator associated with h_κ. Then, for the normalized weight function $W^B_{\kappa,\mu}$ in Definition 8.1.1,*

$$\mathbf{P}_n(W^B_{\kappa,\mu};x,y)$$
$$= \frac{n+\lambda_{\kappa,\mu}}{\lambda_{\kappa,\mu}}c_\mu V_\kappa\left[\int_{-1}^1 C_n^{\lambda_{\kappa,\mu}}\left(\langle x,\cdot\rangle + t\sqrt{1-\|x\|^2}\sqrt{1-\|y\|^2}\right)(1-t^2)^{\mu-1}dt\right](y),$$

where $c_\mu = [B(\frac{1}{2},\mu)]^{-1}$.

Proof The weight function $h_{\kappa,\mu}$ in (8.1.1) is invariant under the group $W\times\mathbb{Z}_2$. Its intertwining operator, denoted by $V_{\kappa,\mu}$, satisfies the relation $V_{\kappa,\mu} = V_\kappa \times V_\mu^{\mathbb{Z}_2}$, where V_κ is the intertwining operator associated with h_κ and $V_\mu^{\mathbb{Z}_2}$ is the intertwining operator associated with $h_\mu(y)=|y_{d+1}|^\mu$. Hence, it follows from the closed formula for $V_\mu^{\mathbb{Z}_2}$ in Theorem 7.5.4 that

$$V_{\kappa,\mu}f(x,x_{d+1}) = c_\mu\int_{-1}^1 V_\kappa[f(\cdot,x_{d+1}t)](x)(1+t)(1-t^2)^{\mu-1}dt.$$

Using the fact that $\mathbf{P}_n(W^B_{\kappa,\mu})$ is the kernel corresponding to h-harmonics that are even in y_{d+1}, as in Theorem 4.2.8, the formula follows from Corollary 7.3.2. □

8.1.3 Orthogonal polynomials for $\mathbb{Z}_2^d$-invariant weight functions

In this subsection we consider the orthogonal polynomials associated with the weight function $W^B_{\kappa,\mu}$ defined in (8.1.5), which are invariant under $\mathbb{Z}_2^d$. Here

$$\lambda_\kappa = |\kappa| + \frac{d-1}{2} \qquad \text{and} \qquad \lambda_{\kappa,\mu} = \lambda_\kappa + \mu.$$

We start with an orthonormal basis for $\mathscr{V}_n^d(W^B_{\kappa,\mu})$ derived from the basis of h-harmonics in Theorem 7.5.2. Use the same notation as in Section 5.2: associated with $x=(x_1,\ldots,x_d) \in \mathbb{R}^d$ define $\mathbf{x}_j = (x_1,\ldots,x_j)$ for $1 \le j \le d$ and $\mathbf{x}_0 = 0$. Associated with $\kappa = (\kappa_1,\ldots,\kappa_d)$ define $\kappa^j = (\kappa_j,\ldots,\kappa_d)$, and similarly for $\alpha \in \mathbb{N}_0^d$.

Theorem 8.1.11 *For $d \ge 2$, $\alpha \in \mathbb{N}_0^d$ and $|\alpha| = n$, define*

$$P^n_\alpha(W^B_{\kappa,\mu};x) = (h^B_{\alpha,n})^{-1} \prod_{j=1}^d \left[(1-\|\mathbf{x}_{j-1}\|^2)^{\alpha_j/2} C^{(\mu+\lambda_j,\kappa_j)}_{\alpha_j}\left(\frac{x_j}{\sqrt{1-\|\mathbf{x}_{j-1}\|^2}} \right) \right],$$

where $|\kappa^j| = \kappa_j + \cdots + \kappa_d$, $\lambda_j = |\alpha^{j+1}| + |\kappa^{j+1}| + \frac{d-j}{2}$ and

$$\left[h^B_{\alpha,n}\right]^2 = \frac{1}{(|\kappa| + \frac{d+1}{2})_n} \prod_{j=1}^d h^{(\mu+\lambda_j,\kappa_j)}_{\alpha_j} (\kappa_j+\lambda_j)_{\alpha_j},$$

$h_n^{(\lambda,\mu)}$ being the normalization constant of $C_n^{(\lambda,\mu)}$. Then $\{P^n_\alpha(W^B_{\kappa,\mu}) : |\alpha| = n\}$ is an orthonormal basis of $V_n(W^B_{\kappa,\mu})$ for $W^B_{\kappa,\mu}$ in (8.1.5).

Proof Using the correspondence in (4.2.2), the statement follows from Theorem 7.5.2 if we replace d by $d+1$ and take $\kappa_{d+1} = \mu$; we need only consider $\alpha_{d+1} = 0$. □

In particular, when $\kappa = 0$ this basis reduces to the basis in Proposition 5.2.2 for classical orthogonal polynomials.

The correspondence in Theorem 4.2.4 can also be used for defining monic orthogonal polynomials. Since $x_{d+1}^2 = 1 - \|x\|^2$ for $(x, x_{d+1}) \in S^d$, the monic h-harmonics $\widetilde{R}_\alpha(y)$ in Definition 7.5.8, with d replaced by $d+1$, become monic polynomials in x if $\alpha = (\alpha_1,\ldots,\alpha_d,0)$ and $y = r(x,x_{d+1})$ by the correspondence (4.2.2). This suggests the following definition:

Definition 8.1.12 For $b \in \mathbb{R}^d$ such that $\max_j |b_j| \le 1$, define polynomials $\widetilde{R}^B_\alpha(x)$, $\alpha \in \mathbb{N}_0^d$ and $x \in B^d$ by

$$c_\kappa \int_{[-1,1]^d} \frac{1}{[1-2z(b,x,t)+\|b\|^2]^{\lambda_\kappa}} \prod_{i=1}^d (1+t_i)(1-t_i^2)^{\kappa_i-1}\,\mathrm{d}t = \sum_{\alpha\in\mathbb{N}_0^d} b^\alpha \widetilde{R}^B_\alpha(x),$$

where $z(b,x,t) = (b_1x_1t_1 + \cdots + b_dx_dt_d)$.

The polynomials $\widetilde{R}^B_\alpha$ with $|\alpha| = n$ form a basis of the subspace of orthogonal polynomials of degree n with respect to W^B_κ. It is given by an explicit formula, as follows.

Proposition 8.1.13 *For $\alpha \in \mathbb{N}^d_0$, let $\beta = \alpha - \lfloor \frac{\alpha+1}{2} \rfloor$. Then*

$$\widetilde{R}^B_\alpha(x) = \frac{2^{|\alpha|}(\lambda_{\kappa,\mu})_{|\alpha|}}{\alpha!} \frac{(\frac{1}{2})_{\alpha-\beta}}{(\kappa+\frac{1}{2})_{\alpha-\beta}} R^B_\alpha(x),$$

where $R_\alpha(x)$ is defined by

$$R_\alpha(x) = x^\alpha F_B\left(-\beta, -\alpha+\beta-\kappa+\tfrac{1}{2}; -|\alpha| - \lambda_{\kappa,\mu}+1; \frac{1}{x_1^2}, \ldots, \frac{1}{x_d^2}\right).$$

In particular, $R^B_\alpha(x) = x^\alpha - Q_\alpha(x)$, $Q_\alpha \in \Pi^d_{n-1}$, is the monic orthogonal polynomial with respect to W^B_κ on B^d.

Proof Setting $b_{d+1} = 0$ and $\|x\| = 1$ in the generating function in Definition 7.5.8 shows that the generating function for $\widetilde{R}^B_\alpha$ is the same as that for $\widetilde{R}_{(\beta,0)}(x)$. Consequently $\widetilde{R}^B_\alpha(x) = \widetilde{R}_{(\alpha,0)}(x, x_{d+1})$ for $(x, x_{d+1}) \in S^d$. Since $\widetilde{R}_{(\alpha,0)}(x, x_{d+1})$ is even in its $(d+1)$th variable, the correspondence in (4.2.2) shows that $\widetilde{R}^B_\alpha$ is orthogonal and that its properties can be derived from those of $\widetilde{R}_\alpha$. □

In particular, if $\kappa_i = 0$ for $i = 1, \ldots, d$ and $\kappa_{d+1} = \mu$, so that W^B_κ becomes the classical weight function $(1 - \|x\|^2)^{\mu-1/2}$, then the limit relation (1.5.1) shows that in the limit the generating function becomes the generating function (5.2.5), so that R^B_α coincides with Appell's orthogonal polynomials V_α.

The L^2 norm of R^B_α can also be deduced from the correspondence (4.2.2). Recall that $w^B_{\kappa,\mu}$ is the normalization constant of the weight function $W^B_{\kappa,\mu}$ in (8.1.5). We denote the normalized L^2 norm with respect to $W^B_{\kappa,\mu}$ by

$$\|f\|_{2,B} = \left(w^B_{\kappa,\mu} \int_{B^d} |f(x)|^2 W^B_{\kappa,\mu}(x)\,\mathrm{d}x\right)^{1/2}.$$

As in the case of h-spherical harmonics, the polynomial $x^\alpha - R^B_\alpha = Q^\alpha$ is the best approximation to x^α from the polynomial space Π^d_{n-1} in the L^2 norm. A restatement is the first part of the following theorem.

Theorem 8.1.14 *The polynomial R^B_α has the smallest $\|f\|_{2,B}$ norm among all polynomials of the form $x^\alpha - P(x)$, $P \in \Pi^d_{n-1}$. Furthermore, for $\alpha \in \mathbb{N}^d_0$,*

$$\|R^B_\alpha\|^2_{2,B} = \frac{2\lambda_\kappa \prod_{i=1}^d (\kappa_i + \frac{1}{2})_{\alpha_i}}{(\lambda_{\kappa,\mu})_{|\alpha|}} \int_0^1 \prod_{i=1}^d \frac{C^{(1/2,\kappa_i)}_{\alpha_i}(t)}{k^{(1/2,\kappa_i)}_{\alpha_i}} t^{|\alpha|+2\lambda_{\kappa,\mu}-1}\,\mathrm{d}t.$$

Proof Since the monic orthogonal polynomial R^B_α is related to the h-harmonic polynomial $R_{(\alpha,0)}$ in Subsection 7.5.3 by the formula $R^B_\alpha(x) = R_{(\alpha,0)}(x, x_{d+1})$,

$(x, x_{d+1}) \in S^d$, it follows readily from Lemma 4.2.3 that the norm of R_α can be deduced from that of $R_{(\alpha,0)}$ in Theorem 7.5.13 directly. □

For the classical weight function $W_\mu(x) = (1-\|x\|^2)^{\mu-1/2}$, the norm of R_α can be expressed as the integral of the product Legendre polynomials $P_n(t) = C_n^{1/2}(t)$. Equivalently, as the best approximation in the L^2 norm, it gives the following:

Corollary 8.1.15 *Let $\lambda_\mu = \mu + \frac{d-1}{2} > 0$ and $n = |\alpha|$ for $\alpha \in \mathbb{N}_0^d$. For the classical weight function $W_\mu(x) = (1-\|x\|^2)^{\mu-1/2}$ on B^d,*

$$\min_{Q \in \Pi_{n-1}^d} \|x^\alpha - Q(x)\|_{2,B}^2 = \frac{\lambda_\mu\, \alpha!}{2^{n-1}(\lambda_\mu)_n} \int_0^1 \prod_{i=1}^d P_{\alpha_i}(t) t^{n+2\lambda_\mu-1}\, dt.$$

Proof Set $\kappa_i = 0$ for $1 \le i \le d$ and $\mu = \kappa_{d+1}$ in the formula in Theorem 8.1.14. The stated formula follows from $(1)_{2n} = 2^{2n}(\frac{1}{2})_n(1)_n$, $n! = (1)_n$ and the fact that $C_m^{(1/2,0)}(t) = C_m^{1/2}(t) = P_m(t)$. □

8.1.4 Reproducing kernel for $\mathbb{Z}_2^d$-invariant weight functions

In the case of the space $\mathscr{V}_n^d(W_{\kappa,\mu}^B)$ associated with $\mathbb{Z}_2^d$, we can use the explicit formula for V in Theorem 7.5.4 to give an explicit formula for the reproducing kernel.

Theorem 8.1.16 *For the normalized weight function $W_{\kappa,\mu}^B$ in* (8.1.5) *associated with the abelian group $\mathbb{Z}_2^d$,*

$$\begin{aligned}
&\mathbf{P}_n(W_{\kappa,\mu}^B; x, y) \\
&\quad = \frac{n+\lambda_{\kappa,\mu}}{\lambda_{\kappa,\mu}} \int_{-1}^1 \int_{[-1,1]^d} C_n^{\lambda_{\kappa,\mu}}\left(t_1x_1y_1 + \cdots + t_dx_dy_d + s\sqrt{1-\|x\|^2}\sqrt{1-\|y\|^2}\right) \\
&\qquad \times \prod_{i=1}^d c_{\kappa_i}(1+t_i)(1-t_i^2)^{\kappa_i-1} dt\; c_\mu\, (1-s^2)^{\mu-1} ds.
\end{aligned}$$

Proof This is a consequence of Theorem 4.2.8 and the explicit formula in Theorem 7.5.5, in which we again replace d by $d+1$ and let $\kappa_{d+1} = \mu$. □

In particular, taking the limit $\kappa_i \to 0$ for $i = 1, \ldots, d$ and using (1.5.1) repeatedly, the above theorem becomes Theorem 5.2.8, which was stated without proof in Section 5.2.

In the case of $\mathbb{Z}_2^d$, we can also derive an analogue of the Funk–Hecke formula for $W_{\kappa,\mu}^B$ on the ball. Let us first recall that the intertwining operator associated with the weight function $h_{\kappa,\mu}(x) = \prod_{i=1}^d |x_i|^{\kappa_i} |x_{d+1}|^\mu$ is given by

$$Vf(x) = \int_{[-1,1]^{d+1}} f(t_1x_1, \ldots, t_{d+1}x_{d+1}) \prod_{i=1}^{d+1} c_{\kappa_i}(1+t_i)(1-t_i^2)^{\kappa_i-1} dt,$$

with $\mu = \kappa_{d+1}$. Associated with this operator, define

$$V_B f(x) = \int_{[-1,1]^d} f(t_1 x_1, \dots, t_d x_d) \prod_{i=1}^{d} c_{\kappa_i} (1+t_i)(1-t_i^2)^{\kappa_i - 1} \mathrm{d}t.$$

Then the Funk–Hecke formula on the ball B^d is given by

Theorem 8.1.17 *Let $W^B_{\kappa,\mu}$ be defined as in (8.1.5). Let f be a continuous function on $[-1,1]$ and $P_n \in \mathscr{V}_n^d(W^B_{\kappa,\mu})$. Then*

$$w^B_{\kappa,\mu} \int_{B^d} V_B f(\langle x, \cdot \rangle)(y) P_n(y) W^B_{\kappa,\mu}(y)\,\mathrm{d}y = \tau_n(f) P_n(x), \qquad \|x\| = 1,$$

where $\tau_n(f)$ is the same as in Theorem 7.3.4 with λ_κ replaced by $\lambda_{\kappa,\mu}$.

Proof From the definition of $V_B f$ and the definition of Vf with $\kappa_{d+1} = \mu$, it follows that

$$V_B f(\langle x, \cdot \rangle)(y) = Vf(\langle (x, x_{d+1}), \cdot \rangle)(y, 0) = Vf(\langle (x,0), \cdot \rangle)(y, y_{d+1}),$$

where we take $x \in B^d$ and $x_{d+1} = \sqrt{1 - \|x\|^2}$ so that $(x, x_{d+1}) \in S^d$. In the following we also write $y \in B^d$ and $(y, y_{d+1}) \in S^d$. Then, using the relation between P_n and the h-harmonic Y_n in (4.2.2) and the integration formula in Lemma 4.2.3, it follows that

$$\begin{aligned}
&\int_{B^d} V_B f(\langle x, \cdot \rangle)(y) P_k^n(y) W^B_{\kappa,\mu}(y)\,\mathrm{d}y \\
&\quad = \int_{S^d} Vf(\langle (x,0), \cdot \rangle)(y, y_{d+1}) Y_n(y, y_{d+1}) h_\kappa^2(y, y_{d+1})\,\mathrm{d}\omega \\
&\quad = \tau_n(f) Y_n(x, 0) = \tau_n(f) P_n(x),
\end{aligned}$$

where the second equality follows from the Funk–Hecke formula of Theorem 7.3.4, which requires that $\|(x,0)\| = \|x\| = 1$. □

The most interesting case of the formula occurs when $\kappa = 0$, that is, in the case of the classical weight function W^B_μ on B^d. Indeed, taking the limit $\kappa_i \to 0$, we have that $V_B = id$ for W^B_μ. Hence, we have the following corollary, in which $\lambda_\mu = \mu + \frac{d-1}{2}$.

Corollary 8.1.18 *Let f be a continuous function on $[-1,1]$. For $W^B_\mu(x) = (1 - \|x\|^2)^{\mu - 1/2}$, let $P_n \in \mathscr{V}_n^d(W^B_\mu)$. Then*

$$w^B_\mu \int_{B^d} f(\langle x, y \rangle) P_n(y) W^B_\mu(y)\,\mathrm{d}y = \tau_n(f) P_n(x), \qquad \|x\| = 1,$$

where $\tau_n(f)$ is the same as in Theorem 7.3.4 but with λ_κ replaced by λ_μ.

An interesting consequence of this corollary and Theorem 5.2.8 is the following proposition.

Proposition 8.1.19 *Let η satisfy $\|\eta\| = 1$; then $C_n^{\lambda_\mu}(\langle x, \eta\rangle)$ is an element of $\mathscr{V}_n^d(W_\mu^B)$. Further, if ξ also satisfies $\|\xi\| = 1$ then*

$$\int_{B^d} C_n^{\lambda_\mu}(\langle x, \xi\rangle) C_n^{\lambda_\mu}(\langle x, \eta\rangle) W_\mu^B(x)\,\mathrm{d}x = \frac{\lambda_\mu}{n+\lambda_\mu} C_n^{\lambda_\mu}(\langle \eta, \xi\rangle).$$

Proof Taking $y = \eta$ in the formula of Theorem 5.2.8 gives

$$\mathbf{P}_n(W_\mu^B; x, \eta) = \frac{n+\lambda_\mu}{\lambda_\mu} C_n^{\lambda_\mu}(\langle x, \eta\rangle),$$

which is an element of $\mathscr{V}_n^d(W_\mu^B)$ since $\mathbf{P}_n(W_\mu^B)$ is the reproducing kernel. The displayed equation follows from choosing $f(x) = C_n^{\lambda_\mu}(\langle x, \xi\rangle)$ and $P_n(x) = C_n^{\lambda_\mu}(\langle x, \eta\rangle)$ in Corollary 8.1.18. The constant is

$$\tau_n = \left[C_n^{\lambda_\mu}(1)\right]^{-1} \int_{-1}^{1} \left[C_n^{\lambda_\mu}(t)\right]^2 w_{\lambda_\mu}(t)\,\mathrm{d}t,$$

which can be shown from the structure constants for C_n^λ in Subsection 1.4.3 to equal $\lambda_\mu/(n+\lambda_\mu)$. □

One may ask how to choose a set $\{\eta_i\} \subset S^d$ such that the polynomials $C_n^{\lambda_m u}(\langle x, \eta_i\rangle)$ form a basis for $\mathscr{V}_n^d(W_\mu^B)$ for a given n.

Proposition 8.1.20 *The set $\{C_n^{\lambda_\mu}(\langle x, \eta_i\rangle) : |\eta_i| = 1, 1 \le i \le r_n^d\}$ is a basis for $\mathscr{V}_n^d(W_\mu^B)$ if the matrix $A_n^\mu = \left(C_n^{\lambda_\mu}(\langle \eta_i, \eta_j\rangle)\right)_{i,j=1}^{r_n^d}$ is nonsingular.*

Proof This set is a basis for $\mathscr{V}_n(W_\mu^B)$ if and only if $\{C_n^{\lambda_\mu}(\langle x, \eta\rangle)\}$ is linearly independent. If A_n^μ is nonsingular then $\sum_k c_k C_n^{\lambda_\mu}(\langle x, \eta_i\rangle) = 0$, obtained upon setting $x = \eta_j$, implies that $c_k = 0$ for all k. □

Choosing η_i such that A_n^μ is nonsingular has a strong connection with the distribution of points on the sphere, which is a difficult problem. One may further ask whether it is possible to choose η_i such that the basis is orthonormal. Proposition 8.1.20 shows that the polynomials $C_n^{\lambda_\mu}(\langle x, \eta_i\rangle)$, $1 \le i \le N$, are mutually orthogonal if and only if $C_n^{\lambda_\mu}(\langle \eta_i, \eta_j\rangle) = 0$ for every pair of i, j with $i \ne j$, in other words, if the $\langle \eta_i, \eta_j\rangle$ are zeros of the Gegenbauer polynomial $C_n^{\lambda_\mu}(t)$. While it is unlikely that such a system of points exists in S^d for $d > 2$, the following special result in $d = 2$ is interesting. Recall that $U_n(t)$ denotes the Chebychev polynomial of the second kind, which is the Gegenbauer polynomial of index 1; see Subsection 1.4.3.

Proposition 8.1.21 *For the Lebesgue measure on B^2, an orthonormal basis for the space $\mathscr{V}_n^2$ is given by the polynomials*

$$U_k^n(x_1,x_2) = \frac{1}{\sqrt{\pi}} U_n\Big(x_1 \cos\frac{k\pi}{n+1} + x_2 \sin\frac{k\pi}{n+1}\Big), \qquad 0 \le k \le n.$$

Proof Recall that $U_n(t) = \sin(n+1)\theta/\sin\theta$ with $t = \cos\theta$; the zeros of U_n are $t_k = \cos k\pi/(n+1)$ for $k = 1,\ldots,n$. Let $\eta_i = (\cos\phi_i, \sin\phi_i)$ with $\phi_i = i\pi/(n+1)$ for $i = 0,1,\ldots,n$. Then $\langle \eta_i, \eta_j\rangle = \cos(\phi_i - \phi_j)$ and $\phi_i - \phi_j = \phi_{i-j}$ for $i > j$. Hence, by Proposition 8.1.19, $\{U_n(\langle x, \eta_i\rangle)\}$ is orthogonal with respect to the weight function $W_{1/2}(x) = 1/\pi$. □

The same argument shows that there does not exist a system of points $\{\eta_i : 1 \le i \le n+1\}$ on S^1 such that the polynomials $C_n^{\mu+1/2}(\langle x, \eta_i\rangle)$ are orthonormal for $\mu \ne \frac{1}{2}$.

8.2 Generalized Classical Orthogonal Polynomials on the Simplex

As was shown in Section 4.4, the orthogonal polynomials on the simplex are closely related to those on the sphere and on the ball. The classical orthogonal polynomials on the simplex were discussed in Section 5.3.

8.2.1 Weight function and differential–difference equation

Here we consider orthogonal polynomials with respect to weight functions defined below.

Definition 8.2.1 Let h_κ be a reflection-invariant weight function and assume that h_κ is also invariant under $\mathbb{Z}_2^d$. Define a weight function on T^d by

$$W_{\kappa,\mu}^T(x) := h_\kappa^2(\sqrt{x_1},\ldots,\sqrt{x_d})(1-|x|)^{\mu-1/2}/\sqrt{x_1\ldots x_d}\,, \quad \mu > -\tfrac{1}{2}.$$

We call polynomials that are orthogonal to $W_{\kappa,\mu}^T$ *generalized classical orthogonal polynomials on* T^d.

In Definition 4.4.2 the weight function $W^T = W_{\kappa,\mu}^T$ corresponds to the weight function $W_{\kappa,\mu}^B$ in Definition 8.1.1. Let the normalization constant of $W_{\kappa,\mu}^T$ be denoted by $w_{\kappa,\mu}^T$. Then Lemma 4.4.1 implies that $w_{\kappa,\mu}^T = w_{\kappa,\mu}^B$.

The requirement that h_κ is also $\mathbb{Z}_2^d$-invariant implies that the reflection group G, under which h_κ is defined to be invariant, is a semi-product of another reflection group, G_0, and $\mathbb{Z}_2^d$. Essentially, in the indecomposable case this limits G to two classes, $\mathbb{Z}_2^d$ itself and the hyperoctahedral group. We list the corresponding weight functions below.

The abelian group $\mathbb{Z}_2^d$ The weight function is

$$W_{\kappa,\mu}^T(x;\mathbb{Z}_2^d) = x_1^{\kappa_1-1/2}\cdots x_d^{\kappa_d-1/2}(1-|x|)^{\mu-1/2}, \qquad (8.2.1)$$

where $\kappa_i \ge 0$ and $\mu \ge 0$. The orthogonal polynomials associated with $W^T_{\kappa,\mu}$ are the classical orthogonal polynomials in Section 5.3. This weight function corresponds to $W^B_{\kappa,\mu}$ in (8.1.5).

The hyperoctahedral group The weight function is

$$W^T_{\kappa,\mu}(x) = \prod_{i=1}^{d} x_i^{\kappa'-1/2} \prod_{1\le i<j\le d} |x_i - x_j|^{\kappa}(1-|x|)^{\mu-1/2}, \qquad (8.2.2)$$

where $\kappa' \ge 0$ and $\kappa \ge 0$. This weight function corresponds to $W^B_{\kappa,\mu}(x)$ in (8.1.9).

The relation described in Theorem 4.4.4 allows us to derive differential–difference equations for orthogonal polynomials on T^d. Denote the orthogonal polynomials with respect to $W^T_{\kappa,\mu}$ by $P^n_\alpha(W^T_{\kappa,\mu};x)$. By (4.4.1) these polynomials correspond to the orthogonal polynomials $P^{2n}_\alpha(W^B_{\kappa,\mu};x)$ of degree $2n$ on B^d, which are even in each of their variables. Making the change of variables $x_i \mapsto \sqrt{z_i}$ gives

$$\frac{\partial}{\partial x_i} = 2\sqrt{z_i}\frac{\partial}{\partial z_i} \qquad \text{and} \qquad \frac{\partial^2}{\partial x_i^2} = 2\left(\frac{\partial}{\partial z_i} + 2z_i \frac{\partial^2}{\partial z_i^2}\right).$$

Using these formulae and the explicit formula for the h-Laplacian given in Subsection 8.1.1, we can derive a differential–difference equation for the orthogonal polynomials on T^d from the equation in Theorem 8.1.3.

In the case of $\mathbb{Z}_2^d$, the Dunkl operator $\mathscr{D}_j$, see (7.5.2), takes a particularly simple form; it becomes a purely differential operator when acting on functions that are even in each of their variables. Hence, upon making the change of variables $x_i \mapsto \sqrt{x_i}$ and using the explicit formula for Δ_h in (7.5.3), we can derive from the equation in Theorem 8.1.3 a differential equation for orthogonal polynomials with respect to $W^T_{\kappa,\mu}$.

Theorem 8.2.2 *The classical orthogonal polynomials in $\mathscr{V}^d_n(W^T_{\kappa,\mu})$ satisfy the partial differential equation*

$$\sum_{i=1}^{d} x_i(1-x_i)\frac{\partial^2 P}{\partial x_i^2} - 2\sum_{1\le i<j\le d} x_i x_j \frac{\partial^2 P}{\partial x_i \partial x_j} + \sum_{i=1}^{d}\left[\left(\kappa_i + \tfrac{1}{2}\right) - \left(|\kappa| + \mu + \tfrac{d+1}{2}\right)x_i\right]\frac{\partial P}{\partial x_i} = \lambda_n P, \qquad (8.2.3)$$

where $\lambda_n = -n(n + |\kappa| + \mu + \frac{d-1}{2})$ and $\kappa_i \ge 0$, $\mu \ge 0$ for $1 \le i \le d$.

This is the same as the equation, (5.3.4), satisfied by the classical orthogonal polynomials on the simplex. Although the above deduction requires $\kappa_i \ge 0$ and $\mu \ge 0$, the analytic continuation shows that it holds for all $\kappa_i > -\frac{1}{2}$ and $\mu > -\frac{1}{2}$.

The above derivation of (8.2.3) explains the role played by the symmetry of T^d. In the case of the hyperoctahedral group, we use the explicit formula for Δ_h given in (8.1.10). The fact that the equation in Theorem 8.1.3 is applied to functions that are even in each variable allows us to drop the first difference part in Δ_h and combine the other two difference parts. The result is as follows.

Theorem 8.2.3 *The orthogonal polynomials in $\mathscr{V}_n^d(W^T_{\kappa,\mu})$ associated with the hyperoctahedral group satisfy the differential–difference equation*

$$\begin{aligned}\Bigg[&\sum_{i=1}^{d} x_i(1-x_i)\frac{\partial^2}{\partial x_i^2} - 2\sum_{1\le i<j\le d} x_i x_j \frac{\partial^2}{\partial x_i \partial x_j}\\ &+\sum_{i=1}^{d}\left(\kappa'+1-2(2\gamma_\kappa+2\mu+d)x_i\right)\frac{\partial}{\partial x_i}\\ &+\kappa\sum_{1\le i<j\le d}\frac{1}{x_i-x_j}\left(2\left(x_i\frac{\partial}{\partial x_i}-x_j\frac{\partial}{\partial x_j}\right)-\frac{x_i+x_j}{x_i-x_j}(1-\sigma_{i,j})\right)\Bigg]P\\ &\qquad = -n(n+\gamma_\kappa+\mu+\tfrac{d-1}{2})P,\end{aligned}$$

where $\gamma_\kappa = \binom{d}{2}\kappa + d\kappa'$.

8.2.2 Orthogonal basis and reproducing kernel

According to Theorem 4.4.6 the orthogonal polynomials associated with $W^T_{\kappa,\mu}$ are related to the h-harmonics associated with $h_{\kappa,\mu}$ in (8.1.1) as shown in Proposition 4.4.8. In the case of the classical weight function $W^T_{\kappa,\mu}$ in (8.2.1) the orthonormal basis derived from Theorems 4.4.4 and 8.1.11 turns out to be exactly the set of the classical orthogonal polynomials given in Section 5.3. There is another way of obtaining an orthonormal basis, based on the following formula for changing variables.

Lemma 8.2.4 *Let f be integrable over T^d. Then*

$$\int_{T^d} f(x)\,\mathrm{d}x = \int_0^1 s^{d-1}\int_{|u|=1} f(su)\,\mathrm{d}u\,\mathrm{d}s.$$

The formula is obtained by setting $x = su$, with $0\le s\le 1$ and $|u|=1$; these can be called the ℓ^1-radial coordinates of the simplex. Let us denote by $\{Q^m_\beta\}_{|\beta|=m}$ a sequence of orthonormal homogeneous polynomials of degree m associated with the weight function $h^2(\sqrt{u_1},\ldots,\sqrt{u_d})/\sqrt{u_1\ldots u_d}$ on the simplex in homogeneous coordinates, as in Proposition 4.4.8.

Proposition 8.2.5 *Let Q^m_β be defined as above. Then the polynomials*

$$P^n_{\beta,m}(x) = b_{m,n}p_{n-m}^{(\gamma_\kappa+2m+(d-2)/2,\mu-1/2)}(2s-1)Q^m_\beta(x)$$

for $0 \le m \le n$ *and* $|\beta| = m$, *where*

$$[b_{m,n}]^2 = (\gamma_\kappa + \tfrac{d}{2})_{2m}/(\gamma_\kappa + \tfrac{d}{2} + \mu + \tfrac{1}{2})_{2m},$$

form an orthonormal basis of $\mathscr{V}_n^d(W_{\kappa,\mu}^T)$ *and the* $P_{\beta,m}^n$ *are homogeneous in the homogeneous coordinates of* T^d.

Proof For brevity, let $\lambda = \gamma + \frac{d-2}{2}$. It follows from Lemma 8.2.4 that

$$\int_{T^d} P_{\beta,m}^n(x) P_{\beta',m'}^n(x) h_\kappa^2(\sqrt{x_1},\dots,\sqrt{x_d})(1-|x|)^{\mu-1/2} \frac{dx}{\sqrt{x_1\cdots x_d}}$$
$$= c\int_0^1 p_{n-m}^{(\lambda+2m,\mu-1/2)}(2s-1)\, p_{n-m'}^{(\lambda+2m',\mu-1/2)}(2s-1)\, s^{\lambda+m+m'}(1-s)^{\mu-1/2}\, ds$$
$$\times \int_{|u|=1} Q_\beta^m(u) Q_{\beta'}^{m'}(u) h_\kappa^2(\sqrt{u_1},\dots,\sqrt{u_d}) \frac{du}{\sqrt{u_1\cdots u_d}},$$

where $c = b_{m,n} b_{m',n}$, from which the pairwise orthogonality follows from that of Q_β^m and of the Jacobi polynomials. It follows from Lemma 8.2.4 that the normalization constant $w_{\kappa,\mu}^T$ of $W_{\kappa,\mu}^T$ is also given by

$$\frac{1}{w_{\kappa,\mu}^T} = \int_{T^d} W_{\kappa,\mu}(x)\, dx$$
$$= \tfrac{1}{2} B\left(\lambda+1, \mu+\tfrac{1}{2}\right) \int_{|u|=1} h_\kappa^2(\sqrt{u_1},\dots,\sqrt{u_d}) \frac{du}{\sqrt{u_1\cdots u_d}}.$$

Hence, multiplying the integral of $[P_{\beta,m}^n(x)]^2$ by $w_{\kappa,\mu}^T$, we can find $b_{m,n}$ from the equation $1 = b_{m,n}^2 B(\lambda+1, \mu+\frac{1}{2})/B(\lambda+2m+1, \mu+\frac{1}{2})$. □

Using the relation to orthogonal polynomials on B^d and Theorem 4.4.4, an orthonormal basis of $\mathscr{V}_n^d(W_{\kappa,\mu}^T)$ can be derived from those given in Proposition 8.1.5, in which $\lambda_\kappa = \gamma_\kappa + \frac{d-1}{2}$.

Proposition 8.2.6 *For* $0 \le j \le 2n$ *let* $\{Y_{\nu,2n-2j}^h\}$ *denote an orthonormal basis of* $\mathscr{H}_{2n-2j}^d(h_\kappa^2, \mathbb{Z}_2^d)$; *then the polynomials*

$$P_{\beta,j}^n(W_{\kappa,\mu}^T; x) = [c_{j,n}^T]^{-1} P_j^{(\mu-1/2, n-2j+\lambda_\kappa-1/2)}(2|x|-1)$$
$$\times Y_{\nu,2n-2j}^h(\sqrt{x_1},\dots,\sqrt{x_d})$$

form an orthonormal basis of $\mathscr{V}_n^d(W_{\kappa,\mu}^T)$, *where* $c_{j,n}^T = c_{j,2n}^B$.

The relation between the orthogonal polynomials on B^d and those on T^d also allows us to derive an explicit formula for the reproducing kernel $\mathbf{P}_n(W_{\kappa,\mu}^T)$ of $\mathscr{V}_n^d(W_{\kappa,\mu}^T)$. Let us introduce the notation $\{x\}^{1/2} = (\sqrt{x_1},\dots,\sqrt{x_d})$ for $x \in T^d$.

Theorem 8.2.7 *Let* V_κ *be the intertwining operator associated with* h_κ *and* $\lambda = \gamma_\kappa + \mu + \frac{d-1}{2}$. *Then*

$$\mathbf{P}_n(W^T_{\kappa,\mu};x,y)$$
$$= \frac{2n+\lambda}{\lambda} c_\mu V_\kappa \Big[\int_{-1}^{1} C^\lambda_{2n}\big(\langle \{x\}^{1/2}, \cdot\rangle + t\sqrt{1-|x|}\sqrt{1-|y|}\,\big)(1-t^2)^{\mu-1}\,\mathrm{d}t \Big](\{y\}^{1/2}).$$

Proof Using Theorem 4.4.4, we see that the reproducing kernel $\mathbf{P}_n(W^T_{\kappa,\mu})$ is related to the reproducing kernel $\mathbf{P}_n(W^B_{\kappa,\mu})$ on the unit ball by

$$\mathbf{P}_n(W^T_{\kappa,\mu};x,y) = \frac{1}{2^d} \sum_{\varepsilon\in\mathbb{Z}_2^d} \mathbf{P}_{2n}\big(W^B_{\kappa,\mu};(\varepsilon_1\sqrt{x_1},\ldots,\varepsilon_d\sqrt{x_d}\,),\{y\}^{1/2}\,\big).$$

In fact, since the reproducing kernel on the left-hand side is unique, it is sufficient to show that the right-hand side has the reproducing property; that is, if we denote the right-hand side temporarily by $Q_n(x,y)$ then

$$\int_{T^d} P(y)Q_n(x,y)W^T_{\kappa,\mu}(y)\,\mathrm{d}y = P(x)$$

for any polynomial $P \in \mathcal{V}^d_n(W^T_{\kappa,\mu})$. We can take $P = P^n_{\beta,j}(W^T_{\kappa,\mu})$ in Proposition 8.2.6. Upon making the change of variables $y_i \mapsto y_i^2$, $1\le i\le d$, and using Lemma 4.4.1 it is easily seen that this equation follows from the reproducing property of $\mathbf{P}_{2n}(W^B_{\kappa,\mu})$. Hence, we can use Theorem 8.1.10 to get an explicit formula for $\mathbf{P}_n(W^T_{\kappa,\mu})$. Since the weight function is invariant under both the reflection group and $\mathbb{Z}_2^d$, and V_κ commutes with the action of the group, it follows that $R(\varepsilon)V_\kappa = V_\kappa R(\varepsilon)$ for $\varepsilon\in\mathbb{Z}_2^d$, where $R(\varepsilon)f(x) = f(x\varepsilon)$. Therefore the summation over $\mathbb{Z}_2^d$ does not appear in the final formula. □

In the case of $\mathbb{Z}_2^d$, we can use the formula for the intertwining operator $V = V_\kappa$ in Theorem 7.5.4 to write down an explicit formula for the reproducing kernel. The result is a closed formula for the classical orthogonal polynomials on T^d.

Theorem 8.2.8 *For the classical orthogonal polynomials associated with the normalized weight function $W^T_{\kappa,\mu}$ in* (8.2.1) *with $\lambda_{\kappa,\mu} = |\kappa|+\mu+\frac{d-1}{2}$,*

$$\mathbf{P}_n(W^T_{\kappa,\mu},x,y)$$
$$= \frac{2n+\lambda_{\kappa,\mu}}{\lambda_{\kappa,\mu}} \int_{-1}^{1}\int_{[-1,1]^d} C^{\lambda_{\kappa,\mu}}_{2n}\,(z(x,y,t,s)) \prod_{i=1}^{d} c_{\kappa_i}(1-t_i^2)^{\kappa_i-1}\mathrm{d}t\,(1-s^2)^{\mu-1}\mathrm{d}s, \tag{8.2.4}$$

where $z(x,y,t,s) = \sqrt{x_1y_1}\,t_1+\cdots+\sqrt{x_dy_d}\,t_d + s\sqrt{1-|x|}\sqrt{1-|y|}$, and if a particular κ_i or μ is zero then the formula holds under the limit relation (1.5.1).

Using the fact that $C^\lambda_{2n}(t) = [(\lambda)_n/(\frac{1}{2})_n]P_n^{\lambda-1/2,-1/2}(2t^2-1)$ and setting $\mu = \kappa_{d+1}$, the identity (8.2.4) becomes (5.3.5) and proves the latter. Some applications of this identity in the summability of orthogonal expansions will be given in Chapter 9.

8.2.3 Monic orthogonal polynomials

The correspondence between the orthogonal polynomials on the simplex and those on the sphere and on the ball can be used to define monic orthogonal polynomials on the simplex. In this subsection we let $\mu = \kappa_{d+1}$ and write $W^T_{\kappa,\mu}$ as W^T_κ. In particular, $\lambda_{\kappa,\mu}$ becomes λ_κ defined by

$$\lambda_k = |\kappa| + \tfrac{d-1}{2}, \qquad |\kappa| = \kappa_1 + \cdots + \kappa_{d+1}.$$

Since the simplex T^d is symmetric in $(x_1,\ldots,x_d,x_{d+1})$, $x_{d+1} = 1-|x|$, we use the homogeneous coordinates $X := (x_1,\ldots,x_d,x_{d+1})$. For the monic h-harmonics defined in Definition 7.5.8, with d replaced by $d+1$ the polynomial $\widetilde{R}_{2\alpha}$ is even in each of its variables, and this corresponds to, using (4.4.1), monic orthogonal polynomials $\widetilde{R}^T_\alpha$ in $\mathscr{V}^d_n(W^T_\kappa)$ in the homogeneous coordinates X. This leads to the following definition.

Definition 8.2.9 For $b \in \mathbb{R}^d$ with $\max_j |b_j| \le 1$ and $\alpha \in \mathbb{N}_0^{d+1}$, define polynomials $\widetilde{R}^T_\alpha(x)$ by

$$c_\kappa \int_{[-1,1]^{d+1}} \frac{1}{[1-2(b_1x_1t_1+\cdots+b_{d+1}x_{d+1}t_{d+1})+\|b\|^2]^{\lambda_\kappa}} \times \prod_{i=1}^{d+1}(1-t_i^2)^{\kappa_i-1}\,\mathrm{d}t = \sum_{\alpha\in\mathbb{N}_0^{d+1}} b^{2\alpha}\widetilde{R}^T_\alpha(x), \qquad x\in T^d.$$

The main properties of $\widetilde{R}^T_\alpha$ are given in the following proposition.

Proposition 8.2.10 *For each $\alpha \in \mathbb{N}_0^{d+1}$ with $|\alpha| = n$, the polynomials*

$$\widetilde{R}^T_\alpha(x) = \frac{2^{2|\alpha|}(\lambda_\kappa)_{2|\alpha|}}{(2\alpha)!}\frac{(\frac12)_\alpha}{(\kappa+\frac12)_\alpha}R^T_\alpha(x),$$

where

$$R^T_\alpha(x) = X^\alpha F_B\Big(-\alpha, -\alpha-\kappa+\tfrac12; -2|\alpha|-\lambda_\kappa+1; \frac{1}{x_1},\ldots,\frac{1}{x_{d+1}}\Big)$$
$$= (-1)^n\frac{(\kappa+\frac12)_\alpha}{(n+|\kappa|+\frac{d-1}{2})_n}F_A\Big(|\alpha|+|\kappa|+\tfrac{d-1}{2}, -\alpha; \kappa+\tfrac12; X\Big)$$

are polynomials orthogonal with respect to W^T_κ on the simplex T^d. Moreover, $R^T_\alpha(x) = X^\alpha - Q_\alpha(x)$, where Q_α is a polynomial of degree at most $n-1$ and $\{R^T_\alpha, \alpha = (\alpha',0), |\alpha| = n\}$ is a basis for $\mathscr{V}^d_n(W^T_\kappa)$.

Proof We return to the generating function for the h-harmonics $\widetilde{R}_\alpha$ in Definition 7.5.8, with d replaced by $d+1$. The explicit formula for $\widetilde{R}_\alpha(x)$ shows that it is even in each variable only if each α_i is even. Let $\varepsilon \in \{-1,1\}^{d+1}$. Then $\widetilde{R}_\alpha(x\varepsilon) = \widetilde{R}_\alpha(\varepsilon_1x_1,\ldots,\varepsilon_{d+1}x_{d+1}) = \varepsilon^\alpha\widetilde{R}_\alpha(x)$. It follows that on the one hand

$$\sum_{\beta\in\mathbb{N}_0^{d+1}} b^{2\beta}\widetilde{R}_{2\beta}(x) = \frac{1}{2^{d+1}}\sum_{\alpha\in\mathbb{N}_0^{d+1}} b^{\alpha}\sum_{\varepsilon\in\{-1,1\}^{d+1}}\widetilde{R}_{\alpha}(x\varepsilon).$$

On the other hand, using the explicit formula for V_κ, the generating function gives

$$\frac{1}{2^{d+1}}\sum_{\varepsilon\in\{-1,1\}^{d+1}}\sum_{\alpha\in\mathbb{N}_0^{d+1}} b^{\alpha}\widetilde{R}_{\alpha}(x\varepsilon)$$

$$= c_\kappa\int_{[-1,1]^{d+1}}\sum_{\varepsilon\in\{-1,1\}^{d+1}}\frac{\prod_{i=1}^{d+1}(1+t_i)(1-t_i^2)^{\kappa_i-1}}{[1-2(b_1x_1t_1\varepsilon_1+\cdots+b_{d+1}x_{d+1}t_{d+1}\varepsilon_{d+1})+\|b\|^2]^{\lambda_\kappa}}\,\mathrm{d}t$$

for $\|x\|=1$. Making the change of variables $t_i\mapsto t_i\varepsilon_i$ in the integral and using the fact that $\sum_\varepsilon\prod_{i=1}^{d+1}(1+\varepsilon_it_i)=2^{d+1}$, we see that the generating function for $\widetilde{R}_{2\beta}(x)$ agrees with the generating function for $\widetilde{R}_{\beta}^{T}(x_1^2,\ldots,x_{d+1}^2)$ in Definition 8.2.9. Consequently, the formulae for R_α^T follow from the corresponding ones for $R_{2\alpha}$ in Propositions 7.5.10 and 7.5.11, both with d replaced by $d+1$. The polynomial R_α^T is homogeneous in X. Using the correspondence between the orthogonal polynomials on S^d and on T^d, we see that the R_α^T are orthogonal with respect to W_κ^T. If $\alpha_{d+1}=0$ then $R_\alpha^T(x)=x^\alpha-Q_\alpha$, which proves the last assertion of the proposition. □

In the case $\alpha_{d+1}=0$, the explicit formula of R_α^T shows that $R_{(\alpha,0)}^T(x)=x^\alpha-Q_\alpha(x)$, so that it agrees with the monic orthogonal polynomial V_α^T in Section 5.3. Setting $b_{d+1}=0$ in Definition 8.2.9 gives the generating function of $R_{(\alpha,0)}^T = V_\alpha^T(x)$.

In the case of a simplex, the polynomial R_α^T is the orthogonal projection of $X^\alpha = x_1^{\alpha_1}\cdots x_d^{\alpha_d}(1-|x|)^{\alpha_{d+1}}$. We compute its L^2 norm. Let

$$\|f\|_{2,T} := \left(w_\kappa^T\int_{T^d}|f(x)|^2W_\kappa^T(x)\,\mathrm{d}x\right)^{1/2}.$$

Theorem 8.2.11 *Let $\beta\in\mathbb{N}_0^{d+1}$. The polynomial R_β^T has the smallest $\|\cdot\|_{2,T}$ norm among all polynomials of the form $X^\beta-P$, $P\in\Pi_{|\beta|-1}^d$, and the norm is given by*

$$\|R_\beta\|_{2,T} = \frac{\lambda_\kappa\left(\kappa+\frac{1}{2}\right)_{2\beta}}{(\lambda_\kappa)_{2|\beta|}}\sum_\gamma\frac{(-\beta)_\gamma\left(-\beta-\kappa+\frac{1}{2}\right)_\gamma}{\left(-2\beta-\kappa+\frac{1}{2}\right)_\gamma\gamma!(2|\beta|-|\gamma|+\lambda_\kappa)}$$

$$= \frac{\lambda_\kappa\beta!\left(\kappa+\frac{1}{2}\right)_\beta}{(\lambda_\kappa)_{2|\beta|}}\int_0^1\prod_{i=1}^{d+1}P_{\beta_i}^{(0,\kappa_i-1/2)}(2r-1)r^{|\beta|+\lambda_\kappa-1}\,\mathrm{d}r.$$

Proof By (4.4.4) the norm of R_α^T is related to the norm of the h-harmonic R_α on S^d. Since $R_\alpha^T(x_1^2,\ldots,x_{d+1}^2)=R_{2\alpha}(x_1,\ldots,x_{d+1})$, the norm of R_α^T can be derived from the norm of $R_{2\alpha}$. Using the fact that $C_{2\beta_i}^{(1/2,\kappa_i)}(t)=P_{\beta_i}^{(0,\kappa_i-1/2)}(2t^2-1)$, the norm $\|R^T_\alpha\|_{2,T}$ follows from making the change of variable $t^2 \mapsto r$ in the integral in Theorem 7.5.13 and replacing d by $d+1$. □

In particular, if $\beta_{d+1} = 0$ then the norm of $R_{(\beta,0)}(x)$ is the smallest norm among all polynomials of the form $x^\beta - P$, $P \in \Pi^d_{n-1}$.

Corollary 8.2.12 *Let $\alpha \in \mathbb{N}^d_0$ and $n = |\alpha|$. Then*

$$\inf_{Q\in\Pi^d_{n-1}} \|x^\alpha - Q(x)\|^2_{2,T} = \frac{\lambda_\kappa \alpha! \prod_{i=1}^d \left(\kappa_i + \frac{1}{2}\right)_{\alpha_i}}{(\lambda_\kappa)_{2|\alpha|}} \times \int_0^1 \prod_{i=1}^d P^{(0,\kappa_i-1/2)}_{\alpha_i}(2r-1) r^{|\alpha|+\lambda_\kappa-1}\, \mathrm{d}r.$$

We note that this is a highly nontrivial identity. In fact, even the positivity of the integral on the right-hand side is not obvious.

8.3 Generalized Hermite Polynomials

The multiple Hermite polynomials are discussed in Subection 5.1.3.

Definition 8.3.1 Let h_κ be a reflection-invariant weight function. The generalized Hermite weight function is defined by

$$W^H_\kappa(x) = h^2_\kappa(x)\mathrm{e}^{-\|x\|^2}, \qquad x \in \mathbb{R}^d.$$

Again recall that c'_h is the normalization constant of h^2_κ over S^{d-1}. Using polar coordinates, it follows that the normalization constant of W^H_κ is given by

$$w^H_\kappa = \left(\int_{\mathbb{R}^d} W^H_\kappa(x)\,\mathrm{d}x\right)^{-1} = \frac{2c'_h}{\sigma_{d-1}\Gamma(\gamma_\kappa + \frac{d}{2})}.$$

If $\kappa = 0$ or W is the orthogonal group then $h_\kappa(x) = 1$ and the weight function $W^H_\kappa(x)$ reduces to the classical Hermite weight function $\mathrm{e}^{-\|x\|^2}$.

An orthonormal basis for the generalized Hermite weight function can be given in terms of h-harmonics and the polynomials H^λ_n on $\mathbb{R}$ in Subsection 1.5.1. We call it the spherical polar Hermite polynomial basis. Denote by $\widetilde{H}^\lambda_n$ the normalized H^λ_n and again let $\lambda_\kappa = \gamma_\kappa + \frac{d-1}{2}$.

Proposition 8.3.2 *For $0 \le 2j \le n$, let $\{Y^h_{\nu,n-2j}\}$ denote an orthonormal basis of $\mathcal{H}^d_{n-2j}(h^2_\kappa)$; then the polynomials*

$$P^n_{\nu,j}(W^H_\kappa; x) = [c^H_{j,n}]^{-1} \widetilde{H}^{n-2j+\lambda_\kappa}_{2j}(\|x\|) Y^h_{\nu,n-2j}(x)$$

form an orthonormal basis of $\mathcal{V}^d_n(W^H_\kappa)$, where $[c^H_{j,n}]^2 = (\lambda_\kappa + \frac{1}{2})_{n-2j}$.

In the case where h_κ is invariant under $\mathbb{Z}_2^d$, an orthonormal basis is given by the product of the generalized Hermite polynomials. We call it the Cartesian Hermite polynomial basis.

Proposition 8.3.3 *For $W_{\kappa,\mu}^H(x) = w_{\kappa,\mu}^H \prod_{i=1}^d |x_i|^{2\kappa_i} e^{-\|x\|^2}$, an orthonormal basis of $\mathscr{V}_n^d(W_\kappa^H)$ is given by $H_\alpha^\kappa(x) = \widetilde{H}_{\alpha_1}^{\kappa_1}(x_1)\cdots\widetilde{H}_{\alpha_d}^{\kappa_d}(x_d)$ for $\alpha \in \mathbb{N}_0^d$ and $|\alpha| = n$.*

In particular, specializing to $\kappa = 0$ leads us back to the classical multiple Hermite polynomials.

There is a limit relation between the polynomials in $\mathscr{V}_n^d(W_\kappa^H)$ and the generalized classical orthogonal polynomials on B^d.

Theorem 8.3.4 *For $P_{\nu,j}^n(W_{\kappa,\mu}^B)$ in Proposition 8.1.5 and $P_{\nu,j}^n(W_\kappa^H)$ in Proposition 8.3.2,*

$$\lim_{\mu\to\infty} P_{\nu,j}^n\left(W_{\kappa,\mu}^B; x/\sqrt{\mu}\right) = P_{\nu,j}^n(W_\kappa^H; x).$$

Proof First, using the definition of the generalized Gegenbauer polynomials, we can rewrite $P_{\nu,j}^n(W_{\kappa,\mu}^B)$ as

$$P_{\beta,j}^n(W_{\kappa,\mu}^B; x) = [b_{j,n}^B]^{-1}\widetilde{C}_{2j}^{(\mu,n-2j+\lambda_\kappa)}(\|x\|)Y_{\nu,n-2j}^h(x),$$

where $\widetilde{C}_{2j}{}^{(\mu,\tau)}$ are the orthonormal polynomials and

$$[b_{j,n}^B]^{-1} = \frac{(\lambda_\kappa+\frac{1}{2})_{n-2j}}{(\lambda_\kappa+\mu+1)_{n-2j}}.$$

Using the limit relation in Proposition 1.5.7 and the normalization constants of both H_n^λ and $C_n^{(\lambda,\mu)}$, we obtain

$$\widetilde{H}_n^\lambda(x) = \lim_{\mu\to\infty} \widetilde{C}_n^{(\mu,\lambda)}(x/\sqrt{\mu}). \tag{8.3.1}$$

Since h-harmonics are homogeneous, $Y_{\nu,j}^h(x/\sqrt{\mu}) = \mu^{-j/2}Y_{\nu,j}^h(x)$. It can also be verified that $\mu^{-(n-2j)/2}b_{j,n}^B \to c_{j,n}^H$ as $\mu \to \infty$. The stated result follows from these relations and the explicit formulae for the bases. □

In the case where h_κ is invariant under $\mathbb{Z}_2^d$, a similar limit relation holds for the other orthonormal basis, see Theorem 8.1.1. The proof uses (8.3.1) and will be left to the reader.

Theorem 8.3.5 *For $P_\alpha^n(W_{\kappa,\mu}^B)$ in Theorem* 8.1.11,

$$\lim_{\mu\to\infty} P_\alpha^n\left(W_{\kappa,\mu}^B; x/\sqrt{\mu}\right) = \widetilde{H}_{\alpha_1}^{\kappa_1}(x_1)\cdots\widetilde{H}_{\alpha_d}^{\kappa_d}(x_d).$$

As a consequence of the limiting relation, we can derive a differential–difference equation for the polynomials in $\mathscr{V}_n^d(W_\kappa^H)$ from the equation satisfied by the classical orthogonal polynomials on B^d.

Theorem 8.3.6 *The polynomials in $\mathscr{V}_n^d(W_\kappa^H)$ satisfy the differential–difference equation*

$$(\Delta_h - 2\langle x, \nabla\rangle)P = -2nP. \tag{8.3.2}$$

Proof Since the polynomials $P_{\nu,j}^n(W_\kappa^H)$ form a basis for $\mathscr{V}_n^d(W_\kappa^H)$, it suffices to prove that these polynomials satisfy the stated equation. The proof is based on the limit relation in Theorem 8.3.4. The $P_{\nu,j}^n(W_{\kappa,\mu}^B)$ satisfy the equation in Theorem 8.1.3. Making the change of variables $x_i \mapsto x_i/\sqrt{\mu}$, it follows that Δ_h becomes $\mu\Delta_h$ and the equation in Theorem 8.1.3 becomes

$$\left[\mu\Delta_h - \sum_{i=1}^d \sum_{j=1}^d x_i x_j \partial_i \partial_j - (2\gamma + 2\mu + d)\langle x, \nabla\rangle\right] P(x/\sqrt{\mu})$$
$$= -n(n+d+2\gamma+2\mu-1)P(x/\sqrt{\mu}).$$

Taking $P = P_{\nu,j}^n(W_{\kappa,\mu}^B)$ in the above equation, dividing the equation by μ and letting $\mu \to \infty$, we conclude that $P_{\nu,j}^n(W_\kappa^H)$ satisfies the stated equation. □

Several special cases are of interest. First, if $h_\kappa(x) = 1$ then Δ_h reduces to the ordinary Laplacian and equation (8.3.2) becomes the partial differential equation (5.1.7) satisfied by the multiple Hermite polynomials. Two other interesting cases are the following.

The symmetric group The weight function is

$$W_\kappa^H(x) = \prod_{1 \le i < j \le d} |x_i - x_j|^{2\kappa} e^{-\|x\|^2}. \tag{8.3.3}$$

The hyperoctahedral group The weight function is

$$W_\kappa^B(x) = \prod_{i=1}^d |x_i|^{2\kappa'} \prod_{1 \le i \le j \le d} |x_i^2 - x_j^2|^\kappa e^{-\|x\|^2}. \tag{8.3.4}$$

These weight functions should be compared with those in (8.1.7) and (8.1.9). In the case of the hyperoctahedral group, equation (8.3.2) is related to the Calogero–Sutherland model in mathematical physics; see the discussion in Section 11.6.

The limiting relation also allows us to derive other properties of the generalized Hermite polynomials, for example, the following Mehler-type formula.

Theorem 8.3.7 *Let $\{P_\alpha^n(W_\kappa^H)\}$, $|\alpha| = n$ and $n \in \mathbb{N}_0$, denote an orthonormal basis with respect to W_κ^H on $\mathbb{R}^d$. Then, for $0 < z < 1$ and $x, y \in \mathbb{R}^d$,*

$$\sum_{n=0}^{\infty}\sum_{|\alpha|=n} P_\alpha^n(W_\kappa^H;x)P_\alpha^n(W_\kappa^H;y)z^n$$
$$= \frac{1}{(1-z^2)^{\gamma_\kappa+d/2}}\exp\left[-\frac{z^2(\|x\|^2+\|y\|^2)}{1-z^2}\right]V_\kappa\left[\exp\left(\frac{2z\langle x,\cdot\rangle}{1-z^2}\right)\right](y).$$

Proof The inner sum over α of the stated formula is the reproducing kernel of $\mathscr{V}_n^d(W_\kappa^H)$, which is independent of the particular basis. Hence, we can work with the basis $P_{\beta,j}^n(W_\kappa^H)$ in Proposition 8.3.2. Recall that $\lambda = \lambda_{\kappa,\mu} = \gamma+\mu+\frac{d-1}{2}$. Then the explicit formula for the reproducing kernel $\mathbf{P}_n(W_{\kappa,\mu}^B)$ in Theorem 8.1.10 can be written as

$$\mathbf{P}_n(W_{\kappa,\mu}^B;x,y) = \frac{n+\lambda}{\lambda}c_\mu\int_{-1}^{1} C_n^\lambda(z)V_\kappa[E_\mu(x,y,\cdot,z)](y)(1-z^2)^{\lambda-1/2}\,\mathrm{d}z,$$

where $z = \langle x,\cdot\rangle + t\sqrt{1-\|x\|^2}\sqrt{1-\|y\|^2}$ and E_μ is defined by

$$E_\mu(x,y,u,v)$$
$$= \left(1-\frac{(v-\langle x,u\rangle)^2}{(1-\|x\|^2)(1-\|y\|^2)}\right)^{\mu-1}\frac{1}{\sqrt{1-\|x\|^2}\sqrt{1-\|y\|^2}(1-v^2)^{\lambda-1/2}}$$

if v satisfies $(v-\langle x,u\rangle)^2 \le (1-\|x\|^2)(1-\|y\|^2)$ and $E_\mu(x,y,u,v)=0$ otherwise. In particular, expanding $V_\kappa[E_\mu(x,y,\cdot,z)]$ as a Gegenbauer series in $C_n^\lambda(z)$, the above formula gives the coefficients of the expansion. Taking into consideration the normalization constants,

$$\frac{c_{\lambda+1/2}}{c_\mu}\sum_{n=0}^{\infty}\frac{\Gamma(2\lambda)\Gamma(n+1)}{\Gamma(n+2\lambda)}\mathbf{P}_n(W_{\kappa,\mu}^B;x,y)C_n^\lambda(z) = V_\kappa[E_\mu(x,y,(\cdot),z)].$$

Replace x and y by $x/\sqrt{\mu}$ and $y/\sqrt{\mu}$, respectively, in the above equation. From the fact that $c_{\lambda+1/2}/c_\mu \to 1$ as $\mu\to\infty$ and that $\mu^{-n}C_n^\mu(z)\to 2^n z^n/n!$, we can use the limit relation in Theorem 8.3.4 to conclude that the left-hand side of the formula displayed above converges to the left-hand side of the stated equation. Moreover, from the definition of E_μ we can write

$$E_\mu\left(\frac{x}{\sqrt{\mu}},\frac{y}{\sqrt{\mu}},\frac{u}{\sqrt{\mu}},z\right)$$
$$= (1-z^2)^{-\gamma-d/2}\left(1-\frac{\|x\|^2}{\mu}\right)^{-\mu+1}$$
$$\times\left(1-\frac{\|y\|^2}{\mu}\right)^{-\mu+1}\left[1-\frac{\|x\|^2+\|y\|^2-2z\langle x,u\rangle}{\mu(1-z^2)}+O(\mu^{-2})\right]^{\mu-1},$$

which implies, upon taking limit $\mu\to\infty$, that

$$\lim_{\mu\to\infty}E_\mu\left(\frac{x}{\sqrt{\mu}},\frac{y}{\sqrt{\mu}},\frac{u}{\sqrt{\mu}},z\right)$$
$$= \frac{1}{(1-z^2)^{\gamma+d/2}}\exp\left[-\frac{z^2(\|x\|^2+\|y\|^2)}{1-z^2}\right]\exp\left(\frac{2z\langle x,u\rangle}{1-z^2}\right).$$

Since V_κ is a bounded linear operator, we can take the limit inside V_κ. Hence, using

$$V_\kappa\left[E_\mu\left(\frac{x}{\sqrt{\mu}},\frac{y}{\sqrt{\mu}},\cdot,z\right)\right]\left(\frac{y}{\sqrt{\mu}}\right)=V_\kappa\left[E_\mu\left(\frac{x}{\sqrt{\mu}},\frac{y}{\sqrt{\mu}},\frac{(\cdot)}{\sqrt{\mu}},z\right)\right](y),$$

it follows that $V_\kappa[E_\mu]$ converges to the right-hand side of the stated formula, which concludes the proof. □

In particular, if $h_\kappa(x)=1$ then $V_\kappa = id$ and the formula reduces to

$$\sum_{n=0}^{\infty}\sum_{|\alpha|=n}\widetilde{H}_\alpha^\kappa(x)\widetilde{H}_\alpha^\kappa(y)z^n=\prod_{i=1}^{d}\frac{1}{(1-z^2)^{1/2}}\exp\left[\frac{2x_iy_iz-z^2(x_i^2+y_i^2)}{1-z^2}\right],$$

which is a product of the classical Mehler formula.

In the case of the Cartesian Hermite polynomials H_α^κ, we can derive an explicit generating function. Let us define $b_\alpha^\kappa = b_{\alpha_1}^{\kappa_1}\cdots b_{\alpha_d}^{\kappa_d}$, where b_n^λ is the normalization constant of H_n^λ, which was defined in Subsection 1.5.1.

Theorem 8.3.8 *For all $x\in B^d$ and $y\in\mathbb{R}^d$,*

$$\sum_{|\alpha|=n}\frac{H_\alpha^\kappa(y)}{b_\alpha^\kappa}x^\alpha = \frac{1}{\sqrt{\pi}2^n n!}\int_{-\infty}^{\infty}\int_{[-1,1]^d}H_n(z(x,y,t,s))$$
$$\times\prod_{i=1}^{d}c_{\kappa_i}(1+t_i)(1-t_i^2)^{\kappa_i-1}\mathrm{d}t\,\mathrm{e}^{-s^2}\mathrm{d}s,$$

where $z(x,y,t,s)=t_1x_1y_1+\cdots+t_dx_dy_d+s\sqrt{1-\|x\|^2}$.

Proof We use the limit relation in Theorem 8.3.5 and the relation

$$\lim_{\mu\to\infty}\mu^{-n/2}\widetilde{C}_n^{(\mu,\lambda)}(x)=2^n[b_n^\lambda]^{-1/2}x^n,$$

which can be verified by using Definition 1.5.5 of $C_n^{(\lambda,\mu)}(x)$, the series expansion in terms of the ${}_2F_1$ formula for the Jacobi polynomial, and the formula for the structure constants. Using the formula for $P_\alpha^n(W_{\kappa,\mu}^B)$ in Theorem 8.1.11 and the fact that $h_{\alpha,n}^B\to 1$ as $\mu\to\infty$, we obtain

$$\lim_{\mu\to\infty}\mu^{-n/2}P_\alpha^n(W_{\kappa,\mu}^B;x)=2^n[b_\alpha^\kappa]^{-1/2}x^\alpha,$$

which implies, together with Theorem 8.3.5, the limit relation

$$\lim_{\mu\to\infty}\mu^{-n/2}\sum_{|\alpha|=n}P_\alpha^n(W_{\kappa,\mu}^B;x)P_\alpha^n(W_{\kappa,\mu}^B;y/\sqrt{\mu})=2^n\sum_{|\alpha|=n}H_\alpha^\kappa(y)\frac{x^\alpha}{b_\alpha^\kappa}.$$

The sum on the left-hand side is $\mathbf{P}_n(W_{\kappa,\mu}^B,x,y/\sqrt{\mu})$, which satisfies the explicit formula in Theorem 8.1.16. Now, multiply the right-hand side of the explicit formula by $\mu^{-n/2}$, replace y by $y/\sqrt{\mu}$ and then make the change of variables $t \mapsto s/\sqrt{\mu}$; its limit as $\mu \to \infty$ becomes, using Proposition 1.5.7, the right-hand side of the stated formula. □

In the case $\kappa = 0$, Theorem 8.3.8 yields the following formula for the classical Hermite polynomials:

$$\sum_{|\alpha|=n} \frac{H_\alpha(y)}{\alpha!} x^\alpha = \frac{1}{n!\sqrt{\pi}} \int_{-\infty}^{\infty} H_n(x_1 y_1 + \cdots + x_d y_d + s\sqrt{1-\|x\|^2}) e^{-s^2} ds$$

for all $x \in B^d$ and $y \in \mathbb{R}^d$. Furthermore, it is the extension of the following well-known formula for the classical Hermite polynomials (Erdélyi *et al.* [1953, Vol. II, p. 288]):

$$\sum_{|\alpha|=n} \frac{H_\alpha(y)}{\alpha!} x^\alpha = \frac{1}{n!} H_n(x_1 y_1 + \cdots + x_d y_d), \qquad \|x\| = 1.$$

In general, in the case $\|x\| = 1$ we have

Corollary 8.3.9 *For* $\|x\| = 1$ *and* $y \in \mathbb{R}^d$,

$$\sum_{|\alpha|=n} \frac{H_\alpha^\kappa(y)}{b_\alpha^\kappa} x^\alpha = \frac{1}{2^n n!} \int_{[-1,1]^d} H_n(t_1 x_1 y_1 + \cdots + t_d x_d y_d)$$
$$\times \prod_{i=1}^d c_{\kappa_i} (1+t_i)(1-t_i^2)^{\kappa_i - 1} dt.$$

8.4 Generalized Laguerre Polynomials

Much as for the relation between the orthogonal polynomials on T^d and those on B^d (see Section 4.4), there is a relation between the orthogonal polynomials on $\mathbb{R}_+^d$ and those on $\mathbb{R}^d$. In fact, since

$$\int_{\mathbb{R}^d} f(y_1^2, \ldots, y_d^2)\, dy = \int_{\mathbb{R}_+^d} f(x_1, \ldots, x_d) \frac{dx}{\sqrt{x_1 \cdots x_d}}$$

in analogy to Lemma 4.4.1, the analogue of Theorem 4.4.4 holds for $W^H(x) = W(x_1^2, \ldots, x_d^2)$ and $W^L(y) = W(y_1, \ldots, y_d)/\sqrt{y_1 \cdots y_d}$. Let $\mathscr{V}_{2n}(W^H, \mathbb{Z}_2^d)$ be the collection of all elements in $\mathscr{V}_{2n}(W^H)$ that are invariant under $\mathbb{Z}_2^d$.

Theorem 8.4.1 *Let* W^H *and* W^L *be defined as above. Then the relation* (4.4.1) *defines a one-to-one correspondence between an orthonormal basis of* $\mathscr{V}_{2n}(W^H, \mathbb{Z}_2^d)$ *and an orthonormal basis of* $\mathscr{V}_n^d(W^L)$.

Hence, we can study the Laguerre polynomials via the Hermite polynomials that are invariant under $\mathbb{Z}_2^d$. We are mainly interested in the case where H is a reflection-invariant weight function.

Definition 8.4.2 Let h_κ be a reflection-invariant weight function and assume that h_κ is also invariant under $\mathbb{Z}_2^d$. Define a weight function on $\mathbb{R}_+^d$ by

$$W_\kappa^L(x) = h_\kappa^2(\sqrt{x_1}, \ldots, \sqrt{x_d})\mathrm{e}^{-|x|}/\sqrt{x_1 \cdots x_d}, \qquad x \in \mathbb{R}_+^d.$$

The polynomials orthogonal to W_κ^L are called generalized Laguerre polynomials.

An orthonormal basis can be given in terms of $\mathbb{Z}_2^d$-invariant general Hermite polynomials; see Proposition 8.3.2.

Proposition 8.4.3 *For $0 \le j \le n$ let $\{Y_{\nu,2n-2j}^h\}$ denote an orthonormal basis of $\mathscr{H}_{2n-2j}^d(h_\kappa^2)$; then the polynomials*

$$P_{\nu,j}^n(W_\kappa^L;x) = [c_{j,n}^L]^{-1}\widetilde{H}_{2j}^{2n-2j+\lambda_\kappa}(\sqrt{|x|})Y_{\nu,2n-2j}^h(\sqrt{x_1}, \ldots, \sqrt{x_d}),$$

where $c_{j,n}^L = c_{j,n}^H$ from Proposition 8.3.2, form an orthonormal basis of $\mathscr{V}_n^d(W_\kappa^L)$.

Just as in the case of the orthogonal polynomials on the simplex, in the indecomposable case the requirement that h_κ is also invariant under $\mathbb{Z}_2^d$ limits us to essentially two possibilities:

The abelian group $\mathbb{Z}_2^d$ The weight function is

$$W_\kappa^L(x;\mathbb{Z}_2^d) = x_1^{\kappa_1} \cdots x_d^{\kappa_d}\mathrm{e}^{-|x|}. \tag{8.4.1}$$

The orthogonal polynomials associated with W_κ^L are the classical orthogonal polynomials from Subsection 5.1.4.

The hyperoctahedral group The weight function is

$$W_\kappa^L(x) = \prod_{i=1}^d x_i^{\kappa'} \prod_{1 \le i<j \le d} |x_i - x_j|^\kappa \mathrm{e}^{-|x|}. \tag{8.4.2}$$

Note that the parameters κ in the two weight functions above are chosen so that the notation is in line with the usual multiple Laguerre polynomials. They are related to $W_{\kappa+1/2}^H(x) = \prod_{i=1}^d |x_i|^{2\kappa_i+1}\mathrm{e}^{-\|x\|^2}$ and to $W_{\kappa+1/2}^H(x)$ in (8.3.4), respectively, where we define $\kappa + \frac{1}{2} = (\kappa_1 + \frac{1}{2}, \ldots, \kappa_d + \frac{1}{2})$.

Another approach to the generalized Laguerre polynomials is to treat them as the limit of the orthogonal polynomials on the simplex. Indeed, from the explicit formulae for orthonormal bases in Proposition 8.2.6, the limit relation (8.3.1) leads to the following.

Theorem 8.4.4 *For the polynomials $P_{\nu,j}^n(W_{\kappa,\mu}^T)$ defined in Proposition 8.2.5 and the polynomials $P_{\nu,j}^n(W_\kappa^L)$ defined in Proposition 8.4.3,*

$$\lim_{\mu \to \infty} P_{\nu,j}^n(W_{\kappa+1/2,\mu}^T;x/\mu) = P_{\nu,j}^n(W_\kappa^L;x).$$

In the case where h_κ is invariant under $\mathbb{Z}_2^d$, the multiple Laguerre polynomials are also the limit of the classical orthogonal polynomials on T^d. The verification of the following limit is left to the reader.

Theorem 8.4.5 *For the classical orthogonal polynomials $P_\alpha^n(W_{\kappa+1/2}^T)$ defined in Proposition 5.3.1 with $\mu = \kappa_{d+1}$,*

$$\lim_{\mu\to\infty} P_\alpha(W_{\kappa+1/2,\mu}^T; x/\mu) = \widetilde{L}_{\alpha_1}^{\kappa_1}(x_1)\cdots\widetilde{L}_{\alpha_d}^{\kappa_d}(x_d).$$

From the relation in Theorem 8.4.1, we can derive the differential–difference equations satisfied by the generalized Laguerre polynomials from those satisfied by the generalized Hermite polynomials, in the same way that the equation satisfied by the orthogonal polynomials on T^d is derived. We can also derive these equations from the limit relation in Theorem 8.4.4 in a similar way to the equation satisfied by the generalized Hermite polynomials. Using either method, we have a proof of the fact that the multiple Laguerre polynomials satisfy the partial differential equation (5.1.9). Moreover, we have

Theorem 8.4.6 *The orthogonal polynomials of $\mathcal{V}_n^d(W_\kappa^L)$ associated with the hyperoctahedral group satisfy the differential–difference equation*

$$\left[\sum_{i=1}^d \left(x_i\frac{\partial^2}{\partial x_i^2} + (\kappa' + 1 - x_i)\frac{\partial}{\partial x_i}\right)\right.$$
$$\left. + \kappa \sum_{1\le i<j\le d} \frac{1}{x_i - x_j}\left(2\left(x_i\frac{\partial}{\partial x_i} - x_j\frac{\partial}{\partial x_j}\right) - \frac{x_i + x_j}{x_i - x_j}[1-(ij)]\right)\right]P = -nP.$$

As in the Hermite case, this equation is an example of the Calogero–Sutherland model in physics; see the discussion in Section 11.6.

There is another way of obtaining an orthonormal basis for the generalized Laguerre polynomials, using orthogonal polynomials on the simplex. In fact, as in Lemma 8.2.4, the following formula holds:

$$\int_{\mathbb{R}_+^d} f(x)\,dx = \int_0^\infty s^{d-1}\int_{|u|=1} f(su)\,du\,ds \tag{8.4.3}$$

for an integrable function f on $\mathbb{R}_+^d$. Denote by $\{R_\beta^m\}_{|\beta|=m}$ a sequence of orthonormal homogeneous polynomials of degree m in the homogeneous coordinates from Proposition 4.4.8 associated with the weight function $W(u) = h^2(\sqrt{u_1},\ldots,\sqrt{u_d})/\sqrt{u_1\cdots u_d}$ on the simplex.

Proposition 8.4.7 *Let the R_β^m be defined as above. Then an orthonormal basis of $\mathcal{V}_n^d(W_\kappa^L)$ consists of*

$$P_{\beta,m}^n(x) = b_{m,n}L_{n-m}^{2m+\lambda_\kappa}(s)R_\beta^m(x), \qquad 0\le m\le n, \quad |\beta| = m,$$

where $x = su$, $|u| = 1$ and $[b_{m,n}]^2 = \Gamma(\lambda_\kappa + 2m + \frac{1}{2})/\Gamma(\lambda_\kappa + \frac{1}{2})$.

The verification of this basis is left to the reader.

From the limit relation in Theorem 8.4.4, we can also derive a Mehler-type formula for the Laguerre polynomials.

Theorem 8.4.8 *Let $\{P_\alpha^n(W_\kappa^L) : |\alpha| = n,\ n \in \mathbb{N}_0\}$, denote an orthonormal basis with respect to W_κ^L on $\mathbb{R}^d$. Then, for $0 < z < 1$ and $x, y \in \mathbb{R}_+^d$,*

$$\sum_{n=0}^{\infty} \sum_{|\alpha|=n} P_\alpha^n(W_\kappa^L; x) P_\alpha^n(W_\kappa^L; y) z^n$$
$$= \frac{1}{(1-z)^{\gamma_\kappa + d/2}} \exp\Big[-\frac{z(|x|+|y|)}{1-z}\Big] V_\kappa\Big[\exp\Big(\frac{2\sqrt{z}\langle\{x\}^{1/2},\cdot\rangle}{1-z}\Big)\Big](\{y\}^{1/2}).$$

Proof The proof follows the same lines as that of Theorem 8.3.7, using the explicit formula for the reproducing kernel of $\mathcal{V}_n^d(W_{\kappa,\mu}^T)$ in Theorem 8.2.7. We will merely point out the differences. Let K_μ be defined from the formula for the reproducing kernel in Theorem 8.2.7, in a similar way to the kernel $V_\kappa[E_\mu(x, y, \cdot, z)](y)$ in the proof of Theorem 8.3.7. Since we have C_{2n}^λ in the formula of Theorem 8.2.7, we need to write

$$\widetilde{K}_\mu(x, y, u, z) = \tfrac{1}{2}[K_\mu(x, y, u, z) + K_\mu(x, y, u, -z)]$$

in terms of the Gegenbauer series. Since $\widetilde{K}_\mu$ is even in z, the coefficients of $C_{2n+1}^{(\lambda)}$ in the expansion vanish. In this way $\widetilde{K}_\mu$ can be written as an infinite sum in z^{2n} with coefficients given by $\mathbf{P}_n(W_{\kappa,\mu}^T; x, y)$. The rest of the proof follows the proof of Theorem 8.3.7. We leave the details to the reader. □

In the case of the classical multiple Laguerre polynomials, the Mehler formula can be written out explicitly, using the explicit formula for the intertwining operator in Theorem 7.5.4. Moreover, in terms of the modified Bessel function I_τ of order τ,

$$I_\tau(x) = \frac{1}{\sqrt{\pi}\,\Gamma(\tau + \frac{1}{2})} \Big(\frac{x}{2}\Big)^\tau \int_{-1}^{1} \mathrm{e}^{-xt}(1-t^2)^{\tau-1/2}\, \mathrm{d}t.$$

This can be written as $I_\tau(x) = \mathrm{e}^{-\tau\pi\mathrm{i}/2} J_\tau(x\mathrm{e}^{\pi\mathrm{i}/2})$, where J_τ is the standard Bessel function of order τ. The formula then becomes

$$\sum_{n=0}^{\infty} \sum_{|\alpha|=n} \widetilde{L}_\alpha^\kappa(x) \widetilde{L}_\alpha^\kappa(x) z^n$$
$$= \prod_{i=1}^{d} \frac{\Gamma(\kappa_i + 1)}{1-z} \exp\Big[-\frac{z(x_i + y_i)}{1-z}\Big] (x_i y_i z)^{-\kappa_i/2} I_{\kappa_i}\Big(\frac{2\sqrt{z x_i y_i}}{1-z}\Big),$$

which is the product over i of the classical Mehler formula for the Laguerre polynomials of one variable (see, for example, Erdélyi *et al.* [1953, Vol. II, p. 189]).

8.5 Notes

The partial differential equations satisfied by the classical orthogonal polynomials were derived using the explicit formula for the biorthonormal basis in Sections 5.2 and 5.3; see Chapter XII of Erdélyi *et al.* [1953]. The derivation of the differential–difference equation satisfied by the generalized classical orthogonal polynomials was given in Xu [2001b].

The U_α^μ basis defined by the Rodrigues-type formula is not mutually orthogonal. There is a biorthogonal basis defined by the projection operator $\mathrm{proj}_n(W_{\kappa,\mu}^B)$ onto $\mathscr{V}_n(W_{\kappa,\mu}^B)$: for $\alpha \in \mathbb{N}_0^d$, define

$$R_\alpha(x) := \mathrm{proj}_n(W_{\kappa,\mu}^B; V_\kappa\{\cdot\}^\alpha, x),$$

where V_κ is the intertwining operator associated with h_κ. Then $\{U_\alpha^\mu : |\alpha| = n\}$ and $\{R_\beta : |\beta| = n\}$ are biorthogonal with respect to $W_{\kappa,\mu}^B$ on B^d. These bases were defined and studied in Xu [2005d], which contains further properties of U_a^μ. The monic orthogonal polynomials in the case of $\mathbb{Z}_2^d$ were studied in Xu [2005c].

The explicit formulae for the reproducing kernel in Theorems 8.1.10 and 8.2.8 were useful for a number of problems. The special case of the formula in Theorem 5.2.8 was first derived by summing up the product of orthogonal polynomials in Proposition 5.2.1, and making use of the product formula of Gegenbauer polynomials, in Xu [1999a]; it appeared first in Xu [1996a]. The connection to h-harmonics was essential in the derivation of Theorem 8.2.8 in Xu [1998d]. They have been used for studying the summability of orthogonal expansions, see the discussion in the next chapter, and also in constructing cubature formulae in Xu [2000c].

The orthonormal basis in Proposition 8.1.21 was first discovered in Logan and Shepp [1975] in connection with the Radon transform. The derivation from the Funk–Hecke formula and further discussions were given in Xu [2000b]. This basis was used in Bojanov and Petrova [1998] in connection with a numerical integration scheme. More recently, it was used in Xu [2006a] to derive a reconstruction algorithm for computerized tomography. In this connection, orthogonal polynomials on a cylinder were studied in Wade [2010].

The relation between h-harmonics and orthogonal polynomials on the simplex was used in Dunkl [1984b] in the case $d = 2$ with a product weight function. The general classical orthogonal polynomials were studied in Xu [2001b, 2005a].

In the case of symmetric groups, the differential–difference equation for the generalized Hermite polynomials is essentially the Schrödinger equation for a type of Calogero–Sutherland model. See the discussion in Section 11.6. The equation for the general reflection group was derived in Rösler [1998] using a different method. The derivation as the limit of the h-Laplacian is in Xu [2001b].

Hermite introduced biorthonormal polynomials for the classical Hermite weight function, which were subsequently studied by many authors; see Appell and de Fériet [1926] and Erdélyi *et al.* [1953, Vol. II, Chapter XII]. Making use of the intertwining operator V, such a basis can be defined for the generalized Hermite polynomials associated with a reflection group, from which follows another proof of the Mehler-type formula for the generalized Hermite polynomials; see Rösler [1998]. Our proof follows Xu [2001b]. In the case where h_κ is associated with the symmetric groups, the Mehler-type formula was proved in Baker and Forrester [1997a]. For the classical Mehler formula, see Erdélyi *et al.* [1953, Vol. II, p. 194], for example.

9

Summability of Orthogonal Expansions

The basics of Fourier orthogonal expansion were discussed in Section 3.6. In the present chapter various convergence results for orthogonal expansions are discussed. We start with a result on a fairly general system of orthogonal polynomials, then consider the convergence of partial-sum operators and Cesàro means and move on to the summability of the orthogonal expansions for h-harmonics and the generalized classical orthogonal polynomials.

9.1 General Results on Orthogonal Expansions

We start with a general result about the convergence of the partial sum of an orthogonal expansion, under a condition on the behavior of the Christoffel function defined in (3.6.8). In the second subsection we discuss the basics of Cesàro means.

9.1.1 Uniform convergence of partial sums

Recall that $\mathscr{M}$ was defined in Subsection 3.2.3 as the set of nonnegative Borel measures on $\mathbb{R}^d$ with moments of all orders. Let $\mu \in \mathscr{M}$ satisfy the condition (3.2.17) in Theorem 3.2.17 and let it have support set $\Omega \in \mathbb{R}^d$. For $f \in L^2(\mathrm{d}\mu)$, let $E_n(\mathrm{d}\mu;f)_2$ be the error of the best $L^2(\mathrm{d}\mu)$ approximation from Π_n^d; that is,

$$E_n(\mathrm{d}\mu;f)_2 = \inf_{P\in\Pi_n^d} \int_\Omega |f(x) - P(x)|^2 \,\mathrm{d}\mu(x).$$

Note that $E_n(\mathrm{d}\mu,f)_2 \to 0$ as $n \to \infty$ according to Theorem 3.2.18. Let $S_n(\mathrm{d}\mu;f)$ denote the partial sum of the Fourier orthogonal expansion with respect to the measure $\mathrm{d}\mu$, as defined in (3.6.3), and let $\Lambda_n(\mathrm{d}\mu;x)$ denote the Christoffel function as defined in (3.6.8).

Lemma 9.1.1 *Let $\mu \in \mathscr{M}$ satisfy* (3.2.17). *If, for $x \in \mathbb{R}^d$,*

$$\sum_{m=0}^{\infty} \frac{E_{2^m}(\mathrm{d}\mu;f)_2}{\sqrt{\Lambda_{2^m+1}(\mathrm{d}\mu,x)}} < \infty, \tag{9.1.1}$$

then $S_n(\mathrm{d}\mu;f)$ converges at the point x. If (9.1.1) *holds uniformly on a set E in Ω then $S_n(\mathrm{d}\mu)$ converges uniformly on E.*

Proof Since $S_n(\mathrm{d}\mu;f)$ is the best $L^2(\mathrm{d}\mu)$ approximation of $f \in L^2(\mathrm{d}\mu)$ from Π_n^d, using (3.6.3) and the Parseval identity in Theorem 3.6.8 as well as the orthogonality,

$$\begin{aligned} E_n(\mathrm{d}\mu;f)_2^2 &= \int_{\mathbb{R}^d} [f - S_n(\mathrm{d}\mu;f)]^2 \,\mathrm{d}\mu \\ &= \int_{\mathbb{R}^d} \left(\sum_{k=n+1}^{\infty} \mathbf{a}_k^{\mathrm{T}}(f)\mathbb{P}_k(x) \right)^2 \mathrm{d}\mu = \sum_{k=n+1}^{\infty} \|\mathbf{a}_k(f)\|_2^2. \end{aligned}$$

Therefore, using the Cauchy–Schwarz inequality, we obtain

$$\begin{aligned} \left(\sum_{k=2^m+1}^{2^{m+1}} |\mathbf{a}_k^{\mathrm{T}}(f)\mathbb{P}_k(x)| \right)^2 &\le \sum_{k=2^m+1}^{2^{m+1}} \|\mathbf{a}_k^{\mathrm{T}}(f)\|_2^2 \sum_{k=2^m+1}^{2^{m+1}} \|\mathbb{P}_k(x)\|_2^2 \\ &\le \sum_{k=2^m+1}^{\infty} \|\mathbf{a}_k^{\mathrm{T}}(f)\|_2^2 \sum_{k=0}^{2^{m+1}} \|\mathbb{P}_k(x)\|_2^2 \\ &= [E_{2^m}(\mathrm{d}\mu;f)_2]^2 [\Lambda_{2^{m+1}}(\mathrm{d}\mu;x)]^{-1}. \end{aligned}$$

Taking the square root and summing over m proves the stated result. □

Proposition 9.1.2 *Let $\mu \in \mathscr{M}$ satisfy* (3.2.17). *If*

$$[n^d \Lambda_n(\mathrm{d}\mu;x)]^{-1} \le M^2 \tag{9.1.2}$$

holds uniformly on a set E with some constant $M > 0$ then $S_n(\mathrm{d}\mu;f)$ is uniformly and absolutely convergent on E for every $f \in L^2(\mathrm{d}\mu)$ such that

$$\sum_{k=1}^{\infty} E_k(\mathrm{d}\mu;f)_2 k^{(d-2)/2} < \infty. \tag{9.1.3}$$

Proof Since $E_n(\mathrm{d}\mu;f)_2$ is nonincreasing,

$$\begin{aligned} \frac{E_{2^m}(\mathrm{d}\mu;f)_2}{\sqrt{\Lambda_{2^m+1}(\mathrm{d}\mu,x)}} &\le M 2^{(m+1)d/2} \frac{1}{2^{m-1}} \sum_{k=2^{m-1}+1}^{2^m} E_k(\mathrm{d}\mu;f)_2 \\ &= M 2^{d/2+1} 2^{m(d-2)/2} \sum_{k=2^{m-1}+1}^{2^m} E_k(\mathrm{d}\mu;f)_2 \\ &\le c \sum_{k=2^{m-1}+1}^{2^m} E_k(\mathrm{d}\mu;f)_2 k^{(d-2)/2}. \end{aligned}$$

Summing over m, the stated result follows from Lemma 9.1.1. □

Corollary 9.1.3 *Let $\mu \in \mathscr{M}$ and suppose that μ has a compact support set Ω in $\mathbb{R}^d$. Suppose that* (9.1.2) *holds uniformly on a subset $E \subset \Omega$. If $f \in C^{[d/2]}(\Omega)$ and each of its $[\frac{d}{2}]$th derivatives satisfies*

$$|D^{[d/2]}f(x) - D^{[d/2]}f(y)| \le c\|x-y\|_2^{\beta},$$

where, for odd d, $\beta > \frac{1}{2}$ and, for even d, $\beta > 0$ then $S_n(\mathrm{d}\mu; f)$ converges uniformly and absolutely to f on E.

Proof For f satisfying the above assumptions, a standard result in approximation theory (see, for example, Lorentz [1986, p. 90]) shows that there exists $P \in \Pi_n^d$ such that

$$E_n(\mathrm{d}\mu, f)_2 \le \|f - P\|_\infty \le cn^{-([d/2]+\beta)}.$$

Using this estimate, our assumption on β implies that (9.1.3) holds. □

Since there are $\binom{n+d}{d} = O(n^d)$ terms in the sum of $[\Lambda_n(\mathrm{d}\mu;x)]^{-1} = \mathbf{K}_n(x,x)$, the condition (9.1.2) appears to be a reasonable assumption. We will show that this condition often holds. In the following, if $\mathrm{d}\mu$ is given by a weight function $\mathrm{d}\mu(x) = W(x)\,\mathrm{d}x$ then we write $\Lambda_n(W;x)$. The notation $A \sim B$ means that there are two constants c_1 and c_2 such that $c_1 \le A/B \le c_2$.

Proposition 9.1.4 *Let W, W_0 and W_1 be weight functions defined on $\mathbb{R}^d$. If $c_1 W_0(x) \le W(x) \le c_2 W_1(x)$ for some positive constant $c_2 > c_1 > 0$ then $c_1 \Lambda_n(W_0;x) \le \Lambda_n(W;x) \le c_2 \Lambda_n(W_1;x)$.*

Proof By Theorem 3.6.6, Λ_n satisfies the property

$$\Lambda_n(W;x) = \min_{P(x)=1, P\in\Pi_n^d} \int_{\mathbb{R}^d} |P(y)|^2 W(y)\,\mathrm{d}y;$$

the stated inequality follows as an easy consequence. □

We will establish (9.1.2) for the classical-type weight functions. The proposition allows us an extension to more general classes of weight functions.

Proposition 9.1.5 *Let $W(x) = \prod_{i=1}^d w_i(x_i)$, where w_i are weight functions defined on $[-1,1]$. Then $\Lambda_n(W;x) \sim n^{-d}$ for all x in the interior of $[-1,1]^d$.*

Proof Let $p_m(w_i)$ denote orthonormal polynomials with respect to w_i. The orthonormal polynomials with respect to W are $P_\alpha^n(x) = p_{\alpha_1}(w_1;x_1)\cdots p_{\alpha_d}(w_d;x_d)$. Let $\lambda_n(w_i;t)$ be the Christoffel function associated with the weight function w_i on $[-1,1]$. It is known that $\lambda_n(w_i;t) \sim n^{-1}$ for $t \in (-1,1)$; see for example Nevai [1986]. Hence, on the one hand,

$$\Lambda_n(W;x) = \left(\sum_{m=0}^{n} \sum_{|\alpha|=m} [P_\alpha^n(x)]^2 \right)^{-1}$$
$$\geq \left(\sum_{k_1=0}^{n} \cdots \sum_{k_d=0}^{n} p_{k_1}^2(w_1;x_1) \cdots p_{k_d}^2(w_d;x_d) \right)^{-1}$$
$$= \prod_{i=1}^{d} \lambda_n(w_i;x_i) \sim n^{-d}.$$

On the other hand, since $\{\alpha : |\alpha| \leq n\}$ contains $\{\alpha : 0 \leq \alpha_i \leq [n/d]\}$, the inequality can be reversed to conclude that $\Lambda_n(W;x) \leq \prod_{i=1}^d \lambda_{[n/d]} (w_i;x_i) \sim n^{-d}$. □

Proposition 9.1.6 *For the weight function $W = W_{\kappa,\mu}^B$ in (8.1.5) on B^d or the weight function $W = W_{\kappa,\mu}^T$ in (8.2.1) on T^d, $[n^d \Lambda_n(W;x)]^{-1} \leq c$ for all x in the interior of B^d or T^d, respectively.*

Proof We prove only the case of T^d; the case of B^d is similar. In the formula for the reproducing kernel in Theorem 8.2.8, let $\kappa_i \to 0$ and $\mu \to 0$ and use the formula (1.5.1) repeatedly to obtain

$$P_n(W_{0,0}^T;x,x) = \frac{2n + \frac{d-1}{2}}{2^d(d-1)} \sum_{\varepsilon \in \mathbb{Z}_2^{d+1}} C_{2n}^{(d-1)/2}(x_1\varepsilon_1 + \cdots + \varepsilon_{d+1}x_{d+1}),$$

where $x_{d+1} = 1 - x_1 - \cdots - x_d$. Hence, using the value of $C_n^{(d-1)/2}(1) = (d-1)_n/n! \sim n^{d-2}$ and the fact that $C_n^{(d-1)/2}(x) \sim n^{(d-3)/2}$ (see, for example, Szegő [1975, p. 196]), we have $P_n(W_{0,0}^T;x,x) \sim n^{d-1}$. Consequently, we obtain $\Lambda_n(W_{0,0}^T;x) \sim n^{-d}$ since $\Lambda_n(W_{0,0}^T;x) = [K_n(W_{0,0}^T;x,x)]^{-1}$. Now, if all the parameters κ_i and μ are even integers then the polynomial q defined by $q^2(x) = W_{\kappa,\mu}^T(x)/W_{0,0}^T(x)$ is a polynomial of degree $m = \frac{1}{2}(|\kappa| + \mu)$. Using the property of Λ_n in Theorem 3.6.6, we obtain

$$\Lambda_n(W_{\kappa,\mu}^T;x) = \min_{P(x)=1, P\in\Pi_n^d} \int_{T^d} [P(y)]^2 W_{\kappa,\mu}^T(y)\, dy$$
$$= q^2(x) \min_{P(x)=1, P\in\Pi_n^d} \int_{T^d} |P(y)q(y)/q(x)|^2 W_0(y)\, dy$$
$$\geq q^2(x)\Lambda_{n+m}(W_{0,0};x).$$

This shows that $n^{-d}\Lambda_n(W_{\kappa,\mu}^T;x) = O(1)$ holds for every x in the interior of T^d when all parameters are even integers. For the general parameters, Proposition 9.1.4 can be used to finish the proof. □

In fact the estimate $\Lambda_n(W;x) \sim n^{-d}$ holds for $W_{\kappa,\mu}^B$ and $W_{\kappa,\mu}^T$. Furthermore, the asymptotics of $\Lambda_n(W;x)$ are known in some cases; see the notes at the end of the chapter.

9.1.2 Cesàro means of the orthogonal expansion

As in classical Fourier analysis, the convergence of the partial-sum operator for the Fourier orthogonal expansion requires that the function is smooth. For functions that are merely continuous, it is necessary to consider summability methods. A common method is the Cesàro (C,δ)-mean.

Definition 9.1.7 Let $\{c_n\}_{n=0}^{\infty}$ be a given sequence. For $\delta > 0$, the Cesàro (C,δ)-means are defined by

$$s_n^\delta = \sum_{k=0}^{n} \frac{(-n)_k}{(-n-\delta)_k} c_k.$$

The sequence $\{c_n\}$ is (C,δ)-summable, by Cesàro's method of order δ, to s if s_n^δ converges to s as $n \longrightarrow \infty$.

It is a well-known fact that if s_n converges to s then s_n^δ converges to s. The reverse, however, is not true. For a sequence that does not converge, s_n^δ may still converge for some δ. We note the following properties of s_n^δ (see, for example, Chapter III of Zygmund [1959]).

1. If s_n is the nth partial sum of the series $\sum_{k=0}^{\infty} c_k$ then
$$s_n^\delta = \frac{\delta}{\delta+n} \sum_{k=0}^{n} \frac{(-n)_k}{(1-\delta-n)_k} s_k.$$
2. If $s_k = 1$ for all k then $s_n^\delta = 1$ for all n.
3. If s_n^δ converges to s then s_n^σ converges to s for all $\sigma > \delta$.

For a weight function W and a function f, denote the nth partial sum of the Fourier orthogonal expansion of f by $S_n(W;f)$, which is defined as in (3.6.3). Its Cesàro (C,δ)-means are denoted by $S_n^\delta(W;f)$ and are defined as the (C,δ)-means of the orthogonal expansion $\sum_{n=0}^{\infty} \mathbf{a}_n^{\mathrm{T}}(f)\mathbb{P}_n$. Since $S_n(W;1) = 1$, property (2) shows that $S_n^\delta(W;1) = 1$.

One important property of the partial-sum operator $S_n(W;f)$ is that it is a projection operator onto Π_n^d; that is, $S_n(W;p) = p$ for any polynomial p of degree n. This property does not hold for the Cesàro means. However, even if $S_n(W;f)$ does not converge, the operator $\sigma_n(W;f) = 2S_{2n-1}(W;f) - S_{n-1}(W;f)$ may converge and, moreover, $\sigma_n(p) = p$ holds for all $p \in \Pi_{n-1}^d$. In classical Fourier analysis, the $\sigma_n(f)$ are called the de la Vallée–Poussin means. The following definition gives an analogue that uses Cesàro means of higher order.

Definition 9.1.8 For an integer $m \geq 1$, define

$$\sigma_n^m(W;f) = \frac{1}{n^m} \sum_{j=0}^{m} (2^j n)_m \prod_{i=0, i\neq j}^{m} \frac{1}{2^j - 2^i} S_{2^j n-1}^m(W;f).$$

Note that when $m = 1$, σ_n^m is precisely the de la Vallée–Poussin mean σ_n. The main properties of the means σ_n^m are as follows.

Theorem 9.1.9 *The means $\sigma_n^m(W;f)$ satisfy the following properties: $\sigma_n^m(W;p) = p$ for all $p \in \Pi_{n-1}^d$ and the $\sigma_n^m(W;\cdot)$ are uniformly bounded in n whenever the $S_n^m(W;\cdot)$ are uniformly bounded.*

Proof The proof will show how the formula for $\sigma_n^m(W;f)$ is determined. Consider

$$\sum_{j=0}^m a_j S_{2^j n-1}^m(W;f) = \sum_{j=0}^m \frac{a_j}{\binom{2^j n+m-1}{m}} \sum_{k=0}^{2^j(n-1)} \binom{2^j n-k+m-2}{2^j n-1-k} S_k(W;f).$$

We will choose a_j so that the terms involving $S_k(W;f)$ become 0 for $0 \le k \le n-1$ and

$$\sigma_n^m(W;f) = \sum_{j=1}^m a_j^* \sum_{k=n}^{2^j n-1} \binom{2^j n-k+m-2}{2^j n-1-k} S_k(W;f),$$

where $a_j^* = a_j / \binom{2^j n+m-1}{m}$; this will show that $\sigma_n^m(W;p) = p$ for all $p \in \Pi_{n-1}^d$. That is, together with $\sigma_n^m(W;1) = 1$, we choose a_j such that

$$\sum_{j=0}^m a_j^* \binom{2^j n-k+m-2}{2^j n-1-k} = 0, \qquad k = 0,1,\ldots,n-1 \quad \text{and} \quad \sum_{j=0}^m a_j = 1.$$

Since m is an integer, it follows that

$$\binom{2^j n-k+m-2}{2^j n-1-k} = \frac{(2^j n-k+m-2)\cdots(2^j n-k)}{(m-1)!} = \sum_{s=0}^{m-1} p_s(2^j n) k^s,$$

where the $p_s(t)$ are polynomials of degree $m-1-s$. The leading term of $p_s(t)$ is a constant multiple of t^{m-1-s}. Similarly,

$$\binom{2^j n+m-1}{m} = \frac{(2^j n)^m}{m!} + \text{lower-degree terms}.$$

Hence, it suffices to choose a_j^* such that

$$\sum_{j=0}^m a_j^* 2^{js} = 0, \qquad s = 0,1,\ldots,m-1 \quad \text{and} \quad \sum_{j=0}^m 2^{jm} a_j^* = \frac{m!}{n^m}.$$

To find such a_j^*, consider the Lagrange interpolation polynomial L_m of degree m defined by

$$L_m(f;t) = \sum_{j=0}^{m-1} f(2^j)\ell_j(t) \qquad \text{with} \qquad \ell_j(t) = \prod_{i=0, i\ne j}^{m-1} \frac{t-2^i}{2^j-2^i},$$

which satisfies $L_m(f;2^j) = f(2^j)$ for $j = 0,1,\ldots,m-1$, and

$$f(t) - L_m(f;t) = \frac{f^{(m)}(\xi)}{m!} \prod_{i=0}^{m-1} (t-2^i), \qquad \xi \in (1, 2^{n-1}).$$

By the uniqueness of the interpolating polynomial, $L_m(f) = f$ whenever f is a polynomial of degree at most $m-1$. Set $f(t) = t^s$ with $s = m, m-1, \ldots, 0$ in the above formulae and let $t = 2^m$; then

$$2^{m^2} - \sum_{j=0}^{m-1} 2^{jm}\ell_j(2^m) = \prod_{i=0}^{m-1}(2^m - 2^i) \qquad \text{and} \qquad 2^{sm} - \sum_{j=0}^{m-1} 2^{sj}\ell_j(2^m) = 0.$$

Hence, the choice $a_m^* = m!/[n^m \prod_{j=0}^{m-1}(2^m - 2^j)]$ and $a_j^* = -\ell_j(2^m)a_m^*$ for $j = 0, 1, \ldots, m-1$ gives the desired result. The definition of ℓ_j leads to the equation $a_j^* = m!/[n^m \prod_{i=0, i\neq j}^{m}(2^j - 2^i)]$, which agrees with the definition of $\sigma_n^m(W; f)$.

Since

$$n^{-m}\binom{2^j n + m}{m} = \prod_{i=1}^{m}(2^j + i/n) \le (2^j + 1)^m,$$

it follows that

$$\sum_{j=0}^{m} |a_j| \le \sum_{j=0}^{m} \left[(2^j + 1)^m / \prod_{i=0, i\neq j}^{m} |2^j - 2^i|\right] = A_m,$$

which is independent of n. Consequently, $|\sigma_n^m(W; f, x)| \le A_m \sup_n |S_n^m(W; f, x)|$. □

Let the weight function W be defined on the domain Ω. For a continuous function f on the domain Ω, define the best approximation of f from the space Π_n^d of polynomials of degree at most n by

$$E_n(f)_\infty = \min_{P \in \Pi_n^d} \|f - P\| = \min_{P \in \Pi_n^d} \max_{x \in \Omega} |f(x) - P(x)|.$$

Corollary 9.1.10 *Suppose that the Cesàro means $S_n^m(W; f)$ converge uniformly to a continuous function f. Then*

$$|\sigma_n^m(W; f) - f(x)| \le B_m E_n(f),$$

where B_m is a constant independent of n.

Proof Since $S_n^m(W; f)$ converges uniformly, there is a constant c_m independent of n such that $\|S_n^m(W; f)\| \le c_m \|f\|$ for all n. Let P be a polynomial of degree n such that $E_n(f) = \|f - P\|$. Then, since $\sigma_n^m(W; P) = P$, the stated result follows from the triangle inequality

$$\|\sigma_n^m(W; f) - f\| \le A_m \sup_n \|S_n^m(W; f - P)\| + \|f - P\| \le (A_m c_m + 1)E_n(f),$$

where $A_m = \sum_{j=0}^m \left[(2^j + 1)^m / \prod_{i=0, i\neq j}^m |2^j - 2^i|\right]$ as in the proof of Theorem 9.1.9. □

9.2 Orthogonal Expansion on the Sphere

Any $f \in L^2(h_\kappa^2 \mathrm{d}\omega)$ can be expanded into an h-harmonic series. Let $\{S_{\alpha,n}\}$ be an orthonormal basis of $\mathcal{H}_n^d(h_\kappa^2)$. For f in $L^2(h_\kappa^2\,\mathrm{d}\omega)$, its Fourier expansion in terms of $\{S_{\alpha,n}\}$ is given by

$$f \sim \sum_{n=0}^{\infty}\sum_{\alpha} a_\alpha^n(f) S_{\alpha,n}, \qquad \text{where} \qquad a_\alpha^n(f) = \int_{S^{d-1}} f S_{\alpha,n} h_\kappa^2\,\mathrm{d}\omega.$$

For the ordinary harmonics, this is usually called the Laplace series. The nth component of this expansion is written, see (7.3.1), as an integral with reproducing kernel $P_n(h_\kappa^2;x,y)$:

$$S_n(h_\kappa^2;f,x) = c_h' \int_{S^{d-1}} f(y) P_n(h_\kappa^2;x,y) h_\kappa^2(y)\,\mathrm{d}\omega.$$

It should be emphasized that the nth component is independent of the particular choice of orthogonal basis. Many results on orthogonal expansions for weight functions whose support set has a nonempty interior, such as those in Section 9.1, hold also for h-harmonic expansions. For example, a standard L^2 argument shows that:

Proposition 9.2.1 *If $f \in L^2(h_\kappa^2\,\mathrm{d}\omega)$ then $S_n(h_\kappa^2;f)$ converges to f in $L^2(h_\kappa^2\,\mathrm{d}\omega)$.*

For uniform convergence or L^p-convergence for $p \neq 2$, it is necessary to consider a summability method. It turns out that the formula for $P_n(h_\kappa^2;x,y)$ in Corollary 7.3.2 allows us to show that the summability of h-harmonics can be reduced to that of Gegenbauer polynomials.

In the following, we denote by $L^p(h_\kappa^2\,\mathrm{d}\omega)$ the space of Lebesgue-measurable functions f defined on S^{d-1} for which the norm

$$\|f\|_{p,h} = \left(c_h' \int_{S^{d-1}} |f(x)|^p h_\kappa^2\,\mathrm{d}\omega\right)^{1/p}$$

is finite. When $p = \infty$, we take the space as $C(S^{d-1})$, the space of continuous functions with uniform norm $\|f\|_\infty$. The essential ingredient is the explicit formula for the reproducing kernel in Corollary 7.3.2. Denote by w_λ the normalized weight function

$$w_\lambda(t) = B(\lambda + \tfrac{1}{2}, \tfrac{1}{2})^{-1}(1-t^2)^{\lambda-1/2}, \qquad t \in [-1,1],$$

whose orthogonal polynomials are the Gegenbauer polynomials (Subsection 1.4.3). A function $f \in L^2(w_\lambda, [-1,1])$ can be expressed as a Gegenbauer expansion:

$$f \sim \sum_{k=0}^{\infty} b_k(f) \widetilde{C}_n^\lambda \qquad \text{with} \qquad b_k(f) := \int_{-1}^{1} f(t) \widetilde{C}_n^\lambda(t) w_\lambda(t)\,\mathrm{d}t,$$

where $\widetilde{C}_n^\lambda$ denotes the orthonormal Gegenbauer polynomial. A summability method for the Gegenbauer expansion is a sum in the form $\phi_n(f,w_\lambda;t) = \sum_{k=0}^n c_{k,n} b_k(f) \widetilde{C}_k^\lambda(t)$, where the $c_{k,n}$ are given constants such that $\sum_{k=0}^n c_{k,n} \to 1$ as $n \to \infty$. Evidently the sum can be written as an integral:

$$\phi_n(f,w_\lambda;t) = \int_{-1}^1 p_n^\phi(w_\lambda;s,t) f(s) w_\lambda(s)\,\mathrm{d}s,$$

with $p_n^\phi(w_\lambda;s,t) = \sum_{k=0}^n c_{k,n} \widetilde{C}_k^\lambda(s) \widetilde{C}_k^\lambda(t)$.

Proposition 9.2.2 *For $p = 1$ and $p = \infty$, the operator norm is given by*

$$\sup_{\|f\|_p \le 1} \|\phi_n(f,w_\lambda)\|_p = \max_{t\in[-1,1]} \int_{-1}^1 |p_n^\phi(w_\lambda;s,t)| w_\lambda(s)\,\mathrm{d}s,$$

where $\|g\|$ is the $L^1(w_\lambda,[-1,1])$ norm of g.

Proof An elementary inequality shows that the left-hand side is no greater than the right-hand side. For $p = \infty$ and fixed t, the choice $f(s) = \operatorname{sign}[p_n^\phi(s,t)]$ is then used to show that the equality holds. For $p = 1$, consider a sequence $s_n(f)$ defined by $s_n(f,t) = \int_{-1}^1 f(u) q_n(u,t) w_\lambda(u)\,\mathrm{d}u$ which converges uniformly to f whenever f is continuous on $[-1,1]$ (for example, the Cesàro means of sufficiently high order). Then choose $f_N(s) = q_N(s,t^*)$, where t^* is a point for which the maximum in the right-hand side is attained, such that

$$\phi_n(f_N,w_\lambda;t) = \int_{-1}^1 q_N(s,t^*) p_n^\phi(w_\lambda;s,t) w_\lambda(s)\,\mathrm{d}s = s_N(p_n^\phi(w_\lambda;\cdot,t);t^*),$$

which converges to $p_n^\phi(w_\lambda;t^*,t)$ as $N \to \infty$. The dominated convergence theorem shows that $\|\phi_n(f_N,w_\lambda)\| \to \int_{-1}^1 |p_n^\phi(w_\lambda;t,t^*)| w_\lambda(t)\,\mathrm{d}t$ as $N \to \infty$, which gives the stated identity for $p = 1$. □

Theorem 9.2.3 *Let $\lambda = \gamma_\kappa + \frac{d-2}{2}$. If $\phi_n(\cdot,w_\lambda;t) = \sum_{k=0}^n c_{k,n} b_k(\cdot) \widetilde{C}_k^\lambda(t)$ defines a bounded operator on $L^p(w_\lambda,[-1,1])$ for $1 \le p \le \infty$ then $\Phi_n(\cdot) = \sum_{k=0}^n c_{k,n} S_k(\cdot,h_\kappa^2)$ defines a bounded operator on $L^p(h_\kappa^2\,\mathrm{d}\omega)$. More precisely, if*

$$\left(\int_{-1}^1 |\phi_n(f,w_\lambda;t)|^p w_\lambda(t)\,\mathrm{d}t\right)^{1/p} \le C\left(\int_{-1}^1 |f(t)|^p w_\lambda(t)\mathrm{d}t\right)^{1/p}$$

for $f \in L^p(w_\lambda,[-1,1])$, where C is a constant independent of f and n, then

$$\left(\int_{S^{d-1}} |\Phi_n(g;x)|^p h_\kappa^2(x)\,\mathrm{d}\omega\right)^{1/p} \le C\left(\int_{S^{d-1}} |g(x)|^p h_\kappa^2(x)\,\mathrm{d}\omega\right)^{1/p}$$

for $g \in L^p(h_\kappa^2\,\mathrm{d}\omega)$ with the same constant. In particular, if $\phi_n(f,w_\lambda)$ converges to f in $L^p(w_\lambda,[-1,1])$ then the means $\Phi_n(g)$ converge to g in $L^p(h_\kappa^2\,\mathrm{d}\omega)$.

Proof Like ϕ_n, the means Φ_n can be written as integrals:

$$\Phi_n(g;x) = c_h' \int_{S^{d-1}} P_n^{\Phi}(x,y) g(y) h_\kappa^2(y)\, d\omega$$

with $P_n^{\Phi}(x,y) = \sum_{k=0}^{n} c_{k,n} P_k(h_\kappa^2; x,y)$. Since $\tilde{C}_k^\lambda(t)\tilde{C}_k^\lambda(1) = [(n+\lambda)/\lambda] C_k^\lambda(t)$, it follows from the explicit formula for $P_n(h_\kappa^2;x,y)$ given by Corollary 7.3.2 and (7.3.1) that

$$P_n^{\Phi}(x,y) = V\left[p_n^{\phi}(w_\lambda; \langle x,\cdot\rangle, 1)\right](y).$$

Since V is positive,

$$\left|V\left[p_n^{\phi}(w_\lambda; \langle x,\cdot\rangle, 1)\right](y)\right| \le V\left[\left|p_n^{\phi}(w_\lambda; \langle x,\cdot\rangle, 1)\right|\right](y).$$

Consequently, for $p = \infty$ and $p = 1$, applying Corollary 7.4.5 gives

$$\begin{aligned}
\|\Phi_n(g)\|_{p,h} &\le \|g\|_{p,h} c_h' \int_{S^{d-1}} \left|V\left[p_n^{\phi}(w_\lambda; \langle x,\cdot\rangle, 1)\right](y)\right| h_\kappa^2(y)\, d\omega(y) \\
&\le \|g\|_{p,h} \int_{-1}^{1} |p_n^{\phi}(w_\lambda; 1, t)| w_\lambda(t)\, dt \\
&\le C\|g\|_{p,h}
\end{aligned}$$

by Proposition 9.2.2. The case $1 < p < \infty$ follows from the Riesz interpolation theorem; see Rudin [1991]. □

Note that the proof in fact shows that the convergence of the Φ_n means depends only on the convergence of ϕ_n at the point $t = 1$.

Let $S_n^\delta(h_\kappa^2; f)$ denote the nth Cesàro (C,δ) mean of the h-harmonic expansion. It can be written as an integral,

$$S_n^\delta(h_\kappa^2; f, x) = c_h' \int_{S^{d-1}} f(y) P_n^\delta(h_\kappa; x,y) h_\kappa^2(y)\, d\omega,$$

where $P_n^\delta(h_\kappa; x,y)$ is the Cesàro (C,δ)-mean of the sequence of reproducing kernels as in Definition 9.1.7. As an immediate corollary of Theorem 9.2.3, we have

Corollary 9.2.4 *Let $f \in L^p(h_\kappa^2\, d\omega)$, $1 \le p < \infty$, or $f \in C(S^{d-1})$. Then the h-harmonic expansion of f with respect to h_κ^2 is (C,δ)-summable in $L^p(h_\kappa^2\, d\omega)$ or $C(S^{d-1})$ provided that $\delta_\kappa > \gamma_\kappa + \frac{d-2}{2}$.*

Proof The (C,δ)-means of the Gegenbauer expansion with respect to w_λ converge if and only if $\delta > \lambda$; see Szegő [1975, p. 246, Theorem 9.1.3]. □

Corollary 9.2.5 *The (C,δ)-means of the h-harmonic expansion with respect to h_κ^2 define a positive linear operator provided that $\delta \ge 2\gamma_\kappa + d - 1$.*

Proof According to an inequality due to Kogbetliantz [1924], see also Askey [1975, p. 71], the (C,δ)-kernel of the Gegenbauer expansion with respect to w_λ is positive if $\delta \ge 2\lambda+1$. □

Note that for $\kappa = 0$ these results reduce to the classical results of ordinary harmonics. In that case, the condition $\delta > \frac{d-2}{2}$ is the so-called critical index (see, for example, Stein and Weiss [1971]), and both corollaries are sharp.

We conclude this section with a result on the expansion of the function $Vf(\langle x,y\rangle)$ in h-harmonics. It comes as an application of the Funk–Hecke formula in Theorem 7.3.4. Let us denote by $s_n^\delta(w_\lambda;\,)$ the Cesàro (C,δ)-means of the Gegenbauer expansion with respect to w_λ.

Proposition 9.2.6 *Let f be a continuous function on $[-1,1]$. Then*

$$S_n^\delta(h_\kappa^2;Vf(\langle x,\cdot\rangle),y) = V[s_n^\delta(w_{\gamma_\kappa+(d-2)/2};f,\langle x,\cdot\rangle)](y).$$

Proof Let $\lambda = \gamma_\kappa + \frac{d-2}{2}$. For each fixed $x \in S^{d-1}$, $P_n(h_\kappa^2;x,y)$ is an element in $\mathscr{H}_n^d(h_\kappa^2)$ and it follows from the Funk–Hecke formula that

$$\begin{aligned} S_n(h_\kappa^2;Vf(\langle x,\cdot\rangle),y) &= \int_{S^{d-1}} Vf(\langle x,u\rangle)P_n(h_\kappa^2;y,u)h_\kappa^2(u)\,\mathrm{d}\omega(u) \\ &= \tau_n(f)\frac{n+\lambda}{\lambda}V\left[C_n^\lambda(\langle x,\cdot\rangle)\right](y) \end{aligned}$$

by Corollary 7.3.2. Hence, by the formula for $\tau_n(f)$ in Theorem 7.3.4,

$$S_n(h_\kappa^2;Vf(\langle x,\cdot\rangle),y) = \int_{-1}^1 f(t)\widetilde{C}_n^\lambda(t)w_\lambda(t)\mathrm{d}t\,V[\widetilde{C}_n^\lambda(\langle x,\cdot\rangle)](y).$$

Taking the Cesàro means of this identity gives the stated formula. □

9.3 Orthogonal Expansion on the Ball

For a weight function W defined on the set Ω of $\mathbb{R}^d$, denote by $L^p(W,\Omega)$ the weighted L^p space with norm defined by

$$\|f\|_{W,p} = \left(\int_\Omega |f(x)|^p W(x)\,\mathrm{d}x\right)^{1/p}$$

for $1 \le p < \infty$. When $p = \infty$, take the space as $C(\Omega)$, the space of continuous functions on Ω with uniform norm.

Recall that $\mathscr{V}_n^d(W)$ denotes the space of orthogonal polynomials of degree n with respect to W and $\mathbf{P}_n(W;x,y)$ denotes the reproducing kernel of $\mathscr{V}_n^d(W)$. Denote by $\mathbf{P}_n(f;x)$ the nth component of the orthogonal expansion which has $\mathbf{P}_n(W;x,y)$ as the integral kernel. Let H be an admissible function defined on $\mathbb{R}^{d+m+1}$, as in Definition 4.3.1. Denote the weighted L^p space by $L^p(H,S^{d+m})$ to emphasize the dependence on S^{d+m}. The orthogonal polynomials with respect to the weight function W_H^m defined in (4.3.1) are related to the orthogonal polynomials with respect to $H\,\mathrm{d}\omega_{d+m}$ on S^{d+m}, as discussed in Section 4.3.

Theorem 9.3.1 *Let H be an admissible weight function and W_m^H be defined as in (4.3.1). Let p be fixed, $1 \le p \le \infty$. If the means $\Phi_n(\cdot) = \sum_{k=0}^n c_{k,n} S_k(\cdot)$ define bounded operators on $L^p(H, S^{d+m})$ then the means $\Psi_n(\cdot) = \sum_{k=0}^n c_{k,n} \mathbf{P}_k(\cdot)$ define bounded operators on $L^p(W_H^m, B^d)$. More precisely, if*

$$\left(\int_{S^{d+m}} |\Phi_n(F;x)|^p H(x)\,\mathrm{d}\omega\right)^{1/p} \le C\left(\int_{S^{d+m}} |F(x)|^p H(x)\,\mathrm{d}\omega\right)^{1/p}$$

for $F \in L^p(H, S^{d+m})$, where C is a constant independent of F and n, then

$$\left(\int_{B^d} |\Psi_n(f;x)|^p W_H^m(x)\,\mathrm{d}x\right)^{1/p} \le C\left(\int_{B^d} |f(x)|^p H(x)\,\mathrm{d}x\right)^{1/p}$$

for $f \in L^p(W_H^m, B^d)$. In particular, if the means $\Phi_n(F)$ converge to F in $L^p(H, S^{d+m})$ then the means $\Psi_n(f)$ converge to f in $L^p(W_H^m, B^d)$.

Proof For $f \in L^p(W_H^m, B^d)$, define a function F on S^{d+m} by $F(x) = f(\mathbf{x}_1)$, where $x = (\mathbf{x}_1, \mathbf{x}_2) \in S^{d+m}$, $\mathbf{x}_1 \in B^d$. By Lemma 4.3.2 it follows that

$$\int_{S^{d+m}} |F(x)|^p H(x)\,\mathrm{d}\omega = C\int_{B^d} |f(x)|^p W_H^m(x)\,\mathrm{d}x;$$

hence, $F \in L^p(H, S^{d+m})$. Using Theorem 4.3.5 and the notation $y = (\mathbf{y}_1, \mathbf{y}_2) \in S^{d+m}$ and $\mathbf{y}_2 = |\mathbf{y}_2|\eta$, $\eta \in S^m$, the sums $P_n(F)$ and $\mathbf{P}_n(f)$ are seen to be related as follows:

$$\begin{aligned}
\mathbf{P}_n(f;\mathbf{x}_1) &= \int_{B^d} \mathbf{P}_n(W_H^m; \mathbf{x}_1, \mathbf{y}_1) f(\mathbf{y}_1) W_H^m(\mathbf{y}_1)\,\mathrm{d}\mathbf{y}_1 \\
&= \int_{B^d} \left[\int_{S^m} P_n\left(H; x, \left(\mathbf{y}_1, \sqrt{1-|\mathbf{y}_1|^2}\eta\right)\right) H_2(\eta)\,\mathrm{d}\omega(\eta)\right] f(\mathbf{y}_1) W_H^m(\mathbf{y}_1)\,\mathrm{d}\mathbf{y}_1 \\
&= \int_{S^{d+m}} P_n(H;x,y) F(y) H(y)\,\mathrm{d}y = P_n(F;x),
\end{aligned}$$

where we have used Lemma 4.3.2. Using the same lemma again,

$$\begin{aligned}
\int_{B^d} |\Phi_n(f;\mathbf{x}_1)|^p W_H^m(\mathbf{x}_1)\,\mathrm{d}\mathbf{x}_1 &= \int_{S^{d+m}} |\Phi_n(f;\mathbf{x}_1)|^p H(x)\,\mathrm{d}\omega(x) \\
&= \int_{S^{d+m}} |\Psi_n(F;x)|^p H(x)\,\mathrm{d}\omega(x).
\end{aligned}$$

Hence, the boundedness of the last integral can be used to conclude that

$$\begin{aligned}
\int_{B^d} |\Phi_n(f;\mathbf{x}_1)|^p W_H^m(\mathbf{x}_1)\,\mathrm{d}\mathbf{x}_1 &\le C^p \int_{S^{d+m}} |F(x)|^p H(x)\,\mathrm{d}\omega \\
&= C^p \int_{B^d} |f(x)|^p W_H^m(x)\,\mathrm{d}x.
\end{aligned}$$

This completes the proof for $1 \le p < \infty$. In the case $p = \infty$, the norm becomes the uniform norm and the result follows readily from the fact that $\mathbf{P}_n(f;\mathbf{x}_1) = P_n(F;x)$. This completes the proof for $m > 0$. In the case $m = 0$, we use Theorem 4.2.8 instead of Theorem 4.3.5. □

Theorem 9.3.1 shows that, in order to study the summability of the Fourier orthogonal expansion on the unit ball, we need only study the summability on the unit sphere. In particular, in the limiting case $m = 0$ we can use Theorem 9.2.3 with d replaced by $d+1$ to obtain the following result for the reflection-invariant weight function $W^B_{\kappa,\mu}$ defined in Definition 8.1.1.

Theorem 9.3.2 *Let $\lambda = \gamma_\kappa + \frac{d-2}{2}$. If the expression $\phi_n(\cdot) = \sum_{k=0}^n c_{k,n} b_k(\cdot) \widetilde{C}_k^\lambda$ defines bounded operators on $L^p(w_\lambda, [-1,1])$, $1 \le p \le \infty$, then the expression $\Psi_n(\cdot) = \sum_{k=0}^n c_{k,n} \mathbf{P}_k(\cdot)$ defines bounded operators on $L^p(W^B_{\kappa,\mu}, B^d)$. More precisely, if*

$$\left(\int_{-1}^1 |\phi_n(f;t)|^p w_\lambda(t)\,\mathrm{d}t\right)^{1/p} \le C\left(\int_{-1}^1 |f(t)|^p w_\lambda(t)\,\mathrm{d}t\right)^{1/p}$$

for $f \in L^p(w_\lambda, [-1,1])$, where C is a constant independent of f and n, then

$$\left(\int_{B^d} |\Psi_n(g;x)|^p W^B_{\kappa,\mu}(x)\,\mathrm{d}x\right)^{1/p} \le C\left(\int_{B^d} |g(x)|^p W^B_{\kappa,\mu}(x)\,\mathrm{d}x\right)^{1/p}$$

for $g \in L^p(W^B_{\kappa,\mu}, B^d)$ with the same constant. In particular, if the means $\phi_n(f)$ converge to f in $L^p(w_\lambda, [-1,1])$, then the means $\Phi_n(g)$ converge to g in $L^p(W^B_{\kappa,\mu}, B^d)$.

For $\delta > 0$, let $S_n^\delta(f; W^B_{\kappa,\mu})$ denote the Cesàro (C,δ) means of the orthogonal expansion with respect to $W^B_{\kappa,\mu}$. As a corollary of the above theorem and the results for Gegenbauer expansions, we can state analogues of Corollaries 9.2.4 and 9.2.5.

Corollary 9.3.3 *Let $W^B_{\kappa,\mu}$ be as in Definition 8.1.1 with $\mu \ge 0$ and let $\lambda_{\kappa,\mu} := \gamma_\kappa + \mu + \frac{d-1}{2}$. Let $f \in L^p(W^B_{\kappa,\mu}, B^d)$. Then the orthogonal expansion of f with respect to $W^B_{\kappa,\mu}$ is (C,δ)-summable in $L^p(W^B_{\kappa,\mu}, B^d)$, $1 \le p \le \infty$, provided that $\delta > \lambda_{\kappa,\mu}$. Moreover, the (C,δ)-means define a positive linear operator if $\delta \ge 2\lambda_{\kappa,\mu} + 1$.*

For the classical weight function $W^B_\mu(x) = (1 - \|x\|^2)^{\mu-1/2}$ $(\kappa = 0)$, the condition for (C,δ)-summability in Corollary 9.3.3 is sharp.

Theorem 9.3.4 *The orthogonal expansion of every continuous function f with respect to W_μ, with $\mu \ge 0$, is uniformly (C,δ)-summable on B^d if and only if $\delta > \mu + \frac{d-1}{2}$.*

Proof We only need to prove the necessary condition. Let $\|x\| = 1$. Let $\mathbf{K}_n^\delta(W_\mu^B)$ denote the (C,δ)-means of the reproducing kernel of the orthogonal expansion. We need to show that

$$\mathscr{I}_n^\mu(x) = \int_{B^d} \left| \mathbf{K}_n^\delta(W_\mu^B; x, y) \right| (1 - \|y\|^2)^{\mu - 1/2} \, dy$$

is unbounded when $\delta = \mu + \frac{d-1}{2}$. By Theorem 5.2.8, $P_n(W_\mu^B; x, y) = \widetilde{C}_n^{\mu+(d-1)/2}(1)\widetilde{C}_n^{\mu+(d-1)/2}(\langle x, y\rangle)$, where we use the normalized Gegenbauer polynomials. Therefore $\mathbf{K}_n^\delta(W_\mu^B; x, y) = K_n^\delta(w_{\mu+(d-1)/2}; 1, \langle x, y\rangle)$, where $K_n^\delta(w_\lambda)$ denotes the (C,δ) means of the Gegenbauer expansion with respect to w_λ. Using the polar coordinates and Lemma 7.4.4,

$$\mathscr{I}_n^\mu = \sigma_{d-2} \int_0^1 r^{d-1} \int_{-1}^1 \left| K_n^\delta(w_{\mu+(d-1)/2}; rs, 1) \right| (1 - s^2)^{(d-3)/2} \, ds \, (1 - r^2)^{\mu - 1/2} \, dr.$$

Making the change of variables $s \mapsto t/r$ and exchanging the order of integration, use of a beta integral shows that

$$\begin{aligned} \mathscr{I}_n^\mu &= \sigma_{d-2} \int_{-1}^1 \left| K_n^\delta(w_{\mu+(d-1)/2}; t, 1) \right| \int_{|t|}^1 (r^2 - t^2)^{(d-3)/2} \, r(1 - r^2)^{\mu - 1/2} \, dr \, dt \\ &= \sigma_{d-2} A_\mu \int_{-1}^1 \left| K_n^\delta(w_{\mu+(d-1)/2}; t, 1) \right| (1 - t^2)^{\mu + (d-2)/2} \, dt \end{aligned}$$

where A_μ is a constant. Therefore, the (C,δ)-summability of the orthogonal expansion with respect to W_μ^B on the boundary of B^d is equivalent to the (C,δ)-summability of the Gegenbauer expansion with index $\mu + \frac{d-1}{2}$ at the point $x = 1$. The desired result follows from Szegő [1975, Theorem 9.1.3, p. 246], where the result is stated for the Jacobi expansion; a shift of $\frac{1}{2}$ on the index is necessary for the Gegenbauer expansion. □

The proof in fact shows that the maximum of the (C,δ)-means with respect to W_μ^B is attained on the boundary of the unit ball.

Since the weight function on the sphere requires all parameters to be nonnegative, we need to assume $\mu \ge 0$ in the above results. The necessary part of the last theorem, however, holds for all $\mu > -\frac{1}{2}$.

For the weight function $W_{\kappa,\mu}^B$ defined in (8.1.5), further results about orthogonal expansions can be derived. Let $S_n^\delta(W_{\kappa,\mu}^B; f)$ denote the Cesàro (C,δ)-means of the orthogonal expansion with respect to $W_{\kappa,\mu}^B$ and let $\mathbf{P}(W_{\kappa,\mu}^B; f)$ denote the nth component of the expansion as in (3.6.2). Recall that $s_n^\delta(w; f)$ denotes the nth (C,δ)-mean of the orthogonal expansion with respect to w. Let $w_{a,b}(t) = |t|^{2a}(1 - t^2)^{b-1/2}$ denote the generalized Gegenbauer weight function.

Proposition 9.3.5 *Let* $W_{\kappa,\mu}^B(x) = \prod_{i=1}^d |x_i|^{\kappa_i}(1 - \|x\|^2)^{\mu - 1/2}$. *Let* f_0 *be an even function defined on* $[-1, 1]$ *and* $f(x) = f_0(\|x\|)$; *then*

$$S_n^\delta(W_{\kappa,\mu}^B; f, x) = s_n^\delta(w_{|\kappa|+(d-1)/2,\mu}; f_0, \|x\|).$$

Proof The formula for the intertwining operator V in Theorem 7.5.4 shows that, for $y = sy'$ with $|y'| = 1$, $V[g(\langle x, \cdot\rangle)](y) = V[g(\langle sx, \cdot\rangle)](y')$. Hence, using polar coordinates, the formulae for the reproducing kernel in Theorem 8.1.10 and Corollary 7.4.5, we obtain, with $\lambda = |\kappa| + \frac{d-1}{2}$,

$$\begin{aligned}
\mathbf{P}_n(W_{\kappa,\mu}^B; f, x) &= w_{\kappa,\mu}^B c_\mu \frac{n+\lambda+\mu}{\lambda+\mu} \int_0^1 f_0(s) s^{d-1} (1-s^2)^{\mu-1/2} \\
&\quad \times \int_{S^{d-1}} V\Big(\int_{-1}^1 C_n^{\lambda+\mu}\Big(\langle sx, \cdot\rangle + \sqrt{1-\|x\|^2}\sqrt{1-s^2}\,t\Big)(1-t^2)^{\mu-1}\mathrm{d}t\Big)(y') \\
&\quad \times h_\kappa^2(y')\,\mathrm{d}\omega(y')\,\mathrm{d}s \\
&= c_\mu w_{\kappa,\mu}^B \frac{n+\lambda+\mu}{\lambda+\mu} B_\kappa \int_0^1 f_0(s) s^{d-1}(1-s^2)^{\mu-1/2} \\
&\quad \times \int_{-1}^1 \int_{-1}^1 C_n^{\lambda+\mu}\Big(s\|x\|u + \sqrt{1-s^2}\sqrt{1-\|x\|^2}\,t\Big)(1-t^2)^{\mu-1}\mathrm{d}t \\
&\quad \times (1-u^2)^{\lambda-1}\mathrm{d}u\,\mathrm{d}s.
\end{aligned}$$

Upon using Corollary 7.5.6 we conclude that

$$\begin{aligned}
\mathbf{P}_n(W_{\kappa,\mu}^B; f, x) &= c\,\widetilde{C}_n^{(\mu,\lambda)}(\|x\|) \int_0^1 \Big[\widetilde{C}_n^{(\mu,\lambda)}(s) + \widetilde{C}_n^{(\mu,\lambda)}(-s)\Big] f_0(|s|) \\
&\qquad\qquad \times s^{d-1}(1-s^2)^{\mu-1/2}\,\mathrm{d}s \\
&= c\,\widetilde{C}_n^{(\mu,\lambda)}(\|x\|) \int_{-1}^1 \widetilde{C}_n^{(\mu,\lambda)}(s) f_0(s) |s|^{d-1} (1-s^2)^{\mu-1/2}\,\mathrm{d}s,
\end{aligned}$$

where $\{\widetilde{C}_n^{(\mu,\lambda)}(t)\}$ are the normalized generalized Gegenbauer polynomials and c is the normalization constant of $w_{\lambda,\mu}$, as can be seen by setting $n = 0$. Taking Cesàro (C, δ)-means proves the stated result. □

Proposition 9.3.5 states that the partial sum of a radial function is also a radial function. It holds, in particular, for the classical weight function W_μ^B. For W_μ^B, there is another class of functions that are preserved in such a manner. A function f defined on $\mathbb{R}^d$ is called a *ridge function* if $f(x) = f_0(\langle x, y\rangle)$ for some $f_0 : \mathbb{R} \mapsto \mathbb{R}$ and $y \in \mathbb{R}^d$. The next proposition states that the partial sum of a ridge function is also a ridge function.

Proposition 9.3.6 *Let $f : \mathbb{R} \mapsto \mathbb{R}$. For $\|y\| = 1$ and $n \geq 0$,*

$$S_n^\delta(W_\mu^B; f(\langle x, \cdot\rangle), y) = s_n^\delta(w_{\mu+(d-1)/2}; f, \langle x, y\rangle).$$

In fact, using the the Funk–Hecke formula in Theorem 8.1.17 and following the proof of Proposition 9.2.6, such a result can be established, for functions of the form $V_B f(\langle x, \cdot\rangle)$, for the weight function $W^B_{\kappa,\mu}$ in Proposition 9.3.5. The most interesting case is the result for the classical weight function. The proof is left to the reader.

9.4 Orthogonal Expansion on the Simplex

Although the structures of the orthogonal polynomials and the reproducing kernel on T^d follow from those on B^d, the summability of orthogonal expansions on T^d does not follow in general as a consequence of the summability on B^d. To state the result, let us denote by $w^{(a,b)}(t) = (1-t)^a(1+t)^b$ the Jacobi weight function and by $p_n^{(a,b)}$ the orthonormal Jacobi polynomials with respect to $w^{(a,b)}$. In the following theorem, let $\mathbf{P}_n(f)$ denote the nth component of the orthogonal expansion with respect to $W^T_{\kappa,\mu}$.

Theorem 9.4.1 *Let* $\lambda = \gamma_\kappa + \mu + \frac{d-2}{2}$. *If* $\phi_n(\cdot) = \sum_{k=0}^n c_{k,n} b_k(\cdot) p_k^{(\lambda,-1/2)}$ *defines bounded operators on* $L^p(w^{(\lambda,-1/2)},[-1,1])$, $1 \le p \le \infty$, *then the means* $\Phi_n(\cdot) = \sum_{k=0}^n c_{k,n}\mathbf{P}_k(\cdot)$ *define bounded operators on* $L^p(W^T_{\kappa,\mu},T^d)$. *More precisely, if*

$$\Big(\int_{-1}^1 |\phi_n(f;t)|^p w^{(\lambda,-1/2)}(t)\,\mathrm{d}t\Big)^{1/p} \le C\Big(\int_{-1}^1 |f(t)|^p w^{(\lambda,-1/2)}(t)\,\mathrm{d}t\Big)^{1/p}$$

for $f \in L^p(w^{(\lambda,-1/2)},[-1,1])$, *where* C *is a constant independent of* f *and* n, *then*

$$\Big(\int_{T^d} |\Phi_n(g;x)|^p W^T_{\kappa,\mu}(x)\,\mathrm{d}x\Big)^{1/p} \le C\Big(\int_{T^d} |g(x)|^p W^T_{\kappa,\mu}(x)\,\mathrm{d}x\Big)^{1/p}$$

for $g \in L^p(W^T_{\kappa,\mu},T^d)$ *with the same constant. In particular, if the means* $\phi_n(f)$ *converge to* f *in* $L^p(w^{(\lambda,-1/2)},[-1,1])$ *then the means* $\Phi_n(g)$ *converge to* g *in* $L^p(W^T_{\kappa,\mu},T^d)$.

Proof We follow the proof of Theorem 9.2.3. The means Φ_n can be written as integrals,

$$\Phi_n(g;x) = w^T_{\kappa,\mu}\int_{T^d}\mathbf{P}_n^{\Phi}(x,y)g(y)W^T_{\kappa,\mu}(y)\,\mathrm{d}y,$$

with $\mathbf{P}_n^{\Phi}(x,y) = \sum_{k=0}^n c_{k,n}\mathbf{P}_k(W^T_{\kappa,\mu};x,y)$. Likewise, write $\phi_n(f)$ as an integral with kernel $p_n^{\phi}(w^{\lambda,-1/2};s,t) = \sum_{k=0}^n c_{k,n}p_k^{(\lambda,-1/2)}(s)p_k^{(\lambda,-1/2)}(t)$. Thus, following the proof of Theorem 9.2.3, the essential part is to show that

$$\mathscr{I}_n(\Phi;x) = \int_{T^d}|\mathbf{P}_n^{\Phi}(x,y)|W^T_{\kappa,\mu}(y)\,\mathrm{d}y = \int_{S^d}|\mathbf{P}_n^{\Phi}(x,\{y\}^2)|h^2_{\kappa,\mu}(y')\,\mathrm{d}\omega(y')$$

with $y' = (y, y_{d+1})$ uniformly bounded for $x \in T^d$, where the second equality follows from Lemmas 4.4.1 and 4.2.3. Let V be the intertwining operator associated with $h_{\kappa,\mu}$ and the reflection group $W = W_0 \times \mathbb{Z}_2$. From the proof of Theorem

8.1.10, we see that the formula for the reproducing kernel $\mathbf{P}_k(W^T_{\kappa,\mu})$ in Theorem 8.2.7 can be written as

$$\mathbf{P}_k(W^T_{\kappa,\mu};x,y) = \frac{2n+\lambda+\frac{1}{2}}{\lambda+\frac{1}{2}} V\left[C^{\lambda+1/2}_{2n}(\langle\{x'\}^{1/2},\cdot\rangle)\right](\{y'\}^{1/2}),$$

where $x' = (x, x_{d+1})$ and $|x'| = 1$. Using the relation between the Jacobi and Gegenbauer polynomials (Definition 1.5.5 with $\mu = 0$), we have

$$\frac{2n+\lambda+\frac{1}{2}}{\lambda+\frac{1}{2}} C^{\lambda+1/2}_{2n}(t) = p_n^{(\lambda,-1/2)}(1) p_n^{(\lambda,-1/2)}(2t^2-1).$$

Hence the kernel can be written in terms of Jacobi polynomials as

$$\mathbf{P}^\Phi_n(x,\{y\}^2) = 2^{-d} \sum_{\varepsilon\in\mathbb{Z}_2^d} V\left[p^\phi_n(w^{(\lambda,-1/2)};1,2\langle\cdot,\mathbf{z}(\varepsilon)\rangle^2-1)\right](y')$$

where $\mathbf{z}(\varepsilon) = \{x'\}^{1/2}\varepsilon$. We observe that $|\mathbf{z}(\varepsilon)|^2 = |x'| = 1$. Hence, since V is a positive operator, we can use Corollary 7.4.5 to conclude that

$$\begin{aligned}
\mathscr{I}_n(\Phi;x) &\le 2^{-d} \sum_{\varepsilon\in\mathbb{Z}_2^d} \int_{S^d} V\left[|p^\phi_n(w^{(\lambda,-1/2)};1,2\langle\cdot,\mathbf{z}(\varepsilon)\rangle^2-1)|\right](y') h^2_\kappa(y')\,d\omega \\
&= B_\kappa \int_{-1}^1 |p^\phi_n(w^{(\lambda,-1/2)};1,2t^2-1)|(1-t^2)^\lambda\,dt \\
&= \int_{-1}^1 |p^\phi_n(w^{(\lambda,-1/2)};1,u)| w^{(\lambda,-1/2)}(u)\,du,
\end{aligned}$$

where the last step follows from a change of variables, and the constant in front of the integral becomes 1 (on setting $n = 0$). This completes the proof. □

As a corollary of the above theorem and the results for the Jacobi expansion, we can state an analogue of Corollary 9.3.3.

Corollary 9.4.2 *Let $W^T_{\kappa,\mu}$ be as in Definition 8.2.1 with $\mu,\kappa \ge 0$ and $f \in L^p(W^T_{\kappa,\mu},T^d)$. Then the orthogonal expansion of f with respect to $W^T_{\kappa,\mu}$ is (C,δ)-summable in $L^p(W^T_{\kappa,\mu},T^d)$, $1 \le p \le \infty$, provided that $\delta > \gamma_\kappa + \mu + \frac{d-1}{2}$. Moreover, the (C,δ)-means are positive linear operators if $\delta \ge 2\gamma_\kappa + 2\mu + d$.*

In particular, these results apply to the classical orthogonal polynomials on T^d. In this case, it should be pointed out that the condition $\delta > \gamma_\kappa + \mu + \frac{d-1}{2}$ is sharp only when at least one $\kappa_i = 0$. For sharp results and further discussion, see the notes at the end of the chapter.

For $\delta > 0$, let $S^\delta_n(W^T_{\kappa,\mu};f)$ denote the Cesàro (C,δ)-means of the orthogonal expansion with respect to $W^T_{\kappa,\mu}$. Let us also point out the following analogue of Proposition 9.3.5.

Proposition 9.4.3 *Let $w^{(a,b)}(t) = (1-t)^a(1+t)^b$ be the Jacobi weight function. Let $f(x) = f_0(2|x|-1)$. Then*

$$S_n^\delta(W_{\kappa,\mu}^T; f, x) = s_n^\delta(w^{(\gamma_\kappa+(d-2)/2,\mu-1/2)}; f_0, 2|x|-1).$$

Proof Let $P_{\beta,m}^n$ denote the orthonormal basis in Proposition 8.2.5. By definition,

$$\mathbf{P}_n(W_{\kappa,\mu}^T; x, y) = \sum_{m=0}^{n} \sum_{|\beta|=m} P_{\beta,m}(x) P_{\beta,m}(y).$$

Since $P_{\beta,m}^n(x) = b_{m,n} p_{n-m}^{(\gamma+2m+(d-2)/2,\mu-1/2)}(2s-1) R_\beta^m(x)$ and R_β^m is homogeneous in the homogeneous coordinates of T^d, using the ℓ^1-radial coordinates for the simplex, $x = su$, we obtain

$$\begin{aligned}
&w_{\kappa,\mu}^T \int_{T^d} f(x) \mathbf{P}_n(W_{\kappa,\mu}^T; x, y) W_{\kappa,\mu}^T(y)\, \mathrm{d}y \\
&= w_{\kappa,\mu}^T \int_0^1 f_0(2s-1) s^{\gamma+d-1} \\
&\quad \times (1-s)^{\mu-1/2} \int_{|u|=1} \mathbf{P}_n(W_{\kappa,\mu}^T; x, su) h(\sqrt{u_1}, \ldots, \sqrt{u_d}) \frac{\mathrm{d}u}{\sqrt{u_1 \cdots u_d}}\, \mathrm{d}s \\
&= c \int_0^1 f_0(2s-1) p_n^{(\gamma+(d-2)/2,\mu-1/2)}(2t-1) p_n^{(\gamma+(d-2)/2,\mu-1/2)}(2s-1) \\
&\quad \times s^{\gamma+d-1}(1-s)^{\mu-1/2}\, \mathrm{d}s,
\end{aligned}$$

where $t = |x|$ and $c = B(\gamma + \frac{d}{2}, \mu + \frac{1}{2})^{-1}$. □

9.5 Orthogonal Expansion of Laguerre and Hermite Polynomials

For the multiple Laguerre polynomials with respect to the weight function $W_\kappa^L(x) = x^\kappa \mathrm{e}^{-|x|}$ with $x^\kappa = \prod_{i=1}^d |x_i|^{\kappa_i}$, we start with

Theorem 9.5.1 *If f is continuous at the origin and satisfies*

$$\int_{x \in \mathbb{R}_+^d, |x| \ge 1} |f(x)| x^\kappa |x|^{-\delta-1/2} \mathrm{e}^{-|x|/2}\, \mathrm{d}x < \infty$$

then the Cesàro (C,δ)-means of the multiple Laguerre expansion of f converge at the origin if and only if $\delta > |\kappa| + d - \frac{1}{2}$.

Proof The generating function of the Laguerre polynomials (Subsection 1.4.2) can be written as

$$\sum_{n=0}^{\infty} \widetilde{L}_n^\alpha(0) \widetilde{L}_n^\alpha(x) r^n = (1-r)^{-\alpha-1} \mathrm{e}^{-xr/(1-r)}, \qquad |r| < 1.$$

Summing the above formula gives a generating function for $\mathbf{P}_n(W_\kappa^L;x,0)$:

$$\sum_{n=0}^{\infty} \mathbf{P}_n(W_\kappa^L;x,0)r^n = \sum_{n=0}^{\infty} \sum_{|\mathbf{k}|=n} \widetilde{L}_{\mathbf{k}}^{\kappa}(0)\widetilde{L}_{\mathbf{k}}^{\kappa}(x)r^n$$
$$= (1-r)^{-|\kappa|-d} e^{-|x|r/(1-r)},$$

which implies that $\mathbf{P}_n(W_\kappa^L;x,0) = L_n^{|\kappa|+d-1}(|x|)$. Multiplying the last expression by the power series of $(1-r)^{-\delta-1}$ shows that

$$\mathbf{P}_n^{\delta}(W_\kappa^L;x,0) = \frac{n!}{(\delta+1)_n} L_n^{|\kappa|+\delta+d}(|x|).$$

Therefore, we conclude that

$$S_n^{\delta}(W_\kappa^L;f,0) = \frac{n!}{(\delta+1)_n} w_\kappa^L \int_{\mathbb{R}_+^d} f(x) L_n^{|\kappa|+d+\delta}(|x|) W_\kappa^L(x)\, \mathrm{d}x.$$

Using ℓ^1 radial coordinates as in (8.4.3), it follows that

$$S_n^{\delta}(W_\kappa^L;f,0) = w_\kappa^L \frac{n!}{(\delta+1)_n} \int_0^\infty \left(\int_{|y|=1} f(ry) y^\kappa \mathrm{d}y \right)$$
$$\times L_n^{|\kappa|+d+\delta}(r) r^{|\kappa|+d-1} e^{-r} \mathrm{d}r.$$

The right-hand side is the (C,δ)-mean of the Laguerre expansion of a one-variable function. Indeed, define $F(r) = c_d \int_{|y|=1} f(ry) y^\kappa \mathrm{d}y$, where the constant $c_d = \Gamma(|\kappa|+d)/\prod_{i=1}^d \Gamma(\kappa_i+1)$; then the above equation can be written as

$$S_n^{\delta}(W_\kappa^L;f,0) = s^{\delta}(w_{|\alpha|+d-1};F,0),$$

where $w_a(t) = t^a e^{-t}$ is the usual Laguerre weight function for one variable. Note that $c_d = 1/\int_{|y|=1} y^\kappa \mathrm{d}y$, so that F is just the average of f over the simplex $\{y : |y|=1\}$. This allows us to use the summability theorems for the Laguerre expansion of one variable. The desired result follows from Szegő [1975, p. 247, Theorem 9.1.7]. The condition of that theorem is verified as follows:

$$\int_1^\infty |F(r)| r^{|\kappa|+d-1-\delta-1/2} e^{-r/2} \mathrm{d}r$$
$$\le c_d \int_1^\infty \int_{|y|=1} |f(ry) y^\kappa| \mathrm{d}y \, r^{|\kappa|+d-\delta-3/2} e^{-r/2} \mathrm{d}r$$
$$\le c_d \int_{x\in\mathbb{R}_+^d, |x|\ge 1} |f(x) x^\kappa| \, |x|^{-\delta-1/2} e^{-|x|/2} \mathrm{d}x,$$

which is bounded under the given condition; it is evident that F is continuous at $r=0$ if f is continuous at the origin. □

The result can be extended to $\mathbb{R}_+$ by using a convolution structure of the Laguerre expansion on $\mathbb{R}_+$. The convolution is motivated by the product formula of Laguerre polynomials,

$$L_n^\lambda(x)L_n^\lambda(y) = \frac{\Gamma(n+\lambda+1)2^\lambda}{\Gamma(n+1)\sqrt{2\pi}} \int_0^\pi L_n^\lambda(x+y+2\sqrt{xy}\cos\theta)e^{-\sqrt{xy}\cos\theta} \times j_{\lambda-1/2}(\sqrt{xy}\sin\theta)\sin^{2\lambda}\theta\, d\theta,$$

where j_μ is the Bessel function of fractional order (see, for example, Abramowitz and Stegun [1970]). For a function f on $\mathbb{R}_+$ define the Laguerre translation operator $T_x^\lambda f$ by

$$T_x^\lambda f(y) = \frac{\Gamma(\lambda+1)2^\lambda}{\sqrt{2\pi}} \int_0^\pi f(x+y+2\sqrt{xy}\cos\theta)e^{-\sqrt{xy}\cos\theta} \times j_{\lambda-1/2}(\sqrt{xy}\sin\theta)\sin^{2\lambda}\theta\, d\theta.$$

The product formula implies that $L_n^\lambda(x)L_n^\lambda(y) = L_n^\lambda(0)T_x^\lambda L_n^\lambda(y)$. The convolution of f and g is defined by

$$(f*g)(x) = \int_0^\infty f(y)T_x^\lambda g(y)y^\lambda e^{-y}\, dy.$$

It satisfies the following property due to Görlich and Markett [1982].

Lemma 9.5.2 *Let $\lambda \ge 0$ and $1 \le p \le \infty$. Then, for $f \in L^p(w_\lambda, \mathbb{R}_+)$ and $g \in L^1(w_\lambda, \mathbb{R}_+)$,*

$$\|f*g\|_{p,w_\lambda} \le \|f\|_{p,w_\lambda}\|g\|_{1,w_\lambda}.$$

For the proof we refer to the paper of Görlich and Markett or to the monograph by Thangavelu [1993, p. 139]. The convolution structure can be extended to multiple Laguerre expansions and used to prove the following result.

Theorem 9.5.3 *Let $\kappa_i \ge 0$, $1 \le i \le d$, and $1 \le p \le \infty$. The Cesàro (C,δ)-means of the multiple Laguerre expansion are uniformly bounded as operators on $L^p(W_\kappa, \mathbb{R}_+^d)$ if $\delta > |\kappa| + d - \frac{1}{2}$. Moreover, for $p = 1$ and ∞, the (C,δ)-means converge in the $L^p(W_\kappa, \mathbb{R}_+^d)$ norm if and only if $\delta > |\kappa| + d - \frac{1}{2}$.*

Proof The product formula of the Laguerre polynomials and the definition of the reproducing kernel gives

$$\mathbf{P}_n^\delta(W_\kappa; x, y) = T_{x_1}^{\kappa_1} \cdots T_{x_d}^{\kappa_d} P_n^\delta(W_\kappa; 0, y)$$

where T_{x_i} acts on the variable y_i. Therefore, it follows that

$$S_n^\delta(W_\kappa; f, x) = \int_{\mathbb{R}_+^d} f(y)T_{x_1}^{\kappa_1} \cdots T_{x_d}^{\kappa_d}\mathbf{P}_n^\delta(W_\kappa; 0, \mathbf{y})W_\kappa(\mathbf{y})\, d\mathbf{y},$$

which can be written as a d-fold convolution in an obvious way. Therefore, applying the inequality in Lemma 9.5.2 d times gives

$$\|S_n^\delta(W_\kappa; f)\|_{p,W_\kappa} \le \|\mathbf{P}_n^\delta(W_\kappa; 0, \cdot)\|_{1,W_\kappa}\|f\|_{p,W_\kappa}.$$

The norm of $\mathbf{P}_n^\delta(W_\kappa, 0, \mathbf{x})$ is bounded if and only if $\delta > |\kappa| + d - \frac{1}{2}$ by the Theorem 9.5.1. The convergence follows from the standard density argument. □

The following result states that the expansion of an ℓ^1-radial function is also an ℓ^1-radial function.

Proposition 9.5.4 *Let $w_a(t) = t^a \mathrm{e}^{-t}$ denote the Laguerre weight function. Let f_0 be a function defined on $\mathbb{R}_+$ and $f(x) = f_0(|x|)$. Then, for the multiple Laguerre expansion,*

$$S_n(W_\kappa^L; f, x) = s_n^\delta(w_{|\kappa|+d-1}; f_0, |x|).$$

Proof The proof is based on Proposition 8.4.7. Using (8.4.3) and the orthogonality of R_β^m, we obtain

$$\begin{aligned}
&\int_{\mathbb{R}_+^d} f_0(|y|) \mathbf{P}_n(W_\kappa^L; x, y) y^\kappa \mathrm{e}^{-|y|} \, \mathrm{d}y \\
&\quad = \int_0^\infty f(r) r^{|\kappa|+d-1} \mathrm{e}^{-r} \int_{|u|=1} \mathbf{P}_n(W_\kappa; x, ru) u^\kappa \, \mathrm{d}u \, \mathrm{d}r \\
&\quad = c \int_0^\infty f(r) r^{|\kappa|+d-1} \mathrm{e}^{-r} \widetilde{L}_n^{|\kappa|+d-1}(r) \, \mathrm{d}r \, \widetilde{L}_n^{|\kappa|+d-1}(|x|),
\end{aligned}$$

where the constant c can be determined by setting $n = 0$. Taking the Cesàro means gives the desired result. □

For the summability of the multiple Hermite expansion, there is no special boundary point for $\mathbb{R}^d$ or convolution structure. In this case the summability often involves techniques from classical Fourier analysis.

Proposition 9.5.5 *Let W_κ^H be as in Definition 8.3.1. Let $f_\mu(x) = f(\sqrt{\mu}x)$. Then, for each $x \in \mathbb{R}^d$,*

$$\lim_{\mu \to \infty} S_n^\delta(W_{\kappa,\mu}^B; f_\mu, x/\sqrt{\mu}) = S_n^\delta(W_\kappa^H; f, x).$$

Proof We look at the Fourier coefficient of f_μ. Making the change of variables $x \mapsto y/\sqrt{\mu}$ leads to

$$\begin{aligned}
a_\alpha^n(W_{\kappa,\mu}^B; f) &= w_{\kappa,\mu}^B \int_{B^d} f(\sqrt{\mu}x) P_\alpha^n(W_{\kappa,\mu}^B; x) W_{\kappa,\mu}^B(x) \, \mathrm{d}x \\
&= w_{\kappa,\mu}^B \mu^{-(\gamma+d)/2} \int_{\mathbb{R}^d} f(y) \chi_\mu(y) P_\alpha^n(W_{\kappa,\mu}^B; y/\sqrt{\mu}) h_\kappa^2(y) (1 - \|y\|^2/\mu)^{\mu-1/2} \, \mathrm{d}y,
\end{aligned}$$

where χ_μ is the characteristic function of the set $\{y : \|y\| \le \sqrt{\mu}\}$. Using the fact that $w_{\kappa,\mu}^B \mu^{-(\gamma+d)/2} \to w_\kappa^H$ as $\mu \to \infty$ and Theorem 8.3.4, the dominated convergence theorem shows that

$$\lim_{\mu \to \infty} a_\alpha^n(W_{\kappa,\mu}^B; f) = w_\kappa^H \int_{\mathbb{R}^d} f(y) P_\alpha^n(W_\kappa^H; y) h_\kappa^2(y) \mathrm{e}^{-\|y\|^2} \mathrm{d}y,$$

which is the Fourier coefficient $a^n_\alpha(W^H_\kappa; f)$; the stated equation follows from Theorem 8.3.4. □

A similar result can be seen to hold for the Laguerre expansion by using the limit relation in Theorem 8.4.4. We leave it to the reader. As an application of this limit relation we obtain

Proposition 9.5.6 *Let* $w_a(t) = |t|^{2a}\mathrm{e}^{-t^2}$. *Let* f_0 *be an even function on* $\mathbb{R}$ *and* $f(x) = f_0(\|x\|)$. *Then, for the orthogonal expansion with respect to* $W^H_\kappa(x) = \prod_{i=1}^d x_i^{2\kappa_i}\mathrm{e}^{-\|x\|^2}$,

$$S_n^\delta(W^H_\kappa; f, x) = s_n^\delta(w_{|\kappa|+(d-1)/2}; f_0, \|x\|).$$

Proof In the identity of Proposition 9.3.5 set $f(x) = f(\sqrt{\mu}x)$ and replace x by $x/\sqrt{\mu}$; then take the limit $\mu \to \infty$. The left-hand side is the limit in Proposition 9.5.5, while the right-hand side is the same limit as that for $d = 1$. □

Furthermore, for the classical multiple Hermite polynomials there is an analogue of Proposition 9.3.6 for ridge functions.

Proposition 9.5.7 *Let* $W^H(x) = \mathrm{e}^{-\|x\|^2}$ *and* $w(t) = \mathrm{e}^{-t^2}$. *Let* $f : \mathbb{R} \mapsto \mathbb{R}$. *For* $\|y\| = 1$ *and* $n \geq 0$,

$$S_n^\delta(W^H; f(\langle x, \cdot\rangle), y) = s_n^\delta(w; f, \langle x, y\rangle).$$

The proof follows by taking the limit $\mu \to \infty$ in Proposition 9.3.6.

Corollary 9.5.8 *Let* $f(x) = f_0(\|x\|)$ *be as in Proposition 9.5.6. Let* $f \in L^p(W^H; \mathbb{R}^d)$, $1 \leq p < \infty$. *Then* $S_n^\delta(W^H; f)$ *converges to* f *in the* $L^p(W^H; \mathbb{R}^d)$ *norm if* $\delta > \frac{d-1}{2}$.

Proof Using polar coordinates, we obtain

$$\pi^{-d/2}\int_{\mathbb{R}^d}|S_n^\delta(W^H; f, x)|^p \mathrm{e}^{-\|x\|^2}\,\mathrm{d}x = \pi^{-d/2}\sigma_{d-1}\int_0^\infty s_n^\delta(w_{(d-1)/2}; f_0, r) r^{d-1}\mathrm{e}^{-r^2}\,\mathrm{d}r.$$

Hence, the stated result follows from the result for the Hermite expansion on the real line; see Thangavelu [1993]. □

For general results on the summability of multiple Hermite expansion, we again refer to Thangavelu [1993].

9.6 Multiple Jacobi Expansion

Multiple Jacobi polynomials are orthogonal with respect to the weight function $W_{\alpha,\beta}(x) = \prod_{i=1}^{d}(1-x_i)^{\alpha_i}(1+x_i)^{\beta_i}$; see Subsection 5.1.1. Recall that $p_n^{(a,b)}(t)$ denotes an orthonormal Jacobi polynomial with respect to $(1-t)^a(1+t)^b$. There is a convolution structure for the Jacobi polynomials, discovered by Gasper [1972].

Lemma 9.6.1 *Let $\alpha,\beta > -1$. There is an integral representation of the form*

$$p_n^{(\alpha,\beta)}(x)p_n^{(\alpha,\beta)}(y) = p_n^{(\alpha,\beta)}(1)\int_{-1}^{1} p_n^{(\alpha,\beta)}(t)\,\mathrm{d}\mu_{x,y}^{(\alpha,\beta)}(t), \qquad n \ge 0,$$

where the real Borel measure $\mathrm{d}\mu_{x,y}^{(\alpha,\beta)}$ on $[-1,1]$ satisfies

$$\int_{-1}^{1} |\mathrm{d}\mu_{x,y}^{(\alpha,\beta)}(t)| \le M, \qquad -1 < x,y < 1,$$

for some constant M independent of x,y if and only if $\alpha \ge \beta$ and $\alpha+\beta \ge -1$. Moreover, the measures are nonnegative, that is, $\mathrm{d}\mu_{x,y}^{(\alpha,\beta)}(t) \ge 0$ if and only if $\beta \ge -\frac{1}{2}$ or $\alpha+\beta \ge 0$.

The product in Lemma 9.6.1 gives rise to a convolution structure of multiple Jacobi polynomials, which allows us to reduce the summability on $[-1,1]^d$ to the summability at the point $\varepsilon = (1,\ldots,1)$, just as in the Laguerre case. The Jacobi polynomials have a generating function (Bailey [1935, p. 102, Ex. 19])

$$\begin{aligned} G^{(\alpha,\beta)}(r;x) &= \sum_{k=0}^{\infty} p_k^{(\alpha,\beta)}(1)p_k^{(\alpha,\beta)}(x)r^n \\ &= \frac{1-r}{(1+r)^{\alpha+\beta+2}}\, {}_2F_1\left(\begin{matrix}\frac{\alpha+\beta+2}{2}, \frac{\alpha+\beta+3}{2} \\ \beta+1\end{matrix}; \frac{2r(1+x)}{(1+r)^2}\right), \qquad 0 \le r < 1. \end{aligned}$$

Multiplying this formula with different variables gives a generating function for the multiple Jacobi polynomials,

$$\sum_{n=0}^{\infty} r^n \sum_{|\mathbf{k}|=n} P_{\mathbf{k}}^{(\alpha,\beta)}(x)P_{\mathbf{k}}^{(\alpha,\beta)}(\mathbf{1}) = \prod_{i=1}^{d} G^{(\alpha_i,\beta_i)}(r;x_i) := G_d^{(\alpha,\beta)}(r;x),$$

where $\mathbf{1} = (1,1,\ldots,1)$. Further multiplication, by

$$(1-r)^{-\delta-1} = \sum_{n=0}^{\infty} \frac{(\delta+1)_n r^n}{n!},$$

gives

$$\sum_{n=0}^{\infty} \frac{(\delta+1)_n}{n!} \mathbf{K}_{n,d}^{\delta}(W_{\alpha,\beta};x,\mathbf{1})r^n = (1-r)^{-\delta-1}G_d^{(\alpha,\beta)}(r;\mathbf{1}) \tag{9.6.1}$$

for the Cesàro (C,δ)-means of the multiple Jacobi polynomials.

Theorem 9.6.2 *Let $\alpha_j \ge -\frac{1}{2}$ and $\beta_j \ge -\frac{1}{2}$ for $1 \le j \le d$. Then the Cesàro (C,δ)-means $S^{\delta}_{n,d}(W_{\alpha,\beta};f)$ of the product Jacobi expansion define a positive approximate identity on $C([-1,1])^d$ if $\delta \ge \sum_{i=1}^d(\alpha_i+\beta_i)+3d-1$; moreover, the order of summability is best possible in the sense that (C,δ)-means are not positive for $0<\delta<\sum_{i=1}^d(\alpha_i+\beta_i)+3d-1$.*

Proof Let $\rho(\alpha)=\sum_{i=1}^d \alpha_i$; if all $\alpha_i \ge 0$ then $\rho(\alpha)=|\alpha|$. Lemma 9.6.1 shows that it is sufficient to prove that, for $\alpha_j \ge \beta_j$, the kernel $K^{\delta}_{n,d}(W_{\alpha,\beta};x,\mathbf{1}) \ge 0$ if and only if $\delta \ge \rho(\alpha)+\rho(\beta)+3d-1$. For $d=1$, the $(C,\alpha+\beta+2)$-means $K^{\alpha+\beta+2}_{n,1}(w_{\alpha,\beta};x,1)$ are nonnegative for $-1\le x\le 1$, as proved in Gasper [1977]. Hence, by (9.6.1) with $d=1$, the function $(1-r)^{-\alpha-\beta-3}G^{(\alpha,\beta)}(r;x)$ is a *completely monotone function* of r, that is, a function whose power series has all nonnegative coefficients. Since multiplication is closed in the space of completely monotone functions, it follows that

$$(1-r)^{-\rho(\alpha)-\rho(\beta)-3d}G_d^{(\alpha,\beta)}(r;\mathbf{x})=\prod_{j=1}^d(1-r)^{-\alpha_j-\beta_j-3}G_d^{(\alpha_j,\beta_j)}(r;x_j)$$

is a completely monotone function. Consequently, by (9.6.1) we conclude that the means $K^{\rho(\alpha)+\rho(\beta)+3d-1}_{n,d}(W_{\alpha,\beta};\mathbf{x},\mathbf{1})\ge 0$. We now prove that the order of summation cannot be improved. If the (C,δ_0)-means are positive then (C,δ)-means are positive for $\delta\ge\delta_0$. Hence, it suffices to show that the $(C,\rho(\alpha)+\rho(\beta)+3d-1-\sigma)$-means of the kernel are not positive for $0<\sigma<1$. From the generating function and the fact that ${}_2F_1(a,b;c;0)=1$, we conclude that, for $\delta=\rho(\alpha)+\rho(\beta)+3d-2$,

$$\begin{aligned}(1-r)^{-\delta-1}G_d^{(\alpha,\beta)}(r;-\mathbf{1}) &= (1-r)(1-r^2)^{-\rho(\alpha)-\rho(\beta)-2d}\\ &=\sum_{k=0}^{\infty}\frac{(\rho(\alpha)+\rho(\beta)+2d)_k}{k!}\left(r^{2k}-r^{2k+1}\right).\end{aligned}$$

Hence, setting

$$A_k=\frac{(\rho(\alpha)+\rho(\beta)+3d-1)_k}{k!}K^{\rho(\alpha)+\rho(\beta)+3d-2}_{k,d}(W_{\alpha,\beta};-\mathbf{1},\mathbf{1})$$

and comparing with (9.6.1), we conclude that

$$A_{2k}=-A_{2k+1}=\frac{(\rho(\alpha)+\rho(\beta)+2d)_k}{k!}\ge 0.$$

Therefore, it follows that

$$\begin{aligned}&\frac{(\rho(\alpha)+\rho(\beta)+3d-\sigma+2)_{2n-1}}{(2n-1)!}K^{\rho(\alpha)+\rho(\beta)+3d-1-\sigma}_{2n+1,d}(W_{\alpha,\beta};-\mathbf{1},\mathbf{1})\\ &\quad=\sum_{k=0}^{2n+1}\frac{(-\sigma)_{2n+1-k}}{(2n+1-k)!}A_k=-\sigma\sum_{k=0}^{n}\frac{1}{2n-2k+1}\frac{(1-\sigma)_{2n-2k}}{(2n-2k)!}A_{2k}.\end{aligned}$$

Since $0<\sigma<1$, we conclude that the $(C,\rho(\alpha)+\rho(\beta)+3d-1-\sigma)$-means are not positive. □

The positivity of these Cesàro means implies their convergence, since it shows that the uniform norm of $S_n^\delta(W_{\alpha,\beta};f)$ is 1. The explicit formula for the reproducing kernel is known only in one special case, namely, the case when $\alpha_i = \beta_i = -\frac{1}{2}$ for all i; see Xu [1995] or Berens and Xu [1996]. This formula, given below, is nonetheless of interest and shows a character different from the explicit formulae for the reproducing kernel on the ball and on the simplex. Denote the weight function by $W_0(x)$, that is,

$$W_0(x) = (1-x_1^2)^{-1/2}\cdots(1-x_d^2)^{-1/2}, \qquad x \in [-1,1]^d.$$

We also need the notion of divided difference. The divided difference of a function $f:\mathbb{R}\to\mathbb{R}$ at the pairwise distinct points $x_0, x_1, \dots, x_n$ in $\mathbb{R}$, is defined inductively by

$$[x_0]f = f(x_0) \qquad \text{and} \qquad [x_0,\dots,x_n]f = \frac{[x_0,\dots,x_{n-1}]f - [x_1,\dots,x_n]f}{x_0 - x_n}.$$

The difference is a symmetric function of the coordinates of the points.

Theorem 9.6.3 *Let* $\mathbf{1} = (1,1,\dots 1)$. *Then*

$$\mathbf{P}_n(W_0;x,\mathbf{1}) = [x_1,\dots,x_d]\,G_n$$

with

$$G_n(t) = (-1)^{[(d+1)/2]}2(1-t^2)^{(d-1)/2}\begin{cases} T_n(t) & \text{for } d \text{ even},\\ U_{n-1}(t) & \text{for } d \text{ odd}.\end{cases}$$

Proof The generating function of the Chebyshev polynomials T_n and U_n, given in Proposition 1.4.12, implies that

$$\frac{(1-r^2)^d}{\prod_{i=1}^d(1-2rx_i+r^2)} = \sum_{n=0}^{\infty}\mathbf{P}(W_0;x,\mathbf{1})r^n.$$

The left-hand side can be expanded as a power series using the formula

$$[x_1,\dots,x_d]\frac{1}{a-b(\cdot)} = \frac{b^{d-1}}{\prod_{i=1}^d(a-bx_i)},$$

which can be proved by induction on the number of variables; the result is

$$\begin{aligned}\frac{(1-r^2)^d}{\prod_{i=1}^d(1-2rx_i+r^2)} &= \frac{(1-r^2)^d}{(2r)^{d-1}}[x_1,\dots,x_d]\frac{1}{1-2r(\cdot)+r^2}\\ &= \frac{(1-r^2)^d}{(2r)^{d-1}}[x_1,\dots,x_d]\sum_{n=d-1}^{\infty}U_n(\cdot)r^n\\ &= 2^{1-d}[x_1,\dots,x_d]\sum_{n=0}^{\infty}r^n\sum_{k=0}^{d}\binom{d}{k}(-1)^kU_{n-2k+d-1}(\cdot),\end{aligned}$$

where the second equation uses the generating function of the Chebyshev polynomials of the second kind, given in Definition 1.4.10, and the third equation uses the fact that $[x_1,\dots,x_d]p = 0$ whenever p is a polynomial of degree at most $d-2$. Using $U_{m+1}(t) = \sin m\theta/\sin\theta$ and $\sin m\theta = (\mathrm{e}^{im\theta} - \mathrm{e}^{-im\theta})/(2\mathrm{i})$, with $t = \cos\theta$, as well as the binomial theorem, we obtain

$$\begin{aligned}
&\sum_{k=0}^{d}\binom{d}{k}(-1)^k U_{n-2k+d-1}(t)\\
&= \frac{1}{\sin\theta}\sum_{k=0}^{d}\binom{d}{k}(-1)^k(\sin\theta)^{n-2k+d}\\
&= \frac{1}{2\mathrm{i}\sin\theta}\left[\mathrm{e}^{\mathrm{i}(n+d)\theta}(1-\mathrm{e}^{-2\mathrm{i}\theta})^d + \mathrm{e}^{-\mathrm{i}(n+d)\theta}(1-\mathrm{e}^{2\mathrm{i}\theta})^d\right]\\
&= (2\mathrm{i})^d(\sin\theta)^{d-1}\frac{\mathrm{e}^{\mathrm{i}n\theta}(-1)^d - \mathrm{e}^{-\mathrm{i}n\theta}}{2\mathrm{i}} = 2^{d-1}G_n(t),
\end{aligned}$$

from which the stated formula follows on comparing the coefficients of r^n in the two power expansions. □

Proposition 9.6.4 *Define* $x = \cos\Theta = (\cos\theta_1,\dots,\cos\theta_d)$ *and* $y = \cos\Phi = (\cos\phi_1,\dots,\cos\phi_d)$. *For* $\tau\in\mathbb{Z}_2^d$, *denote by* $\Phi+\tau\Theta$ *the vector that has components* $\phi_i+\tau_i\theta_i$. *Then*

$$\mathbf{P}_n(W_0;x,y) = \sum_{\tau\in\mathbb{Z}_2^d}\mathbf{P}_n(W_0;\cos(\Phi+\tau\Theta),\mathbf{1}).$$

Proof With respect to the normalized weight function $\pi^{-1}(1-t^2)^{-1/2}$, the orthonormal Chebyshev polynomials are $\widetilde{T}_0(t) = 1$ and $\widetilde{T}_n(t) = \sqrt{2}\cos n\theta$. Hence

$$\mathbf{P}_n(W_0;\mathbf{x},\mathbf{y}) = 2^d\sum_{|\alpha|=n}{}'\cos\alpha_1\theta_1\cos\alpha_1\phi_1\cdots\cos\alpha_d\theta_d\cos\alpha_d\phi_d,$$

where $\alpha\in\mathbb{N}_0^d$ and the notation $\sum'$ means that whenever $\alpha_i = 0$ the term containing $\cos\alpha_i\theta_i$ is halved. Then the stated formula is seen to be merely a consequence of the addition formula for the cosine function. □

The multiple Chebyshev expansion is related to the multiple Fourier series on $\mathbb{T}^d$. In fact, $\mathbf{P}_n(W_0;x,\mathbf{1})$ is the Dirichlet kernel in the following sense:

$$\mathbf{P}_n(W_0;x,\mathbf{1}) = \sum_{|\alpha|=n}\mathrm{e}^{\mathrm{i}\langle\alpha,\Theta\rangle}, \qquad x = \cos\Theta.$$

The corresponding summability of the multiple Fourier series is called ℓ_1 summability; it has a completely different character to the usual spherical summability (in which the summation of the kernel is taken over multi-indices in $\{\alpha\in\mathbb{N}_0^d : \|\alpha\| = n\}$); see, for example, Stein and Weiss [1971], Podkorytov [1981] or Berens and Xu [1997].

9.7 Notes

Section 9.1 Proposition 9.1.2 and its proof are extensions of the one-variable results; see Freud [1966, p. 139]. The properties of best polynomial approximations can be found in books on approximation theory, for example, Lorentz [1986], Cheney [1998] or DeVore and Lorentz [1993].

For $d = 1$, the asymptotics of the Christoffel function are known for general weight functions. It was proved in Máté, Nevai and Totik [1991] that

$$\lim_{n\to\infty} n\lambda_n(\mathrm{d}\mu, x) = \pi\mu'(x)\sqrt{1-x^2} \qquad \text{for a.e.} \quad x \in [-1,1]$$

for measures μ belonging to the Szegő class (that is, $\log \mu'(\cos t) \in L^1[0,2\pi]$). For orthogonal polynomials of several variables, we may conjecture that an analogous result holds:

$$\lim_{n\to\infty} \binom{n+d}{d} \Lambda_n(W;x) = \frac{W(x)}{W_0(x)},$$

where W_0 is the analogue of the normalized Chebyshev weight function for the given domain. This limit relation was established for several classical-type weight functions on the cube $[-1,1]^d$ with $W_0(x) = \pi^{-d}(1-x_1^2)^{-1/2}\cdots(1-x_d^2)^{-1/2}$, on the ball B^d with $W_0(x) = w^B(1-\|x\|^2)^{-1/2}$ and on the simplex T^d with $W_0(x) = w^T x_1^{-1/2}\cdots x_d^{-1/2}(1-|x|)^{-1/2}$, where w^B and w^T are normalization constants, so that W_0 has unit integral on the corresponding domain; see Xu [1995], Bos [1994] and Xu [1996a, b], The above limit was also studied for a central symmetric weight function in Bos, Della Vecchia and Mastroianni [1998]. More recently, fairly general results on the asymptotics of the Christoffel function were established in Kroó and Lubinsky [2013a, b]; their results were motivated by the universality limit, a concept originating in random matrix theory.

Section 9.2 Kogbetliantz's inequality on the positivity of the (C,δ)-means of Gegenbauer polynomials is a special case of a much more general positivity result on Jacobi polynomials due to Askey and to Gasper (see, for example, Askey [1975] and Gasper [1977]). These inequalities also imply other positivity results for h-harmonics.

For the weight function $h_\kappa(x) = \prod_{i=1}^d |x_i|^{\kappa_i}$, which is invariant under $\mathbb{Z}_2^d$, the result in Corollary 9.2.4 can be improved as follows.

Theorem 9.7.1 *The (C,δ)-means of the h-harmonic expansion of every continuous function for $h_\kappa(x) = \prod_{i=1}^d |x_i|^{\kappa_i}$ converge uniformly to f if and only if $\delta > \frac{d-2}{2} + |\kappa| - \min_{1\le i\le d}\kappa_i$.*

A further result shows that the great circles defined by the intersection of S^{d-1} and the coordinate planes form boundaries on S^{d-1}. Define

$$S^{d-1}_{\text{int}} = S^{d-1} \Big\backslash \bigcup_{i=1}^{d} \{x \in S^{d-1} : x_i = 0\},$$

which is the interior region bounded by these boundaries on S^{d-1}. Then, for continuous functions, $S_n^\delta(h_\kappa^2, f; x)$ converges to $f(x)$ for every $x \in S^{d-1}_{\text{int}}$ provided that $\delta > \frac{d-2}{2}$, independently of κ. The proof of these results is in Li and Xu [2003], which uses rather involved sharp estimates of the Cesàro means of the reproducing kernels given in Theorem 7.5.5. For L^p convergence with $1 < p < \infty$, a sharp critical index for the convergence of $S_n^\delta(h_\kappa^2, f)$ was established in Dai and Xu [2009b].

Section 9.3 The summability of the expansion in classical orthogonal polynomials on the ball can be traced back to the work of Chen, Koschmieder, and several others, where the assumption that 2μ is an integer was imposed; see Section 12.7 of Erdélyi *et al.* [1953]. Theorem 9.3.4 was proved in Xu [1999a]. The general relation between summability of orthogonal expansion on the sphere and on the ball is studied in Xu [2001a]. For the weight function $W^B_{\kappa,\mu}(x) = \prod_{i=1}^d |x_i|^{\kappa_i}(1-\|x\|^2)^{\mu-1/2}$, which is invariant under $\mathbb{Z}_2^d$, Corollary 9.3.3 can be improved to a sharp result: $S_n^\delta(W^B_{\kappa,\mu}, f)$ converges to f uniformly for continuous functions f if and only if $\delta > \frac{d-1}{2} + |\kappa| - \min_{1\le i\le d+1} \kappa_i$, with $\kappa_{d+1} = \mu$ as in Theorem 9.7.1; further, pointwise convergence holds for x inside B^d and x not on one of the coordinate hyperplanes, provided that $\delta > \frac{d-1}{2}$. The proof uses Theorem 9.3.1.

Section 9.4 A result similar to that in Theorem 9.7.1 also holds for classical orthogonal polynomials on the simplex. The proof requires a sharp estimate for the Cesàro means of the reproducing kernels given in Theorem 8.2.7. The estimate is similar to that for the kernel in the case of $\mathbb{Z}_2^d$-invariant h_κ^2, but there are additional difficulties. A partial result appeared in Li and Xu [2003]; a complete analogue to the result on the ball was proved in Dai and Xu [2009a], that is, that $S_n^\delta(W^T_{\kappa,\mu}, f)$ converges to f uniformly for continuous functions f if and only if $\delta > \frac{d-1}{2} + |\kappa| - \min_{1\le i\le d+1} \kappa_i$ with $\kappa_{d+1} = \mu$. The pointwise convergence in the interior of T^d holds if $\delta > \frac{d-1}{2}$. Furthermore, a sharp result for convergence in L^p, $1 < p < \infty$, was established in Dai and Xu [2009b].

Section 9.5 For Laguerre expansions, there are several different forms of summability, depending on the L^2 spaces under consideration. For example, considering the Laguerre functions $\mathscr{L}_n^\alpha(x) = L_n^\alpha(x)e^{-x/2}x^{-\alpha/2}$ and their other varieties, $\mathscr{L}_n^\alpha(\frac{1}{2}x^2)x^{-\alpha}$ or $\mathscr{L}_n^\alpha(x^2)(2x)^{1/2}$, one can form orthogonal systems in $L^2(\mathrm{d}x, \mathbb{R}_+)$ and $L^2(x^{2\alpha+1}\mathrm{d}x, \mathbb{R}_+)$. Consequently, there have been at least four types of Laguerre expansion on $\mathbb{R}_+$ studied in the literature and each has its extension in the multiple setting; see Thangavelu [1992] for the definitions. We considered only orthogonal polynomials in $L^2(x^\alpha\,\mathrm{d}x; \mathbb{R}_+)$, following Xu [2000d]. For results in the multiple Hermite and Laguerre expansions, see the monograph by Thangavelu [1993].

Section 9.6 For multiple Jacobi expansions the order of δ can be reduced if we consider only convergence without positivity. However, although their product nature leads to a convolution structure, which allows us to reduce the summability to the point $x = \mathbf{1}$, convergence at the point $\mathbf{1}$ no longer follows from the product of the results for one variable.

Theorem 9.7.2 *Let $\alpha_j, \beta_j \geq -\frac{1}{2}$. The Cesàro (C,δ) means of the multiple Jacobi expansion with respect to $W_{\alpha,\beta}$ are uniformly convergent in the norm of $C([-1,1]^d)$ provided that $\delta > \sum_{j=1}^{d} \max\{\alpha_j, \beta_j\} + \frac{d}{2}$.*

Similar results also hold for the case where $\alpha_j > -1$, $\beta_j > -1$ and $\alpha_j + \beta_j \geq -1$, $1 \leq j \leq d$, with a properly modified condition on δ. In particular, if $\alpha_j = \beta_j = -\frac{1}{2}$ then convergence holds for $\delta > 0$. This is the order of the ℓ_1 summability of the multiple Fourier series. The proof of these results was given in Li and Xu [2000], which uses elaborate estimates of the kernel function that are written in terms of the integral of the Poisson kernel of the Jacobi polynomials.

Further results The concise formulae for the reproducing kernels with respect to the h_κ^2 associated with $\mathbb{Z}_2^d$ on the sphere, $W_{\kappa,\mu}^B$ on the unit ball and $W_{\kappa,\mu}^T$ on the simplex open the way for in-depth study of various topics in approximation theory and harmonic analysis. Many recent advances in this direction are summed up in the book Dai and Xu [2013]. These kernels are also useful in the construction of highly localized bases, called needlets, a name coined in Narcowich, Petrushev, and Ward [2006], where such bases were constructed on the unit sphere. These bases were constructed and studied by Petrushev and Xu [2008a] for W_μ^B on the ball; by Ivanov, Petrushev and Xu [2010], [2012] for $W_{\kappa,\mu}^T$ on the simplex and on product domains, respectively; by Kerkyacharian, Petrushev, Picard and Xu [2009] for the Laguerre weight on $\mathbb{R}_+^d$; and by Petrushev and Xu [2008b] for the Hermite weight on $\mathbb{R}^d$.

The Cesàro-summability of orthogonal expansions on a cylindrical domain was studied by Wade [2011].

10

Orthogonal Polynomials Associated with Symmetric Groups

In this chapter we consider analysis associated with symmetric groups. The differential–difference operators for these groups, called type A in Weyl group nomenclature, are crucial in this theory. The techniques tend to be algebraic, relying on methods from combinatorics and linear algebra. Nevertheless the chapter culminates in explicit evaluations of norm formulae and integrals of the Macdonald–Mehta–Selberg type. These integrals involve the weight function $\prod_{1\leq i<j\leq d}\left|x_i-x_j\right|^{2\kappa}$ on the torus and the weight function on $\mathbb{R}^d$ equipped with the Gaussian measure. The fundamental objects are a commuting set of self-adjoint operators and the associated eigenfunction decomposition. The simultaneous eigenfunctions are certain homogeneous polynomials, called nonsymmetric Jack polynomials. The Jack polynomials are a family of parameterized symmetric polynomials, which have been studied mostly in combinatorial settings.

The fact that the symmetric group is generated by transpositions of adjacent entries will frequently be used in proofs; for example, it suffices to prove invariance under adjacent transpositions to show group invariance. Two bases of polynomials will be used, not only the usual monomial basis but also the p-basis; these are polynomials, defined by a generating function, which have convenient transformation formulae for the differential–difference operators. Also, they provide expressions for the nonsymmetric Jack polynomials which are independent of the number of trailing zeros of the label $\alpha\in\mathbb{N}_0^d$. The chapter concludes with expressions for the type-A exponential-type kernel and the intertwining operators and an algorithm for the nonsymmetric Jack polynomials labeled by partitions. The algorithm is easily implementable in a symbolic computation system. There is a brief discussion of the associated Hermite-type polynomials.

10.1 Partitions, Compositions and Orderings

In this chapter we will be concerned with the action of the symmetric group S_d on $\mathbb{R}^d$ and on polynomials. Some orderings of $\mathbb{N}_0^d$ are fundamental and are especially relevant to the idea of expressing a permutation as a product of adjacent transpositions.

Consider the elements of S_d as functions on $\{1,2,\dots,d\}$. For $x \in \mathbb{R}^d$ and $w \in S_d$ let $(xw)_i = x_{w(i)}$ for $1 \le i \le d$; extend this action to polynomials by writing $wf(x) = f(xw)$. This has the effect that monomials transform to monomials, $w(x^\alpha) = x^{w\alpha}$, where $(w\alpha)_i = \alpha_{w^{-1}(i)}$ for $\alpha \in \mathbb{N}_0^d$. (Consider x as a row vector α as a column vector and w as a permutation matrix, with 1s at the $(w(j),j)$ entries.) The reflections in S_d are transpositions interchanging x_i and x_j, denoted by (i,j) for $i \neq j$. The term *composition* is a synonym for multi-index; it is commonly used in algebraic combinatorics. The basis polynomials (some orthogonal) that will be considered have composition labels and are contained in subspaces invariant under S_d. In such a subspace the "standard" composition label will be taken to be the associated partition. We recall the notation ε_i for the unit label: $(\varepsilon_i)_j = \delta_{i,j}$. This will be used in calculations of quantities such as $\alpha - \varepsilon_i$ (when $\alpha \in \mathbb{N}_0^d$ and $\alpha_i \ge 1$).

Definition 10.1.1 A partition (with no more than d parts) is a composition $\lambda \in \mathbb{N}_0^d$ such that $\lambda_1 \ge \lambda_2 \ge \cdots \ge \lambda_d$. The set of all such partitions is denoted by $\mathbb{N}_0^{d,P}$. For any $\alpha \in \mathbb{N}_0^d$, let α^+ be the unique partition such that $\alpha^+ = w\alpha$ for some $w \in S_d$. The length of a composition is $\ell(\alpha) := \max\{i : \alpha_i > 0\}$.

There is a total order on compositions, namely, lexicographic order ("lex order" for short) (see also Section 3.1), defined as follows: $\alpha \succ_L \beta$ means that $\alpha_i = \beta_i$ for $1 \le i < m$ and $\alpha_m > \beta_m$ for some $m \le d$ ($\alpha, \beta \in \mathbb{N}_0^d$); but there is a partial order better suited for our purposes, the dominance order.

Definition 10.1.2 For $\alpha, \beta \in \mathbb{N}_0^d$, say, α dominates β, that is, $\alpha \succeq \beta$, when $\sum_{i=1}^j \alpha_i \ge \sum_{i=1}^j \beta_i$ for $1 \le j \le d$. Also, $\alpha \succ \beta$ means $\alpha \succeq \beta$ with $\alpha \neq \beta$. Further, $\alpha \rhd \beta$ means that $|\alpha| = |\beta|$ and either $\alpha^+ \succ \beta^+$ or $\alpha^+ = \beta^+$ and $\alpha \succ \beta$.

Clearly the relations $\succeq$ and $\unrhd$ are partial orderings ($\alpha \unrhd \beta$ means $\alpha \rhd \beta$ or $\alpha = \beta$). Dominance has the nice property of "reverse invariance": for $\alpha \in \mathbb{N}_0^d$ let $\alpha^R = (\alpha_d, \alpha_{d-1}, \dots, \alpha_1)$; then, for $\alpha, \beta \in \mathbb{N}_0^d$, $|\alpha| = |\beta|$ and $\alpha \succeq \beta$ implies that $\beta^R \succeq \alpha^R$. In our applications dominance order will be used to compare compositions α, β having $\alpha^+ = \beta^+$ and partitions λ, μ having $|\lambda| = |\mu|$. The following lemma shows the effect of two basic operations.

Lemma 10.1.3 *Suppose that $\alpha \in \mathbb{N}_0^d$ and $\lambda \in \mathbb{N}_0^{d,P}$; then*

(i) *if $\alpha_i > \alpha_j$ and $i < j$ then $\alpha \succ (i,j)\alpha$;*
(ii) *if $\lambda_i > \lambda_j + 1$ (implying $i < j$) then $\lambda \succ \mu^+$, where $\mu_i = \lambda_i - 1, \mu_j = \lambda_j + 1$ and $\mu_k = \lambda_k$ for $k \neq i, j$;*

(iii) $\alpha^{+} \succeq \alpha$;

(iv) *if* $\lambda_i > \lambda_j + 1$ *and* $1 \le s < \lambda_i - \lambda_j$ *then* $\lambda \succ \left(\mu^{(s)}\right)^{+}$ *where* $\mu_i^{(s)} = \lambda_i - s$, $\mu_j^{(s)} = \lambda_j + s$ *and* $\mu_k^{(s)} = \lambda_k$ *for* $k \ne i, j$ *(that is,* $\mu^{(s)} = \lambda - s(\varepsilon_i - \varepsilon_j)$*).*

Proof For part (i) let $\beta = (i,j)\,\alpha$; then $\sum_{k=1}^{m} \alpha_k = \sum_{k=1}^{m} \beta_k + (\alpha_i - \alpha_j)$ for $i \le m < j$, otherwise $\sum_{k=1}^{m} \alpha_j = \sum_{k=1}^{m} \beta_j$. For part (ii) assume (by relabeling if necessary) that $\lambda_i > \lambda_{i+1} \ge \cdots \ge \lambda_{j-1} > \lambda_j \ge \lambda_{j+1}$ (possibly $j = i+1$) and that now $\mu_k = \lambda_k$ for all k except i, for which $\mu_i = \lambda_i - 1$ and j, for which $\mu_j = \lambda_j + 1$. By construction $\mu_i \ge \mu_{i+1}$ and $\mu_{j-1} \ge \mu_j$, so $\mu = \mu^{+}$ (this is the effect of the relabeling) and clearly $\lambda \succ \mu^{+}$. For part (iii) note that α^{+} can be obtained by applying a finite sequence of adjacent transpositions (apply $(i, i+1)$ when $\alpha_i < \alpha_{i+1}$).

For part (iv), note that $\left(\mu^{(\lambda_i - \lambda_j - s)}\right)^{+} = \left(\mu^{(s)}\right)^{+}$. Part (ii) shows that $\lambda \succ \left(\mu^{(1)}\right)^{+} \succ \left(\mu^{(2)}\right)^{+} \succ \cdots \succ \left(\mu^{(t)}\right)^{+}$ for $t = [\frac{1}{2}(\lambda_i - \lambda_j)]$, and this proves part (iv). □

10.2 Commuting Self-Adjoint Operators

The differential–difference operators of Chapter 4 for the group S_d allow one parameter, denoted by κ, and have the formula, for $f \in \Pi^d$ and $1 \le i \le d$,

$$\mathscr{D}_i f(x) = \frac{\partial f(x)}{\partial x_i} + \kappa \sum_{j=1, j \ne i}^{d} \frac{f(x) - f(x(i,j))}{x_i - x_j},$$

often abbreviated as $\partial_i + \kappa \sum_{j \ne i} \left[1 - (i,j)\right] / (x_i - x_j)$.

Lemma 10.2.1 *The following commutants hold for* $1 \le i, j \le d$*:*

(i) $x_i \mathscr{D}_i f(x) - \mathscr{D}_i x_i f(x) = -f(x) - \kappa \sum_{j \ne i} (i,j) f(x)$;

(ii) $x_j \mathscr{D}_i f(x) - \mathscr{D}_i x_j f(x) = \kappa (i,j) f(x)$ *for* $i \ne j$.

Proof In Proposition 6.4.10 set $u = \varepsilon_i, t = \varepsilon_j$; then

$$x_j \mathscr{D}_i f(x) = \mathscr{D}_i x_j f(x) - \langle \varepsilon_j, \varepsilon_i \rangle f(x) - \kappa \sum_{k \ne i} \langle \varepsilon_j, \varepsilon_i - \varepsilon_k \rangle (i,k) f(x).$$

Only the positive roots $v = \pm(\varepsilon_i - \varepsilon_k)$ appear in the sum (the inner product is the usual one on $\mathbb{R}^d$). □

Lemma 10.2.2 *For* $m, n \in \mathbb{N}_0$,

$$\frac{x_1^m x_2^n - x_1^n x_2^m}{x_1 - x_2} = \operatorname{sign}(m-n) \sum_{i=\min\{m,n\}}^{\max\{m,n\}-1} x_1^{m+n-1-i} x_2^i.$$

We will describe several inner products $\langle \cdot, \cdot \rangle$ on Π^d with the following properties; such inner products are called permissible.

1. if $p \in \mathscr{P}_n^d$ and $q \in \mathscr{P}_m^d$ then $n \neq m$ implies that $\langle p, q \rangle = 0$;
2. for $p, q \in \Pi^d$ and $w \in S_d$, $\langle wp, wq \rangle = \langle p, q \rangle$;
3. for each i, the operator $\mathscr{D}_i x_i$ is self-adjoint, that is, $\langle \mathscr{D}_i (x_i p(x)), q(x) \rangle = \langle p(x), \mathscr{D}_i (x_i q(x)) \rangle$.

One such inner product has already been defined, in Definition 7.2.2, namely $\langle p, q \rangle = p(\mathscr{D}_1, \mathscr{D}_2, \dots, \mathscr{D}_d) q(x)|_{x=0}$. Here $\mathscr{D}_i^* = x_i$ (a multiplication factor) and so $\mathscr{D}_i x_i$ is self-adjoint for each i. These operators do not commute, but a simple modification provides sufficient commuting self-adjoint operators that the decomposition of Π^d into joint eigenfunctions provides an orthogonal basis.

Recall that for two linear operators A, B, operating on the same linear space, the commutant is $[A, B] = AB - BA$.

Lemma 10.2.3 *For $i \neq j$, $[\mathscr{D}_i x_i, \mathscr{D}_j x_j] = \kappa(\mathscr{D}_i x_i - \mathscr{D}_j x_j)(i, j)$; moreover, $[\mathscr{D}_i x_i, \mathscr{D}_j x_j - \kappa(i, j)] = 0$.*

Proof By part (ii) of Lemma 10.2.1, $x_j \mathscr{D}_i = \mathscr{D}_i x_j + \kappa(i, j)$. Since $\mathscr{D}_i \mathscr{D}_j = \mathscr{D}_j \mathscr{D}_i$, we have

$$\begin{aligned} \mathscr{D}_i x_i \mathscr{D}_j x_j - \mathscr{D}_j x_j \mathscr{D}_i x_i &= \mathscr{D}_i [\mathscr{D}_j x_i + \kappa(i, j)] x_j - \mathscr{D}_j (\mathscr{D}_i x_j + \kappa(i, j)) x_i \\ &= \kappa(\mathscr{D}_i x_i - \mathscr{D}_j x_j)(i, j). \end{aligned}$$

Also, $\kappa(\mathscr{D}_i x_i - \mathscr{D}_j x_j)(i, j) = \kappa[\mathscr{D}_i x_i (i, j) - (i, j) \mathscr{D}_i x_i]$. □

Definition 10.2.4 For $1 \le i \le d$ define the self-adjoint operators $\mathscr{U}_i$ on Π^d by

$$\mathscr{U}_i = \mathscr{D}_i x_i + \kappa - \kappa \sum_{j=1}^{i-1} (j, i).$$

Note that each $\mathscr{U}_i$ preserves the degree of homogeneity; that is, it maps $\mathscr{P}_n^d$ to $\mathscr{P}_n^d$ for each n. The transpositions are self-adjoint because $\langle (i, j)p, q \rangle = \langle p, (i, j)^{-1} q \rangle = \langle p, (i, j) q \rangle$ by the hypothesis on the inner product.

Theorem 10.2.5 *For $1 \le i < j \le d$, $\mathscr{U}_i \mathscr{U}_j = \mathscr{U}_j \mathscr{U}_i$.*

Proof Write $\mathscr{U}_i = \mathscr{D}_i x_i + \kappa - \kappa A$ and $\mathscr{U}_j = \mathscr{D}_j x_j + \kappa - \kappa(i, j) - \kappa B$, where $A = \sum_{k<i} (k, i)$ and $B = \sum_{k<j, k \neq i} (k, j)$. Then $[\mathscr{U}_i, \mathscr{U}_j] = [\mathscr{D}_i x_i, \mathscr{D}_j x_j - \kappa(i, j)] - \kappa[\mathscr{D}_i x_i, B] - \kappa[A, \mathscr{D}_j x_j] + \kappa^2 [A, B + (i, j)]$. The first term is zero by Lemma 10.2.3, and the middle two are zero because the transpositions in A, B individually

commute with the other terms; for example $\mathscr{D}_i x_i (k,j) = (k,j)\mathscr{D}_i x_i$ for $k \neq i$. Further,

$$\begin{aligned}[A, B+(i,j)] &= \sum_{k=1}^{i-1}\sum_{s=1}^{j-1}[(k,i),(s,j)] \\ &= \sum_{k=1}^{i-1}\left([(k,i)(k,j)-(i,j)(k,i)]-[(k,j)(k,i)-(k,i)(i,j)]\right).\end{aligned}$$

Each bracketed term is zero since $(k,i)(k,j) = (i,j)(k,i)$ in the sense of multiplication in the group S_d. □

We consider the action of $\mathscr{U}_i$ on a monomial x^α. For any i, by Lemma 10.2.2 we have

$$\mathscr{D}_i x_i x^\alpha = (\alpha_i+1)x^\alpha + \kappa\sum_{\beta\in A} x^\beta - \kappa\sum_{\beta\in B} x^\beta$$

where

$$\begin{aligned} A &= \bigcup_{j\neq i}\{\alpha + k(\varepsilon_j - \varepsilon_i) : (\alpha_i \geq \alpha_j), 0 \leq k \leq \alpha_i - \alpha_j\}, \\ B &= \bigcup_{j\neq i}\{\alpha + k(\varepsilon_i - \varepsilon_j) : (\alpha_i < \alpha_j - 1), 1 \leq k \leq \alpha_j - \alpha_i - 1\}.\end{aligned}$$

The combined coefficient of x^α is $(\alpha_i+1) + \kappa\#\{j : \alpha_j \leq \alpha_i, j \neq i\}$. Monomials with coefficient κ are of the form x^β, where

$$\beta = (\alpha_1, \ldots, \alpha_i - k, \ldots, \alpha_j + k, \ldots) \qquad \text{for } 1 \leq k \leq \alpha_i - \alpha_j;$$

in each case $\alpha^+ \succ \beta^+$, by part (iv) of Lemma 10.1.3, except for $\beta = (i,j)\alpha$ (for $k = \alpha_i - \alpha_j$). Monomials with coefficient $-\kappa$ have the form x^β, where $\beta = (\alpha_1, \ldots, \alpha_i + k, \ldots, \alpha_j - k, \ldots)$ for $1 \leq k \leq \alpha_j - \alpha_i - 1$. Again by Lemma 10.1.3, $\alpha^+ \succ \beta^+$. When the remaining terms of $\mathscr{U}_i$ are added, the effect is to remove the terms $\kappa(j,i)x^\alpha$ with $j < i$ and $\alpha_j < \alpha_i$, for which $(j,i)\alpha \succ \alpha$. Thus

$$\mathscr{U}_i x^\alpha = (\kappa(d - \#\{j : \alpha_j > \alpha_i\} - \#\{j : j < i \text{ and } \alpha_j = \alpha_i\}) + \alpha_i + 1)x^\alpha + q_{\alpha,i},$$

where $q_{\alpha,i}$ is a sum of terms $\pm\kappa x^\beta$ with $\alpha \rhd \beta$. This shows that $\mathscr{U}_i$ is represented by a triangular matrix on the monomial basis (triangularity with respect to the partial order $\trianglerighteq$). It remains to study the simultaneous eigenfunctions, called nonsymmetric Jack polynomials, in detail.

10.3 The Dual Polynomial Basis

There is a basis of homogeneous polynomials which has relatively tractable behavior under the actions of $\mathscr{D}_i$ and $\mathscr{U}_i$ and also provides an interesting inner product. These polynomials are products of analogues of x_i^n.

Definition 10.3.1 For $1 \le i \le d$ define the homogeneous polynomials $p_n(x_i;x)$, $n \in \mathbb{N}_0$, by logarithmic differentiation applied to the generating function:

$$\sum_{n=0}^{\infty} p_n(x_i;x)\, y^n = (1-x_i y)^{-1} \prod_{j=1}^{d} (1-x_j y)^{-\kappa},$$

valid for $|y| < \min_j (1/|x_j|)$.

We will show that $\mathscr{D}_i p_n(x_i;x) = (d\kappa + n)\, p_{n-1}(x_i;x)$ and $\mathscr{D}_j p_n(x_i;x) = 0$ for $j \neq i$, using an adaptation of logarithmic differentiation. Let $f_0 = (1-x_i y)^{-1}$ and $f_1 = \prod_{j=1}^{d} (1-x_j y)^{-\kappa}$ (the latter is S_d-invariant); then

$$\begin{aligned}\frac{\mathscr{D}_i(f_0 f_1)}{f_0 f_1} - \frac{y^2 \partial (f_0 f_1)/\partial y}{f_0 f_1} &= \frac{(1+\kappa)y}{1-x_i y} + \kappa \sum_{j \neq i} \frac{1}{x_i - x_j}\left(1 - \frac{1-x_i y}{1-x_j y}\right) \\ &\quad - \frac{y^2 x_i}{1-x_i y} - \kappa \sum_{j=1}^{d} \frac{y^2 x_j}{1-x_j y} \\ &= (1+\kappa) y + \kappa \sum_{j \neq i} \frac{y - y^2 x_j}{1 - x_j y} = (d\kappa + 1) y.\end{aligned}$$

In the generating function,

$$\sum_{n=0}^{\infty} \mathscr{D}_i p_n(x_i;x)\, y^n = \sum_{n=0}^{\infty} p_n(x_i;x)\,(n + d\kappa + 1)\, y^{n+1},$$

as claimed. For $j \neq i$,

$$\frac{\mathscr{D}_j(f_0 f_1)}{f_0 f_1} = \kappa \frac{y}{1-x_j y} + \kappa \frac{1}{x_j - x_i}\left(1 - \frac{1-x_i y}{1-x_j y}\right) = 0.$$

These polynomials are multiplied together (over i) to form a basis.

Definition 10.3.2 For $\alpha \in \mathbb{N}_0^d$ let $p_\alpha(x) = \prod_{i=1}^{d} p_{\alpha_i}(x_i;x)$; alternatively, set

$$\sum_{\alpha \in \mathbb{N}_0^d} p_\alpha(x)\, y^\alpha = \prod_{i=1}^{d} \left((1-x_i y_i)^{-1} \prod_{j=1}^{d} (1-x_j y_i)^{-\kappa} \right),$$

valid for $\max_i |y_i| < \min_i (1/|x_i|)$.

The polynomials $p_\alpha(x)$ transform like monomials under the action of S_d; indeed, if $w \in S_d$ then $w p_\alpha(x) = p_\alpha(xw) = p_{w\alpha}(x)$; this can be proved using the generating function.

Denote the generating function by $F(x,y)$. The value of $\mathscr{D}_i p_\alpha$ will be derived from the equation

$$\frac{\mathscr{D}_i F}{F} = \frac{y_i^2}{F} \frac{\partial F}{\partial y_i} + (d\kappa + 1) y_i + \kappa \sum_{j \neq i} \frac{y_i y_j \left[F - (y_j, y_i) F\right]}{(y_i - y_j) F},$$

where (y_i, y_j) denotes the interchange of the variables y_i, y_j in F. A typical term in the difference part has the form

$$\frac{1}{x_i - x_j}\left(1 - \frac{(1-x_iy_i)(1-x_jy_j)}{(1-x_iy_j)(1-x_jy_i)}\right) = \frac{y_i - y_j}{(1-x_iy_j)(1-x_jy_i)}.$$

Using similar methods as for $p_n(x_i;x)$ we obtain

$$\begin{aligned}
&\frac{1}{F}\left(\mathscr{D}_i F - y_i^2 \frac{\partial F}{\partial y_i}\right)\\
&\quad = \frac{y_i - y_i^2 x_i}{1 - x_i y_i} + \kappa \sum_{j=1}^{d}\left(\frac{y_j}{1-x_iy_j} - \frac{y_i^2 x_j}{1-x_jy_i}\right) + \kappa \sum_{j\neq i} \frac{y_i - y_j}{(1-x_iy_j)(1-x_jy_i)}\\
&\quad = (d\kappa+1)y_i + \kappa\sum_{j\neq i}\left[\frac{y_i-y_j}{(1-x_iy_j)(1-x_jy_i)} + \frac{y_j}{1-x_iy_j} - \frac{y_i}{1-x_jy_i}\right]\\
&\quad = (d\kappa+1)y_i + \kappa\sum_{j\neq i}\frac{y_iy_j(x_i-x_j)}{(1-x_iy_j)(1-x_jy_i)};
\end{aligned}$$

the sum in the last line equals

$$\kappa\sum_{j\neq i}\frac{y_iy_j\,(F-(y_j,y_i)F)}{(y_i-y_j)F}.$$

As before, if the superscript (x) refers to the variable being acted upon then $\mathscr{D}_i^{(x)} x_i F(x,y) = \mathscr{D}_i^{(y)} y_i F(x,y)$ for each i. Indeed,

$$\begin{aligned}
\frac{1}{F}\mathscr{D}_i^{(x)} x_i F &= 1 + \frac{x_iy_i}{1-x_iy_i} + \kappa\sum_{j=1}^{d}\frac{x_iy_j}{1-x_iy_j}\\
&\quad + \kappa\sum_{j\neq i}\frac{1}{x_i-x_j}\left(x_i - \frac{x_j(1-x_iy_i)(1-x_jy_j)}{(1-x_iy_j)(1-x_jy_i)}\right)\\
&= 1 + \frac{(1+\kappa)x_iy_i}{1-x_iy_i} + \kappa\sum_{j\neq i}\frac{1-x_jy_j}{(1-x_iy_j)(1-x_jy_i)},
\end{aligned}$$

which is symmetric in x, y (as is F, of course).

Proposition 10.3.3 *For $\alpha \in \mathbb{N}_0^d$ and $1 \le i \le d$, $\mathscr{D}_i p_\alpha = 0$ if $\alpha_i = 0$, else*

$$\mathscr{D}_i p_\alpha = [\kappa(1 + \#\{j : \alpha_j < \alpha_i\}) + \alpha_i]\, p_{\alpha-\varepsilon_i} + \kappa\sum_{\beta\in A} p_\beta - \kappa\sum_{\beta\in B} p_\beta,$$

where

$$A = \bigcup_{j\neq i}\{\alpha + k\varepsilon_i - (k+1)\varepsilon_j : \max\{0, \alpha_j - \alpha_i\} \le k \le \alpha_j - 1\},$$

$$B = \bigcup_{j\neq i}\{\alpha - (k+1)\varepsilon_i + k\varepsilon_j : \max\{1, \alpha_i - \alpha_j\} \le k \le \alpha_i - 1\}.$$

Proof We have that $\mathscr{D}_i p_\alpha$ is the coefficient of y^α in

$$\left(y_i^2 \frac{\partial}{\partial y_i} + (d\kappa + 1)\, y_i\right) F + \kappa \sum_{j \neq i} \frac{y_i y_j}{(y_i - y_j)} \left[F - (y_j, y_i) F\right].$$

The first term contributes $(d\kappa + \alpha_i)\, p_{\alpha - \varepsilon_i}$. For any $j \neq i$ let $\beta \in \mathbb{N}_0^d$ satisfy $\beta_k = \alpha_k$ for $k \neq i, j$. Then p_β appears in the sum with coefficient κ when there is a solution to $\beta_i - s = \alpha_i$, $\beta_j + 1 + s = \alpha_j$, $0 \leq s \leq \beta_i - \beta_j - 1$ (this follows from Lemma 10.2.2), that is, for $0 \leq \beta_j \leq \min\{\alpha_i - 1, \alpha_j - 1\}$ and $\beta_i + \beta_j = \alpha_i + \alpha_j - 1$, and p_β appears in the sum with coefficient $-\kappa$ when there is a solution to the same equations but with β_i and β_j interchanged. In the case $\alpha_j \geq \alpha_i$ the values $\beta_i = \alpha_i - 1$, $\beta_j = \alpha_j$ appear (so that $\beta = \gamma$). These are deleted from the set B and combined to give the coefficient of p_γ; the total is $\kappa(d - \#\{j : j \neq i \text{ and } \alpha_j \geq \alpha_i\})$. □

A *raising operator* for the polynomials p_α is useful, as follows.

Definition 10.3.4 For $\alpha \in \mathbb{N}_0^d$ and $1 \leq i \leq d$ let $\rho_i p_\alpha = p_{\alpha + \varepsilon_i}$, and extend by linearity to $\operatorname{span}\{p_\beta : \beta \in \mathbb{N}_0^d\}$.

Proposition 10.3.5 *For $1 \leq i \leq d$ the operator $\mathscr{D}_i \rho_i$ satisfies*

$$\mathscr{D}_i \rho_i p_\alpha = [\kappa(d - \#\{j : \alpha_j > \alpha_i\}) + \alpha_i + 1]\, p_\alpha + \kappa \sum_{\beta \in A} p_\beta - \kappa \sum_{\beta \in B} p_\beta,$$

where $\alpha \in \mathbb{N}_0^d$ and

$$A = \bigcup_{j \neq i} \{\alpha + k(\varepsilon_i - \varepsilon_j) : \max\{1, \alpha_j - \alpha_i\} \leq k \leq \alpha_j\},$$

$$B = \bigcup_{j \neq i} \{\alpha + k(\varepsilon_j - \varepsilon_i) : \max\{1, \alpha_i - \alpha_j + 1\} \leq k \leq \alpha_i\}.$$

Proof This is a consequence of Proposition 10.3.3 with α replaced by $\alpha + \varepsilon_i$. The sets A, B used before are reformulated, setting $\beta_j = \alpha_j - k$ in A and $\beta_i = \alpha_i - k$ in B. □

Corollary 10.3.6 *For $\alpha \in \mathbb{N}_0^d$ and $1 \leq i \leq d$,*

$$\mathscr{D}_i \rho_i p_\alpha = [\kappa(d - \#\{j : \alpha_j > \alpha_i\}) + \alpha_i + 1]\, p_\alpha + \kappa \sum_{\alpha_j > \alpha_i} p_{(i,j)\alpha} + \kappa q_{\alpha,i},$$

where $q_{\alpha,i}$ is a sum of terms of the form $\pm p_\beta$ with $|\beta| = |\alpha|$ and $\beta^+ \succ \alpha^+$ and $\alpha_j = 0$ implies that $\beta_j = 0$.

Proof Suppose that $\beta, \gamma \in \mathbb{N}_0^d$ with $|\beta| = |\gamma|$ and $\beta_k = \gamma_k$ for all $k \neq i, j$ (for any i, j with $i \neq j$); then part (iv) of Lemma 10.1.3 implies that $\beta^+ \succ \gamma^+$ if and only if $\min\{\beta_i, \beta_j\} < \min\{\gamma_i, \gamma_j\} \leq \max\{\gamma_i, \gamma_j\} < \max\{\beta_i, \beta_j\}$; of course the latter inequality is redundant since $\beta_i+\beta_j=\gamma_i+\gamma_j$. Then any $\beta\in A\cup B$ (the sets in Proposition 10.3.5) satisfies the condition $\min\{\beta_i,\beta_j\}<\min\{\alpha_i,\alpha_j\}$ for some j; the exceptions $\beta=(i,j)\,\alpha$ occur in A when $\alpha_i\le\alpha_j-1$. □

It remains to show that $\{p_\alpha:\alpha\in\mathbb{N}_0^d\}$ is actually a basis for Π^d for all but a discrete set of negative values of κ; further, $\mathscr{D}_i\rho_i=\mathscr{D}_ix_i+\kappa$, as one might already suspect from the coefficient of p_α in $\mathscr{D}_i\rho_ip_\alpha$. The following is valid for $\operatorname{span}\{p_\alpha\}\subseteq\Pi^d$, which in fact is an apparent restriction, only.

Proposition 10.3.7 *For* $1\le i\le d$, $\mathscr{D}_i\rho_i=\mathscr{D}_ix_i+\kappa$.

Proof The generating function for $p_n(x_i;x)$ is $(1-x_iy)^{-1}\prod_{j=1}^{d}(1-x_jy)^{-\kappa}$. Let $\prod_{j=1}^{d}(1-x_jy)^{-\kappa}=\sum_{n=0}^{\infty}\pi_n(x)y^n$, where π_n is a homogeneous S_d-invariant polynomial; thus

$$p_{n+1}(x_i;x)=\sum_{j=0}^{n+1}x_i^j\pi_{n+1-j}(x)=\pi_{n+1}(x)+x_ip_n(x_i;x).$$

Also, $\partial\pi_{n+1}(x)/\partial x_i=\kappa p_n(x_i;x)$ because

$$\sum_{n=0}^{\infty}\frac{\partial}{\partial x_i}\pi_n(x)y^n=\frac{\kappa y}{1-x_iy}\prod_{j=1}^{d}(1-x_jy)^{-\kappa}.$$

Let $\alpha\in\mathbb{N}_0^d$ and write $p_\alpha(x)=p_{\alpha_i}(x_i;x)g(x)$, where $g(x)=\prod_{j\ne i}p_{\alpha_j}(x_j;x)$. Then

$$\begin{aligned}\mathscr{D}_i\rho_ip_\alpha-\mathscr{D}_ix_ip_\alpha&=\mathscr{D}_i\left[\left(p_{1+\alpha_i}(x_i;x)-x_ip_{\alpha_i}(x_i;x)\right)g(x)\right]\\&=\mathscr{D}_i\left[\pi_{1+\alpha_i}(x)g(x)\right]\\&=\left(\frac{\partial}{\partial x_i}\pi_{1+\alpha_i}(x)\right)g(x)+\pi_{1+\alpha_i}(x)\mathscr{D}_ig(x)\\&=\kappa p_{\alpha_i}(x_i;x)g(x)=\kappa p_\alpha(x).\end{aligned}$$

This uses the product rule for $\mathscr{D}_i$ with one factor invariant and the fact that $\mathscr{D}_ig(x)=0$ (Proposition 10.3.3). □

Proposition 10.3.8 *For* $1\le i\le d$ *and* $\kappa\ge0$, *the linear operator* $\mathscr{D}_i\rho_i$ *is one to one on* $\operatorname{span}\{p_\alpha:\alpha\in\mathbb{N}_0^d\}$.

Proof Because $\mathscr{D}_i\rho_i=(1,i)\,\mathscr{D}_1\rho_1(1,i)$ it suffices to show that $\mathscr{D}_1\rho_1$ is one to one. The effect of $\mathscr{D}_1\rho_1$ on $\{p_\alpha:\alpha\in\mathbb{N}_0^d,|\alpha|=n\}$ respects the partial order $\trianglelefteq$ because $\mathscr{D}_1\rho_1p_\alpha=[\kappa(d-\#\{j:\alpha_j>\alpha_1\})+\alpha_1+1]\,p_\alpha+\kappa\sum_{\alpha_j>\alpha_1}p_{(1,j)\alpha}+\kappa q_{\alpha,1}$ (Corollary 10.3.6), where each term p_β in $q_{\alpha,1}$ satisfies $\beta^+\succ\alpha^+$ and $(1,j)\,\alpha\succ\alpha$ when $\alpha_j>\alpha_1$ (part (i) of Lemma 10.1.3). □

To show that $\{p_\alpha : \alpha \in \mathbb{N}_0^d\}$ is a basis we let

$$\mathscr{V}_n^m = \operatorname{span}\{p_\alpha : \alpha \in \mathbb{N}_0^d, |\alpha| = n, \alpha_i = 0 \text{ for } i > m\}$$

and demonstrate that $\dim \mathscr{V}_n^m = \dim \mathscr{P}_n^m$, where $\mathscr{P}_n^m$ are the homogeneous polynomials in m variables, by a double induction argument for $1 \leq m \leq d$.

Proposition 10.3.9 *For $n \in \mathbb{N}_0$, $1 \leq m \leq d$ and $\kappa \geq 0$,*

$$\dim \mathscr{V}_n^m = \binom{m+n-1}{n} \qquad \textit{with} \qquad \mathscr{V}_n^m = \{p \in \mathscr{P}_n^d : \mathscr{D}_i p = 0, \textit{for } i > m\}.$$

Proof The boundary values are $\dim \mathscr{V}_0^m = 1$ and

$$\dim \mathscr{V}_n^1 = \dim [c p_n(x_1; x) : c \in \mathbb{R}] = 1;$$

the latter holds because $p_n(1;(1,0,\ldots)) = (\kappa+1)_n/n! \neq 0$. Suppose that $\dim \mathscr{V}_n^m = \binom{m+n-1}{n}$ for all pairs (m,n) with $m+n = s$ and $1 \leq m \leq d$; this implies that the associated $\{p_\alpha\}$ are linearly independent. For $m < d$ and $m+n = s$ we have

$$\begin{aligned}\mathscr{V}_n^{m+1} &= \operatorname{span}\{p_\alpha : |\alpha| = n, \alpha_i = 0 \text{ for } i > m+1, \alpha_{m+1} \geq 1\} + \mathscr{V}_n^m \\ &= \rho_{m+1} \mathscr{V}_{n-1}^{m+1} + \mathscr{V}_n^m.\end{aligned}$$

This is a direct sum because $\mathscr{D}_{m+1}\rho_{m+1}$ is one to one on $\mathscr{V}_{n-1}^{m+1}$ and $\mathscr{D}_{m+1}\mathscr{V}_n^m = \{0\}$, which shows that $\dim \mathscr{V}_n^{m+1} = \dim \mathscr{V}_{n-1}^{m+1} + \dim \mathscr{V}_n^m = \binom{m+n-1}{n-1} + \binom{m+n-1}{n} = \binom{m+n}{n}$ by the inductive hypothesis. Further, $\mathscr{V}_n^m = \{p \in \mathscr{V}_n^{m+1} : \mathscr{D}_{m+1} p = 0\}$; together with $\mathscr{V}_n^1 \subset \mathscr{V}_n^2 \subset \cdots \subset \mathscr{V}_n^d$ this proves the remaining statement. □

We now discuss the matrices of $\mathscr{U}_i$ with respect to the basis $\{p_\alpha : \alpha \in \mathbb{N}_0^d\}$; their triangularity is opposite to that of the basis $\{x^\alpha : \alpha \in \mathbb{N}_0^d\}$. The eigenvalues appear in so many calculations that they deserve a formal definition. There is an associated *rank* function r.

Definition 10.3.10 For $\alpha \in \mathbb{N}_0^d$ and $1 \leq i \leq d$ let

$$\begin{aligned}r_\alpha(i) &= \#\{j : \alpha_j > \alpha_i\} + \#\{j : 1 \leq j \leq i, \alpha_j = \alpha_i\}, \\ \xi_i(\alpha) &= (d - r_\alpha(i) + 1)\kappa + \alpha_i + 1.\end{aligned}$$

For a partition $\lambda \in \mathbb{N}_0^{d,P}$, one has $r_\lambda(i) = i$ and $\xi_i(\lambda) = \kappa(d-i+1) + \lambda_i + 1$. Thus r_α is a permutation of $\{1,\ldots,N\}$ for any $\alpha \in \mathbb{N}_0^d$.

Proposition 10.3.11 *For $\alpha \in \mathbb{N}_0^d$ and $1 \leq i \leq d$, $\mathscr{U}_i p_\alpha = \xi_i(a) p_\alpha + \kappa q_{\alpha,i}$, where $q_{\alpha,i}$ is a sum of $\pm p_\beta$ with $\beta \rhd \alpha$; also, if $\alpha_j = 0$ for each $j \geq m$ then $\beta_j = 0$ for $j \geq m$, with $2 \leq m < d$.*

Proof By Corollary 10.3.6 and Proposition 10.3.7,

$$\begin{aligned}\mathscr{U}_i p_\alpha &= \left[\kappa\left(d-\#\{j:\alpha_j>\alpha_i\}\right)+\alpha_i+1\right]p_\alpha \\ &\quad +\kappa\sum_{\alpha_j>\alpha_i} p_{(i,j)\alpha}-\kappa\sum_{j<i} p_{(i,j)\alpha}+\kappa q_{\alpha,i} \\ &= \xi_i(a)p_\alpha+\kappa\sum\{p_{(i,j)\alpha}: j>i;\alpha_j>\alpha_i\} \\ &\quad -\kappa\sum\{p_{(i,j)\alpha}: j<i;\alpha_j<\alpha_i\}+\kappa q_{\alpha,i},\end{aligned}$$

where $q_{\alpha,i}$ is a sum of $\pm p_\beta$ with $\beta^+\succ\alpha^+$ and $|\beta|=|\alpha|$. Also, $(i,j)\alpha\succ\alpha$ in the two cases $j>i,\alpha_j>\alpha_i$ and $j<i,\alpha_j<\alpha_i$; that is, $(i,j)\alpha\rhd\alpha$. □

Corollary 10.3.12 *For $\alpha\in\mathbb{N}_0^d$ and $2\le m<d$, if $\alpha_j=0$ for all $j\ge m$ then $\mathscr{U}_i p_\alpha=[\kappa(d-i+1)+1]\,p_\alpha=\xi_i(\alpha)p_\alpha$ for all $i\ge m$.*

Proof Suppose that α satisfies the hypothesis and $i\ge m$; then, by Proposition 10.3.5,

$$\mathscr{D}_i\rho_i p_\alpha=[\kappa(d-\#\{j:\alpha_j>0\})+1]\,p_\alpha+\kappa\sum_{\alpha_j>0}(i,j)\,p_\alpha,$$

and $\kappa\sum_{j<i}(i,j)\,p_\alpha=\kappa\sum_{\alpha_j>0}(i,j)\,p_\alpha+\kappa\,[\#\{j:j<i;\alpha_i=0\}]\,p_\alpha$. Thus $\mathscr{U}_i p_\alpha=[\kappa(d-\#\{j:\alpha_j>0\}-\#\{j:j<i;\alpha_j=0\})+1]p_\alpha=[\kappa(d-i+1)+1]p_\alpha$. □

The set of values $\{\xi_i(\alpha)\}$ determines α^+, provided that $\kappa>0$: arrange the values in decreasing order, so that if $\alpha_i>\alpha_j$ then $\xi_i(\alpha)>\xi_j(\alpha)$ and if $\alpha_i=\alpha_j$ and $i<j$ then $\xi_i(\alpha)-\xi_j(\alpha)=m\kappa$ for some $m=1,2,\ldots$ It is now easy to see that the list $\{\xi_1(\alpha),\ldots,\xi_d(\alpha)\}$ determines α and thus a uniquely defined set of joint eigenvectors of $\{\mathscr{U}_i\}$. The following are the nonsymmetric Jack polynomials normalized to be monic in the p-basis.

Theorem 10.3.13 *For $\alpha\in\mathbb{N}_0^d$ let ζ_α be the unique polynomial of the form*

$$\zeta_\alpha=p_\alpha+\sum\{B(\beta,\alpha)p_\beta:\beta\rhd\alpha\}$$

which satisfies $\mathscr{U}_i\zeta_\alpha=\xi_i(\alpha)\zeta_\alpha$ for $1\le i\le d$. The coefficients $B(\beta,\alpha)$ are in $\mathbb{Q}(\kappa)$ and do not depend on the number d of variables provided that $\alpha_j=0$ for all $j>m$ for some m and $d\ge m$. For any permissible inner product, $\alpha\ne\beta$ implies that $\langle\zeta_\alpha,\zeta_\beta\rangle=0$.

Proof The existence of the polynomials in the stated form follows from the self-adjointness of the operators $\mathscr{U}_i$ and their triangular matrix action on the p-basis ordered by $\rhd$. Suppose that $\alpha_j=0$ for all $j>m$, with some fixed m. Because each $\mathscr{U}_i$ with $i\le m$ has span$\{p_\beta:\beta_j=0$ for all $j>m\}$ as an invariant subspace, this space contains the joint eigenvectors of $\{\mathscr{U}_i:1\le i\le m\}$. In the action of $\mathscr{U}_i$ the value of d appears only on the diagonal, so that $\mathscr{U}_i-(d\kappa)1$ has the same eigenvectors as $\mathscr{U}_i$ and the eigenvectors have coefficients that are rational in $\mathbb{Q}(\kappa)$ with respect to the p-basis. By Corollary 10.3.12 any polynomial in $\operatorname{span}\{p_\beta : \beta_j = 0 \text{ for all } j > m\}$ is an eigenfunction of $\mathscr{U}_i$ with eigenvalue $\kappa(d-i+1)+1$ when $i > m$ (as was shown, this agrees with $\xi_i(\alpha)$ when $\alpha_j = 0$ for all $j > m$).

If $\alpha, \beta \in \mathbb{N}_0^d$ and $\alpha \neq \beta$ then $\xi_i(\alpha) \neq \xi_i(\beta)$ for some i; thus $\langle \zeta_\alpha, \zeta_\beta \rangle = 0$ for any inner product for which $\mathscr{U}_i$ is self-adjoint. □

10.4 S_d-Invariant Subspaces

Using only the properties of permissible inner products (that they are S_d-invariant and that each $\mathscr{D}_i\rho_i$ or $\mathscr{D}_i x_i$ is self-adjoint), many useful relations can be established on the subspaces $\operatorname{span}\{\zeta_\alpha : \alpha^+ = \lambda\}$ for any partition $\lambda \in \mathbb{N}_0^{d,P}$. We will show that $\langle \zeta_\alpha, \zeta_\alpha \rangle$ is a multiple of $\langle \zeta_\lambda, \zeta_\lambda \rangle$ which is independent of the choice of inner product. Of course, $\langle \zeta_\alpha, \zeta_\beta \rangle = 0$ if $\alpha \neq \beta$ because $\xi_i(\alpha) \neq \xi_i(\beta)$ for some i. It is convenient to have a basis which transforms like x^α or p_α (where $(xw)^\alpha = x^{w\alpha}$ for $w \in S_d$), and this is provided by the orbit of ζ_λ. In fact the action of S_d on the eigenfunctions ζ_α is not conveniently expressible, but adjacent transpositions $(i, i+1)$ not only have elegant formulations but are crucial in the development.

Lemma 10.4.1 *For $1 \leq i \leq d$ the commutant $[\mathscr{U}_i, (j, j+1)] = 0$ for $j > i$ or $j < i-1$ and $(i, i+1)\,\mathscr{U}_i = \mathscr{U}_{i+1}\,(i, i+1) + \kappa$.*

Proof First, $[\mathscr{D}_i\rho_i, (j, j+1)] = 0$ if $j < i-1$ or $j > i$. Also, $(j,k)(m,j)(j,k) = (m,k)$ provided that j,k,m are all distinct. This shows that $[\sum_{k<i}(k,i), (j, j+1)] = 0$ under the same conditions on j, when $j < i-1$ the key fact is $(j, j+1)\{(j,i) + (j+1,i)\} = \{(j+1,i) + (j,i)\}(j, j+1)$. For $i = j$, we have $(i, i+1)\,\mathscr{U}_i = \mathscr{D}_{i+1}\rho_{i+1}(i, i+1) - \kappa\sum_{m<i}(m, i+1)(i, i+1) = \mathscr{U}_{i+1}(i, i+1) + \kappa(i, i+1)(i, i+1)$. □

Proposition 10.4.2 *For $\alpha \in \mathbb{N}_0^d$, if $\alpha_i = \alpha_{i+1}$ for some $i < d$ then $(i, i+1)\,\zeta_\alpha = \zeta_\alpha$; if $\lambda \in \mathbb{N}_0^{d,P}$ and $w\lambda = \lambda$ then $w\zeta_\lambda = \zeta_\lambda$ with $w \in S_d$.*

Proof Let $g = (i, i+1)\,\zeta_\alpha - \zeta_\alpha$. By the lemma, $\mathscr{U}_j g = \xi_j(\alpha) g$ when $j < i$ or $j > i+1$, while $\mathscr{U}_{i+1} g = \xi_i(\alpha)\,(i, i+1)\,\zeta_\alpha - (\xi_{i+1}(\alpha) + \kappa)\,\zeta_\alpha = \xi_i(\alpha)\, g$ and similarly $\mathscr{U}_i g = \xi_{i+1}(\alpha)(i, i+1)\zeta_\alpha - (\xi_i(\alpha) - \kappa)\zeta_\alpha = \xi_{i+1}(\alpha) g$. Consequently, either $g = 0$ or it is a joint eigenfunction of the $\mathscr{U}_j$ with eigenvalues $(\xi_1(\alpha), \ldots, \xi_{i+1}(\alpha), \xi_i(\alpha), \ldots)$. This is impossible by the definition of ξ_j.

For a partition λ, the permutations which fix λ are generated by adjacent transpositions (for example, suppose that $\lambda_{j-1} > \lambda_j = \cdots = \lambda_k > \lambda_{k+1}$ for some $0 < j < k < d$; then $(i, i+1)\,\lambda = \lambda$ for $j \leq i < k$). Thus $w\lambda = \lambda$ implies $w\zeta_\lambda = \zeta_\lambda$. □

With this, one can define a "monomial"-type basis for $\operatorname{span}\{\zeta_\alpha : \alpha^+ = \lambda\}$ for each $\lambda \in \mathbb{N}_0^{d,P}$ (the term monomial refers to the actions $wx^\alpha = x^{w\alpha}$ or $wp_\alpha = p_{w\alpha}$ for $w \in S_d$ and $\alpha \in \mathbb{N}_0^d$).

Definition 10.4.3 For $\lambda \in \mathbb{N}_0^{d,P}$ let $\omega_\lambda = \zeta_\lambda$ and for $\alpha \in \mathbb{N}_0^d$ with $\alpha^+ = \lambda$ let $\omega_\alpha = w\omega_\lambda$, where $w\lambda = \alpha$.

Proposition 10.4.2 shows that ω_α is well defined, for if $w_1\lambda = \alpha = w_2\lambda$ then $w_2^{-1}w_1\lambda = \lambda$ and $w_2\omega_\lambda = w_2\left(w_2^{-1}w_1\omega_\lambda\right) = w_1\omega_\lambda$. We now collect some properties of ω_α.

Proposition 10.4.4 *For $\alpha \in \mathbb{N}_0^d$ and $w \in S_d$ the following hold:*

(i) $w\omega_\alpha = \omega_{w\alpha}$;
(ii) $\omega_\alpha = p_\alpha + \sum\{B(\beta,\alpha)\,p_\beta : \beta^+ \succ \alpha^+, \alpha_i = 0 \text{ implies } \beta_i = 0\}$, *where the coefficients $B(\beta,\alpha) \in \mathbb{Q}(\kappa)$ and are independent of d;*
(iii) $\mathscr{D}_i\rho_i\omega_\alpha = [\kappa(d - \#\{j : \alpha_j > \alpha_i\}) + \alpha_i + 1]\,\omega_\alpha + \kappa\sum_{\alpha_j > \alpha_i}\omega_{(i,j)\alpha}$ *for* $1 \le i \le d$.

Proof Let $\lambda = \alpha^+$. Part (i) follows from the basic property that $w\lambda = \lambda$ implies $w\omega_\lambda = \omega_\lambda$. Theorem 10.3.13 shows that ω_λ satisfies part (ii) because λ is maximal for the dominance ordering $\succeq$ on $\{\beta : \beta^+ = \lambda\}$, that is, $\beta \succeq \lambda$ and $\beta^+ = \lambda$ implies that $\beta = \lambda$. The equation in (ii) transforms under $w \in S_d$ with $w\lambda = \alpha$ to the claimed expression.

The fact that $\omega_\lambda = \zeta_\lambda$ is an eigenfunction of $\mathscr{U}_i$ implies that

$$\begin{aligned}\mathscr{D}_i\rho_i\omega_\lambda &= [\kappa(d-i+1) + \lambda_i + 1]\,\omega_\lambda + \kappa\sum_{j<i}(i,j)\,\omega_\lambda\\ &= [\kappa(d - \#\{j : \lambda_j > \lambda_i\}) + \lambda_i + 1]\,\omega_\lambda + \kappa\sum_{\lambda_j > \lambda_i}\omega_{(i,j)\lambda},\end{aligned}$$

where the terms in the sum in the first line which correspond to values of j satisfying $j < i$ and $\lambda_j = \lambda_i$ have been added into the first part of the second line. Again apply w to this formula with $w\lambda = \alpha$ (the index i is changed also, but this does not matter because i is arbitrary). □

A few more details will establish that $\operatorname{span}\{\omega_\alpha : \alpha^+ = \lambda\} = \operatorname{span}\{\zeta_\alpha : \alpha^+ = \lambda\}$. The important thing is to give an explicit method for calculating ζ_α from ζ_λ, for $\alpha^+ = \lambda$. Adjacent transpositions are the key.

Proposition 10.4.5 *Suppose that $\alpha \in \mathbb{N}_0^d$ and $\alpha_i > \alpha_{i+1}$; then, for $c = \kappa[\xi_i(\alpha) - \xi_{i+1}(\alpha)]^{-1}$ and $\sigma = (i, i+1)$,*

$$\zeta_{\sigma\alpha} = \sigma\zeta_\alpha - c\zeta_\alpha, \qquad \sigma\zeta_{\sigma\alpha} = (1 - c^2)\zeta_\alpha - c\zeta_{\sigma\alpha}$$

and $\operatorname{span}\{\zeta_\alpha, \zeta_{\sigma\alpha}\}$ *is invariant under* σ.

Proof Note that $\xi_i(\alpha) = \xi_{i+1}(\sigma\alpha), \xi_{i+1}(\alpha) = \xi_i(\sigma\alpha)$ and $\xi_i(\alpha) - \xi_{i+1}(\alpha) \geq \alpha_i - \alpha_{i+1} + \kappa$; thus $0 < c < 1$. Let $g = \sigma\zeta_\alpha - c\zeta_\alpha$; then $\mathscr{U}_j g = \xi_j(\alpha) g = \xi_j(\sigma\alpha) g$ for $j < i$ or $j > i+1$ (by Lemma 10.4.1). Further, $\mathscr{U}_i g = \xi_{i+1}(\alpha) g$ and $\mathscr{U}_{i+1} g = \xi_i(\alpha) g$, because $\sigma \mathscr{U}_i \sigma = \mathscr{U}_{i+1} + \kappa\sigma$. This shows that g is a scalar multiple of $\zeta_{\sigma\alpha}$, having the same eigenvalues. The coefficient of $p_{\sigma\alpha}$ in g is 1 (coming from $\sigma\zeta_\alpha$ because $p_{\sigma\alpha}$ does not appear in the expansion of ζ_α since $\alpha \succ \sigma\alpha$).

Thus $\sigma\zeta_\alpha = c\zeta_\alpha + \zeta_{\sigma\alpha}$. To find $\sigma\zeta_{\sigma\alpha}$, note that $\sigma\zeta_{\sigma\alpha} = \zeta_\alpha - c\sigma\zeta_\alpha = (1-c^2)\zeta_\alpha - c\zeta_{\sigma\alpha}$. □

Now, for $\lambda \in \mathbb{N}_0^{d,P}$ let

$$E_\lambda = \text{span}\{\omega_\alpha : \alpha^+ = \lambda\}.$$

Since E_λ is closed under the action of S_d, the proposition shows that $\text{span}\{\zeta_\alpha : \alpha^+ = \lambda\} = E_\lambda$. The set $\{\zeta_\alpha\}$ forms an orthogonal basis and E_λ is invariant under each $\mathscr{D}_i\rho_i, 1 \leq i \leq d$. The next task is to determine the structural constants $\langle \zeta_\alpha, \zeta_\alpha \rangle$ and $\zeta_\alpha(\mathbf{1}^d)$ in terms of $\langle \zeta_\lambda, \zeta_\lambda \rangle$ and $\zeta_\lambda(\mathbf{1}^d)$, respectively.

Proposition 10.4.6 *Suppose that $\alpha \in \mathbb{N}_0^d$ and $\alpha_i > \alpha_{i+1}, \sigma = (i, i+1)$, and $c = \kappa[\xi_i(\alpha) - \xi_{i+1}(\alpha)]^{-1}$. Then:*

(i) $\langle \zeta_{\sigma\alpha}, \zeta_{\sigma\alpha} \rangle = (1-c^2) \langle \zeta_\alpha, \zeta_\alpha \rangle$;
(ii) $\zeta_{\sigma\alpha}(\mathbf{1}^d) = (1-c)\zeta_\alpha(\mathbf{1}^d)$;
(iii) $f_0 = (1+c)\zeta_\alpha + \zeta_{\sigma\alpha}$ *satisfies* $\sigma f_0 = f_0$;
(iv) $f_1 = (1-c)\zeta_\alpha - \zeta_{\sigma\alpha}$ *satisfies* $\sigma f_1 = -f_1$.

Proof Because $\langle \zeta_\alpha, \zeta_{\sigma\alpha} \rangle = 0$ and σ is an isometry we have

$$\langle \zeta_\alpha, \zeta_\alpha \rangle = \langle \sigma\zeta_\alpha, \sigma\zeta_\alpha \rangle = c^2 \langle \zeta_\alpha, \zeta_\alpha \rangle + \langle \zeta_{\sigma\alpha}, \zeta_{\sigma\alpha} \rangle;$$

thus $\langle \zeta_{\sigma\alpha}, \zeta_{\sigma\alpha} \rangle = (1-c^2) \langle \zeta_\alpha, \zeta_\alpha \rangle$. The map of evaluation at $\mathbf{1}^d$ is invariant under σ; thus $\zeta_{\sigma\alpha}(\mathbf{1}^d) = \sigma\zeta_\alpha(\mathbf{1}^d) - c\zeta_\alpha(\mathbf{1}^d) = (1-c)\zeta_\alpha(\mathbf{1}^d)$. Parts (iii) and (iv) are trivial calculations. □

Aiming to use induction on adjacent transpositions, we introduce a function which measures (heuristically speaking) how much α is out of order.

Definition 10.4.7 Functions on $\alpha \in \mathbb{N}_0^d$ for $1 \leq i, j \leq d$ and $\varepsilon = \pm$ are given by:

(i) $\tau_{i,j}^\varepsilon(\alpha) = 1 + \dfrac{\varepsilon\kappa}{\xi_j(\alpha) - \xi_i(\alpha)}$ if $i < j$ and $\alpha_i < \alpha_j$, else $\tau_{i,j}^\varepsilon(\alpha) = 1$,
(ii) $\mathscr{E}_\varepsilon(\alpha) = \prod_{i<j} \tau_{i,j}^\varepsilon(\alpha)$ $\quad$ ($\mathscr{E}_\varepsilon(\lambda) = 1$ for $\lambda \in \mathbb{N}_0^{d,P}$).

Theorem 10.4.8 *For $\alpha \in \mathbb{N}_0^d$ and $\lambda = \alpha^+$,*

(i) $\langle \zeta_\alpha, \zeta_\alpha \rangle = \mathscr{E}_+(\alpha)\mathscr{E}_-(\alpha) \langle \zeta_\lambda, \zeta_\lambda \rangle$,
(ii) $\zeta_\alpha(\mathbf{1}^d) = \mathscr{E}_-(\alpha)\zeta_\lambda(\mathbf{1}^d)$.

Proof First, suppose that $\beta^+ = \lambda$, $\beta_i > \beta_{i+1}$ and $\sigma = (i, i+1)$; then

$$\mathscr{E}_\varepsilon(\sigma\beta)/\mathscr{E}_\varepsilon(\beta) = 1 + \varepsilon\kappa[\xi_i(\beta) - \xi_{i+1}(\beta)]^{-1}.$$

Indeed, $\tau^\varepsilon_{k,j}(\sigma\beta) = \tau^\varepsilon_{k,j}(\beta)$ for all values of $j,k \neq i, i+1$. Also, $\tau^\varepsilon_{i,j}(\sigma\beta) = \tau^\varepsilon_{i+1,j}(\beta)$ and $\tau^\varepsilon_{i+1,j}(\sigma\beta) = \tau^\varepsilon_{i,j}(\beta)$ for $j > i+1$; $\tau^\varepsilon_{j,i}(\sigma\beta) = \tau^\varepsilon_{j,i+1}(\beta)$ and $\tau^\varepsilon_{j,i+1}(\sigma\beta) = \tau^\varepsilon_{j,i}(\beta)$ for $j < i$. The claim is now proven from the special values $\tau^\varepsilon_{i,i+1}(\beta) = 1$ and $\tau^\varepsilon_{i,i+1}(\sigma\beta) = 1 + \varepsilon\kappa[\xi_i(\beta) - \xi_{i+1}(\beta)]^{-1}$ (recall that $\xi_i(\beta) = \xi_{i+1}(\sigma\beta)$ and $\xi_{i+1}(\beta) = \xi_i(\sigma\beta)$).

By part (i) of Proposition 10.4.6,

$$\frac{\langle \zeta_{\sigma\beta}, \zeta_{\sigma\beta}\rangle}{\langle \zeta_\beta, \zeta_\beta\rangle} = 1 - c^2 = \frac{\mathscr{E}_+(\sigma\beta)\mathscr{E}_-(\sigma\beta)}{\mathscr{E}_+(\beta)\mathscr{E}_-(\beta)}$$

(recall that $c = \kappa[\xi_i(\beta) - \xi_{i+1}(\beta)]^{-1}$. Similarly,

$$\frac{\zeta_{\sigma\beta}(\mathbf{1}^d)}{\zeta_\beta(\mathbf{1}^d)} = 1 - c = \frac{\mathscr{E}_-(\sigma\beta)}{\mathscr{E}_-(\beta)}.$$

Since α is obtained from λ by a sequence of adjacent transpositions satisfying this hypothesis, the proof is complete. □

The last results of this section concern the unique symmetric and skew-symmetric polynomials in E_λ (which exist if $\lambda_1 > \cdots > \lambda_d$). Recall that $\lambda^R = (\lambda_d, \lambda_{d-1}, \ldots, \lambda_1)$; the set of values $\{\xi_i(\lambda^R)\}$ is the same as $\{\xi_i(\lambda)\}$ but not necessarily in reverse order. For example, suppose that $\lambda_{j-1} > \lambda_j = \cdots = \lambda_k > \lambda_{k+1}$ for some $0 < j < k < d$; then the list $\{\xi_i(\lambda) : i = j, \ldots, k\}$ agrees with $\{\xi_{d-k-i+1}(\lambda^R) : i = j, \ldots, k\}$ (in that order). Thus

$$\mathscr{E}_\varepsilon(\lambda^R) = \prod\left\{1 + \frac{\varepsilon\kappa}{\xi_i(\lambda) - \xi_j(\lambda)} : \lambda_i > \lambda_j\right\}.$$

Definition 10.4.9 For $\lambda \in \mathbb{N}_0^{d,P}$ let

$$j_\lambda = \mathscr{E}_+(\lambda^R) \sum_{\alpha^+ = \lambda} \frac{1}{\mathscr{E}_+(\alpha)} \zeta_\alpha$$

and, if $\lambda_1 > \cdots > \lambda_d$, let

$$a_\lambda = \mathscr{E}_-(\lambda^R) \sum_{w \in S_d} \frac{\operatorname{sign}(w)}{\mathscr{E}_-(w\lambda)} \zeta_{w\lambda}.$$

Recall that $\operatorname{sign}(w) = (-1)^m$, where w can be expressed as a product of m transpositions. For $\lambda \in \mathbb{N}_0^{d,P}$ we use the notation $\#S_d(\lambda) = \#\{\alpha \in \mathbb{N}_0^d : \alpha^+ = \lambda\}$ for the cardinality of the S_d-orbit of λ.

Theorem 10.4.10 *For $\lambda \in \mathbb{N}_0^{d,P}$, the polynomial j_λ has the following properties:*

(i) $j_\lambda = \sum_{\alpha^+ = \lambda} \omega_\alpha$;
(ii) $w j_\lambda = j_\lambda$ *for each* $w \in S_d$;

(iii) $j_\lambda(\mathbf{1}^d) = \#S_d(\lambda)\,\zeta_\lambda(\mathbf{1}^d)$;
(iv) $\langle j_\lambda, j_\lambda\rangle = \#S_d(\lambda)\,\mathscr{E}_+(\lambda^R)\,\langle\zeta_\lambda,\zeta_\lambda\rangle$.

Proof It suffices to show that $(i,i+1)j_\lambda = j_\lambda$ for any $i<d$. Indeed, with $\sigma = (i,i+1)$,

$$j_\lambda = \mathscr{E}_+(\lambda^R)\Big[\sum\Big\{\frac{\zeta_\alpha}{\mathscr{E}_+(\alpha)} : \alpha^+ = \lambda, \alpha_i = \alpha_{i+1}\Big\}$$
$$+\sum\Big\{\frac{\zeta_\alpha}{\mathscr{E}_+(\alpha)} + \frac{\zeta_{\sigma\alpha}}{\mathscr{E}_+(\sigma\alpha)} : \alpha^+ = \lambda, \alpha_i > \alpha_{i+1}\Big\}\Big].$$

Each term in the first sum is invariant under σ. By part (iii) of Proposition 10.4.6, each term in the second sum is invariant because

$$\mathscr{E}_+(\sigma\alpha)/\mathscr{E}_+(\alpha) = 1+\kappa[\xi_i(\alpha)-\xi_{i+1}(\alpha)]^{-1}.$$

Thus j_λ is invariant under S_d and is a scalar multiple of $\sum_{\alpha^+=\lambda}\omega_\alpha$. Since λ^R is minimal for $\succeq$ in $\{\alpha : \alpha^+ = \lambda\}$ and, by the triangularity for ζ_α in the p-basis, the coefficient of p_{λ^R} in j_λ is 1, part (i) holds. Part (iii) is an immediate consequence of (i).

By the group invariance of the inner product,

$$\langle j_\lambda, j_\lambda\rangle = \sum_{\alpha^+=\lambda}\langle j_\lambda,\omega_\alpha\rangle = \#\{\alpha : \alpha^+=\lambda\}\,\langle j_\lambda,\zeta_\lambda\rangle$$
$$= \#S_d(\lambda)\,\mathscr{E}_+(\lambda^R)\,\langle\zeta_\lambda,\zeta_\lambda\rangle.$$

Note that $\omega_\lambda = \zeta_\lambda$ and $\mathscr{E}_+(\lambda) = 1$. □

Theorem 10.4.11 *For $\lambda \in \mathbb{N}_0^{d,P}$ with $\lambda_1 > \cdots > \lambda_d$, the polynomial a_λ has the following properties:*

(i) $a_\lambda = \sum_{w\in S_d}\operatorname{sign}(w)\,\omega_{w\lambda}$;
(ii) $wa_\lambda = \operatorname{sign}(w)a_\lambda$ *for each* $w \in S_d$;
(iii) $\langle a_\lambda, a_\lambda\rangle = d!\,\mathscr{E}_-(\lambda^R)\,\langle\zeta_\lambda,\zeta_\lambda\rangle$.

Proof The proof is similar to the previous one. To show part (ii) it suffices to show that $\sigma a_\lambda = -a_\lambda$ for $\sigma = (i,i+1)$ and for each $i<d$. As before, each term in the sum is skew under σ:

$$a_\lambda = \mathscr{E}_-(\lambda^R)\sum\Big\{\operatorname{sign}(w)\Big[\frac{\zeta_{w\lambda}}{\mathscr{E}_-(w\lambda)} - \frac{\zeta_{\sigma w\lambda}}{\mathscr{E}_-(\sigma w\lambda)}\Big] : w\in S_d, (w\lambda)_i > (w\lambda)_{i+1}\Big\}.$$

Both a_λ and $\sum_{w\in S_d}\operatorname{sign}(w)\,\omega_{w\lambda}$ have the skew property and coefficient $\operatorname{sign}(w)$ for p_{λ^R}, where $w\lambda = \lambda^R$. Thus part (i) holds. Finally,

$$\langle a_\lambda, a_\lambda\rangle = \sum_{w_1\in S_d}\sum_{w_2\in S_d}\operatorname{sign}(w_1)\operatorname{sign}(w_2)\,\langle\omega_{w_1\lambda},\omega_{w_2\lambda}\rangle$$
$$= d!\sum_{w\in S_d}\operatorname{sign}(w)\,\langle\omega_\lambda,\omega_{w\lambda}\rangle = d!\,\langle\zeta_\lambda,a_\lambda\rangle$$
$$= d!\,\mathscr{E}_-(\lambda^R)\,\langle\zeta_\lambda,\zeta_\lambda\rangle,$$

and this proves part (iii) (in the first line we change the variable of summation, letting $w_2 = w_1 w$). □

The polynomial j_λ is a scalar multiple of the Jack polynomial $J_\lambda(x; 1/\kappa)$. The actual multiplying factor will be discussed later.

10.5 Degree-Changing Recurrences

In this section the emphasis will be on the transition from $\zeta_{\lambda-\varepsilon_m}$ to ζ_λ, where $\lambda \in \mathbb{N}_0^{d,P}$ and $\lambda_m > \lambda_{m+1} = 0$ for some $m \le d$. By use of the defining properties we will determine the ratios $\langle \zeta_\lambda, \zeta_\lambda \rangle / \langle \zeta_{\lambda-\varepsilon_m}, \zeta_{\lambda-\varepsilon_m} \rangle$ for each of three permissible inner products. This leads to norm and $\mathbf{1}^d$ evaluation formulae in terms of hook-length products. A certain cyclic shift is a key tool.

Definition 10.5.1 For $1 < m \le d$, let $\theta_m = (1,2)(2,3)\cdots(m-1,m) \in S_d$, so that $\theta_m \lambda = (\lambda_m, \lambda_1, \lambda_2, \ldots, \lambda_{m-1}, \lambda_{m+1}, \ldots)$ for $\lambda \in \mathbb{N}_0^{d,P}$.

Suppose that $\lambda \in \mathbb{N}_0^{d,P}$ satisfies $m = \ell(\lambda)$, and let

$$\widetilde{\lambda} = (\lambda_m - 1, \lambda_1, \lambda_2, \ldots, \lambda_{m-1}, 0, \ldots) = \theta_m(\lambda - \varepsilon_m).$$

We will show that $\mathscr{D}_m \zeta_\lambda = [(d-m+1)\kappa + \lambda_m]\,\theta_m^{-1} \zeta_{\widetilde{\lambda}}$, which leads to the desired recurrences. The idea is to identify $\mathscr{D}_m \zeta_\lambda$ as a joint eigenfunction of the operators $\theta_m^{-1} \mathscr{U}_i \theta_m$ and match up the coefficients of the $p_{\widetilde{\lambda}}$.

Lemma 10.5.2 *Suppose that $\lambda \in \mathbb{N}_0^{d,P}$ and $m = \ell(\lambda)$; then the following hold:*

(i) $x_m \mathscr{D}_m \zeta_\lambda = [\kappa(d-m) + \lambda_m]\,\zeta_\lambda - \kappa \sum_{j>m} (m,j)\,\zeta_\lambda$;

(ii) $x_m \mathscr{D}_m \zeta_\lambda(\mathbf{1}^d) = \lambda_m \zeta_\lambda(\mathbf{1}^d)$;

(iii) $\langle x_m \mathscr{D}_m \zeta_\lambda, \zeta_\lambda \rangle = \dfrac{\lambda_m}{\kappa + \lambda_m}[\kappa(d-m+1) + \lambda_m]\langle \zeta_\lambda, \zeta_\lambda \rangle$;

(iv) $$\langle x_m \mathscr{D}_m \zeta_\lambda, x_m \mathscr{D}_m \zeta_\lambda \rangle = \frac{\lambda_m}{\kappa + \lambda_m}[\kappa(d-m+1) + \lambda_m] \times [\kappa(d-m) + \lambda_m]\langle \zeta_\lambda, \zeta_\lambda \rangle.$$

Proof By part (i) of Lemma 10.2.1,

$$\begin{aligned} x_m \mathscr{D}_m \zeta_\lambda &= \Big(\mathscr{D}_m x_m - 1 - \kappa \sum_{j \ne m} (m,j) \Big) \zeta_\lambda \\ &= (\mathscr{U}_m - \kappa - 1)\zeta_\lambda - \kappa \sum_{j>m} (m,j)\zeta_\lambda \\ &= [\xi_m(\lambda) - \kappa - 1]\zeta_\lambda - \kappa \sum_{j>m} (m,j)\zeta_\lambda \\ &= [\kappa(d-m) + \lambda_m]\zeta_\lambda - \kappa \sum_{j>m} (m,j)\zeta_\lambda. \end{aligned}$$

This proves parts (i) and (ii). Next, observe that $\langle\zeta_\lambda,(m,j)\zeta_\lambda\rangle = \langle\zeta_\lambda,(m,m+1)\zeta_\lambda\rangle$ for $j > m$; by Proposition 10.4.5, $\sigma\zeta_\lambda = \zeta_{\sigma\lambda} + c\zeta_\lambda$ where $\sigma = (m,m+1)$ and $c = \kappa[\xi_m(\lambda) - \xi_{m+1}(\lambda)]^{-1} = \kappa(\kappa+\lambda_m)^{-1}$ (recall that $\xi_i(\lambda) = \kappa(d-i+1) + \lambda_i + 1$ for each $\lambda \in \mathbb{N}_0^{d,P}$). Thus

$$\begin{aligned}\langle x_m\mathcal{D}_m\zeta_\lambda,\zeta_\lambda\rangle &= [\kappa(d-m)+\lambda_m]\langle\zeta_\lambda,\zeta_\lambda\rangle - \kappa(d-m)\langle\zeta_\lambda,(m,j)\zeta_\lambda\rangle\\ &= \left([\kappa(d-m)+\lambda_m] - \frac{\kappa^2(d-m)}{\kappa+\lambda_m}\right)\langle\zeta_\lambda,\zeta_\lambda\rangle\\ &= \frac{\lambda_m}{\kappa+\lambda_m}[\kappa(d-m+1)+\lambda_m]\langle\zeta_\lambda,\zeta_\lambda\rangle.\end{aligned}$$

For part (iv), let $a = \kappa(d-m)+\lambda_m$; the group invariance properties of the inner product are now used repeatedly:

$$\begin{aligned}&\langle x_m\mathcal{D}_m\zeta_\lambda, x_m\mathcal{D}_m\zeta_\lambda\rangle\\ &= a^2\langle\zeta_\lambda,\zeta_\lambda\rangle - 2a\kappa\sum_{j>m}\langle\zeta_\lambda,(m,j)\zeta_\lambda\rangle\\ &\quad + \kappa^2\Big(\sum_{j>m}\langle(m,j)\zeta_\lambda,(m,j)\zeta_\lambda\rangle + 2\sum_{m<j<k}\langle(m,j)\zeta_\lambda,(m,k)\zeta_\lambda\rangle\Big)\\ &= \left(a^2 - 2ac\kappa(d-m) + \kappa^2(d-m)(1+c(d-m-1))\right)\langle\zeta_\lambda,\zeta_\lambda\rangle\\ &= \frac{\lambda_m}{\kappa+\lambda_m}[\kappa(d-m)+\lambda_m][\kappa(d-m+1)+\lambda_m]\langle\zeta_\lambda,\zeta_\lambda\rangle.\end{aligned}$$

□

Next, we consider the behavior of $\mathcal{D}_m\zeta_\lambda$ under $\mathcal{D}_i\rho_i$, with the aim of identifying it as a transform (θ_m^{-1}) of a joint eigenfunction. The cyclic shift θ_m satisfies $\theta_m^{-1}(1) = m$ and $\theta_m^{-1}(i) = i-1$ for $2 \le i \le m$. Recall the commutation relations for $\mathcal{D}_i\rho_i, (i,j)$ and $w \in S_d$:

1. $w\mathcal{D}_i\rho_i = \mathcal{D}_{w(i)}\rho_{w(i)}w$;
2. $w\left(w^{-1}(i), w^{-1}(j)\right) = (i,j)w$.

Lemma 10.5.3 *Suppose that $\lambda \in \mathbb{N}_0^{d,P}$ and $m = \ell(\lambda)$; then*

(i) $\mathcal{D}_i\rho_i\mathcal{D}_m\zeta_\lambda = \left(\xi_i(\lambda) + \kappa[\sum_{j<i}(j,i) + (i,m)]\right)\mathcal{D}_m\zeta_\lambda$ *for $i < m$;*
(ii) $\mathcal{D}_m\rho_m\mathcal{D}_m\zeta_\lambda = [\xi_m(\lambda) - 1]\mathcal{D}_m\zeta_\lambda$*;*
(iii) $\mathcal{U}_i\theta_m\mathcal{D}_m\zeta_\lambda = [\kappa(d-i+1)+1]\theta_m\mathcal{D}_m\zeta_\lambda$ *for $i > m$;*
(iv) $\mathcal{U}_i\theta_m\mathcal{D}_m\zeta_\lambda = \xi_{i-1}(\lambda)\theta_m\mathcal{D}_m\zeta_\lambda$ *for $1 < i \le m$;*
(v) $\mathcal{U}_1\theta_m\mathcal{D}_m\zeta_\lambda = [\xi_m(\lambda) - 1]\theta_m\mathcal{D}_m\zeta_\lambda$.

Proof By Proposition 10.3.7 and Lemma 10.2.1, $\mathcal{D}_i\rho_i\mathcal{D}_m = \kappa\mathcal{D}_m + \mathcal{D}_i[\mathcal{D}_m x_i + \kappa(i,m)]$ for $i < m$; thus

$$\begin{aligned}\mathcal{D}_i\rho_i\mathcal{D}_m\zeta_\lambda &= \mathcal{D}_m\mathcal{D}_i\rho_i\zeta_\lambda + \kappa(i,m)\mathcal{D}_m\zeta_\lambda\\ &= \mathcal{D}_m\Big(\xi_i(\lambda) + \kappa\sum_{j<i}(j,i)\Big)\zeta_\lambda + \kappa(i,m)\mathcal{D}_m\zeta_\lambda,\end{aligned}$$

which proves part (i), since $\mathscr{D}_i(i,m) = (i,m)\mathscr{D}_m$ and $\mathscr{D}_m(j,i) = (j,i)\mathscr{D}_m$ for $j < i < m$. Next, $\mathscr{D}_m\rho_m\mathscr{D}_m = \kappa\mathscr{D}_m + \mathscr{D}_m[\mathscr{D}_m x_m - 1 - \kappa\sum_{j\neq m}(j,m)]$, so that

$$\begin{aligned}\mathscr{D}_m\rho_m\mathscr{D}_m\zeta_\lambda &= \mathscr{D}_m\Big(\mathscr{U}_m - 1 - \kappa\sum_{j>m}(j,m)\Big)\zeta_\lambda\\ &= [\xi_m(\lambda) - 1]\,\mathscr{D}_m\zeta_\lambda - \kappa\sum_{j>m}(j,m)\,\mathscr{D}_j\zeta_\lambda.\end{aligned}$$

However, $\mathscr{D}_j\zeta_\lambda = 0$ for $j > m$ because the expansion of ζ_λ in the p-basis (see Definition 10.3.2 and Proposition 10.3.9) contains no p_β with $\beta_j > 0$. Also, $\mathscr{D}_m\zeta_\lambda$ has the latter property, which shows that $\mathscr{U}_i\mathscr{D}_m\zeta_\lambda = [\kappa(d-i+1)+1]\mathscr{D}_m\zeta_\lambda$ for each $i > m$ (Corollary 10.3.12). Thus part (iii) follows from $\theta_m\mathscr{U}_i = \mathscr{U}_i\theta_m$ for $i > m$.

Apply θ_m to both sides of the equation in (i):

$$\Big(\mathscr{D}_i\rho_i - \kappa\Big[\sum_{j<i}(j,i) + (i,m)\Big]\Big)\mathscr{D}_m\zeta_\lambda = \xi_i(\lambda)\mathscr{D}_m\zeta_\lambda,$$

and obtain equation (iv) with index $i+1$ (so that $2 \le i+1 \le m$). A similar operation on equation (ii) proves equation (v). □

Consider the multi-index $\widetilde{\lambda} = (\lambda_m - 1, \lambda_1, \lambda_2, \ldots, \lambda_{m-1}, 0, \ldots)$; clearly the eigenvalues are given by $\xi_1(\widetilde{\lambda}) = [\kappa(d-m) + \lambda_m] = \xi_m(\lambda) - 1$ and $\xi_i(\widetilde{\lambda}) = \xi_{i-1}(\lambda)$ for $2 \le i \le m$, because $\lambda_m - 1 < \lambda_{m-1}$. The remaining eigenvalues are trivially $\xi_i(\widetilde{\lambda}) = \xi_i(\lambda) = \kappa(d-i+1)+1$ for $i > m$. Thus $\theta_m\mathscr{D}_m\zeta_\lambda$ is a multiple of $\zeta_{\widetilde{\lambda}}$; the value of the multiple can be determined from the coefficient of $p_{\widetilde{\lambda}}$ in $\theta_m\mathscr{D}_m\zeta_\lambda$, that is, the coefficient of $p_{\lambda-\varepsilon_m}$ in $\mathscr{D}_m\zeta_\lambda$. As usual, this calculation depends on the dominance ordering.

Lemma 10.5.4 *Suppose that $\lambda \in \mathbb{N}_0^{d,P}$, $m = \ell(\lambda)$ and $p_{\lambda-\varepsilon_m}$ appears with a nonzero coefficient in $\mathscr{D}_m p_\alpha$; then one of the following holds:*

(i) $\alpha = \lambda$, *with coefficient* $\kappa(d-m+1) + \lambda_m$;
(ii) $\alpha = \lambda + k(\varepsilon_j - \varepsilon_m)$, *where* $j > m$, $1 \le k < \lambda_m$, *with coefficient* κ;
(iii) $\alpha = \lambda + k(\varepsilon_m - \varepsilon_j)$, *where* $j < m$, $1 \le k \le \lambda_j - \lambda_m$, *with coefficient* $-\kappa$.

Proof By Proposition 10.3.3, when $\alpha = \lambda$ the coefficient of $p_{\lambda-\varepsilon_i}$ in $\mathscr{D}_m p_\alpha$ is $\kappa(1 + \#\{j : \lambda_j < \lambda_m\}) + \lambda_m = \kappa(d-m+1) + \lambda_m$. The two other possibilities correspond to sets A, B with α differing from λ in two entries. In the case of set A, for some $j \ne m$ we have $\alpha_j = \lambda_j + k, \alpha_m = \lambda_m - k$ and $\lambda_j \le \lambda_m - k - 1$. Since λ_m is the smallest nonzero part of λ this implies that $j > m$ and $\lambda_j = 0$. In the case of set B, for some $j \ne m$ we have $\alpha_j = \lambda_j - k, \alpha_m = \lambda_m + k$ and $\lambda_m \le \lambda_j - k < \lambda_j$ (and $\lambda_j > \lambda_m$ implies $j < m$). □

Theorem 10.5.5 *Suppose that* $\lambda \in \mathbb{N}_0^{d,P}$ *and* $m = \ell(\lambda)$*; then*

$$\mathscr{D}_m \zeta_\lambda = [\kappa(d-m+1) + \lambda_m]\, \theta_m^{-1} \zeta_{\widetilde{\lambda}}.$$

Proof In Lemma 10.5.3 it was established that $\mathscr{D}_m \zeta_\lambda$ is a scalar multiple of $\theta_m^{-1}\zeta_{\widetilde{\lambda}}$. The multiple equals the coefficient of $p_{\lambda-\varepsilon_m}$ in $\mathscr{D}_m \zeta_\lambda$ (because the coefficient of $p_{\widetilde{\lambda}}$ in $\zeta_{\widetilde{\lambda}}$ is 1). But $p_{\lambda-\varepsilon_m}$ only arises from $\mathscr{D}_m p_\lambda$ in the expansion of $\mathscr{D}_m \zeta_\lambda$, by Lemma 10.5.4: case (i) is the desired coefficient; case (ii) cannot occur because if p_β appears in the expansion of ζ_λ then $\beta_j = 0$ for all $j > m$; case (iii) cannot occur because $\lambda \succ \lambda + k(\varepsilon_m - \varepsilon_j)$ for $j < m, 1 \le k \le \lambda_j - \lambda_m$. □

Because θ_m is an isometry we can now assert that

$$\langle \mathscr{D}_m \zeta_\lambda, \mathscr{D}_m \zeta_\lambda \rangle = [\kappa(d-m+1) + \lambda_m]^2\, \mathscr{E}_+(\widetilde{\lambda}) \mathscr{E}_-(\widetilde{\lambda}) \langle \zeta_{\lambda-\varepsilon_m}, \zeta_{\lambda-\varepsilon_m} \rangle$$

and

$$\mathscr{D}_m \zeta_\lambda(\mathbf{1}^d) = [\kappa(d-m+1) + \lambda_m]\, \mathscr{E}_-(\widetilde{\lambda}) \zeta_{\lambda-\varepsilon_m}(\mathbf{1}^d).$$

10.6 Norm Formulae

10.6.1 Hook-length products and the pairing norm

The quantities (functions on $\mathbb{N}_0^{d,P}$) which have the required recurrence behavior (see Lemma 10.5.2 and Theorem 10.5.5) are of two kinds: the easy one is the generalized Pochhammer symbol, and the harder one comes from hook length products.

Definition 10.6.1 For a partition $\lambda \in \mathbb{N}_0^{d,P}$ and implicit parameter κ, the generalized Pochhammer symbol is defined by

$$(t)_\lambda = \prod_{i=1}^{d} (t - (i-1)\kappa)_{\lambda_i}.$$

For example, in the situation described above,

$$\frac{(d\kappa+1)_\lambda}{(d\kappa+1)_{\lambda-\varepsilon_m}} = \frac{(\kappa(d-m+1)+1)_{\lambda_m}}{(\kappa(d-m+1)+1)_{\lambda_m-1}} = \kappa(d-m+1) + \lambda_m.$$

Next, we calculate $\mathscr{E}_\varepsilon(\widetilde{\lambda})$: indeed,

$$\begin{aligned}\mathscr{E}_\varepsilon(\widetilde{\lambda}) &= \prod_{j=2}^{m}\Big(1 + \frac{\varepsilon\kappa}{\xi_j(\widetilde{\lambda}) - \xi_1(\widetilde{\lambda})}\Big) = \prod_{j=2}^{m}\Big(1 + \frac{\varepsilon\kappa}{\xi_{j-1}(\lambda) - [\xi_m(\lambda) - 1]}\Big)\\ &= \prod_{j=1}^{m-1}\Big(\frac{\kappa(m-j+\varepsilon 1) + \lambda_j - \lambda_m + 1}{\kappa(m-j) + \lambda_j - \lambda_m + 1}\Big).\end{aligned}$$

Recall that $\xi_j(\lambda) = \kappa(d-j+1)+\lambda_j+1$. It turns out that the appropriate device to handle this factor is the hook-length product of a tableau or of a composition. Here we note that a *tableau* is a function whose domain is a Ferrers diagram and whose values are numbers or algebraic expressions. The *Ferrers diagram* of $\alpha \in \mathbb{N}_0^d$ is the set $\{(i,j) : 1 \le i \le d, 1 \le j \le \alpha_i\}$: for each *node* (i,j) with $1 \le j \le \alpha_i$ there are two special subsets, the *arm*

$$\{(i,l) : j < l \le \alpha_i\}$$

and the *leg*

$$\{(l,j) : l > i, j \le \alpha_l \le \alpha_i\} \cup \{(l,j) : l < i, j \le \alpha_l + 1 \le \alpha_i\}$$

(when $j = \alpha_l + 1$ the point is outside the diagram). The node itself, the arm and the leg make up the *hook*. The definition of hooks for compositions is from [160, p. 15]. The cardinality of the leg is called the leg length, formalized by the following:

Definition 10.6.2 For $\alpha \in \mathbb{N}_0^d, 1 \le i \le d$ and $1 \le j \le \alpha_i$ the leg length is

$$L(\alpha;i,j) := \#\{l : l > i, j \le \alpha_l \le \alpha_i\} + \#\{l : l < i, j \le \alpha_l + 1 \le \alpha_i\}.$$

For $t \in \mathbb{Q}(\kappa)$ the *hook length* and the hook-length product for α are given by

$$h(\alpha,t;i,j) = \alpha_i - j + t + \kappa L(\alpha;i,j)$$

$$h(\alpha,t) = \prod_{i=1}^{\ell(\alpha)} \prod_{j=1}^{\alpha_i} h(\alpha,t;i,j).$$

For the special case $\kappa = 1$ and $\alpha \in \mathbb{N}_0^{d,P}$, this goes back to the beginnings of the representation theory for the symmetric group S_d (due to Schur and to Young in the early twentieth century); for arbitrary parameter values the concept is due to Stanley (1989), with a more recent modification by Knop and Sahi (1997).

Here is an example. Below, we give the tableau (Ferrers diagram) for $\lambda = (5,3,2,2)$ with the factors of the hook-length product $h(\lambda,t)$ entered in each cell:

$4+t+3\kappa$	$3+t+3\kappa$	$2+t+\kappa$	$1+t$	t
$2+t+2\kappa$	$1+t+2\kappa$	t		
$1+t+\kappa$	$t+\kappa$			
$1+t$	t			

Compare this with the analogous diagram for $\alpha = (2,3,2,5)$:

$1+t+\kappa$	$t+\kappa$			
$2+t+2\kappa$	$1+t+2\kappa$	$1+t$		
$1+t$	t			
$4+t+3\kappa$	$3+t+3\kappa$	$2+t+3\kappa$	$1+t+\kappa$	t

There is an important relation between $h(\alpha,t)$, for the values $t = 1, \kappa+1$, and $\mathscr{E}_\varepsilon(\alpha)$. We will use the relation $\xi_i(\alpha) - \xi_j(\alpha) = \kappa[r_\alpha(j) - r_\alpha(i)] + \alpha_i - \alpha_j$:

Lemma 10.6.3 *For $\alpha \in \mathbb{N}_0^d$, we have*

$$h(\alpha, \kappa+1) = h\left(\alpha^+, \kappa+1\right) \mathscr{E}_+(\alpha)$$

and

$$h(\alpha, 1) = \frac{h(\alpha^+, 1)}{\mathscr{E}_-(\alpha)}.$$

Proof We will use induction on adjacent transpositions. The statements are true for $\alpha = \alpha^+$. Fix α^+ and suppose that $\alpha_i > \alpha_{i+1}$ for some i. Let $\sigma = (i, i+1)$. Consider the ratio $h(\sigma\alpha, t)/h(\alpha, t)$. The only node whose hook length changes (in the sense of an interchange of rows i and $i+1$ of the Ferrers diagram) is $(i, \alpha_{i+1}+1)$. Explicitly, $h(\sigma\alpha, t; s, j) = h(\alpha, t; s, j)$ for $s \neq i, i+1$ and $1 \leq j \leq \alpha_s$, $h(\sigma\alpha, t; i, j) = h(\alpha, t; i+1, j)$ for $1 \leq j \leq \alpha_{i+1}$ and $h(\sigma\alpha, t; i+1, j) = h(\alpha, t; i, j)$ for $1 \leq j \leq \alpha_i$ except for $j = \alpha_{i+1}+1$. Thus $h(\sigma\alpha, t)/h(\alpha, t) = h(\sigma\alpha, t; i+1, \alpha_{i+1}+1)/h(\alpha, t; i, \alpha_{i+1}+1)$. Note that $L(\sigma\alpha; i+1, \alpha_{i+1}+1) = L(\alpha; i, \alpha_{i+1}+1)+1$ (since the node $(i, \alpha_{i+1}+1)$ is adjoined to the leg). Let

$$\begin{aligned}
E_1 &= \{s : s \leq i, \alpha_s \geq \alpha_i\} \cup \{s : s > i, \alpha_s > \alpha_i\}, \\
E_2 &= \{s : s \leq i+1, \alpha_s \geq \alpha_{i+1}\} \cup \{s : s > i+1, \alpha_s > \alpha_{i+1}\};
\end{aligned}$$

thus, by definition, $r_\alpha(i) = \#E_1$ and $r_\alpha(i+1) = \#E_2$. Now, $E_1 \subset E_2$ so that $r_\alpha(i+1) - r_\alpha(i) = \#(E_2 \backslash E_1)$ and

$$E_2 \backslash E_1 = \{s : s < i, \alpha_i > \alpha_s \geq \alpha_{i+1}\} \cup \{i\} \cup \{s : s > i+1, \alpha_i \geq \alpha_s > \alpha_{i+1}\}.$$

This shows that $\#(E_2 \backslash E_1) = 1 + L(\alpha; i, \alpha_{i+1}+1)$ and

$$\begin{aligned}
h(\alpha, t; i, \alpha_{i+1}+1) &= \kappa[r_\alpha(i+1) - r_\alpha(i) - 1] + t + \alpha_i - \alpha_{i+1} - 1, \\
h(\sigma\alpha, t; i+1, \alpha_{i+1}+1) &= \kappa[r_\alpha(i+1) - r_\alpha(i)] + t + \alpha_i - \alpha_{i+1} - 1.
\end{aligned}$$

Thus

$$\begin{aligned}
\frac{h(\sigma\alpha, \kappa+1; i+1, \alpha_{i+1}+1)}{h(\alpha, \kappa+1; i, \alpha_{i+1}+1)} &= \frac{\kappa[r_\alpha(i+1) - r_\alpha(i) + 1] + \alpha_i - \alpha_{i+1}}{\kappa(r_\alpha(i+1) - r_\alpha(i)) + \alpha_i - \alpha_{i+1}} \\
&= 1 + \frac{\kappa}{\kappa[r_\alpha(i+1) - r_\alpha(i)] + \alpha_i - \alpha_{i+1}} \\
&= \frac{\mathscr{E}_+(\sigma\alpha)}{\mathscr{E}_+(\alpha)}
\end{aligned}$$

(the last equation is proven in Theorem 10.4.8) and

$$\begin{aligned}
\frac{h(\sigma\alpha, 1; i+1, \alpha_{i+1}+1)}{h(\alpha, 1; i, \alpha_{i+1}+1)} &= \frac{\kappa[r_\alpha(i+1) - r_\alpha(i)] + \alpha_i - \alpha_{i+1}}{\kappa[r_\alpha(i+1) - r_\alpha(i) - 1] + \alpha_i - \alpha_{i+1}} \\
&= \left(1 - \frac{\kappa}{\kappa[r_\alpha(i+1) - r_\alpha(i)] + \alpha_i - \alpha_{i+1}}\right)^{-1} \\
&= \frac{\mathscr{E}_-(\alpha)}{\mathscr{E}_-(\sigma\alpha)}.
\end{aligned}$$

Thus $h(\alpha,\kappa+1)$ and $h(\alpha^+,\kappa+1)\mathscr{E}_+(\alpha)$ have the same transformation properties under adjacent transpositions and hence are equal. Similarly, $h(\alpha,1) = h(\alpha^+,1)/\mathscr{E}_-(\alpha)$. $\square$

Next we consider the ratio $h(\lambda,t)/h(\lambda-\varepsilon_m,t)$. Each cell above (m,λ_m) changes (the leg length decreases by 1) and each entry in row m changes. Thus

$$\frac{h(\lambda,t)}{h(\lambda-\varepsilon_m,t)} = \prod_{i=1}^{m-1} \frac{\lambda_i-\lambda_m+t+\kappa(m-i)}{\lambda_i-\lambda_m+t+\kappa(m-i-1)}\,(\lambda_m-1+t).$$

The following relates this to $\mathscr{E}_\varepsilon(\widetilde{\lambda})$.

Lemma 10.6.4 *Suppose that $\lambda\in\mathbb{N}_0^{d,P}$ and $\widetilde{\lambda}=(\lambda_m-1,\lambda_1,\ldots,\lambda_{m-1},0,\ldots)$ with $m=\ell(\lambda)$; then*

$$(\kappa+\lambda_m)\mathscr{E}_+(\widetilde{\lambda}) = \frac{h(\lambda,\kappa+1)}{h(\lambda-\varepsilon_m,\kappa+1)} \qquad \textit{and} \qquad \frac{1}{\lambda_m}\mathscr{E}_-(\widetilde{\lambda}) = \frac{h(\lambda-\varepsilon_m,1)}{h(\lambda,1)}.$$

Proof For the first equation,

$$\begin{aligned}\mathscr{E}_+(\widetilde{\lambda}) &= \prod_{i=1}^{m-1}\left(\frac{\kappa(m-i+1)+\lambda_i-\lambda_m+1}{\kappa(m-i)+\lambda_i-\lambda_m+1}\right)\\ &= \frac{h(\lambda,\kappa+1)}{h(\lambda-\varepsilon_m,\kappa+1)(\kappa+\lambda_m)}\end{aligned}$$

by use of the above formula for $h(\lambda,t)/h(\lambda-\varepsilon_m,t)$ with $t=\kappa+1$. For the second equation,

$$\mathscr{E}_-(\widetilde{\lambda}) = \prod_{i=1}^{m-1}\left(\frac{\kappa(m-i-1)+\lambda_i-\lambda_m+1}{\kappa(m-i)+\lambda_i-\lambda_m+1}\right) = \lambda_m\frac{h(\lambda-\varepsilon_m,1)}{h(\lambda,1)}$$

by the same formula with $t=1$. $\square$

The easiest consequence is the formula for $\zeta_\lambda(\mathbf{1}^d)$; the proof of Lemma 10.6.4 illustrates the ideas that will be used for the norm calculations.

Proposition 10.6.5 *For $\alpha\in\mathbb{N}_0^d$ and $\lambda=\alpha^+\in\mathbb{N}_0^{d,P}$,*

$$\zeta_\alpha(\mathbf{1}^d) = \mathscr{E}_-(\alpha)\frac{(d\kappa+1)_\lambda}{h(\lambda,1)} = \frac{(d\kappa+1)_\lambda}{h(\alpha,1)}.$$

Proof From Theorem 10.4.8 we have $\zeta_\alpha(\mathbf{1}^d)=\mathscr{E}_-(\alpha)\zeta_\lambda(\mathbf{1}^d)$. Suppose that $m=\ell(\lambda)$; then, by part (ii) of Lemma 10.5.2 and Theorem 10.5.5, $\zeta_\lambda(\mathbf{1}^d) = \lambda_m^{-1}\mathscr{D}_m\zeta_\lambda(\mathbf{1}^d) = \lambda_m^{-1}[\kappa(d-m+1)+\lambda_m]\mathscr{E}_-(\widetilde{\lambda})\zeta_{\lambda-\varepsilon_m}(\mathbf{1}^d)$; thus

$$\frac{\zeta_\lambda(\mathbf{1}^d)}{\zeta_{\lambda-\varepsilon_m}(\mathbf{1}^d)} = [\kappa(d-m+1)+\lambda_m]\frac{h(\lambda-\varepsilon_m,1)}{h(\lambda,1)}.$$

Clearly $\zeta_{\mathbf{0}}\left(\mathbf{1}^d\right)=1$ (the trivial $\mathbf{0}=(0,\ldots,0)\in\mathbb{N}_0^d$) and

$$\frac{(d\kappa+1)_\lambda}{(d\kappa+1)_{\lambda-\varepsilon_m}}=\kappa(d-m+1)+\lambda_m.$$

An obvious inductive argument finishes the proof. □

We now specialize to the permissible inner product $\langle f,g\rangle_h=f(\mathscr{D})\,g(x)\,|_{x=0}$. In this case $\mathscr{D}_m$ is the adjoint of multiplication by x_m, and we use part (iii) of Lemma 10.5.2.

Theorem 10.6.6 *For $\alpha\in\mathbb{N}_0^d$ and $\lambda=\alpha^+\in\mathbb{N}_0^{d,P}$,*

$$\langle\zeta_\alpha,\zeta_\alpha\rangle_h=\mathscr{E}_+(\alpha)\,\mathscr{E}_-(\alpha)\,(d\kappa+1)_\lambda\,\frac{h(\lambda,\kappa+1)}{h(\lambda,1)}.$$

Proof From Theorem 10.4.8, $\langle\zeta_\alpha,\zeta_\alpha\rangle_h=\mathscr{E}_+(\alpha)\,\mathscr{E}_-(\alpha)\,\langle\zeta_\lambda,\zeta_\lambda\rangle_h$. From the definition of the inner product, $\langle x_m\mathscr{D}_m\zeta_\lambda,\zeta_\lambda\rangle_h=\langle\mathscr{D}_m\zeta_\lambda,\mathscr{D}_m\zeta_\lambda\rangle_h$. Suppose that $m=\ell(\lambda)$; then, again by part (iii) of Lemma 10.5.2 and Theorem 10.5.5,

$$\begin{aligned}\langle\zeta_\lambda,\zeta_\lambda\rangle_h&=\frac{\kappa+\lambda_m}{\lambda_m\left[\kappa(d-m+1)+\lambda_m\right]}\langle\mathscr{D}_m\zeta_\lambda,\mathscr{D}_m\zeta_\lambda\rangle_h\\&=\frac{\kappa+\lambda_m}{\lambda_m}\left[\kappa(d-m+1)+\lambda_m\right]\left\langle\zeta_{\widetilde{\lambda}},\zeta_{\widetilde{\lambda}}\right\rangle_h\\&=\frac{\kappa+\lambda_m}{\lambda_m}\left[\kappa(d-m+1)+\lambda_m\right]\mathscr{E}_+(\widetilde{\lambda})\mathscr{E}_-(\widetilde{\lambda})\left\langle\zeta_{\lambda-\varepsilon_m},\zeta_{\lambda-\varepsilon_m}\right\rangle_h.\end{aligned}$$

Thus

$$\frac{\langle\zeta_\lambda,\zeta_\lambda\rangle_h}{\left\langle\zeta_{\lambda-\varepsilon_m},\zeta_{\lambda-\varepsilon_m}\right\rangle_h}=\frac{(d\kappa+1)_\lambda\,h(\lambda,\kappa+1)\,h(\lambda-\varepsilon_m,1)}{(d\kappa+1)_{\lambda-\varepsilon_m}\,h(\lambda-\varepsilon_m,\kappa+1)\,h(\lambda,1)}.$$

Induction finishes the proof. □

Corollary 10.6.7 *For $\alpha\in\mathbb{N}_0^d$,*

$$\langle\zeta_\alpha,\zeta_\alpha\rangle_h=(d\kappa+1)_{\alpha^+}\,\frac{h(\alpha,\kappa+1)}{h(\alpha,1)}.$$

10.6.2 The biorthogonal-type norm

Recall the generating function

$$F(x,y)=\sum_{\alpha\in\mathbb{N}_0^d}p_\alpha(x)\,y^\alpha=\prod_{i=1}^d(1-x_iy_i)^{-1}\prod_{j,k=1}^d(1-x_jy_k)^{\ \kappa}.$$

Because F is symmetric in x,y, the matrix A given by

$$\sum_{\alpha\in\mathbb{N}_0^d}p_\alpha(x)\,y^\alpha=\sum_{n=0}^\infty\sum_{|\alpha|=|\beta|=n}A_{\beta\alpha}x^\beta y^\alpha,$$

expressing $p_\alpha(x)$ in terms of x^β, is symmetric (consider the matrix A, which is locally finite, as a direct sum of ordinary matrices, one for each degree $n = |\alpha| = |\beta| \in \mathbb{N}$). The matrix is invertible and thus we can define a symmetric bilinear form on Π^d (in the following sum the coefficients have only finitely many nonzero values):

$$\left\langle \sum_{\alpha\in\mathbb{N}_0^d} a_\alpha x^\alpha, \sum_{\beta\in\mathbb{N}_0^d} b_\beta x^\beta \right\rangle_p = \sum_{\alpha,\beta} a_\alpha b_\beta \left(A^{-1}\right)_{\alpha\beta},$$

that is,

$$\left\langle p_\alpha, x^\beta \right\rangle_p = \delta_{\alpha,\beta}.$$

From the transformation properties of p_α it follows that $\langle wf, wg\rangle_p = \langle f,g\rangle_p$ for $f,g \in \Pi^d$ and $w \in S_d$. We showed that $\mathscr{D}_i^{(x)} x_i F(x,y) = \mathscr{D}_i^{(y)} y_i F(x,y)$, which implies that $\mathscr{D}_i^{(x)} x_i$ is self-adjoint with respect to the form defined above ($1 \le i \le d$). To see this, let $\mathscr{D}_i^{(x)} x_i x^\alpha = \sum_\gamma B_{\gamma\alpha} x^\gamma$; then $BA = AB^{\mathrm{T}}$ (T denotes the transpose). Thus $\langle\cdot,\cdot\rangle_p$ is a permissible inner product. To compute $\langle\zeta_\lambda,\zeta_\lambda\rangle_p$ we use the fact that $\langle x_i f, \rho_i g\rangle_p = \langle f,g\rangle_p$ for $f,g \in \Pi^d$, $1 \le i \le d$.

Lemma 10.6.8 *Suppose that $\lambda \in \mathbb{N}_0^{d,P}$ and $m = \ell(\lambda)$; then*

$$\rho_m \mathscr{D}_m \zeta_\lambda = [\kappa(d-m+1) + \lambda_m]\,\zeta_\lambda + f_0,$$

where $\mathscr{D}_m f_0 = 0$ and $\langle w\zeta_\lambda, f_0\rangle = 0$ for any $w \in S_d$.

Proof Let $f_0 = \rho_m \mathscr{D}_m \zeta_\lambda - [\kappa(d-m+1)+\lambda_m]\,\zeta_\lambda$; by part (ii) of Lemma 10.5.3, $\mathscr{D}_m f_0 = 0$. Further, the basis elements p_β which appear in the expansions of $\mathscr{D}_m \zeta_\lambda$ and ζ_λ satisfy $\beta_j = 0$ for all $j > m$. By Proposition 10.3.9, if p_β appears in f_0 then $\beta_j = 0$ for all $j > m$. The expansion of f_0 in the orthogonal basis $\{\zeta_\beta\}$ involves only compositions β with the same property, $\beta_j = 0$ for all $j > m$. Thus $\langle f_0, g\rangle = 0$ for any $g \in E_\lambda$ (note that $\lambda_m > 0$). Since E_λ is closed under the action of S_d, this shows that $\langle f_0, w\zeta_\lambda\rangle = 0$ for all $w \in S_d$. □

The statement in Lemma 10.6.8 applies to any permissible inner product.

Theorem 10.6.9 *For $\alpha \in \mathbb{N}^d$,*

$$\langle\zeta_\alpha,\zeta_\alpha\rangle_p = \frac{h(\alpha,\kappa+1)}{h(\alpha,1)}, \qquad \lambda = \alpha^+ \in \mathbb{N}_0^{d,P}.$$

Proof Let $\lambda = \alpha^+ \in \mathbb{N}_0^{d,P}$. From Theorem 10.4.8 we have

$$\langle\zeta_\alpha,\zeta_\alpha\rangle_p = \mathscr{E}_+(\alpha)\mathscr{E}_-(\alpha)\langle\zeta_\lambda,\zeta_\lambda\rangle_p.$$

From the definition of the inner product,

$$\langle x_m \mathscr{D}_m \zeta_\lambda, \rho_m \mathscr{D}_m \zeta_\lambda \rangle_p = \langle \mathscr{D}_m \zeta_\lambda, \mathscr{D}_m \zeta_\lambda \rangle_p.$$

Suppose that $m = \ell(\lambda)$; then, by parts (i) and (iii) of Lemma 10.5.2 and Lemma 10.6.8,

$$\begin{aligned}\langle \mathscr{D}_m \zeta_\lambda, \mathscr{D}_m \zeta_\lambda \rangle_p &= \langle x_m \mathscr{D}_m \zeta_\lambda, [\kappa(d-m+1)+\lambda_m]\,\zeta_\lambda + f_m \rangle_p \\ &= [\kappa(d-m+1)+\lambda_m]\,\langle x_m \mathscr{D}_m \zeta_\lambda, \zeta_\lambda \rangle_p \\ &\quad + [(d-m)\kappa + \lambda_m]\,\langle \zeta_\lambda, f_m \rangle_p - \kappa \sum_{j>m} \langle (m,j)\,\zeta_\lambda, f_m \rangle_p \\ &= \frac{\lambda_m}{\kappa+\lambda_m}\,[\kappa(d-m+1)+\lambda_m]^2\,\langle \zeta_\lambda, \zeta_\lambda \rangle_p.\end{aligned}$$

But, as before (by Theorem 10.5.5),

$$\begin{aligned}\langle \mathscr{D}_m \zeta_\lambda, \mathscr{D}_m \zeta_\lambda \rangle_p &= [\kappa(d-m+1)+\lambda_m]^2 \left\langle \zeta_{\widetilde{\lambda}}, \zeta_{\widetilde{\lambda}} \right\rangle_p \\ &= [\kappa(d-m+1)+\lambda_m]^2\, \mathscr{E}_+(\widetilde{\lambda})\mathscr{E}_-(\widetilde{\lambda}) \langle \zeta_{\lambda-\varepsilon_m}, \zeta_{\lambda-\varepsilon_m} \rangle_p\end{aligned}$$

and so

$$\frac{\langle \zeta_\lambda, \zeta_\lambda \rangle_p}{\left\langle \zeta_{\lambda-\varepsilon_m}, \zeta_{\lambda-\varepsilon_m} \right\rangle_p} = \frac{\kappa+\lambda_m}{\lambda_m}\mathscr{E}_+(\widetilde{\lambda})\mathscr{E}_-(\widetilde{\lambda}) = \frac{h(\lambda,\kappa+1)\,h(\lambda-\varepsilon_m,1)}{h(\lambda-\varepsilon_m,\kappa+1)\,h(\lambda,1)}.$$

As before, induction on $n = |\lambda|$ finishes the proof ($\langle 1,1 \rangle_p = 1$). □

Considering the generating function in Definition 10.3.2 as a reproducing kernel for $\langle \cdot,\cdot \rangle_p$ establishes the following.

Corollary 10.6.10 *For $\kappa > 0, \max_i |x_i| < 1, \max_i |y_i| < 1$, the following expansion holds:*

$$\prod_{i=1}^{d}(1-x_i y_i)^{-1} \prod_{j,k=1}^{d}(1-x_j y_k)^{-\kappa} = \sum_{\alpha\in\mathbb{N}_0^d} \frac{h(\alpha,1)}{h(\alpha,\kappa+1)}\zeta_\alpha(x)\zeta_\alpha(y).$$

Later we will consider the effect of this corollary on the determination of the intertwining operator.

10.6.3 The torus inner product

By considering the variables x_i as coordinates on the *complex d-torus*, one can define an L^2 structure which provides a permissible inner product. Algebraically this leads to 'constant-term' formulae: the problem is to determine the constant term in a Laurent polynomial (in $x_1, x_1^{-1}, x_2, x_2^{-1}, \ldots$) when $\kappa \in \mathbb{N}$.

Definition 10.6.11 The d-torus is given by

$$\mathbb{T}^d = \{x : x_j = \exp(\mathrm{i}\,\theta_j); -\pi < \theta_j \le \pi; 1 \le j \le d\}.$$

The standard measure m on $\mathbb{T}^d$ is

$$\mathrm{d}m(x) = (2\pi)^{-d}\,\mathrm{d}\theta_1 \cdots \mathrm{d}\theta_d$$

and the conjugation operator on polynomials (with real coefficients) is

$$g^*(x) = g\left(x_1^{-1}, x_2^{-1}, \ldots, x_d^{-1}\right), \qquad g \in \Pi^d.$$

The weight function is a power of

$$k(x) = \prod_{1 \le i < j \le d} (x_i - x_j)\left(x_i^{-1} - x_j^{-1}\right);$$

thus $k(x) \ge 0$ since $(x_i - x_j)\left(x_i^{-1} - x_j^{-1}\right) = \left[2 \sin \frac{1}{2}(\theta_i - \theta_j)\right]^2$ for $x \in \mathbb{T}^d$. For now we do not give the explicit value of the normalizing constant c_κ, where $c_\kappa^{-1} = \int_{\mathbb{T}^d} k^\kappa \,\mathrm{d}m$. The associated inner product is

$$\langle f, g \rangle_{\mathbb{T}} = c_\kappa \int_{\mathbb{T}^d} f g^* k^\kappa \,\mathrm{d}m, \qquad f, g \in \Pi^d.$$

For real polynomials $\langle f, g \rangle_{\mathbb{T}} = \langle g, f \rangle_{\mathbb{T}}$, since $k^\kappa \,\mathrm{d}m$ is invariant under inversion, that is, $x_j \to x_j^{-1}$ for all j. Also, $\langle wf, wg \rangle_{\mathbb{T}} = \langle f, g \rangle_{\mathbb{T}}$ for each $w \in S_d$ by the invariance of k.

The fact that $k(x)$ is homogeneous of degree 0 implies that homogeneous polynomials of different degrees are orthogonal for $\langle \cdot, \cdot \rangle_{\mathbb{T}}$, when $\kappa > 0$. The next theorem completes the proof that $\langle \cdot, \cdot \rangle_{\mathbb{T}}$ is a permissible inner product. Write

$$\partial_i := \frac{\partial}{\partial x_i} \qquad \text{and} \qquad \delta_i := \sum_{j \ne i} \frac{1}{x_i - x_j}[1 - (i, j)]$$

in the next proof.

Theorem 10.6.12 *For $\kappa > 0$, polynomials $f, g \in \Pi^d$ and $1 \le i \le d$,*

$$\langle \mathscr{D}_i x_i f, g \rangle_{\mathbb{T}} = \langle f, \mathscr{D}_i x_i g \rangle_{\mathbb{T}};$$

if f, g are homogeneous of different degrees then $\langle f, g \rangle_{\mathbb{T}} = 0$.

Proof Since $(\partial/\partial\theta_i) f(x) = \mathrm{i}\exp(\mathrm{i}\theta_i)\,\partial_i f(x) = \mathrm{i}x_i \partial_i f(x)$, integration by parts shows that $\int_{\mathbb{T}^d} x_i \partial_i h(x)\,\mathrm{d}m(x) = 0$ for any periodic differentiable h. Also, note that $x_i \partial_i g^*(x) = -x_i^{-1}(\partial_i g)^* = -(x_i \partial_i g)^*$. Thus the Euler operator $\psi = \sum_{i=1}^d x_i \partial_i$ is self-adjoint for $\langle \cdot, \cdot \rangle_{\mathbb{T}}$ (observe that $\psi(k^\kappa) = \kappa k^{\kappa-1} \psi k = 0$ for any $\kappa > 0$). Suppose that f, g are homogeneous of degrees m, n, respectively; then $m \langle f, g \rangle_{\mathbb{T}} = \langle \psi f, g \rangle_{\mathbb{T}} = \langle \psi f, \psi g \rangle_{\mathbb{T}} = n \langle f, g \rangle_{\mathbb{T}}$; so $m \ne n$ implies $\langle f, g \rangle_{\mathbb{T}} = 0$. Next,

$$\int_{\mathbb{T}^d} \left(\mathscr{D}_i x_i f(x)\, g^*(x) - f(x)\left[\mathscr{D}_i x_i g(x)\right]^*\right) k(x)^\kappa \, \mathrm{d}m(x)$$
$$= \int_{\mathbb{T}^d} \left(f g^* + x_i (\partial_i f)\, g^* - f g^* + x_i f \partial_i g^* + x_i f g^* \kappa \frac{1}{k}\partial_i k\right) k^\kappa \, \mathrm{d}m(x)$$
$$+ \kappa \int_{\mathbb{T}^d} \left(-x_i f g^* \frac{1}{k}\partial_i k + (\delta_i x_i f)\, g^* - f\,(\delta_i x_i g)^*\right) k^\kappa \, \mathrm{d}m(x).$$

This equation results from integration by parts, and the first part is clearly zero. The integrand of the second part equals

$$\kappa \sum_{j \neq i} \Big[- f g^* x_i \Big(\frac{1}{x_i - x_j} + \frac{-x_i^{-2}}{x_i^{-1} - x_j^{-1}} \Big)$$
$$+ \frac{x_i f - x_j (i,j) f}{x_i - x_j} g^* - f \frac{x_i^{-1} g^* - x_j^{-1} (i,j) g^*}{x_i^{-1} - x_j^{-1}} \Big]$$
$$= \kappa \sum_{j \neq i} \frac{1}{x_i - x_j} \big[- f g^* (x_i + x_j) + x_i f g^* - x_j g^* (i,j) f$$
$$+ x_j f g^* - x_i f (i,j) g^* \big].$$

For each $j \neq i$ the corresponding term is skew under the transposition (i,j) and the measure $k^\kappa \, \mathrm{d}m$ is invariant, and so the second part of the integral is also zero, proving the theorem. □

Theorem 10.6.13 *For $\alpha \in \mathbb{N}_0^d$,*

$$\langle \zeta_\alpha, \zeta_\alpha \rangle_{\mathbb{T}} = \frac{(d\kappa + 1)_\lambda \, h(\alpha, \kappa + 1)}{((d-1)\kappa + 1)_\lambda \, h(\alpha, 1)}.$$

Proof From Theorem 10.4.8 we have $\langle \zeta_\alpha, \zeta_\alpha \rangle_{\mathbb{T}} = \mathscr{E}_+(\alpha)\mathscr{E}_-(\alpha) \langle \zeta_\lambda, \zeta_\lambda \rangle_{\mathbb{T}}$. By the definition of the inner product $\langle x_m \mathscr{D}_m \zeta_\lambda, x_m \mathscr{D}_m \zeta_\lambda \rangle_{\mathbb{T}} = \langle \mathscr{D}_m \zeta_\lambda, \mathscr{D}_m \zeta_\lambda \rangle_{\mathbb{T}}$ (that is, multiplication by x_i is an isometry). Suppose that $m = \ell(\lambda)$; then by part (iv) of Lemma 10.5.2 and Theorem 10.5.5

$$\langle \zeta_\lambda, \zeta_\lambda \rangle_{\mathbb{T}} = \frac{\kappa + \lambda_m}{\lambda_m [\kappa(d - m + 1) + \lambda_m][\kappa(d-m) + \lambda_m]} \langle \mathscr{D}_m \zeta_\lambda, \mathscr{D}_m \zeta_\lambda \rangle_{\mathbb{T}}$$
$$= \frac{(\kappa + \lambda_m)\,[\kappa(d-m+1) + \lambda_m]}{\lambda_m\,[\kappa(d-m) + \lambda_m]} \left\langle \zeta_{\widetilde{\lambda}}, \zeta_{\widetilde{\lambda}} \right\rangle_{\mathbb{T}}$$
$$= \frac{(\kappa + \lambda_m)\,[\kappa(d-m+1) + \lambda_m]}{\lambda_m\,[\kappa(d-m) + \lambda_m]} \mathscr{E}_+(\widetilde{\lambda})\mathscr{E}_-(\widetilde{\lambda}) \langle \zeta_{\lambda - \varepsilon_m}, \zeta_{\lambda - \varepsilon_m} \rangle_{\mathbb{T}}.$$

Thus

$$\frac{\langle \zeta_\lambda, \zeta_\lambda \rangle_{\mathbb{T}}}{\left\langle \zeta_{\lambda - \varepsilon_m}, \zeta_{\lambda - \varepsilon_m} \right\rangle_{\mathbb{T}}} = \frac{(d\kappa + 1)_\lambda \, ((d-1)\kappa + 1)_{\lambda - \varepsilon_m} \, h(\lambda, \kappa + 1)\, h(\lambda - \varepsilon_m, 1)}{(d\kappa + 1)_{\lambda - \varepsilon_m} \, ((d-1)\kappa + 1)_\lambda \, h(\lambda - \varepsilon_m, \kappa + 1)\, h(\lambda, 1)}.$$

Induction on $n = |\lambda|$ finishes the proof. □

10.6.4 Monic polynomials

Another normalization of nonsymmetric Jack polynomials is such that they are monic in the x-basis; the polynomials are then denoted by ζ^x_α. For $\alpha \in \mathbb{N}_0^d$,

$$\zeta^x_\alpha(x) = x^\alpha + \sum\{B(\beta,\alpha)x^\beta : \alpha \rhd \beta\},$$

with coefficients in $\mathbb{Q}(\kappa)$. Note that the triangularity in the x-basis is in the opposite direction to that in the p-basis. Since this polynomial has the same eigenvalues for $\{\mathscr{U}_i\}$ as ζ_α, it must be a scalar multiple of the latter. The value of this scalar multiple can be obtained from the fact that $\langle\zeta_\alpha, \zeta^x_\alpha\rangle_p = 1$ (since $\langle p_\alpha, x^\alpha\rangle_p = 1$ and the other terms appearing in $\zeta_\alpha, \zeta^x_\alpha$ are pairwise orthogonal, owing to the $\rhd$-ordering). Let $\zeta^x_\alpha = c_\alpha\zeta_\alpha$; then $1 = c_\alpha\langle\zeta_\alpha, \zeta_\alpha\rangle_p$ and thus

$$c_\alpha = \frac{h(\alpha,1)}{h(\alpha,\kappa+1)}.$$

The formulae for norms and value at $\mathbf{1}^d$ imply the following (where $\lambda = \alpha^+$):

1. $\zeta^x_\alpha(\mathbf{1}^d) = \dfrac{(d\kappa+1)_\lambda}{h(\alpha,\kappa+1)}$;
2. $\langle\zeta^x_\alpha, \zeta^x_\alpha\rangle_h = (d\kappa+1)_\lambda \dfrac{h(\alpha,1)}{h(\alpha,\kappa+1)}$;
3. $\langle\zeta^x_\alpha, \zeta^x_\alpha\rangle_{\mathbb{T}} = \dfrac{(d\kappa+1)_\lambda}{((d-1)\kappa+1)_\lambda}\dfrac{h(\alpha,1)}{h(\alpha,\kappa+1)}$;
4. $\langle\zeta^x_\alpha, \zeta^x_\alpha\rangle_p = \dfrac{h(\alpha,1)}{h(\alpha,\kappa+1)}$.

10.6.5 Normalizing constants

The alternating polynomial for S_d, namely

$$a(x) = \prod_{1\le i<j\le d}(x_i - x_j),$$

has several important properties relevant to these calculations. For an inner product related to an L^2 structure, the value of $\langle a,a\rangle$ shows the effect of changing κ to $\kappa+1$. Also, a is h-harmonic because $\Delta_h a$ is a skew polynomial of degree $\binom{d}{2} - 2$, hence 0. Let

$$\delta = (d-1, d-2, \ldots, 1, 0) \in \mathbb{N}_0^{d,P};$$

then a is the unique skew element of E_δ (and $a = a_\delta$, cf. Definition 10.4.9). The previous results will thus give the values of $\langle a,a\rangle$ in the permissible inner products.

Theorem 10.6.14 *The alternating polynomial satisfies*

$$a(x) = \sum_{w\in S_d}\operatorname{sign}(w)x^{w\delta} = \sum_{w\in S_d}\operatorname{sign}(w)p_{w\delta} = \sum_{w\in S_d}\operatorname{sign}(w)\omega_{w\delta}.$$

Proof The first equation for a is nothing other than the Vandermonde determinant. By Proposition 10.4.4, $\omega_\delta = p_\delta + \sum\{B(\beta,\delta)\, p_\beta : \beta^+ \succ \delta; |\beta| = |\delta|\}$. But $\beta^+ \succ \delta$ and $|\beta| = |\delta| = \binom{d}{2}$ implies that $\beta_j = \beta_k$ for some $j \neq k$; in turn this implies that $\sum_{w\in S_d} \operatorname{sign}(w) p_{w\beta} = 0$. Thus $\sum_{w\in S_d} \operatorname{sign}(w)\omega_{w\delta} = \sum_{w\in S_d} \operatorname{sign}(w) p_{w\delta}$. To compute this we use a determinant formula of Cauchy: for $1 \le i,j \le d$ let $A_{ij} = (1 - x_i y_j)^{-1}$; then

$$\det A = \frac{a(x)\,a(y)}{\prod_{i,j=1}^{d}(1-x_i y_j)}.$$

To prove this, use row operations to put zeros in the first column (below the first row): let $B_{1j} = A_{1j}$ and $B_{ij} = A_{ij} - A_{i1}A_{1j}/A_{11}$ for $2 \le i \le d$ and $1 \le j \le d$; thus $B_{i1} = 0$ and

$$B_{ij} = \frac{(x_1 - x_i)(y_1 - y_j)}{(1 - x_1 y_j)(1 - x_i y_1)(1 - x_i y_j)}.$$

Extracting the common factors from each row and column shows that

$$\det A = \det B = \frac{\prod_{i=2}^{d}(x_1 - x_i)\prod_{j=2}^{d}(y_1 - y_j)}{(1 - x_1 y_1)\prod_{i=2}^{d}(1 - x_i y_1)\prod_{j=2}^{d}(1 - x_1 y_j)} \det(A_{ij})_{i,j=2}^{d},$$

which proves the formula inductively.

Now apply the skew operator to the generating function for $\{p_\alpha\}$:

$$\begin{aligned}
&\sum_{w\in S_d} \operatorname{sign}(w) \sum_{\alpha\in\mathbb{N}_0^d} p_{w\alpha}(x)\, y^\alpha \\
&\quad = \sum_{w\in S_d} \operatorname{sign}(w) \prod_{i=1}^{d} [1 - (wx)_i y_i]^{-1} \prod_{j,k=1}^{d} (1 - x_j y_k)^{-\kappa} \\
&\quad = a(x)\,a(y) \prod_{j,k=1}^{d} (1 - x_j y_k)^{-\kappa - 1}.
\end{aligned}$$

This holds by the Cauchy formula, since the second part of the product is S_d-invariant. The coefficient of y^δ in $a(y)$ is 1 since the rest of the product does not contribute to this exponent, and thus $\sum_{w\in S_d} \operatorname{sign}(w) p_{w\delta}(x) = a(x)$. □

Proposition 10.6.15 *In the pairing norm,* $\langle a,a\rangle_p = d!$*; also,*

$$\mathscr{E}_-(\delta^R)\, h(\delta,\kappa+1)/h(\delta,1) = 1.$$

Proof By the norm formula in Theorem 10.4.11,

$$\langle a,a\rangle_p = d!\,\mathscr{E}_-(\delta^R)\,\langle \zeta_\delta, \zeta_\delta\rangle_p = d!\,\mathscr{E}_-(\delta^R) h(\delta,\kappa+1)/h(\delta,1);$$

but $\langle a,a\rangle_p = \sum_{w_1,w_2\in S_d} \operatorname{sign}(w_1 w_2) \left\langle p_{w_1\delta}(x), x^{w_2\delta}\right\rangle_p = d!$, by definition. □

The second part of the proposition can be proved by a direct calculation showing that

$$\mathscr{E}_-\left(\delta^R\right) = \prod_{1\le i<j\le d} \frac{(j-i-1)\,\kappa+(j-i)}{(j-i)\,(\kappa+1)} = \frac{h(\delta,1)}{h(\delta,\kappa+1)}.$$

Theorem 10.6.16 *The torus norm for a is* $\langle a,a\rangle_{\mathbb{T}} = d!\dfrac{(d\kappa+1)_\delta}{((d-1)\,\kappa+1)_\delta}$, *and for* $\kappa>0$ *the normalizing constant is*

$$c_\kappa = \left(\int_{\mathbb{T}^d} k^\kappa\,\mathrm{d}m\right)^{-1} = \frac{\Gamma(\kappa+1)^d}{\Gamma(\kappa d+1)}$$

for $\kappa\in\mathbb{N}, c_\kappa = (\kappa!)^d/(\kappa d)!$.

Proof By the norm formula and Proposition 10.6.15,

$$\langle a,a\rangle_{\mathbb{T}} = d!\mathscr{E}_-\left(\delta^R\right)\frac{(d\kappa+1)_\delta}{((d-1)\,\kappa+1)_\delta}\frac{h(\delta,\kappa+1)}{h(\delta,1)} = d!\frac{(d\kappa+1)_\delta}{((d-1)\,\kappa+1)_\delta}.$$

Further, $c_{\kappa+1}^{-1} = \int_{\mathbb{T}^d} aa^* k^\kappa\,\mathrm{d}m = c_\kappa^{-1}\langle a,a\rangle_{\mathbb{T}}$ and so

$$\frac{c_\kappa}{c_{\kappa+1}} = d!\frac{(d\kappa+1)_\delta}{((d-1)\,\kappa+1)_\delta} = \frac{(d\kappa+1)_d}{(\kappa+1)^d}.$$

Analytic function theory and the asymptotics for the gamma function are used to complete the proof. Since $k\ge 0$, the function

$$\phi(\kappa) = \frac{\Gamma(\kappa+1)^d}{\Gamma(\kappa d+1)}\int_{\mathbb{T}^d} k^\kappa\,\mathrm{d}m$$

is analytic on $\{\kappa : \operatorname{Re}\kappa > 0\}$; additionally, the formula for $c_\kappa/c_{\kappa+1}$ shows that $\phi(\kappa+1) = \phi(\kappa)$. Analytic continuation shows that ϕ is the restriction of an entire function that is periodic and of period 1. We will show that $\phi(\kappa) = O\left(|\kappa|^{(d-1)/2}\right)$, implying that ϕ is a polynomial and periodic, hence constant; of course, $\phi(0)=1$. Restrict κ to a strip, say $\{\kappa : 1\le\operatorname{Re}\kappa\le 2\}$. On this strip $|\int_{\mathbb{T}^d} k^\kappa\,\mathrm{d}m| \le \int_{\mathbb{T}^d} k^{\operatorname{Re}\kappa}\,\mathrm{d}m \le M_1$ for some $M_1>0$. By the Gauss multiplication formula, we have

$$\frac{\Gamma(\kappa+1)^d}{\Gamma(\kappa d+1)} = (2\pi)^{(d-1)/2} d^{-d\kappa-1/2}\prod_{j=1}^{d-1}\frac{\Gamma(\kappa+1)}{\Gamma(\kappa+j/d)}.$$

The asymptotic formula $\Gamma(\kappa+a)/\Gamma(\kappa+b) = \kappa^{a-b}\left(1+\mathscr{O}(\kappa^{-1})\right)$, valid on $\operatorname{Re}\kappa\ge 1$, applied to each factor yields

$$\left|\frac{\Gamma(\kappa+1)^d}{\Gamma(\kappa d+1)}\right| = M_2\left|d^{-d\kappa}\right||\kappa|^{(d-1)/2}\left[1+O\left(\frac{1}{\kappa}\right)\right].$$

Extend this bound by periodicity. Hence $\phi=1$. □

Recall that, for $f,g \in \Pi^d$ (Theorem 7.2.7),

$$\langle f,g\rangle_h = (2\pi)^{-d/2} b_\kappa \int_{\mathbb{R}^d} \left(\mathrm{e}^{-\Delta_h/2} f\right)\left(\mathrm{e}^{-\Delta_h/2} g\right) |a(x)|^{2\kappa} \mathrm{e}^{-\|x\|^2/2}\,\mathrm{d}x,$$

where the normalizing constant b_κ satisfies $\langle 1,1\rangle_h = 1$. Since $\mathrm{e}^{-\Delta_h/2} a = a$, the following holds.

Theorem 10.6.17 *For $\kappa > 0$,*

$$\langle a,a\rangle_h = d!\,(d\kappa+1)_\delta \qquad \textit{and} \qquad b_\kappa = \prod_{j=2}^{d} \frac{\Gamma(\kappa+1)}{\Gamma(j\kappa+1)}.$$

Proof Using the method of the previous proof,

$$\langle a,a\rangle_h = d!\mathscr{E}_-(\delta^R) \frac{(d\kappa+1)_\delta\, h(\delta,\kappa+1)}{h(\delta,1)} = d!\,(d\kappa+1)_\delta.$$

Further,

$$b_{\kappa+1}^{-1} = (2\pi)^{-d/2} \int_{\mathbb{R}^d} \prod_{i<j} |x_i - x_j|^{2\kappa+2} \mathrm{e}^{-\|x\|^2/2}\,\mathrm{d}x = b_\kappa^{-1} \langle a,a\rangle_h.$$

Also, $d!\,(d\kappa+1)_\delta = d!\prod_{j=2}^d (j\kappa+1)_{j-1} = (\kappa+1)^{1-d} \prod_{j=2}^d (j\kappa+1)_j$; thus

$$\prod_{j=2}^{d} \frac{\Gamma(j\kappa+1)}{\Gamma(\kappa+1)} d!\,(d\kappa+1)_\delta = \prod_{j=2}^{d} \frac{\Gamma(j\kappa+j+1)}{\Gamma(\kappa+2)}.$$

We will use the same technique as for the torus, first converting the integral to one over the sphere S^{d-1}, a compact set. Let $\gamma = \frac{1}{2}\kappa d(d-1)$, the degree of h. In spherical polar coordinates,

$$\begin{aligned}(2\pi)^{-d/2} &\int_{\mathbb{R}^d} \prod_{i<j} |x_i - x_j|^{2\kappa} \mathrm{e}^{-\|x\|^2/2}\,\mathrm{d}x \\ &= 2^\gamma \sigma_d \frac{\Gamma\left(\frac{d}{2}+\gamma\right)}{\Gamma\left(\frac{d}{2}\right)} \int_{S^{d-1}} \prod_{i<j} |x_i - x_j|^{2\kappa} \mathrm{d}\omega(x);\end{aligned}$$

the normalizing constant (see (4.1.3)) satisfies $\sigma_{d-1} = \int_{S^{d-1}} \mathrm{d}\omega = 1$. Define the analytic function

$$\begin{aligned}\phi(\kappa) = \Gamma\left(\tfrac{d}{2} + \tfrac{d(d-1)}{2}\kappa\right) \prod_{j=2}^{d} \Bigg[&\frac{\Gamma(\kappa+1)}{\Gamma(j\kappa+1)} \frac{2^{\kappa d(d-1)/2}}{\Gamma(\frac{d}{2})\sigma_{d-1}} \\ &\times \int_{S^{d-1}} \prod_{i<j} |x_i - x_j|^{2\kappa} \mathrm{d}\omega(x) \Bigg].\end{aligned}$$

The formula for $b_\kappa / b_{\kappa+1}$ shows that $\phi(\kappa+1) = \phi(\kappa)$. By analytic continuation, ϕ is entire and periodic. On the strip $\{\kappa : 1 \le \operatorname{Re}\kappa \le 2\}$ the factors of ϕ excluding the gamma functions in κ are uniformly bounded. From the asymptotic formulae (valid in $-\pi < \arg z < \pi$ and $a, b > 0$),

$$\log\Gamma(z) = \left(z - \tfrac{1}{2}\right)\log z - z + \tfrac{1}{2}\log(2\pi) + \mathscr{O}\left(\frac{1}{|z|}\right),$$

$$\log\Gamma(az+b) = \left(az + b - \tfrac{1}{2}\right)(\log z + \log a) - az + \tfrac{1}{2}\log(2\pi) + \mathscr{O}\left(\frac{1}{|z|}\right);$$

the second formula is obtained from the first by the use of $\log(az+b) = \log az + b/(az) + \mathscr{O}\left(|z|^{-2}\right)$ (see Lebedev [1972], p. 10). Apply this to the first part of $\log\phi$:

$$\begin{aligned}
&\log\Gamma\left(\tfrac{d}{2} + \tfrac{d(d-1)}{2}\kappa\right) + (d-1)\log\Gamma(\kappa+1) - \sum_{j=2}^{d}\log\Gamma(j\kappa+1) \\
&= \tfrac{d-1}{2}(1 + d\kappa)\log\tfrac{d(d-1)}{2} - \sum_{j=2}^{d}\left(j\kappa + \tfrac{1}{2}\right)\log j + \tfrac{1}{2}\log(2\pi) \\
&\quad + \left(\tfrac{d-1}{2} + \tfrac{d(d-1)}{2}\kappa + (d-1)\left(\kappa + \tfrac{1}{2}\right) - \sum_{j=2}^{d}\left(j\kappa + \tfrac{1}{2}\right)\right)\log\kappa \\
&\quad + \left(-\tfrac{d(d-1)}{2} - (d-1) + \sum_{j=2}^{d} j\right)\kappa + O\left(|\kappa|^{-1}\right) \\
&= C_1 + C_2\kappa + \tfrac{d-1}{2}\log\kappa + O\left(|\kappa|^{-1}\right),
\end{aligned}$$

where C_1 and C_2 are real constants depending only on d. Consequently, $\phi(\kappa) = O\left(|\kappa|^{(d-1)/2}\right)$ in the strip $\{\kappa : 1 \le \operatorname{Re}\kappa \le 2\}$. By periodicity the same bound applies for all κ, implying that ϕ is a periodic polynomial and hence constant. This proves the formula for b_κ. □

10.7 Symmetric Functions and Jack Polynomials

In the present context, a symmetric function is an element of the abstract ring of polynomials in infinitely many indeterminates $e_0 = 1, e_1, e_2, \ldots$ with coefficients in some field (an extension of $\mathbb{Q}$). Specialized to $\mathbb{R}^d$, the generators e_i become the elementary symmetric functions of degree i (and $e_j = 0$ for $j > d$). For fixed $\alpha \in \mathbb{N}_0^d$ a polynomial p_α can be expressed as a polynomial in $x_1, \ldots, x_d$ and the indeterminates e_i (involving an arbitrary unbounded number of variables) as follows. Let $n \ge \max_i \alpha_i$; then p_α is the coefficient of y^α in

$$\prod_{i=1}^{d}(1 - x_i y_i)^{-1}\prod_{j=1}^{d}\left[1 - e_1 y_j + e_2 y_j^2 + \cdots + (-1)^n e_n y_j^n\right]^{-\kappa}.$$

When e_i is evaluated at $x \in \mathbb{R}^d$ the original formula is recovered. As remarked before, the polynomials ζ_β have expressions in the p-basis with coefficients independent of the number d of underlying variables provided that $d \geq m$, where $\beta_j = 0$ for any $j > m$. Thus the mixed $\{x_j, e_i\}$ expression for p_α provides a similar one for ζ_β which is now meaningful for any number ($\geq m$) of variables.

The Jack polynomials are parametrized by $1/\kappa$ and the original formulae (Stanley [1989]) use two hook-length products h^*, h_*, called upper and lower, respectively. In terms of our notation:

Definition 10.7.1 For a partition $\lambda \in \mathbb{N}_0^{d,P}$, with $m = \ell(\lambda)$ and parameter κ, the upper and lower hook-length products are

$$h^*(\lambda) = \kappa^{-|\lambda|} h(\lambda, 1) = \prod_{i=1}^{m} \prod_{j=1}^{\lambda_i} \left[\kappa^{-1}(\lambda_i - j + 1) + \#\{k : k > i; j \leq \lambda_k\}\right],$$

$$h_*(\lambda) = \kappa^{-|\lambda|} h(\lambda, \kappa) = \prod_{i=1}^{m} \prod_{j=1}^{\lambda_i} \left[\kappa^{-1}(\lambda_i - j) + 1 + \#\{k : k > i; j \leq \lambda_k\}\right].$$

There is a relationship between $h\left(\lambda^R, \kappa+1\right)$ and $h(\lambda, \kappa)$, which is needed in the application of our formulae. The quantity $\#S_d(\lambda) = \#\{\alpha : \alpha^+ = \lambda\} = \dim E_\lambda$ appears because of the symmetrization.

Let $m_j(\lambda) = \#\{i : 1 \leq i \leq d, \lambda_i = j\}$ (zero for $j > \lambda_1$) for $j \geq 0$. Then $\#S_d(\lambda) = d!/\prod_{j=0}^{\lambda_1} m_j(\lambda)!$.

Lemma 10.7.2 *For a partition* $\lambda \in \mathbb{N}_0^{d,P}$,

$$h\left(\lambda^R, \kappa+1\right) = \#S_d(\lambda) \frac{(d\kappa+1)_\lambda}{(d\kappa)_\lambda} h(\lambda, \kappa).$$

Proof We will evaluate the ratios $(d\kappa+1)_\lambda / (d\kappa)_\lambda$ and $h\left(\lambda^R, \kappa+1\right)/h(\lambda, \kappa)$. Clearly,

$$\begin{aligned}\frac{(d\kappa+1)_\lambda}{(d\kappa)_\lambda} &= \frac{1}{\kappa^d d!} \prod\nolimits_{i=1}^{d} \left[(d+1-i)\kappa + \lambda_i\right] \\ &= \frac{[d - \ell(\lambda)]!}{\kappa^{\ell(d)} d!} \prod\nolimits_{i=1}^{\ell(\lambda)} \left[(d+1-i)\kappa + \lambda_i\right].\end{aligned}$$

If the multiplicity of a particular λ_i is 1 (in λ) then

$$L\left(\lambda^R; d+1-i, j+1\right) = L(\lambda; i, j)$$

for $1 \leq j < \lambda_i$. To account for multiplicities, let w be the inverse of r_{λ^R}, that is, $r_{\lambda^R}(w(i)) = i$ for $1 \leq i \leq d$. Then $\lambda^R_{w(i)} = \lambda_i$ (for example, suppose that $\lambda_1 > \lambda_2 = \lambda_3 > \lambda_4$; then $w(1) = d, w(2) = d-2, w(3) = d-1$). As in the multiplicity-1 case we have

$$L\left(\lambda^R; w(i), j+1\right) = L(\lambda; i, j), \qquad 1 \leq j < \lambda_i.$$

Then $h\left(\lambda^R, \kappa+1; w(i), j+1\right) = [\lambda_i - j + \kappa(L(\lambda; i, j) + 1)]$, which is the same as $h(\lambda, \kappa; i, j)$ for $1 \le j < \lambda_i$. Thus

$$\frac{h\left(\lambda^R, \kappa+1\right)}{h(\lambda, \kappa)} = \prod_{i=1}^{\ell(\lambda)} \frac{h\left(\lambda^R, \kappa+1; w(i), 1\right)}{h(\lambda, \kappa; i, \lambda_i)}.$$

By the construction of w, $h\left(\lambda^R, \kappa+1; w(i), 1\right) = (\lambda_i + \kappa(d+1-i))$. Suppose that λ_i has multiplicity m (that is, $\lambda_{i-1} > \lambda_i = \cdots = \lambda_{i+m-1} > \lambda_{i+m}$); it then follows that $L(\lambda; i+l-1, \lambda_i) = m - l$ for $1 \le l \le m$ and

$$\prod_{l=1}^{m} h(\lambda, \kappa; i+l-1, \lambda_i) = \prod_{l=1}^{m} [(m-l+1)\kappa] = \kappa^m m!.$$

Thus

$$\frac{h\left(\lambda^R, \kappa+1\right)}{h(\lambda, \kappa)} = \frac{\prod_{i=1}^{\ell(\lambda)} [\lambda_i + \kappa(d+1-i)]}{\kappa^{\ell(\lambda)} \prod_{j=1}^{\lambda_1} m_\lambda(j)!}.$$

Combining the ratios, we find that

$$h\left(\lambda^R, \kappa+1\right) = \frac{(d\kappa+1)_\lambda}{(d\kappa)_\lambda} \frac{d!}{[d-\ell(\lambda)]! \prod_{j=1}^{\lambda_1} m_\lambda(j)!} h(\lambda, \kappa).$$

Since $m_\lambda(0) = d - \ell(\lambda)$, this completes the proof. □

There is another way to express the commuting operators, formulated by Cherednik [1991] for $f \in \Pi^d$:

$$X_i f = \frac{1}{\kappa} x_i \frac{\partial f}{\partial x_i} + x_i \sum_{j<i} \frac{f - (i,j)f}{x_i - x_j} + \sum_{j>i} \frac{x_j [f - (i,j)f]}{x_i - x_j} + (1-i)f;$$

it is easy to check that $\kappa X_i f - \mathscr{D}_i x_i f = -\kappa \sum_{j<i} (i,j)f - [1 + \kappa(d-1)]f$, so that $\kappa X_i f = \mathscr{U}_i f - (1+\kappa d)f$. Another definition of nonsymmetric Jack polynomials is as the simultaneous eigenfunctions of $\{X_i\}$, normalized to be monic in the x-basis. Another notation for these in the literature is $E_\alpha\left(x; \kappa^{-1}\right)$; this is the same polynomial as ζ_α^x.

Previously, we defined symmetric (that is, S_d-invariant) elements j_λ of the spaces E_λ for each $\lambda \in \mathbb{N}_0^{d,P}$. An alternative approach to j_λ is to regard them as the simultaneous eigenfunctions of d algebraically independent symmetric differential operators. These eigenfunctions can be taken as restrictions of elementary symmetric functions of $\{\mathscr{U}_i\}$; subsequently to the definition we will prove their S_d-invariance.

Definition 10.7.3 For $0 \le i \le d$, let $\mathscr{T}_i$ be the linear operator on Π^d defined by (for indeterminate t)

$$\sum_{i=0}^{d} t^i \mathscr{T}_i = \prod_{j=1}^{d} (1 + t\mathscr{U}_j).$$

The definition is unambiguous since $\{\mathscr{U}_i\}$ is a commutative set.

Theorem 10.7.4 *For $1 \le i \le d$ and $w \in S_d$, $w\mathcal{T}_i = \mathcal{T}_i w$. Further, $\mathcal{T}_i$ coincides with a differential operator when restricted to Π^W (the symmetric polynomials).*

Proof It suffices to show that $(j,j+1)\mathcal{T}_i = \mathcal{T}_i(j,j+1)$ for $1 \le j < d$. Let $\sigma = (j,j+1)$. By Lemma 10.4.1 $\sigma\mathcal{U}_k = \mathcal{U}_k\sigma$ for each $k \ne j, j+1$, so it remains to prove that $\sigma(1+t\mathcal{U}_j)(1+t\mathcal{U}_{j+1})\sigma = (1+t\mathcal{U}_j)(1+t\mathcal{U}_{j+1})$. Again by Lemma 10.4.1, $\sigma(\mathcal{U}_j+\mathcal{U}_{j+1})\sigma = (\mathcal{U}_{j+1}+\kappa\sigma)+(\mathcal{U}_j-\kappa\sigma)$, implying that $\mathcal{U}_j+\mathcal{U}_{j+1}$ commutes with σ. Also, $\sigma\mathcal{U}_j\mathcal{U}_{j+1} = (\mathcal{U}_{j+1}\sigma+\kappa)\mathcal{U}_{j+1} = \mathcal{U}_{j+1}(\mathcal{U}_j\sigma-\kappa)+\kappa\mathcal{U}_{j+1} = \mathcal{U}_j\mathcal{U}_{j+1}\sigma$. Thus $\sigma\mathcal{T}_i = \mathcal{T}_i\sigma$.

This shows that $q \in \Pi^W$ implies $\mathcal{T}_i q \in \Pi^W$. The same proof as that used for Theorem 6.7.10 (which depended on the Leibniz rule) can be used to show that $\mathcal{T}_i$ is consistent with a differential operator of degree $\le i$ on Π^W, provided that $ad(q)^2\,\mathcal{U}_k = 0$ for each $q \in \Pi^W$ and $1 \le k \le d$. But, for any $f \in \Pi^d$, we have $q\mathcal{U}_k f = \mathcal{U}_k(qf) - \left(x_k\partial q/\partial x_k\right)f$, so that $ad(q)\mathcal{U}_k = -x_k\partial q/\partial x_k$ (a multiplier) and $[ad(q)]^2\mathcal{U}_k = 0$. □

The eigenvalue of $\mathcal{T}_i$ acting on ζ_λ is the coefficient of t^i in $\prod_{j=1}^d [1+t\xi_j(\lambda)]$; define

$$\sum_{i=0}^{d} v_i(\lambda)t^i = \prod_{j=1}^{d} (1+t[\kappa(d-i+1)+\lambda_i+1]).$$

The values $\{v_i(\lambda) : 1 \le i \le d\}$ determine $\lambda \in \mathbb{N}_0^{d,P}$ uniquely. For any $w \in S_d$, $w\zeta_\lambda$ is an eigenfunction of $\mathcal{T}_i$ with eigenvalue $v_i(\lambda)$. The following is now obvious.

Theorem 10.7.5 *For each $\lambda \in \mathbb{N}_0^{d,P}$ there is a unique (up to scalar multiplication) symmetric polynomial q such that $\mathcal{T}_i q = v_i(\lambda)q$, for each i with $1 \le i \le d$ and $q = bj_\lambda$ for some $b \in \mathbb{Q}(\kappa)$.*

The *Jack polynomials* $J_\lambda(x;\kappa^{-1})$ (in d variables, thus labeled by partitions with no more than d nonzero parts) were defined to be an orthogonal family for an algebraically defined inner product on symmetric functions (see Jack [1970/71], Stanley [1989] and Macdonald [1995]) and for the torus (Beerends and Opdam [1993]). The polynomials with $\kappa = \frac{1}{2}$ are called zonal spherical functions in statistics (James and Constantine [1974]). Bergeron and Garsia [1992] showed that these are spherical functions on a certain finite homogeneous space. Beginning with a known coefficient, on the one hand we can express J_λ in terms of j_λ, indeed the coefficient of x^λ in $J_\lambda(x;\kappa^{-1})$ is $h_*(\lambda) = \kappa^{-|\lambda|}h(\lambda,\kappa)$. On the other hand,

$$j_\lambda = \mathcal{E}_+(\lambda^R)\sum_{\alpha^+=\lambda}\frac{1}{\mathcal{E}_+(\alpha)}\zeta_\alpha,$$

and the expansion of ζ_α in the x-basis shows that x^λ only occurs in ζ_λ. But $E_\lambda(x;\kappa^{-1})$ is monic in x^λ and thus the coefficient of x^λ in j_λ is $h(\lambda^R,\kappa+1)/h(\lambda,1)$.

Proposition 10.7.6 *For $\lambda \in \mathbb{N}_0^{d,P}$,*

$$J_\lambda\left(\cdot;\kappa^{-1}\right) = \frac{(d\kappa)_\lambda\, h(\lambda,1)}{(d\kappa+1)_\lambda\, \kappa^{|\lambda|}\#S_d(\lambda)}\, j_\lambda.$$

Proof Let $J_\lambda\left(\cdot;\kappa^{-1}\right) = b j_\lambda$ and match up the coefficients of x^λ. This implies that

$$b = \frac{h(\lambda,\kappa)\, h(\lambda,1)}{k^{|\lambda|} h(\lambda^R,\kappa+1)},$$

and Lemma 10.7.2 finishes the calculation. □

Corollary 10.7.7 *For $\lambda \in \mathbb{N}_0^{d,P}$ the following hold:*

(i) $J_\lambda(\mathbf{1}^d;\kappa^{-1}) = \kappa^{-|\lambda|}(d\kappa)_\lambda$;

(ii) $\langle J_\lambda(\cdot;\kappa^{-1}), J_\lambda(\cdot;\kappa^{-1})\rangle_h = (d\kappa)_\lambda\, h_*(\lambda)\, h^*(\lambda)$;

(iii) $\langle J_\lambda(\cdot;\kappa^{-1}), J_\lambda(\cdot;\kappa^{-1})\rangle_p = \dfrac{(d\kappa)_\lambda}{(d\kappa+1)_\lambda} h_*(\lambda)\, h^*(\lambda)$;

(iv) $\langle J_\lambda(\cdot;\kappa^{-1}), J_\lambda(\cdot;\kappa^{-1})\rangle_{\mathbb{T}} = \dfrac{(d\kappa)_\lambda}{((d-1)\kappa+1)_\lambda} h_*(\lambda)\, h^*(\lambda)$.

Proof This involves straightforward calculations using the norm formulae from the previous section, Theorem 10.4.10 and Lemma 10.7.2. □

For their intrinsic interest as well as for applications to Calogero–Sutherland models we derive the explicit forms of the two operators, $\sum_{i=1}^d \mathscr{U}_i$ and $\sum_{i=1}^d \mathscr{U}_i^2$ (actually, we will use a convenient shift of the second). Recall the notation $\partial_i = \partial/\partial x_i$ for $1 \le i \le d$.

Proposition 10.7.8 *We have $\sum_{i=1}^d \mathscr{U}_i = \sum_{i=1}^d x_i\partial_i + d + \frac{1}{2}\kappa d(d+1)$.*

Proof For $1 \le i \le d$, by Lemma 10.2.1

$$\begin{aligned}\mathscr{U}_i &= x_i\mathscr{D}_i + \kappa + 1 + \kappa\sum_{j\neq i}(i,j) - \kappa\sum_{j<i}(i,j)\\ &= x_i\mathscr{D}_i + \kappa + 1 + \kappa\sum_{j>i}(i,j)\end{aligned}$$

and

$$\begin{aligned}\sum_{i=1}^d \mathscr{U}_i &= \sum_{i=1}^d x_i\partial_i + \kappa\sum_{1\le i<j\le d}[1-(i,j)] + \kappa\sum_{1\le i<j\le d}(i,j) + d(1+\kappa)\\ &= \sum_{i=1}^d x_i\partial_i + d(1+\kappa) + \tfrac{1}{2}\kappa d(d-1).\end{aligned}$$

We used here the general formula for $\sum_{i=1}^d x_i\mathscr{D}_i$ (see Proposition 6.5.3). □

Theorem 10.7.9 *An invariant operator commuting with each $\mathcal{U}_j$ is given by*

$$\sum_{i=1}^{d}(\mathcal{U}_i - 1 - \kappa d)^2$$
$$= \sum_{i=1}^{d}(x_i\partial_i)^2 + 2\kappa \sum_{1\le i<j\le d} x_i x_j \left[\frac{\partial_i - \partial_j}{x_i - x_j} - \frac{1-(i,j)}{(x_i-x_j)^2}\right] + \tfrac{1}{6}\kappa^2 d(d-1)(2d-1).$$

Proof For $i \ne j$, let $\phi_{ij} = [1-(i,j)]/(x_i - x_j)$, an operator on polynomials. From the proof of the preceding proposition, $\mathcal{U}_i - 1 - \kappa d = x_i\partial_i + \kappa\sum_{j\ne i} x_i\phi_{ij} - \kappa(d-1) + \kappa\sum_{j>i}(i,j)$, and we write

$$\sum_{i=1}^{d}(\mathcal{U}_i - 1 - \kappa d)^2 = \sum_{i=1}^{d}(x_i\partial_i)^2 + \kappa(E_1+E_2) + \kappa^2(E_3+E_4)$$
$$- 2\kappa(d-1)\sum_{i=1}^{d}(\mathcal{U}_i - 1 - \kappa) + \kappa^2 d(d-1)^2,$$

where

$$E_1 = \sum_{i\ne j}(x_i\partial_i x_i\phi_{ij} + x_i\phi_{ij}x_i\partial_i),$$
$$E_2 = \sum_{i<j}[x_i\partial_i(i,j) + (i,j)x_i\partial_i] = \sum_{i\ne j}(i,j)x_i\partial_i,$$
$$E_3 = \sum_{i\ne j}\left(\sum_{k\ne i} x_i\phi_{ij}x_i\phi_{ik} + \sum_{k>i}[x_i\phi_{ij}(i,k) + (i,k)x_i\phi_{ij}]\right),$$
$$E_4 = \sum_{i=1}^{d-1}\sum_{j>i}\sum_{k>i}(i,j)(i,k)$$
$$= \sum_{i=1}^{d-1}\left(d - i + \sum_{i<j<k}[(i,j)(i,k) + (i,k)(i,j)]\right).$$

The argument depends on the careful grouping of terms and permuting the indices of summation. Fix a pair (i,j) with $i<j$ and combine the corresponding terms in E_1+E_2 to obtain

$$x_i\frac{1-(i,j)}{x_i-x_j} + x_i^2\frac{\partial_i - \partial_i(i,j)}{x_i-x_j} - x_i^2\frac{1-(i,j)}{(x_i-x_j)^2}$$
$$+ x_i\frac{x_i\partial_i - (i,j)x_i\partial_i}{x_i-x_j} + (i,j)x_i\partial_i + x_j\frac{1-(i,j)}{x_j-x_i} + x_j^2\frac{\partial_j - \partial_j(i,j)}{x_j-x_i}$$
$$- x_j^2\frac{1-(i,j)}{(x_i-x_j)^2} + x_j\frac{x_j\partial_j - (i,j)x_j\partial_j}{x_j-x_i} + (i,j)x_j\partial_j;$$

the second line comes from interchanging i and j. The coefficient of $(i,j)\partial_i$ is $-x_ix_j + x_j^2/(x_i-x_j) + x_j = 0$, and similarly the coefficient of $(i,j)\partial_j$ is zero. The nonvanishing terms are

$$2\frac{x_i^2\partial_i - x_j^2\partial_j}{x_i - x_j} - 2\frac{x_ix_j[1-(i,j)]}{(x_i-x_j)^2}$$
$$= 2x_ix_j\left[\frac{\partial_i - \partial_j}{x_i - x_j} - \frac{1-(i,j)}{(x_i-x_j)^2}\right] + 2(x_i\partial_i + x_j\partial_j).$$

Summing the last term over all $i < j$ results in $2(d-1)\sum_{i=1}^{d} x_i\partial_i$.

There are terms in E_3 with only two distinct indices: for any pair $\{i,j\}$ with $i < j$ one has

$$x_i\phi_{ij}x_i\phi_{ij} + x_j\phi_{ji}x_j\phi_{ji} + x_i\phi_{ij}(i,j) + (i,j)x_i\phi_{ij} = [1-(i,j)] - [1-(i,j)] = 0,$$

because $(i,j)\phi_{ij} = \phi_{ij}$ so that

$$x_i\phi_{ij}x_i\phi_{ij} = x_i/x_i - x_j(x_i\phi_{ij} - x_j\phi_{ij}) = x_i\phi_{ij},$$

which combines with $x_j\phi_{ji}x_j\phi_{ji}$ to produce $1-(i,j)$. Also,

$$x_i\phi_{ij}(i,j) + (i,j)x_i\phi_{ij} = x_i(i,j) - 1/x_i - x_j + x_j(i,j) - 1/x_j - x_i = -[1-(i,j)].$$

For any triple (i,j,k) (all distinct, and the order matters), the first part of E_3 (in the second term interchange j and k) contributes

$$\frac{x_i^2[1-(i,k)]}{(x_i-x_j)(x_i-x_k)} - \frac{x_ix_k[(i,k)-(i,k)(i,j)]}{(x_i-x_k)(x_k-x_j)}$$

and the second part of E_3 contributes (but only for $k > i$)

$$\frac{x_i[p(i,k)-(i,j)(i,k)]}{x_i - x_j} + \frac{x_k[(i,k)-(i,k)(i,j)]}{x_k - x_j};$$

however, this is symmetric in i,k since $(k,j)(k,i) = (i,k)(i,j)$ (interchanging i and k in the first term). Fix a triple (p,q,s) with $p < q$, and then the transposition (p,q) appears in the triples $(p,s,q),(q,s,p)$ with coefficient

$$\frac{-x_p^2}{(x_p-x_s)(x_p-x_q)} + \frac{-x_q^2}{(x_q-x_s)(x_q-x_p)} - \frac{x_px_q}{(x_p-x_q)(x_q-x_s)}$$
$$- \frac{x_px_q}{(x_q-x_p)(x_p-x_s)} + \frac{x_p}{x_p-x_s} + \frac{x_q}{x_q-x_s} = 0.$$

Finally we consider the coefficient of a given 3-cycle (of permutations), say, $(p,q)(p,s) = (q,s)(q,p) = (s,p)(s,q)$, which arises as follows:

$$\frac{x_px_q}{(x_p-x_q)(x_q-x_s)} - \frac{x_q}{x_q-x_s} + \frac{x_qx_s}{(x_q-x_s)(x_s-x_p)} - \frac{x_s}{x_s-x_p}$$
$$+ \frac{x_sx_p}{(x_s-x_p)(x_p-x_q)} - \frac{x_p}{x_p-x_q} = -1.$$

This is canceled by the corresponding 3-cycle in E_4. It remains to find the coefficient of 1 (the identity operator) in E_3 and E_4. The six permutations of any triple $p < q < s$ correspond to the six permutations of $x_p^2/(x_p-x_q)(x_p-x_s)$, and these terms add to 2. Thus E_3 and E_4 contribute $2\binom{d}{3} + \sum_{i=1}^{d-1}(d-i) = \frac{1}{6}d(d-1)(2d-1)$. Further,

$$2\kappa(d-1)\sum_{i=1}^{d}[x_i\partial_i - (\mathscr{U}_i - 1 - \kappa)] + \kappa^2 d(d-1)^2 = 0$$

and this finishes the proof. □

Corollary 10.7.10 *For $\lambda \in \mathbb{N}_0^{d,P}$ and $\alpha^+ = \lambda$,*

$$\left[\sum_{i=1}^{d}(x_i\partial_i)^2 + 2\kappa\sum_{i<j}x_ix_j\left(\frac{\partial_i - \partial_j}{x_i - x_j} - \frac{1-(i,j)}{(x_i-x_j)^2}\right)\right]\zeta_\alpha = \sum_{i=1}^{d}\lambda_i[\lambda_i - 2\kappa(i-1)]\zeta_\alpha.$$

Proof The operator is $\sum_{i=1}^{d}(\mathscr{U}_i - 1 - \kappa d)^2 - \frac{1}{6}\kappa^2 d(d-1)(2d-1)$. Because of its invariance, the eigenvalue for ζ_α is the same as that for ζ_λ, namely, $\sum_{i=1}^{d}[\kappa(d-i+1) + \lambda_i - \kappa d]^2 - \kappa^2\sum_{i=1}^{d}(i-1)^2$. □

The difference part of the operator vanishes for symmetric polynomials.

Corollary 10.7.11 *For $\lambda \in \mathbb{N}_0^{d,P}$,*

$$\left(\sum_{i=1}^{d}(x_i\partial_i)^2 + 2\kappa\sum_{i<j}x_ix_j\frac{\partial_i - \partial_j}{x_i - x_j}\right)j_\lambda = \sum_{i=1}^{d}\lambda_i[\lambda_i - 2\kappa(i-1)]j_\lambda.$$

This is the second-order differential operator whose eigenfunctions are Jack polynomials.

10.8 Miscellaneous Topics

Recall the exponential-type kernel $K(x,y)$, defined for all $x,y \in \mathbb{R}^d$, with the property that $\mathscr{D}_i^{(x)}K(x,y) = y_iK(x,y)$ for $1 \le i \le d$. The component $K_n(x,y)$ that is homogeneous of degree n in x (and in y) has the reproducing property for the inner product $\langle\cdot,\cdot\rangle_h$ on $\mathscr{P}_n^d$, that is, $K_n\left(x, \mathscr{D}^{(y)}\right)q(y) = q(x)$ for $q \in \mathscr{P}_n^d$. So, the orthogonal basis can be used to obtain the following:

$$K_n(x,y) = \sum_{|\alpha|=n}\frac{h(\alpha,1)}{h(\alpha,\kappa+1)(d\kappa+1)_{\alpha^+}}\zeta_\alpha(x)\zeta_\alpha(y).$$

The intertwining map V can also be expressed in these terms. Recall its defining properties: $V\mathscr{P}_n^d \subseteq \mathscr{P}_n^d, V1 = 1$ and $\mathscr{D}_iVq(x) = V\partial q(x)/\partial x_i$ for every $q \in \Pi^d, 1 \le i \le d$. The homogeneous component of degree n of V^{-1} is given by

$$V^{-1}q(x) = \frac{1}{n!}\sum_{|\alpha|=n}\binom{n}{\alpha}x^\alpha\left(\mathscr{D}^{(y)}\right)^\alpha q(y).$$

Define a linear map ξ on polynomials by $\xi p_\alpha = x^\alpha/\alpha!$; then $\xi^{-1}V^{-1}q(x) = \sum_{|\alpha|=n} p_\alpha(x)(\mathscr{D}^{(y)})^\alpha q(y) = \langle F_n(x,\cdot),q\rangle_h$, the pairing of the degree-n homogeneous part of $F(x,y)$ (the generating function for p_α and the reproducing kernel for $\langle\cdot,\cdot\rangle_p$) with q in the h inner product (see Definition 7.2.2).

Write $q = \sum_{|\beta|=n} b_\beta \zeta_\beta$ (expanding in the ζ-basis); then, with $|\alpha| = n = |\beta|$,

$$\langle F_n(x,\cdot),q\rangle_h = \sum_{\alpha,\beta} \frac{1}{\langle \zeta_\alpha,\zeta_\alpha\rangle_p} \zeta_\alpha(x) b_\beta \langle \zeta_\alpha,\zeta_\beta\rangle_h = \sum_\alpha (d\kappa+1)_{\alpha^+} b_\alpha \zeta_\alpha(x).$$

Thus $\xi^{-1}V^{-1}$ acts as the scalar $(d\kappa+1)_\lambda 1$ on the space E_λ; hence $V\xi = (d\kappa+1)_\lambda^{-1} 1$ on E_λ. From the expansion

$$\zeta_\alpha = p_\alpha + \sum\{B_{\beta\alpha} p_\beta : \beta \rhd \alpha\},$$

the matrix $B_{\beta,\alpha}$ can be defined for all $\alpha,\beta \in \mathbb{N}_0^d$ with $|\alpha| = n = |\beta|$, for fixed n, such that $B_{\alpha,\alpha} = 1$ and $B_{\beta\alpha} = 0$ unless $\beta \unrhd \alpha$. The inverse of the matrix has the same triangularity property, so that

$$p_\alpha = \zeta_\alpha + \sum\left\{ \left(B^{-1}\right)_{\beta\alpha} \zeta_\beta : \beta \rhd \alpha \right\}$$

and thus

$$Vx^\alpha = \alpha! V\xi p_\alpha = \frac{\alpha!}{(d\kappa+1)_{\alpha^+}} \zeta_\alpha + \sum\left\{ \frac{\alpha!}{(d\kappa+1)_{\beta^+}} (B^{-1})_{\beta\alpha} \zeta_\beta : \beta \rhd \alpha \right\}.$$

Next, we consider algorithms for p_α and ζ_α. Since p_α is a product of polynomials $p_n(x_i;x)$ only these need to be computed. Further, we showed in the proof of Proposition 10.3.7 that $p_n(x_i;x) = \pi_n(x) + x_i p_{n-1}(x_i;x)$ (and also that $\partial \pi_{n+1}(x)/\partial x_i = \kappa p_n(x_i;x)$), where π_n is generated by

$$\sum_{n=0}^\infty \pi_n(x) t^n = \prod_{i=1}^d (1-x_i t)^{-\kappa} = \left(\sum_{j=0}^d (-1)^j e_j t^j \right)^{-\kappa},$$

in terms of the elementary symmetric polynomials in x ($e_0 = 1, e_1 = x_1 + x_2 + \cdots$, $e_2 = x_1x_2 + \cdots, \ldots, e_d = x_1x_2\cdots x_d$).

Proposition 10.8.1 *For $\kappa > 0$, the polynomials π_n are given by the recurrence (with $\pi_j = 0$ for $j < 0$)*

$$\pi_n = \frac{1}{n} \sum_{i=1}^d (-1)^{i-1} (n - i + i\kappa) e_i \pi_{n-i}.$$

Proof Let $g = \sum_{j=0}^d (-1)^j e_j t^j$ and differentiate both sides of the generating equation:

$$\sum_{n=1}^\infty n\pi_n(x) t^n = t\frac{\partial}{\partial t} g^{-\kappa} = -\kappa \left(\sum_{j=1}^d (-1)^j j e_j t^j \right) g^{-\kappa-1};$$

thus

$$\sum_{n=1}^{\infty} n\pi_n(x)t^n \sum_{j=0}^{d} (-1)^j e_j t^j = -\kappa \sum_{j=1}^{d} (-1)^j j e_j t^j \sum_{n=0}^{\infty} \pi_n(x)t^n.$$

The coefficient of t^m in this equation is

$$\sum_{j=0}^{d} (-1)^j (m-j) e_j \pi_{m-j} = -\kappa \sum_{j=1}^{d} (-1)^j j e_j \pi_{m-j},$$

which yields $m\pi_m = -\sum_{j=1}^{d} (-1)^j (m-j+\kappa j) e_j \pi_{m-j}$. □

It is an easy exercise to show that π_n is the sum over distinct $\gamma \in \mathbb{N}_0^d$, with $\gamma_1 + 2\gamma_2 + \cdots + d\gamma_d = n$, of $(\kappa)_{|\gamma|}(-1)^{n-|\gamma|}\prod_{j=1}^{d}\left(e_j^{\gamma_j}/\gamma_j!\right)$.

The monic polynomials ζ_α^x can be computed algorithmically by means of a kind of Yang–Baxter graph. This is a directed graph, and each node of the graph consists of a triple $(\alpha, \xi(\alpha), \zeta_\alpha^x)$ where $\alpha \in \mathbb{N}_0^d$ and $\xi(\alpha)$ denotes the vector $(\xi_i(\alpha))_{i=1}^d$. The latter is called the *spectral vector* because it contains the $\{\mathscr{U}_i\}$-eigenvalues. The node can be labeled by α for brevity. The adjacent transpositions are fundamental in this application and are denoted by $s_i = (i, i+1)$, $1 \le i < d$. One type of edge in the graph comes from the following:

Proposition 10.8.2 *Suppose that $\alpha \in \mathbb{N}_0^d$ and $\alpha_i < \alpha_{i+1}$; then*

$$\left(1-c^2\right)\zeta_\alpha^x = s_i\zeta_{s_i\alpha}^x - c\zeta_{s_i\alpha}^x,$$

where

$$c = \frac{\kappa}{\xi_{i+1}(\alpha) - \xi_i(\alpha)}\,\zeta_{s_i\alpha}^x = s_i\zeta_\alpha^x + c\zeta_\alpha^x$$

and the spectral vector $\xi(s_i\alpha) = s_i\xi(\alpha)$.

Proof The proof is essentially the same as that for Proposition 10.4.5. In the monic case the leading term of $s_i\zeta_\alpha^x + c\zeta_\alpha^x$ is $x^{s_i\alpha}$, because every monomial x^β in ζ_α^x satisfies $\beta \trianglelefteq \alpha \triangleleft s_i\alpha$. □

The other type of edge is called an affine step and corresponds to the map $\Phi\alpha = (\alpha_2, \alpha_3, \ldots, \alpha_d, \alpha_1 + 1)$ for $\alpha \in \mathbb{N}_0^d$.

Lemma 10.8.3 *For $\alpha \in \mathbb{N}_0^d$ the spectral vector $\xi(\Phi\alpha) = \Phi\xi(\alpha)$.*

Proof For $2 \le j \le d$, let $\varepsilon_j = 1$ if $\alpha_1 \ge \alpha_j$ (equivalently, $\alpha_1 + 1 > \alpha_j$) and $\varepsilon_j = 0$ otherwise. Then, for $1 \le i < d$,

$$\begin{aligned} r_{\Phi\alpha}(i) &= \varepsilon_{i+1} + \#\{l : 2 < l < i+1, \alpha_l \ge \alpha_{i+1}\} \\ &\quad + \#\{l : i+1 < l \le d, \alpha_l > \alpha_{i+1}\} \\ &= r_\alpha(i+1); \end{aligned}$$

see Definition 10.3.10. Thus $\xi_{\Phi\alpha}(i) = \kappa[d - r_\alpha(i+1)+1] + \alpha_{i+1} + 1 = \xi_\alpha(i+1)$. Similarly, $r_{\Phi\alpha}(d) = r_\alpha(1)$ and $\xi_{\Phi\alpha}(d) = \xi_\alpha(1)+1$. □

From the commutation relations in Lemma 10.2.1 we obtain

$$\begin{aligned} \mathscr{U}_i x_d f &= x_d\,[\mathscr{U}_i - \kappa\,(i,d)]\,f, \qquad 1 \le i < d, \\ \mathscr{U}_d x_d f &= x_d\,(1 + \mathscr{D}_d x_d)\,f. \end{aligned}$$

Let $\theta_d = s_1 s_2 \dots s_{d-1}$; thus $\theta_d\,(d) = 1$ and $\theta_d\,(i) = i+1$ for $1 \le i < d$ (a cyclic shift). Then, by the use of $\mathscr{U}_{i+1} = s_i \mathscr{U}_i s_i - \kappa s_i$ and $s_j \mathscr{U}_i s_j = s_j \mathscr{U}_i s_j$ for $j < i-1$ or $j > i$,

$$\begin{aligned} \mathscr{U}_i x_d f &= x_d\left(\theta_d^{-1}\mathscr{U}_{i+1}\theta_d\right) f, \qquad 1 \le i < d, \\ \mathscr{U}_d x_d f &= x_d\left(1 + \theta_d^{-1}\mathscr{U}_1\theta_d\right) f. \end{aligned}$$

If f satisfies $\mathscr{U}_i f = \lambda_i f$ for $1 \le i \le d$ then $\mathscr{U}_i\left(x_d\theta_d^{-1}f\right) = \lambda_{i+1}\left(x_d\theta_d^{-1}f\right)$ for $1 \le i < d$ and $\mathscr{U}_d\left(x_d\theta_d^{-1}f\right) = (\lambda_1 + 1)\left(x_d\theta_d^{-1}f\right)$. Note that $\theta_d^{-1}f\,(x) = f\left(x\theta_d^{-1}\right) = f\,(x_d, x_1, \dots, x_{d-1})$.

Proposition 10.8.4 *Suppose that $\alpha \in \mathbb{N}_0^d$; then*

$$\zeta^x_{\Phi\alpha}\,(x) = x_d\,\zeta^x_\alpha\,(x_d, x_1, \dots, x_{d-1}).$$

Proof The above argument shows that $x_d\zeta^x_\alpha\left(x\theta_d^{-1}\right)$ has spectral vector $\xi\,(\Phi\alpha)$. The coefficient of $x^{\Phi\alpha}$ in $x_d\zeta^x_\alpha\left(x\theta_d^{-1}\right)$ is the same as the coefficient of x^α in ζ^x_α. □

The Yang–Baxter graph is interpreted as an algorithm as follows:

1. the base point is $\left(0^d, ((d-i+1)\,\kappa+1)_{i=1}^d, 1\right)$,
2. the node $(\alpha, \xi\,(\alpha), f)$ is joined to $(\Phi\alpha, \Phi\xi\,(\alpha), x_d f\,(x_d, x_1, \dots, x_{d-1}))$ by a directed edge labeled S;
3. if $\alpha_i < \alpha_{i+1}$ then the node $(\alpha, \xi\,(\alpha), f)$ is joined to
$$\left(s_i\alpha, s_i\xi\,(\alpha), \left(s_i + \frac{\kappa}{\xi_{i+1}\,(\alpha) - \xi_i\,(\alpha)}\right) f\right)$$
by a directed edge labeled T_i.

A path in the graph is a sequence of connected edges (respecting the directions) joining the base point to another node. By Propositions 10.8.4 and 10.8.2 the end result of the path is a triple $(\alpha, \xi\,(\alpha), \zeta^x_\alpha)$.

For example, here is a path to the node labeled $(0,1,0,2)$: $(0,0,0,0) \xrightarrow{S} (0,0,0,1) \xrightarrow{T_3} (0,0,1,0) \xrightarrow{T_2} (0,1,0,0) \xrightarrow{S} (1,0,0,1) \xrightarrow{T_3} (1,0,1,0) \xrightarrow{S} (0,1,0,2)$.

All paths with the same end point have the same length. To show, this introduce the function $\tau(t) = \frac{1}{2}(|t| + |t+1| - 1)$ for $t \in \mathbb{Z}$; then $\tau(t) = t$ for $t \geq 0$ and $= -t-1$ for $t \leq -1$, also $\tau(-t-1) = \tau(t)$. For $\alpha \in \mathbb{N}_0^d$, define $\tau'(\alpha) = \sum_{1\leq i<j\leq d} \tau(\alpha_i - \alpha_j)$.

Proposition 10.8.5 *For $\alpha \in \mathbb{N}_0^d$, the number of edges in a path joining the base point to the node $(\alpha, \xi(\alpha), \zeta_\alpha^x)$ is $\tau'(\alpha) + |\alpha|$.*

Proof Argue by induction on the length. The induction starts with $\tau'(0^d) + |0^d| = 0$. Suppose that the claim is true for some α. Consider the step to $\Phi\alpha$. It suffices to show that $\tau'(\Phi\alpha) = \tau'(\alpha)$, since $|\Phi\alpha| = |\alpha| + 1$. Indeed,

$$\begin{aligned}\tau'(\Phi\alpha) - \tau'(\alpha) &= \sum_{i=1}^{d-1} \tau(\alpha_{i+1} - (\alpha_1 + 1)) - \sum_{i=2}^{d} \tau(\alpha_1 - \alpha_i)\\ &= \sum_{i=2}^{d} [\tau(-(\alpha_1 - \alpha_i) - 1) - \tau(\alpha_1 - \alpha_i)] = 0.\end{aligned}$$

Now, suppose that $\alpha_i < \alpha_{i+1}$ (that is, $\alpha_i - \alpha_{i+1} \leq -1$); then $|s_i\alpha| = |\alpha|$ and

$$\begin{aligned}\tau'(s_i\alpha) - \tau'(\alpha) &= \tau(\alpha_{i+1} - \alpha_i) - \tau(\alpha_i - \alpha_{i+1})\\ &= (\alpha_{i+1} - \alpha_i) - (\alpha_{i+1} - \alpha_i - 1)\\ &= 1.\end{aligned}$$

This completes the induction. □

In the above example

$$\tau'((0,1,0,2)) = \tau(-1) + \tau(-2) + \tau(1) + \tau(-1) + \tau(-2) = 3.$$

Recall that $e^{-\Delta_h/2}$ is an isometric map on polynomials from the inner product $\langle\cdot,\cdot\rangle_h$ to $\mathscr{H} = L^2\left(\mathbb{R}^d; b_\kappa \prod_{i<j} |x_i - x_j|^{2\kappa} e^{-\|x\|^2/2}\, dx\right)$. The images of $\{\zeta_\alpha : \alpha \in \mathbb{N}_0^d\}$ under $e^{-\Delta_h/2}$ form the orthogonal basis of nonsymmetric Jack–Hermite polynomials. By determining $e^{-\Delta_h/2}\mathscr{U}_i e^{\Delta_h/2}$, we will exhibit commuting self-adjoint operators on $\mathscr{H}$ whose simultaneous eigenfunctions are exactly these polynomials.

Proposition 10.8.6 *For $1 \leq i \leq d$, the operator*

$$\mathscr{U}_i^{\mathscr{H}} - e^{-\Delta_h/2}\mathscr{U}_i e^{\Delta_h/2} - \mathscr{D}_i x_i + \kappa - \mathscr{D}_i^2 - \kappa \sum_{j<i} (j,i)$$

is self-adjoint on $\mathscr{H}$.

Proof The self-adjoint property is a consequence of the isometric transformation. Since Δ_h commutes with each transposition (j,i), it suffices to show that

$$\mathrm{e}^{-\Delta_h/2}\mathscr{D}_i x_i = \left(\mathscr{D}_i x_i - \mathscr{D}_i^2\right)\mathrm{e}^{-\Delta_h/2}.$$

Use the relation (Lemma 7.1.10) $\Delta_h x_i = x_i\Delta_h + 2\mathscr{D}_i$ to show inductively that

$$\Delta_h^n \mathscr{D}_i x_i - \mathscr{D}_i x_i \Delta_h^n = 2n\mathscr{D}_i^2\Delta_h^{n-1};$$

now apply Δ_h to both sides of the equation to obtain

$$\begin{aligned}\Delta_h^{n+1}\mathscr{D}_i x_i - \Delta_h\mathscr{D}_i x_i\Delta_h^n &= 2n\mathscr{D}_i^2\Delta_h^n\\ &= \Delta_h^{n+1}\mathscr{D}_i x_i - \mathscr{D}_i x_i\Delta_h^{n+1} - 2\mathscr{D}_i^2\Delta_h^n.\end{aligned}$$

Multiply the equation for n by $\left(-\frac{1}{2}\right)^n/n!$ and sum over $n = 0,1,2,\ldots$; the result is $\mathrm{e}^{-\Delta_h/2}\mathscr{D}_i x_i - \mathscr{D}_i x_i \mathrm{e}^{-\Delta_h/2} = -\mathscr{D}_i^2\mathrm{e}^{-\Delta_h/2}$. □

The nonsymmetric Jack–Hermite polynomials $\{\mathrm{e}^{-\Delta_h/2}\zeta_\alpha\}$ are not homogeneous; the norms in $\mathscr{H}$ are the same as the $\langle\cdot,\cdot\rangle_h$ norms for $\{\zeta_\alpha\}$.

10.9 Notes

The forerunners of the present study of orthogonal polynomials with finite group symmetry were the Jack polynomials and also the Jacobi polynomials of Heckman [1987], Heckman and Opdam [1987] and Opdam [1988]. These are trigonometric polynomials periodic on the root lattice of a Weyl group and are eigenfunctions of invariant (under the corresponding Weyl group) differential operators (see, for example, Section 5.5 for an example in the type-A category). Beerends and Opdam [1993] found an explicit relationship between Jack polynomials and the Heckman–Opdam Jacobi polynomials of type A. Okounkov and Olshanski [1998] studied the asymptotics of Jack polynomials as $d \to \infty$. It seems clear now that the theory of orthogonal polynomials with respect to invariant weight functions requires the use of differential–difference operators. Differential operators do not suffice. After Dunkl's [1989a] construction of rational differential–difference operators, Heckman 1991a] defined trigonometric ones; the connection between the two is related to the change of variables $x_j = \exp(\mathrm{i}\,\theta_j)$, at least for S_d. Heckman's operators are self-adjoint but not commuting; later, Cherednik [1991] found a modification providing commutativity. Lapointe and Vinet [1996] introduced the explicit form of $\mathscr{U}_i$ used in this chapter to generate Jack polynomials by a Rodrigues-type formula. Opdam [1995] constructed orthogonal decompositions for polynomials associated with Weyl groups in terms of commuting self-adjoint differential–difference operators. Several details and formulae in this chapter are special cases of Opdam's results; because we considered only symmetric groups, the formulae are a good deal more accessible.

Dunkl [1998b] introduced the p-basis, partly for the purpose of analyzing the intertwining operator. The algorithm for ζ_λ in terms of $\zeta_{\lambda-\varepsilon_m}$ (when $m = \ell(\lambda)$) was taken from this paper. Sahi [1996] computed the norms of the nonsymmetric Jack polynomials for the inner product $\langle\cdot,\cdot\rangle_p$, that is, the formal biorthogonality $\langle p_\alpha, x^\beta\rangle_p = \delta_{\alpha\beta}$. Knop and Sahi [1997] developed a hook-length product associated with tableaux of compositions (the row lengths were not in monotone order). Baker and Forrester, in a series of papers ([1997a,b, 1998]) studied nonsymmetric Jack polynomials for their application to Calogero–Moser systems (see Chapter 11) and also exponential-type kernels. Baker and Forrester and also Lassalle [1991a] studied the Hermite polynomials of type A. Some techniques involving the p-basis and the symmetric and skew-symmetric polynomials appeared in Dunkl [1998a] and in Baker, Dunkl and Forrester [2000]. The evaluation of the pairing $\prod_{i<j}(\mathscr{D}_i - \mathscr{D}_j)\prod_{i<j}(x_i - x_j)$, used here in norm calculations, was used previously in Dunkl and Hanlon [1998].

Proposition 10.8.5 was proved in Dunkl and Luque [2011, Proposition 2.13]. Etingof [2010] gave a proof for the explicit evaluation integrals of Macdonald–Mehta–Selberg type for all the irreducible groups except type B_N.

11

Orthogonal Polynomials Associated with Octahedral Groups, and Applications

11.1 Introduction

The adjoining of sign changes to the symmetric group produces the hyperoctahedral group. Many techniques and results from the previous chapter can be adapted to this group by considering only functions that are even in each variable. A second parameter κ' is associated with the conjugacy class of sign changes. The main part of the chapter begins with a description of the differential–difference operators for these groups and their effect on polynomials of arbitrary parity (odd in some variables, even in the others). As in the type-A case there is a fundamental set of first-order commuting self-adjoint operators, and their eigenfunctions are expressed in terms of nonsymmetric Jack polynomials. The normalizing constant for the Hermite polynomials, that is, the Macdonald–Mehta–Selberg integral, is computed by the use of a recurrence relation and analytic-function techniques. There is a generalization of binomial coefficients for the nonsymmetric Jack polynomials which can be used for the calculation of the Hermite polynomials. Although no closed form is as yet available for these coefficients, we present an algorithmic scheme for obtaining specific desired values (by symbolic computation). Calogero and Sutherland were the first to study nontrivial examples of many-body quantum models and to show their complete integrability. These systems concern identical particles in a one-dimensional space, the line or the circle. The corresponding models have been extended to include nonsymmetric wave functions by allowing the exchange of spin values between two particles. We give a concise description of the Schrödinger equations for the models and of the construction of wave functions and commuting operators, using type A and type B operators. Type A operators belong to the symmetric groups; see Section 6.2 for a classification of reflection groups. The chapter concludes with notes on the research literature in current mathematics and physics journals.

11.2 Operators of Type B

The Coxeter groups of type B are the symmetry groups of the hypercubes $\{(\pm 1, \pm 1, \ldots, \pm 1)\} \subset \mathbb{R}^d$ and of the hyperoctahedra $\{\pm \varepsilon_i : 1 \le i \le d\}$. For given d, the corresponding group is denoted W_d and is of order $2^d d!$. For compatibility with the previous chapter we use κ, κ' for the values of the multiplicity function, with κ assigned to the class of roots $\{\pm\varepsilon_i \pm \varepsilon_j : 1 \le i < j \le d\}$ and κ' assigned to the class of roots $\{\pm\varepsilon_i : 1 \le i \le d\}$. For $i \ne j$ the roots $\varepsilon_i - \varepsilon_j$, $\varepsilon_i + \varepsilon_j$ correspond to reflections σ_{ij}, τ_{ij} respectively, where, for $x \in \mathbb{R}^d$,

$$x\sigma_{ij} = (x_1, \ldots, \overset{i}{x_j}, \ldots, \overset{j}{x_i}, \ldots, x_d),$$
$$x\tau_{ij} = (x_1, \ldots, -\overset{i}{x_j}, \ldots, -\overset{j}{x_i}, \ldots, x_d).$$

As before, the superscripts above a symbol refer to its position in the list of elements. For $1 \le i \le d$ the root ε_i corresponds to the reflection (sign change)

$$x\sigma_i = (x_1, \ldots, -\overset{i}{x_i}, \ldots, x_d).$$

Note that $\sigma_i\sigma_{ij}\sigma_i = \sigma_j\sigma_{ij}\sigma_j = \tau_{ij}$. The associated Dunkl operators are as follows.

Definition 11.2.1 For $1 \le i \le d$ and $f \in \Pi^d$ the type-B operator $\mathscr{D}_i$ is given by

$$\mathscr{D}_i f(x) = \frac{\partial}{\partial x_i} f(x) + \kappa' \frac{f(x) - f(x\sigma_i)}{x_i} + \kappa \sum_{j \ne i} \left(\frac{f(x) - f(x\sigma_{ij})}{x_i - x_j} + \frac{f(x) - f(x\tau_{ij})}{x_i + x_j} \right).$$

Most of the work in constructing orthogonal polynomials was done in the previous chapter. The results transfer, with these conventions: the superscript A indicates the symmetric-group operators and, for $x \in \mathbb{R}^d$, let

$$\begin{aligned} y = {} & (y_1, \ldots, y_d) = (x_1^2, \ldots, x_d^2), \\ \mathscr{D}_i^A g(y) = {} & \frac{\partial}{\partial y_i} g(y) + \kappa \sum_{j \ne i} \frac{g(y) - (i,j)g(y)}{y_i - y_j}, \\ \mathscr{U}_i^A g(y) = {} & \mathscr{D}_i^A \rho_i g(y) - \kappa \sum_{j<i} (i,j) g(y); \end{aligned}$$

note that both σ_{ij} and τ_{ij} induce the transposition (i,j) on y and recall that $\mathscr{D}_i^A \rho_i = \mathscr{D}_i^A y_i + \kappa$. To keep the numerous inner products distinct, we use $\langle \cdot, \cdot \rangle_B$ for the h inner product, that is, for $f, g \in \Pi^d$ let

$$\langle f, g \rangle_B = f(\mathscr{D}_1, \ldots, \mathscr{D}_d) g(x_1, \ldots, x_d)|_{x=0}.$$

As in the type A case, we construct commuting self-adjoint operators whose simultaneous eigenfunctions can be expressed in terms of the nonsymmetric Jack polynomials. If two polynomials have opposite parities in the same variables then, by a simple argument, they are orthogonal.

Definition 11.2.2 For any subset $E \subseteq \{i : 1 \le i \le d\}$, let

$$x_E = \prod_{i\in E} x_i.$$

Suppose that g_1, g_2 are polynomials in y and E, F are sets with $E \ne F$; then, for any suitable inner product (specifically, one that is invariant under the sign-change group $\mathbb{Z}_2^d$), one has $\langle x_E g_1(y), x_F g_2(y)\rangle = 0$. Applying $\mathscr{D}_i$ to a polynomial $x_E g(y)$ gives two qualitatively different formulae depending on whether $i \in E$.

Proposition 11.2.3 *Let $g(y)$ be a polynomial and $E \subseteq \{j : 1 \le j \le d\}$. Then, for $1 \le i \le d$, if $i \notin E$,*

$$\mathscr{D}_i x_E g(y) = 2x_i x_E \mathscr{D}_i^A g(y);$$

if $i \in E$,

$$\mathscr{D}_i x_E g(y) = 2\frac{x_E}{x_i}\left((\kappa' - \tfrac{1}{2} + \mathscr{D}_i^A y_i) g(y) - \kappa \sum_{j\in E, j\ne i} (i,j) g(y) \right).$$

Proof We use the product rule, Proposition 6.4.12, specialized as follows:

$$\begin{aligned}\mathscr{D}_i x_E g(y) &= x_E \mathscr{D}_i g(y) + g(y)\frac{\partial x_E}{\partial x_i} + \kappa' g(y) \frac{x_E - \sigma_i x_E}{x_i} \\ &\quad + \kappa \sum_{j\ne i} (i,j) g(y) \Big(\frac{x_E - \sigma_{ij} x_E}{x_i - x_j} + \frac{x_E - \tau_{ij} x_E}{x_i + x_j} \Big).\end{aligned}$$

Considering the first term, we obtain

$$\begin{aligned}\mathscr{D}_i g(y) &= 2x_i \frac{\partial}{\partial y_i} g(y) + \kappa \sum_{j\ne i} [g(y) - (i,j)g(y)] \Big(\frac{1}{x_i - x_j} + \frac{1}{x_i + x_j} \Big) \\ &= 2x_i \left(\frac{\partial}{\partial y_i} g(y) + \kappa \sum_{j\ne i} \frac{g(y) - (i,j)g(y)}{y_i - y_j} \right) = 2x_i \mathscr{D}_i^A g(y).\end{aligned}$$

Next, if $i \notin E$ then $x_E - \sigma_i x_E = 0$ and $x_E - \sigma_{ij} x_E = 0 = x_E - \tau_{ij} x_E$ for $j \notin E$. If $i \notin E$ and $j \in E$ then

$$\frac{x_E - \sigma_{ij} x_E}{x_i - x_j} + \frac{x_E - \tau_{ij} x_E}{x_i + x_j} = \frac{x_E}{x_j} \left(\frac{x_j - x_i}{x_i - x_j} + \frac{x_j + x_i}{x_i + x_j} \right) = 0.$$

This proves the first formula.

If $i \in E$ then $x_E - \sigma_i x_E = 2x_E$ and $x_E - \sigma_{ij} x_E = 0 = x_E - \tau_{ij} x_E$ for $j \in E$, while for $j \notin E$ we have

$$\frac{x_E - \sigma_{ij} x_E}{x_i - x_j} + \frac{x_E - \tau_{ij} x_E}{x_i + x_j} = \frac{x_E}{x_i} \left(\frac{x_i - x_j}{x_i - x_j} + \frac{x_i + x_j}{x_i + x_j} \right) = 2\frac{x_E}{x_i}.$$

Thus, for $i \in E$,

$$\mathscr{D}_i x_E g(y) = 2\frac{x_E}{x_i}\Big(y_i \mathscr{D}_i^A g(y) + (\tfrac{1}{2}+\kappa')g(y) + \kappa \sum\{(i,j)g(y) : j \notin E\}\Big).$$

The commutation $y_i \mathscr{D}_i^A g(y) = \mathscr{D}_i^A y_i g(y) - g(y) - \kappa \sum_{j\neq i}(i,j)g(y)$ established in Lemma 10.2.1 finishes the proof. □

Lemma 11.2.4 *For $i \neq j$, the commutant*

$$[\mathscr{D}_i x_i, \mathscr{D}_j x_j] = \kappa(\mathscr{D}_i x_i - \mathscr{D}_j x_j)(\sigma_{ij}+\tau_{ij}) = \kappa(\mathscr{D}_i x_i, \sigma_{ij}+\tau_{ij}).$$

Proof By Proposition 6.4.10, $\mathscr{D}_j x_i - x_i \mathscr{D}_j = -\kappa(\sigma_{ij} - \tau_{ij})$ (as operators on Π^d); thus

$$\begin{aligned}
&\mathscr{D}_i x_i\, \mathscr{D}_j x_j - \mathscr{D}_j x_j \mathscr{D}_i x_i \\
&\quad = \mathscr{D}_i[\mathscr{D}_j x_i + \kappa(\sigma_{ij}-\tau_{ij})]x_j - \mathscr{D}_j[\mathscr{D}_i x_j + \kappa(\sigma_{ij}-\tau_{ij})]x_i \\
&\quad = \kappa\mathscr{D}_i(\sigma_{ij}-\tau_{ij})x_j - \kappa\mathscr{D}_j(\sigma_{ij}-\tau_{ij})x_i \\
&\quad = \kappa(\mathscr{D}_i x_i - \mathscr{D}_j x_j)(\sigma_{ij}+\tau_{ij}),
\end{aligned}$$

because $\sigma_{ij}x_j = x_i\sigma_{ij}$ and $\tau_{ij}x_j = -x_i\tau_{ij}$. The remaining equation follows from $(\sigma_{ij}+\tau_{ij})\mathscr{D}_i x_i = \mathscr{D}_j x_j(\sigma_{ij}+\tau_{ij})$. □

Definition 11.2.5 For $1 \leq i \leq d$, the self-adjoint (in $\langle\cdot,\cdot\rangle_B$) operators $\mathscr{U}_i$ are given by

$$\mathscr{U}_i = \mathscr{D}_i x_i - \kappa \sum_{j<i}(\sigma_{ij}+\tau_{ij}).$$

Theorem 11.2.6 *For $1 \leq i,j \leq d$, $\mathscr{U}_i\mathscr{U}_j = \mathscr{U}_j\mathscr{U}_i$.*

Proof Assume that $i<j$ and let $\psi_{rs} = \sigma_{rs}+\tau_{rs}$ for $r \neq s$. Further, let $\mathscr{U}_i = \mathscr{D}_i x_i - \kappa A$, $\mathscr{U}_j = \mathscr{D}_j x_j - \kappa\psi_{ij} - \kappa B$, where $A = \sum_{k<i}\psi_{ki}$ and $B = \sum_{k<j, k\neq i}\psi_{kj}$. Then

$$[\mathscr{U}_i, \mathscr{U}_j] = [\mathscr{D}_i x_i, \mathscr{D}_j x_j - \kappa\psi_{ij}] - \kappa[\mathscr{D}_i x_i, B] - \kappa[A, \mathscr{D}_j x_j] + \kappa^2[A, \psi_{ij}+B].$$

The first term is zero by Lemma 11.2.4 and the next two are zero by the transformation properties of $\mathscr{D}_i x_i$ and $\mathscr{D}_j x_j$. The last term reduces to

$$\kappa^2 \sum_{k=1}^{i-1}([\psi_{ki}, \psi_{kj}] + [\psi_{ki}, \psi_{ij}]) = \kappa^2 \sum_{k=1}^{i-1}\big((\psi_{ki}\psi_{kj} - \psi_{ij}\psi_{ki}) + (\psi_{ki}\psi_{ij} - \psi_{kj}\psi_{ki})\big).$$

Each bracketed term is zero; to see this, start with the known (from S_d) relation $\sigma_{ki}\sigma_{kj} - \sigma_{ij}\sigma_{ki} = 0$ and conjugate it by $\sigma_j, \sigma_i, \sigma_k$ to obtain $\sigma_{ki}\tau_{kj} - \tau_{ij}\sigma_{ki} = 0$, $\tau_{ki}\sigma_{kj} - \tau_{ij}\tau_{ki} = 0$, $\tau_{ki}\tau_{kj} - \sigma_{ij}\tau_{ki} = 0$ respectively. (Replacing i,k by k,i shows that the second term is zero.) □

11.3 Polynomial Eigenfunctions of Type B

The nonsymmetric Jack polynomials immediately provide simultaneous eigenfunctions which are even in each variable.

Proposition 11.3.1 *For $\alpha \in \mathbb{N}_0^d$ and $1 \le i \le d$,*

$$\mathscr{U}_i \zeta_\alpha(y) = \left[\xi_i(\alpha) - \kappa + \kappa' - \tfrac{1}{2}\right] \zeta_\alpha(y).$$

Proof By Proposition 11.2.3,

$$\begin{aligned}
\mathscr{D}_i x_i \zeta_\alpha(y) - \kappa \sum_{j<i} (\sigma_{ij} + \tau_{ij}) \zeta_\alpha(y) \\
&= 2(\kappa' - \tfrac{1}{2} + \mathscr{D}_i^A y_i)\zeta_\alpha(y) - 2\kappa \sum_{j<i} (i,j) \zeta_\alpha(y) \\
&= 2(\kappa' - \kappa - \tfrac{1}{2} + \mathscr{U}_i^A)\zeta_\alpha(y) \\
&= 2(\xi_i(\alpha) - \kappa + \kappa' - \tfrac{1}{2})\zeta_\alpha(y).
\end{aligned}$$

Recall that $\mathscr{D}_i^A y_i = \mathscr{D}_i^A \rho_i - \kappa$. □

For mixed parity, the simplest structure involves polynomials of the form $x_1 x_2 \cdots x_k g(y)$; this is a corollary, given below, of the following.

Lemma 11.3.2 *Let $E \subseteq \{j : 1 \le j \le d\}$; then, for $i \notin E$,*

$$\mathscr{U}_i x_E g(y) = 2x_E \left((\mathscr{U}_i^A - \kappa + \kappa' - \tfrac{1}{2}) - \kappa \sum_{j>i, j \in E} (i,j) \right) g(y),$$

and, for $i \in E$,

$$\mathscr{U}_i x_E g(y) = 2x_E \left(\mathscr{D}_i^A y_i - \kappa \sum_{j<i, j \in E} (i,j) \right) g(y).$$

Proof For any $j \ne i$ note that $(\sigma_{ij} + \tau_{ij}) x_E g(y) = 0$ if $\#(\{i,j\} \cap E) = 1$, otherwise $(\sigma_{ij} + \tau_{ij}) x_E g(y) = 2x_E (i,j) g(y)$. For $i \notin E$, by the second part of Proposition 11.2.3,

$$\mathscr{U}_i x_E g(y) = 2x_E \left((\mathscr{D}_i^A y_i + \kappa' - \tfrac{1}{2}) - \kappa \sum_{j \in E} (i,j) \right) g(y) - 2\kappa x_E \sum_{j<i, j \notin E} (i,j) g(y),$$

which proves the first part. For $i \in E$, note that $x_i x_E = x_E y_i / x_i$ and, by the first part of Proposition 11.2.3,

$$\begin{aligned}
\mathscr{U}_i x_E g(y) &= \mathscr{D}_i x_i x_E g(y) - 2\kappa x_E \sum_{j<i, j \in E} (i,j) g(y) \\
&= 2x_E \left(\mathscr{D}_i^A y_i g(y) - \kappa \sum_{j<i, j \in E} (i,j) g(y) \right).
\end{aligned}$$

This completes the proof. □

Corollary 11.3.3 *For $1 \le k \le d$ and $\alpha \in \mathbb{N}_0^d$, let $E = \{j : 1 \le j \le k\}$; then, for $1 \le i \le k$,*

$$\mathscr{U}_i x_E \zeta_\alpha(y) = 2[\xi_i(\alpha) - \kappa] x_E \zeta_\alpha(y)$$

and, for $k < i \le d$,

$$\mathscr{U}_i x_E \zeta_\alpha(y) = 2[\xi_i(\alpha) - \kappa + \kappa' - \tfrac{1}{2}] x_E \zeta_\alpha(y).$$

It is clear from S_d theory that $\{x_E \zeta_\alpha(y) : \alpha \in \mathbb{N}_0^d\}$ is a complete set of simultaneous eigenfunctions for $\{\mathscr{U}_i\}$ of this parity type. To handle arbitrary subsets E one uses a permutation which preserves the relative order of the indices corresponding to even and odd parities, respectively. For given k, we consider permutations $w \in S_d$ with certain properties; the set $w(E) = \{w(1), w(2), \dots, w(k)\}$ will be the set of indices having odd parities, that is, $wx_E = \prod_{i=1}^k x_{w(i)}$.

Proposition 11.3.4 *For $1 \le k < d$, let $E = \{j : 1 \le j \le k\}, \alpha \in \mathbb{N}_0^d$, and let $w \in S_d$ with the property that $w(i) < w(j)$ whenever $1 \le i < j \le k$ or $k+1 \le i < j \le d$; then, for $1 \le i \le k$,*

$$\mathscr{U}_{w(i)} w x_E \zeta_\alpha(y) = 2[\xi_i(\alpha) - \kappa] w x_E \zeta_\alpha(y)$$

and, for $k < i \le d$,

$$\mathscr{U}_{w(i)} w x_E \zeta_\alpha(y) = 2[\xi_i(\alpha) - \kappa + \kappa' - \tfrac{1}{2}] w x_E \zeta_\alpha(y).$$

Proof The transformation properties $w\mathscr{D}_i^A y_i = \mathscr{D}_{w(i)}^A y_{w(i)} w$ and

$$(r,s)w = w(w^{-1}(r), w^{-1}(s)), \qquad \text{for } r \neq s$$

will be used. When $1 \le i \le k$, by Lemma 11.3.2 we have

$$\mathscr{U}_{w(i)} w x_E \zeta_\alpha(y) = 2x_{w(E)} w \left(\mathscr{D}_i^A y_i - \kappa \sum_{j<w(i), j\in w(E)} (w^{-1}(j), i) \right) \zeta_\alpha(y)$$

but the set $\{w^{-1}(j) : j < w(i), j \in w(E)\}$ equals $\{r : 1 \le r < i\}$ by the construction of w. Thus $\mathscr{U}_{w(i)} w x_E \zeta_\alpha(y) = 2x_{w(E)} w(\mathscr{U}_i^A - \kappa)\zeta_\alpha(y) = 2[\xi_i(\alpha) - \kappa] w x_E \zeta_\alpha(y)$.

When $k < i \le d$, we express the first part of Lemma 11.3.2 as

$$\begin{aligned}\mathscr{U}_{w(i)} w x_E \zeta_\alpha(y) = 2x_{w(E)} w \Bigg(& \mathscr{D}_i^A y_i + \kappa' - \tfrac{1}{2} \\ & - \kappa \sum_{j\in w(E)} (w^{-1}(j), i) - \kappa \sum_{j<w(i), j\notin w(E)} (w^{-1}(j), i) \Bigg) \zeta_\alpha(y) \\ = {} & 2x_{w(E)} w(\mathscr{U}_i^A - \kappa + \kappa' - \tfrac{1}{2})\zeta_\alpha(y).\end{aligned}$$

The set $\{r : 1 \le r < i\}$ equals $E \cup \{r : k < r < i\}$ and, again by the construction of w, $\{r : k < r < i\} = \{w^{-1}(j) : j < w(i), j \notin w(E)\}$. □

To restate the conclusion: the space of polynomials which are odd in the variables $x_{w(i)}$, $1 \le i \le k$, is the span of the $\{\mathscr{U}_i\}$ simultaneous eigenfunctions $wx_E \zeta_\alpha(y)$.

To conclude this section, we compute the $\langle \cdot, \cdot \rangle_B$ norms of $x_E \zeta_\alpha(y)$ for $E = \{r : 1 \le r \le k\}$. By group invariance the norm of $wx_E \zeta_\alpha(y)$ has the same value. First the S_d-type results will be used for the all-even case $\zeta_\alpha(y)$; this depends on the use of permissible (in the sense of Section 10.3) inner products.

Proposition 11.3.5 *The inner product $\langle \cdot, \cdot \rangle_B$ restricted to polynomials in $y = (x_1^2, \ldots, x_d^2)$ is permissible.*

Proof Indeed, for $f, g \in \Pi^d$ the value of this inner product is given by

$$\langle f, g \rangle_B = f(\mathscr{D}_1^2, \ldots, \mathscr{D}_d^2) g(x_1^2, \ldots, x_d^2)|_{x=0}.$$

The invariance $\langle wf, wg \rangle_B = \langle f, g \rangle_B$ for $w \in S_d$ is obvious. By the second part of Proposition 11.2.3, $\mathscr{D}_i^A y_i f(y) = (\frac{1}{2}\mathscr{D}_i x_i + \frac{1}{2} - \kappa') f(y)$; the right-hand side is clearly a self-adjoint operator in $\langle \cdot, \cdot \rangle_B$. □

By Theorem 10.4.8 the following holds.

Corollary 11.3.6 *For $\alpha \in \mathbb{N}_0^d$ and $\lambda = \alpha^+$,*

$$\langle \zeta_\alpha(y), \zeta_\alpha(y) \rangle_B = \mathscr{E}_+(\alpha)\mathscr{E}_-(\alpha) \langle \zeta_\lambda(y), \zeta_\lambda(y) \rangle_B.$$

As in Section 10.6, the calculation of $\langle \zeta_\lambda(y), \zeta_\lambda(y) \rangle_B$ for $\lambda \in \mathbb{N}_0^{d,P}$ depends on a recurrence relation involving $\mathscr{D}_m^A \zeta_\lambda$, where $m = \ell(\lambda)$. We recall some key details from Section 10.6. For $1 < m \le d$ let $\theta_m = (1,2)(2,3)\cdots(m-1,m) \in S_d$. When $m = \ell(\lambda)$, let $\widetilde{\lambda} = (\lambda_m - 1, \lambda_1, \lambda_2, \ldots, \lambda_{m-1}, 0 \ldots)$; then $\mathscr{D}_m^A \zeta_\lambda = [\kappa(d - m + 1) + \lambda_m]\theta_m^{-1}\zeta_{\widetilde{\lambda}}$. This fact will again be used for norm calculations. For such λ, for conciseness in calculations set

$$a_\lambda = (d-m)\kappa + \kappa' + \lambda_m - \tfrac{1}{2},$$
$$b_\lambda = (d-m+1)\kappa + \lambda_m.$$

Lemma 11.3.7 *Suppose that $\lambda \in \mathbb{N}_0^{d,P}$ and $m = \ell(\lambda)$. Then*

$$\mathscr{D}_m^2 \zeta_\lambda(y) = 4a_\lambda \mathscr{D}_m^A \zeta_\lambda(y) = 4a_\lambda b_\lambda \theta_m^{-1} \zeta_{\widetilde{\lambda}}(y).$$

Proof By Proposition 11.2.3,

$$\mathscr{D}_m^2 \zeta_\lambda(y) = \mathscr{D}_m[2x_m \mathscr{D}_m^A \zeta_\lambda(y)] = 4[(\kappa' - \tfrac{1}{2})\mathscr{D}_m^A \zeta_\lambda(y) + \mathscr{D}_m^A y_m \mathscr{D}_m^A \zeta_\lambda(y)].$$

By Lemma 10.5.3, $\mathscr{D}_m^A y_m \mathscr{D}_m^A \zeta_\lambda(y) = [\xi_m(\lambda) - \kappa - 1]\mathscr{D}_m^A \zeta_\lambda(y)$ because $\mathscr{D}_m^A y_m = \mathscr{D}_m^A \rho_m - \kappa$. Since $\xi_m(\lambda) = (d-m+1)\kappa + \lambda_m + 1$, we obtain

$$\mathscr{D}_m^2 \zeta_\lambda(y) = 4[\kappa' - \tfrac{1}{2} + \xi_m(\lambda) - \kappa - 1]\mathscr{D}_m^A \zeta_\lambda(y) = 4a_\lambda \mathscr{D}_m^A \zeta_\lambda(y).$$

Finally, $\mathscr{D}_m^A \zeta_\lambda(y) = b_\lambda \theta_m^{-1} \zeta_{\widetilde{\lambda}}(y)$. □

Theorem 11.3.8 *Suppose that $\lambda \in \mathbb{N}_0^{d,P}$. Then*

$$\langle \zeta_\lambda(y), \zeta_\lambda(y) \rangle_B = 2^{2|\lambda|}(d\kappa+1)_\lambda((d-1)\kappa+\kappa'+\tfrac{1}{2})_\lambda \frac{h(\lambda,\kappa+1)}{h(\lambda,1)}.$$

Proof Suppose that $m = \ell(\lambda)$. By the defining property of $\langle \cdot,\cdot \rangle_B$ we have on the one hand

$$\begin{aligned}
\left\langle \mathscr{D}_m^2 \zeta_\lambda(y), \mathscr{D}_m^2 \zeta_\lambda(y) \right\rangle_B &= \left\langle x_m^2 \mathscr{D}_m^2 \zeta_\lambda(y), \zeta_\lambda(y) \right\rangle_B \\
&= 4a_\lambda \left\langle x_m^2 \mathscr{D}_m^A \zeta_\lambda(y), \zeta_\lambda(y) \right\rangle_B \\
&= 4a_\lambda \left\langle y_m \mathscr{D}_m^A \zeta_\lambda(y), \zeta_\lambda(y) \right\rangle_B \\
&= 4a_\lambda b_\lambda \frac{\lambda_m}{\kappa+\lambda_m} \langle \zeta_\lambda(y), \zeta_\lambda(y) \rangle_B,
\end{aligned}$$

because $\left\langle y_m \mathscr{D}_m^A \zeta_\lambda(y), \zeta_\lambda(y) \right\rangle_B = [\lambda_m/(\kappa+\lambda_m)] b_\lambda \langle \zeta_\lambda(y), \zeta_\lambda(y) \rangle_B$ by part (iii) of Lemma 10.5.2. On the other hand,

$$\begin{aligned}
\left\langle \mathscr{D}_m^2 \zeta_\lambda(y), \mathscr{D}_m^2 \zeta_\lambda(y) \right\rangle_B &= (4a_\lambda b_\lambda)^2 \left\langle \theta_m^{-1} \zeta_{\widetilde{\lambda}}(y), \theta_m^{-1} \zeta_{\widetilde{\lambda}}(y) \right\rangle_B \\
&= (4a_\lambda b_\lambda)^2 \left\langle \zeta_{\widetilde{\lambda}}(y), \zeta_{\widetilde{\lambda}}(y) \right\rangle_B \\
&= (4a_\lambda b_\lambda)^2 \mathscr{E}_+(\widetilde{\lambda}) \mathscr{E}_-(\widetilde{\lambda}) \left\langle \zeta_{\lambda-\varepsilon_m}(y), \zeta_{\lambda-\varepsilon_m}(y) \right\rangle_B.
\end{aligned}$$

Combining the two displayed equations shows that

$$\langle \zeta_\lambda(y), \zeta_\lambda(y) \rangle_B = 4a_\lambda b_\lambda \frac{\kappa+\lambda_m}{\lambda_m} \mathscr{E}_+(\widetilde{\lambda}) \mathscr{E}_-(\widetilde{\lambda}) \left\langle \zeta_{\lambda-\varepsilon_m}(y), \zeta_{\lambda-\varepsilon_m}(y) \right\rangle_B.$$

From Lemma 10.6.4,

$$\frac{\kappa+\lambda_m}{\lambda_m} \mathscr{E}_+(\widetilde{\lambda}) \mathscr{E}_-(\widetilde{\lambda}) = \frac{h(\lambda,\kappa+1) h(\lambda-\varepsilon_m,1)}{h(\lambda-\varepsilon_m,\kappa+1) h(\lambda,1)}.$$

Moreover,

$$\frac{(d\kappa+1)_\lambda}{(d\kappa+1)_{\lambda-\varepsilon_m}} = (d-m+1)\kappa+\lambda_m = b_\lambda$$

and

$$\frac{((d-1)\kappa+\kappa'+\frac{1}{2})_\lambda}{((d-1)\kappa+\kappa'+\frac{1}{2})_{\lambda-\varepsilon_m}} = (d-m)\kappa+\kappa'+\lambda_m-\frac{1}{2} = a_\lambda.$$

The proof is completed by induction on $|\lambda|$. □

When $\kappa = 0 = \kappa'$ these formulae reduce to the trivial identity $\left\langle x^\lambda, x^\lambda \right\rangle = \prod_{i=1}^d \lambda_i!$, because $2^{2n}(1)_n(\frac{1}{2})_n = (2n)!$ for $n \in \mathbb{N}_0$. It remains to compute the quantity $\langle x_E \zeta_\alpha(y), x_E \zeta_\alpha(y) \rangle_B$ for $E = \{j : 1 \le j \le k\}$ and $\alpha \in \mathbb{N}_0^d$. The following is needed to express the result. Let $\lfloor r \rfloor$ denote the largest integer $\le r$.

Definition 11.3.9 For $\beta \in \mathbb{N}_0^d$ let $e(\beta), o(\beta) \in \mathbb{N}_0^d$ with $e(\beta)_i = \lfloor \beta_i/2 \rfloor$ and $o(\beta)_i = \beta_i - e(\beta)_i$ for $1 \le i \le d$.

(Roughly, $e(\beta)$ and $o(\beta)$ denote the even and odd components of β.) We start with $\beta \in \mathbb{N}_0^d$ such that β_i is odd exactly when $1 \le i \le k$ and consider the $\{\mathscr{U}_i\}$ simultaneous eigenfunctions indexed by β, which are given by $x_E \zeta_\alpha(y)$ where $\alpha = e(\beta)$.

Theorem 11.3.10 *Suppose that* $1 \le k \le d$ *and* $\beta \in \mathbb{N}_0^d$ *satisfies the condition that* β_i *is odd for* $1 \le i \le k$ *and* β_i *is even for* $k < i$*, and let* $E = \{j : 1 \le j \le k\}$*. Then*

$$\begin{aligned}&\langle x_E \zeta_{e(\beta)}(y), x_E \zeta_{e(\beta)}(y) \rangle_B \\ &\quad = 2^{|\beta|} (d\kappa + 1)_{e(\beta)^+} ((d-1)\kappa + \kappa' + \tfrac{1}{2})_{o(\beta)^+} \frac{h(e(\beta), \kappa+1)}{h(e(\beta), 1)}.\end{aligned}$$

Proof For any $\alpha \in \mathbb{N}_0^d$ and $1 \le m \le k$, by Proposition 11.2.3 we have

$$\begin{aligned}\mathscr{D}_m x_1 x_2 \cdots x_m \zeta_\alpha(y) &= 2 x_1 x_2 \cdots x_{m-1} \left(\kappa' - \tfrac{1}{2} + \mathscr{D}_m^A y_m - \kappa \sum_{j<m} (j,m) \right) \zeta_\alpha(y) \\ &= 2 x_1 x_2 \cdots x_{m-1} (\kappa' - \kappa - \tfrac{1}{2} + \mathscr{U}_m^A) \zeta_\alpha(y) \\ &= 2[\xi_m(\alpha) - \kappa + \kappa' - \tfrac{1}{2}] x_1 x_2 \cdots x_{m-1} \zeta_\alpha(y).\end{aligned}$$

Let $\alpha = e(\beta)$; then, using this formula inductively, we obtain

$$\begin{aligned}\langle x_E \zeta_{e(\beta)}(y), x_E \zeta_{e(\beta)}(y) \rangle_B &= \langle \mathscr{D}_1 \cdots \mathscr{D}_k x_E \zeta_{e(\beta)}(y), \zeta_{e(\beta)}(y) \rangle_B \\ &= 2^k \prod_{i=1}^k [\xi_i(\alpha) - \kappa + \kappa' - \tfrac{1}{2}] \langle \zeta_{e(\beta)}(y), \zeta_{e(\beta)}(y) \rangle_B .\end{aligned}$$

The last inner product has already been evaluated (and $2^{2|e(\beta)|} 2^k = 2^{|\beta|}$), so it remains to show that

$$\prod_{i=1}^k [\xi_i(\alpha) - \kappa + \kappa' - \tfrac{1}{2}] ((d-1)\kappa + \kappa' + \tfrac{1}{2})_{e(\beta)^+} = ((d-1)\kappa + \kappa' + \tfrac{1}{2})_{o(\beta)^+}.$$

Let $\alpha = e(\beta)$ and $\lambda = \alpha^+$ and let $w \in S_d$ be such that $\alpha_i = \lambda_{w(i)}$ and if $\alpha_i = \alpha_j$ for some $i < j$ then $w(i) < w(j)$. This implies that $\mu = o(\beta)^+$ satisfies $o(\beta)_i = \mu_{w(i)}$, specifically that $\mu_{w(i)} = \alpha_i + 1$ for $1 \le i \le k$ and $\mu_{w(i)} = \alpha_i$ for $k < i$. This shows that $\mu \in \mathbb{N}_0^{d,P}$; indeed by construction $w(i) < w(j)$ implies that $\lambda_{w(i)} \ge \lambda_{w(j)}$ and $\mu_{w(i)} \ge \mu_{w(j)}$. The latter could only be negated if $\alpha_i = \lambda_{w(i)} = \mu_{w(i)}$ and $\mu_{w(j)} = \alpha_j + 1$ (so that $j \le k < i$), but $\alpha_j + 1 > \alpha_i \ge \alpha_j$ implies that $\alpha_i = \alpha_j$ and $j < i$ implies that $w(j) < w(i)$, a contradiction.

For $1 \le i \le k$, we have $\xi_i(\alpha) = \kappa[d - \#\{j : \alpha_j > \alpha_i\} - \#\{j : j < i; \alpha_j = \alpha_i\}] + \alpha_i + 1 = [d - w(i) + 1]\kappa + \lambda_{w(i)} + 1 = [d - w(i) + 1]\kappa + \mu_{w(i)}$. Thus

$$\begin{aligned}
&\prod_{i=1}^{k}[\xi_i(\alpha) - \kappa + \kappa' - \tfrac{1}{2}]((d-1)\kappa + \kappa' + \tfrac{1}{2})_\lambda \\
&\quad = \prod_{i=1}^{k}\left([d - w(i)]\kappa + \kappa' + \mu_{w(i)} - \tfrac{1}{2}\right)\prod_{i=1}^{d}\left([d - w(i)]\kappa + \kappa' + \tfrac{1}{2}\right)_{\lambda_{w(i)}} \\
&\quad = \prod_{i=1}^{k}\left([d - w(i)]\kappa + \kappa' + \tfrac{1}{2}\right)_{\mu_{w(i)}}\prod_{i=k+1}^{d}\left([d - w(i)]\kappa + \kappa' + \tfrac{1}{2}\right)_{\lambda_{w(i)}} \\
&\quad = ((d-1)\kappa + \kappa' + \tfrac{1}{2})_\mu.
\end{aligned}$$

This completes the proof. □

Example 11.3.11 To illustrate the construction of w in the above proof, let $\beta = (7,5,7,2,6,8)$; then $\alpha = e(\beta) = (3,2,3,1,3,4)$, $\lambda = \alpha^+ = (4,3,3,3,2,1)$, $\mu = o(\beta)^+ = (4,4,4,3,3,1)$ and

$$w = \begin{pmatrix} 1 & 2 & 3 & 4 & 5 & 6 \\ 2 & 5 & 3 & 6 & 4 & 1 \end{pmatrix},$$

using the standard notation for permutations.

We turn now to the Selberg–Macdonald integral and use the same techniques as in the type-A case. The alternating polynomial for S_d, namely

$$a_B(x) = \prod_{1 \le i < j \le d}(x_i^2 - x_j^2) = a(y),$$

plays a key role in these calculations. For the type-B inner product the value of $\langle a_B, a_B\rangle_B$ shows the effect of changing κ to $\kappa+1$. Also, a_B is h-harmonic because $\Delta_h a_B$ is (under the action of σ_{ij} or τ_{ij}) a skew polynomial of degree $2\binom{d}{2} - 2$, hence 0 (note that $\Delta_h = \sum_{i=1}^{d}\mathscr{D}_i^2$). As before, let

$$\delta = (d-1, d-2, \ldots, 1, 0) \in \mathbb{N}_0^{d,P};$$

then a_B is the unique skew element of span$\{w\zeta_\delta : w \in W_d\}$ and $a = a_\delta$ in Definition 10.4.9). Thus the previous results will give the values of $\langle a_B, a_B\rangle_B$.

Recall that, for $f, g \in \Pi^d$ (Theorem 7.2.7),

$$\begin{aligned}
\langle f, g\rangle_B &= (2\pi)^{-d/2} b(\kappa, \kappa') \\
&\quad \times \int_{\mathbb{R}^d}\left(\mathrm{e}^{-\Delta_h/2} f\right)\left(\mathrm{e}^{-\Delta_h/2} g\right)\prod_{i=1}^{d}|x_i|^{2\kappa'}|a_B(x)|^{2\kappa}\mathrm{e}^{-\|x\|^2/2}\,\mathrm{d}x,
\end{aligned}$$

where the normalizing constant $b(\kappa, \kappa')$ satisfies $\langle 1, 1\rangle_B = 1$. Since $\mathrm{e}^{-\Delta_h/2}a_B = a_B$ the following holds (note that $2|\delta| = d(d-1)$).

Theorem 11.3.12 *We have* $\langle a_B, a_B\rangle_B = 2^{d(d-1)}d!\,(d\kappa+1)_\delta\,((d-1)\kappa+\kappa'+\frac{1}{2})_\delta$ *and, for* $\kappa,\kappa'\geq 0$,

$$b(\kappa,\kappa')^{-1} = (2\pi)^{-d/2}\int_{\mathbb{R}^d}\prod_{i=1}^{d}|x_i|^{2\kappa'}\prod_{1\leq i<j\leq d}|x_i^2-x_j^2|^{2\kappa}\mathrm{e}^{-\|x\|^2/2}\,\mathrm{d}x$$
$$= 2^{d[(d-1)\kappa+\kappa']}\frac{\Gamma(\kappa'+\frac{1}{2})}{\pi^{d/2}\Gamma(\kappa+1)^{d-1}}\prod_{j=2}^{d}\Gamma(j\kappa+1)\Gamma((j-1)\kappa+\kappa'+\tfrac{1}{2}).$$

Proof Using the same method as in the proof of Theorem 10.6.17, and by Theorem 11.3.8,

$$\langle a_B,a_B\rangle_B = 2^{d(d-1)}d!\mathscr{E}_-(\delta^R)(d\kappa+1)_\delta((d-1)\kappa+\kappa'+\tfrac{1}{2})_\delta\frac{h(\delta,\kappa+1)}{h(\delta,1)}$$
$$= 2^{d(d-1)}d!\,(d\kappa+1)_\delta\,((d-1)\kappa+\kappa'+\tfrac{1}{2})_\delta.$$

Further,

$$b(\kappa+1,\kappa')^{-1} = (2\pi)^{-d/2}\int_{\mathbb{R}^d}\prod_{i=1}^{d}|x_i|^{2\kappa'}\prod_{i<j}\left|x_i^2-x_j^2\right|^{2\kappa+2}\mathrm{e}^{-\|x\|^2/2}\,\mathrm{d}x$$
$$= b(\kappa,\kappa')^{-1}\langle a_B,a_B\rangle_B.$$

From the proof of Theorem 10.6.17 we have

$$\prod_{j=2}^{d}\frac{\Gamma(j\kappa+1)}{\Gamma(\kappa+1)}d!\,(d\kappa+1)_\delta = \prod_{j=2}^{d}\frac{\Gamma(j\kappa+j+1)}{\Gamma(\kappa+2)}.$$

Also,

$$\prod_{j=2}^{d}\Gamma((j-1)\kappa+\kappa'+\tfrac{1}{2})((d-1)\kappa+\kappa'+\tfrac{1}{2})_\delta = \prod_{j=2}^{d}\Gamma((j-1)(\kappa+1)+\kappa'+\tfrac{1}{2})$$

and

$$b(0,\kappa')^{-1} = (2\pi)^{-d/2}\int_{\mathbb{R}^d}\prod_{i=1}^{d}|x_i|^{2\kappa'}\mathrm{e}^{-\|x\|^2/2}\,\mathrm{d}x = \left(2^{\kappa'}\frac{\Gamma(\kappa'+\frac{1}{2})}{\Gamma(\frac{1}{2})}\right)^d,$$

by an elementary calculation. We will use the same technique as in the type A case, first converting the integral to one over the sphere S^{d-1}, a compact set. Let $\gamma = \kappa d(d-1)+\kappa' d$. Using spherical polar coordinates,

$$(2\pi)^{-d/2}\int_{\mathbb{R}^d}\prod_{i=1}^{d}|x_i|^{2\kappa'}\prod_{i<j}\left|x_i^2-x_j^2\right|^{2\kappa}\mathrm{e}^{-\|x\|^2/2}\,\mathrm{d}x$$
$$= 2^\gamma\sigma_{d-1}^{-1}\frac{\Gamma(\frac{d}{2}+\gamma)}{\Gamma(\frac{d}{2})}\int_{S^{d-1}}\prod_{i=1}^{d}|x_i|^{2\kappa'}\prod_{i<j}\left|x_i^2-x_j^2\right|^{2\kappa}\mathrm{d}\omega(x);$$

the normalizing constant (see (4.1.3)) satisfies $\sigma_{d-1}^{-1}\int_{S^{d-1}} \mathrm{d}\omega = 1$. Fix $\kappa' \geq 0$ and define the analytic function

$$\phi(\kappa) = \Gamma\left(d[(d-1)\kappa + \kappa' + \tfrac{1}{2}]\right)\prod_{j=1}^{d}\left(\frac{\Gamma(\kappa+1)\Gamma(\frac{1}{2})}{\Gamma(j\kappa+1)\Gamma((j-1)\kappa+\kappa'+\frac{1}{2})}\right.$$
$$\left.\times\frac{\sigma_{d-1}^{-1}}{\Gamma\left(\frac{d}{2}\right)}\int_{S^{d-1}}\prod_{i=1}^{d}|x_i|^{2\kappa'}\prod_{i<j}\left|x_i^2-x_j^2\right|^{2\kappa}\mathrm{d}\omega(x)\right).$$

The formula for $b(\kappa,\kappa')/b(\kappa+1,\kappa')$ shows that $\phi(\kappa+1) = \phi(\kappa)$ and also, that $\phi(0) = 1$. By analytic continuation ϕ is entire and periodic. On the strip $\{\kappa : 1 \leq \operatorname{Re}\kappa \leq 2\}$ the factors of ϕ in the second line are uniformly bounded. Apply the asymptotic formula (from the proof of Theorem 10.6.17), valid in $-\pi < \arg\kappa < \pi$ and $a, b > 0$,

$$\log\Gamma(a\kappa+b) = a\kappa(\log\kappa+\log a-1)$$
$$+\left(b-\tfrac{1}{2}\right)(\log\kappa+\log a)+\tfrac{1}{2}\log(2\pi)+O\left(|z|^{-1}\right)$$

to the first line of the expression for $\phi(\kappa)$:

$$\log\Gamma\left(d[(d-1)\kappa+\kappa'+\tfrac{1}{2}]\right)+(d-1)\log\Gamma(\kappa+1)$$
$$-\sum_{j=2}^{d}\log\Gamma(j\kappa+1)+d\log\Gamma(\tfrac{1}{2})-\sum_{j=0}^{d-1}\log\Gamma(j\kappa+\kappa'+\tfrac{1}{2})$$
$$=\left[d(d-1)\kappa+d\kappa'+\tfrac{d-1}{2}\right]\log\left[d(d-1)\right]-\sum_{j=2}^{d}(j\kappa+\tfrac{1}{2})\log j$$
$$-\sum_{j=1}^{d-1}(j\kappa+\kappa')\log j-\tfrac{d-2}{2}\log(2\pi)-\log\Gamma(\kappa'+\tfrac{1}{2})+\tfrac{d}{2}\log\pi$$
$$+\kappa\left(d(d-1)+(d-1)-\sum_{j=2}^{d}j-\sum_{j=1}^{d-1}j\right)(\log\kappa-1)$$
$$+\left(d\kappa'+\tfrac{d-1}{2}+\tfrac{d-1}{2}-\tfrac{d-1}{2}-(d-1)\kappa'\right)\log\kappa+O\left(|\kappa|^{-1}\right)$$
$$=C_1(d,\kappa')+C_2(d)\kappa+(\kappa'+\tfrac{d-1}{2})\log\kappa+O\left(|\kappa|^{-1}\right),$$

where C_1, C_2 are real constants depending on d, κ'. Therefore, in the strip $\{\kappa : 1 \leq \operatorname{Re}\kappa \leq 2\}$, we have $\phi(\kappa) = O(|\kappa|^{\kappa'+(d-1)/2})$. By periodicity the same bound applies for all κ, implying that ϕ is a periodic polynomial and hence constant. This proves the formula for $b(\kappa,\kappa')$. □

Note that the most important part of the above asymptotic calculation was to show that the coefficient of $\kappa\log\kappa$ is zero. The value of the integral in Theorem 11.3.12 is the type B case of a general result for root systems of Macdonald [1982]; it was conjectured by Mehta [1991] yet turned out to be a limiting case of Selberg's integral (an older result, Selberg [1944]). The proof given here could be said to be almost purely algebraic but, of course, it does use asymptotics for the gamma function and some elementary entire-function theory.

11.4 Generalized Binomial Coefficients

The use of generalized binomial coefficients exploits the facts that the vector $(1,1,\ldots,1)\in\mathbb{R}^d$ is invariant under S^d, or equivalently, orthogonal to all the roots $\varepsilon_i-\varepsilon_j$, $i<j$, and that $\sum_{i=1}^d \mathcal{D}_i^A=\sum_{i=1}^d \partial/\partial y_i$. For consistency, we will continue to use y as the variable involved in type-A actions. Baker and Forrester [1998] introduced the concept, motivated by the binomial coefficients previously defined in connection with the Jack polynomials.

Definition 11.4.1 For $\alpha,\beta\in\mathbb{N}_0^d$ and parameter $\kappa\geq 0$, the binomial coefficient $\binom{\alpha}{\beta}_\kappa$ is defined by the expansion

$$\frac{\zeta_\alpha(y+\mathbf{1}^d)}{\zeta_\alpha(\mathbf{1}^d)}=\sum_\beta\binom{\alpha}{\beta}_\kappa\frac{\zeta_\beta(y)}{\zeta_\beta(\mathbf{1}^d)}.$$

We will show that $\binom{\alpha}{\beta}_\kappa=0$ unless $\alpha^+\supseteq\beta^+$ (this ordering on partitions means the inclusion of Ferrers diagrams, that is, $\alpha_i^+\geq\beta_i^+$ for $1\leq i\leq d$). Also, there is a recurrence relation which needs only the values of $\binom{\alpha}{\beta}_\kappa$ at $|\beta|=|\alpha|-1$ at the start (however, even these are complicated to obtain; no general formula is as yet available). Ordinary calculus and homogeneity properties provide some elementary identities. Of course, when $\kappa=0$ the polynomials ζ_α reduce to y^α and $\binom{\alpha}{\beta}_0=\prod_{i=1}^d\binom{\alpha_i}{\beta_i}$, a product of ordinary binomial coefficients.

Proposition 11.4.2 *For $\alpha\in\mathbb{N}_0^d$ and $\kappa\geq 0, s\in\mathbb{R}, j\in\mathbb{N}_0$,*

$$\frac{\zeta_\alpha(y+s\mathbf{1}^d)}{\zeta_\alpha(\mathbf{1}^d)}=\sum_\beta\binom{\alpha}{\beta}_\kappa s^{|\alpha|-|\beta|}\frac{\zeta_\beta(y)}{\zeta_\beta(\mathbf{1}^d)}$$

and

$$\left(\sum_{i=1}^d\mathcal{D}_i^A\right)^j\zeta_\alpha(y)=j!\sum_{|\beta|=|\alpha|-j}\binom{\alpha}{\beta}_\kappa\frac{\zeta_\alpha(\mathbf{1}^d)}{\zeta_\beta(\mathbf{1}^d)}\zeta_\beta(y).$$

Proof For the first part, suppose $s\neq 0$; then $s^{-|\alpha|}\zeta_\alpha(y+s\mathbf{1}^d)=\zeta_\alpha(s^{-1}y+\mathbf{1}^d)$. In Definition 11.4.1 replace y by $s^{-1}y$ and observe that ζ_β is homogeneous of degree $|\beta|$. Consider the first equation as a Taylor's series in s at $s=0$; then

$$\frac{\partial}{\partial s}=\sum_{i=1}^d\frac{\partial}{\partial y_i}=\sum_{i=1}^d\mathcal{D}_i^A.$$

□

Proposition 11.4.3 *For $\alpha, \beta \in \mathbb{N}_0^d, \kappa \geq 0$ and $|\beta| < m < |\alpha|$,*

$$\binom{\alpha}{\beta}_\kappa \binom{|\alpha| - |\beta|}{m - |\beta|} = \sum_{|\gamma|=m} \binom{\alpha}{\gamma}_\kappa \binom{\gamma}{\beta}_\kappa.$$

Proof For arbitrary $s, t \in \mathbf{R}$,

$$\begin{aligned}
\sum_\beta \binom{\alpha}{\beta}_\kappa (s+t)^{|\alpha|-|\beta|} \frac{\zeta_\beta(y)}{\zeta_\beta(\mathbf{1}^d)} &= \frac{\zeta_\alpha(y + (s+t)\mathbf{1}^d)}{\zeta_\alpha(\mathbf{1}^d)} \\
&= \sum_\gamma \binom{\alpha}{\gamma}_\kappa t^{|\alpha|-|\gamma|} \frac{\zeta_\gamma(y + s\mathbf{1}^d)}{\zeta_\gamma(\mathbf{1}^d)} \\
&= \sum_\gamma \sum_\beta \binom{\alpha}{\gamma}_\kappa \binom{\gamma}{\beta}_\kappa t^{|\alpha|-|\gamma|} s^{|\gamma|-|\beta|} \frac{\zeta_\beta(y)}{\zeta_\beta(\mathbf{1}^d)}.
\end{aligned}$$

The coefficient of $s^{m-|\beta|} t^{|\alpha|-m}$ in the equations gives the claimed formula. □

This shows that the investigation of $\binom{\alpha}{\beta}_\kappa$ begins by setting $|\beta| = |\alpha| - 1$, that is, by using the expansion of $\sum_{i=1}^d \mathscr{D}_i^A \zeta_\alpha$. The relevance of the binomial coefficients lies in the computation of the type B Hermite polynomials $\mathrm{e}^{-\Delta_h/2} x_E \zeta_\lambda(y)$ (where $E = \{j : 1 \leq j \leq k\}$ for some k).

Proposition 11.4.4 *For $\alpha \in \mathbb{N}_0^d$ and $\kappa \geq 0, j \in \mathbb{N}_0$,*

$$\left(\sum_{i=1}^d y_i\right)^j \zeta_\alpha(y) = j! \sum_{|\beta|=|\alpha|+j} \binom{\beta}{\alpha}_\kappa \frac{\mathscr{E}_+(\alpha) h(\alpha^+, \kappa+1)}{\mathscr{E}_+(\beta) h(\beta^+, \kappa+1)} \zeta_\beta(y).$$

Proof The adjoint of multiplication by $(\sum_{i=1}^d y_i)^j$ in the h inner product is $(\sum_{i=1}^d \mathscr{D}_i^A)^j$. In an abstract sense, if T is a linear operator with matrix $M_{\alpha\beta}$, where $T\zeta_\alpha = \sum_\beta M_{\beta\alpha} \zeta_\beta$, then the adjoint T^* has matrix

$$M^*_{\alpha\beta} = M_{\beta\alpha} \frac{\langle \zeta_\beta, \zeta_\beta \rangle_h}{\langle \zeta_\alpha, \zeta_\alpha \rangle_h}.$$

By Proposition 10.6.5 and Theorem 10.6.6, $\zeta_\gamma(\mathbf{1}^d) = \mathscr{E}_-(\gamma)(d\kappa+1)_{\gamma^+}/h(\gamma^+, 1)$ and $\langle \zeta_\gamma, \zeta_\gamma \rangle_h = \mathscr{E}_+(\gamma)\mathscr{E}_-(\gamma)(d\kappa+1)_{\gamma^+} h(\gamma^+, \kappa+1)/h(\gamma^+, 1)$ for any $\gamma \in \mathbb{N}_0^d$. □

In the B inner product the adjoint of multiplication by $(\sum_{i=1}^d y_i)^j = (\sum_{i=1}^d x_i^2)^j$ is Δ_h^j. For $\gamma \in \mathbb{N}_0^d$ and $0 \leq k \leq d$, let

$$E_k = \{j : 1 \leq j \leq k\},$$

$$\eta_k = \sum_{i=1}^k \varepsilon_i = (1, 1, \ldots, 1, 0, \ldots, 0).$$

Proposition 11.4.5 *For $\alpha \in \mathbb{N}_0^d$ and $\kappa \geq 0$, $j \in \mathbb{N}_0$, $0 \leq k \leq d$,*

$$(4^j j!)^{-1}\Delta_h^j x_{E_k}\zeta_\alpha(y) = \sum_{|\beta|=|\alpha|-j}\left[\binom{\alpha}{\beta}_\kappa \frac{(d\kappa+1)_{\alpha^+}((d-1)\kappa+\kappa'+\frac{1}{2})_{(\alpha+\eta_k)^+}h(\beta,1)}{(d\kappa+1)_{\beta^+}((d-1)\kappa+\kappa'+\frac{1}{2})_{(\beta+\eta_k)^+}h(\alpha,1)} x_{E_k}\zeta_\beta(y)\right].$$

Proof We use the adjoint formula from Proposition 11.4.4. From Theorem 11.3.10, for any $\gamma \in \mathbb{N}_0^d$,

$$\langle x_{E_k}\zeta_\gamma(y), x_{E_k}\zeta_\gamma(y)\rangle_B = 2^{2|\gamma|+k}(d\kappa+1)_{\gamma^+}((d-1)\kappa+\kappa'+\tfrac{1}{2})_{(\gamma+\eta_k)^+}\frac{h(\gamma,\kappa+1)}{h(\gamma,1)}.$$

This proves the formula. □

Corollary 11.4.6 *For $\alpha \in \mathbb{N}_0^d$ and $|\alpha| = m$, the lowest-degree term in the expression $e^{-\Delta_h/2}x_{E_k}\zeta_\alpha(y)$ is*

$$\begin{aligned}\frac{(-\frac{1}{2}\Delta_h)^m}{m!}x_{E_k}\zeta_\alpha(y) &= (-2)^m(d\kappa+1)_{\alpha^+}\frac{((d-1)\kappa+\kappa'+\frac{1}{2})_{(\alpha+\eta_k)^+}}{((d-1)\kappa+\kappa'+\frac{1}{2})_{\eta_k}}\frac{\mathcal{E}_-(\alpha)}{h(\alpha^+,1)}x_{E_k} \\ &= (-2)^m\frac{((d-1)\kappa+\kappa'+\frac{1}{2})_{(\alpha+\eta_k)^+}}{\prod_{i=1}^k[(d-i)\kappa+\kappa'+\frac{1}{2}]}x_{E_k}\zeta_\alpha(\mathbf{1}^d).\end{aligned}$$

Proof It is clear from the definition that $\binom{\alpha}{\mathbf{0}}_\kappa = 1$, where $\mathbf{0} \in \mathbb{N}_0^d$. Further, $\zeta_\alpha(\mathbf{1}^d) = \mathcal{E}_-(\alpha)(d\kappa+1)_{\alpha^+}/h(\alpha^+,1)$. □

We turn to the problem posed by $\sum_{i=1}^d \mathcal{D}_i^A\zeta_\alpha$. Recall from Chapter 8 the definition of the invariant subspace E_λ associated with $\lambda \in \mathbb{N}_0^{d,P}$, namely, $E_\lambda = \operatorname{span}\{w\zeta_\lambda : w \in S_d\} = \operatorname{span}\{\zeta_\alpha : \alpha^+ = \lambda\}$. This space is invariant under S_d and the operators $\mathcal{D}_i^A y_i$. Also recall, from Section 10.8, the intertwining operator V, with $\mathcal{D}_i^A V = V\partial/\partial y_i$, and the utility operator ξ, where $\xi p_\alpha = y^\alpha/\alpha!$.

Theorem 11.4.7 *For $\lambda \in \mathbb{N}_0^{d,P}$, if $q(y) \in E_\lambda$ and $1 \leq i \leq d$ then*

$$\mathcal{D}_i^A q = \sum_{\lambda_j > \lambda_{j+1}} ((d-j+1)\kappa+\lambda_j)q_j,$$

where $q_j \in E_{\lambda-\varepsilon_j}$ and $\mathcal{D}_i^A\rho_i q_j = (\xi_j(\lambda)-1)q_j$, for j such that $\lambda_j > \lambda_{j+1}$.

Proof Expand q in the p-basis; then $q = \rho_i g + q_0$, where $\mathcal{D}_i^A q_0 = 0$. This is obtained by setting q_0 as the part of the expansion in which all appearing p_α

have $\alpha_i = 0$ (and in $\rho_i g$ each term p_α has $\alpha_i \geq 1$, by the definition of ρ_i). Expand g as a sum $\Sigma_\mu g_\mu$ of projections onto E_μ with $|\mu| = |\lambda| - 1$, that is,

$$q = \rho_i \sum_\mu g_\mu + q_0, \qquad \mathscr{D}_i^A q = \mathscr{D}_i^A \rho_i \sum_\mu g_\mu.$$

Apply $V\xi$ to both sides of the first equation. It was shown in Section 10.8 that $V\xi$ is scalar on each E_μ with eigenvalue $1/(d\kappa+1)_\mu$. Also, $(\partial/\partial y_i)\xi p_\alpha(y) = 0$ if $\alpha_i = 0$ and $(\partial/\partial y_i)\xi p_\alpha(y) = \xi p_{\alpha-\varepsilon_i}$ if $\alpha_i \geq 1$. Thus $[1/(d\kappa+1)_\lambda]\mathscr{D}_i^A q = \mathscr{D}_i^A V\xi q = V(\partial/\partial y_i)\xi(\rho_i \Sigma_\mu g_\mu + q_0) = V\xi \Sigma_\mu g_\mu = \Sigma_\mu [1/(d\kappa+1)_\mu] g_\mu$, and so

$$\mathscr{D}_i^A q = \sum_\mu \frac{(d\kappa+1)_\lambda}{(d\kappa+1)_\mu} g_\mu = \sum_\mu \mathscr{D}_i^A \rho_i g_\mu.$$

Since each E_μ is invariant under $\mathscr{D}_i^A \rho_i$, this shows that g_μ is an eigenfunction of $\mathscr{D}_i^A \rho_i$ with eigenvalue $(d\kappa+1)_\lambda/(d\kappa+1)_\mu$. The eigenvalues of $\mathscr{D}_i^A \rho_i = (1,i)\mathscr{D}_1^A \rho_1 (1,i)$ are the same as those of $\mathscr{U}_1^A = \mathscr{D}_1^A \rho_1$, namely, $\xi_1(\alpha)$ for $\alpha \in \mathbb{N}_0^d$; additionally, each eigenvalue is a linear polynomial in d (and for fixed $\zeta_\alpha \in E_\mu$, that is, $\alpha^+ = \mu$, the coefficients of ζ_α in the p-basis are independent of d, where $\alpha_s = 0$ for all $s > k$ and $d \geq k$), so we conclude that $(d\kappa+1)_\mu$ is a factor of $(d\kappa+1)_\lambda$ as a polynomial in d. This implies that $\mu_j \leq \lambda_j$ for each j. Combine this with the restriction that $\mu \in \mathbb{N}_0^{d,P}$ and $|\mu| = |\lambda| - 1$ to show that $\mu = \lambda - \varepsilon_j$ for some j with $\lambda_j > \lambda_{j+1}$; this includes the possibility that $\lambda_d > 0$. Also,

$$\frac{(d\kappa+1)_\lambda}{(d\kappa+1)_{\lambda-\varepsilon_j}} = (d-j+1)\kappa + \lambda_j = \xi_j(\lambda) - 1.$$

□

Corollary 11.4.8 *For $\alpha, \beta \in \mathbb{N}_0^d$ and $\kappa \geq 0$, $\binom{\alpha}{\beta}_\kappa \neq 0$ implies that $\alpha^+ \supseteq \beta^+$.*

Proof Theorem 11.4.7 shows that $\binom{\alpha}{\beta}_\kappa \neq 0$ and $|\beta| = |\alpha| - 1$ implies that $\beta^+ = \alpha^+ - \varepsilon_j$ for some j with $\alpha_j^+ > \alpha_{j+1}^+$, that is, $\alpha^+ \supseteq \beta^+$. Propositions 11.4.2 and 11.4.3 and an inductive argument finish the proof. □

Next we consider a stronger statement that can be made about the expression $\Sigma_{i=1}^d \mathscr{D}_i^A \zeta_\lambda$ for $\lambda \in \mathbb{N}_0^{d,P}$. The following is a slight modification of Lemma 10.5.3.

Lemma 11.4.9 *For $\alpha \in \mathbb{N}_0^d$ and $1 \leq i, k \leq d$ the following hold:*
if $i < k$ then

$$\Big(\mathscr{D}_i^A \rho_i - \kappa \sum_{j<i} (j,i) - \kappa(i,k)\Big) \mathscr{D}_k^A \zeta_\alpha = \xi_i(\alpha) \mathscr{D}_k^A \zeta_\alpha;$$

if $i > k$ then

$$\mathscr{U}_i^A \mathscr{D}_k^A \zeta_\alpha = \xi_i(\alpha) \mathscr{D}_k^A \zeta_\alpha + \kappa(k,i) \mathscr{D}_i^A \zeta_\alpha.$$

Further,

$$\mathscr{D}_k^A \rho_k \mathscr{D}_k^A \zeta_\alpha = [\xi_k(\alpha) - 1]\mathscr{D}_k^A \zeta_\alpha - \kappa \sum_{j>k} (k,j)\mathscr{D}_j^A \zeta_\alpha.$$

Proof The case $i < k$ is part (i) of Lemma 10.5.3. For $i > k$, we have

$$\mathscr{D}_i^A \rho_i \mathscr{D}_k^A \zeta_\alpha = \mathscr{D}_k^A \mathscr{D}_i^A \rho_i \zeta_\alpha + \kappa(i,k)\mathscr{D}_k^A \zeta_\alpha$$

and

$$-\kappa \sum_{j<i} (j,i)\mathscr{D}_k^A \zeta_\alpha = -\kappa \mathscr{D}_k^A \sum_{j<i, j\neq k} (j,i)\zeta_\alpha - \kappa(k,i)\mathscr{D}_k^A \zeta_\alpha.$$

Add the two equations to obtain

$$\begin{aligned}\mathscr{U}_i^A \mathscr{D}_k^A \zeta_\alpha &= \mathscr{D}_k^A \left(\mathscr{D}_i^A \rho_i - \kappa \sum_{j<i, j\neq k} (j,i) \right) \zeta_\alpha \\ &= \mathscr{D}_k^A \mathscr{U}_i^A \zeta_\alpha + \kappa \mathscr{D}_k^A (k,i)\zeta_\alpha.\end{aligned}$$

This proves the second part.

Finally, $\mathscr{D}_k^A \rho_k \mathscr{D}_k^A \zeta_\alpha = \mathscr{D}_k^A [\mathscr{D}_k^A \rho_k - \kappa \sum_{j\neq k}(j,k) - 1]\zeta_\alpha = \mathscr{D}_k^A (\mathscr{U}_k^A - 1)\zeta_\alpha - \kappa \mathscr{D}_k^A \sum_{j>k}(j,k)\zeta_\alpha$ and $\mathscr{D}_k^A (j,k) = (j,k)\mathscr{D}_j^A$. □

It is convenient in the following to have a notation for projections. For $\mu \in \mathbb{N}_0^{d,P}$, denote the orthogonal projection of $f(y) \in \Pi^d$ onto E_μ by $\pi(\mu)f$, so that

$$f = \sum_\mu \pi(\mu) f \qquad \text{and} \qquad \pi(\mu) f \in E_\mu.$$

The formulae in Lemma 11.4.9 can be applied to the projections $\pi(\lambda - \varepsilon_j)$ of $\mathscr{D}_k^A \zeta_\alpha$, for $\lambda = \alpha^+$ and $\lambda_j > \lambda_{j+1}$, because $E_{\lambda - \varepsilon_j}$ is invariant under $\mathscr{D}_i^A \rho_i$ and elements of S_d.

Proposition 11.4.10 *For $\lambda \in \mathbb{N}_0^{d,P}$, if $1 \le j < k$ and $\lambda_j > \lambda_{j+1}$ then we have $\pi(\lambda - \varepsilon_j)\mathscr{D}_k^A \zeta_\lambda = 0$.*

Proof Let $g = \pi(\lambda - \varepsilon_j)\mathscr{D}_k^A \zeta_\lambda$. By Lemma 11.4.9, $[\mathscr{D}_i^A \rho_i - \kappa \sum_{j<i}(j,i) - \kappa(i,k)]g = \xi_i(\lambda) g$ for all $i < k$. The operators on the left-hand side can be expressed as $w^{-1}\mathscr{U}_{i+1}^A w$ for the cyclic permutation $w = (1,2,\ldots,k) \in S_d$, with eigenvalues drawn from $\{\xi_s(\lambda - \varepsilon_j) : 1 \le s \le d\}$ on the space $E_{\lambda - \varepsilon_j}$. These eigenvalues are $\xi_1(\lambda), \xi_2(\lambda), \ldots, \xi_j(\lambda) - 1, \ldots, \xi_k(\lambda), \ldots$; $\xi_j(\lambda)$ does not appear in this list, hence $g = 0$. □

Note that this argument uses the fact that λ is a partition in a crucial way.

Lemma 11.4.11 *For $\lambda \in \mathbb{N}_0^{d,P}$, $\lambda_j > \lambda_{j+1}$ and $k \le j$, let $g_k = \pi(\lambda - \varepsilon_j)\mathscr{D}_k^A \zeta_\lambda$. Then*

$$\left(\mathscr{D}_i^A\rho_i - \kappa\sum_{s<i}(s,i) - \kappa(i,k)\right)g_k = \xi_i(\lambda)g_k \qquad \textit{for } i<k;$$
$$\mathscr{D}_k^A\rho_k g_k = [\xi_k(\lambda)-1]g_k - \kappa\sum_{i=k+1}^{j}(k,i)g_i;$$
$$\mathscr{U}_i^A g_k = \xi_i(\lambda)g_k + \kappa(k,i)g_i \qquad \textit{for } k<i\le j;$$
$$\mathscr{U}_i^A g_k = \xi_i(\lambda)g_k \qquad \textit{for } i>j.$$

Proof This proceeds by an application of the projection $\pi(\lambda-\varepsilon_j)$ to each formula in Lemma 11.4.9 in combination with the fact that $\pi(\lambda-\varepsilon_j)\mathscr{D}_i^A\zeta_\lambda = 0$ for $j<i$ (Proposition 11.4.10). □

As in Definition 10.5.1, let $\theta_j = (1,2)(2,3)\cdots(j-1,j)\in S_d$. The following has almost the same proof as Theorem 10.5.5.

Proposition 11.4.12 *For $\lambda\in\mathbb{N}_0^{d,P}$ and $\lambda_j>\lambda_{j+1}$, set*

$$\widetilde{\lambda} = (\lambda_j-1,\lambda_1,\ldots,\lambda_{j-1},\lambda_{j+1},\ldots);$$

then

$$\pi(\lambda-\varepsilon_j)\mathscr{D}_j^A\zeta_\lambda = [(d-j+1)\kappa+\lambda_j]\theta_j^{-1}\zeta_{\widetilde{\lambda}}.$$

Proof Let $g=\pi(\lambda-\varepsilon_j)\mathscr{D}_j^A\zeta_\lambda$. By Lemma 11.4.11, g satisfies the following:

$$\left(\mathscr{D}_i^A\rho_i - \kappa\sum_{s<i}(s,i) - \kappa(i,k)\right)g = \xi_i(\lambda)g, \qquad \text{for } i<j;$$
$$\mathscr{D}_j^A\rho_j g = [\xi_j(\lambda)-1]g,$$
$$\mathscr{U}_i^A g = \xi_i(\lambda)g \qquad \text{for } i>j.$$

Just as in Lemma 10.5.3, these equations show that $\theta_j g$ is a scalar multiple of $\zeta_{\widetilde{\lambda}}$, where the multiple is the coefficient of $p_{\lambda-\varepsilon_j}$ in $\mathscr{D}_j^A\zeta_\lambda$. Use Lemma 10.5.4 (the hypothesis $m=\ell(\lambda)$ can be replaced by $\lambda_j>\lambda_{j+1}$), which implies that $p_{\lambda-\varepsilon_j}$ appears with a nonzero coefficient in $\mathscr{D}_j^A p_\alpha$ if one of the following holds: (1) $\alpha=\lambda$, with coefficient $(d-j+1)\kappa+\lambda_j$; (2) $\alpha=\lambda+k(\varepsilon_j-\varepsilon_i)$ with coefficient $-\kappa$ where $1\le k\le\lambda_i-\lambda_j$ (this implies that $i>j$ and $\lambda\succ\lambda+k(\varepsilon_j-\varepsilon_i)$ or, more precisely, for $k<\lambda_i-\lambda_j, \lambda\succ[\lambda+k(\varepsilon_j-\varepsilon_i)]^+$ and for $k=\lambda_i-\lambda_j, \lambda\succ(i,j)\lambda = \lambda+k(\varepsilon_j-\varepsilon_i)$, so this case cannot occur); (3) $\alpha=\lambda+k(\varepsilon_i-\varepsilon_j)$ with coefficient κ and $1\le k\le\lambda_j-\lambda_i-1$, thus $\lambda_i\le\lambda_j-2$ and $i>j$, but this case cannot occur either, because $\lambda\succ\lambda+k(\varepsilon_i-\varepsilon_j)$ when $i>j$. □

The part of Lemma 11.4.11 dealing with $\mathscr{D}_k^A\rho_k g_k$ actually contains more information than appears at first glance. It was shown in Theorem 11.4.7 that $\pi(\lambda-\varepsilon_j)\mathscr{D}_k^A\zeta_\lambda$ is an eigenfunction of $\mathscr{D}_k^A\rho_k$ with eigenvalue $\xi_j(\lambda)-1$. Thus the lemma implies that $[\xi_k(\lambda) - \xi_j(\lambda)]g_k = \kappa \sum_{i=k+1}^{j}(k,i)g_i$; but this provides an algorithm for computing $g_{j-1}, g_{j-2}, \ldots, g_1$ from g_j, which was obtained in Proposition 11.4.12.

Algorithm 11.4.13 For $\lambda \in \mathbb{N}_0^{d,P}$ and $\lambda_j > \lambda_{j+1}$,

$$\pi(\lambda - \varepsilon_j) \sum_{i=1}^{d} \mathscr{D}_i^A \zeta_\lambda = [(d-j+1)\kappa + \lambda_j]\phi_{j-1}$$

where

$$\begin{aligned} \phi_{1-j} &= \zeta_{\lambda-\varepsilon_j}, \\ \phi_{1-i} &= \Big((i,i+1) - \frac{\kappa}{\xi_i(\lambda) - \xi_j(\lambda) + 1}\Big)\phi_{-i} \qquad \text{for } i = j-1, j-2, \ldots, 1, \\ \phi_i &= \Big((i,i+1) + \frac{\kappa}{\xi_i(\lambda) - \xi_j(\lambda)}\Big)\phi_{i-1} \qquad \text{for } i = 1, 2, \ldots, j-1. \end{aligned}$$

Proof Let $g_i = [(d-j+1)\kappa + \lambda_j]^{-1}\pi(\lambda - \varepsilon_j)\mathscr{D}_i^A \zeta_\lambda$; then $g_i = 0$ for $i > j$ (by Proposition 11.4.10), so we require to show that $\phi_{j-1} = \sum_{i=1}^{j} g_i$. Note that these polynomials have been rescaled from those in Lemma 11.4.11, but the same equations hold. For convenience let $b_i = \xi_i(\lambda) - \xi_j(\lambda) = \lambda_i - \lambda_j + \kappa(j-i), 1 \le i < j$. Then the lemma shows that

$$b_i g_i = \kappa \sum_{s=i+1}^{j} (i,s) g_s.$$

By Proposition 11.4.12,

$$g_j = \Big(1 - \frac{\kappa(1,j)}{b_1 + 1}\Big)\Big(1 - \frac{\kappa(2,j)}{b_2+1}\Big) \cdots \Big(1 - \frac{\kappa(j-1,j)}{b_{j-1}+1}\Big)\zeta_{\lambda-\varepsilon_j}.$$

Next we show that

$$\begin{aligned} g_i &= \frac{\kappa}{b_i}\Big(1 + \frac{\kappa}{b_{j-1}}(j-1,j)\Big) \cdots \Big(1 + \frac{\kappa}{b_{i+1}}(i+1,j)\Big)(i,j)g_j, \\ \sum_{s=i}^{j} g_s &= \Big(1 + \frac{\kappa}{b_{j-1}}(j-1,j)\Big)\Big(1 + \frac{\kappa}{b_{j-2}}(j-2,j)\Big) \cdots \Big(1 + \frac{\kappa}{b_i}(i,j)\Big) g_j, \end{aligned}$$

arguing inductively for $i = j-1, j-2, \ldots, 1$. The first step, $i = j-1$, follows immediately from Lemma 11.4.11. Suppose that the formulae are valid for $g_{i+1}, \ldots, g_j$; then

$$\begin{aligned} g_i &= \frac{\kappa}{b_i} \sum_{s=i+1}^{j} (i,s) g_s \\ &= \frac{\kappa}{b_i} \sum_{s=i+1}^{j} \Big(1 + \frac{\kappa}{b_{j-1}}(j-1,j)\Big) \cdots \Big(1 + \frac{\kappa}{b_{s+1}}(s+1,j)\Big)\frac{\kappa}{b_s}(i,s)(s,j)g_j \end{aligned}$$

$$
\begin{aligned}
&= \frac{\kappa}{b_i}\sum_{s=i+1}^{j}\left(1+\frac{\kappa}{b_{j-1}}(j-1,j)\right)\cdots\left(1+\frac{\kappa}{b_{s+1}}(s+1,j)\right)\frac{\kappa}{b_s}(s,j)(i,j)g_j \\
&= \frac{\kappa}{b_i}\left(1+\frac{\kappa}{b_{j-1}}(j-1,j)\right)\cdots\left(1+\frac{\kappa}{b_{i+1}}(i+1,j)\right)(i,j)g_j.
\end{aligned}
$$

This uses $(s,i)(s,j)=(s,j)(i,j)$; the last step follows using simple algebra (for example, $(1+t_1)(1+t_2)(1+t_3)=1+t_1+(1+t_1)t_2+(1+t_1)(1+t_2)t_3$). The formula for $\sum_{s=i}^{j} g_s$ follows easily from that for g_i.

To obtain the stated formulae, replace g_j by $\theta_j^{-1}(\theta_j g_j)$. Then

$$
\theta_j g_j = \left((1,2)-\frac{\kappa}{b_1+1}\right)\left((2,3)-\frac{\kappa}{b_2+1}\right)\cdots\left((j-1,j)-\frac{\kappa}{b_{j-1}+1}\right)\zeta_{\lambda-\varepsilon_j},
$$

and the remaining factors are transformed as follows:

$$
\begin{aligned}
&\left(1+\frac{\kappa}{b_{j-1}}(j-1,j)\right)\left(1+\frac{\kappa}{b_{j-2}}(j-2,j)\right)\cdots\left(1+\frac{\kappa}{b_1}(1,j)\right) \\
&\times (j-1,j)\ldots(2,3)(1,2) \\
&\quad = \left((j-1,j)+\frac{\kappa}{b_{j-1}}\right)\left((j-2,j-1)+\frac{\kappa}{b_{j-2}}\right)\cdots\left((1,2)+\frac{\kappa}{b_1}\right).
\end{aligned}
$$

This completes the proof. □

To compute the expansion of $\pi(\lambda-\varepsilon_j)\sum_{i=1}^{d}\mathscr{D}_i^A\zeta_\lambda$ in terms of ζ_α with $\alpha^+ = \lambda-\varepsilon_j$, the algorithm can be considered to have the starting point $\phi_0=\zeta_{\tilde{\lambda}}$. Each step involves an adjacent transposition which has a relatively simple action on each ζ_α (recall Proposition 10.4.5). Indeed, let $\sigma=(i,i+1)$. Then: $\sigma\zeta_\alpha=\zeta_\alpha$ if $\alpha_i=\alpha_{i+1}$; $\sigma\zeta_\alpha=\zeta_{\sigma\alpha}+c\zeta_\alpha$ if $\alpha_i>\alpha_{i+1}$; $\sigma\zeta_\alpha=(1-c^2)\zeta_{\sigma\alpha}+c\zeta_\alpha$ if $\alpha_i<\alpha_{i+1}$, where $c=\kappa[\xi_i(\alpha)-\xi_{i+1}(\alpha)]^{-1}$. So, in general (with distinct parts of λ), the expansion consists of 2^{j-1} distinct ζ_α.

11.5 Hermite Polynomials of Type B

In this section we examine the application of the operator $\exp(-u\Delta_h)$ to produce orthogonal polynomials for weights of Gaussian type, namely,

$$
w(x)^2 e^{-v\|x\|^2}, \qquad x\in\mathbb{R}^d
$$

where

$$
w(x)=\prod_{i=1}^{d}|x_i|^{\kappa'}\prod_{j<k}\left|x_j^2-x_k^2\right|^{\kappa}
$$

and $uv=\frac{1}{4}$, $v>0$. In the following let $\gamma=\deg w(x)=d\kappa'+d(d-1)\kappa$.

Lemma 11.5.1 *Suppose that $f\in\mathscr{P}_m^d$, $g\in\mathscr{P}_n^d$ for some $m,n\in\mathbb{N}_0, v>0$ and $u=1/(4v)$; then*

$$\langle f,g\rangle_B = (2v)^{\gamma+(m+n)/2}\left(\frac{v}{\pi}\right)^{d/2} b\left(\kappa,\kappa'\right) \times \int_{\mathbb{R}^d}\left(e^{-u\Delta_h}f\right)\left(e^{-u\Delta_h}g\right)w(x)^2 e^{-v\|x\|^2}\,dx.$$

Proof In the formula to be proved, with $v=\frac{1}{2}$, make the change of variables $x=(2v)^{1/2}z$, that is, set $x_i=(2v)^{1/2}z_i$ for each i. Then $dx=(2v)^{d/2}\,dz$, and the weight function contributes the factor $(2v)^\gamma$. For $j\le\frac{n}{2}$ let $f_j(x)=\Delta_h^j f(x)$; then $e^{-\Delta_h/2}f(x)=\sum_j\frac{1}{j!}(-\frac{1}{2})^j f_j(x)$, which transforms to

$$\sum_j\frac{1}{j!}\left(\frac{-1}{2}\right)^j(2v)^{n/2-j}f_j(z)=(2v)^{n/2}\sum_j\frac{1}{j!}\left(\frac{-1}{4v}\right)^j f_j(z)=(2v)^{n/2}e^{-u\Delta_h}f.$$

A similar calculation for g finishes the proof. □

Of course, it is known that $\langle f,g\rangle_B=0$ for $m\ne n$.

As in Section 10.9, the commuting self-adjoint operators $\mathcal{U}_i$ can be transformed to commuting self-adjoint operators on $L^2(\mathbb{R}^d;w^2e^{-v\|x\|^2}dx)$. For $v>0$ and $u=1/(4v), 1\le i\le d$, let

$$\mathcal{U}_i^v=e^{-u\Delta_h}\mathcal{U}_i e^{u\Delta_h}.$$

Proposition 11.5.2 *For* $1\le i\le d$,

$$\mathcal{U}_i^v=\mathcal{D}_i x_i-2u\mathcal{D}_i^2-\kappa\sum_{j<i}(\sigma_{ji}+\tau_{ji}).$$

Proof Because Δ_h commutes with each $w\in W_d$ it is clear that $e^{-u\Delta_h}(\sigma_{ji}+\tau_{ji})e^{u\Delta_h}=\sigma_{ji}+\tau_{ji}$ for $j<i$. From the commutation $\Delta_h x_i-x_i\Delta_h=2\mathcal{D}_i$, we argue inductively that $\Delta_h^n x_i-x_i\Delta_h^n=2n\mathcal{D}_i\Delta_h^{n-1}$ for $n\in\mathbb{N}_0$. By summing this equation (multiplying it by $(-u)^n/n!$) we obtain

$$e^{-u\Delta_h}x_i=(x_i-2u\mathcal{D}_i)e^{-u\Delta_h}.$$

Finally, apply $\mathcal{D}_i$ to the left-hand side of this equation to prove the stated formula. □

The construction of an orthogonal basis labeled by $\mathbb{N}_0^d$ in terms of the nonsymmetric Jack polynomials proceeds as follows. For $\alpha\in\mathbb{N}_0^d$,

1. let $E=\{i:\alpha_i \text{ is odd}\}$;
2. if E is empty let $w=1$, otherwise let $k=\#(E), E_k=\{i:1\le i\le k\}$ and let w be the unique permutation ($\in S_d$) such that $w(E_k)=E$ and $w(i)<w(j)$ whenever $1\le i<j\le k$ or $k+1\le i<j\le d$;
3. let $\beta=w^{-1}(\alpha)$, that is, $\beta_i=\alpha_{w(i)}$ for each i;
4. let $H_\alpha(x;\kappa,\kappa',v)=e^{-u\Delta_h}wx_{E_k}\zeta_{e(\beta)}(y)$ (recall that $e(\beta)_i=\lfloor\frac{1}{2}\beta_i\rfloor$).

By Propositions 11.3.4 and 11.5.2, $H_\alpha(x)$ is a $\{\mathscr{U}_i^\nu\}$ simultaneous eigenfunction such that

$$\mathscr{U}_{w(i)}^\nu H_\alpha(x) = 2[\xi_i(\beta) - \kappa]H_\alpha(x) \qquad \text{for } 1 \le i \le k,$$
$$\mathscr{U}_{w(i)}^\nu H_\alpha(x) = 2[\xi_i(\beta) - \kappa + \kappa' - \tfrac{1}{2}]H_\alpha(x) \qquad \text{for } k < i \le d.$$

Since w acts isometrically, we have

$$\begin{aligned}
&(2\nu)^\gamma \left(\frac{\nu}{\pi}\right)^{d/2} b\left(\kappa,\kappa'\right) \int_{\mathbb{R}^d} H_\alpha(x;\kappa,\kappa',\nu)^2 w^2(x) \mathrm{e}^{-\nu\|x\|^2}\mathrm{d}x \\
&= (2\nu)^{-|\alpha|} \left\langle x_{E_k}\zeta_{e(\beta)}(y), x_{E_k}\zeta_{e(\beta)}(y)\right\rangle_B \\
&= \nu^{-|\alpha|}(d\kappa+1)_{e(\alpha)^+}((d-1)\kappa+\kappa'+\tfrac{1}{2})_{o(\alpha)^+} \\
&\quad \times \mathscr{E}_+(e(\beta))\mathscr{E}_-(e(\beta))\frac{h(e(\alpha)^+,\kappa+1)}{h(e(\alpha)^+,1)}.
\end{aligned}$$

The substitutions $e(\beta)^+ = e(\alpha)^+, o(\beta)^+ = o(\alpha)^+$ are valid since $\alpha = w(\beta)$. The most common specializations of ν are $\nu = \frac{1}{2}$ and $\nu = 1$ (when the polynomials $H_\alpha(x)$ reduce to the products of ordinary Hermite polynomials for $\kappa = 0 = \kappa'$).

11.6 Calogero–Sutherland Systems

We begin with the time-independent Schrödinger wave equation for d particles in a one-dimensional space: particle i has mass m_i, coordinate x_i and is subject to an external potential $V_i(x_i)$ and the interaction between particles i,j is given by $V_{ij}(x_i - x_j), i < j$. The force acting on i due to j is $F_{ij} = -(\partial/\partial x_i)V_{ij}(x_i - x_j)$, which equals $-F_{ji} = -(\partial/\partial x_j)V_{ij}(x_i - x_j)$ according to Newton's third law. The Hamiltonian of the system is

$$\mathscr{H} = -\frac{\hbar^2}{2}\sum_{i=1}^d \frac{1}{m_i}\left(\frac{\partial}{\partial x_i}\right)^2 + \sum_{i=1}^d V_i(x_i) + \sum_{1\le i<j\le d} V_{ij}(x_i - x_j),$$

and the Schrödinger equation is

$$\mathscr{H}\psi(x) = E\psi(x),$$

where the eigenvalue E is the energy level associated with the wave function ψ. The quantum mechanical interpretation of ψ is that $|\psi(x)|^2$ is the probability density function for the location of the particles (which is independent of time by hypothesis). This interpretation requires that the integral of $|\psi(x)|^2$ over the state space must be 1 (this is one reason for the importance of computing L^2-norms of orthogonal polynomials!). Planck's constant is used in the form $\hbar = h/(2\pi)$. We will consider only systems for which the Hamiltonian is invariant under the action of the symmetric group S_d and the interactions are multiples of $\left|x_i - x_j\right|^{-2}$ (called $1/r^2$ interactions, where r denotes the distance between two particles). For systems on the line $\mathbb{R}$ the external potential which we will consider is that corresponding to harmonic confinement (producing a simple harmonic oscillator). In the mathematical analysis of $\mathscr{H}$ it is customary to change the scales and units in such a way that the equations have as few parameters as possible. The term *complete integrability* refers to the existence of d independent and commuting differential operators, one of which is $\mathscr{H}$. The relevance to orthogonal polynomials is that Calogero–Sutherland systems with $1/r^2$ interactions have wave functions of the form $p(x)\psi_0(x)$, where p is a polynomial and $\psi_0(x)$ is the base state (the state of lowest energy). This situation can be illustrated by the harmonic oscillator.

11.6.1 The simple harmonic oscillator

Consider a particle of mass m moving on a line subject to the restoring force $F(x) = -kx$ or, equivalently, with potential $V(x) = kx^2/2$, $x \in \mathbb{R}$. The Schrödinger wave equation is

$$-\frac{\hbar^2}{2m}\frac{\mathrm{d}^2}{\mathrm{d}x^2}\psi(x) + \frac{kx^2}{2}\psi(x) = E\psi(x).$$

Write the equation as

$$\mathscr{H}_1\psi(x) = -\frac{\mathrm{d}^2}{\mathrm{d}x^2}\psi(x) + v^2x^2\psi(x) = \lambda\psi(x),$$

where v is a positive constant and λ denotes the rescaled energy level. Ignoring the question of normalization for now, let the base state be $\psi_0(x) = \exp(-vx^2/2)$; then by an easy calculation it can be shown that

$$\mathscr{H}_1(p(x)\psi_0(x)) = \psi_0(x)[-(\mathrm{d}/\mathrm{d}x)^2 + 2xv\mathrm{d}/\mathrm{d}x + v]p(x).$$

For each $n \in \mathbb{N}_0$ let $p(x) = H_n(\sqrt{v}x)$; then by using the corresponding differential equation for the Hermite polynomials, Subsection 1.4.1, namely $-H_n''(x) + 2xH_n'(x) = 2nH_n(x)$, we obtain

$$\mathscr{H}_1[H_n(\sqrt{v}x)\psi_0(x)] = (2n+1)vH_n(\sqrt{v}x)\psi_0(x).$$

This shows that the energy levels are evenly spaced and

$$\{(2^n n!\sqrt{\pi/v})^{-1/2}H_n(\sqrt{v}x)\exp(-vx^2/2) : n \in \mathbb{N}_0\}$$

is a complete set of wave functions, an orthonormal basis for $L^2(\mathbb{R})$.

The machinery of Section 11.5 for the case $d = 1, \kappa' = 0$ asserts that the appropriately scaled Hermite polynomials are obtained from $\exp\{[-u(\mathrm{d}/\mathrm{d}x)^2]x^n\}$ (with $u = 1/(4v), n \in \mathbb{N}_0$); indeed

$$\exp\left\{-\frac{1}{4v}\left(\frac{\mathrm{d}}{\mathrm{d}x}\right)^2 x^n\right\} = 2^{-n}v^{-n/2}H_n(\sqrt{v}x),$$

using the formula in Proposition 1.4.3.

11.6.2 Root systems and the Laplacian

The following calculation is useful for any root system related to a Calogero model. Recall the notation of Section 4.4, with a multiplicity function κ_v defined on a root system R.

Lemma 11.6.1 *Let $\psi(x) = \prod_{v\in R_+} |\langle x,v\rangle|^{\kappa_v}$ and let f be a smooth function on $\mathbb{R}^d$; then*

$$\psi(x)^{-1}\Delta[\psi(x)f(x)]$$
$$= \Delta f(x) + 2\sum_{v\in R_+} \kappa_v \frac{\langle \nabla f(x), v\rangle}{\langle x,v\rangle} + \sum_{v\in R_+} \frac{\kappa_v(\kappa_v - 1)\|v\|^2}{\langle x,v\rangle^2} f(x).$$

Proof From the product rule for Δ we have

$$\psi^{-1}\Delta\psi = \Delta + 2\sum_{i=1}^{d} \frac{1}{\psi}\frac{\partial\psi}{\partial x_i}\frac{\partial}{\partial x_i} + \frac{1}{\psi}\Delta\psi.$$

The middle term follows from

$$\frac{1}{\psi}\frac{\partial\psi}{\partial x_i} = \sum_{v\in R_+} \kappa_v \frac{v_i}{\langle x,v\rangle}.$$

It remains to compute $\Delta\psi/\psi$. Indeed,

$$\frac{1}{\psi}\left(\frac{\partial}{\partial x_i}\right)^2 \psi = \left(\sum_{v\in R_+} \kappa_v \frac{v_i}{\langle x,v\rangle}\right)^2 - \sum_{v\in R_+} \kappa_v \frac{v_i^2}{\langle x,v\rangle^2}$$

and

$$\frac{1}{\psi}\Delta\psi = -\sum_{v\in R_+} \kappa_v \frac{\langle v,v\rangle}{\langle x,v\rangle^2} + \sum_{u,v\in R_+} \kappa_u\kappa_v \frac{\langle u,v\rangle}{\langle x,u\rangle\langle x,v\rangle}.$$

In the second sum the subset of terms for which $\sigma_u\sigma_v = w$ for a fixed rotation $\neq 1$ adds to zero, by Lemma 6.4.6. This leaves the part with $u = v$ and thus $\Delta\psi/\psi = \sum_{v\in R_+} \left[\kappa_v(\kappa_v - 1)\langle v,v\rangle/\langle x,v\rangle^2\right]$. The proof is complete. □

Lemma 11.6.1 will be used here just for operator types A and B.

11.6.3 Type A models on the line

Suppose that there are d identical particles on the line $\mathbb{R}$, each subject to an external potential v^2x^2 (the simple harmonic oscillator potential), with $1/r^2$ interactions and with parameter κ. The Hamiltonian is

$$\mathscr{H} = -\Delta + v^2\|x\|^2 + 2\kappa \sum_{1\le i<j\le d} \frac{\kappa - 1}{(x_i - x_j)^2}.$$

The base state is $\psi_0(x) = \exp(-v\|x\|^2/2)\prod_{i<j}|x_i - x_j|^\kappa$ for $x \in \mathbb{R}^d$. Let $a(x) = \prod_{i<j}(x_i - x_j)$. To facilitate the computations we will introduce two commutation

rules (the second is the type A case of Lemma 11.6.1; $f(x)$ is a smooth function on $x \in \mathbb{R}^d$). The commutation rules are as follows:

$$\exp(v\|x\|^2/2)\Delta(\exp(-v\|x\|^2/2)f(x)) = \Big(\Delta - 2v\sum_{i=1}^d x_i\frac{\partial}{\partial x_i} + v(v\|x\|^2 - d)\Big)f(x), \tag{11.6.1}$$

$$|a(x)|^{-\kappa}\Delta[|a(x)|^{\kappa}f(x)] = \left(\Delta + 2\kappa\sum_{i<j}\frac{1}{x_i - x_j}\left(\frac{\partial}{\partial x_i} - \frac{\partial}{\partial x_j}\right) + 2\kappa(\kappa-1)\sum_{i<j}\frac{1}{(x_i-x_j)^2}\right)f(x).$$

Combining the two equation (and noting that $\sum_{i=1}^d x_i(\partial/\partial x_i)\,|a(x)|^{\kappa} = [\frac{1}{2}\kappa d(d-1)\,|a(x)|^{\kappa}]$, we obtain

$$\psi_0(x)^{-1}\mathscr{H}[\psi_0(x)f(x)] = \Bigg[-\Delta - 2\kappa\sum_{1\le i<j\le d}\frac{1}{x_i - x_j}\left(\frac{\partial}{\partial x_i} - \frac{\partial}{\partial x_j}\right) + 2v\sum_{i=1}^d x_i\frac{\partial}{\partial x_i} + dv[1+\kappa(d-1)]\Bigg]f(x).$$

We recall the type-A Laplacian

$$\Delta_h = \sum_{i=1}^d(\mathscr{D}_i^A)^2 = \Delta + 2\kappa\sum_{1\le i<j\le d}\left[\frac{1}{x_i - x_j}\left(\frac{\partial}{\partial x_i} - \frac{\partial}{\partial x_j}\right) - \frac{1-(i,j)}{(x_i-x_j)^2}\right],$$

and observe its close relationship to $\psi_0^{-1}\mathscr{H}\psi_0$; for symmetric polynomials the terms $1-(i,j)$ vanish. Further (Proposition 6.5.3),

$$\sum_{i=1}^d x_i\mathscr{D}_i^A f(x) = \sum_{i=1}^d x_i\frac{\partial}{\partial x_i}f(x) + \kappa\sum_{i<j}[1-(i,j)]f(x).$$

It was shown (Section 10.8) that $e^{-u\Delta_h}\mathscr{U}_i^A e^{u\Delta_h} = \mathscr{U}_i^A - 2u(\mathscr{D}_i^A)^2$, so that, with $uv = \frac{1}{4}$,

$$2ve^{-u\Delta_h}\sum_{i=1}^d\mathscr{U}_i^A e^{u\Delta_h} = -\Delta_h + 2v\sum_{i=1}^d\mathscr{U}_i^A = -\Delta_h + 2v\left(\sum_{i=1}^d x_i\frac{\partial}{\partial x_i} + d + \frac{\kappa}{2}d(d+1)\right).$$

Thus, when restricted to symmetric polynomials the conjugate of the Hamiltonian can be expressed as

$$\psi_0^{-1}\mathscr{H}\psi_0 = 2ve^{-u\Delta_h}\sum_{i=1}^d\mathscr{U}_i^A e^{u\Delta_h} - dv(1+2\kappa).$$

For any partition $\lambda \in \mathbb{N}_0^{d,P}$ there is an invariant eigenfunction j_λ of $\sum_{i=1}^d\mathscr{U}_i^A$ with eigenvalue $\sum_{i=1}^d\xi_i(\lambda) = d + \frac{\kappa}{2}d(d+1) + |\lambda|$ (where $|\lambda| = \sum_i\lambda_i$). Further, in Definition 10.7.3 we constructed invariant differential operators commuting with $\sum_{i=1}^{d}\mathscr{U}_i^A$ by restricting $\mathscr{T}_j$ to symmetric polynomials, where $\sum_{j=1}^{d}t^j\mathscr{T}_j = \prod_{i=1}^{d}(1+t\,\mathscr{U}_i^A)$ (with a formal variable t). Thus $\mathscr{T}_j' = \mathrm{e}^{-u\Delta_h}\mathscr{T}_j\mathrm{e}^{u\Delta_h}$ is the elementary symmetric polynomial of degree j in $\{\mathrm{e}^{-u\Delta_h}\mathscr{U}_i^A\mathrm{e}^{u\Delta_h} : 1\le i\le d\}$; each $\mathrm{e}^{-u\Delta_h}\mathscr{U}_i^A\mathrm{e}^{u\Delta_h}$ is a second-order differential–difference operator and $\mathscr{T}_j'$ maps symmetric polynomials onto themselves. By the same proof as that of Theorem 10.7.4, $\mathscr{T}_j'$ when restricted to symmetric polynomials acts as a differential operator of degree $2j$. This shows the complete integrability of $\mathscr{H}$; each operator $\psi_0\mathscr{T}_j'\psi_0^{-1}$ commutes with it.

Given $\lambda\in\mathbb{N}_0^{d,P}$, let $\psi_\lambda = (\mathrm{e}^{-u\Delta_h}j_\lambda)\psi_0$; then $\mathscr{H}\psi_\lambda = v[2|\lambda|+d+\kappa d(d-1)]\psi_\lambda$. This provides a complete orthogonal decomposition of the symmetric wave functions for indistinguishable particles.

Next we consider a simple modification to the Hamiltonian to allow nonsymmetric wave functions. The physical interpretation is that each particle has a two-valued spin, which can be exchanged with another particle. Refer to the expression for Δ_h, and add

$$2\kappa\sum_{i<j}\frac{1-(i,j)}{(x_i-x_j)^2}$$

to the potential. The modified Hamiltonian is

$$\mathscr{H}' = -\Delta+v^2\|x\|^2+2\kappa\sum_{1\le i<j\le d}\frac{\kappa-(i,j)}{(x_i-x_j)^2}.$$

This operator has the same symmetric eigenfunctions as $\mathscr{H}$ and in addition has eigenfunctions $(\mathrm{e}^{-u\Delta_h}\zeta_\alpha)\psi_0$, where ζ_α is the nonsymmetric Jack polynomial for $\alpha\in\mathbb{N}_0^d$; the eigenvalues are given by $v[2|\alpha|+d+\kappa d(d-1)]$. The transpositions (i,j) are called exchange operators in the physics context, since they exchange the spin values of two particles.

This establishes the complete integrability of the linear Calogero–Moser system with harmonic confinement, $1/r^2$ interactions and exchange terms. The wave functions involve Hermite polynomials of type A. The normalizations can be determined from the results in Section 10.7.

11.6.4 Type *A* models on the circle

Consider d identical particles located on the unit circle at polar coordinates $(\theta_1,\theta_2,\dots,\theta_d)$, with no external potential and inverse-square law $(1/r^2)$ interactions. The chordal distance between particles j and k is $2\left|\sin[\frac{1}{2}(\theta_j-\theta_k)]\right|$. The unique nature of the $1/r^2$ interaction is illustrated by the identity

$$\frac{1}{4\sin^2\frac{1}{2}\theta} = \sum_{n\in\mathbb{Z}}\frac{1}{(\theta-2\pi n)^2},$$

which shows that the particles can be considered to be on a line. They can be considered as the countable union of copies of one particle; these copies are located in 2π-periodically repeating positions on the line and they all interact according to the $1/r^2$ law. The coupling constant in the potential is written as $2\kappa(\kappa-1)$, and $\kappa > 1$. The Hamiltonian is

$$\mathscr{H}_\theta = -\sum_{i=1}^{d}\left(\frac{\partial}{\partial\theta_i}\right)^2 + \frac{\kappa}{2}\sum_{1\le i<j\le d}\frac{\kappa-1}{\sin^2[\frac{1}{2}(\theta_i-\theta_j)]}.$$

Now change the coordinate system to $x_j = \exp(\mathrm{i}\theta_j)$ for $1\le j\le d$; this transforms trigonometric polynomials to Laurent polynomials in x (polynomials in $x_i^{\pm 1}$), and we have $-(\partial/\partial\theta_i)^2 = (x_i\partial_i)^2$. Further, $\sin^2[\frac{1}{2}(\theta_i-\theta_j] = \frac{1}{4}(x_i-x_j)(x_i^{-1}-x_j^{-1})$ for $i\ne j$. In the x-coordinate system the Hamiltonian is

$$\mathscr{H} = \sum_{i=1}^{d}(x_i\partial_i)^2 - 2\kappa(\kappa-1)\sum_{1\le i<j\le d}\frac{x_ix_j}{(x_i-x_j)^2}.$$

The (non-normalized) base state is

$$\begin{aligned}\psi_0(x) &= \prod_{1\le i<j\le d}\left|(x_i-x_j)(x_i^{-1}-x_j^{-1})\right|^{\kappa/2}\\ &= \prod_{1\le i<j\le d}|x_i-x_j|^{\kappa}\prod_{i=1}^{d}|x_i|^{-\kappa(d-1)/2}.\end{aligned}$$

Even though $|x_i|=1$ on the unit torus, the function ψ_0 is to be interpreted as a positively homogeneous function of degree 0 on $\mathbb{C}^d$.

Proposition 11.6.2 *For any smooth function f on $\mathbb{C}^d$,*

$$\begin{aligned}&\psi_0(x)^{-1}\sum_{i=1}^{d}(x_i\partial_i)^2[\psi_0(x)f(x)]\\ &\quad= \sum_{i=1}^{d}(x_i\partial_i)^2 f(x) + 2\kappa\sum_{1\le i<j\le d}x_ix_j\left(\frac{\partial_i-\partial_j}{x_i-x_j} - \frac{\kappa-1}{(x_i-x_j)^2}\right)f(x)\\ &\quad\quad+\kappa(d-1)\sum_{i=1}^{d}x_i\partial_i f(x) + \frac{\kappa^2}{12}d(d^2-1)f(x).\end{aligned}$$

Proof Define the operator $\delta_i = \psi_0^{-1}x_i\partial_i\psi_0$; then, by logarithmic differentiation,

$$\delta_i = x_i\partial_i + \frac{\kappa}{2}\sum_{j\ne i}\frac{x_i+x_j}{x_i-x_j}$$

and so

$$\begin{aligned}\delta_i^2 &= (x_i\partial_i)^2 + 2\kappa\sum_{j\ne i}\frac{x_ix_j\partial_i}{x_i-x_j} + \kappa(d-1)x_i\partial_i\\ &\quad -\kappa\sum_{j\ne i}\frac{x_ix_j}{(x_i-x_j)^2} + \frac{\kappa^2}{4}\left(\sum_{j\ne i}\frac{x_i+x_j}{x_i-x_j}\right)^2.\end{aligned}$$

Now sum this expression over $1 \le i \le d$. Expand the squared-sum term as follows:

$$\sum_{i=1}^{d} \sum_{j,k \neq i} \left(\frac{x_i^2}{(x_i - x_j)(x_i - x_k)} + \frac{x_i x_j + x_i x_k + x_j x_k}{(x_i - x_j)(x_i - x_k)} \right).$$

When $j = k$, the sum contributes $d(d-1) + 8\sum_{i<j} x_i x_j/(x_i - x_j)^2$ (each pair $\{i, j\}$ with $i < j$ appears twice in the triple sum, and $(x_i + x_j)^2 = (x_i - x_j)^2 + 4x_i x_j$). When i, j, k are all distinct, the first terms each contributes $2\binom{d}{3}$ and the second terms cancel out (the calculations are similar to those in the proof of Theorem 10.7.9). Finally,

$$\frac{\kappa^2}{4}\left(d(d-1) + \frac{1}{3} d(d-1)(d-2)\right) = \tfrac{1}{12}\kappa^2 d(d^2 - 1).$$

□

Using Proposition 10.7.8 and Theorem 10.7.9 the transformed Hamiltonian can be expressed in terms of $\{\mathscr{U}_i^A\}$. Indeed,

$$\begin{aligned}
&\sum_{i=1}^{d} \left[\mathscr{U}_i^A - 1 - \tfrac{1}{2}\kappa(d+1)\right)^2 \\
&\quad = \sum_{i=1}^{d} (x_i \partial_i)^2 + 2\kappa \sum_{1 \le i < j \le d} x_i x_j \left(\frac{\partial_i - \partial_j}{x_i - x_j} - \frac{1 - (i,j)}{(x_i - x_j)^2} \right) \\
&\quad\quad + \kappa(d-1) \sum_{i=1}^{d} x_i \partial_i + \tfrac{1}{12}\kappa^2 d(d^2 - 1).
\end{aligned}$$

To show this, write

$$\begin{aligned}
&\sum_{i=1}^{d} [\mathscr{U}_i^A - 1 - \tfrac{1}{2}\kappa(d+1)]^2 \\
&\quad = \sum_{i=1}^{d} (\mathscr{U}_i^A - 1 - \kappa d)^2 + \kappa(d-1) \sum_{i=1}^{d} (\mathscr{U}_i^A - 1 - \kappa d) + d[\tfrac{1}{2}\kappa(d-1)]^2
\end{aligned}$$

and use Proposition 10.7.8. As for the line model, the potential can be modified by exchange terms (interchanging spins of two particles). The resulting spin Hamiltonian is

$$\mathscr{H}' = \sum_{i=1}^{d} (x_i \partial_i)^2 - 2\kappa \sum_{1 \le i < j \le d} \frac{x_i x_j}{(x_i - x_j)^2} [\kappa - (i,j)]$$

and, for a smooth function $f(x)$,

$$\psi_0(x)^{-1} \mathscr{H}' [\psi_0(x) f(x)] = \sum_{i=1}^{d} \left[\mathscr{U}_i^A - 1 - \tfrac{1}{2}\kappa(d+1)\right]^2 f(x).$$

This immediately leads to a complete eigenfunction decomposition for $\mathscr{H}'$ in terms of nonsymmetric Jack polynomials. Let $e_d = \prod_{i=1}^{d} x_i$ (where e_d is an elementary symmetric polynomial); then $\mathcal{U}_i^A[e_d^m f(x)] = e_d^m(\mathcal{U}_i^A + m)f(x)$ for any $m \in \mathbb{Z}$ (see Proposition 6.4.12) and the eigenvalue is shifted by m.

Lemma 11.6.3 *Let $\alpha \in \mathbb{N}_0^d$ and $m \in \mathbb{N}_0$; then $e_d^m \zeta_\alpha$ is a scalar multiple of ζ_β, where $\beta = \alpha + m\mathbf{1}^d$.*

Proof Both $e_d^m \zeta_\alpha$ and ζ_β are eigenfunctions of $\mathcal{U}_i^A$ with eigenvalue $\xi_i(\alpha) + m$, for $1 \le i \le d$. □

Corollary 11.6.4 *The Laurent polynomials $e_d^m \zeta_\alpha$ with $m \in \mathbb{Z}$ and $\alpha \in \mathbb{N}_0^d$ with* $\min_i \alpha_i = 0$ *form a basis for the set of all Laurent polynomials.*

Proof For any Laurent polynomial $g(x)$ there exists $m \in \mathbb{Z}$ such that $e_d^m g(x)$ is a polynomial in $x_1, \dots, x_d$. Lemma 11.6.3 shows that there are no linear dependences in the set. □

This leads to the determination of the energy levels.

Theorem 11.6.5 *Let $\lambda \in \mathbb{N}_0^{d,P}, \lambda_d = 0, m \in \mathbb{Z}$ and $\alpha^+ = \lambda$; then*

$$\psi(x) = \psi_0(x) e_d^m \zeta_\alpha(x)$$

is a wave function for $\mathcal{H}'$, and

$$\mathcal{H}'\psi = \sum_{i=1}^{d} \Big[\lambda_i + m + \tfrac{1}{2}\kappa(d+1-2i)\Big]^2 \psi.$$

The normalization constant for ψ was found in Theorems 10.6.13 and 10.6.16. The eigenvalues can also be expressed as $\sum_{i=1}^d (\lambda_i + m)[\lambda_i + m + \kappa(d+1-2i)] + \frac{1}{2}\kappa^2 d(d^2-1)$. The complete integrability of $\mathcal{H}$ (as an invariant differential operator) is a consequence of Theorem 10.7.4

11.6.5 Type B models on the line

Suppose that there are d identical particles on the line at coordinates $x_1, x_2, \dots, x_d$, subject to an external harmonic confinement potential, with $1/r^2$ interactions between both the particles and their mirror images; as well, there is a barrier at the origin, and spin can be exchanged between particles. The Hamiltonian with exchange terms for this spin type-B Calogero model is

$$\begin{aligned}\mathcal{H} = & -\Delta + v^2\|x\|^2 + \sum_{i=1}^{d} \frac{\kappa'(\kappa' - \sigma_i)}{x_i^2} \\ & + 2\kappa \sum_{1 \le i < j \le d} \left(\frac{\kappa - \sigma_{ij}}{(x_i - x_j)^2} + \frac{\kappa - \tau_{ij}}{(x_i + x_j)^2} \right),\end{aligned}$$

where v, κ, κ' are positive parameters. The base state is

$$\psi_0(x) = \mathrm{e}^{-v\|x\|^2/2} \prod_{1\le i<j\le d} \left|x_i^2 - x_j^2\right|^{\kappa} \prod_{i=1}^{d} |x_i|^{\kappa'}.$$

By Lemma 11.6.1 specialized to type-B models, and the commutation for $\mathrm{e}^{-v\|x\|^2/2}$ from equation (11.6.1), we have

$$\begin{aligned}\psi_0(x)^{-1}\mathscr{H}\psi_0(x) &= -\Delta - \kappa' \sum_{i=1}^{d}\left(\frac{2}{x_i}\partial_i - \frac{1-\sigma_i}{x_i^2}\right)\\ &\quad - 2\kappa \sum_{1\le i<j\le d}\left(\frac{\partial_i - \partial_j}{x_i - x_j} - \frac{1-\sigma_{ij}}{(x_i-x_j)^2} + \frac{\partial_i+\partial_j}{x_i+x_j} - \frac{1-\tau_{ij}}{(x_i+x_j)^2}\right)\\ &\quad + 2v\sum_{i=1}^{d} x_i\partial_i + v[d + 2\kappa d(d-1) + 2\kappa' d]\\ &= -\Delta_h + 2v\sum_{i=1}^{d} x_i\partial_i + vd[1 + 2\kappa(d-1) + 2\kappa'].\end{aligned}$$

Using Propositions 6.4.10 and 6.5.3 we obtain

$$\begin{aligned}&\sum_{i=1}^{d}\mathscr{D}_i x_i - \kappa \sum_{1\le i<j\le d}(\sigma_{ij}+\tau_{ij})\\ &\qquad = \sum_{i=1}^{d} x_i\partial_i + d[1+\kappa'+\kappa(d-1)] + \kappa'\sum_{i=1}^{d}\sigma_i.\end{aligned}$$

Let $u = 1/(4v)$. Then, by Proposition 11.5.2,

$$2v\mathrm{e}^{-u\Delta_h}\sum_{i=1}^{d}\mathscr{U}_i\mathrm{e}^{u\Delta_h} = -\Delta_h + 2v\sum_{i=1}^{d}x_i\partial_i + 2vd[1+\kappa'+\kappa(d-1)] + 2v\kappa'\sum_{i=1}^{d}\sigma_i.$$

Thus

$$\psi_0(x)^{-1}\mathscr{H}\psi_0(x) = v\mathrm{e}^{-u\Delta_h}\left(2\sum_{i=1}^{d}\mathscr{U}_i - d - 2\kappa'\sum_{i=1}^{d}\sigma_i\right)\mathrm{e}^{u\Delta_h}.$$

Apply this operator to the Hermite polynomial $H_\alpha(x;\kappa,\kappa',v)$, $\alpha \in \mathbb{N}_0^d$, constructed in Section 11.5; thus

$$\mathscr{H}(\psi_0 H_\alpha) = 2v(|\alpha| + d[\tfrac{1}{2} + (d-1)\kappa + \kappa'])\psi_0 H_\alpha.$$

The normalization constants were obtained in Section 11.5. The type-B version of Theorem 10.7.4 is needed to establish the complete integrability of this quantum model. As in Definition 10.7.3, define the operators $\mathscr{T}_j$ by

$$\sum_{i=1}^{d} t^i \mathscr{T}_i = \prod_{j=1}^{d} (1 + t\mathscr{U}_j).$$

Theorem 11.6.6 *For $1 \le i \le d$, the operators $\mathscr{T}_i$ commute with each $w \in W_d$.*

Proof We need only consider the reflections in W_d. Clearly $\sigma_j \mathscr{U}_i = \mathscr{U}_i \sigma_j$ for each j. It now suffices to consider adjacent transpositions $\sigma_{j,j+1}$, since $\sigma_i \sigma_{ij} \sigma_i = \tau_{ij}$. Fix j and let $\sigma = \sigma_{j,j+1}$; then $\sigma \mathscr{U}_i = \mathscr{U}_i \sigma$ whenever $i \ne j, j+1$. Also, $\sigma \mathscr{U}_j = \mathscr{U}_{j+1}\sigma + \kappa(\sigma + \tau_{j,j+1})$. It is now clear that $\sigma(\mathscr{U}_j + \mathscr{U}_{j+1})\sigma = \mathscr{U}_j + \mathscr{U}_{j+1}$ and $\sigma(\mathscr{U}_j \mathscr{U}_{j+1})\sigma = \mathscr{U}_j \mathscr{U}_{j+1}$ (note that $\sigma \tau_{ij} \sigma = \tau_{i,j+1}$ for $i \ne j, j+1$). □

The operators $\psi_0 \mathrm{e}^{-u\Delta_h} \mathscr{T}_i \mathrm{e}^{u\Delta_h} \psi_0^{-1}$ commute with the Hamiltonian $\mathscr{H}$. When restricted to W_d-invariant functions they all coincide with invariant differential operators.

11.7 Notes

Symmetries of the octahedral type have appeared in analysis for decades, if not centuries. Notably, the quantum mechanical descriptions of atoms in certain crystals are given as spherical harmonics with octahedral symmetry (once called "cubical harmonics"). Dunkl [1984c, 1988] studied the type-B weight functions but only for invariant polynomials, using what is now known as the restriction of the h-Laplacian. Later, in Dunkl [1999c] most of the details for Sections 9.2 and 9.3 were worked out. The derivation of the Macdonald–Mehta–Selberg integral in Section 9.3 is essentially based on the technique in Opdam [1991], where Opdam proved Macdonald's conjectures about integrals related to the crystallographic root systems and situated on the torus (more precisely, Euclidean space modulo the root lattice). The original proof for the type-B integral used a limiting argument on Selberg's integral (see Askey [1980]). For the importance of Selberg's integral and its numerous applications, see the recent survey by Forrester and Warnaar [2008]. Generalized binomial coefficients for nonsymmetric Jack polynomials were introduced by Baker and Forrester [1998], but Algorithm 11.4.13 is new. There is a connection between these coefficients and the theory of shifted, or nonhomogeneous, Jack polynomials. The Hermite polynomials, which are called Laguerre polynomials by some authors in cases where the parities are the same for all variables, have appeared in several articles. The earlier articles, by Lassalle [1991a, b, c], Yan [1992] and Baker and Forrester [1997a], dealt with the invariant cases only. Later, when the nonsymmetric Jack polynomials became more widely known, the more general Hermite polynomials were studied, again by Baker and Forrester [1998], Rösler and Voit [1998] and Ujino and Wadati [1995, 1997], who called them "Hi-Jack" polynomials. The idea of using $e^{-u\Delta_h}$ to generate bases for Gaussian-type weight functions appeared in Dunkl [1991].

The study of Calogero–Sutherland systems has been carried on by many authors. We refer to the book *Calogero–Moser–Sutherland models*, edited by van Diejen and Vinet [2000] for a more comprehensive and up-to-date bibliography. We limited our discussion to the role of the type-A and type-B polynomials in the wave functions and did not attempt to explain the physics of the theory or physical objects which can be usefully modeled by these systems. Calogero [1971] and Sutherland [1971, 1972] proved that inverse square potentials for identical particles in a one-dimensional space produce exactly solvable models. Before their work, only the Dirac-delta interaction (collisions) model had been solved. Around the same time, the Jack polynomials were constructed (Jack [1970/71], Stanley [1989]). Eventually they found their way into the Sutherland systems, and Lapointe and Vinet [1996] used physics concepts such as annihilation and creation operators to prove new results on the Jack polynomials and also applied them to the wave function problem. Other authors whose papers have a significant component dealing with the use of orthogonal polynomials are Kakei [1998], Ujino and Wadati [1999], Nishino, Ujino and Wadati [1999] and Uglov [2000]. The type-B model on the line with spin terms was first studied by Yamamoto [1995]. Kato and Yamamoto [1998] used group-invariance methods to evaluate correlation integrals for these models.

Genest, Ismail, Vinet and Zhedanov [2013] analyzed in great detail the Hamiltonians associated with the differential–difference operators for $\mathbb{Z}_2^2$ with weight function $|x_1|^{\kappa_1}|x_2|^{\kappa_2}$. This work was continued by Genest, Vinet and Zhedanov [2013]. The generalized Hermite polynomials appear in their analyses of the wave functions.

The theory of nonsymmetric Jack polynomials has been extended to the family of complex reflection groups named $G(n,1,d)$, that is, the group of $d\times d$ permutation matrices whose nonzero entries are the nth roots of unity (so $G(1,1,d)$ is the same as the symmetric group S_d and $G(2,1,d)$ is the same as W_d) in Dunkl and Opdam [2003].

References

Abramowitz, M. and Stegun, I. (1970). *Handbook of Mathematical Functions*, Dover, New York.

Agahanov, C. A. (1965). A method of constructing orthogonal polynomials of two variables for a certain class of weight functions (in Russian), *Vestnik Leningrad Univ.* **20**, no. 19, 5–10.

Akhiezer, N. I. (1965). *The Classical Moment Problem and Some Related Questions in Analysis*, Hafner, New York.

Aktaş, R. and Xu, Y. (2013) Sobolev orthogonal polynomials on a simplex, *Int. Math. Res. Not.*, **13**, 3087–3131.

de Álvarez, M., Fernández, L., Pérez, T. E. and Piñar, M. A. (2009). A matrix Rodrigues formula for classical orthogonal polynomials in two variables, *J. Approx. Theory* **157**, 32–52.

Andrews, G. E., Askey, R. and Roy, R. (1999). *Special Functions*, Encyclopedia of Mathematics and its Applications 71, Cambridge University Press, Cambridge.

Aomoto, K. (1987). Jacobi polynomials associated with Selberg integrals, *SIAM J. Math. Anal.* **18**, no. 2, 545–549.

Appell, P. and de Fériet, J. K. (1926). *Fonctions hypergéométriques et hypersphériques, polynomes d'Hermite*, Gauthier-Villars, Paris.

Area, I., Godoy, E., Ronveaux, A. and Zarzo, A. (2012). Bivariate second-order linear partial differential equations and orthogonal polynomial solutions. *J. Math. Anal. Appl.* **387**, 1188–1208.

Askey, R. (1975). *Orthogonal Polynomials and Special Functions*, Regional Conference Series in Applied Mathemathics 21, SIAM, Philadelphia.

Askey, R. (1980). Some basic hypergeometric extensions of integrals of Selberg and Andrews, *SIAM J. Math. Anal.* **11**, 938–951.

Atkinson, K. and Han, W. (2012). *Spherical Harmonics and Approximations on the Unit Sphere: An Introduction*, Lecture Notes in Mathematics 2044, Springer, Heidelberg.

Atkinson, K. and Hansen, O. (2005). Solving the nonlinear Poisson equation on the unit disk, *J. Integral Eq. Appl.* **17**, 223–241.

Axler, S., Bourdon, P. and Ramey, W. (1992). *Harmonic Function Theory*, Springer, New York.

Bacry, H. (1984). Generalized Chebyshev polynomials and characters of $GL(\mathbf{N},\mathbf{C})$ and $SL(\mathbf{N},\mathbf{C})$, in *Group Theoretical Methods in Physics*, pp. 483–485, Lecture Notes in Physics, 201, Springer-Verlag, Berlin.

Badkov, V. (1974). Convergence in the mean and almost everywhere of Fourier series in polynomials orthogonal on an interval, *Math. USSR-Sb.* **24**, 223–256.

Bailey, W. N. (1935). *Generalized Hypergeometric Series*, Cambridge University Press, Cambridge.

Baker, T. H., Dunkl, C. F. and Forrester, P. J. (2000). Polynomial eigenfunctions of the Calogero–Sutherland–Moser models with exchange terms, in *Calogero–Sutherland–Moser Models*, pp. 37–51, CRM Series in Mathematical Physics, Springer, New York.

Baker, T. H. and Forrester, P. J. (1997a). The Calogero–Sutherland model and generalized classical polynomials, *Comm. Math. Phys.* **188**, 175–216.

Baker, T. H. and Forrester, P. J. (1997b). The Calogero–Sutherland model and polynomials with prescribed symmetry, *Nuclear Phys.* B **492**, 682–716.

Baker, T. H. and Forrester, P. J. (1998). Nonsymmetric Jack polynomials and integral kernels, *Duke Math. J.* **95**, 1–50.

Barrio, R., Peña, J. M. and Sauer, T. (2010). Three term recurrence for the evaluation of multivariate orthogonal polynomials, *J. Approx. Theory* **162**, 407–420.

Beerends, R. J. (1991). Chebyshev polynomials in several variables and the radial part of the Laplace–Beltrami operator, *Trans. Amer. Math. Soc.* **328**, 779–814.

Beerends, R. J. and Opdam, E. M. (1993). Certain hypergeometric series related to the root system *BC*, *Trans. Amer. Math. Soc.* **339**, 581–609.

Berens, H., Schmid, H. and Xu, Y. (1995a). On two-dimensional definite orthogonal systems and on lower bound for the number of associated cubature formulas, *SIAM J. Math. Anal.* **26**, 468–487.

Berens, H., Schmid, H. and Xu, Y. (1995b). Multivariate Gaussian cubature formula, *Arch. Math.* **64**, 26–32.

Berens, H. and Xu, Y. (1996). Fejér means for multivariate Fourier series, *Math. Z.* **221**, 449–465.

Berens, H. and Xu, Y. (1997). $\ell-1$ summability for multivariate Fourier integrals and positivity, *Math. Proc. Cambridge Phil. Soc.* **122**, 149–172.

Berg, C. (1987). The multidimensional moment problem and semigroups, in *Moments in Mathematics*, pp. 110–124, Proceedings of Symposia in Applied Mathematics 37, American Mathemathical Society, Providence, RI.

Berg, C., Christensen, J. P. R. and Ressel, P. (1984). *Harmonic Analysis on Semigroups, Theory of Positive Definite and Related Functions*, Graduate Texts in Mathematics 100, Springer, New York–Berlin.

Berg, C. and Thill, M. (1991). Rotation invariant moment problems, *Acta Math.* **167**, 207–227.

Bergeron, N. and Garsia, A. M. (1992). Zonal polynomials and domino tableaux, *Discrete Math.* **99**, 3–15.

Bertran, M. (1975). Note on orthogonal polynomials in v-variables, *SIAM. J. Math. Anal.* **6**, 250–257.

Bojanov, B. and Petrova, G. (1998). Numerical integration over a disc, a new Gaussian quadrature formula, *Numer. Math.* **80**, 39–50.

Bos, L. (1994). Asymptotics for the Christoffel function for Jacobi like weights on a ball in $\mathbb{R}^m$, *New Zealand J. Math.* **23**, 99–109.

Bos, L., Della Vecchia, B. and Mastroianni, G. (1998). On the asymptotics of Christoffel functions for centrally symmetric weights functions on the ball in $\mathbb{R}^n$, *Rend. del Circolo Mat. di Palermo*, **52**, 277–290.

Bracciali, C. F., Delgado, A. M., Fernández, L., Pérez, T. E. and Piñar, M. A. (2010). New steps on Sobolev orthogonality in two variables, *J. Comput. Appl. Math.* **235**, 916–926.

Braaksma, B. L. J. and Meulenbeld, B. (1968) Jacobi polynomials as spherical harmonics, *Indag. Math.* **30**, 384–389.

Buchstaber, V., Felder, G. and Veselov, A. (1994). Elliptic Dunkl operators, root systems, and functional equations, *Duke Math. J.* **76**, 885–891.

Calogero, F. (1971). Solution of the one-dimensional N-body problems with quadratic and/or inversely quadratic pair potentials, *J. Math. Phys.* **12**, 419–436.

zu Castell, W., Filbir, F. and Xu, Y. (2009). Cesàro means of Jacobi expansions on the parabolic biangle, *J. Approx. Theory* **159**, 167–179.

Cheney, E. W. (1998). *Introduction to Approximation Theory*, reprint of the 2nd edn (1982), AMS Chelsea, Providence, RI.

Cherednik, I. (1991). A unification of Knizhnik–Zamolodchikov and Dunkl operators via affine Hecke algebras, *Invent. Math.* **106**, 411–431.

Chevalley, C. (1955). Invariants of finite groups generated by reflections, *Amer. J. Math.* **77**, 777–782.

Chihara, T. S. (1978). *An Introduction to Orthogonal Polynomials*, Mathematics and its Applications 13, Gordon and Breach, New York.

Cichoń, D., Stochel, J. and Szafraniec, F. H. (2005). Three term recurrence relation modulo ideal and orthogonality of polynomials of several variables, *J. Approx. Theory* **134**, 11–64.

Connett, W. C. and Schwartz, A. L. (1995). Continuous 2-variable polynomial hypergroups, in *Applications of Hypergroups and Related Measure Algebras* (Seattle, WA, 1993), pp. 89–109, Contemporary Mathematics 183, American Mathematical Society, Providence, RI.

van der Corput, J. G. and Schaake, G. (1935). Ungleichungen für Polynome und trigonometrische Polynome, *Compositio Math.* **2**, 321–361.

Coxeter, H. S. M. (1935). The complete enumeration of finite groups of the form $R_i^2 = \left(R_iR_j\right)^{k_{ij}} = 1$, *J. London Math. Soc.* **10**, 21–25.

Coxeter, H. S. M. (1973). *Regular Polytopes*, 3rd edn, Dover, New York.

Coxeter, H. S. M. and Moser, W. O. J. (1965). *Generators and Relations for Discrete groups*, 2nd edn, Springer, Berlin–New York.

Dai, F. and Wang, H. (2010). A transference theorem for the Dunkl transform and its applications, *J. Funct. Anal.* **258**, 4052–4074.

Dai, F. and Xu, Y. (2009a). Cesàro means of orthogonal expansions in several variables, *Const. Approx.* **29**, 129–155.

Dai, F. and Xu, Y. (2009b). Boundedness of projection operators and Cesàro means in weighted L^p space on the unit sphere, *Trans. Amer. Math. Soc.* **361**, 3189–3221.

Dai, F. and Xu, Y. (2013). *Approximation Theory and Harmonic Analysis on Spheres and Balls*, Springer, Berlin–NewYork.

Debiard, A. and Gaveau, B. (1987). Analysis on root systems, *Canad. J. Math.* **39**, 1281–1404.

Delgado, A. M., Fernández, L, Pérez, T. E., Piñar, M. A. and Xu, Y. (2010). Orthogonal polynomials in several variables for measures with mass points, *Numer. Algorithm* **55**, 245–264.

Delgado, A., Geronimo, J. S., Iliev, P. and Marcellán, F. (2006). Two variable orthogonal polynomials and structured matrices, *SIAM J. Matrix Anal. Appl.* **28**, 118–147.

Delgado, A. M., Geronimo, J. S., Iliev, P. and Xu, Y. (2009). On a two variable class of Bernstein–Szego measures, *Constr. Approx.* **30**, 71–91.

DeVore, R. A. and Lorentz, G. G. (1993). Constructive approximation, Grundlehren der Mathematischen Wissenschaften 303, Springer, Berlin.

van Diejen, J. F. (1999). Properties of some families of hypergeometric orthogonal polynomials in several variables, *Trans. Amer. Math. Soc.* **351**, no. 1, 233–270.

van Diejen, J. F. and Vinet, L. (2000). *Calogero–Sutherland–Moser models*, CRM Series in Mathematical Physics, Springer, New York.

Dieudonné, J. (1980). *Special Functions and Linear Representations of Lie Groups*, CBMS Regional Conference Series in Mathematics 42, American Mathematical Society, Providence, RI.

Dijksma, A. and Koornwinder, T. H. (1971). Spherical harmonics and the product of two Jacobi polynomials, *Indag. Math.* **33**, 191–196.

Dubiner, M. (1991). Spectral methods on triangles and other domains, *J. Sci. Comput.* **6**, 345–390.

Dunkl, C. F. (1981). Cube group invariant spherical harmonics and Krawtchouk polynomials, *Math. Z.* **92**, 57–71.

Dunkl, C. F. (1982). An additional theorem for Heisenberg harmonics, in *Proc. Conf. on Harmonic Analysis in Honor of Antoni Zygmund*, pp. 688–705, Wadsworth International, Belmont, CA.

Dunkl, C. F. (1984a). The Poisson kernel for Heisenberg polynomials on the disk, *Math. Z.* **187**, 527–547.

Dunkl, C. F. (1984b). Orthogonal polynomials with symmetry of order three, *Canad. J. Math.* **36**, 685–717.

Dunkl, C. F. (1984c). Orthogonal polynomials on the sphere with octahedral symmetry, *Trans. Amer. Math. Soc.* **282**, 555–575.

Dunkl, C. F. (1985). Orthogonal polynomials and a Dirichlet problem related to the Hilbert transform, *Indag. Math.* **47**, 147–171.

Dunkl, C. F. (1986). Boundary value problems for harmonic functions on the Heisenberg group, *Canad. J. Math.* **38**, 478–512.

Dunkl, C. F. (1987). Orthogonal polynomials on the hexagon, *SIAM J. Appl. Math.* **47**, 343–351.

Dunkl, C. F. (1988). Reflection groups and orthogonal polynomials on the sphere, *Math. Z.* **197**, 33–60.

Dunkl, C. F. (1989a). Differential–difference operators associated to reflection groups, *Trans. Amer. Math. Soc.* **311**, 167–183.

Dunkl, C. F. (1989b). Poisson and Cauchy kernels for orthogonal polynomials with dihedral symmetry, *J. Math. Anal. Appl.* **143**, 459–470.

Dunkl, C. F. (1990). Operators commuting with Coxeter group actions on polynomials, in *Invariant Theory and Tableaux* (Minneapolis, MN, 1988), pp. 107–117, IMA Mathematics and Applications 19, Springer, New York.

Dunkl, C. F. (1991). Integral kernels with reflection group invariance, *Canad. J. Math.* **43**, 1213–1227.

Dunkl, C. F. (1992). Hankel transforms associated to finite reflection groups, in *Hypergeometric Functions on Domains of Positivity, Jack Polynomials, and Applications* (Tampa, FL, 1991), pp. 123–138, Contemporary Mathematics 138, American Mathematical Society, Providence, RI.

Dunkl, C. F. (1995). Intertwining operators associated to the group S_3, *Trans. Amer. Math. Soc.* **347**, 3347–3374.

Dunkl, C. F. (1998a). Orthogonal polynomials of types A and B and related Calogero models, *Comm. Math. Phys.* **197**, 451–487.

Dunkl, C. F. (1998b). Intertwining operators and polynomials associated with the symmetric group, *Monatsh. Math.* **126**, 181–209.

Dunkl, C. F. (1999a). Computing with differential–difference operators, *J. Symbolic Comput.* **28**, 819–826.

Dunkl, C. F. (1999b). Planar harmonic polynomials of type B, *J. Phys. A: Math. Gen.* **32**, 8095–8110.

Dunkl, C. F. (1999c). Intertwining operators of type B_N, in *Algebraic Methods and q-Special Functions* (Montréal, QC, 1996), pp. 119–134, CRM Proceedings Lecture Notes 22, American Mathematical Society, Providence, RI.

Dunkl, C. F. (2007). An intertwining operator for the group B2, *Glasgow Math. J.* **49**, 291–319.

Dunkl, C. F., de Jeu, M. F. E. and Opdam, E. M. (1994). Singular polynomials for finite reflection groups, *Trans. Amer. Math. Soc.* **346**, 237–256.

Dunkl, C. F. and Hanlon, P. (1998). Integrals of polynomials associated with tableaux and the Garsia–Haiman conjecture. *Math. Z.* **228**, 537–567.

Dunkl, C. F. and Luque, J.-G. (2011). Vector-valued Jack polynomials from scratch, *SIGMA* **7**, 026, 48 pp.

Dunkl, C. F. and Opdam, E. M. (2003). Dunkl operators for complex reflection groups, *Proc. London Math. Soc.* **86**, 70–108.

Dunkl, C. F. and Ramirez, D. E. (1971). *Topics in Harmonic Analysis*, Appleton-Century Mathematics Series, Appleton-Century-Crofts (Meredith Corporation), New York.

Dunn, K. B. and Lidl, R. (1982). Generalizations of the classical Chebyshev polynomials to polynomials in two variables, *Czechoslovak Math. J.* **32**, 516–528.

Eier, R. and Lidl, R. (1974). Tschebyscheffpolynome in einer und zwei Variablen, *Abh . Math. Sem. Univ. Hamburg* **41**, 17-27.

Eier, R. and Lidl, R. (1982). A class of orthogonal polynomials in k variables, *Math. Ann.* **260**, 93–99.

Eier, R., Lidl, R. and Dunn, K. B. (1981). Differential equations for generalized Chebyshev polynomials, *Rend. Mat.* **1**, no. 7, 633–646.

Engelis, G. K. (1974). Certain two-dimensional analogues of the classical orthogonal polynomials (in Russian), *Latvian Mathematical Yearbook* **15**, 169–202, 235, Izdat. Zinatne, Riga.

Engels, H. (1980). *Numerical Quadrature and Cubature*, Academic Press, New York.

Erdélyi, A. (1965). Axially symmetric potentials and fractional integration, *SIAM J.* **13**, 216–228.

Erdélyi, A., Magnus, W., Oberhettinger, F. and Tricomi, F. G. (1953). *Higher Transcendental Functions*, McGraw-Hill, New York.

Etingof, E. (2010). A uniform proof of the Macdonald–Mehta–Opdam identity for finite Coxeter groups, *Math. Res. Lett.* **17**, 277–284.

Exton, H. (1976). *Multiple Hypergeometric Functions and Applications*, Halsted, New York.

Fackerell E. D. and Littler, R. A. (1974). Polynomials biorthogonal to Appell's polynomials, *Bull. Austral. Math. Soc.* **11**, 181–195.

Farouki, R. T., Goodman, T. N. T. and Sauer, T. (2003). Construction of orthogonal bases for polynomials in Bernstein form on triangular and simplex domains, *Comput. Aided Geom. Design* **20**, 209–230.

Fernández, L., Pérez, T. E. and Piñar, M. A. (2005). Classical orthogonal polynomials in two variables: a matrix approach, *Numer. Algorithms* **39**, 131–142.

Fernández, L., Pérez, T. E. and Piñar, M. A. (2011). Orthogonal polynomials in two variables as solutions of higher order partial differential equations, *J. Approx. Theory* **163**, 84–97.

Folland, G. B. (1975). Spherical harmonic expansion of the Poisson–Szegő kernel for the ball, *Proc. Amer. Math. Soc.* **47**, 401–408.

Forrester, P. J. and Warnaar, S. O. (2008). The importance of the Selberg integral *Bull. Amer. Math. Soc.* **45**, 489–534.

Freud, G. (1966). *Orthogonal Polynomials*, Pergamon, New York.

Fuglede, B. (1983). The multidimensional moment problem, *Expos. Math.* **1**, 47–65.

Gasper, G. (1972). Banach algebras for Jacobi series and positivity of a kernel, *Ann. Math.* **95**, 261–280.

Gasper, G. (1977). Positive sums of the classical orthogonal polynomials, *SIAM J. Math. Anal.* **8**, 423–447.

Gasper, G. (1981). Orthogonality of certain functions with respect to complex valued weights, *Canad. J. Math.* **33**, 1261–1270.

Gasper, G. and Rahman, M. (1990). *Basic Hypergeometric Series*, Encyclopedia of Mathematics and its Applications 35, Cambridge University Press, Cambridge.

Gekhtman, M. I. and Kalyuzhny, A. A. (1994). On the orthogonal polynomials in several variables, *Integr. Eq. Oper. Theory* **19**, 404–418.

Genest, V., Ismail, M., Vinet, L. and Zhedanov, A. (2013). The Dunkl oscillator in the plane I: superintegrability, separated wavefunctions and overlap coefficients, *J. Phys. A: Math. Theor.* **46**, 145 201.

Genest, V., Vinet, L. and Zhedanov, A. (2013). The singular and the 2:1 anisotropic Dunkl oscillators in the plane, *J. Phys. A: Math. Theor.* **46**, 325 201.

Ghanmi, A. (2008). A class of generalized complex Hermite polynomials, *J. Math. Anal. Appl.* **340**, 1395–1406.

Ghanmi, A. (2013). Operational formulae for the complex Hermite polynomials $H_{p,q}(z,\bar{z})$, *Integral Transf. Special Func.* **24**, 884–895.

Görlich, E. and Markett, C. (1982). A convolution structure for Laguerre series, *Indag. Math.* **14**, 161–171.

Griffiths, R. (1979). A transition density expansion for a multi-allele diffusion model. *Adv. Appl. Prob.* **11**, 310–325.

Griffiths, R. and Spanò, D. (2011). Multivariate Jacobi and Laguerre polynomials, infinite-dimensional extensions, and their probabilistic connections with multivariate Hahn and Meixner polynomials, *Bernoulli* **17**, 1095–1125.

Griffiths, R. and Spanò, D. (2013). Orthogonal polynomial kernels and canonical correlations for Dirichlet measures, *Bernoulli* **19**, 548–598.

Groemer, H. (1996). *Geometric Applications of Fourier Series and Spherical Harmonics*, Cambridge University Press, New York.

Grove, L. C. (1974). The characters of the hecatonicosahedroidal group, *J. Reine Angew. Math.* **265**, 160–169.

Grove, L. C. and Benson, C. T. (1985). *Finite Reflection Groups*, 2nd edn, Graduate Texts in Mathematics 99, Springer, Berlin–New York.

Grundmann, A. and Möller, H. M. (1978). Invariant integration formulas for the n-simplex by combinatorial methods, *SIAM. J. Numer. Anal.* **15**, 282–290.

Haviland, E. K. (1935). On the momentum problem for distributions in more than one dimension, I, II, *Amer. J. Math.* **57**, 562–568; **58**, 164–168.

Heckman, G. J. (1987). Root systems and hypergeometric functions II, *Compositio Math.* **64**, 353–373.

Heckman, G. J. (1991a). A remark on the Dunkl differential–difference operators, in *Harmonic Analysis on Reductive Groups* (Brunswick, ME, 1989), pp. 181–191, Progress in Mathematics 101, Birkhäuser, Boston, MA.

Heckman, G. J. (1991b). An elementary approach to the hypergeometric shift operators of Opdam, *Invent. Math.* **103**, 341–350.

Heckman, G. J. and Opdam, E. M. (1987). Root systems and hypergeometric functions, I, *Compositio Math.* **64**, 329–352.

Helgason, S. (1984). *Groups and Geometric Analysis*, Academic Press, New York.

Higgins, J. R. (1977). *Completeness and Basis Properties of Sets of Special Functions*, Cambridge University Press, Cambridge.

Hoffman, M. E. and Withers, W. D. (1988). Generalized Chebyshev polynomials associated with affine Weyl groups, *Trans. Amer. Math. Soc.* **308**, 91–104.

Horn, R. A. and Johnson, C. R. (1985). *Matrix Analysis*, Cambridge University Press, Cambridge.

Hua, L. K. (1963). *Harmonic Analysis of Functions of Several Complex Variables in the Classical Domains*, Translations of Mathematical Monographs 6, American Mathematical Society, Providence, RI.

Humphreys, J. E. (1990). *Reflection Groups and Coxeter Groups*, Cambridge Studies in Advanced Mathematics 29, Cambridge University Press, Cambridge.

Ignatenko, V. F. (1984). On invariants of finite groups generated by reflections, *Math. USSR Sb.* **48**, 551–563.

Ikeda, M. (1967). On spherical functions for the unitary group, I, II, III, *Mem. Fac. Engrg Hiroshima Univ.* **3**, 17–75.

Iliev, P. (2011). Krall–Jacobi commutative algebras of partial differential operators, *J. Math. Pures Appl.* **96**, no. 9, 446–461.

Intissar, A. and Intissar, A. (2006). Spectral properties of the Cauchy transform on $L^2(\mathbb{C}; e^{|z|^2} d\lambda)$, *J. Math. Anal. Appl.* **313**, 400–418.

Ismail, M. (2005). *Classical and Quantum Orthogonal Polynomials in One Variable*, Encyclopedia of Mathematics and its Applications 71, Cambridge University Press, Cambridge.

Ismail, M. (2013). Analytic properties of complex Hermite polynomials. *Trans. Amer. Math. Soc.*, to appear.

Ismail, M. and P. Simeonov (2013). Complex Hermite polynomials: their combinatorics and integral operators. *Proc. Amer. Math. Soc.*, to appear.

Itô, K. (1952). Complex multiple Wiener integral, *Japan J. Math.* **22**, 63–86.

Ivanov, K., Petrushev, P. and Xu, Y. (2010). Sub-exponentially localized kernels and frames induced by orthogonal expansions. *Math. Z.* **264**, 361–397.

Ivanov, K., Petrushev, P. and Xu, Y. (2012). Decomposition of spaces of distributions induced by tensor product bases, *J. Funct. Anal.* **263**, 1147–1197.

Jack, H. (1970/71). A class of symmetric polynomials with a parameter, *Proc. Roy. Soc. Edinburgh Sect. A.* **69**, 1–18.

Jackson, D. (1936). Formal properties of orthogonal polynomials in two variables, *Duke Math. J.* **2**, 423–434.

James, A. T. and Constantine, A. G. (1974). Generalized Jacobi polynomials as spherical functions of the Grassmann manifold, *Proc. London Math. Soc.* **29**, 174–192.

de Jeu, M. F. E. (1993). The Dunkl transform, *Invent. Math.* **113**, 147–162.

Kakei, S. (1998). Intertwining operators for a degenerate double affine Hecke algebra and multivariable orthogonal polynomials, *J. Math. Phys.* **39**, 4993–5006.

Kalnins, E. G., Miller, W., Jr and Tratnik, M. V. (1991). Families of orthogonal and biorthogonal polynomials on the n-sphere, *SIAM J. Math. Anal.* **22**, 272–294.

Kanjin, Y. (1985). Banach algebra related to disk polynomials, *Tôhoku Math. J.* **37**, 395–404.

Karlin S. and McGregor, J. (1962). Determinants of orthogonal polynomials, *Bull. Amer. Math. Soc.* **68**, 204–209.

Karlin, S. and McGregor, J. (1975). Some properties of determinants of orthogonal polynomials, in *Theory and Application of Special Functions*, ed. R. A. Askey, pp. 521–550, Academic Press, New York.

Kato, Y. and Yamamoto, T. (1998). Jack polynomials with prescribed symmetry and hole propagator of spin Calogero–Sutherland model, *J. Phys. A*, **31**, 9171–9184.

Kerkyacharian, G., Petrushev, P., Picard, D. and Xu, Y. (2009). Decomposition of Triebel–Lizorkin and Besov spaces in the context of Laguerre expansions, *J. Funct. Anal.* **256**, 1137–1188.

Kim, Y. J., Kwon, K. H. and Lee, J. K. (1998). Partial differential equations having orthogonal polynomial solutions, *J. Comput. Appl. Math.* **99**, 239–253.

Knop, F. and Sahi, S (1997). A recursion and a combinatorial formula for Jack polynomials. *Invent. Math.* **128**, 9–22.

Koelink, E. (1996). Eight lectures on quantum groups and q-special functions, *Rev. Colombiana Mat.* **30**, 93–180.

Kogbetliantz, E. (1924). Recherches sur la sommabilité des séries ultrasphériques par la méthode des moyennes arithmétiques, *J. Math. Pures Appl.* **3**, no. 9, 107–187.

Koornwinder, T. H. (1974a). Orthogonal polynomials in two variables which are eigenfunctions of two algebraically independent partial differential operators, I, II, *Proc. Kon. Akad. v. Wet., Amsterdam* **36**, 48–66.

Koornwinder, T. H. (1974b). Orthogonal polynomials in two variables which are eigenfunctions of two algebraically independent partial differential operators, III, IV, *Proc. Kon. Akad. v. Wet., Amsterdam* **36**, 357–381.

Koornwinder, T. H. (1975). Two-variable analogues of the classical orthogonal polynomials, in *Theory and Applications of Special Functions*, ed. R. A. Askey, pp. 435–495, Academic Press, New York.

Koornwinder, T. H. (1977). Yet another proof of the addition formula for Jacobi polynomials, *J. Math. Anal. Appl.* **61**, 136–141.

Koornwinder, T. H. (1992). Askey–Wilson polynomials for root systems of type *BC*, in *Hypergeometric Functions on Domains of Positivity, Jack Polynomials, and Applications*, pp. 189–204, Contemporary Mathematics 138, American Mathematical Society, Providence, RI.

Koornwinder T. H. and Schwartz, A. L. (1997). Product formulas and associated hypergroups for orthogonal polynomials on the simplex and on a parabolic biangle, *Constr. Approx.* **13**, 537–567.

Koornwinder, T. H. and Sprinkhuizen-Kuyper, I. (1978). Generalized power series expansions for a class of orthogonal polynomials in two variables, *SIAM J. Math. Anal.* **9**, 457–483.

Kowalski, M. A. (1982a). The recursion formulas for orthogonal polynomials in n variables, *SIAM J. Math. Anal.* **13**, 309–315.

Kowalski, M. A. (1982b). Orthogonality and recursion formulas for polynomials in n variables, *SIAM J. Math. Anal.* **13**, 316–323.

Krall, H. L. and Sheffer, I. M. (1967). Orthogonal polynomials in two variables, *Ann. Mat. Pura Appl.* **76**, no. 4, 325–376.

Kroó, A. and Lubinsky, D. (2013a). Christoffel functions and universality on the boundary of the ball, *Acta Math. Hungarica* **140**, 117–133.

Kroó, A. and Lubinsky, D. (2013b). Christoffel functions and universality in the bulk for multivariate orthogonal polynomials, *Can. J. Math.* **65**, 600–620.

Kwon, K. H., Lee, J. K. and Littlejohn, L. L. (2001). Orthogonal polynomial eigenfunctions of second-order partial differential equations, *Trans. Amer. Math. Soc.* **353**, 3629–3647.

Lapointe, L. and Vinet, L. (1996). Exact operator solution of the Calogero–Sutherland model, *Comm. Math. Phys.* **178**, 425–452.

Laporte, O. (1948). Polyhedral harmonics, *Z. Naturforschung* **3a**, 447–456.

Larcher, H. (1959). Notes on orthogonal polynomials in two variables. *Proc. Amer. Math. Soc.* **10**, 417–423.

Lassalle, M. (1991a). Polynômes de Hermite généralisé (in French) *C. R. Acad. Sci. Paris Sér. I Math.* **313**, 579–582.

Lassalle, M. (1991b). Polynômes de Jacobi généralisé (in French) *C. R. Acad. Sci. Paris Sér. I Math.* **312**, 425–428.

Lassalle, M. (1991c). Polynômes de Laguerre généralisé (in French) *C. R. Acad. Sci. Paris Sér. I Math.* **312**, 725–728.

Lasserre, J. (2012). The existence of Gaussian cubature formulas, *J. Approx. Theory*, **164**, 572–585.

Lebedev, N. N. (1972). *Special Functions and Their Applications*, revised edn, Dover, New York.

Lee J. K. and Littlejohn, L. L. (2006). Sobolev orthogonal polynomials in two variables and second order partial differential equations, *J. Math. Anal. Appl.* **322**, 1001–1017.

Li, H., Sun, J. and Xu, Y. (2008). Discrete Fourier analysis, cubature and interpolation on a hexagon and a triangle, *SIAM J. Numer. Anal.* **46**, 1653–1681.

Li, H. and Xu, Y. (2010). Discrete Fourier analysis on fundamental domain and simplex of A_d lattice in d-variables, *J. Fourier Anal. Appl.* **16**, 383–433.

Li, Zh.-K. and Xu, Y. (2000). Summability of product Jacobi expansions, *J. Approx. Theory* **104**, 287–301.

Li, Zh.-K. and Xu, Y. (2003). Summability of orthogonal expansions, *J. Approx. Theory*, **122**, 267–333.

Lidl, R. (1975). Tschebyscheff polynome in mehreren variabelen, *J. Reine Angew. Math.* **273**, 178–198.

Littlejohn, L. L. (1988). Orthogonal polynomial solutions to ordinary and partial differential equations, in *Orthogonal Polynomials and Their Applications* (Segovia, 1986), Lecture Notes in Mathematics 1329, 98–124, Springer, Berlin.

Logan, B. and Shepp, I. (1975). Optimal reconstruction of a function from its projections, *Duke Math. J.* **42**, 649–659.

Lorentz, G. G. (1986). *Approximation of Functions*, 2nd edn, Chelsea, New York.

Lyskova, A. S. (1991). Orthogonal polynomials in several variables, *Dokl. Akad. Nauk SSSR* **316**, 1301–1306; translation in *Soviet Math. Dokl.* **43** (1991), 264–268.

Macdonald, I. G. (1982). Some conjectures for root systems, *SIAM, J. Math. Anal.* **13**, 988–1007.

Macdonald, I. G. (1995). *Symmetric Functions and Hall Polynomials*, 2nd edn, Oxford University Press, New York.

Macdonald, I. G. (1998). *Symmetric Functions and Orthogonal Polynomials*, University Lecture Series 12, American Mathematical Society, Providence, RI.

Maiorov, V. E. (1999). On best approximation by ridge functions, *J. Approx. Theory* **99**, 68–94.

Markett, C. (1982). Mean Cesàro summability of Laguerre expansions and norm estimates with shift parameter, *Anal. Math.* **8**, 19–37.

Marr, R. (1974). On the reconstruction of a function on a circular domain from a sampling of its line integrals, *J. Math. Anal. Appl.*, **45**, 357–374.

Máté, A., Nevai, P. and Totik, V. (1991). Szegő's extremum problem on the unit circle, *Ann. Math.* **134**, 433–453.

Mehta, M. L. (1991). *Random Matrices*, 2nd edn, Academic Press, Boston, MA.

Möller, H. M. (1973). Polynomideale und Kubaturformeln, Thesis, University of Dortmund.

Möller, H. M. (1976). Kubaturformeln mit minimaler Knotenzahl, *Numer. Math.* **25**, 185–200.

Moody, R. and Patera, J. (2011). Cubature formulae for orthogonal polynomials in terms of elements of finite order of compact simple Lie groups, *Adv. Appl. Math.* **47**, 509–535.

Morrow, C. R. and Patterson, T. N. L. (1978). Construction of algebraic cubature rules using polynomial ideal theory, *SIAM J. Numer. Anal.* **15**, 953–976.

Müller, C. (1997). *Analysis of Spherical Symmetries in Euclidean Spaces*, Springer, New York.

Mysovskikh, I. P. (1970). A multidimensional analog of quadrature formula of Gaussian type and the generalized problem of Radon (in Russian), *Vopr. Vychisl. i Prikl. Mat., Tashkent* **38**, 55–69.

Mysovskikh, I. P. (1976). Numerical characteristics of orthogonal polynomials in two variables, *Vestnik Leningrad Univ. Math.* **3**, 323–332.

Mysovskikh, I. P. (1981). *Interpolatory Cubature Formulas*, Nauka, Moscow.

Narcowich, F. J., Petrushev, P, and Ward, J. D. (2006) Localized tight frames on spheres, *SIAM J. Math. Anal.* **38**, 574–594.

Nelson, E. (1959). Analytic vectors, *Ann. Math.* **70**, no. 2, 572–615.

Nevai, P. (1979). Orthogonal polynomials, *Mem. Amer. Math. Soc.* **18**, no. 213.

Nevai, P. (1986). Géza Freud, orthogonal polynomials and Christoffel functions. A case study, *J. Approx. Theory* **48**, 3–167.

Nesterenko, M., Patera, J., Szajewska, M. and Tereszkiewicz, A. (2010). Orthogonal polynomials of compact simple Lie groups: branching rules for polynomials, *J. Phys. A* **43**, no. 49, 495 207, 27 pp.

Nishino, A., Ujino, H. and Wadati, M. (1999). Rodrigues formula for the nonsymmetric multivariable Laguerre polynomial, *J. Phys. Soc. Japan* **68**, 797–802.

Noumi, M. (1996). Macdonald's symmetric polynomials as zonal spherical functions on some quantum homogeneous spaces, *Adv. Math.* **123**, 16–77.

Nussbaum, A. E. (1966). Quasi-analytic vectors, *Ark. Mat.* **6**, 179–191.

Okounkov, A. and Olshanski, G. (1998). Asymptotics of Jack polynomials as the number of variables goes to infinity, *Int. Math. Res. Not.* **13**, 641–682.

Opdam, E. M. (1988). Root systems and hypergeometric functions, III, IV, *Compositio Math.* **67**, 21–49, 191–209.

Opdam, E. M. (1991). Some applications of hypergeometric shift operators, *Invent. Math.* **98**, 1–18.

Opdam, E. M. (1995). Harmonic analysis for certain representations of graded Hecke algebras, *Acta Math.* **175**, 75–121.

Pérez, T. E., Piñar, M. A. and Xu, Y. (2013). Weighted Sobolev orthogonal polynomials on the unit ball, *J. Approx. Theory*, **171**, 84–104.

Petrushev, P. (1999). Approximation by ridge functions and neural networks, *SIAM J. Math. Anal.* **30**, 155–189.

Petrushev, P. and Xu, Y. (2008a). Localized polynomial frames on the ball, *Constr. Approx.* **27**, 121–148.

Petrushev, P. and Xu, Y. (2008b). Decomposition of spaces of distributions induced by Hermite expansions, *J. Fourier Anal. Appl.* **14**, 372–414.

Piñar, M. and Xu, Y. (2009). Orthogonal polynomials and partial differential equations on the unit ball, *Proc. Amer. Math. Soc.* **137**, 2979–2987.

Podkorytov, A. M. (1981). Summation of multiple Fourier series over polyhedra, *Vestnik Leningrad Univ. Math.* **13**, 69–77.

Proriol, J. (1957). Sur une famille de polynomes à deux variables orthogonaux dans un triangle, *C. R. Acad. Sci. Paris* **245**, 2459–2461.

Putinar, M. (1997). A dilation theory approach to cubature formulas, *Expos. Math.* **15**, 183–192.

Putinar, M. (2000). A dilation theory approach to cubature formulas II, *Math. Nach.* **211**, 159–175.

Putinar, M. and Vasilescu, F. (1999). Solving moment problems by dimensional extension, *Ann. Math.* **149**, 1087–1107.

Radon, J. (1948). Zur mechanischen Kubatur, *Monatsh. Math.* **52**, 286–300.

Reznick, B. (2000). Some concrete aspects of Hilbert's 17th problem, in *Real Algebraic Geometry and Ordered Structures*, pp. 251–272, Contemporary Mathematics 253, American Mathematical Society, Providence, RI.

Ricci, P. E. (1978). Chebyshev polynomials in several variables (in Italian), *Rend. Mat.* **11**, no. 6, 295–327.

Riesz, F. and Sz. Nagy, B. (1955). *Functional Analysis*, Ungar, New York.

Rosengren, H. (1998). Multilinear Hankel forms of higher order and orthogonal polynomials, *Math. Scand.* **82**, 53–88.

Rosengren, H. (1999). Multivariable orthogonal polynomials and coupling coefficients for discrete series representations, *SIAM J. Math. Anal.* **30**, 232–272.

Rosier, M. (1995). Biorthogonal decompositions of the Radon transform, *Numer. Math.* **72**, 263–283.

Rösler, M. (1998). Generalized Hermite polynomials and the heat equation for Dunkl operators, *Comm. Math. Phys.* **192**, 519–542.

Rösler, M. (1999). Positivity of Dunkl's intertwining operator, *Duke Math. J.*, **98**, 445–463.

Rösler, M. and Voit, M. (1998). Markov processes related with Dunkl operators. *Adv. Appl. Math.* **21**, 575–643.

Rozenblyum, A. V. and Rozenblyum, L. V. (1982). A class of orthogonal polynomials of several variables, *Differential Eq.* **18**, 1288–1293.

Rozenblyum, A. V. and Rozenblyum, L. V. (1986). Orthogonal polynomials of several variables, which are connected with representations of groups of Euclidean motions, *Differential Eq.* **22**, 1366–1375.

Rudin, W. (1991). *Functional Analysis*, 2nd edn., International Series in Pure and Applied Mathematics, McGraw-Hill, New York.

Sahi, S. (1996). A new scalar product for nonsymmetric Jack polynomials, *Int. Math. Res. Not.* **20**, 997–1004.

Schmid, H. J. (1978). On cubature formulae with a minimum number of knots, *Numer. Math.* **31**, 281–297.

Schmid, H. J. (1995). Two-dimensional minimal cubature formulas and matrix equations, *SIAM J. Matrix Anal. Appl.* **16**, 898–921.

Schmid, H. J. and Xu, Y. (1994). On bivariate Gaussian cubature formula, *Proc. Amer. Math. Soc.* **122**, 833–842.

Schmüdgen, K. (1990). *Unbounded Operator Algebras and Representation Theory*, Birkhäuser, Boston, MA.

Selberg, A. (1944). Remarks on a multiple integral (in Norwegian), *Norsk Mat. Tidsskr.* **26**, 71–78.

Shephard, G. C. and Todd, J. A. (1954). Finite unitary reflection groups, *Canad. J. Math.* **6**, 274–304.

Shishkin, A. D. (1997). Some properties of special classes of orthogonal polynomials in two variables, *Integr. Transform. Spec. Funct.* **5**, 261–272.

Shohat, J. and Tamarkin, J. (1943). *The Problem of Moments*, Mathematics Surveys 1, American Mathematical Society, Providence, RI.

Sprinkhuizen-Kuyper, I. (1976). Orthogonal polynomials in two variables. A further analysis of the polynomials orthogonal over a region bounded by two lines and a parabola, *SIAM J. Math. Anal.* **7**, 501–518.

Stanley, R. P. (1989). Some combinatorial properties of Jack symmetric functions, *Adv. Math.* **77**, 76–115.

Stein, E. M. (1970). *Singular Integrals and Differentiability Properties of Functions*, Princeton University Press, Princeton, NJ.

Stein, E. M. and Weiss, G. (1971). *Introduction to Fourier Analysis on Euclidean Spaces*, Princeton University Press, Princeton, NJ.

Stokman, J. V. (1997). Multivariable big and little q-Jacobi polynomials, *SIAM J. Math. Anal.* **28**, 452–480.

Stokman, J. V. and Koornwinder, T. H. (1997). Limit transitions for BC type multivariable orthogonal polynomials, *Canad. J. Math.* **49**, 373–404.

Stroud, A. (1971). *Approximate Calculation of Multiple Integrals*, Prentice-Hall, Englewood Cliffs, NJ.

Suetin, P. K. (1999). *Orthogonal Polynomials in Two Variables*, translated from the 1988 Russian original by E. V. Pankratiev, Gordon and Breach, Amsterdam.

Sutherland, B. (1971). Exact results for a quantum many-body problem in one dimension, *Phys. Rev. A* **4**, 2019–2021.

Sutherland, B. (1972). Exact results for a quantum many-body problem in one dimension, II, *Phys. Rev. A* **5**, 1372–1376.

Szegő, G. (1975). *Orthogonal Polynomials*, 4th edn, American Mathematical Society Colloquium Publication 23, American Mathematical Society, Providence, RI.

Thangavelu, S. (1992). Transplantaion, summability and multipliers for multiple Laguerre expansions, *Tôhoku Math. J.* **44**, 279–298.

Thangavelu, S. (1993). *Lectures on Hermite and Laguerre Expansions*, Princeton University Press, Princeton, NJ.

Thangavelu, S. and Xu, Y. (2005). Convolution operator and maximal function for the Dunkl transform, *J. Anal. Math.* **97**, 25–55.

Tratnik, M.V. (1991a). Some multivariable orthogonal polynomials of the Askey tableau – continuous families, *J. Math. Phys.* **32**, 2065–2073.

Tratnik, M.V. (1991b). Some multivariable orthogonal polynomials of the Askey tableau – discrete families, *J. Math. Phys.* **32**, 2337–2342.

Uglov, D. (2000). Yangian Gelfand–Zetlin bases, gl_N-Jack polynomials, and computation of dynamical correlation functions in the spin Calogero–Sutherland model, in *Calogero–Sutherland–Moser Models*, pp. 485–495, CRM Series in Mathematical Physics, Springer, New York.

Ujino, H. and Wadati, M. (1995). Orthogonal symmetric polynomials associated with the quantum Calogero model, *J. Phys. Soc. Japan* **64**, 2703–2706.

Ujino, H. and Wadati, M. (1997). Orthogonality of the Hi-Jack polynomials associated with the Calogero model, *J. Phys. Soc. Japan* **66**, 345–350.

Ujino, H. and Wadati, M. (1999). Rodrigues formula for the nonsymmetric multivariable Hermite polynomial, *J. Phys. Soc. Japan* **68**, 391–395.

Verlinden, P. and Cools, R. (1992). On cubature formulae of degree $4k + 1$ attaining Möller's lower bound for integrals with circular symmetry, *Numer. Math.* **61**, 395–407.

Vilenkin, N. J. (1968). *Special Functions and the Theory of Group Representations*, American Mathematical Society Translation of Mathematics Monographs 22, American Mathematical Society, Providence, RI.

Vilenkin, N. J. and Klimyk, A. U. (1991a). *Representation of Lie Groups and Special Functions. Vol. 1. Simplest Lie Groups, Special Functions and Integral Transforms*, Mathematics and its Applications (Soviet Series) 72, Kluwer, Dordrecht.

Vilenkin, N. J. and Klimyk, A. U. (1991b). *Representation of Lie Groups and Special Functions. Vol. 2. Class I Representations, Special Functions, and Integral Transforms*, Mathematics and its Applications (Soviet Series) 72, Kluwer, Dordrecht, Netherlands.

Vilenkin, N. J. and Klimyk, A. U. (1991c). *Representation of Lie Groups and Special Functions. Vol. 3. Classical and Quantum Groups and Special Functions*, Mathematics and its Applications (Soviet Series) 72, Kluwer, Dordrecht, Netherlands.

Vilenkin, N. J. and Klimyk, A. U. (1995). *Representation of Lie Groups and Special Functions. Recent Advances*, Mathematics and its Applications 316, Kluwer, Dordrecht, Netherlands.

Volkmer, H. (1999). Expansions in products of Heine–Stieltjes polynomials, *Constr. Approx.* **15**, 467–480.

Vretare, L. (1984). Formulas for elementary spherical functions and generalized Jacobi polynomials, *SIAM. J. Math. Anal.* **15**, 805–833.

Wade, J. (2010). A discretized Fourier orthogonal expansion in orthogonal polynomials on a cylinder, *J. Approx. Theory* **162**, 1545–1576.

Wade, J. (2011). Cesàro summability of Fourier orthogonal expansions on the cylinder, *J. Math. Anal. Appl.* **402** (2013), 446–452.

Waldron, S. (2006). On the Bernstein–Bézier form of Jacobi polynomials on a simplex, *J. Approx. Theory* **140**, 86–99.

Waldron, S. (2008). Orthogonal polynomials on the disc, *J. Approx. Theory* **150**, 117–131.

Waldron, S. (2009). Continuous and discrete tight frames of orthogonal polynomials for a radially symmetric weight. *Constr. Approx.* **30**, 3352.

Wünsche, A. (2005). Generalized Zernike or disc polynomials, *J. Comp. Appl. Math.* **174**, 135–163.

Xu, Y. (1992). Gaussian cubature and bivariable polynomial interpolation, *Math. Comput.* **59**, 547–555.

Xu, Y. (1993a). On multivariate orthogonal polynomials, *SIAM J. Math. Anal.* **24**, 783–794.

Xu, Y. (1993b). Unbounded commuting operators and multivariate orthogonal polynomials, *Proc. Amer. Math. Soc.* **119**, 1223–1231.

Xu, Y. (1994a). Multivariate orthogonal polynomials and operator theory, *Trans. Amer. Math. Soc.* **343**, 193–202.

Xu, Y. (1994b). Block Jacobi matrices and zeros of multivariate orthogonal polynomials, *Trans. Amer. Math. Soc.* **342**, 855–866.

Xu, Y. (1994c). Recurrence formulas for multivariate orthogonal polynomials, *Math. Comput.* **62**, 687–702.

Xu, Y. (1994d). On zeros of multivariate quasi-orthogonal polynomials and Gaussian cubature formulae, *SIAM J. Math. Anal.* **25**, 991–1001.

Xu, Y. (1994e). Solutions of three-term relations in several variables, *Proc. Amer. Math. Soc.* **122**, 151–155.

Xu, Y. (1994f). *Common Zeros of Polynomials in Several Variables and Higher Dimensional Quadrature*, Pitman Research Notes in Mathematics Series 312, Longman, Harlow.

Xu, Y. (1995). Christoffel functions and Fourier series for multivariate orthogonal polynomials, *J. Approx. Theory* **82**, 205–239.

Xu, Y. (1996a). Asymptotics for orthogonal polynomials and Christoffel functions on a ball, *Meth. Anal. Appl.* **3**, 257–272.

Xu, Y. (1996b). Lagrange interpolation on Chebyshev points of two variables, *J. Approx. Theory* **87**, 220–238.

Xu, Y. (1997a). On orthogonal polynomials in several variables, in *Special Functions, q-Series and Related Topics*, pp. 247–270, The Fields Institute for Research in Mathematical Sciences, Communications Series 14, American Mathematical Society, Providence, RI.

Xu, Y. (1997b). Orthogonal polynomials for a family of product weight functions on the spheres, *Canad. J. Math.* **49**, 175–192.

Xu, Y. (1997c). Integration of the intertwining operator for h-harmonic polynomials associated to reflection groups, *Proc. Amer. Math. Soc.* **125**, 2963–2973.

Xu, Y. (1998a). Intertwining operator and h-harmonics associated with reflection groups, *Canad. J. Math.* **50**, 193–209.

Xu, Y. (1998b). Orthogonal polynomials and cubature formulae on spheres and on balls, *SIAM J. Math. Anal.* **29**, 779–793.

Xu, Y. (1998c). Orthogonal polynomials and cubature formulae on spheres and on simplices, *Meth. Anal. Appl.* **5**, 169–184.

Xu, Y. (1998d). Summability of Fourier orthogonal series for Jacobi weight functions on the simplex in $\mathbb{R}^d$, *Proc. Amer. Math. Soc.* **126**, 3027–3036.

Xu, Y. (1999a). Summability of Fourier orthogonal series for Jacobi weight on a ball in $\mathbb{R}^d$, *Trans. Amer. Math. Soc.* **351**, 2439–2458.

Xu, Y. (1999b). Aymptotics of the Christoffel functions on a simplex in $\mathbb{R}^d$, *J. Approx. Theory* **99**, 122–133.

Xu, Y. (1999c). Cubature formulae and polynomial ideals, *Adv. Appl. Math.* **23**, 211–233.

Xu, Y. (2000a). Harmonic polynomials associated with reflection groups, *Canad. Math. Bull.* **43**, 496–507.

Xu, Y. (2000b). Funk–Hecke formula for orthogonal polynomials on spheres and on balls, *Bull. London Math. Soc.* **32**, 447–457.

Xu, Y. (2000c). Constructing cubature formulae by the method of reproducing kernel, *Numer. Math.* **85**, 155–173.

Xu, Y. (2000d). A note on summability of Laguerre expansions, *Proc. Amer. Math. Soc.* **128**, 3571–3578.

Xu, Y. (2000e). A product formula for Jacobi polynomials, in *Special Functions* (Hong Kong, 1999), pp. 423–430, World Science, River Edge, NJ.

Xu, Y. (2001a). Orthogonal polynomials and summability in Fourier orthogonal series on spheres and on balls, *Math. Proc. Cambridge Phil. Soc.* **31** (2001), 139–155.

Xu, Y. (2001b). Orthogonal polynomials on the ball and the simplex for weight functions with reflection symmetries, *Constr. Approx.* **17**, 383–412.

Xu, Y. (2004). On discrete orthogonal polynomials of several variables, *Adv. Appl. Math.* **33**, 615–632.

Xu, Y. (2005a). Weighted approximation of functions on the unit sphere, *Constr. Approx.* **21**, 1–28.

Xu, Y. (2005b). Second order difference equations and discrete orthogonal polynomials of two variables, *Int. Math. Res. Not.* **8**, 449–475.

Xu, Y. (2005c). Monomial orthogonal polynomials of several variables, *J. Approx. Theory*, **133**, 1–37.

Xu, Y. (2005d). Rodrigues type formula for orthogonal polynomials on the unit ball, *Proc. Amer. Math. Soc.* **133**, 1965–1976.

Xu, Y. (2006a). A direct approach to the reconstruction of images from Radon projections, *Adv. Applied Math.* **36**, 388–420.

Xu, Y. (2006b). A family of Sobolev orthogonal polynomials on the unit ball, *J. Approx. Theory* **138**, 232–241.

Xu, Y. (2006c). Analysis on the unit ball and on the simplex, *Electron. Trans. Numer. Anal.*, **25**, 284–301.

Xu, Y. (2007). Reconstruction from Radon projections and orthogonal expansion on a ball, *J. Phys. A: Math. Theor.* **40**, 7239–7253.

Xu, Y. (2008). Sobolev orthogonal polynomials defined via gradient on the unit ball, *J. Approx. Theory* **152**, 52–65.

Xu, Y. (2010). Fourier series and approximation on hexagonal and triangular domains, *Const. Approx.* **31**, 115–138.

Xu, Y. (2012). Orthogonal polynomials and expansions for a family of weight functions in two variables. *Constr. Approx.* **36**, 161–190.

Xu, Y. (2013). Complex vs. real orthogonal polynomials of two variables. *Integral Transforms Spec. Funct.*, to appear. arXiv:1307.7819.

Yamamoto, T. (1995). Multicomponent Calogero model of B_N-type confined in a harmonic potential, *Phys. Lett. A* **208**, 293–302.

Yan, Z. M. (1992). Generalized hypergeometric functions and Laguerre polynomials in two variables, in *Hypergeometric Functions on Domains of Positivity, Jack Polynomials, and Applications*, pp. 239–259, Contemporary Mathematics 138, American Mathematical Society, Providence, RI.

Zernike, F. and Brinkman, H. C. (1935). Hypersphärishe Funktionen und die in sphärischen Bereichen orthogonalen Polynome, *Proc. Kon. Akad. v. Wet., Amsterdam* **38**, 161–170.

Zygmund, A. (1959). *Trigonometric Series*, Cambridge University Press, Cambridge.

Author Index

Symbol Index

Subject Index